Climate, Ticks and Disease

CABI Climate Change Series

Climate change is a major environmental challenge to the world today, with significant threats to ecosystems, food security, water resources and economic stability overall. In order to understand and research ways to alleviate the effects of climate change, scientists need access to information that not only provides an overview of and background to the field, but also keeps them up to date with the latest research findings.

This series addresses many topics relating to climate change, including strategies to develop sustainable systems that minimize impact on climate and/or mitigate the effects of human activity on climate change. Coverage will encompass all areas of environmental and agricultural sciences. Aimed at researchers, upper-level students and policy makers, titles in the series provide international coverage of topics related to climate change, including both a synthesis of facts and discussions of future research perspectives and possible solutions.

Titles Available

1. Climate Change and Crop Production
   Edited by Matthew P. Reynolds
2. Crop Stress Management and Climate Change
   Edited by José L. Araus and Gustavo A. Slafer
3. Temperature Adaptation in a Changing Climate: Nature at Risk
   Edited by Kenneth Storey and Karen Tanino
4. Plant Genetic Resources and Climate Change
   Edited by Michael Jackson, Brian Ford-Lloyd and Martin Parry
5. Climate Change Impact and Adaptation in Agricultural Systems
   Edited by Jürg Fuhrer and Peter J. Gregory
6. Livestock Production and Climate Change
   Edited by Pradeep K. Malik, Raghavendra Bhatta, Junichi Takahashi, Richard A. Kohn and Cadaba S. Prasad
7. Climate Change and Insect Pests
   Edited by Christer Björkman and Pekka Niemelä
8. Climate Change and Agricultural Water Management in Developing Countries
   Edited by Chu Thai Hoanh, Vladimir Smakhtin and Robyn Johnston
9. Climate Change Challenges and Adaptations at Farm-level: Case Studies from Asia and Africa
   Edited by Naveen Prakash Singh, Cynthia Bantilan, Kattarkandi Byjesh and Swamikannu Nedumaran
10. Climate Change Impacts on Urban Pests
    Edited by Partho Dhang
11. Climate Change Impacts and Sustainability: Ecosystems of Tanzania
    Edited by Pius Z. Yanda, Claude G. Mung'ong'o and Edmund B. Mabhuye
12. Climate, Ticks and Disease
    Edited by Pat Nuttall

Climate, Ticks and Disease

Edited by

Pat Nuttall

University of Oxford, Oxford, UK

CABI

CABI is a trading name of CAB International

CABI
Nosworthy Way
Wallingford
Oxfordshire OX10 8DE
UK

Tel: +44 (0)1491 832111
Fax: +44 (0)1491 833508
E-mail: info@cabi.org
Website: www.cabi.org

CABI
WeWork
One Lincoln Street
24th Floor
Boston, MA 02111
USA

Tel: +1 (617)682-9015
E-mail: cabi-nao@cabi.org

A catalogue record for this book is available from the British Library, London, UK.

Library of Congress Control Number: 2021939387

References to Internet websites (URLs) were accurate at the time of writing.

ISBN-13: 978 1 78924 963 7 (hardback)
978 1 78924 964 4 (ePDF)
978 1 78924 965 1 (ePub)

DOI: 10.1079/9781789249637.0000

Commissioning Editor: David Hemming
Editorial Assistant: Ali Thompson
Production Editor: Shankari Wilford

Typeset by SPi, Pondicherry, India
Printed and bound in the USA by Integrated Books International, Dulles, Virginia

Contents

Contributors **xi**

Preface **xix**

Section 1: Climate

1.1. Climate and the tick ecosystem

eo1: Future Climate of Africa 1
Rosalind Cornforth and Heather Plumpton

eo2: Vegetation–Climate Interactions: Into the Tick Zone 8
Deborah Hemming, James P. Duffy, Neil Kaye and Ilya Maclean

1.2. Modelling climate change impacts

eo3: Climate Change and Lyme Disease 18
Peter Cox, Chris Huntingford, Morgan Sparey and Pat Nuttall

eo4: How to Model the Impact of Climate Change on Vector-Borne Diseases? 26
Cyril Caminade

eo5: Challenges of Modelling and Projecting Tick Distributions 32
William Wint and Neil Alexander

eo6: Considerations for Predicting Climate Change Implications on Future Spatial Distribution Ranges of Ticks 38
Ram Raghavan and Roman Ganta

1.3. Synopsis: Climate

Synopsis: Climate 44
Pat Nuttall

Section 2: Ticks

2.1. Climate impacts on tick physiology

eo7: Can the Impact of Climate Change on the Tick Microbiome Bring a New Epidemiological Landscape to Tick-Borne Diseases? 46
Alejandro Cabezas-Cruz

eo8: Climate Influence on Tick Neurobiology 50
Ladislav Šimo

eo9: The Impact of Climate Change on Tick Host-Seeking Behaviour 56
Isobel Ronai

eo10: Expected Transitions in Ticks and Their Heritable Endosymbionts Under Environmental Changes 61
Yuval Gottlieb and Olivier Duron

eo11: Drought and Tick Dynamics During Climate Change 67
Joshua Benoit and Kennan Oyen

eo12: Climate Influences on Reproduction and Immunity in the Soft Tick, *Ornithodoros moubata* 74
DeMar Taylor and Mari Ogihara

eo13: Climate Change and Ticks: Measuring Impacts 84
Richard Wall and Saeed Alasmari

2.2. Climate impacts on tick populations

eo14: Scandinavia and Ticks in a Changing Climate 90
Lene Jung Kjær and René Bødker

eo15: Birds, Ticks and Climate Change 96
Tom Kelly, John Healy and Neil Coughlan

eo16: How Tick Vectors are Coping with Global Warming 110
André Wilke, John Beier, Domenico Otranto and Giovanni Benelli

eo17: Possible Direct and Human-Mediated Impact of Climate Change on Tick Populations in Turkey 115
Sırrı Kar and Aysen Gargili Keles

eo18: Climate Change Alone Cannot Explain Altered Tick Distribution Across Europe: A Spotlight on Endemic and Invasive Tick Species 125
Frederic Stachurski, Nathalie Boulanger, Adrien Blisnick, Laurence Vial and Sarah Bonnet

eo19: Climate and Management Effects on Tick–Game Animal Dynamics 132
Andrew Hoodless and Rufus Sage

eo20: Climate-Driven Livestock Management Shifts and Tick Populations 139
Georgia Titcomb

eo21: Potential Impacts of Climate Change on Medically Important Tick Species in North America 145
Geoffrey Lynn, Sukanya Narasimhan and Erol Fikrig

2.3. Climate impacts on tick species

eo22: Climate Change and Tick Evolution: Lessons from the Past 152
Ben Mans

eo23: *Amblyomma* Ticks and Future Climates 159
Stefan Vilges de Oliveira

eo24: Climate Impacts on *Dermacentor reticulatus* Tick Population Dynamics and Range 166
Zbigniew Zając

eo25: Changes Expected in *Ixodes ricinus* Temporal and Spatial Distribution in Europe 171
Annapaola Rizzoli

eo26: Range Expansion of *Ixodes scapularis* in the USA 176
Durland Fish

eo27: Distribution, Seasonal Occurrence and Biological Characteristics of *Haemaphysalis longicornis*, a Vector of Bovine Piroplasmosis in Japan 183
Rika Umemiya-Shirafuji

eo28: Climate and Vector Potential of Medically Important North American Ticks 188
Lars Eisen and Rebecca Eisen

eo29: The Impact of Climate Change on the Biology of the Cattle Tick, *Rhipicephalus microplus*: Current Knowledge and Gaps to be Filled 193
Isabel Santos, Matias Szabó, Alexandre Rodrigues Caetano and Maristela Martins de Camargo

2.4. Climate impacts on vector capacity

eo30: Climate Impacts on the Vector Capacity of Tropical and Temperate Populations of the Brown Dog Tick, *Rhipicephalus sanguineus* sensu lato 209
Gervasio Bechara and Gustavo Sanches

eo31: Argasidae: Distribution and Vectorial Capacity in a Changing Global Environment 216
Serhii Filatov and Ryan Rego

eo32: Effects of Climate Change on Babesiosis Vectors 223
Jeremy Gray

2.5. Synopsis: Ticks

Synopsis: Ticks 228
Pat Nuttall

Section 3: Disease

3.1. Vector–host–pathogen triangle

eo33: Conflict and Cooperation in Tick–Host–Pathogen Interactions Contribute to Increased Tick Fitness and Survival 232
José de la Fuente and Margarita Villar

eo34: Climate, Ticks and Pathogens: Gaps and Caveats 240
Agustín Estrada-Peña, Natalia Fernández-Ruiz and José de la Fuente

eo35: Climate and Prediction of Tick-Borne Diseases Facing the Complexity of the Pathogen–Tick–Host Triad at Northern Latitudes 247
Atle Mysterud

eo36: Is the Clock 'Ticking' for Climate Change? 253
Ernest Gould and Xavier de Lamballerie

eo37: Climate Instability and Emerging Tick-Borne Disease 259
Janet Foley

eo38: Co-infections of Ticks 264
Martin Pfeffer, Nina Król and Anna Obiegala

eo39: Impact of Climate Change on Co-feeding Transmission 270
Jianhong Wu and Xue Zhang

eo40: Human Behaviour Trumps Entomological Risk 276
Sam Telford

eo41: It's All in the Timing: Effect of Tick Phenology on Pathogen Transmission Dynamics 283
Maria Diuk-Wasser

eo42: *Anaplasma* Species' Novel Tick–Host–Pathogen Relationships and Effects of Climate Change 293
Daniel Sonenshine

eo43: Zoonotic Potential in the Genera *Anaplasma* and *Ehrlichia* 300
Snorre Stuen

eo44: Tick Vectors, Tick-Borne Diseases and Climate Change 307
Abdallah Samy, Abdelghafar Alkishe, Tatjana Pustahija and Townsend Peterson

eo45: Climate and Other Global Factors at the Zoonotic Interface in America: Influence on Diseases Caused by Tick-Borne Pathogens 318
Ulrike Munderloh and Timothy Kurtti

eo46: Microclimatic Conditions and RNA Viruses in Ticks 326
Yan-Jang Huang, Stephen Higgs and Dana Vanlandingham

3.2. Vector-borne infections of humans

eo47: Climate, Ticks and Tick-Borne Encephalitis in Central Europe 331
Vlasta Danielová and Milan Daniel

eo48: Tick-Borne Viral Haemorrhagic Fever Infections 341
Dennis Bente

eo49: Climate Impact on Lyme Borreliosis and Its Causative Agents 349
Natalie Rudenko, Libor Grubhoffer and Maryna Golovchenko

eo50: Climate Change and Tick-Borne Encephalitis in the Greater Alpine Region 354
Franz Rubel

eo51: The Expansion of Japanese Spotted Fever and the Complex Group of Spotted Fever Group Rickettsiae in Japan 360
Megumi Sato, Reiko Arai and Marcello Otake Sato

eo52: Spatiotemporal and Demographic Patterns of Transmission of Kyasanur Forest Disease Virus in India 365
Neha Pandey and Sunit K. Singh

eo53: Argasid Ticks, Relapsing Fever and a Changing Climate 370
Mike Teglas

eo54: The Potential Effects of Climate Change on Lyme Borreliosis in East-Central Europe 375
Attila Trájer

eo55: Epidemiology of Severe Fever with Thrombocytopaenia Syndrome in China 382
Wei Liu, Li-Qun Fang and Hao Li

eo56: Climate Change and Debilitating Symptom Complexes Attributed to Ticks in Australia 391
Mona Dehhaghi, Hamed Kazemi Shariat Panahi, Richard Schloeffel, Bernard Hudson, Benjamin Ruiwen and Gilles Guillemin

eo57: Effect of Climate Change on Mosquito-Borne Pathogens 400
Michael Turell

3.3. Vector-borne infections of domesticated animals

eo58: *Ornithodoros* Tick Vectors and African Swine Fever Virus 405
Armanda Bastos and Fernando Boinas

eo59: Tick-Borne Diseases of Livestock in the UK 413
Nicholas Johnson and L. Paul Phipps

eo60: Impact of Climate Change on Tick-Borne Diseases of Livestock in Pakistan – Looking Ahead 418
Abdul Ghafar, Robin Gasser and Abdul Jabbar

eo61: The Emergence of Tick-Borne Diseases in Domestic Animals in Australia 424
Charlotte Oskam, Isobel Ronai and Peter Irwin

3.4. Vector-borne infections in different regions

eo62: Tick-Borne Infections in Central Europe 430
Maria Kazimirova

eo63: Impact of Climate Change on Ticks and Tick-Borne Infections in Russia 438
Ed Korenberg

eo64: Is Climate Change Affecting Ticks and Tick-Borne Diseases in Taiwan? 444
Chi-Chien Kuo

eo65: Ticks and Tick-Borne Pathogens in the Caribbean Region in the Context of Climate Change 449
Dasiel Obregón Alvarez, Roxanne Albertha Charles and Agustín Estrada-Peña

eo66: The Strange Case of Tick-Borne Viruses in Turkey 455
Sırrı Kar and Koray Ergünay

eo67: Melting, Melting Pot – Climate Change and Its Impact on Ticks and Tick-Borne Pathogens in the Arctic 460
Jiří Černý, Jana Elsterová and Lauren Culler

eo68: Ticks and Tick-Borne Diseases in the Middle East 469
Ali Reza Chavshin and Seyyed Javad Seyyed-Zadeh

eo69: The Emergence of Ticks and Tick-Borne Diseases in the USA 477
Charles Hart and Saravanan Thangamani

eo70: Role of Climate and Other Factors in Determining the Dynamics of Tick and Tick-Transmitted Pathogen Populations and Distribution in Western, Central and Eastern Africa 486
Naftaly Githaka, Esther Kanduma and Richard Bishop

eo71: Ticks and Tick-Borne Pathogens in China 492
Hao Li, Li-Qun Fang and Wei Liu

eo72: Tick-Borne Rickettsioses in Africa 500
Agatha Kolo

eo73: Climate and the Emergence of Tick-Borne Disease in Canada 507
Patrick Leighton, Robbin Lindsay and Nicholas Ogden

eo74: Climate Change Impacts on *Ixodes ricinus* Ticks in Scotland and Implications for Lyme Disease Risk 513
Lucy Gilbert

eo75: Possible Impact of Climate and Environmental Change on Ticks and Tick-Borne Disease in England 518
Jolyon Medlock and Kayleigh Hansford

eo76: Climate Change, Ticks and Tick-Borne Pathogens in Northern Europe 528
John Pettersson

eo77: Tick and Tick-Borne Disease Circulation in a Changing Marine Ecosystem 532
Karen D. McCoy

3.5. Synopsis: Disease
Synopsis: Disease 541
Pat Nuttall

Section 4: Final Synopsis and Future Predictions
Final Synopsis and Future Predictions 548
Pat Nuttall

Index **553**

Contributors

Alasmari, Saeed, School of Biological Sciences, Life Sciences Building, University of Bristol, Bristol, UK. Email: sa16081@bristol.ac.uk

Alexander, Neil, Environmental Research Group Oxford Ltd, c/o Dept Zoology, University of Oxford, 11a Mansfield Road, Oxford OX1 3SZ, UK. Email: neil.s.alexander@gmail.com

Alkishe, Abdelghafar, Biodiversity Institute, University of Kansas, Lawrence, Kansas, USA and Zoology Department, Faculty of Science, University of Tripoli, Tripoli, Libya. Email: al.keesh@ku.edu

Arai, Reiko, Niigata Prefectural Institute of Public Health and Environmental Sciences, 314-1 Sowa, Nishi-ku, Niigata City, 950-2144 Japan. Email: arai.reiko@pref.niigata.lg.jp

Bastos, Armanda, Department of Zoology and Entomology, Faculty of Natural and Agricultural Sciences, University of Pretoria, Pretoria, South Africa. Email: adbastos@zoology.up.ac.za

Bechara, Gervasio Henrique, Graduate Program in Animal Science, School of Life Sciences, Pontifical Catholic University of Paraná, Rua Imaculada Conceição 1155, Bairro Prado Velho, 80215-901 Curitiba-PR, Brazil. Email: gervasio.bechara@pucpr.br

Beier, John C., Department of Public Health, Miller School of Medicine, University of Miami, Miami, Florida, USA. Email: jbeier@miami.edu

Benelli, Giovanni, Department of Agriculture, Food and Environment, University of Pisa, Pisa, Italy. Email: giovanni.benelli@unipi.it

Benoit, Joshua, Department of Biological Sciences, University of Cincinnati, 711E Rieveschl Hall, 318 College Drive, Cincinnati, OH 45221-0006, USA. Email: joshua.benoit@uc.edu

Bente, Dennis A., Microbiology & Immunology Department, University of Texas Medical Branch, Galveston National Laboratory, 301 University Blvd, Route 0610, Galveston, TX 77555-1019, USA. Email: dabente@utmb.edu

Bishop, Richard P., Department of Veterinary Microbiology and Pathology, Washington State University, Pullman, Washington, USA. Email: richard.bishop2@wsu.edu

Blisnick, Adrien, Institut Pasteur, Department of Virology, Arboviruses and Insect Vectors, Paris, France. Email: adrien.blisnick@pasteur.fr

Bødker, René, Section for Animal Welfare and Disease Control, Department of Veterinary and Animal Sciences, Faculty of Health and Medical Sciences, University of Copenhagen, Copenhagen, Denmark. Email: rebo@sund.ku.dk

Boinas, Fernando, Centro de Investigação Interdisciplinar em Sanidade Animal, Faculdade de Medicina Veterinária, Universidade de Lisboa, Avenida da Universidade Técnica, 1300-477 Lisboa, Portugal. Email: fboinas@fmv.ulisboa.pt

Bonnet, Sarah I., INRAE, Département Santé Animale, Centre Val de Loire – Tours, 37380 Nouzilly, France. Email: sarah.bonnet@inrae.fr

Boulanger, Nathalie, UR 7290, Virulence bactérienne précoce: groupe Borrelia, Fédération de médecine translationnelle, Université de Strasbourg, France; and Centre National de Référence Borrelia, France. Email: nboulanger@unistra.fr

Cabezas-Cruz, Alejandro, UMR BIPAR, Laboratoire de Santé Animale, ANSES, INRAE, École Nationale Vétérinaire d'Alfort, UMR BIPAR, Laboratoire de Santé Animale, Maisons-Alfort, F-94700, France. Email: alejandro.cabezas@vet-alfort.fr

Caetano, Alexandre Rodrigues, Embrapa Recursos Geneticos e Biotecnologia, Parque Estacao Biologica, Final Av. W/5 Norte C.P. 02372, Brasilia, Distrito Federal CEP-70770-917, Brazil. Email: alexandre.caetano@embrapa.br

Caminade, Cyril, Department of Livestock and One Health, Institute of Infection, Veterinary and Ecological Sciences, University of Liverpool, Liverpool, UK. Email: cyril.caminade@liverpool.ac.uk

Černý, Jiří, Centre of Animal Infectious Diseases, Faculty of Tropical AgriSciences, Czech University of Life Sciences, Kamýcká 129, 165 00 Prague 6 – Suchdol, Czech Republic. Email: jiricerny@ftz.czu.cz

Charles, Roxanne Albertha, School of Veterinary Medicine, University of the West Indies, St. Augustine, Trinidad and Tobago. Email: roxanne.charles@sta.uwi.edu

Chavshin, Ali Reza, Department of Medical Entomology and Vector Control, School of Public Health, Urmia University of Medical Sciences, Urmia, Iran. Email: chavshin@gmail.com

Cornforth, Rosalind J., Walker Institute, University of Reading, Reading, UK. Email: r.j.cornforth@reading.ac.uk

Coughlan, Neil E., Department of Zoology, School of Biological, Earth and Environmental Sciences, University College Cork, Republic of Ireland. Email: neil.coughlan.zoology@gmail.com

Cox, Peter, College of Engineering, Mathematics and Physical Science, University of Exeter, Exeter, UK. Email: p.m.cox@exeter.ac.uk

Culler, Lauren E., Institute of Arctic Studies, Dartmouth College, Hanover, New Hampshire, USA. Email: lauren.e.culler@dartmouth.edu

Daniel, Milan, National Institute of Public Health, Prague, Czech Republic. Email: midaniel@seznam.cz

Danielová, Vlasta, National Institute of Public Health, Prague, Czech Republic. Email: vdanielova@seznam.cz

Dehhaghi, Mona, EMC Building, Macquarie University, Macquarie Park, NSW 2109, Australia; and PANDIS.org. Email: mona.aghdaee@mq.edu.au

Diuk-Wasser, Maria A., Ecology, Evolution and Environmental Biology, Columbia University, 1200 Amsterdam Avenue, New York, NY 10027, USA. Email: mad2256@columbia.edu

Duffy, James, P., Environment and Sustainability Institute, University of Exeter, Penryn, Cornwall, UK. Email: j.duffy@exeter.ac.uk

Duron, Olivier, MiVEGEC (Maladies Infectieuses et Vecteurs: Ecologie, Génétique, Evolution et Contrôle), Centre National de la Recherche Scientifique (CNRS), Institut pour la Recherche et le Développement (IRD), Université de Montpellier (UM), Montpellier, France; and CREES (Centre de Recherche en Écologie et Évolution de la Santé), Montpellier, France. Email: olivier.dupont@ird.fr

Eisen, Lars, Division of Vector-Borne Diseases, Centers for Disease Control and Prevention, Fort Collins, CO 80521, USA. Email: evp4@cdc.gov

Eisen, Rebecca J., Division of Vector-Borne Diseases, Centers for Disease Control and Prevention, Fort Collins, CO 80521, USA. Email: dyn2@cdc.gov

Elsterová, Jana, Institute of Parasitology, Biology Centre, Czech Academy of Sciences, České Budějovice, Czech Republic; and Veterinary Research Institute, Brno, Czech Republic. Email: elsterova@paru.cas.cz

Ergünay, Koray, Department of Medical Microbiology, Faculty of Medicine, Hacettepe University, Ankara, Turkey. Email: ekoray@hacettepe.edu.tr

Estrada-Peña, Agustín, Department of Animal Health, Faculty of Veterinary Medicine, University of Zaragoza, Spain; and Emerging Zoonoses Research Group, Instituto Agroalimentario de Aragón (IA2), Zaragoza, Spain. Email: aestrada@unizar.es

Fang, Li-Qun, Beijing Institute of Microbiology and Epidemiology, Beijing, 100071, People's Republic of China. Email: fanglq@bmi.ac.cn

Fernández-Ruiz, Natalia, Department of Animal Health, Faculty of Veterinary Medicine, University of Zaragoza, Zaragoza, Spain. Email: natalia.ferru@gmail.com

Fikrig, Erol, Yale School of Medicine, New Haven, CT 06510, USA. Email: erol.fikrig@yale.edu

Filatov, Serhii, National Scientific Centre, Institute of Experimental and Clinical Veterinary Medicine, Kharkiv – 61023, Ukraine. Email: filatovmidge@gmail.com

Fish, Durland, Department of Epidemiology of Microbial Disease, Yale School of Public Health, New Haven, Connecticut, USA. Email: durland.fish@yale.edu

Foley, Janet, University of California, 1320 Tupper Hall, Davis, CA 95616, USA. Email: jefoley@ucdavis.edu

de la Fuente, José de la, SaBio, Instituto de Investigación en Recursos Cinegéticos IREC-CSIC-UCLM-JCCM, Ronda de Toledo s/n, 13005 Ciudad Real, Spain; and Department of Veterinary Pathobiology, Center for Veterinary Health Sciences, Oklahoma State University, Stillwater, OK 74078, USA. Email: jose_delafuente@yahoo.com

Ganta, Roman R., Department of Diagnostic Medicine/Pathobiology, College of Veterinary Medicine, Kansas State University, Manhattan, KS 66506, USA. Email: rganta@vet.k-state.edu

Gargili Keles, Aysen, Health Sciences Faculty, Marmara University, Istanbul, Turkey. Email: agargili@yahoo.com

Gasser, Robin Beat, Department of Veterinary Biosciences, Melbourne Veterinary School, Faculty of Veterinary and Agricultural Sciences, The University of Melbourne, 250 Princes Highway, Werribee, VIC 3030, Australia. Email: robinbg@unimelb.edu.au

Ghafar, Abdul, Department of Veterinary Biosciences, Melbourne Veterinary School, Faculty of Veterinary and Agricultural Sciences, The University of Melbourne, 250 Princes Highway, Werribee, VIC 3030, Australia. Email: abdul.ghafar@unimelb.edu.au

Gilbert, Lucy, Institute of Biodiversity, Animal Health and Comparative Medicine, Graham Kerr Building, University Avenue, University of Glasgow, Glasgow G12 8QQ, UK. Email: lucy.gilbert@glasgow.ac.uk

Githaka, Naftaly, International Livestock Research Institute (ILRI), PO Box 30709-00100, Nairobi, Kenya. Email: n.githaka@cgiar.org

Golovchenko, Maryna, Biology Centre, Czech Academy of Sciences, Institute of Parasitology, Branisovska 31, 37005 České Budějovice, Czech Republic. Email: marina@paru.cas.cz

Gottlieb, Yuval, Koret School of Veterinary Medicine, The Robert H. Smith Faculty of Agriculture, Food and Environment, The Hebrew University of Jerusalem, Rehovot, Israel. Email: gottlieb.yuval@mail.huji.ac.il

Gould, Ernest, UMR IRD 190, Inserm 1207, Unité des Virus Émergents, University of Aix-Marseille, Marseille, France. Email: eagvirology@gmail.com

Gray, Jeremy, UCD School of Biology and Environmental Science, University College Dublin, Dublin, Republic of Ireland. Email: jeremy.gray@ucd.ie

Grubhoffer, Libor, Biology Centre, Czech Academy of Sciences, Institute of Parasitology, Branisovska 31, 37005 České Budějovice, Czech Republic; and University of South Bohemia, Branisovska 31, 37005 České Budějovice, Czech Republic. Email: libor.grubhoffer@bc.cas.cz

Guillemin, Gilles J., Neuroinflammation Group, Faculty of Medicine and Health Sciences, Macquarie University, Macquarie Park, NSW 2019, Australia; and PANDIS.org. Email: gilles.guillemin@mq.edu.au

Hansford, Kayleigh M., Medical Entomology Group, Public Health England, Porton Down, Salisbury, UK. Email: kayleigh.hansford@phe.gov.uk

Hart, Charles, Department of Microbiology and Immunology, SUNY Upstate Medical University, Syracuse, New York, USA. Email: hartch@upstate.edu

Healy, John E., Department of Zoology, School of Biological, Earth and Environmental Sciences, University College Cork, Republic of Ireland. Email: e.healy@ucc.ie

Hemming, Deborah, Met Office Hadley Centre, Met Office, Exeter, Devon, UK and Birmingham Institute of Forest Research, University of Birmingham, Birmingham, UK. Email: debbie.hemming@metoffice.gov.uk

Higgs, Stephen, Department of Diagnostic Medicine/Pathobiology, College of Veterinary Medicine and Biosecurity Research Institute, Kansas State University, Manhattan, KS 66506, USA. Email: shiggs@bri.ksu.edu

Hoodless, Andrew, Game & Wildlife Conservation Trust, Burgate Manor, Fordingbridge, UK. Email: ahoodless@gwct.org.uk

Huang, Yan-Jang S., Department of Diagnostic Medicine/Pathobiology, College of Veterinary Medicine and Biosecurity Research Institute, Kansas State University, Manhattan, KS 66506, USA. Email: scott0204@gmail.com

Hudson, Bernard, Department of Microbiology and Infectious Disease, Royal North Shore Hospital, Sydney, New South Wales, Australia; and PANDIS.org. Email: bernard.hudson@sydney.edu.au

Huntingford, Chris, UK Centre for Ecology & Hydrology, Crowmarsh Gifford, Wallingford, UK. Email: chg@ceh.ac.uk

Irwin, Peter, Centre for Biodiversity and One Health, Harry Butler Institute, Murdoch University, Perth, Australia. Email: p.irwin@murdoch.edu.au

Jabbar, Abdul, Department of Veterinary Biosciences, Melbourne Veterinary School, Faculty of Veterinary and Agricultural Sciences, The University of Melbourne, 250 Princes Highway, Werribee, VIC 3030, Australia. Email: jabbara@unimelb.edu.au

Johnson, Nicholas, Arbovirus Research Team, Virology Department, Animal and Plant Health Agency (Weybridge), Woodham Lane, Addlestone KT15 3NB, UK; and Faculty of Health and Medical Science, University of Surrey, Guildford, Surrey GU2 7XH, UK. Email: nick.johnson@apha.gov.uk

Kanduma, Esther, Department of Biochemistry, School of Medicine, University of Nairobi, PO Box 30197-00100, Nairobi, Kenya. Email: ekanduma@uonbi.ac.ke

Kar, Sırrı, Department of Biology, Tekirdağ Namık Kemal University, Tekirdağ, Turkey. Email: sirrikar@yahoo.com

Kaye, Neil, Met Office Hadley Centre, Met Office, Exeter, Devon, UK. Email: neil.kaye@metoffice.gov.uk

Kazimirova, Maria, Institute of Zoology, Slovak Academy of Sciences, Bratislava 84506, Slovakia. Email: maria.kazimirova@savba.sk

Kelly, Thomas C., Department of Zoology, School of Biological, Earth and Environmental Sciences, University College Cork, Republic of Ireland. Email: t.kelly@ucc.ie

Kjær, Lene Jung, Section for Animal Welfare and Disease Control, Department of Veterinary and Animal Sciences, Faculty of Health and Medical Sciences, University of Copenhagen, Copenhagen, Denmark. Email: lenju@sund.ku.dk

Kolo, Agatha Onyemowo, Next Generation Health Cluster, Council for Scientific and Industrial Research, Pretoria, South Africa. Email: afrianimalfriend@yahoo.co.uk

Korenberg, Eduard, Gamaleya National Research Centre for Epidemiology and Microbiology, Moscow, Russia. Email: edkornberg@yandex.ru

Król, Nina, Institute of Animal Hygiene and Veterinary Public Health, University of Leipzig, Leipzig, Germany. Email: nina.krol@vetmed.uni-leipzig.de

Kuo, Chi-Chien, Department of Life Science, National Taiwan Normal University, Taiwan. Email: cckuo@ntnu.edu.tw

Kurtti, Timothy J., Department of Entomology, University of Minnesota, St. Paul, MN 55108, USA. Email: kurtt001@umn.edu

Lamballerie, Xavier de, UMR IRD 190, Inserm 1207, Unité des Virus Émergents, University of Aix-Marseille, Marseille, France. Email: secretariat.u1207@inserm.fr

Leighton, Patrick A., Research Group on Epidemiology of Zoonoses and Public Health (GREZOSP), Faculty of Veterinary Medicine, Université de Montréal, 3200 rue Sicotte, Saint-Hyacinthe, QC, J2S 2M2, Canada. Email: patrick.a.leighton@umontreal.ca

Li, Hao, Beijing Institute of Microbiology and Epidemiology, Beijing 100071, People's Republic of China. Email: liuwei@bmi.ac.cn

Lindsay, L. Robbin, National Microbiology Laboratory, 1015 Arlington Street, Winnipeg, MB, R3E 3M4, Canada; Email: robbin.lindsay@canada.ca

Liu, Wei, Beijing Institute of Microbiology and Epidemiology, Beijing 100071, People's Republic of China. Email: liuwei@bmi.ac.cn

Lynn, Geoffrey E., Yale School of Medicine, New Haven, CT 06510, USA. Email: geoffrey.lynn@yale.edu

Maclean, Ilya, Environment and Sustainability Institute, University of Exeter, Penryn, Cornwall, UK. Email: i.m.d.maclean@exeter.ac.uk

Mans, Ben J., Agricultural Research Council, Onderstepoort Veterinary Research, Private Bag X05, Onderstepoort 0110, Pretoria, South Africa; Department of Veterinary Tropical Diseases, Vector and Vector-Borne Disease Research Programme, University of Pretoria, Private Bag X04, Onderstepoort 0110, Pretoria, South Africa; and Department of Life and Consumer Sciences, University of South Africa, Florida 1709, South Africa. Email: manssb@arc.agric.za

Martins de Camargo, Maristela, Instituto de Ciências Biomédicas, Universidade de São Paulo, Av. Prof. Lineu Prestes 1730, sala 138, São Paulo, São Paulo CEP-05508-900, Brazil. Email: mmcamar@usp.br

McCoy, Karen D., Centre de Recherches en Écologie et en Évolution de la Santé (CREES), MiVEGEC UMR CNRS-IRD-University of Montpellier, Domaine La Valette – 900, rue Jean François BRETON, 34090 Montpellier, France. Email: karen.mccoy@ird.fr

Medlock, Jolyon M., Medical Entomology Group, Public Health England, Porton Down, Salisbury, UK. Email: jolyon.medlock@phe.gov.uk

Munderloh, Ulrike G., Department of Entomology, University of Minnesota, St. Paul, MN 55108, USA. Email: munde001@umn.edu

Mysterud, Atle, Centre for Ecological and Evolutionary Synthesis (CEES), Department of Biosciences, University of Oslo, PO Box 1066 Blindern, NO-0316 Oslo, Norway. Email: atle.mysterud@ibv.uio.no

Narasimhan, Sukanya, Yale School of Medicine, New Haven, CT 06510, USA. Email: sukuyana.narsimhan@yale.edu

Nuttall, Patricia A., University of Oxford, Department of Biology, Mansfield Road, Oxford OX1 3SZ, UK. Email: pat.nuttall@zoo.ox.ac.uk

Obiegala, Anna, Institute of Animal Hygiene and Veterinary Public Health, University of Leipzig, Leipzig, Germany. Email: anna.obiegala@vetmed.uni-leipzig.de

Obregón Alvarez, Dasiel, School of Environmental Sciences, University of Guelph, Guelph, ON, N1G 2W1, Canada. Email: dasieloa@uoguelph.ca

Ogden, Nicholas H., National Microbiology Laboratory at St. Hyacinthe, and Research Group on Epidemiology of Zoonoses and Public Health (GREZOSP), Faculty of Veterinary Medicine, Université de Montréal. Email: nicholas.ogden@canada.ca

Ogihara, Mari H., Institute of Livestock and Grassland Science, National Agriculture and Food Research Organization, Tsukuba, Japan. Email: ogiharam590@affrc.go.jp

Oliveira, Stefan Vilges de, Department of Collective Health, Faculty of Medicine, Federal University of Uberlândia, Uberlândia, Minas Gerais, Brazil. Email: stefanbio@yahoo.com.br

Oskam, Charlotte, Centre for Biodiversity and One Health, Harry Butler Institute, Murdoch University, Perth, Australia. Email: c.oskam@murdoch.edu.au

Otranto, Domenico, Università degli Studi di Bari, Bari, Italy; and Bu-Ali Sina University, Hamedan, Iran. Email: domenico.otranto@uniba.it

Oyen, Kennan J., Department of Biological Sciences, University of Cincinnati, OH 45221-0006, USA. Email: oyenkj@ucmail.uc.edu

Panahi, Hamed Kazemi Shariat, Neuroinflammation Group, Faculty of Medicine and Health Sciences, Macquarie University, Macquarie Park, NSW 2109, Australia. Email: hamed.kazemi-shariat-panahi@students.mq.edu.au

Pandey, Neha, Molecular Biology Unit, Faculty of Medicine, Institute of Medical Sciences, Banaras Hindu University, Varanasi – 221005, India. Email: neha.mhg.bhu15@gmail.com

Peterson, A. Townsend, Biodiversity Institute, University of Kansas, Lawrence, Kansas, USA. Email: town@ku.edu

Pettersson, John H.-O., Marie Bashir Institute for Infectious Diseases and Biosecurity, School of Life and Environmental Sciences and School of Medical Sciences, The University of Sydney, Sydney, NSW 2006, Australia; and Zoonosis Science Centre, Department of Medical Biochemistry and Microbiology, Uppsala University, Uppsala, Sweden. Email: john.pettersson@gmail.com

Pfeffer, Martin, Institute of Animal Hygiene and Veterinary Public Health, University of Leipzig, Leipzig, Germany. Email: pfeffer@vetmed.uni-leipzig.de

Phipps, L. Paul, Arbovirus Research Team, Virology Department, Animal and Plant Health Agency (Weybridge), Woodham Lane, Addlestone KT15 3NB, UK. Email: paul.phipps@apha.gov.uk

Plumpton, Heather J., Walker Institute, University of Reading, Reading, UK. Email: h.j.plumpton@reading.ac.uk

Pustahija, Tatjana, Institute of Public Health of Vojvodina, Futoska 121, Novi Sad 21000, Republic of Serbia. Email: tatjana.pustahija@izjzv.org.rs

Raghavan, Ram, Veterinary Pathobiology and Public Health, College of Veterinary Medicine and School of Public Health, University of Missouri, Columbia, MO 65211, USA. Email: raghavanrk@missouri.edu

Rego, Ryan O.M., Institute of Parasitology, Biology Centre, Czech Academy of Sciences, České Budějovice 37005, Czech Republic; and Faculty of Science, University of South Bohemia, České Budějovice 37005, Czech Republic. Email: ryanrego@paru.cas.cz

Rizzoli, Annapaola, Department of Biodiversity and Molecular Ecology, Research and Innovation Centre, Fondazione Edmund Mach, San Michele all'Adige (TN), Italy. Email: annapaola.rizzoli@fmach.it

Ronai, Isobel, School of Life and Environmental Sciences, The University of Sydney, Sydney, Australia. Email: isobel.ronai@sydney.edu.au

Rubel, Franz, University of Veterinary Medicine, Vienna, Austria. Email: franz.rubel@vetmeduni.ac.at

Rudenko, Natalie, Biology Centre, Czech Academy of Sciences, Institute of Parasitology, Branisovska 31, 37005 České Budějovice, Czech Republic. Email: natasharudenko@hotmail.com

Ruiwen, Benjamin, Neuroinflammation Group, Faculty of Medicine and Health Sciences, Macquarie University, Macquarie Park, NSW 2109, Australia. Email: benjamin.ruiwen@students.mq.edu.au

Sage, Rufus, Game & Wildlife Conservation Trust, Burgate Manor, Fordingbridge, UK. Email: rsage@gwct.org.uk

Samy, Abdallah M., Entomology Department, Faculty of Science, Ain Shams University, Abbassia, Cairo 11566, Egypt. Email: samy@sci.asu.edu.eg

Sanches, Gustavo Seron, School of Life Sciences, Pontifical Catholic University of Paraná, Rua Imaculada Conceição 1155, Bairro Prado Velho, 80215-901 Curitiba-PR, Brazil. Email: gustavo.seron@pucpr.br

Santos, Isabel Kinney Ferreira, Department of Biochemistry and Immunology, Ribeirão Preto School of Medicine, Universidade de São Paulo, Prédio Central da Medicina, Cidade Universitária, Avenida Bandeirantes 3900, Ribeirão Preto, São Paulo, CEP-14040-900, Brazil. Email: imsantos@fmrp.usp.br

Sato, Marcello Otake, Department of Tropical Medicine and Parasitology, Dokkyo Medical University, 880 Kitakobayashi, Mibu-machi, Shimotsuga-gun, Tochigi, 321-0293 Japan. Email: marcello@dokkyomed.ac.jp

Sato, Megumi, Graduate School of Health Sciences, Niigata University, 2-746 Asahimachi-dori, Chuo-ku, Niigata City, 951-8518 Japan. Email: satomeg@clg.niigata-u.ac.jp

Schloeffel, Richard, Neuroinflammation Group, Faculty of Medicine and Health Sciences, Macquarie University, Macquarie Park, NSW 2109, Australia; and Grove Health Centre, Gordon, New South Wales, Australia. Email: richard.schloeffel@mq.edu.au

Seyyed-Zadeh, Seyyed Javad, Department of Medical Entomology and Vector Control, School of Public Health, Urmia University of Medical Sciences, Urmia, Iran. Email: sjs_68@yahoo.com

Šimo, Ladislav, UMR BIPAR, Laboratoire de Santé Animale, ANSES, INRAE, École Nationale Vétérinaire d'Alfort, UMR BIPAR, Laboratoire de Santé Animale, Maisons-Alfort, F-94700, France. Email: ladislav.simo@vet-alfort.fr

Singh, Sunit Kumar, Molecular Biology Unit, Faculty of Medicine, Institute of Medical Sciences, Banaras Hindu University, Varanasi – 221005, India. Email: sunitsingh2000@gmail.com

Sonenshine, Daniel E., VMBU, Laboratory of Malaria and Vector Research, NIAID, NIH, Rockville, MD 20852, USA. Email: daniel.sonenshine@nih.gov

Sparey, Morgan, College of Engineering, Mathematics and Physical Science, University of Exeter, Exeter, UK. Email: ms959@exeter.ac.uk

Stachurski, Frederic, CIRAD, UMR ASTRE, F-34398 Montpellier, France; and ASTRE, Université de Montpellier, CIRAD, INRAE, Montpellier, France. Email: frederic.stachurski@cirad.fr

Stuen, Snorre, Norwegian University of Life Sciences, Section for Small Ruminant Research, N-4325 Sandnes, Norway. Email: snorre.stuen@nmbu.no

Szabó, Matias Pablo Juan, Hospital Veterinário, Universidade Federal de Uberlândia, Av. Mato Grosso, n° 3289 – Bloco 2 S – Campus Umuarama, Uberlândia, Minas Gerais CEP-38405-314, Brazil. Email: szabo@ufu.br

Taylor, DeMar, Laboratory of Disease Vector Ecology, Faculty of Life and Environmental Sciences, University of Tsukuba, Tsukuba, Japan. Email: taylor.de.mar.ge@u.tsukuba.ac.jp

Teglas, Mike, Department of Agriculture, Veterinary and Rangeland Sciences, University of Nevada, Reno, Nevada, USA. Email: mteglas@unr.edu

Telford III, Sam R., Dept of Infectious Disease and Global Health, Tufts University, 200 Westboro Road, North Grafton, MA 01536, USA. Email: sam.telford@tufts.edu

Thangamani, Saravanan, Department of Microbiology and Immunology, SUNY Upstate Medical University, Syracuse, New York, USA. Email: thangams@upstate.edu

Titcomb, Georgia, Marine Science Institute, University of California, Santa Barbara, California, USA. Email: georgiatitcomb@gmail.com

Trájer, Attila J., Sustainability Solutions Research Lab., University of Pannonia, H-8200, Veszprém, Egyetem utca 10, Hungary. Email: trajer.attila@mk.uni-pannon.hu

Turell, Michael, VectorID LLC, Frederick, Maryland, USA. Email: mturell@erols.com

Umemiya-Shirafuji, Rika, National Research Center for Protozoan Diseases, Obihiro University of Agriculture and Veterinary Medicine, Obihiro, Japan. Email: umemiya@obihiro.ac.jp

Vanlandingham, Dana L., Department of Diagnostic Medicine/Pathobiology, College of Veterinary Medicine and Biosecurity Research Institute, Kansas State University, Manhattan, KS 66506, USA. Email: divanlan@vet.ksu.edu

Vial, Laurence, CIRAD, UMR ASTRE, F-34398 Montpellier, France; and ASTRE, Université de Montpellier, CIRAD, INRAE, Montpellier, France. Email: laurence.vial@cirad.fr

Villar, Margarita, Instituto de Investigación en Recursos Cinegéticos, IREC-CSIC-UCLM-JCCM, Ronda de Toledo s/n, 13005 Ciudad Real, Spain; and Biochemistry Section, Faculty of Science and Chemical Technologies, and Regional Centre for Biomedical Research (CRIB), University of Castilla-La Mancha, 13071 Ciudad Real, Spain. Email: margaritam.villar@uclm.es

Wall, Richard, School of Biological Sciences, Life Sciences Building, University of Bristol, Bristol, UK. Email: richard.wall@bristol.ac.uk

Wilke, André B.B., Department of Public Health, Miller School of Medicine, University of Miami, Miami, Florida, USA. Email: axb1737@miami.edu

Wint, G.R. William, Environmental Research Group Oxford Ltd, c/o Dept Zoology, University of Oxford, 11a Mansfield Road, Oxford OX1 3SZ, UK. Email: william.wint@zoo.ox.ac.uk

Wu, Jianhong, Laboratory for Industrial and Applied Mathematics, York University, Toronto, Ontario, Canada. Email: wujh@yorku.ca

Zając, Zbigniew, Department of Biology and Parasitology, Medical University of Lublin, Radziwiłłowska 11 st., 20-080 Lublin, Poland. Email: zbigniew.zajac@umlub.pl

Zhang, Xue, College of Science, Northeastern University, Shenyang, People's Republic of China. Email: zhangxue@mail.neu.edu.cn

Preface

Carbon dioxide levels in the atmosphere are now 50% higher than in pre-industrial times. This unnaturally high level has not been detected in air bubbles trapped in Antarctic ice during at least the past 800,000 years. According to the Intergovernmental Panel on Climate Change, the average global land surface temperature has increased by 1.5°C since pre-industrial times. The extra heat is driving regional and seasonal temperature extremes, reducing snow cover and sea ice, intensifying heavy rainfall, and changing habitat ranges for plants and animals, expanding some and shrinking others.

Against this background, *Climate, Ticks and Disease* explores the evidence for whether or not global warming is affecting ticks and the many diseases caused by diverse tick-borne pathogens. How best to do this than by asking the relevant experts for their views? In fact, the experts were only too keen to respond even though many had not previously considered how climate change might affect their particular area of expertise; some experts in the climate change field had never previously thought about ticks. The result is a collection of 77 expert opinions arranged in three sections. Section 1. Climate: this sets the scene, including predictions of future climate change, how climate change affects ecosystems, and how to model projections of the spatial distribution of ticks and tick-borne infections under different climate change scenarios. Section 2. Ticks: in this section the focus is on ticks (although tick-borne pathogens creep in) and whether or not changes in climate affect the tick biosphere, from physiology to ecology. Section 3. Disease: here the focus is on the tick–host–pathogen biosphere, ranging from the triangle of tick–host–pathogen molecular interactions to disease ecology in various regions and ecosystems of the world. Each of these three sections ends with a synopsis that aims to give a brief overview of all the expert opinions within the section. The book concludes with Section 4: Final Synopsis and Future Predictions. This synopsis attempts to summarize evidence provided by the experts of tangible impacts of climate change on ticks and tick-borne infections. In constructing their expert opinions, contributors were asked to give their views on what the future might hold. The final synopsis provides a snapshot of their expert thoughts on the future.

I should like to thank Dr Maria Kazimirova (Institute of Zoology, Bratislava, Slovakia) and Professor Rick Ostfeld (Cary Institute of Ecosystem Studies, USA) for their very helpful critiques of the synopses, and all of the contributors who have shared their expert opinions.

Pat Nuttall
March 2021

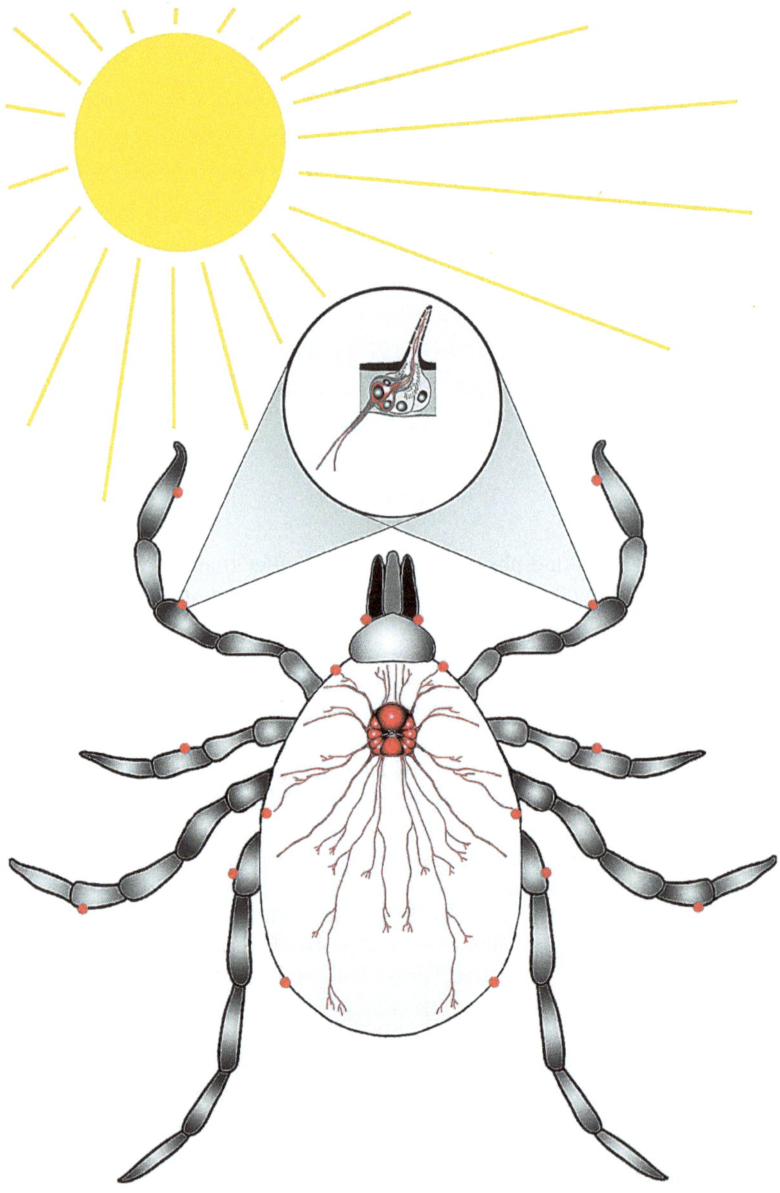

(Image by Ladislav Šimo, French National Research Institute for Agriculture, Food and Environment, Maisons-Alfort, France.)

eo1 Future Climate of Africa

Rosalind Cornforth* **and Heather Plumpton**
University of Reading, Reading, UK

1.1 Expert Opinion

1.1.1 Introduction

Globally, climate change is causing average temperatures to rise and the intensity and frequency of extreme events to increase. However, these changes will not be uniform globally (Sillmann *et al.*, 2013) and will have differential impacts based on population vulnerability. Across Africa, changes to seasonal rainfall and temperature extremes will have significant impacts on society, particularly those strongly dependent on seasonal rains for agriculture. Moreover, these impacts will vary across the region due to a range of local factors. Support for agricultural development programmes, e.g. research into how to meet the challenges of food production in a changing climate (Fig. 1.1), is key in integrating the impacts of climate change into long-term planning. In parallel, support is needed for decision makers to help them work under conditions of uncertainty and help develop ways for local stakeholders to access and assess information as more data and knowledge become available. The contribution of this expert opinion is to provide a synthesis of the current and future climate risks for Africa and the range of impacts in the medium to long term. It also seeks to highlight areas where research and resources might need to be focused, in the future, to limit or avoid negative climate risks to socio-economic development.

1.1.2 What do we understand about the future climate of Africa?

Climate change is projected to continue the current warming trend in Africa, with annual mean temperatures likely to increase beyond 2°C above pre-industrial temperatures by the mid-21st century under a high greenhouse gas emissions scenario (Niang *et al.*, 2014). How the climate will change in different regions over the coming decades, however, remains uncertain.

There is still significant uncertainty in model projections of climate change for Africa, particularly for changes to the rainfall regime. This uncertainty can be a challenge for decision makers designing policies and plans for the future. Some types of uncertainty can never be eliminated fully, and it is important to work with decision makers to support them to work under conditions of uncertainty by considering a range of possible futures (Dessai *et al.*, 2018; Shepherd *et al.*, 2018). However, recent work, particularly through the Future Climate for

*Email: r.j.cornforth@reading.ac.uk

DOI: 10.1079/9781789249637.0001

Fig. 1.1. Harvesting sesame seed in West Africa for the family lunch. (Authors' own photograph.)

Africa (FCFA) (https://futureclimateafrica.org, accessed 26 May 2021) programme, has improved our understanding of African climate and narrowed the range of uncertainty in climate model projections for Africa.

Across Africa, annual and seasonal mean temperatures have increased over the last century, with minimum temperatures increasing more rapidly than maximum temperatures (Niang *et al.*, 2014). Model projections across Africa show mean annual temperatures are expected to continue to rise, particularly in more arid regions, with the rate of increase in minimum temperatures continuing to exceed that of maximum temperatures (Niang *et al.*, 2014). However, this temperature increase is not uniform across the region and depends on factors such as proximity to the ocean. For East Africa, which has the largest proportion of Africa's livestock population, average temperatures have been estimated to increase by 2–3°C by mid-century and 2–5°C by end-of-century (Otieno and Anyah, 2013; Ongoma *et al.*, 2018a). This will be detrimental to the livestock comfort temperature. Under a medium emissions scenario, coastal regions are likely to see the largest increases in the number of warm nights by the end of the century (Ongoma *et al.*, 2018b). Across West and Central Africa, temperature rises are more intense in the driest regions of the Sahel, with maximum temperatures increasing substantially (Diedhiou *et al.*, 2018). The duration of warm spells and the number of heatwaves are also expected to increase, along with the number of warm days, while there is likely to be a reduction in the number of cold extreme events in Africa (Seneviratne *et al.*, 2012).

Projections of future changes in rainfall over Africa are uncertain, with the climate models assessed for the fifth assessment report of the Intergovernmental Panel on Climate Change (IPCC) varying in both whether rainfall will increase or decrease, and by how much (Niang *et al.*, 2014). These projections also suggest there may be increases in heavy rainfall events, but with low confidence (Seneviratne *et al.*, 2012; Niang *et al.*, 2014). More recent work has narrowed the range of uncertainty in model projections and improved our understanding of African climate in three key areas of the precipitation regime: timing of rainy seasons; magnitude of changes to rainfall amount; and changes to the frequency and intensity of extreme rainfall events.

The timing of the rainy seasons in Africa has significant consequences for the large proportion of the population who are dependent on rain-fed agriculture and livestock for their livelihoods (Dunning *et al.*, 2016, 2017). Changes to the timing and intensity of the rainy seasons will impact the length of the growing season and agricultural yields (Vizy and Cook, 2017) and pasture, hence directly affecting the resilience of the population. A continent-scale analysis of changes in wet-season characteristics for Africa under medium and high emission scenarios projects significant regional variation across the continent. In West Africa and the Sahel, the wet season is projected to be delayed by 5–10 days, starting later and finishing later, with no overall change in length of the wet season. In contrast, the later start to the wet season in southern Africa is projected to lead to a shorter wet season with reduced total wet-season rainfall. To the north of the equator, wet-season rainfall is projected to increase with little change in the timing of the rains. In the central equatorial region and the Horn of Africa, where there are two wet seasons a year, the short rainy season is projected to be extended by 8 days on average with higher rainfall amount (Dunning *et al.*, 2018). Factors not considered in these climate models may also impact the seasonal precipitation regime, such as aerosol emissions. Modelling suggests reduced aerosol emissions may delay the timing of the short rainy season in East Africa. However, comparisons across models show significant uncertainties and the need for aerosol emission scenarios to be incorporated into climate models more widely (Scannell *et al.*, 2019). Changes to the timing of the rainy season, which is relied upon heavily for providing water for both crops and livestock (Vetter, 2009; Archer van Garderen, 2011), will also impact on the interaction of vegetation with pests, predators and diseases in the wider ecosystem (Adhikari *et al.*, 2015).

There remain significant uncertainties in whether the total amount of annual and seasonal rainfall will increase or decrease across Africa. However, rainfall intensity is projected to increase across West Africa (see Taylor *et al.*, 2017), southern Africa and Central Africa, with the amount of rainfall per rainy day projected to increase in the wet season. The number of rainy days in the wet season is projected to decrease for West and southern Africa (New *et al.*, 2006), leading to an overall slight decline in total wet-season rainfall. By contrast, Central Africa is projected to experience an increase in total seasonal rainfall due to increasing rain per rainy day and no change in number of rainy days (Dunning *et al.*, 2018).

On top of changes to the seasonal and annual rainfall amounts, the frequency and intensity of wet and dry extremes are critical elements of the precipitation regime with significant impacts on livelihoods and population vulnerability. Model projections of changes to wet and dry extremes have been improved by recent application of convection-permitting models (CPMs) to the African continent, which better simulate the diurnal cycle of tropical convection, vertical cloud structure and coupling between soil and atmosphere in the Sahel (Stratton *et al.*, 2018). The CPM agrees with coarser-resolution models that wet-season rainfall is likely to be less frequent but more intense over much of Africa, even where overall rainfall amount declines. However, the CPM shows larger decreases in rainfall occurrence, and increases in rainfall intensity and extremes, compared with conventional climate models. Flood-inducing events, defined as exceeding 60 mm accumulated rain in 3 h over a 25-km grid scale, are projected to increase from occurring once every 30 years at present to once every 3–4 years in the future. Additionally, dry spells are projected to lengthen significantly during the wet season over the Sahel, Gulf of Guinea and Central Africa. In the Gulf of Guinea, dry spells longer than 10 days are expected to double in frequency in the future (Kendon *et al.*, 2019; Wainwright *et al.*, 2020). These changes in extreme rainfall are larger and more widespread than previously predicted, with implications for disaster risk reduction, food security and infrastructure planning across Africa.

The impact of all these climatic changes across Africa will affect multiple sectors, including food security, vulnerability of the population to disasters, water availability for crops and livestock, public health and infrastructure.

1.1.3 What are the likely impacts on agriculture and people's livelihoods?

These ongoing changes in the climate will have impacts on a range of sectors in the region. Existing stress on water availability is likely to be amplified by climate change (Niang *et al.*, 2014). Estimating the influence of climate change on water resources is, however, limited by uncertainties in future rainfall changes, as well as the influences of non-climatic drivers such as population growth.

Livestock are likely to be affected by projected decreases in rainfall, which may lead to more droughts, impacting drinking-water (Fig. 1.2) and crop production for fodder. Changing weather patterns could expand the distribution of ticks causing animal disease (Niang *et al.*, 2014). Increases in the variability of rainfall, including longer dry spells and more intense rainfall events, are likely to lead to more frequent and intense flood events. Floods may increase livestock mortality due to damage to livestock housing and the potential for increased transmission of water-borne diseases in floodwaters (FAO, IFAD, UNICEF, WFP and WHO, 2018; InterAcademy Partnership, 2018). Furthermore, increases in temperature will lead to increases in heat stress for livestock, leading to reduced welfare, health and productivity (Escarcha *et al.*, 2018). Climate change is expected to affect the incidence and geographical range of diseases, including tick-borne diseases such as theileriosis, through warming temperatures and changes in rainfall

Fig. 1.2. Collecting water from a community borehole in Burkina Faso. (Authors' own photograph.)

(Niang *et al.*, 2014). Climate change is very likely to negatively affect access to safe water, sanitation, food security, access to healthcare and education, and increase the burden of malnutrition (Niang *et al.*, 2014).

1.1.4 Addressing the science gaps to support agricultural decision making

As outlined above, there remain some specific areas where additional climate science is needed in which the direction of forced change in many impact-relevant variables is uncertain. For example, there is still significant uncertainty over the direction and magnitude of precipitation change in East (Wainwright *et al.*, 2019) and West Africa. Further scientific gaps are documented in the latest synthesis report from the IPCC, which highlights outstanding knowledge gaps in both the underlying science on climate change impacts (Hoegh-Guldberg *et al.*, 2018) and our understanding of the pathways to action (de Coninck *et al.*, 2018). More generally, there is a need to characterize risks and opportunities in terms relevant to local stakeholders, to support policy-relevant risk assessment. This includes developing ways of coming up with relevant and plausible climate scenarios and assessment frameworks.

1.2 Future Predictions

Although global aspects of climate change are well understood, there remains great uncertainty at regional and national scales, and traditional probabilistic approaches lead to a 'cascade of uncertainty' (Wilby and Dessai, 2010). As part of the HyCRISTAL project (2015–2021) (https://futureclimateafrica.org/project/hycristal/, accessed 26 May 2021), this uncertainty of the future climate was captured in a set of stories that describe three possible futures for rural East Africa in 2050 (Burgin *et al.*, 2019). These futures do not cover every possible outcome projected but seek to describe some of the resulting impacts that could be experienced in rural areas. Under Future 1, the climate is much wetter and hotter, rainfall intensity and frequency of storms increase, and the long rainy season increases in length. This longer rainy season prevents some staple crops from maturing properly but increases opportunities to grow perennial crops and a wider variety of fruits and vegetables, including crops used for livestock fodder. Under Future 2, the climate is hotter and there is a smaller increase in rainfall intensity and storms, leading to similar problems and opportunities for crop production as in Future 1. Under these slightly wetter conditions, livestock production can increase. Under Future 3, the climate is much hotter and drier with temperatures rising by 3°C on average. Rainfall decreases overall, with shorter rainy seasons. This hotter and drier climate leads to a reduction in the range of perennial crops, including livestock fodder, and other fruits and vegetables that can be grown, and farming communities suffer lack of income and poor nutrition (Burgin *et al.*, 2019).

A similar approach has been applied to model and describe three possible futures for water resources under climate change in West Africa, specifically Burkina Faso and Northern Ghana, in 2050 (Cornforth *et al.*, 2019), and the impacts these futures would have on crop and livestock production. Under Future 1, the climate is much hotter and drier with more erratic rainfall. The lower rainfall leads to low levels of surface water and groundwater, and declines in staple, cash and fodder crops. Livestock mortality increases due to lack of water and fodder, as well as increased water-borne diseases due to people and livestock being concentrated around fewer resources. Under Future 2, groundwater levels are similar to current day, but there is greater variability in the climate with long dry spells and more intense rainfall events. Rain-fed agriculture declines due to risks from the unpredictable nature of the climate. Long dry spells lead to declines in livestock production as under Future 1. Under Future 3, the climate is much wetter, with increases in intensity and frequency of rainfall events and increases in groundwater levels. Rain-fed and irrigated crop and fodder production can increase; however, flooding and waterlogging also increase, although this is spatially variable. Livestock production is supported by availability of water and fodder but suffers increased losses due to damage from flooding (Cornforth *et al.*, 2019).

These alternative possible futures for climate change in East and West Africa illustrate the range of uncertainty from climate projections over the next few decades. This explicit communication of uncertainty in climate projections to decision makers enables more effective adaptation planning. The three possible climate futures offer the potential for planners and policy makers to consider which actions would be beneficial across multiple possible futures and when such actions need to be applied.

However, the impacts of future climate change will vary across a region due to a range of local factors. For example, flooding and waterlogging of soils will be spatially variable and impacts will depend on the types of crops grown, as well as the infrastructure and management options for protection of livestock. Therefore, local-scale studies which incorporate high-resolution (differentiated) information on climate, mapped on to high-resolution (differentiated) information on vulnerability and exposure, will be needed to better assess impacts and consequences at a scale relevant to local decision makers (Young *et al.*, 2021). This approach has been adopted in Uganda to assess the impacts of climate change on sweet potato, an understudied staple crop in East and Central Africa (Ewell, 2002; Young *et al.*, 2020a,b). This approach of working at the local scale, and integrating information from climate, agricultural and social science, produces contextualized assessments of climate risk that are meaningful to local decision makers. Such a systems-based view of risks to livestock, agriculture and wider livelihood resilience enables adaptation planning that acknowledges the interdependence of sectors such as water, agriculture, social protection and health, for a multi-sector and multi-scale approach to climate adaptation (Cornforth *et al.*, 2021).

References

Adhikari, U., Nejadhashemi, A.P. and Woznicki, S.A. (2015) Climate change and eastern Africa: a review of impact on major crops. *Food and Energy Security* 4, 110–132.

Archer van Garderen, E.R.M. (2011) Considering cattle farming in southern Africa under a changing climate. *Weather, Climate and Society* 3, 249–253.

Burgin, L., Walker, G., Cornforth, R., Rowell, D., Marsham, J., *et al.* (2019) *FCFA HyCRISTAL Climate Narrative Rural Infographic and Brief*. Walker Institute, Reading, UK. https://doi.org/10.5281/zenodo.3257288

Cornforth, R.J., Macdonald, D.M., Osbahr, H., Ciampi L., Myers J., *et al.* (2019) *Possible Futures for Groundwater in Burkina Faso under a Changing Climate (Version 1.0)*. Walker Institute, Reading, UK. https://doi.org/10.5281/zenodo.3533108

Cornforth, R.J., Petty, E.C. and Walker, G. (2021) Supporting climate-resilient planning at national and district level: a pathway to multi-stakeholder decision-making in Uganda. In: Conway, D. and Vincent, K. (eds) *Climate Risk in Africa: Adaptation and Resilience*. Palgrave Macmillan, Cham, Switzerland, pp. 131–145.

de Coninck, H., Revi, A., Babiker, M., Bertoldi, P., Buckeridge, M., *et al.* (2018) Strengthening and implementing the global response. In: Masson-Delmotte, V., Zhai, P., Pörtner, H.-O., Roberts, D., Skea, J., *et al.* (eds) *Global Warming of 1.5°C. An IPCC Special Report on the impacts of global warming of 1.5°C above pre-industrial levels and related global greenhouse gas emission pathways, in the context of strengthening the global response to the threat of climate change, sustainable development, and efforts to eradicate poverty*. World Meteorological Organization, Geneva, Switzerland, pp. 313–444.

Dessai, S., Bhave, A., Birch, C., Conway, D., Garcia-Carreras, L., *et al.* (2018) Building narratives to characterise uncertainty in regional climate change through expert elicitation. *Environmental Research Letters* 13, 74005.

Diedhiou, A., Bichet, A., Wartenburger, R., Seneviratne, S.I., Rowell, D.P., *et al.* (2018) Changes in climate extremes over West and Central Africa at 1.5°C and 2°C global warming. *Environmental Research Letters* 13, 65020.

Dunning, C.M., Black, E.C. and Allan, R.P. (2016) The onset and cessation of seasonal rainfall over Africa. *Journal of Geophysical Research: Atmospheres* 121, 11405–11424.

Dunning, C.M., Allan, R.P. and Black, E. (2017) Identification of deficiencies in seasonal rainfall simulated by CMIP5 climate models. *Environmental Research Letters* 12, 114001.

Dunning, C.M., Black, E. and Allan, R.P. (2018) Later wet seasons with more intense rainfall over Africa under future climate change. *Journal of Climate* 31, 9719–9738.

Escarcha, J.F., Lassa, J.A. and Zander, K.K. (2018) Livestock under climate change: a systematic review of impacts and adaptation. *Climate* 6, 54.

Ewell, P.T. (2002) *Sweetpotato Production in Sub-Saharan Africa: Patterns and Key Issues*. International Potato Center (CIP), Nairobi. Available at: https://www.sweetpotatoknowledge.org/files/sweetpotato-production-patterns-key-issues/ (accessed 26 May 2021).

FAO, IFAD, UNICEF, WFP and WHO (2018) *The State of Food Security and Nutrition in the World 2018. Building Climate Resilience for Food Security and Nutrition*. Food and Agriculture Organization of the United Nations, Rome.

Hoegh-Guldberg, O., Jacob, D., Taylor, M., Bindi, M., Brown, S., *et al.* (2018) Impacts of 1.5°C global warming on natural and human systems. In: Masson-Delmotte, V., Zhai, P., Pörtner, H.-O., Roberts, D., Skea, J., *et al.* (eds) *Global Warming of 1.5°C. An IPCC Special Report on the impacts of global warming of 1.5°C above pre-industrial levels and related global greenhouse gas emission pathways, in the context of strengthening the global response to the threat of climate change, sustainable development, and efforts to eradicate poverty*. World Meteorological Organization, Geneva, Switzerland, pp. 175–312.

InterAcademy Partnership (2018) *Opportunities for Future Research and Innovation on Food and Nutrition Security and Agriculture: The InterAcademy Partnership's Global Perspective*. InterAcademy Partnership, Trieste, Italy and Washington, DC. Available at: https://www.interacademies.org/sites/default/files/publication/iap_fnsa_global_web_complete_28nov.pdf (accessed 26 May 2021).

Kendon, E.J., Stratton, R.A., Tucker, S., Marsham, J.H., Berthou, S., *et al.* (2019) Enhanced future changes in wet and dry extremes over Africa at convection-permitting scale. *Nature Communications* 10, 1794.

New, M., Hewitson, B., Stephenson, D.B., Tsiga, A., Kruger, A., *et al.* (2006) Evidence of trends in daily climate extremes over southern and west Africa. *Journal of Geophysical Research D: Atmospheres* 111, D14102.

Niang, I., Ruppel, O.C., Abdrabo, M.A., Essel, A., Lennard, C., *et al.* (2014) Africa. In: Barros, V.R., Field, C.B., Dokken, D.J., Mastrandrea, M.D., Mach, K.J., *et al.* (eds) *Climate Change 2014: Impacts, Adaptation, and Vulnerability. Part B: Regional Aspects. Contribution of Working Group II to the Fifth Assessment Report of the Intergovernmental Panel on Climate Change*. Cambridge University Press, Cambridge and New York, pp. 1199–1265.

Ongoma, V., Chen, H. and Gao, C. (2018a) Projected changes in mean rainfall and temperature over East Africa based on CMIP5 models. *International Journal of Climatology* 38, 1375–1392.

Ongoma, V., Chen, H., Gao, C., Nyongesa, A.M. and Polong, F. (2018b) Future changes in climate extremes over Equatorial East Africa based on CMIP5 multimodel ensemble. *Natural Hazards* 90, 901–920.

Otieno, V.O. and Anyah, R.O. (2013) CMIP5 simulated climate conditions of the Greater Horn of Africa (GHA). Part II: Projected climate. *Climate Dynamics* 41, 2099–2113.

Scannell, C., Booth, B.B.B., Dunstone, N.J., Rowell, D.P., Bernie, D.J., *et al.* (2019) The influence of remote aerosol forcing from industrialized economies on the future evolution of East and West African rainfall. *Journal of Climate* 32, 8335–8354.

Seneviratne, S.I., Nicholls, N., Easterling, D., Goodess, C.M., Kanae, S., *et al.* (2012) Changes in climate extremes and their impacts on the natural physical environment. In: Field, C.B., Barros, V., Stocker, T.F., Qin, D., Dokken, D.J., *et al.* (eds) *Managing the Risks of Extreme Events and Disasters to Advance Climate Change Adaptation. A Special Report of Working Groups I and II of the Intergovernmental Panel on Climate Change (IPCC).* Cambridge University Press, Cambridge and New York, pp. 109–230.

Shepherd, T.G., Boyd, E., Calel, R.A., Chapman, S.C., Dessai, S., *et al.* (2018) Storylines: an alternative approach to representing uncertainty in physical aspects of climate change. *Climatic Change* 151, 555–571.

Sillmann, J., Kharin, V.V., Zhang, X., Zwiers, F.W. and Bronaugh, D. (2013) Climate extremes indices in the CMIP5 multimodel ensemble: Part 1. Model evaluation in the present climate. *Journal of Geophysical Research: Atmospheres* 118, 1716–1733.

Stratton, R.A., Senior, C.A., Vosper, S.B., Folwell, S.S., Boutle, I.A., *et al.* (2018) A Pan-African convection-permitting regional climate simulation with the Met Office Unified Model: CP4-Africa. *Journal of Climate* 31, 3485–3508.

Taylor, C.M., Belušić, D., Guichard, F., Parker, D.J., Vischel, T., *et al.* (2017) Frequency of extreme Sahelian storms tripled since 1982 in satellite observations. *Nature* 544, 475–478.

Vetter, S. (2009) Drought, change and resilience in South Africa's arid and semi-arid rangelands. *South African Journal of Science* 105, 29–33.

Vizy, E.K. and Cook, K.H. (2017) Seasonality of the observed amplified Sahara warming trend and implications for Sahel rainfall. *Journal of Climate* 30, 3073–3094.

Wainwright, C.M., Marsham, J.H., Keane, R.J., Rowell, D.P., Finney, D.L., *et al.* (2019) 'Eastern African Paradox' rainfall decline due to shorter not less intense long rains. *npj Climate and Atmospheric Science* 2, 34. https://doi.org/10.1038/s41612-019-0091-7

Wainwright, C.M., Finney, D.L., Kilavi, M., Black, E. and Marsham, J.H. (2020) Extreme rainfall in East Africa, October 2019–January 2020 and context under future climate change. *Weather* 99, 99.

Wilby, R.L. and Dessai, S. (2010) Robust adaptation to climate change. *Weather* 65, 180–185.

Young, H.R., Cornforth, R.J., Todman, L.C., Miret, J.A., Shepherd, T.G., *et al.* (2020a) Climate change and sweet potato policy briefing note Uganda. *GCRF Sweet Potato Catalyst, Policy Brief No. 1*. Walker Institute, Reading, UK. https://doi.org/10.5281/zenodo.3836642

Young, H.R., Plumpton, H., Miret, J.A., Cornforth, R.J., Petty, C., *et al.* (2020b) Sweet potato production in Uganda in a changing climate: what is the role for fertilisers? *GCRF Sweet Potato Catalyst, Policy Brief No. 2*. Walker Institute, Reading, UK. https://doi.org/10.5281/zenodo.3955098

Young, H., Shepherd, T., Acidri, J., Cornforth, R.J., Petty, C., *et al.* (2021) Storylines for decision-making: climate and food security in Namibia. *Climate and Development* 13, 515–528.

eo2 Vegetation–Climate Interactions: Into the Tick Zone

Deborah Hemming[1]*, James Duffy[2], Neil Kaye[1] and Ilya Maclean[2]
[1]*Met Office Hadley Centre, Exeter, UK and Birmingham Institute of Forest Research, University of Birmingham, Birmingham, UK;* [2]*University of Exeter, Exeter, UK*

2.1 Expert Opinion

2.1.1 Introduction

Plants are the most extensive life form on the planet. They account for around 80% (450 Gt C; 1 Gt C = 10^{15} g carbon) of total biomass, compared with 0.01% for humans (Bar-On *et al.*, 2018), and cover approximately 85% of the ice-free land area (Zhu *et al.*, 2016). Vegetation plays a major role in the Earth's climate system, both influencing and being influenced by climate variations across large spatial and temporal scales. Vegetation also provides many functions and resources that are vital for nearly all other life forms, including: (i) net uptake of carbon dioxide (CO_2) and release of oxygen (O_2); (ii) regulation of major biogeochemical cycles, e.g. water, carbon and nitrogen; (iii) provision of food, resources and medicines; (iv) settings for improved human health and well-being (Karjalainen *et al.*, 2010); and (v) availability of habitats for most of the world's land organisms (Tews *et al.*, 2003).

The close association between vegetation and climate has been recognized for millennia (Woodward, 1987). One of the earliest major studies was by the Greek philosopher Theophrastus (a pupil of Plato and Aristotle), in the third century BC (Entrocassi *et al.*, 2020), who discussed the importance of climate on plant distributions and seasonal changes in his book *Enquiry into plants and minor works on odours and weather signs* (translated by Hort, 1916). Considerable scientific endeavour has since been conducted to understand the fundamental processes that underly these relationships and how they scale from individual plant organs to global vegetation zones (e.g. Budyko, 1974; Brovkin, 2002; McPherson, 2007; Adams, 2010). This has involved a wide range of observations, experiments and models, some highly specific to individual processes, locations or species (e.g. Parmesan *et al.*, 1999; Parmesan and Yohe, 2003; Suttle *et al.*, 2007) and others that consider the role of vegetation–climate interactions within the wider Earth system (e.g. Koster *et al.*, 2006; Hemming *et al.*, 2013).

A key challenge for understanding species and ecosystem responses to global environmental change is bridging the gap in scales between the microclimates in which most organisms live and the meteorological observations and models used to study these responses (Potter *et al.*, 2013). Vegetation is central to this understanding because it has a strong effect on microclimates and provides other functions that influence organism behaviour and survival. This is particularly relevant for ticks as they rely on both vegetation and microclimate conditions for their survival (Medlock *et al.*, 2013).

*Email: debbie.hemming@metoffice.gov.uk

DOI: 10.1079/9781789249637.0002

Here, we give a broad overview of vegetation–climate interactions in the context of scale. We first introduce the importance of scale and pattern for studying ecological processes, then discuss examples of vegetation–climate interactions at different scales from molecular to global. Next, we consider the relevance of these interactions for understanding ecological change, highlighting the significance for ticks. We finish by suggesting how future advances could improve understanding of interactions between organisms and climate, particularly within the tick zone.

2.1.2 The importance of scale and pattern

Scale and pattern are crucial concepts in ecology and central to the understanding of vegetation–climate interactions (Levin, 1992; Chave, 2013). Vegetation and climate interact through exchanges of energy, water, carbon and trace elements, such as iron and zinc (Adams, 2010). The fundamental exchanges take place at very fine spatial and temporal scales within plant canopies or soil layers. These, in turn, aggregate and cascade, non-linearly, to coarser-scale responses that modify the surrounding environment. The patterns of these responses may be detected in the broadest scales, i.e. global (Snyder *et al.*, 2004).

As with most ecological relations, there is no single scale at which vegetation–climate interactions should be studied; different processes and patterns emerge at very different space, time and organizational scales (Levin, 1992). However, it is important that the resolution of observations or models used is appropriate for the scale of the processes or patterns being studied (Chave, 2013) and that the potential uncertainties resulting from non-linearities are understood (Bütikofer *et al.*, 2020). For example, fine-scale microclimate variations at the land surface are extremely important for assessing the risk of species extinction from climate change (Suggitt *et al.*, 2018), but insignificant for classifying broad relationships between climate and global vegetation distribution (Holdridge, 1947). It is also important that observations and models are developed to improve mechanistic modelling, leading to more realistic representation of ecological processes, and bridge across scales in order to better link the primary processes and mechanisms with their larger-scale patterns of influence (Chave, 2013).

2.1.3 Vegetation–climate interactions across scales

Here, we provide examples of how key plant physiological processes are influenced by climate and how these feed back to modify microclimates and larger-scale patterns of change. The primary physiological processes that enable plant growth, i.e. photosynthesis, respiration and transpiration, are closely coupled with microclimate conditions (scales typically < metres and hours) within or surrounding plant organs (leaves, stem, roots). These microclimates are continuously nudged by larger-scale climate and other external drivers, e.g. pollution, but they are also strongly influenced by fine-scale biophysical variations at the land surface, including topography, soil characteristics, and vegetation structure and cover (Brovkin, 2002; Bennie *et al.*, 2014; Suggitt *et al.*, 2017; Maclean *et al.*, 2019). To illustrate how microclimate conditions can interact with primary plant processes, we briefly recap on some of the physiological steps involved and provide examples of how they scale up to large-scale patterns. A more comprehensive review of plant processes and dynamics can be found in Fisher *et al.* (2018) and Taiz *et al.* (2018).

Photosynthesis is the biochemical process whereby photosynthetically active radiation (PAR) (i.e. solar radiation of a specific wave band, 400–700 nm) reaching the leaf surface is absorbed by chlorophyll, the green pigment present inside leaf cells, and some of the energy is used to build carbohydrates from water (H_2O) and CO_2, releasing O_2 as a by-product. Respiration is the opposite process, which breaks down carbohydrates by combining them with O_2, releasing the stored energy for use by the plant, and generating CO_2 and H_2O as by-products. Lastly, transpiration is the process by which H_2O gas is lost from the leaves during photosynthesis and respiration (Taiz *et al.*, 2018). Liquid H_2O in leaves typically originates from the soil via the root and stem system, and at the leaves CO_2,

H_2O, O_2 and other gases are exchanged with the surrounding atmosphere through stomata (pores in leaves). Many steps in these processes are influenced by, and feed back to, the surrounding climate, as briefly discussed below.

The amount of PAR that reaches the leaf surface is reduced and changed by the presence of clouds and aerosols in the atmosphere (Mõttus *et al.*, 2012) and the radiation that penetrates the canopy is modified by the architecture and density of vegetation (Sellers *et al.*, 1997; Yang and Friedl, 2003). Vegetation geometry and density also alter the roughness of the surface and turbulence in the air above it, changing the fluxes between plants and atmosphere (Raupach, 1994) and ultimately the shape of the temperature profile above the canopy (Monin and Obukhov, 1954).

The type, quantity and condition of the vegetation also affect the surface–atmosphere exchanges of energy and gases (McPherson, 2007). The air itself, being almost transparent, absorbs little radiation, whereas vegetation is opaque and thus absorbs radiation. The effect of solar radiation on air temperature is therefore an indirect one: heat is transported via convection from the vegetation to the nearby air. Further away from the vegetation surface the air becomes increasingly turbulent and well-mixed (Monin and Obukhov, 1954). As a result, the finest-scale air temperature variations occur very close to the surface of vegetation as it is here that the thermal mixing of air is at its lowest. Vegetation directly affects the thermal properties of the surface through changes in surface albedo (the average shortwave solar energy reflected at the surface) and emissivity (the effectiveness of a surface to emit energy as thermal radiation). The albedo of vegetated areas is generally considered to be lower than that of regions of barren land. Vegetation also tends to exhibit a relatively high thermal emissivity, due to its water content, leading to more long-wave radiation emitted than in regions with sparse vegetation (Feng *et al.*, 2020).

H_2O fluxes from the leaf surface also influence leaf temperature. Transpiration and evaporation from the leaf surface result in evaporative cooling of the leaf and surrounding atmosphere (Gates, 1968). Leaf water status is influenced by both soil moisture (and associated precipitation) and atmospheric humidity surrounding the leaves, which regulates stomatal conductance and therefore gas exchange through the stomata (Buckley, 2019). However, the rate of transpiration is also determined by the vapour pressure gradient between leaves and the surrounding air. The water potential of leaves, except under strong drought conditions, is generally such that they can be treated as wet surfaces, and the vapour concentration at the evaporating surface is thus similar to the saturated vapour concentration at surface temperature (Monteith, 1958; Campbell and Norman, 2012). As water evaporates from a wet surface, the concentration of H_2O molecules in the gas increases until an equilibrium is reached such that the number of molecules escaping equals the number being recaptured. When the temperature of the leaves increases, the random kinetic energy of the molecules increases and more water escapes. Thus, rates of evapotranspiration are strongly dependent on the temperature of leaves, increasing when leaf temperatures are higher.

The partial pressure of CO_2 in the atmosphere surrounding stomata influences the availability of CO_2 for photosynthesis and other plant functions. Elevated atmospheric CO_2 generally leads to higher rates of photosynthesis and growth, although the scale of response varies considerably by species traits, photosynthetic pathway (i.e. C_3, C_4 or CAM (crassulacean acid metabolism)) and other limiting factors (e.g. water and nutrients) (Taub, 2010). Changes in CO_2 uptake and release by plants also modulate the concentration of atmospheric CO_2, notably in diurnal and annual cycles (Baldocchi, 2008).

The rates of biochemical processes are regulated by enzymes that are directly influenced by temperature (Lin *et al.*, 2013); e.g. Rubisco (ribulose-1,5-bisphosphate carboxylase-oxygenase), the most common enzyme on Earth and the primary one in photosynthesis, and carbonic anhydrase, one of the fastest enzymes in nature, which is involved in the hydration of CO_2 into leaf water (Yakir, 2002). Although species are adapted to different climate conditions, most plant physiological processes take place between 0 and 40°C (Went, 1953). Extreme high and low temperatures reduce the efficiency of these processes and can cause damage to a wide range of plant functions. Combined with daylength, temperature can also advance or delay the phenology (seasonal timing of events) of plants, e.g. bud burst, leaf out or senescence (which can be observed at hemispheric scales) (Lieth, 1974; Schwartz, 1998; Park *et al.*, 2016; Richardson *et al.*, 2018). Phenology also mediates many of the vegetation feedbacks on the climate system (Parmesan, 2007; Richardson *et al.*, 2013).

The lower part of the atmosphere, referred to as the atmospheric or planetary boundary layer (ABL), is directly influenced by the presence of the Earth's surface. Its depth can vary from a few metres to several kilometres, and it responds to surface forcing on timescales of hours or less, depending on surface and meteorological features (Hu, 2015). As spatial and temporal scales increase, the responses of all small-scale plant processes aggregate and net patterns of change may emerge (Thackeray *et al.*, 2016). These can have major large-scale feedbacks, e.g. transpiration from terrestrial vegetation accounts for 60–80% of all evapotranspiration from the land surface and it recycles about 39 ± 10% of global precipitation back to the atmosphere (Schlesinger and Jasechko, 2014). Landscape-scale processes also become important, such as population dynamics, land-use change, fire history, and transport of nutrients, organic matter and pollutants (Wofsy, 2001). The impact of vegetation interactions and changes on the physical climate (e.g. atmospheric circulation, temperature, humidity, clouds, rainfall) and on the atmospheric concentration of biogenic gases (e.g. CO_2, methane, ozone) can be significant at meso scales (i.e. tens to hundreds of kilometres) (Chase *et al.*, 2000; McPherson, 2007) and in some cases (e.g. atmospheric CO_2 concentration) at global and centennial scales (Bonan *et al.*, 1992; Keeling *et al.*, 2005).

The global distribution of plants and animals into distinct biological communities, known as ecosystems (or biomes), is strongly associated with spatial variations in average climate, particularly radiation, temperature and precipitation (von Humboldt and Bonpland, 1805; Woodward, 1987). These broad relationships have been used to classify the world's climate into biologically relevant zones, e.g. the Köppen–Geiger climate classification (Köppen and Geiger, 1936), the Holdridge life zone classification (Holdridge, 1947) and the Walter classification (Walter, 1973). More detailed spatial relationships are also the basis for species distribution models (SDMs) that correlatively relate species occurrence with environmental variables and climate envelope models (CEMs) that use climate observations to predict environmental suitability for different species (Watling *et al.*, 2013). Over long timescales, paleoenvironmental studies have shown that continental-scale shifts in climate and vegetation distribution, or species composition, have coincided over centuries and millennia (e.g. Prentice *et al.*, 1991; Hatté and Schwartz, 2003; Masson-Delmotte *et al.*, 2013).

2.1.4 Relevance for understanding ecological change

Vegetation clearly plays a major role in the Earth's climate system. It also regulates the living conditions of most of the world's organisms and is therefore a key component of ecological change. The study of vegetation–climate interactions and their role in species and ecosystem changes requires a fusion of ecological and meteorological understanding. Historically, these disciplines have tended to approach research from opposing scales. Ecological studies began at small extents, typically less than a square kilometre over timescales of seconds to years, motivated by the desire to understand the responses of individual organisms, species or small/simple ecosystems (Drake *et al.*, 2016; Fulton *et al.*, 2019). In comparison, meteorological studies developed from a requirement to characterize and predict weather and climate around the globe at scales of tens to thousands of square kilometres and hours to decades (WMO, 2014).

In recent decades, observations have expanded significantly and now include harmonized regional- and global-scale networks of ecosystem and meteorological measurements (Hemming *et al.*, 2007; Walters and Scholes, 2017), e.g. FLUXNET (Baldocchi, 2008; Gerken *et al.*, 2019) or the International Long-Term Ecological Research Network (ILTER/LTER) (Dirnböck *et al.*, 2019). The spatial resolution of gridded meteorological data sets has also increased (e.g. Hollis *et al.*, 2019) and the spatio-temporal resolution and variety of products available from satellite remote sensing have grown spectacularly, providing recurring global coverage of a wide range of ecologically and meteorologically relevant data (Melesse *et al.*, 2007; Coniff, 2017). Technological developments in micro-scale monitoring within habitats, and even within organisms (e.g. Sellier *et al.*, 2014; Godyń *et al.*, 2019), have helped advance knowledge of microclimate variations and their role in species and ecosystem responses (Wall *et al.*, 2014; Maclean *et al.*, 2015; Tang and Shao, 2015; Eckmann *et al.*, 2018; Kiselev *et al.*, 2018). Also, major developments in experiments on land–atmosphere exchanges, such as the Global Land–Atmosphere Climate Experiment (GLACE) (Koster *et al.*, 2006), and on controlled environments, such as Free-Air CO_2 Enrichment (FACE), have enabled ecosystem processes and interactions to be observed in the context of projected future changes (Ainsworth and Long, 2004; Hart *et al.*, 2020). Models have developed alongside improvements in observations and experimentation. The scope of ecological models has grown considerably from individual and population scales to communities, ecosystems and even socio-ecological systems that integrate human and natural components (Fulton *et al.*, 2019). Meteorological models have also evolved radically from the early days of numerical weather prediction (Richardson, 1922). Weather forecasting models are now run routinely at sub-kilometre and sub-hourly scales in response to the need to resolve physical processes that are influential on weather timescales, such as cloud convection (Guichard and Couvreux, 2017), air drainage in valleys (Webster *et al.*, 2003), air/sea breezes (Miller *et al.*, 2003) and ocean eddies (Roberts *et al.*, 2016). Also, climate models have evolved into earth system models (ESMs) that integrate atmosphere, ocean, land, ice and biosphere interactions to estimate both physical and biogeochemical processes, particularly the carbon cycle (Heavens *et al.*, 2013; Fisher *et al.*, 2018; Randall *et al.*, 2018).

As observations, experiments and models have advanced, the scales of ecological and meteorological studies have converged (Liu *et al.*, 2006; Bramer *et al.*, 2018). Some of the issues of mismatch and interconnection between ecological and meteorological scales have been resolved (Fulton *et al.*, 2019). For example, most ESMs now include explicit representation of fine-scale physiological processes and ecological dynamics in sub-models referred to as terrestrial biosphere models (TBMs) (Moorcroft *et al.*, 2001; Fisher *et al.*, 2018), enabling large-scale coupled vegetation and climate interactions to be explored under varying environmental forcing scenarios like land-use change and atmospheric CO_2 projections. However, other issues of scale remain, especially at small spatiotemporal scales (Potter *et al.*, 2013).

Understanding and modelling the behaviour of organisms that spend their life cycles within vegetation, such as ticks, is particularly challenging because of the need to account for vegetation dynamics as well as reconcile the mismatch in scales between microclimate conditions experienced by organisms and the resolution of available meteorological observations or models (Bennie *et al.*, 2014; Kearney *et al.*, 2014; Bramer *et al.*, 2018). Many ecological models, including SDMs and CEMs, use statistical relationships between station or gridded meteorological data to predict species and ecosystem responses, and this mismatch in scales can cause significant inaccuracies in the estimates (Trevedi *et al.*, 2008; Randin *et al.*, 2009; Franklin *et al.*, 2013; Slavich *et al.*, 2014; Lenoir *et al.*, 2016). Potter *et al.* (2013) show that the average spatial resolution of SDMs is about 10,000 times larger than the animals studied and 1000 times larger than the plants studied, and this discrepancy is even greater for small organisms. Incorporating sub-grid scale meteorology into ecological models is not simple (Fulton *et al.*, 2019). Boehnke *et al.* (2017) examine this issue for *Ixodes ricinus* ticks within a forest habitat and conclude that to understand the influence of meteorology on tick populations, continuous *in situ* measurements are needed that accurately represent the meteorological conditions within the forest stands. Recent advances in microclimate monitoring and modelling have improved the representation of microclimates in ecological studies (e.g. Morán Cadenas *et al.*, 2007; Maclean *et al.*, 2015; Meineri and Hylander, 2017; Ashcroft, 2018). Further developments are possible using statistical abstractions (Moorcroft *et al.*, 2001), agent-based approaches (DeAngelis and Mooij, 2005; Fulton *et al.*, 2019) and by incorporating microclimate components within existing observation networks and models. Increasing the knowledge of ecological and meteorological interactions across scales will also help to improve understanding and modelling of the mechanistic processes and interdependencies that drive organismal responses to weather and climate.

2.2 Future Predictions

Addressing the major environmental and societal challenges currently posed by climate change, species loss, habitat degradation, pollution and sustainable resource use (and others) will require well-informed ecological understanding. Over recent decades, the temporal and spatial scales of ecological and meteorological studies have converged, and this has enabled major advances in the understanding of vegetation–climate interactions and their relationships with other organisms (as outlined above). Yet there are still considerable uncertainties in predicting species and ecosystem responses to climate and other environmental changes, partly due to persistent mismatches in the scale of some ecological and meteorological data relative to the scale of organisms (Fulton *et al.*, 2019). This is particularly an issue for small organisms, such as ticks, that live in burrows, crevices or leaf litter where microclimates can vary significantly from the climate data available from standard observations and models. Bridging this scale gap will be a key challenge for future research and will require additional microclimate monitoring, further use of high-resolution satellite products, improved methods for downscaling observations and models to microclimate scales, and techniques for transferring and analysing the big data sets that will ensue. Routine interdisciplinary collaborations (including ecologists, engineers, meteorologists, climatologists, statisticians and geographers) will also be essential to ensure the approaches and outputs are scientifically robust (Potter *et al.*, 2013). Although none of this is simple, many of the observations, experiments and models that will be needed are already well-developed. Likewise, innovative analytical methods such as machine learning techniques (Jung *et al.*, 2019), ecological informatics (Boyd and Foody, 2011) and high-volume data transfer streams for making use of the big data ('data sets so large or complex that they challenge established methods for data capture, curation, storage, analysis, and transfer' – Lynch, 2008) are already applied to global ecological concerns (Franklin *et al.*, 2017). It therefore seems feasible that concerted efforts in the near future will be able to bridge the ecological scale gaps that still exist, particularly at small scales. This would be a major step forward for understanding relationships between climate and all living organisms, but especially for small, climate-sensitive species, such as ticks.

Disclaimer

The findings and conclusions of this expert opinion are by the authors and do not necessarily represent the views of the Met Office or the University of Exeter.

References

Adams, J. (2010) *Vegetation–Climate Interaction: How Plants Make the Global Environment*, 2nd edn. Springer, Berlin/Heidelberg, Germany.

Ainsworth, E.A. and Long, S.P. (2004) What have we learned from 15 years of free-air CO_2 enrichment (FACE)? A meta-analytic review of the responses of photosynthesis, canopy properties and plant production to rising CO_2. *New Phytologist* 165, 351–372.

Ashcroft, M.B. (2018) Which is more biased: standardized weather stations or microclimatic sensors? *Ecology and Evolution* 8, 5231–5232.

Baldocchi, D. (2008) 'Breathing' of the terrestrial biosphere: lessons learned from a global network of carbon dioxide flux measurement systems. *Australian Journal of Botany* 56, 1–26.

Bar-On, Y.M., Philips, R. and Milo, R. (2018) The biomass distribution on Earth. *Proceedings of the National Academy of Sciences USA* 115, 6506–6511.

Bennie, J., Wilson, R., Maclean, I.M.D. and Suggitt, A.J. (2014) Seeing the woods for the trees – when is microclimate important in species distribution models? *Global Change Biology* 20, 2699–2700.

Boehnke, D., Gebhardt, R., Petney, T. and Norra, S. (2017) On the complexity of measuring forests microclimate and interpreting its relevance in habitat ecology: the example of *Ixodes ricinus* ticks. *Parasites & Vectors* 10, 549.
Bonan, G., Pollard, D. and Thompson, S. (1992) Effects of boreal forest vegetation on global climate. *Nature* 359, 716–718.
Boyd, D.S. and Foody, G.M. (2011) An overview of recent remote sensing and GIS based research in ecological informatics. *Ecological Informatics* 6, 25–36.
Bramer, I., Anderson, B.J., Bennie, J., Bladon, A.J., Frenne, P.D., *et al.* (2018) Advances in monitoring and modelling climate at ecologically relevant scales. *Advances in Ecological Research* 58, 101–161.
Brovkin, V. (2002) Climate–vegetation interaction. *Journal de Physique IV (Proceedings)* 12, 57–72.
Buckley, T.N. (2019) How do stomata respond to water status? *New Phytologist* 224, 21–36.
Budyko, M.I. (1974) *Climate and Life*. Academic Press, New York.
Bütikofer, L., Anderson, K., Bebber, D.P., Bennie, J.J., Early, R.I. and Maclean, I.M. (2020) The problem of scale in predicting biological responses to climate. *Global Change Biology* 26, 6657–6666.
Campbell, G.S. and Norman, J. (2012) *An Introduction to Environmental Biophysics*. Springer, New York.
Chase, T., Pielke, R. Sr, Kittel, T., Nemani, R.R. and Running, S.W. (2000) Simulated impacts of historical land cover changes on global climate in northern winter. *Climate Dynamics* 16, 93–105.
Chave, J. (2013) The problem of pattern and scale in ecology: what have we learned in 20 years? *Ecology Letters* 16, 4–16.
Coniff, R. (2017) Eyes on nature: how satellite imagery is transforming conservation science. *Yale Environment 360*, 22 June. Available at: https://e360.yale.edu/features/eyes-on-nature-how-satellite-imagery-is-transforming-conservation-science (accessed 26 May 2021).
DeAngelis, D.L. and Mooij, W.M. (2005) Individual-based modeling of ecological and evolutionary processes. *Annual Review of Ecology, Evolution, and Systematics* 36, 147–168.
Dirnböck, T., Haase, P., Mirtl, M., Pauw, J. and Templer, P.H. (2019) Contemporary International Long-Term Ecological Research (ILTER) – from biogeosciences to socio-ecology and biodiversity research. *Regional Environmental Change* 19, 309–311.
Drake, J.E., Tjoelker, M.G., Aspinwall, M.J., Reich, P.B., Barton, C.V.M., *et al.* (2016) Does physiological acclimation to climate warming stabilize the ratio of canopy respiration to photosynthesis? *New Phytologist* 211, 850–863.
Eckmann, T., Morach, A., Hamilton, M., Walker, J., Simpson, L., *et al.* (2018) Measuring and modelling microclimate impacts of *Sequoiadendron giganteum*. *Sustainable Cities and Society* 38, 509–525.
Entrocassi, G.S., Gavilán, R.G. and Sánchez-Mata, D. (2020) *Subtropical Mountain Forests of Las Yungas: Vegetation and Bioclimate. Geobotany Studies (Basics, Methods and Case Studies)*. Springer, Cham, Switzerland.
Feng, H., Ye, S. and Zou, B. (2020) Contribution of vegetation change to the surface radiation budget: a satellite perspective. *Global and Planetary Change* 192, 103225.
Fisher, R.A., Koven, C.D., Anderegg, W.R.L., Christoffersen, B.O. and Dietze, M.C. (2018) Vegetation demographics in Earth System Models: a review of progress and priorities. *Global Change Biology* 24, 35–54.
Franklin, J., Davis, F.W., Ikegami, M., Syphard, A.D., Flint, L.E., *et al.* (2013) Modeling plant species distributions under future climates: how fine scale do climate projections need to be? *Global Change Biology* 19, 473–483.
Franklin, J., Serra-Diaz, J.M., Syphard, A.D. and Regan, H.M. (2017) Big data for forecasting global change impacts on plant communities. *Global Ecology and Biogeography* 26, 6–17.
Fulton, E.A., Blanchard, J.L., Melbourne-Thomas, J., Plaganyi, E.E. and Vivitskaia, J.D. (2019) Where the ecological gaps remain, a modelers' perspective. *Frontiers in Ecology and Evolution* 7, 424.
Gates, D.M. (1968) Transpiration and leaf temperature. *Annual Review of Plant Physiology* 19, 211–238.
Gerken, T., Ruddell, B.L., Yu, R., Stoy, P.C. and Drewry, D.T. (2019) Robust observations of land-to-atmosphere feedbacks using the information flows of FLUXNET. *npj Climate and Atmospheric Sciences* 2, 37.
Godyń, D., Herbut, P. and Angrecka, S. (2019) Measurements of peripheral and deep body temperature in cattle – a review. *Journal of Thermal Biology* 79, 42–49.
Guichard, F. and Couvreux, F. (2017) A short review of numerical cloud-resolving models. *Tellus A: Dynamic Meteorology and Oceanography* 69, 1373578.
Hart, K.M., Curioni, G., Blaen, P., Harper, N.J., Miles, P., *et al.* (2020) Characteristics of free air carbon dioxide enrichment of a northern temperate mature forest. *Global Change Biology* 26, 1023–1037.

Hatté, C. and Schwartz, D. (2003) Reconstruction of paleoclimates by isotopic analysis: what can the fossil isotopic record tell us about the plant life of past environments? *Phytochemistry Reviews* 2, 163–177.

Heavens, N.G., Ward, D.S. and Natalie, M.M. (2013) Studying and projecting climate change with Earth System Models. *Nature Education Knowledge* 4, 4.

Hemming, D.L., Griffiths, H., Loader, N.J., Marca, A.D., Robertson, I., *et al.* (2007) The future of large-scale stable isotope networks. *Terrestrial Ecology* 1, 361–381.

Hemming, D.L., Betts, R.A. and Collins, M. (2013) Sensitivity and uncertainty of modelled terrestrial net primary productivity to doubled CO_2 and associated climate change for a relatively large perturbed physics ensemble. *Agricultural and Forest Meteorology* 170, 79–88.

Holdridge, L.R. (1947) Determination of plant world formations from simple climatic data. *Science* 105, 367–368.

Hollis, D., McCarthy, M.P., Kendon, M., Legg, T. and Simpson, I. (2019) HadUK-Grid – a new UK dataset of gridded climate observations. *Geoscientific Data Journal* 6, 151–159.

Hort, A. (1916) *Enquiry into plants and minor works on odours and weather signs*. By Theophrastus. W. Heinemann, London.

Hu, X.-M. (2015) Air pollution meteorology. In: North, G.R., Pyle, J. and Zhang, F. (eds) *Boundary Layer (Atmospheric) and Air Pollution. Encyclopaedia of Atmospheric Sciences*, 2nd edn. Academic Press, London, pp. 227–236.

Jung, M., Koirala, S., Weber, U., Ichii, K., Gans, F., *et al.* (2019) The FLUXCOM ensemble of global land-atmosphere energy fluxes. *Scientific Data* 6, 74.

Karjalainen, E., Sarjala, T. and Raitio, H. (2010) Promoting human health through forests: overview and major challenges. *Environmental Health and Preventive Medicine* 15, 1–8.

Kearney, M.R., Isaac, A.P. and Porter, W.P. (2014) Microclim: global estimates of hourly microclimate based on long-term monthly climate averages. *Scientific Data* 1, 140006.

Keeling, C.D., Piper, S.C., Bacastow, R.B., Whalen, M., Whorf, T.P., *et al.* (2005) Atmospheric CO_2 and $^{13}CO_2$ exchange with the terrestrial biosphere and oceans from 1978 to 2000: observations and carbon cycle implications. In: Ehleringer, J.R., Cerling, T. and Dearing, M.D. (eds) *A History of Atmospheric CO_2 and Its Effects on Plants, Animals, and Ecosystems. Ecological Studies (Analysis and Synthesis)*, vol. 177. Springer, New York, pp. 83–113.

Kiselev, M.V., Voropay, N.N., Dyukarev, E.A., Kurakov, S.A., Kurakova, P.S. and Makeev, E.A. (2018) Automatic meteorological measuring systems for microclimate monitoring. *IOP Conference Series: Earth and Environmental Science* 190, 012031.

Köppen, W. and Geiger, R. (1936) *Das geographische System der Klimate [The Geographic System of Climates]. Handbuch der Klimatologie [Handbook of Climatology]*, vol. 1. Gebruder Borntraeger, Berlin.

Koster, R.D., Guo, Z., Dirmeyer, P.A., Bonan, G., Chan, E., *et al.* (2006) GLACE: the Global Land–Atmosphere Coupling Experiment. Part 1. Overview. *Journal of Hydrometeorology* 7, 590–610.

Lenoir, J., Hattab, T. and Pierre, G. (2016) Climatic microrefugia under anthropogenic climate change: implications for species redistribution. *Ecography* 40, 253–266.

Levin, S.A. (1992) The problem of pattern and scale in ecology: The Robert H. MacArthur Award Lecture. *Ecology* 73, 1943–1967.

Lieth, H. (1974) *Phenology and Seasonality Modeling*. Springer, Berlin.

Lin, Y.-S., Medlyn, B.E., de Kauwe, M.G. and Ellsworth, D.S. (2013) Biochemical photosynthetic responses to temperature: how do interspecific differences compare with seasonal shifts? *Tree Physiology* 33, 793–806.

Liu, Z., Notaro, M., Kutzbach, J. and Liu, N. (2006) Assessing global vegetation–climate feedbacks from observations. *Journal of Climate* 19, 787–814.

Lynch, C. (2008) How do your data grow? *Nature* 455, 28–29.

Maclean, I.M.D., Hopkins, J.J., Bennie, J., Lawson, C.R. and Wilson, R.J. (2015) Microclimates buffer plant community responses to climate change. *Global Ecology and Biogeography* 24, 1340–1350.

Maclean, I.M.D., Mosedale, J.R. and Bennie, J.J. (2019) Microclima: an R package for modelling meso- and microclimate. *Methods in Ecology and Evolution* 10, 280–290.

Masson-Delmotte, V., Schulz, M., Abe-Ouchi, A., Beer, J., Ganopolski, A., *et al.* (2013) Information from paleoclimate archives. In: Stocker, T.F., Qin, D., Plattner, G.-K., Tignor, M., Allen, S.K., *et al.* (eds) *Climate Change 2013: The Physical Science Basis. Contribution of Working Group I to the Fifth Assessment Report of the Intergovernmental Panel on Climate Change*. Cambridge University Press, Cambridge and New York, pp. 383–464.

McPherson, R.A. (2007) A review of vegetation–atmosphere interactions and their influences on mesoscale phenomena. *Progress in Physical Geography* 31, 261–285.
Medlock, J.M., Hansford, K.M., Bormane, A., Derdakova, M., Estrada-Peña, A., *et al.* (2013) Driving forces for changes in geographical distribution of *Ixodes ricinus* ticks in Europe. *Parasites & Vectors* 6, 1.
Meineri, E. and Hylander, K. (2017) Fine-grain, large-domain climate models based on climate station and comprehensive topographic information improve microrefugia detection. *Ecography* 40, 1003–1013.
Melesse, A.M., Weng, Q., Thenkabail, P.S. and Senay, G.B. (2007) Remote sensing sensors and applications in environmental resources mapping and modelling. *Sensors (Basel)* 7, 3209–3241.
Miller, S.T.K., Keim, B.D., Talbot, R.W. and Mao, H. (2003) Sea breeze: structure, forecasting, and impacts. *Reviews of Geophysics* 41, 1011.
Monin, A.S. and Obukhov, A.M. (1954) Basic laws of turbulent mixing in the surface layer of the atmosphere. *Contributions to the Geophysical Institute of the Academy of Sciences USSR* 151, e187.
Monteith, J.L. (1958) The heat balance of soil beneath crops. In: *UNESCO–Australia Symposium on Arid Zone Climatology with Special Reference to Microclimatology, Canberra, October 1956*. UNESCO, Canberra, pp. 123–128.
Moorcroft, P.R., Hurtt, G.C. and Pacala, S.W. (2001) A method for scaling vegetation dynamics: the ecosystem demography model. *Ecological Monograph* 71, 557–585.
Morán Cadenas, F., Rais, O., Jouda, F., Douet, V., Humair, P.-F., *et al.* (2007) Phenology of *Ixodes ricinus* and infection with *Borrelia burgdorferi sensu lato* along a north- and south-facing altitudinal gradient on Chaumont Mountain, Switzerland. *Journal of Medical Entomology* 44, 683–693.
Mõttus, M., Sulev, M., Baret, F., Lopez-Lozano, R. and Reinart, A. (2012) Photosynthetically active radiation: measurement and modeling. In: Meyers, R.A. (ed.) *Encyclopedia of Sustainability Science and Technology*. Springer, New York. https://doi.org/10.1007/978-1-4419-0851-3_451
Park, T., Ganguly, S., Tommervik, H., Euskirchen, E.S., Hogda, K.A., *et al.* (2016) Changes in growing season duration and productivity of northern vegetation inferred from long-term remote sensing data. *Environmental Research Letters* 11, 084001.
Parmesan, C. (2007) Influences of species, latitudes and methodologies on estimates of phenological response to global warming. *Global Change Biology* 13, 1860–1872.
Parmesan, C. and Yohe, G. (2003) A globally coherent fingerprint of climate change impacts across natural systems. *Nature* 421, 37–42.
Parmesan, C., Ryrholm, N., Stefanescu, C., Hill, J.K., Thomas, C.D., *et al.* (1999) Poleward shifts in geographical ranges of butterfly species associated with regional warming. *Nature* 399, 579–583.
Potter, K.A., Woods, A.H. and Pincebourde, S. (2013) Microclimatic challenges in global change biology. *Global Change Biology* 19, 2932–2939.
Prentice, I.C., Bartlein, P.J. and Webb, T. III (1991) Vegetation and climate change in eastern North America since the last glacial maximum. *Ecology* 72, 2038–2056.
Randall, D.A., Bitz, C.M., Danabasoglu, G., Denning, S., Gent, P., *et al.* (2018) 100 years of Earth System Model development. *Meteorological Monographs* 59, 12.1–12.66.
Randin, C.F., Engler, R., Normand, S., Zappa, M., Zimmermann, N.E., *et al.* (2009) Climate change and plant distribution: local models predict high-elevation persistence. *Global Change Biology* 15, 1557–1569.
Raupach, M. (1994) Simplified expressions for vegetation roughness length and zero plane displacement as functions of canopy height and leaf area index. *Boundary Layer Meteorology* 71, 211–216.
Richardson, L.F. (1922) *Weather Prediction by Numerical Processes*. Cambridge University Press, Boston, Massachusetts.
Richardson, A.D., Keenan, T.F., Migliavacca, M., Ryu, Y., Sonnentag, O. and Toomey, M. (2013) Climate change, phenology, and phenological control of vegetation feedbacks to the climate system. *Agricultural and Forest Meteorology* 169, 156–173.
Richardson, A.D., Hufkens, K., Milliman, T., Aubrecht, D.M., Chen, M., *et al.* (2018) Tracking vegetation phenology across diverse North American biomes using PhenoCam imagery. *Scientific Data* 5, 18028.
Roberts, M.J., Hewitt, H.T., Hyder, P., Ferreira, D., Josey, S.A., *et al.* (2016) Impact of ocean resolution on coupled air–sea fluxes and largescale climate. *Geophysical Research Letters* 43, 10430–10438.
Schlesinger, W.H. and Jasechko, S. (2014) Transpiration in the global water cycle. *Agricultural and Forest Meteorology* 189–190, 115–117.
Schwartz, M.D. (1998) Green-wave phenology. *Nature* 394, 839–840.
Sellers, P., Dickinson, R.E., Randall, D.A., Betts, A.K., Hall, F.G., *et al.* (1997) Modelling the exchanges of energy, water and carbon between continents and the atmosphere. *Science* 275, 502–509.

Sellier, N., Guettier, E. and Staub, C. (2014) A review of methods to measure animal body temperature in precision farming. *American Journal of Agricultural Science and Technology* 2, 74–99.
Slavich, E., Warton, D.I., Ashcroft, M.B., Gollan, J.R. and Ramp, D. (2014) Topoclimate versus macroclimate: how does climate mapping methodology affect species distribution model and climate change projections? *Biodiversity Research* 20, 952–963.
Snyder, P.K., Delire, C. and Foley, J.A. (2004) Evaluating the influence of different vegetation biomes on the global climate. *Climate Dynamics* 23, 279–302.
Suggitt, A.J., Platts, P.J., Barata, I.M., Bennie, J., Burgess, M.D., *et al.* (2017) Conducting robust ecological analyses with climate data. *Oikos* 126, 1533–1541.
Suggitt, A.J., Wilson, R.J., Isaac, N.J.B., Beale, C.M., Auffret, A.G., *et al.* (2018) Extinction risk from climate change is reduced by microclimate buffering. *Nature Climate Change* 8, 713–717.
Suttle, K.B., Thomsen, M.A. and Power, M.E. (2007) Species interactions reverse grassland responses to changing climate. *Science* 315, 640–642.
Taiz, L., Zeiger, E., Moller, I.M. and Murphy, A. (2018) *Fundamentals of Plant Physiology*. Sinauer Associates Inc., Sunderland, Massachusetts.
Tang, L. and Shao, G. (2015) Drone remote sensing for forestry research and practices. *Journal of Forestry Research* 26, 791–797.
Taub, D.R. (2010) Effects of rising atmospheric concentrations of carbon dioxide on plants. *Nature Education Knowledge* 3(10), 21.
Tews, J., Brose, U., Grimm, V., Tielbörger, K., Wichmann, M.C., *et al.* (2003) Animal species diversity driven by habitat heterogeneity/diversity: the importance of keystone structures. *Journal of Biogeography* 31, 79–92.
Thackeray, S.J., Henrys, P.A., Hemming, D., Bell, J.R., Botham, M.S., *et al.* (2016) Phenological sensitivity to climate across taxa and tropic levels. *Nature* 535, 241–245.
Trevedi, M.R., Berry, P.M., Morecroft, M.D. and Dawson, T.P. (2008) Spatial scale affects bioclimate model projections of climate change impacts on mountain plants. *Global Change Biology* 14, 1089–1103.
von Humboldt, A. and Bonpland, A. (1805) *Essai sur la geographie des plantes*. Maxtor, Paris.
Wall, J., Wittemyer, G., Klinkenberg, B. and Douglas-Hamilton, I. (2014) Novel opportunities for wildlife conservation and research with real-time monitoring. *Ecological Applications* 24, 593–601.
Walter, H. (1973) *Vegetation of the Earth*. Springer, London.
Walters, M. and Scholes, R.J. (eds) (2017) *The GEO Handbook on Biodiversity Observation Network*. Springer, Cham, Switzerland.
Watling, J.I., Brandt, L.A., Mazzotti, F.J. and Romañach, S.S. (2013) *Use and Interpretation of Climate Envelope Models: A Practical Guide*. University of Florida, Fort Lauderdale, Florida.
Webster, S., Brown, A.R., Cameron, D.R. and Jones, C.P. (2003) Improvements to the representation of orography in the Met Office Unified Model. *Quarterly Journal of the Royal Meteorological Society* 129, 1989–2010.
Went, F.W. (1953) The effect of temperature on plant growth. *Annual Review of Plant Physiology* 4, 347–362.
Wofsy, S.C. (2001) Where has all the carbon gone? *Science* 292, 2261–2263.
WMO (2014) *Guide to Meteorological Instruments and Methods of Observation (WMO-No. 8)*. World Meteorological Organization, Geneva, Switzerland.
Woodward, F.I. (1987) *Climate and Plant Distribution*. Cambridge University Press, Cambridge.
Yakir, D. (2002) Global enzymes: sphere of influence. *Nature* 416, 795.
Yang, R. and Friedl, M.A. (2003) Modeling the effects of three-dimensional vegetation structure on surface radiation and energy balance in boreal forests. *Journal of Geophysical Research* 108, 8615.
Zhu, Z., Piao, S., Myneni, R.B., Huang, M., Zeng, Z., *et al.* (2016) Greening of the Earth and its drivers. *Nature Climate Change* 6, 791–795.

eo3 Climate Change and Lyme Disease

Peter Cox[1]*, Chris Huntingford[2], Morgan Sparey[1] and Pat Nuttall[3]
[1]*University of Exeter, Exeter, UK;* [2]*UK Centre for Ecology and Hydrology, Wallingford, UK;* [3]*University of Oxford, Oxford, UK*

3.1 Expert Opinion

3.1.1 Introduction

Lyme disease is the most common tick-borne disease in temperate regions (Couper *et al.*, 2021). It results from infection by *Borrelia burgdorferi* bacteria, which are transmitted from birds and mammals to humans by ticks of the genus *Ixodes* (*Ixodes scapularis* in Northeast USA, *Ixodes pacificus* in north-west USA, *Ixodes ricinus* in Europe and *Ixodes persulcatus* in Russia and north-east China) (Stanek *et al.*, 2012). These ticks have a 2- to 6-year life cycle through stages which are sensitive to climate (egg → larva → nymph → adult) (Ogden and Lindsay, 2016). Each life stage can only begin after a tick has had a full blood meal, lasting 2–14 days, from a vertebrate host. Larvae and nymphs typically feed from birds or small mammals such as mice, from which they can pick up *Borrelia* bacteria. Adult ticks feed from larger animals such as deer, sheep and dogs. Lyme disease occurs primarily when infected nymphs feed on passing humans. Adult ticks can also cause Lyme disease, but these are larger and therefore more often detected and removed before a person can be infected (Couper *et al.*, 2021).

Increases have been observed in the incidence of Lyme disease in the USA (especially in New England) (Brownstein *et al.*, 2003), Canada (especially in Ontario) (Cheng *et al.*, 2017), Western Europe (Sykes and Makiello, 2017), Russia (Malkhazova *et al.*, 2014) and China (especially in the eastern provinces) (see Section 3.1.2). Despite important additional driving factors such as land-use and population changes, these increases are broadly consistent with the impacts of climate change on the life cycle of *Ixodes* (Ogden and Lindsay, 2016). Many studies have projected further increases in the incidence of Lyme disease under continuing 21st-century climate change (Ogden *et al.*, 2006, 2014; McPherson *et al.*, 2017; Dumic and Severnini, 2018; Li *et al.*, 2019; Couper *et al.*, 2021). In this expert opinion, we attempt to place these studies in the context of climate projections not for specific greenhouse gas concentration scenarios, but instead for different levels of global warming.

We examine the key climatic driving factors for changes in the incidence of Lyme disease and explore how these depend on the level of global warming. Our aim is to provide some context for the more detailed expert opinions in this book as well as to show how the key climatic driving factors for Lyme disease relate to the Paris targets for climate stabilization. In Section 3.1.2, we summarize some of the key findings in the literature concerning the sensitivity of *Ixodes* life cycles to climate. Section 3.1.3 offers a very brief

*Email: p.m.cox@exeter.ac.uk

DOI: 10.1079/9781789249637.0003

summary of anthropogenic climate change and climate change projections. In Section 3.1.4 we tentatively suggest some simple climate thresholds that are broadly consistent with the current global pattern of Lyme disease hotspots, before exploring the implications for the distribution of Lyme disease under the 2°C limit of the Paris climate agreement in Section 3.2.

3.1.2 Climate effects on the tick vector life cycle

The size of the tick population depends on the number of reproducing adults and is therefore very sensitive to the number of nymphs that make the transition to adulthood. The transition from nymph to adult depends on a nymph being able to consume a blood meal from a vertebrate host. To do this, nymphs place themselves at the tips of leaves in temperate grasslands or woodlands and await a passing animal that they can latch on to and feed from. In the temperate northern hemisphere this 'questing' phase is typically in the late-spring to mid-summer months of May–July.

All else being equal, warmer conditions accelerate the tick life cycle, which tends to increase the number of ticks that reproduce before dying naturally or being killed by predators, disease or climate variability. However, drier conditions have the opposite effect by increasing tick mortality through desiccation, especially in the immature phase (Ginsberg *et al.*, 2017). There is also evidence that dry conditions suppress questing, which also reduces the number of nymphs that can capture the blood meal that will enable them to become a reproducing adult (McClure and Diuk-Wasser, 2019). The impact of climate change on *Borrelia*-infected ticks is therefore likely to be a competition between the amplifying effects of warmer conditions (especially in the winter) and the suppressing effects of summer drying.

Figure 3.1 is a schematic showing areas with higher risk of Lyme disease (hatched areas) on a map showing an aggregated Köppen–Geiger climate classification (Beck *et al.*, 2018). The hatched areas are a qualitative guide derived by 'eye' from various data sources: Lyme disease incidence in states of the USA (CDC, 2019), countries of Europe (with more than 25 annual cases per 100,000 people) (Czupryna *et al.*, 2016; Sykes and Makiello, 2017) and regions of Russia (mean annual cases greater than 200) (Malkhazova *et al.*, 2014); and risk maps for China (Liqun Fang, China, 2021, personal communication). There are, however, significant variations within these large regions (e.g. Lyme disease cases in Sweden are heavily concentrated in the south (eo14: Kjaer and Bødker)). Most of the hotspots are in temperate mid-latitudes classified as a *humid continental, cool summer* climate, and on the boundary of the *sub-Arctic* across much of Eurasia. This boundary

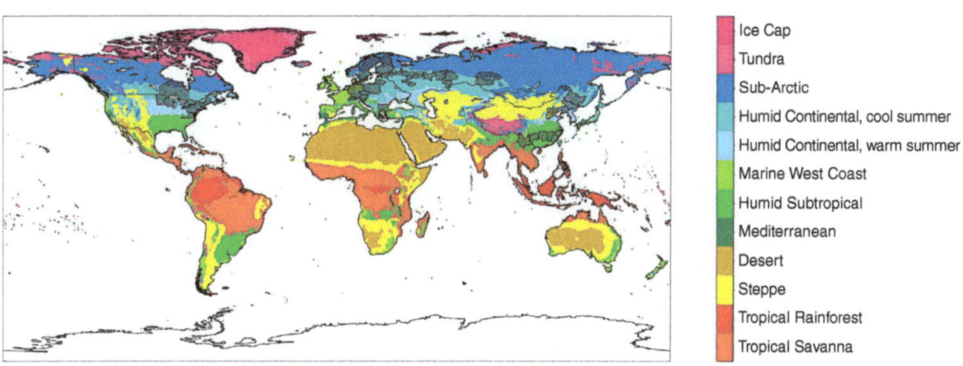

Fig. 3.1. Correspondence of Lyme disease hotspots in Northeast USA, Europe, Russia and China (hatched areas) with Köppen–Geiger climate classifications (colours). Hatching represents states (in the USA), nations (in Europe) or provinces (in Russia, China) where Lyme disease incidence is relatively high (see text for further details). (Authors' own figure.)

will continue to move northwards with global warming and Arctic amplification. There are also significant Lyme disease risks in the *humid subtropical* climates of south China, but these are most likely associated with different species of ticks and *Borrelia* (Sun *et al.*, 2003).

3.1.3 Climate change

Global mean temperature has already changed by more than 1°C since the mid-19th century and this is primarily due to increasing concentrations of greenhouse gases in the atmosphere. Atmospheric carbon dioxide (CO_2) concentrations have risen by almost 50% since pre-industrial times, from around 280 ppmv (parts per million by volume) to 410 ppmv by 2020, as a result of anthropogenic CO_2 emissions from fossil fuel burning, cement production and deforestation. Current global anthropogenic emissions exceed 10 GtC/year (i.e. ten billion tonnes of carbon per year) (Friedlingstein *et al.*, 2020). Over the same period, human activities have also increased the concentrations of other greenhouse gases such as methane (CH_4), which is up by more than 150%, and nitrous oxide (N_2O), which is up by almost 25% since pre-industrial times.

Together, human-induced increases in greenhouse gases have created an energy imbalance at the top of the Earth's atmosphere, which is equivalent now to injecting about 3 W/m^2 into the climate system compared with the pre-industrial climate. Ironically, this greenhouse warming has been partially offset by particulate pollution, especially sulfate aerosols from the burning of sulfurous coal, which reflect sunlight both directly and indirectly by making clouds brighter. The effects of such particulate pollution remain uncertain but are estimated to have offset about 1 W/m^2 of the forcing due to increasing greenhouse gases. The global warming that has been observed can be understood as the response of the climate system to the sum of the positive (warming) forcing from anthropogenic greenhouse gases and the negative (cooling) forcing from anthropogenic aerosols.

Future climate change depends first and foremost on how anthropogenic emissions of greenhouse gases and aerosol precursors change in the future. This in turn depends on the details of regional population change, economic growth and rates of decarbonization (e.g. through a switch away from fossil fuels, energy conservation and leaps of innovation). The authors of the Intergovernmental Panel on Climate Change (IPCC) – specifically Working Group III – deal with these issues by developing alternative storylines for the world which bracket feasible socio-economic futures to the end of the 21st century along with the huge range of atmospheric concentrations of greenhouse gases and aerosols associated with these (Riahi *et al.*, 2017).

However, there is a second major source of uncertainty in climate change projections which we seek to bypass in our analysis here. Even for a given socio-economic scenario, and therefore for given changes in greenhouse gases and aerosols, climate models differ by a factor of more than two in their projections of global mean warming by 2100. The reason for this range is primarily due to differences in the way key climate feedbacks are represented in the different climate models, which affects the response of their modelled temperatures to the net radiative forcing from greenhouse gases and aerosols. The biggest uncertainties still relate to how clouds will respond to climate change and therefore feed back to affect global and regional warming. Indeed, the latest 'CMIP6' climate models, which are improved in many ways, actually produce an even larger range of climate sensitivities as a result of an even larger range of cloud feedbacks (Nijsse *et al.*, 2020).

This might lead the reader to despair that we are just about to say the usual 'can't tell you yet' with regard to future climate change. It is indeed true that the huge range of future greenhouse gas concentrations, and the continuing uncertainty in the climate sensitivity to radiative forcing, combine to give an even larger range of projected climate changes by the end of this century. However, we also know a lot more now about the climate change problem and upon which aspects the different climate models broadly agree. First, climate models agree that global warming is essentially proportional to the total emissions of CO_2 that humankind has emitted since pre-industrial times. Second, although models disagree on the magnitude of global warming, they agree broadly on the patterns of warming at a given level of global warming.

Most notably, all models predict a faster warming of the northern latitudes than the tropics, which is sometimes called 'Arctic amplification', and a faster warming of the land than the sea. The predicted patterns of change in precipitation certainly vary much more between models, but the patterns of temperature-related variables are spatially smoother and more common across models.

We seek to anchor our analysis (see below) to these points of greater certainty. We do that by looking specifically at projected regional climate change at a given level of global warming. This neatly sidesteps the zeroth-order uncertainties associated with future emissions and climate sensitivity, as these primarily affect the level of global warming. However, it is more than just a convenience. It also allows us to focus on the possible impacts of climate change on Lyme disease at levels of global warming that are the focus of international agreements – most notably, the 2°C limit of the Paris Accord.

We utilize an approach developed by two of the co-authors of this expert opinion (Huntingford and Cox, 2000), originally to assess climate impacts on land ecosystems (Huntingford *et al.*, 2010). This approach is called 'pattern scaling' as it involves approximating the projections from individual climate models as the product of a time-varying global warming (which depends on climate sensitivity and the scenario of changing greenhouse gases) and time-invariant spatial patterns (for each climate variable and month). Here we utilize just the spatial patterns from this method, which we multiply by the assumed level of global warming to estimate the regional climate change. Although patterns of near-surface climate change vary somewhat across models, we look specifically at ensemble-mean changes from the CMIP5 generation of climate models (Zelazowski *et al.*, 2018). The ensemble means provide a central estimate of regional climate change at various levels of global warming.

3.1.4 The climate zone for Lyme disease

Figure 3.2 shows the ensemble-mean patterns of projected climate changes at 1°C of global warming, which is essentially the current-day climate. We show changes in annual mean temperature (Fig. 3.2a) and humidity deficit (Fig. 3.2b). As expected, the CMIP5 models project increases in annual mean temperature in the mid- and high-latitude areas that exceed the global mean warming (especially in winter). Such warming is expected to continue to extend Lyme disease hotspot areas to higher altitudes and higher latitudes.

The humidity deficit, Δq, represents the evaporative demand of the atmosphere and depends on both the near-surface air temperature, T_a, and the atmospheric relative humidity, r_h, as: $\Delta q = q_{sat}(T_a)(1-r_h)$. Here $q_{sat}(T_a)$ is the saturated humidity, which depends strongly on the atmospheric temperature T_a such that it increases by about 7% per °C increase in T_a. The CMIP5 models project relatively small changes in near-surface relative humidity, such that the change in humidity deficit is predominantly due to the increase in T_a. As a result, the atmospheric humidity deficit increases appreciably with global warming, especially in the May–July period (Fig. 3.2b). We therefore expect the tick life cycle, which is likely to be accelerated by the warming (Fig. 3.2a), to experience a compensating suppression due to atmospheric drying in the nymph questing season. How these competing effects of climate change will play out in terms of the incidence and regional distribution of Lyme disease is still uncertain but are addressed by other expert opinions in this book.

However, Fig. 3.2c shows a simple attempt to relate the hotspots of Lyme disease as shown in Fig. 3.1 to a minimum annual mean temperature of 3°C and a maximum May–July humidity deficit of 5 g/kg. These thresholds were chosen to be broadly consistent with the climates of Fig. 3.2a and b and the Lyme disease hotspots shown in Figs 3.1 and 3.2c. There is broad agreement on where Lyme disease is most prevalent, especially in Northeast USA, Central Europe and China. The agreement in southern China is especially noteworthy since these hotspots are not in the usual *humid continental, cool summer* climate classification which is typical of other Lyme disease hotspots (Fig. 3.1). The simple threshold model predicts higher prevalence of Lyme disease in France and the UK than observed, perhaps because the tick population in these locations is not limited by climate but instead by land use (e.g. arable agriculture rather than forest and grasslands).

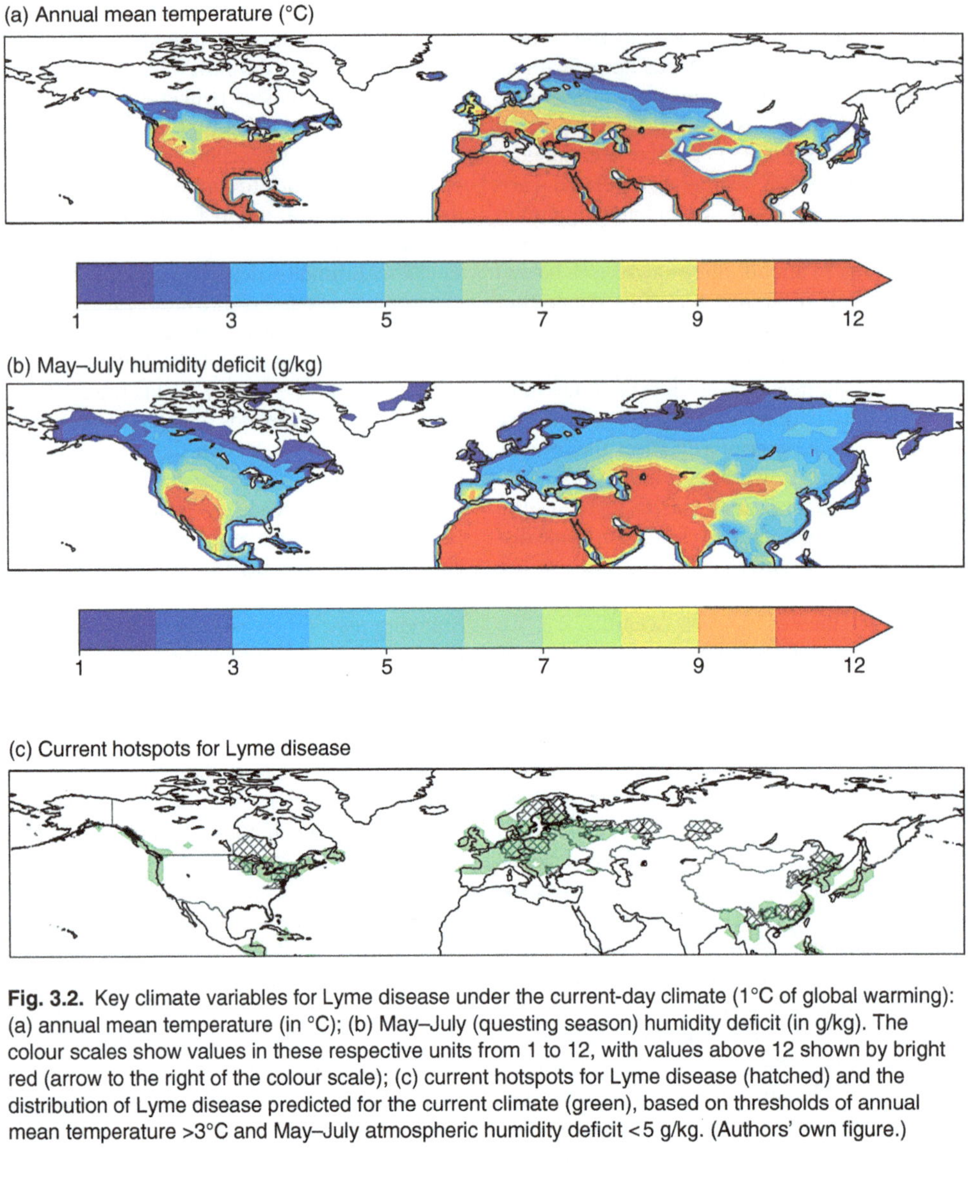

Fig. 3.2. Key climate variables for Lyme disease under the current-day climate (1°C of global warming): (a) annual mean temperature (in °C); (b) May–July (questing season) humidity deficit (in g/kg). The colour scales show values in these respective units from 1 to 12, with values above 12 shown by bright red (arrow to the right of the colour scale); (c) current hotspots for Lyme disease (hatched) and the distribution of Lyme disease predicted for the current climate (green), based on thresholds of annual mean temperature >3°C and May–July atmospheric humidity deficit <5 g/kg. (Authors' own figure.)

3.2 Future Predictions

The success of the climate threshold model in reproducing the large-scale pattern of current Lyme disease gives us some confidence in applying the same model to global warming projections. In particular, we examine the Lyme disease climate zone at 2°C of global warming, which is the upper limit for global warming consistent with the Paris climate agreement.

Figure 3.3 shows the pattern of changes in annual mean temperature (Fig. 3.3a) and May–July humidity deficit (Fig. 3.3b), per °C of global warming, according to the IMOGEN patterns from the CMIP5 models. Polar amplification leads to temperatures over high-latitude land increasing at more than twice the rate of the global mean (Fig. 3.3a). Even in the temperate mid-latitudes where Lyme disease is most prevalent, annual mean temperatures are projected to increase by around 1.5°C per °C of global warming, which will tend to drive the northern edge of Lyme disease to higher latitudes. There is, however, a counteracting effect of increasing humidity deficit (and therefore desiccation of questing ticks) in May–July (Fig. 3.3b). The warming is associated in general with increasing humidity deficit, especially on the most southerly Lyme disease regions where humidity deficit is projected to increase by around 0.5 g/kg per °C of global warming.

In Fig. 3.4 we apply the threshold model for Lyme disease hotspots (Fig. 3.2c) to the patterns of projected climate change at the Paris target of 2°C of global warming. The warming relative to current day (global warming of 1°C) tends to move the Lyme disease zone northwards, into British Columbia, south-eastern Canada, and further into Scandinavia and north-west Russia. However, the increasing atmospheric humidity deficit in the northern hemisphere questing season (May–July) tends to reduce Lyme disease incidence at its current southern edge. As a result, the incidence of Lyme disease in New England, southern China and south-central Europe may actually decline under further global warming, at least according to this very simple climate envelope model.

In summary, Lyme disease, for which ticks (*Ixodes* spp.) act as the vector to transport the infecting bacterium (*B. burgdorferi*) from mice and birds to humans, is increasingly prevalent in New England, Central Europe, Russia and China. The increased incidence of Lyme disease is broadly consistent with the impacts of global warming on the tick life cycle, especially at the northern edge of the temperate mid-latitude regions where Lyme disease is currently most common. However, climate models and simple physics both indicate that global warming will also be associated with a drying of the atmosphere, especially in the summer months when tick nymphs are most sensitive to desiccation. How will these two competing effects play out to affect the overall burden of Lyme disease in the human population? Other

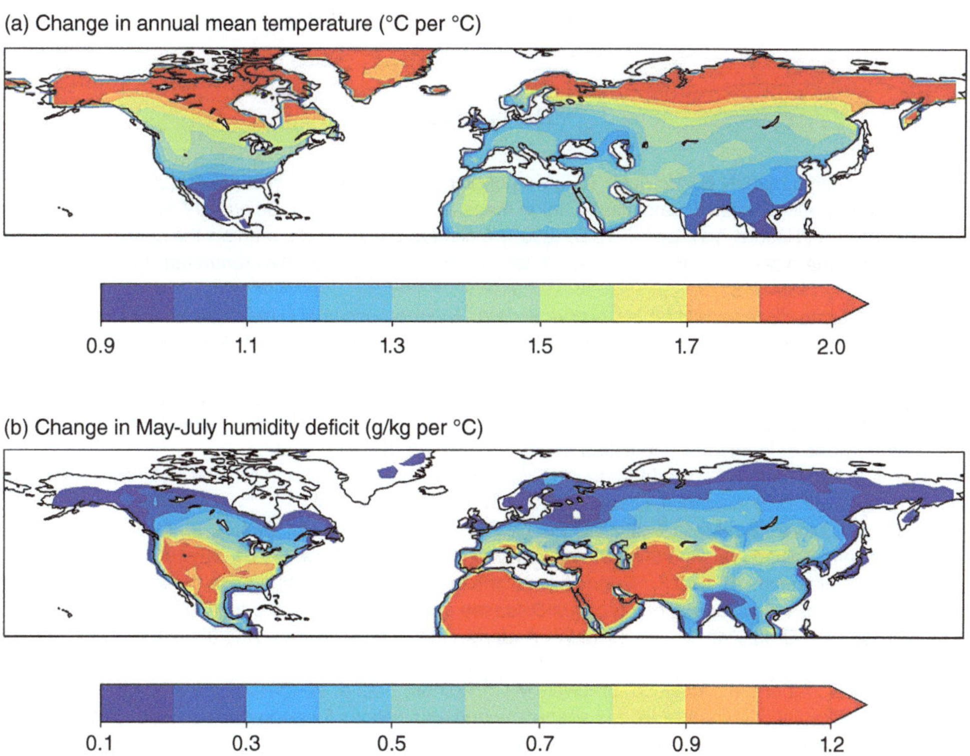

Fig. 3.3. Patterns of change in key climate variables for Lyme disease per °C of global warming: (a) change in annual mean temperature (°C/°C); (b) change in May–July (questing season) humidity deficit (in g/kg per °C). (Authors' own figure.)

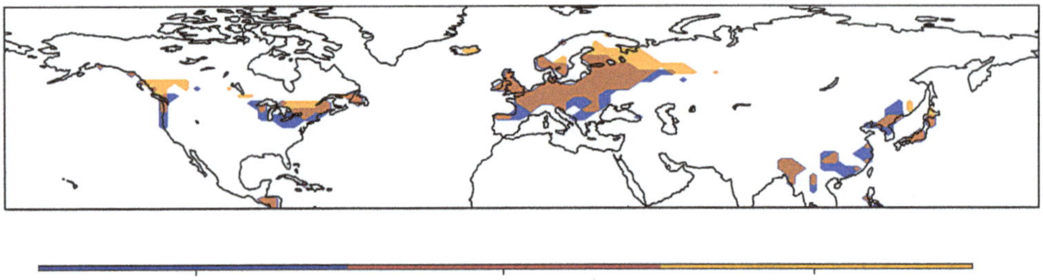

Fig. 3.4. Projected changes in the geographical distribution of Lyme disease hotspots from current climate (1°C of global warming) to the Paris target (2°C of global warming), based on the climate threshold model shown in Fig. 3.2c. (Authors' own figure.)

expert opinions in this book address key issues relevant to answering this question.

Acknowledgements

Peter Cox was supported by the European Research Council ECCLES project (grant agreement number 742472) and two EU Horizon 2020 Research Programme projects: CRESCENDO (grant agreement number 641816) and 4C (grant agreement number 821003). Chris Huntingford was funded by the UKCEH National Capability fund. Morgan Sparey contributed to this expert opinion (especially Fig. 3.1) as part of his Natural Science undergraduate project at the University of Exeter.

References

Beck, H.E., Zimmerman, N.E., McVicar, T.R., Vergopolan, N., Berg, A. and Wood, E.F. (2018) Present and future Köppen–Geiger climate classification maps at 1-km resolution. *Scientific Data* 5, 180214.

Brownstein, J.S., Holford, T.R. and Fish, D. (2003) A climate-based model predicts the spatial distribution of the Lyme disease vector *Ixodes scapularis* in the United States. *Environmental Health Perspectives* 111, 1152–1157.

CDC (2019) *Lyme disease data tables: historical data. Reported cases of Lyme disease by state or locality, 2009–2018*. Centers for Disease Control and Prevention, Atlanta, Georgia. Available at: https://www.cdc.gov/lyme/stats/tables.html (accessed 19 February 2021).

Cheng, A., Chen, D., Woodstock, K., Ogden, N.H., Wu, X. and Wu, J. (2017) Analyzing the potential risk of climate change on Lyme disease in eastern Ontario, Canada using time series remotely sensed temperature data and tick population modelling. *Remote Sensing* 9, 609.

Couper, L.I., MacDonald, A.J. and Mordecai, E.A. (2021) Impact of prior and projected climate change on US Lyme disease incidence. *Global Change Biology* 27, 738–754.

Czupryna, P., Moniuszko-Malinowska, A., Pancewicz, S., Garkowski, A., Gościk, J., *et al.* (2016) Lyme disease in Poland – a serious problem? *Advances in Medical Sciences* 61, 96–100.

Dumic, I. and Severnini, E. (2018) 'Ticking bomb': the impact of climate change on the incidence of Lyme disease. *Canadian Journal of Infectious Diseases and Medical Microbiology* 2018, 5719081.

Friedlingstein, P., O'Sullivan, M., Jones, M.W., Andrew, R.M., Hauck, J., *et al.* (2020) Global carbon budget 2020. *Earth System Science Data* 12, 3269–3340.

Ginsberg, H.S., Albert, M., Acevedo, L., Dyer, M.C., Arsnoe, I.M., *et al.* (2017) Environmental factors affecting survival of immature *Ixodes scapularis* and implications for geographical distribution of Lyme disease: the Climate/Behavior hypothesis. *PLoS ONE* 12, e0168723.

Huntingford, C. and Cox, P.M. (2000) An analogue model to derive additional climate change scenarios from existing GCM simulations. *Climate Dynamics* 16, 575–586.

Huntingford, C., Booth, B.B.B., Sitch, S., Gedney, N., Lowe, J.A., *et al.* (2010) IMOGEN: an intermediate complexity model to evaluate terrestrial impacts of a changing climate. *Geoscientific Model Development* 3, 679–687.

Li, S., Gilbert, L., Vanwambeke, S.O., Yu, J., Purse, B.V. and Harrison, P.A. (2019) Lyme disease risks in Europe under multiple uncertain drivers of change. *Environmental Health Perspectives* 127, 067010.

Malkhazova, S.M., Mironova, V.A., Kotova, T.V., Shartova, N.V. and Orlov, D.S. (2014) Natural-focal diseases: mapping experience in Russia. *International Journal of Health Geographics* 13, 21.

McClure, M. and Diuk-Wasser, M.A. (2019) Climate impacts on blacklegged tick host-seeking behavior. *International Journal for Parasitology* 49, 37–47.

McPherson, M., García-García, A., Cuesta-Valero, F.J., Beltrami, H., Hansen-Ketchum, P., *et al.* (2017) Expansion of the Lyme disease vector *Ixodes scapularis* in Canada inferred from CMIP5 climate projections. *Environmental Health Perspectives* 125, 057008.

Nijsse, F.J.M.M., Cox, P.M. and Williamson, M.S. (2020) Emergent constraints on transient climate response (TCR) and equilibrium climate sensitivity (ECS) from historical warming in CMIP5 and CMIP6 models. *Earth System Dynamics* 11, 737–750.

Ogden, N.H. and Lindsay, L.R. (2016) Effects of climate and climate change on vectors and vector-borne diseases: ticks are different. *Trends in Parasitology* 32, 646–656.

Ogden, N.H., Maarouf, A., Barker, I.K., Bigras-Poulin, M., Lindsay, L.R., *et al.* (2006) Climate change and the potential for range expansion of the Lyme disease vector *Ixodes scapularis* in Canada. *International Journal for Parasitology* 36, 63–70.

Ogden, N.H., Radojevic, M., Wu, X., Duvvuri, V.R., Leighton, P.A. and Wu, J. (2014) Estimated effects of projected climate change on the basic reproductive number of the Lyme disease vector *Ixodes scapularis*. *Environmental Health Perspectives* 122, 631–638.

Riahi, K., van Vuuren, D.P., Kriegler, E., Edmonds, J., O'Neill, B.C., *et al.* (2017) The shared socioeconomic pathways and their energy, land use, and greenhouse gas emissions implications: an overview. *Global Environmental Change* 42, 153–168.

Stanek, G., Wormser, G.P., Gray, J. and Strle, F. (2012) Lyme borreliosis. *The Lancet* 379, 461–473.

Sun, Y., Xu, R. and Cao, W. (2003) *Ixodes sinensis*: competence as a vector to transmit the Lyme disease spirochete *Borrelia garinii*. *Vector-Borne and Zoonotic Diseases* 3, 39–44.

Sykes, R.A. and Makiello, P. (2017) An estimate of Lyme borreliosis incidence in Western Europe. *Journal of Public Health* 39, 74–81.

Zelazowski, P., Huntingford, C., Mercado, L.M. and Schaller, N. (2018) Climate pattern-scaling set for an ensemble of 22 GCMs – adding uncertainty to the IMOGEN version 2.0 impact system. *Geoscientific Model Development* 11, 541–560.

eo4 How to Model the Impact of Climate Change on Vector-Borne Diseases?

Cyril Caminade*
University of Liverpool, Liverpool, UK

4.1 Expert Opinion

4.1.1 Introduction

Vector-borne diseases (VBDs) are transmitted by an ectothermic arthropod or helminth to a human or an animal host. Examples affecting human and animal health are numerous. Malaria, a parasitic disease transmitted by *Anopheles* mosquitoes, still significantly impacts health and economics in Africa. Important arboviruses transmitted by *Aedes* mosquitoes such as dengue, Zika and chikungunya are increasingly plaguing South America and South-east Asia. Lyme disease, a borreliosis transmitted by the bite of an infected ixodid tick to a mammalian host, is also on the rise in temperate regions of Europe and North America. In addition, soil helminths as well as other tick-borne, mosquito-borne and midge-borne pathogens affect livestock production worldwide. Diseases transmitted by sap-sucking insects, such as *Xyllela fastidiosa*, also threaten plant health and therefore agricultural production and food security.

VBDs are climate sensitive, with rainfall and humidity providing suitable conditions for vector breeding sites. As ectothermic vectors do not regulate their internal temperature, air temperature directly affects their development rates, biting rates and the pathogen incubation period in their body. Importantly, as most VBDs affecting humans are zoonotic, and given that vectors are climate sensitive, the 'environment–human health–animal health' nexus has led to the recent development of the One Health framework (Zinsstag *et al.*, 2011).

As temperatures are rising and the hydrological cycle is modified, vectors' distribution, their activity and behaviour, and the potential length of VBD transmission seasons have already been affected worldwide (Caminade *et al.*, 2019). Climate change is expected to impact future vector whereabouts, in conjunction with other critical factors like globalization of goods and services (because vectors are primarily transported over long distances via human activities or migratory animals), migration of populations, socio-economic and human behaviour factors, and drug and insecticide resistance. The future potential development of novel diagnostic tests, new insecticides, vector-control techniques and drugs will ultimately modulate such potential health burdens in the long term. The recent and unprecedented rapid development of efficient vaccines for COVID-19 is a prime example and provides hope for the near-term future. Importantly, vaccines to counter major tick-borne diseases (TBDs) affecting human health, such as Lyme disease, Crimean-Congo haemorrhagic fever

*Email: cyril.caminade@liverpool.ac.uk

DOI: 10.1079/9781789249637.0004

(CCHF), relapsing fever, Rocky Mountain spotted fever and human babesiosis, are currently lacking.

Ticks have four life cycle stages; they evolve from egg to larva to nymph to an adult tick. Depending on the tick species, their life cycle requires a different number of animal hosts to be completed. For most ixodid ticks three intermediate animal host species are usually required, and their life cycle takes about 3 years in total. In temperate regions, larvae will typically feed on small mammals and birds during the warm season. Nymphs then feed on a small rodent when they emerge the following spring. Following a moulting process, a nymph becomes an adult tick the subsequent spring and will seek blood meals from a large ruminant (deer, sheep, cattle) or human host. High-humidity conditions (>85%) are generally required to avoid tick desiccation. In addition, questing for hosts usually occurs at temperatures >7°C (Süss *et al.*, 2008). Because of their climate sensitivity, areas of moderate to heavy rainfall with vegetation that retains humidity favour tick presence. Typical habitats for ixodid tick species include deciduous and coniferous woodlands, moorland, shrubland, rough pasture, forests and urban parks (Medlock *et al.*, 2013).

Recent trends in tick vectors and the diseases they transmit are worrying. In Europe, *Ixodes ricinus*, the sheep tick that commonly transmits Lyme disease, has spread to higher altitudes in the Alps and the Carpathian Mountains (Medlock *et al.*, 2013) and to higher latitudes in Sweden, Norway and Germany (Lindgren *et al.*, 2000). Milder winters, prolonged spring and autumn seasons, increased vegetation cover and the spread of cervids into new regions have been proposed to explain such changes. An increase in Lyme disease and tick-borne encephalitis (TBE) prevalence has been reported for many European countries, however with caveats related to heterogeneity in surveillance means and reporting methods across countries (Sykes and Makiello, 2017). This increase in prevalence is related to several human and socio-economic factors (afforestation, human behaviour, e.g. an increase in green tourism, etc.) as well as increased climatic suitability for ticks. Similar trends have been observed in North America and parts of northern Russia. *Ixodes scapularis*, the deer tick on the east coast, and *Ixodes pacificus*, the main Lyme disease vector over the west coast, have spread further north in the USA during the last decades (Eisen *et al.*, 2016). *I. scapularis* subsequently spread to southern Canadian provinces (Ogden *et al.*, 2009). Prevalence rates of Lyme disease have increased significantly in the USA and Canada over the past decades. Lyme disease is believed to affect about 476,000 Americans every year based on recent insurance data for patients seeking treatment, making Lyme disease the most prevalent VBD in the USA (Schwartz *et al.*, 2021). In addition, other important TBDs such as ehrlichiosis, spotted fever rickettsiosis, babesiosis, tularaemia and Powassan virus disease have all been reported to be on the rise by the Centers for Disease Control and Prevention in the USA.

Given recent observed changes, assessing the spatiotemporal burden of TBD in a changing climate and a more connected world is today a priority for many animal and health specialists worldwide. This expert opinion will provide an overview of mathematical models that have been used to assess the impact of climate change on ticks and TBD, ways forward in terms of improving models for the recent context and broad guidelines for conducting future climate change risk assessment.

4.1.2 A sea of mathematical models

Mathematical models that simulate the risk posed by a particular tick, or a particular TBD, usually fall into three categories. First, statistical models aim to predict the geographic distribution of a particular vector species or disease. This branch of modelling relies on the availability of surveillance data sets or observations (e.g. vector or disease presence/absence) to infer a relationship between environmental, socio-economic or other risk factors (covariates) and the desired outcome. This relationship, or model fit, is then usually extrapolated at larger spatial scale. The importance of each risk factor in explaining the variance of the outcome is usually estimated to better understand the underlying biological mechanisms and provide recommendations to public health experts. Statistical modelling has been favoured by ecologists over the past decades and addresses questions such as: what model will be postulated, how is the model response distributed and are observations independent? Various modelling approaches in that branch can be considered, ranging from simple linear regression models to maximum entropy modelling ('Maxent') to generalized additive models. The quality of model outputs mainly relies on the quality and quantity of observed surveillance data that feed them, the availability and quality of input covariates and the interpretation of biological mechanisms at play to provide robust recommendations. Statistical models are directly derived from observations and may utilize fitting methods which inherently estimate their own uncertainty. Usually, resulting risk maps vary only in space and do not necessarily consider the dynamic relationship between climate variables and disease vectors which tends to follow a seasonal pattern. However, these models can accurately simulate the risk posed by a particular vector or disease at high spatial resolution. Examples of statistical models include early efforts to anticipate the spread of TBE in Europe (Randolph and Rogers, 2000), as well as more recent efforts to map the ecological niches of important tick species in the USA (Wisely and Glass, 2019) and Europe (Williams *et al.*, 2015). Notably, the Maxent method has been widely used to assess the risk posed by insect vectors because of its overall good performance (Elith *et al.*, 2006). More detailed guidelines about the use of statistical models for ecological applications, and their related pros and cons, are provided by Guillera-Arroita *et al.* (2015).

Second, over the past decade, computer scientists have developed a broad range of algorithms that are perfectly suited for prediction, such as neural networks, ensemble of trees and support vector machines, to name a few. These models are referred to as machine learning (ML) models. The ML approach assumes that the data-generating process is extremely complex and unknown, a highly relevant assumption for biological systems, and learns responses by observing inputs to find dominant patterns (Elith *et al.*, 2008). The emphasis of ML is on the ability of a model to do a correct prediction and focuses on what is being predicted and how the success of that prediction should be measured. The uptake of ML models by ecologists has been relatively limited with respect to standard statistical approaches over the past decades, mostly because of the difficulty in interpreting the underlying mechanisms and the complexity of ML models. However, boosted regression trees are now increasingly being utilized to assess the risk posed by vectors and diseases, including recent examples for ticks and TBDs in China (Zhao *et al.*, 2021) and Scandinavia (Kjær *et al.*, 2019).

Finally, mechanistic, or process-based, models make direct assumptions about the biological and environmental mechanisms that drive a particular vector to thrive or disease to be transmitted in a particular location at a given time. The empirical relationship between climate variables and key epidemiological parameters is usually derived directly from available laboratory or field data. Different modelling approaches can be considered ranging from simple climate envelope models to degree-day models to more complex SEIR (susceptible, exposed, infectious, recovered) models that dynamically simulate the life cycle of the vector and the proportions of infected vectors and hosts. The steady state of SEIR models leads to the formulation of R_0, the basic reproduction number. SEIR models for ticks and TBDs are extremely complex and difficult to parameterize given the long time period required for the tick to complete its full life cycle and the number of intermediate animal hosts involved in the process (birds, rodents, ruminants). Process-based models are powerful tools to explore seasonality and potential long-term changes in vector abundance and disease transmission risk, a critical point to assess recent climate change effects and develop early warning systems. The mechanistic modelling framework and variation of the parameter space allow exploring diverse mechanisms at play, such as biodiversity and disease risk amplification/dilution, host-seeking behaviour, ticks and TBD risk importation by birds or co-infection in vectors and hosts. A more detailed review of mechanistic models for ticks and TBD is provided by Ostfeld and Brunner (2015), Lou and Wu (2017) and Norman *et al.* (2016). Importantly, mechanistic models successfully anticipated the spread of *I. scapularis* into southern Canada (Ogden *et al.*, 2005).

Mathematical models for ticks and TBDs are very diverse and some models are better suited to answer particular scientific questions.

However, there are a few common critical points that deserved to be mentioned. First, the quality and quantity of surveillance data to develop and validate them are paramount. On a positive note, surveillance data have improved greatly over the past decade, with projects that now centralize occurrence data sets such as the Global Biodiversity Information Facility or efforts conducted by the US and European centres for disease control that routinely update tick presence data in real time. Importantly, time series of tick abundance are important to investigate seasonality and early impacts of climate change, such as continuous abundance data collected for ixodid ticks by Public Health England in the UK and Institut National de la Recherche Agronomique in France, or other similar initiatives. Second, multi-model comparison exercises using the same inputs should be encouraged because model ensembles always provide more information and allow a more detailed exploration of uncertainties. Finally, given the complexity of TBDs, a One Health multidisciplinary framework – including experts in ecology, veterinary and medical entomology, public health, mathematical modelling, and earth and computer sciences – is key.

4.2 Future Predictions (Brief Modelling Guidelines)

Earth system models are now routinely employed to simulate the impact of climate change on various sectors (e.g. agriculture, water resources, biodiversity, health). Coupled general circulation model and regional climate model simulations are centralized by the Coupled Model Inter-Comparison Project (CMIP), which is now in its sixth phase. All major climate centres contribute to CMIP worldwide, resulting in a large multi-model ensemble available to researchers. Detailed guidelines about the use of multiple climate model ensembles for impact assessment are provided by Semenov and Stratonovitch (2010). I would like to emphasize a few important points here.

1. Climate models still show significant biases; they can overestimate or underestimate some climate variables. Consequently, climate model outputs must be calibrated with respect to climate observations before driving any impact model. This is a critical exercise for TBDs because ticks are sensitive to temperature and humidity thresholds. Various calibration techniques can be applied (Maraun, 2016). Importantly, projects that readily calibrate climate model outputs for the impact community, such as the Inter-Sectoral Impact Model Intercomparison Project in Europe (Warszawski *et al.*, 2014) or the NASA Earth Exchange Global Daily Downscaled Projections in the USA (Thrasher *et al.*, 2013), should be praised and encouraged.

2. Climate model ensembles are multidimensional. Different climate models with different parameterizations can be used, different atmospheric and oceanic initial conditions can be utilized, and different emission scenarios (Representative Concentration Pathways or RCPs) can be considered. RCPs range from the mildest (RCP2.6) to the most extreme (RCP8.5) emission scenario. A detailed climate change risk assessment should ideally include all RCPs and provide an estimation of uncertainties associated with the climate model ensemble and the different impact models (Caminade *et al.*, 2014). This exercise can be computationally expensive, but a subset of climate models can be employed. Climate models should ideally be preselected to represent the full range of uncertainty in the multi-model ensemble (Jones *et al.*, 2019). A detailed assessment of the different sources of uncertainties (related to climate models, impact models and emission scenarios) should also be carried out (Jones *et al.*, 2019). Shared Socio-economic Pathway (SSP) population scenarios can be combined with RCPs to disentangle climate change effects from population changes (Rohat *et al.*, 2020).

3. Long-term changes need to be investigated. The use of 20-year averages to estimate mid- or end-of-century changes is standard. Simulated change in mean seasonal cycles (tick abundance, length of TBD transmission season, etc.) should be estimated to inform end users such as vector control experts and farmers about potential future changes. Ultimately, simulated future trends always need to be compared with recently observed trends, and long-term historical data if available, to provide a better understanding and interpretation of results (past, present and future).

References

Caminade, C., Kovats, S., Rocklov, J., Tompkins, A.M., Morse, A.P., *et al.* (2014) Impact of climate change on global malaria distribution. *Proceedings of the National Academy of Sciences USA* 111, 3286–3291.

Caminade, C., McIntyre, K.M. and Jones, A.E. (2019) Impact of recent and future climate change on vector-borne diseases. *Annals of the New York Academy of Sciences* 1436, 157–173.

Eisen, R.J., Eisen, L. and Beard, C.B. (2016) County-scale distribution of *Ixodes scapularis* and *Ixodes pacificus* (Acari: Ixodidae) in the Continental United States. *Journal of Medical Entomology* 53, 349–386.

Elith J., Graham, C.H., Anderson, R.P., Dudík, M., Ferrier, S., *et al.* (2006) Novel methods improve prediction of species' distributions from occurrence data. *Ecography* 29, 129–151.

Elith, J., Leathwick, J.R and Hastie, T. (2008) A working guide to boosted regression trees. *Journal of Animal Ecology* 77, 802–813.

Guillera-Arroita, G., Lahoz-Monfort, J.J., Elith, J., Gordon, A., Kujala, H., *et al.* (2015) Is my species distribution model fit for purpose? Matching data and models to applications. *Global Ecology and Biogeography* 24, 276–292.

Jones, A.E., Turner, J., Caminade, C., Heath, A.E., Wardeh, M., *et al.* (2019) Bluetongue risk under future climates. *Nature Climate Change* 9, 153–157.

Kjær, L.J., Soleng, A., Edgar, K.S., Lindstedt, H.E.H., Paulsen, K.M., *et al.* (2019) Predicting the spatial abundance of *Ixodes ricinus* ticks in southern Scandinavia using environmental and climatic data. *Scientific Reports* 9, 18144.

Lindgren, E., Talleklint, L. and Polfeldt, T. (2000) Impact of climatic change on the northern latitude limit and population density of the disease-transmitting European tick *Ixodes ricinus*. *Environmental Health Perspectives* 108, 119–123.

Lou, Y. and Wu, J. (2017) Modeling Lyme disease transmission. *Infectious Disease Modelling* 2, 229–243.

Maraun, D. (2016) Bias correcting climate change simulations – a critical review. *Current Climate Change Reports* 2, 211–220.

Medlock, J.M., Hansford, K.M., Bormane, A., Derdakova, M., Estrada-Peña, A., *et al.* (2013) Driving forces for changes in geographical distribution of *Ixodes ricinus* ticks in Europe. *Parasites & Vectors* 6, 1.

Norman, R.A., Worton, A.J. and Gilbert, L. (2016) Past and future perspectives on mathematical models of tick-borne pathogens. *Parasitology* 143, 850–859.

Ogden, N.H., Bigras-Poulin, M., O'Callaghan, C.J., Barker, L.K., Lindsay, L.R., *et al.* (2005) A dynamic population model to investigate effects of climate on geographic range and seasonality of the tick *Ixodes scapularis*. *International Journal for Parasitology* 35, 375–389.

Ogden, N.H., Lindsay, L.R., Morshed, M., Sockett, P.N. and Artsob, H. (2009) The emergence of Lyme disease in Canada. *Canadian Medical Association Journal* 180, 1221–1224.

Ostfeld, R.S. and Brunner, J.L. (2015) Climate change and *Ixodes* tick-borne diseases of humans. *Philosophical Transactions of the Royal Society B* 370, 20140051.

Randolph, S.E. and Rogers, D.J. (2000) Fragile transmission cycles of tick-borne encephalitis virus may be disrupted by predicted climate change. *Philosophical Transactions of the Royal Society B* 267, 1741–1744.

Rohat, G., Monaghan, A., Hayden, M.H., Ryan, S.J., Charriere, E. and Wilhelmi, O.V. (2020) Intersecting vulnerabilities: climatic and demographic contributions to future population exposure to *Aedes*-borne viruses in the United States. *Environmental Research Letters* 15, 084046.

Schwartz, A.M., Kugeler, K.J., Nelson, C.A., Marx, G.E. and Hinckley, A.F. (2021) Use of commercial claims data for evaluating trends in Lyme disease diagnoses, United States, 2010–2018. *Emerging Infectious Diseases* 2, 499–507.

Semenov, M.A. and Stratonovitch, P. (2010) Use of multi-model ensembles from global climate models for assessment of climate change impacts. *Climate Research* 41, 1–14.

Süss, J., Klaus, C., Gerstengarbe, F.-W. and Werner, P.C. (2008) What makes ticks tick? Climate change, ticks, and tick-borne diseases. *Journal of Travel Medicine* 15, 39–45.

Sykes, R.A. and Makiello, P. (2017) An estimate of Lyme borreliosis incidence in Western Europe. *Journal of Public Health* 39, 74–81.

Thrasher, B., Xiong, J., Wang, W., Melton, F., Michaelis, A. and Ramakrishna, N. (2013) Downscaled climate projections suitable for resource management. *Eos* 94, 321–323.

Warszawski, L., Frieler, K., Huber, V., Piontek, F., Serdeczny, O. and Schewe, J. (2014) The Inter-Sectoral Impact Model Intercomparison Project (ISI–MIP): project framework. *Proceedings of the National Academy of Sciences USA* 111, 3228–3232.

Williams, H.W., Cross, D.E., Crump, H.L., Drost, C.J. and Thomas, C.J. (2015) Climate suitability for European ticks: assessing species distribution models against null models and projection under AR5 climate. *Parasites & Vectors* 8, 440.

Wisely, S.M. and Glass, G. (2019) Advancing the science of tick and tick-borne disease surveillance in the United States. *Insects* 10, 361.

Zhao, G.-P., Wang, Y.-X., Fan, Z.-W., Ji, Y., Liu, M.-J., *et al.* (2021) Mapping ticks and tick-borne pathogens in China. *Nature Communications* 12, 1075.

Zinsstag, J., Schelling, E., Waltner-Toews, D. and Tanner, M. (2011) From 'one medicine' to 'one health' and systemic approaches to health and well-being. *Preventive Veterinary Medicine* 101, 148–156.

e05 Challenges of Modelling and Projecting Tick Distributions

William Wint* **and Neil Alexander**
Environmental Research Group Oxford Ltd, Oxford, UK

5.1 Expert Opinion

5.1.1 Introduction

To produce temporal projections of tick distributions, two inputs are needed: present and projected. This may seem to be an unnecessary truism, but the stipulation introduces a number of challenges. This expert opinion looks at the practicalities of modelling both distributions and sets out what data are needed to produce effective (and, it is hoped, reliable) models.

5.1.2 Setting the current baseline

In a world that is changing out of all recognition in a multitude of unexpected ways, it is clearly valuable to be able to predict where human and animal diseases are likely to be a problem in the future. Although such predictions might be most useful (and probably less inaccurate) in the short to medium term, there remains a case to have a long-term perspective in the toolbox, if only to 'guesstimate' whether a currently increasing problem is likely to become a permanent issue or is just a temporary 'blip'.

Other contributions to this book set out the drivers of tick occurrence and provide opinions on where the future will lead us in terms of tick distributions. This one sets out to illustrate what is needed 'behind the scenes' to project long-term changes in where ticks are likely to be found. Of course, what is needed and what is practically available are not necessarily the same thing!

The illustration uses spatial distribution models based on assessing the relationship between the known vector distribution status and the values of a series of driver covariates at a set of representative sample points. The driver values are known for every location in the area of interest and so the established sampled relationship can be used to estimate the vector distribution metric for every location. The methods used to implement these models have been mainstream for many years now and include boosted regression trees (BRT) and random forest, maximum entropy ('Maxent'), various forms of regression and a raft of Bayesian estimations (Breiman, 2001; Elith and Leathwick, 2009; Elith *et al.*, 2006, 2008, 2010). As the process is stochastic, most practitioners use many iterations to produce a mean estimate (Kraemer *et al.*, 2015) and many also combine methods to smooth the idiosyncrasies linked to particular methods, distributions or value ranges (ECDC, 2019).

The examples are derived from work done for the VectorNet project (ECDC, 2018), funded

*Email: william.wint@zoo.ox.ac.uk

DOI: 10.1079/9781789249637.0005

by the European Food Safety Authority and the European Centre for Disease Prevention and Control, which assembles and publishes vector distribution data extracted from the literature (and latterly from coordinated field sampling). A number of tick species are priority targets: *Ixodes ricinus* and *Ixodes persulcatus*, *Hyalomma marginatum*, *Rhipicephalus sanguineus* and *Dermacentor reticulatum*. The focus here is on *I. ricinus*, as the most wide-ranging of these species within Europe, but the principles (and most probably the results) would be similar for all five species.

A first decision to be made is what distribution metric to model. The two main alternatives are presence and absence or abundance. There are arguments as to whether tick abundance is a useful quantity in epidemiological analyses or not (Wint *et al.*, 2018), when it is the infection rate that is a more immediate determinant of disease risk. Leaving these uncertainties aside, there are major practical issues in assembling consistent abundance data. All too often these are not available because different studies look for different life stages or use different methods, frequencies or seasonal coverage. This means combining them into a standardized and coherent metric is difficult or involves convoluted weightings (Wint *et al.*, 2020a,b). As a result, modelling presence and absence, rather than abundance, is much more feasible, although it still relies on having access to a decent set of observed ('training') data. Luckily, in the case of European ticks, the generosity of experts such as Agustin Estrada-Peña has meant that a lot of data extracted from the literature are in the public domain. Although they are not sampled with geography in mind, data from these reports can be aggregated if required to reduce pseudo replication by using an average or maximum value per 1, 5 or 10 km grid.

The availability of standardized training data, especially for abundance, relies not only on standardized methods, but also on a coordinated approach to data sharing. A number of pan-European projects have strongly promoted this (Alexander *et al.*, 2015; COST, 2018; VectorNet – ECDC, 2018) and have emphasized the value of continental analyses to all data contributors, highlighting how it helps improve local models as well as extending coverage. These initiatives have also encouraged authors to provide full and geo-referenced details of sampling effort as well as simple numbers or presence.

Most presence and absence modelling relies on both presence and absence records being available in the training data set (Elith and Leathwick, 2009; ECDC, 2019). There need to be roughly the same number of absences as there are presences, which means there are rarely enough recorded absences as field operatives are not usually encouraged to record zeros. As a result, they need to be generated. Some methods generate them internally (Maxent) and in others the use of pseudo absences is widespread. These are identified using the assumption that the target species is absent where it is not recorded as present. Clearly this is not guaranteed to be true close to recorded presences, so the pseudo absences are often assigned to locations further than a defined distance from the recorded presences. This approach is not based upon anything other than location and does not use any knowledge that might be available about habitat suitability, environmental limits or the like.

This example instead uses suitability masks to define known areas of absence, which can be any distance from known presences, thereby improving the training data accuracy near to known presence records as well as far distant. By incorporating land-use categories, vegetation seasonality and snow coverage, such masks also help to broaden the number of factors used to define target species distributions.

There are, of course, other factors that affect vector distributions that should be incorporated into distribution models: most obviously the hosts on which the vectors rely, in addition to humans. If there are no suitable hosts then there are no vectors. As often as not host distributions are also 'modelled' (Wint *et al.*, 2014; Estrada-Peña and de la Fuente, 2016). It is not always reliable to use these as indicators of absence, as there may well be other hosts that facilitate vector presence; however, they can be used as additional masks to refine the model outputs on the assumption that some hosts are vector amplifiers and so the lack of them significantly reduces the likely number of ticks.

The issues of which driver covariates to use in a spatial model in addition to the habitat masking ones is open to discussion and is largely dictated by what is available. The obvious ones are climate: primarily temperature and rainfall, which in the current example are also complemented by vegetation index, relative humidity, altitude, land-use percentage and human population density. The climate humidity, precipitation and vegetation index data sets are derived directly from satellite imagery or ground stations, and from daily to monthly records for upwards of the last 20 years. This means an appropriate value is difficult to choose; should it be an annual mean, the value for a specific date or month, or for a specific combination of years? As a result, the drivers used here have been subjected to one of the more widely used data reduction techniques, i.e. temporal Fourier processing (Scharlemann *et al.*, 2008), which extracts a limited series of parameters from a long-term time series, to produce biologically relevant indicators of seasonality and variability.

Figure 5.1 shows various models of *I. ricinus* that are ensembles of BRT and random forest outputs based on approximately 7500 training data points, including an equal number of absences inferred from suitability masks. Model A is a simple (probability of presence) model for the present day, based on the principles and data sets outlined above, using climate, land use, vegetation, topography and human population as covariates (Fig. 5.1a). This is a standard covariate data set that has proved successful for a number of vector and host species models (Alexander *et al.*, 2014; Alten *et al.*, 2016; Schaffner *et al.*, 2016; Balenghien *et al.*, 2020).

Model B is the same model with: (i) a suitability mask incorporating land-use categories, a climate envelope and annual seasonality of snow and vegetation cover defined from the literature and VectorNet expert opinion (ECDC, 2019) in grey; and (ii) a host mask derived from three species of deer hosts – red, roe and fallow (Alexander *et al.*, 2014; Wint *et al.*, 2014) – in black (Fig. 5.1b). The masks make a considerable difference: large areas in Scandinavia and northern continental Europe are masked out as being unsuitable in terms of land use and significant parts of west Wales, north-western Russia and the western Iberian Peninsula are masked out because the models of deer hosts suggest that there are not enough deer to amplify the tick populations.

5.2 Future Predictions

Producing 'contemporary' models can be based on the optimum set of covariate driver data, limited only by the accessibility of the information. In recent years this has improved exponentially as global and public-domain data become ever more widely available and at improved spatial and temporal resolution.

Producing projected populations is somewhat more problematic, especially if the desired metric is change, rather than a standalone output. Most projections rely on the logic of extrapolating a contemporary model: i.e. applying the relationships between a target distribution and drivers calculated from present-day distributions to the projected values of the driver covariates for the desired date in the future.

This approach has some inherent faults – most critically that it assumes the relationships between target distribution and the drivers are linear with each driver; that any interactions also remain constant; and that there are no new disruptive factors to override the driver relationships used. None of these assumptions can be relied upon and so projections are, more than is usually the case for stochastic modelling, *caveat emptor*.

A change estimation also means that both models need to be constructed in exactly the same way using the same training data, methods and covariates. If any of these differ for the different dates, then any differences between the models may be the result of the differing covariates, not any changes in the same covariates.

This severely limits the covariates that can be used for modelling and masking to those which are constant (e.g. elevation or daylength) and those for which projections are available. To compound this uncertainty, many projections, especially for climate, are very variable, even within a specified change 'scenario' (more correctly, Representative Concentration Pathways or RCPs that are driven by different assumptions of societal climate change mitigation strategies). This variability can be mitigated to some degree by ensembling a number of global climate models; in this example, the climate projections used are based on an ensemble of the outputs from 17 general circulation models (Messina *et al.*, 2019).

Other projections widely available are for human population densities (Stehfest *et al.*, 2014; van Vuuren *et al.*, 2017; KDP, 2020), but for vegetation indices or land-use categories

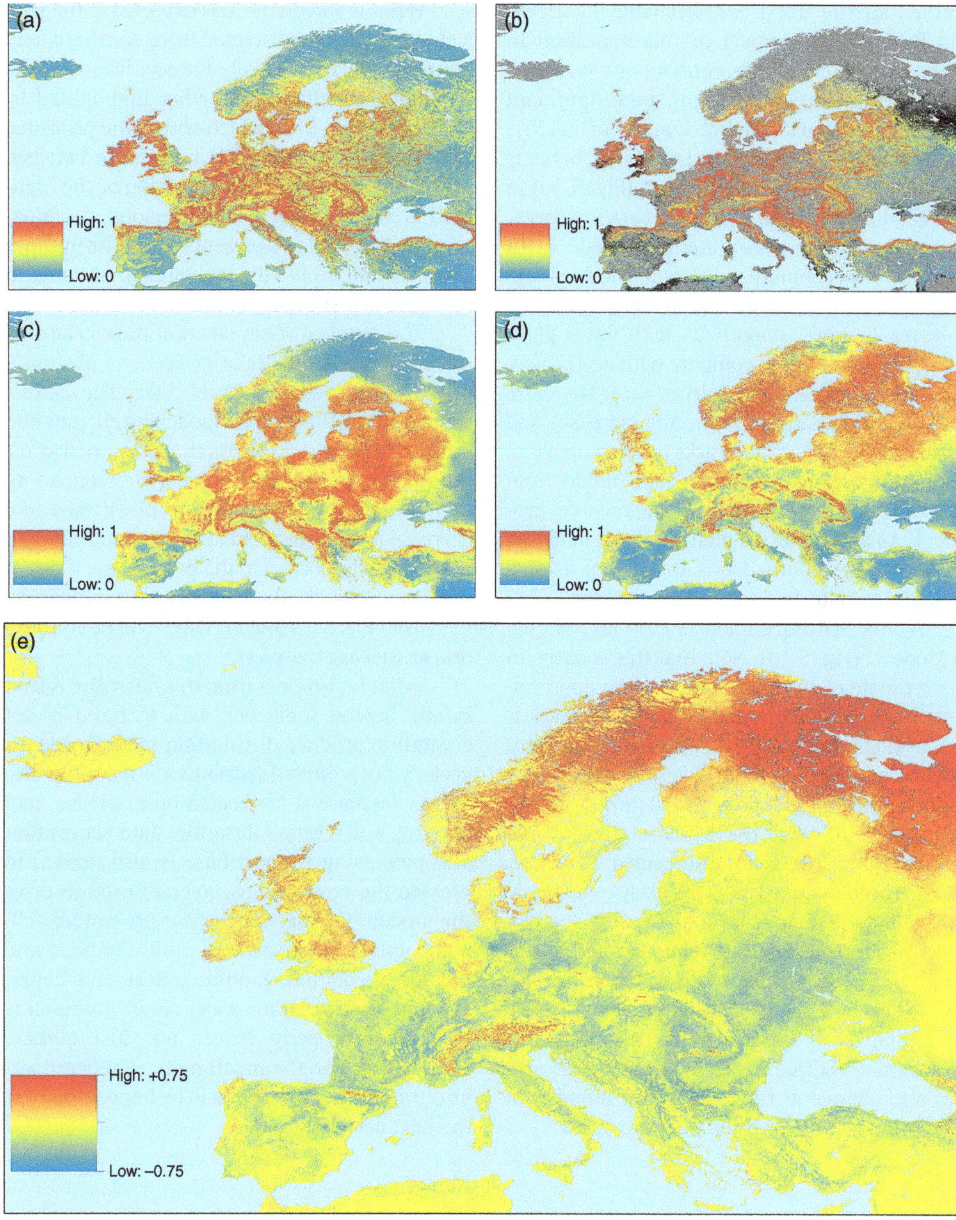

Fig. 5.1. Spatial models of *Ixodes ricinus*. Values are probability of presence: (a) present day, all covariates, unmasked (Model A); (b) present day, all covariates, masked (Model B); (c) present day, projectable covariates (Model C); (d) projected 2080, RCP8.5, projectable covariates (Model D); (e) difference between 2020 and 2080 (2080 – 2020). (Authors' own figure.)

such projections are not yet mature enough to use. This is partly because the projections of land use need to predict changes in spatial distribution – spread or contraction – as well as levels in a particular location and calculating which category spreads to replace another that contracts is very complex.

As a result, projection covariates in this example are limited to human population, temperature, humidity and rainfall, and have been selected for the worst case available, i.e. 2080 and RCP8.5. The impact of this limitation in input covariates on the contemporary model, using the same vector training data input, can be seen by comparing Models C and A in Fig. 5.1. Model C with limited covariates (Fig. 5.1c) is broadly similar to the full model A (Fig. 5.1a) in that the main distribution ranges are northern, central and eastern continental Europe, with distributions becoming patchier to the west and south. The limited covariate model (Fig. 5.1c), however, is more 'smoothed', with fewer gaps. Using the full model to compare with any projections based on the limited driver set is therefore going to exaggerate any projected increase. Add to this the fact that the masks (shown above to be an important refinement) are made from parameters that also cannot be reliably projected. As any projected changes must be based on 'like for like' comparisons, this means using a limited covariate data set and no masking.

A projected distribution for *I. ricinus* is shown in Model D (Fig. 5.1d). Note that this is likely an overestimate of the distribution, as an appropriately projected environmental and host mask is not available to refine the model predictions. What is nevertheless immediately evident is that the tick distribution is projected to move substantially north to cover the whole of the Scandinavian region. This is not unexpected, of course, and has been assumed to be the likely outcome of environmental change for many years. One biological (or ecological) uncertainty is whether this projected expansion of *I. ricinus* will be in any way affected by the fact that some of the new areas are already currently occupied by the tick's congener, *I. persulcatus* (ECDC, 2018). Will this second species also change its range, or will it interfere in some way with the invasion from the south?

What is somewhat less expected is the projected retreat of the vector from southern and Central Europe, especially France, Italy and the Balkans. This change is further highlighted by the map in Fig. 5.1e, which shows the projected change in probability of tick presence between 2020 and 2080. With the exception of the highland areas and some coastal regions, the modelled likelihood of tick presence drops throughout continental Europe (the blue areas), not only in the regions just highlighted.

The number of caveats that have been presented above about these projections obviously begs a question about their reliability. The use of a consistent set of inputs for modelling current and future dates reduces the risk that they are way off the mark, and the geographical coherence into more or less contiguous regions of increase and decrease also suggests that the projected changes are likely to be correct in the broad sense if not in the detail (a bit like the full and reduced covariate models in Fig. 5.1a and c). There will be contraction as well as expansion.

What is also clear from this narrative is that simply having some tick data to hand is not enough to produce distribution models, and especially not projected distributions. A continental vector database is predicated on extensive data sharing, and a very substantial data acquisition and processing support base is also needed to provide the covariate parameters used to drive the models. Furthermore, these are not just climatic, but are more broadly multi-thematic and include vegetation, land-use, host and demographic variables. This wider set of parameters will more properly reflect not just climate change, but also the much more significant environmental change that will be happening over the next decades.

References

Alexander, N.S., Morley, D., Medlock, J., Searle, K. and Wint, W. (2014) A first attempt at modelling roe deer (*Capreolus capreolus*) distributions over Europe. *Open Health Data* 2, e2.

Alexander, N., Allepuz, A., Alten, B., Bøder, R., Bonnet, S., *et al.* (2015) The impact of a decade (2004–2015) of research on vector-borne diseases. In: Lancelot, R., Laurens, S. and Lewer, A.L.A. (eds) *EDEN & EDENext*. CIRAD, Montpellier, France. Available at: http://publications.cirad.fr/une_notice.php?dk=577214 (accessed 27 May 2021).

Alten, B., Versteirt, V., Van Bortel, W., Wint, W. and Alexander, N.S. (2016) VBORNET gap analysis: sand fly vector distribution models utilised to identify areas of potential species distribution in areas lacking records. *Open Health Data* 4, 1.

Balenghien, T., Alexander, N., Arnbórsdóttir, A.L., Bisia, M., Blackwell, A., *et al.* (2020) VectorNet data series 3: Culicoides abundance distribution models for Europe and surrounding regions. *Open Health Data* 7, 2.
Breiman, L. (2001) Random forests. *Machine Learning* 45, 5–32.
COST (2018) Aedes Invasive Mosquitoes. European Cooperation in Science and Technology. Available at: https://www.cost.eu/actions/CA17108/#tabsIName:overview (accessed 27 May 2021).
ECDC (2018) European network for sharing data on the geographic distribution of arthropod vectors, transmitting human and animal disease agents (VectorNet). European Centre for Disease Prevention and Control, Stockholm. Available at: https://www.ecdc.europa.eu/en/about-us/partnerships-and-networks/disease-and-laboratory-networks/vector-net (accessed 20 November 2020).
ECDC (2019) *A Spatial Modelling Method for Vector Surveillance*. European Centre for Disease Prevention and Control, Stockholm.
Elith, J. and Leathwick, J.R. (2009) Species distribution models: ecological explanation and prediction across space and time. *Annual Review of Ecology, Evolution, and Systematics* 40, 677–697.
Elith, J., Graham, C.H., Anderson, R.P., Dudík, M., Ferrier, S., *et al.* (2006) Novel methods improve prediction of species' distributions from occurrence data. *Ecography* 29, 129–151.
Elith, J., Leathwick, J.R., and Hastie, T. (2008) A working guide to boosted regression trees. *Journal of Animal Ecology* 77, 802–813.
Elith J., Kearney, M. and Phillips, S. (2010) The art of modelling range-shifting species. *Methods in Ecology and Evolution*, 1, 330–342.
Estrada-Peña, A. and de la Fuente, J. (2016) Species interactions in occurrence data for a community of tick-transmitted pathogens. *Scientific Data* 3, 160056.
KDP (2020) KNMI Data Platform (KDP). Royal Netherlands Meteorological Institute. Available at: https://dataplatform.knmi.nl (accessed 10 June 2021).
Kraemer, M.U., Sinka, M.E., Duda, K.A., Mylne, A.Q.N., Shearer, F.M., *et al.* (2015) The global distribution of the arbovirus vectors *Aedes aegypti and Ae. albopictus*. *eLife* 4, e08347.
Messina, J.P., Brady, O.J., Golding, N., Kraemer, M.U.G., Wint, G.R.W., *et al.* (2019) The current and future global distribution and population at risk of dengue. *Nature Microbiology* 4, 1508–1515.
Schaffner, F., Versteirt, V., Bortel, W. Van, Zeller, H., Wint, W. and Alexander, N.S. (2016) VBORNET gap analysis: mosquito vector distribution models utilised to identify areas of potential species distribution in areas lacking records. *Open Health Data* 4, e6.
Scharlemann, J.P., Benz, D., Hay, S.I., Purse, B.V., Tatem, A.J., *et al.* (2008) Global data for ecology and epidemiology: a novel algorithm for temporal Fourier processing MODIS data. *PLoS ONE* 3, e1408.
Stehfest, E., van Vuuren, D., Kram, T., Bouwman, L., Alkemade, R., *et al.* (2014) *Integrated Assessment of Global Environmental Change with IMAGE 3.0. Model Description and Policy Applications*. PBL Netherlands Environmental Assessment Agency, The Hague, the Netherlands.
van Vuuren, D.P., Stehfest, E., Gernaat, D.E.H.J., Doelman, J.C., van den Maarten, B., *et al.* (2017) Energy, land-use and greenhouse gas emissions trajectories under a green growth paradigm. *Global Environmental Change* 42, 237–250.
Wint, W., Morley, D., Medlock, J. and Alexander, N. (2014) A first attempt at modelling red deer (*Cervus elaphus*) distributions over Europe. *Open Health Data* 2, e1.
Wint, W., Bülent, A., Balenghien, T., Berriauta, E., Braks, M., *et al.* (2018) The importance of vector abundance and seasonality. *EFSA Supporting Publications* 15, 1491E.
Wint, W., Petric, D., Van Bortel, W., Alexander, N. and Schaffner, F. (2020a) RVF vector spatial distribution models: vector abundance. *EFSA Supporting Publications* 17, 1847E.
Wint, W., Bortel, W.V. and Schaffner, F. (2020b) RVF vector spatial distribution models: probability of presence. *EFSA Supporting Publications* 17, 1800E.

e06 Considerations for Predicting Climate Change Implications on Future Spatial Distribution Ranges of Ticks

Ram Raghavan[1]* and Roman Ganta[2]
[1]*University of Missouri, Columbia, Missouri, USA;* [2]*Kansas State University, Manhattan, Kansas, USA*

6.1 Expert Opinion

Ticks, poikilothermic ectoparasites, live most of their lives exposed to ambient weather conditions. There are other factors involved, but to a large extent adaptation to fluctuations in weather allows ticks to thrive in some environments better than others. As for other arthropods, it has long been known that the distribution of ixodid ticks and their phenology are closely associated with the patterns of weather and that different tick species exhibit natural adaptations to different climates. Growth rates, questing behaviour, survival and reproductive success are only a few of the life-history traits of ticks shown, under field conditions, to be influenced by such weather parameters as temperature, relative humidity and photoperiod (Estrada-Peña *et al.*, 2004; Randolph, 2014). Climate change and consequent changes in the microclimate that are directly experienced by ticks are therefore expected to elicit changes in tick life history and it is logical, as a consequence, to expect changes to tick spatiotemporal distributions as well (Gray *et al.*, 2009; Ostfeld and Brunner, 2015). As has been documented, these changes may render new areas more suitable, potentially leading to an expansion or a shift in distributional range for some tick species (Alkishe *et al.*, 2017; Raghavan *et al.*, 2019a; Fernández-Ruiz and Estrada-Peña, 2021) (Fig. 6.1), as well as some of the currently known distribution ranges of ticks becoming uninhabitable, leading to a contraction in spatial distribution (e.g. Boorgula *et al.*, 2020). There is a notable interest and value in predicting future potential spatial distributions of ticks resulting from ongoing climate change, owing to the medical and veterinary significance of numerous tick species worldwide.

The theory and practice of modelling species spatial distributions of different life forms is a well-developed field in the domain of ecological modelling. Spatial distribution models (SDMs) have been developed and continually refined for several decades, and they have been used ubiquitously for predicting the distributions of a multitude of plant, animal and microbial species (Elith and Leathwick, 2009). With the advent of high-speed computing, contemporary and some of the more popular SDM approaches utilize machine learning algorithms (e.g. Maxent, Genetic Algorithm for Rule-set Production). These approaches rely on the use of species occurrence data in conjunction with environmental data to produce predictive models of a species' spatial distribution. SDMs allow for predictions to be made for different spatiotemporal extents, including predictions of a species' distribution under

*Email: raghavanrk@missouri.edu

DOI: 10.1079/9781789249637.0006

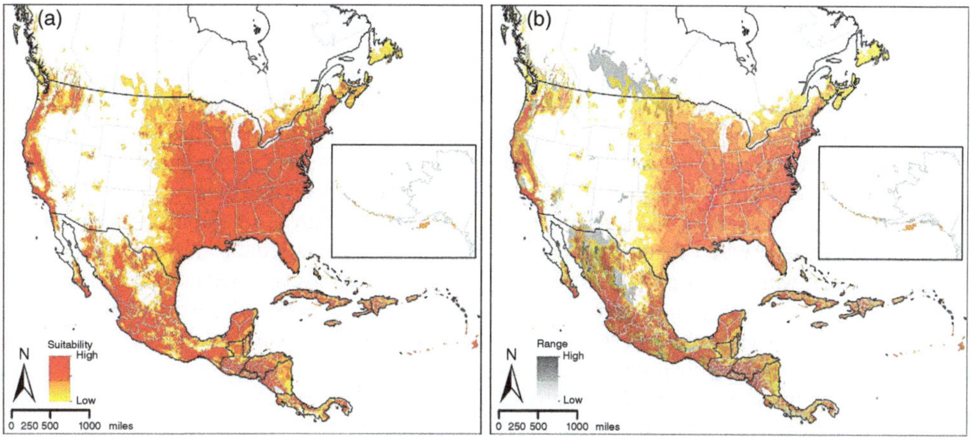

Fig. 6.1. Predicted distribution of bioclimatically suitable areas for *Dermacentor variabilis* (Say) in North America (a) and uncertainty (range = maximum – minimum predicted suitable values) associated with the prediction for 2050 (b). (From Boorgula *et al.*, 2020; CC0 1.0 Universal Public Domain Dedication.)

future climate change scenarios. While the intricacies of machine learning and some of the modelling approaches used in these contemporary approaches are relatively better understood by those who practice in the field of ecological modelling, the tools themselves and some guidance on how to construct SDMs are readily available, which has allowed experts in other scientific domains, including tick biologists and epidemiologists, to experiment with and utilize SDMs. As a result, in recent years, SDMs have been widely used to predict how climate change might affect the future distributions of ticks throughout the world.

Recent studies have shown that there are important considerations when constructing SDMs that are essential to producing reliable models of tick distributions, both in the present time and in the future under climate change scenarios. As it is with any model, some SDMs for ticks are useful and some can be wrong or misleading. Therefore, careful consideration of the methods is necessary in building SDMs, model assumptions, the limitations in predictions and making a careful interpretation of predictions, if possible, supported by field observations.

Maxent, which uses the correlative maximum entropy approach (Phillips and Dudík, 2008), is by far the most widely adapted SDM platform used today for a variety of reasons. We will therefore limit our discussion to this approach alone. Although much of the literature surrounding best practices for SDM uses Maxent as a platform, these recommendations are likely to be useful for other SDM approaches as well. The rationale behind these recommendations have been discussed at length in the disparate literature and are cited here; we encourage those who consider modelling future distribution of ticks to explore them for deeper learning.

Data of geographic locations where ticks are collected, or simply occurrence data in the form of latitude and longitude coordinates for a species of interest, crucially informs the model about the environment in which a given tick species is known to be present currently as well as in the past. Therefore, it is important to include occurrence data from all of a species' known spatial distributional extent. Although it is not feasible to know all the locations in which a species may exist, and to prove absence is difficult, environmental characteristics of the places in which a species has indeed been found will inform the SDMs and by supplying as many unique locations that represent different environmental conditions, a better model prediction can be expected (Kramer-Schadt *et al.*, 2013). Such occurrence data could come from a researcher's own field collections of ticks, but often a good resource is archived specimen data in acarological collections and that is available at natural history museums.

With such occurrence data, one is likely to encounter spatial clustering, as more than one tick or different life stages of the same species from a single vicinity are often collected. Therefore, occurrence data must be rarefied using spatial filters, to avoid effects of spatial bias and over-representation of some environment behind those occurrence points in a model (Kramer-Schadt *et al.*, 2013; Syfert *et al.*, 2013). Care must be taken when rarefying occurrence data that might otherwise explain a unique set of conditions in which a species exists. For instance, occurrences over a fairly homogeneous landscape may use a relatively large spatial filter; for heterogeneous and highly fragmented landscapes, such as hill slopes and other transient zones, a case-by-case evaluation will be more appropriate (Raghavan *et al.*, 2019a). Further, uncertainties associated with occurrence data must be carefully evaluated and those with no uncertainty measurements or excessive uncertainty values must be removed.

Areas in which a species naturally occurs presently may have changed over long periods of time during the species' ecological and evolutionary history. Such areas that have been accessible to a tick over relevant periods of time are ideal for SDM development, testing and comparison (Barve *et al.*, 2011). Based on the rarefied occurrences, an area that is appropriately larger than the currently known extent of a species' distributional range, excluding tick uninhabitable areas, but that would have been otherwise accessible to a species, is recommended to be calculated and used in the constructing of SDMs. This area will be relevant for calibrating the models as well as for testing and validating the SDMs (Barve *et al.*, 2011; Escobar *et al.*, 2014).

Environmental variables to build SDMs for a given species must be chosen in light of their relevance to the life history of that species. For instance, annual temperature and precipitation are recognized as major determinants of many tick species' distribution, and minimum and maximum temperatures may represent a species' tolerance to cold and drought conditions, respectively. For best modelling results, environmental variables must be from the same time period in which the occurrence data were collected and at a spatial resolution that closely matches the resolution (or uncertainty) of species occurrence data. The BIOCLIM (Fick and Hijmans, 2017) variables, which are summary conditions of 19 different bioclimatic conditions, many of which are relevant to tick biology, are commonly used in SDM modelling. MERRAclim, a similar source of bioclimatic data, has some advantages over BIOCLIM data (Vega *et al.*, 2017), one being that the latter uses satellite-based data instead of interpolated summaries of ground-station-derived data.

The strength of relationships between tick species occurrence data and bioclimatic variables needs to be evaluated prior to constructing SDMs. The presence of two or more closely related variables in any model will lead to overfitting, among other potential issues, which can be avoided by selecting more meaningful variable(s) using a correlation estimate. Once variables are selected, they can be used individually, in selected combinations with others as variable sets, or as a single composite layer derived using principal components analysis (PCA). We have found the second option to have advantages over the others as this method allows progressive evaluation of at least two most relevant variable sets, as variable contributions may differ when they are present along with other candidate variables. Each variable set may leave one or more variables from the previous set, based on their model contribution, verified by jackknife plots and other model fitness metrics (Alkishe *et al.*, 2017). With a composite PCA-derived variable, the ability to interpret the significance of individual variables to the SDM is lost.

Maxent allows users to select any one or a combination of different functional forms (i.e. linear, quadratic, threshold, hinge, product, categorical) to describe environmental variables' relationship to the species of interest, and the choice made will affect model outcomes. Similarly, the regularization multiplier, which determines the complexity or generality of the model responses, strongly influences model outcomes. Further, the number and types of environmental variables that are used in building SDMs also influence model predictions. Therefore, a need for careful evaluation of these options has been pointed out (Warren *et al.*, 2014). One way to evaluate such relationships is by constructing several progressively complex models by setting different combinations of environmental variable sets, regularization multiplier and combinations of functional forms for each SDM, a process that can be automated using programming packages such as kuenm (Cobos *et al.*, 2019).

A crucial factor that influences SDM predictions is the predetermination of tolerance towards a model erroneously missing a true habitat or wrongly assigning an unsuitable habitat for a species of interest. Thresholding of models with a predetermined, fixed, allowable omission error rate (e.g. 5%) will allow 5% of records (both occurrence and environmental data) to be omitted, resulting in better-quality models (Peterson *et al.*, 2008; Phillips and Dudík, 2008).

Procedures to evaluate performance of models have been widely discussed. There are some notable issues with using the default model performance metric, the area under the curve (AUC) of the receiver operator characteristic (ROC) curve available in Maxent output (Peterson *et al.*, 2008). It is recommended that models be ranked using different criteria, first based on partial ROC curve values (non-significant models are excluded) and omissions rate (e.g. ≤*x*% are removed), followed by the Akaike Information Criterion (AIC) (least complex being best). Briefly, the partial ROC values determine the significance of a model, omissions rate is an indication of how well the model predicts true occurrences, and AICs indicate model complexity.

The best predictive distribution model(s) thus selected can then be projected into desired future year(s) under different climate scenarios. Careful consideration of two aspects of future climate change data for projecting SDMs is necessary: (i) the general circulation model (GCM); and (ii) the Representative Concentration Pathway (RCP) that is used in constructing the future scenarios (Fig. 6.2). Individual GCMs, particularly when considered in the context of tick SDM, differ (at times, drastically) in the way they project future climate and therefore a good working knowledge of the assumptions made in the creation of these data sets and product details (e.g. resolution, layer depth) is necessary. The RCPs are projections made about greenhouse gas concentration trajectory; some are based on more conservative estimates (e.g. RCP4.5) than others (e.g. RCP8.5) and it would be useful to use both or more scenarios to get a broader assessment. The spatial resolution used for future predictions should ideally use the same resolution used for deriving the present-day distribution SDM.

An issue often encountered in SDM when transferring present-day models to future climate or other regions is the extent to which the projections are purely based on extrapolation (i.e. areas into which the model was transferred that have values outside the range of

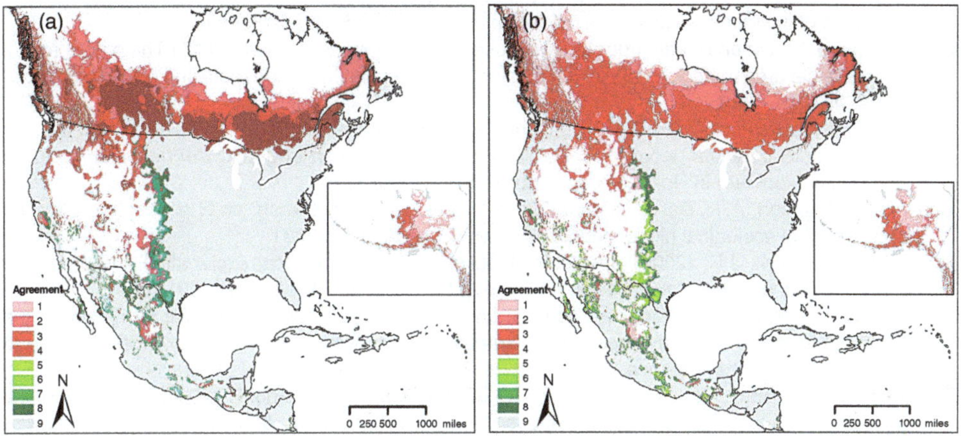

Fig. 6.2. Future potential distribution of *Dermacentor variabilis* (Say) in North America based on two Representative Concentration Pathway (RCP) scenarios, RCP4.5 (a) and RCP8.5 (b), showing agreement between different general circulation models (GCMs) for 2050. Grey areas represent present-day prediction; progressively darker shades of red indicate higher number of GCM agreement and stronger likelihood of future spatial expansion. Darker colours of green indicate higher number of GCM agreement and likelihood of contraction in future spatial distribution. Mobility-oriented parity (MOP) analysis did not reveal any model extrapolative areas. (From Boorgula *et al.*, 2020; CC0 1.0 Universal Public Domain Dedication.)

representative climates in the calibration area), leading to higher uncertainty in highly extrapolative areas (Owens *et al.*, 2013). The mobility-oriented parity (MOP) metric allows calculation of the presence and extent of this problem in SDMs when projected into the future. Models with significant areas with higher MOP values must be rejected and model steps re-evaluated.

6.2 Future Predictions

Future potential spatial distribution of medically important tick species in North America and Europe generally indicates a northward expansion as a result of different climate change scenarios (Alkishe *et al.*, 2017; Raghavan *et al.*, 2019b; Boorgula *et al.*, 2020). These models also indicate a potential reduction in the spatial distribution as depicted in Fig. 6.2. Robustly built SDMs of tick distributions are good indicators of where a tick species can be expected to be found. Yet, they must be interpreted carefully in the context of a given tick species' biology. Abiotic conditions are only part of a potentially large group of influential factors that affect the biology of ticks, including their current and future spatial distributions. It is incorrect to assume that models built purely with summarized abiotic determinants will capture spatial distributions for the present time or in the future under different climate change scenarios. Host availability and host density are important additional factors that affect tick distributions. Indeed, the identification of climatically suitable geographic regions for tick species' distribution is not a definitive indication of a species' presence, as factors such as missing hosts, relative diversity and abundance of hosts during interannual periods will potentially affect tick distribution. Further, it should be remembered that predicting spatial distributions for ticks under future climate change scenarios routinely assumes all else to be the same. In other words, vegetation, tick–host adaptations and any adaptations by the ticks to changing climate are often discounted in the SDMs.

References

Alkishe, A.A., Peterson, A.T. and Samy, A.M. (2017) Climate change influences on the potential geographic distribution of the disease vector tick *Ixodes ricinus*. *PLoS ONE* 12, e0189092.

Barve, N., Barve, V., Jiménez-Valverde, A., Lira-Noriega, A., Maher, S.P., *et al.* (2011) The crucial role of the accessible area in ecological niche modeling and species distribution modeling. *Ecological Modelling* 222, 1810–1819.

Boorgula, G.D., Peterson, A.T., Foley, D.H., Ganta, R.R. and Raghavan, R.K. (2020) Assessing the current and future potential geographic distribution of the American dog tick, *Dermacentor variabilis* (Say) (Acari: Ixodidae) in North America. *PLoS ONE* 15, e0237191.

Cobos, M.E., Peterson, A.T., Barve, N. and Osorio-Olvera, L. (2019) kuenm: an R package for detailed development of ecological niche models using Maxent. *PeerJ* 7, e6281.

Elith, J. and Leathwick, J.R. (2009) Species distribution models: ecological explanation and prediction across space and time. *Annual Review of Ecology, Evolution, and Systematics* 40, 677–697.

Escobar, L.E., Lira-Noriega, A., Medina-Vogel, G. and Peterson, A.T. (2014) Potential for spread of the white-nose fungus (*Pseudogymnoascus destructans*) in the Americas: use of Maxent and NicheA to assure strict model transference. *Geospatial Health* 1, 221–229.

Estrada-Peña, A., Martinez, J.M., Sanchez Acedo, C., Quilez, J. and Del Cacho, E. (2004) Phenology of the tick, *Ixodes ricinus*, in its southern distribution range (central Spain). *Medical and Veterinary Entomology* 18, 387–397.

Fernández-Ruiz, N. and Estrada-Peña, A. (2021) Towards new horizons: climate trends in Europe increase the environmental suitability for permanent populations of *Hyalomma marginatum* (Ixodidae). *Pathogens* 10, 95.

Fick, S.E. and Hijmans, R.J. (2017) WorldClim 2: new 1-km spatial resolution climate surfaces for global land areas. *International Journal of Climatology* 37, 4302–4315.

Gray, J.S., Dautel, H., Estrada-Peña, A., Kahl, O. and Lindgren, E. (2009) Effects of climate change on ticks and tick-borne diseases in Europe. *Interdisciplinary Perspectives on Infectious Diseases* 2009, 593232.

Kramer-Schadt, S., Niedballa, J., Pilgrim, J.D., Schröder, B., Lindenborn, J., *et al.* (2013) The importance of correcting for sampling bias in MaxEnt species distribution models. *Diversity and Distributions* 19, 1366–1379.

Ostfeld, R.S. and Brunner, J.L. (2015) Climate change and *Ixodes* tick-borne diseases of humans. *Philosophical Transactions of the Royal Society B* 370, 20140051.

Owens, H.L., Campbell, L.P., Dornak, L.L., Saupe, E.E., Barve, N., *et al.* (2013) Constraints on interpretation of ecological niche models by limited environmental ranges on calibration areas. *Ecological Modelling* 263, 10–18.

Peterson, A.T., Papeş, M. and Soberón, J. (2008) Rethinking receiver operating characteristic analysis applications in ecological niche modeling. *Ecological Modelling* 213, 63–72.

Phillips, S.J. and Dudík, M. (2008) Modeling of species distributions with Maxent: new extensions and a comprehensive evaluation. *Ecography* 31, 161–175.

Raghavan, R.K., Barker, S.C., Cobos, M.E., Barker, D., Teo, E.J., *et al.* (2019a) Potential spatial distribution of the newly introduced long-horned tick, *Haemaphysalis longicornis* in North America. *Scientific Reports* 9, 498.

Raghavan, R.K., Peterson, A.T., Cobos, M.E., Ganta, R. and Foley, D. (2019b) Current and future distribution of the lone star tick, *Amblyomma americanum* (L.) (Acari: Ixodidae) in North America. *PLoS ONE* 14, e0209082.

Randolph, S.E. (2014) Ecology of non-nidicolous ticks. In: Sonenshine, D.E. and Roe, R.M. (eds) *Biology of Ticks*, vol. 2, 2nd edn. Oxford University Press, New York, pp. 3–38.

Syfert, M.M., Smith, M.J. and Coomes, D.A. (2013) The effects of sampling bias and model complexity on the predictive performance of MaxEnt species distribution models. *PLoS ONE* 8, e55158.

Vega, G.C., Pertierra, L.R. and Olalla-Tárraga, M.Á. (2017) MERRAclim, a high-resolution global dataset of remotely sensed bioclimatic variables for ecological modelling. *Scientific Data* 4, 170078.

Warren, D.L., Wright, A.N., Seifert, S.N. and Shaffer, H.B. (2014) Incorporating model complexity and spatial sampling bias into ecological niche models of climate change risks faced by 90 California vertebrate species of concern. *Diversity and Distributions* 20, 334–343.

Synopsis: Climate

Pat Nuttall*
University of Oxford, Oxford, UK

The first section of this collection of expert opinions considers various aspects of climate change, including predictions of future climate change, how it affects ecosystems and modelling to predict the spatial distribution of ticks and associated diseases.

Future climate change depends on how anthropogenic emissions of greenhouse gases and aerosol precursors change in the future (eo3: Cox *et al.*). Working groups of the Intergovernmental Panel on Climate Change (IPCC) develop alternative scenarios for the world, linking feasible socio-economic futures to the end of the 21st century with the vast range of atmospheric concentrations of greenhouse gases and aerosols associated with these socio-economic futures. There are huge uncertainties, e.g. whether rainfall will increase or decrease in Africa (eo1: Cornforth and Plumpton). Against this background of uncertainty, storylines are created describing alternative outcomes, e.g. the climate is much wetter and hotter, rainfall intensity and frequency of storms increase, and the long rainy season increases in length. This longer rainy season prevents some staple crops from maturing but increases opportunities to grow perennial crops and a wider variety of fruits and vegetables, including crops used for livestock fodder. This explicit communication of uncertainty in climate projections to decision makers enables more effective adaptation planning (eo1: Cornforth and Plumpton).

However, even for given changes in greenhouse gases and aerosols, climate models differ by a factor of >2 in their projections of global mean warming by 2100 (eo3: Cox *et al.*). One way of sidestepping this conundrum is by looking specifically at projected regional climate change at a given level of global warming. Cox *et al.* (eo3) do this using 'pattern scaling' to compare 1°C of global warming (~present day) with the 2°C limit of the Paris Agreement and project changes in the geographical distribution of Lyme borreliosis. The outputs show warming relative to current day moving the Lyme borreliosis zone further northwards while the associated atmospheric humidity deficit reduces the incidence of Lyme borreliosis at its current southern limits.

From a tick perspective, a key challenge for understanding species and ecosystem responses to climate change is bridging the gap in scales between the microclimates in which ticks live and the meteorological observations and models used to study these responses. Vegetation is central to this understanding because it strongly influences microclimates, besides having other functions (eo2: Hemming *et al.*). Although advances in microclimate monitoring and modelling have improved the representation of

*Email: pat.nuttall@zoo.ox.ac.uk

DOI: 10.1079/9781789249637.syn1

microclimates in ecosystems, there is still a major challenge, particularly when ticks live in burrows, crevices or leaf litter where microclimates vary significantly from climate data available from standard observations and models (eo2: Hemming *et al.*).

Despite the challenges of predicting future global warming, let alone how it might affect tick vectors and their habitats, there is an obvious need for accurate models that simulate the risk to humans and domesticated animals of ticks and tick-borne infections. Mathematical models of ticks and associated diseases are diverse and need to be tailored to the questions addressed. Caminade (eo4) makes three points: (i) the quality and quantity of surveillance data to develop and validate models must be good; (ii) model ensembles are desirable as multi-model comparisons using the same inputs allow greater understanding of uncertainties; and (iii) a One Health multidisciplinary framework is key. Often, predictions of how climate change might impact the prevalence and incidence of tick-borne infections are restricted to the expected distribution of the principal vector species. In this case, modelling presence and absence, rather than abundance, is more feasible, although it still relies on having access to a decent set of observed ('training') data (eo5: Wint and Alexander). Applying masks to refine the model outputs, e.g. using a suitability mask (incorporating land-use categories, a climate envelope and annual seasonality) and a host mask (e.g. derived from deer hosts), makes a considerable difference to the model output. For example, a distribution map of *Ixodes ricinus* projected for 2080 with high emissions and the exemplified masks shows the likelihood of tick presence drops throughout much of continental Europe (eo5: Wint and Alexander). Predicting spatial distributions for ticks under future climate change scenarios routinely assumes all else to be the same. In other words, vegetation, tick–host adaptations and tick adaptations to changes in climate are generally overlooked (eo6: Raghavan and Ganta). The next section in this collection of expert opinions considers how tick vectors might adapt to climate change.

e07 Can the Impact of Climate Change on the Tick Microbiome Bring a New Epidemiological Landscape to Tick-Borne Diseases?

Alejandro Cabezas-Cruz*
French National Research Institute for Agriculture, Food and Environment, Maisons-Alfort, France

7.1 Expert Opinion

7.1.1 Introduction

Ticks coexist and interact with tick-borne pathogens, symbionts and commensals, and together they form an ecological unit, the tick holobiont (Díaz-Sánchez *et al.*, 2019). The contributions of the tick microbiota to tick physiology and pathogen life cycle are so relevant that vector capacity cannot be understood without considering tick microbial communities (Narasimhan *et al.*, 2014; Abraham *et al.*, 2017; Díaz-Sánchez *et al.*, 2019). The taxonomic composition of midgut microbiota in different tick species of the genera *Rhipicephalus*, *Ixodes*, *Amblyomma*, *Haemaphysalis*, *Hyalomma*, *Dermacentor*, *Argas* and *Ornithodoros* has been reported (Narasimhan and Fikrig, 2015). Microbiota and pathogens have bidirectional relationships (Narasimhan *et al.*, 2014; Abraham *et al.*, 2017), so that while *Anaplasma phagocytophilum* colonization was reported to reduce the microbial diversity of tick microbiota (Abraham *et al.*, 2017), reduction of microbial diversity by antibiotic treatment decreased *Borrelia burgdorferi* colonization in the tick vector (Narasimhan *et al.*, 2014). Currently, the taxonomic diversity (i.e. number of different operational taxonomic units (OTUs) within a sample) of the tick microbiota is hotly debated; while some studies report a highly diverse microbiota, others report a very poor diversity. On the other hand, the variability of the taxonomic composition (i.e. number of different OTUs per sample) of tick microbiota across samples has been found to be consistently high. This was illustrated in a recent study identifying taxa which demonstrated that all *Ixodes scapularis* samples ($n = 98$) only shared ten bacterial taxa (total 821, 1.2%). This high microbial variability may be due to the multiple factors influencing tick microbiota composition including, but not limited to, tick species, environmental microbes, source of host blood meal, developmental stage and abiotic factors such as temperature (reviewed in Narasimhan and Fikrig, 2015). This expert opinion focuses on the possible impact of climate change on the tick microbiome, with potential consequences for disease ecology. Within the text, 'microbiome' refers to the microorganisms and their genes whereas 'microbiota' only refers to the microbes themselves. 'Holobiont' refers to the close association between host and microbes, that together form a discrete ecological unit.

*Email: alejandro.cabezas@vet-alfort.fr

DOI: 10.1079/9781789249637.0007

7.1.2 The tick microbiome, a redundant source of microbial metabolic pathways

Most of the information available on tick microbiota concerns the taxonomic composition of the microbial communities, while the functional significance of bacterial diversity has remained relatively unexplored. Despite the taxonomic variability observed in tick microbiomes, the strong phylogenetic signal between *Ixodes* ticks and their microbial communities suggests that tick microbiota assemblies are not stochastic (Díaz-Sánchez *et al.*, 2019). Rather, the phylogenetic structure of ixodid tick microbial communities supports the existence of a species-specific tick holobiont with a largely unknown influence on tick fitness and vector competence (Díaz-Sánchez *et al.*, 2019). A hologenome (i.e. the collective genomes of the holobiont) can carry out many different functions arising from an interaction continuum between host and microbiota gene products. For example, the presence of vitamin B synthesis genes in *Francisella* or *Coxiella* symbionts compensates for the shortage of vitamin B in the blood meal (Duron *et al.*, 2018). By complementing a nutritional deficiency in the tick diet, bacteria carrying vitamin B synthesis genes are fundamental for tick survival and their elimination produced anomalies in tick development (Duron *et al.*, 2018). However, vitamin B synthesis genes are not restricted to symbiotic bacteria of the genera *Francisella* or *Coxiella* but are widely distributed throughout several genera of tick microbiota (Obregón *et al.*, 2019). In addition, considering the high diversity of genes and metabolic pathways encoded in the genomes of tick microbiota bacteria, we hypothesize that the contribution of bacteria to tick physiology and survival could extend well beyond vitamin B supplementation (Obregón *et al.*, 2019; Estrada-Peña *et al.*, 2020a). Indeed, the metabolic pathways associated with the tick microbiome include processes as diverse as amino acid, antibiotic, pyrimidine, lipid and amino sugar metabolism (Obregón *et al.*, 2019). Remarkably, metabolic pathway analysis demonstrated that all *I. scapularis* samples ($n = 98$) shared as many as 381 pathways (87.2% of 440 in total) (Estrada-Peña *et al.*, 2020a). Functional analysis of metabolic pathways associated with tick bacterial communities revealed the existence of functional redundancy (i.e. when multiple taxa contribute the same metabolic function) in the tick microbiome (Estrada-Peña *et al.*, 2020a). Our data indicate that the functional microbiome of ticks is indeed highly redundant, with as many as 198 different bacterial taxa contributing to a single pathway (Estrada-Peña *et al.*, 2020b). This suggests that tick microbial communities share a common subset of metabolic pathways, regardless of their taxonomic identity (Estrada-Peña *et al.*, 2020a,b). Altogether, this evidence suggests that: (i) both tick symbionts (i.e. microbes forming a close and long-term biological interaction with the tick) and tick microbiota (i.e. microbes forming a short-term biological interaction with the tick) can complement tick nutritional pathways; and (ii) functional diversity and redundancy enables ticks to continually access stable sources of microbial metabolites regardless of environmental disturbance.

7.1.3 Resistance of the tick microbiome to biotic disturbance

Microbiome resistance can be defined as the degree to which the microbial community withstands change in the face of disturbance. To assess the resistance of tick microbiomes, we measured the extent of changes induced by different biotic factors on the taxonomic and functional profiles of the tick microbiome (Estrada-Peña *et al.*, 2020a,b). The results showed that factors such as anti-tick immunity, pathogen infection, antimicrobial peptides and developmental stage modified the *I. scapularis* microbiota but with limited impact on the functional profiles of the tick microbiome. The rewiring of taxonomic networks after disturbance revealed that the tick microbiota is susceptible to biological disturbance. Moreover, the taxonomic networks that resulted after disruption were less resistant to taxa extinction. However, the factor with the largest effect on bacterial diversity was anti-tick immunity. This was illustrated by altered functional microbiomes due to anti-tick immunity, whereas *A. phagocytophilum* infection and antimicrobial peptide did not alter tick microbiome pathway diversity (Estrada-Peña *et al.*, 2020a).

7.1.4 Impact of temperature on tick microbiota

There are very few studies on the impact of temperature on the taxonomic composition and abundance of tick microbiota. In their study, Thapa *et al.* (2019) quantified the bacterial microbiome of colony-reared *I. scapularis* ticks incubated at different temperatures (i.e. 4, 20, 30 and 37°C) at constant humidity in a controlled laboratory setting. The authors found that colony-reared *I. scapularis* adult ticks incubated for 10 days at 4 or 20°C showed no significant changes in bacterial community composition. The bacterial community was also minimally affected in the first days of incubation at 30 or 37°C. However, bacterial community complexity of the ticks dropped dramatically upon prolonged incubation at 30°C (more than 1 week) and at 37°C (longer than 5 days). These findings suggest that long-term exposure to warmer temperatures eliminates certain groups of bacteria while favouring others. Specifically, Thapa *et al.* (2019) noted a decreased relative abundance of *Pseudomonas* after incubation at 30°C for more than 1 week, and at 37°C for more than 5 days, with a concomitant increase in the relative proportions of *Brevibacterium* and *Streptomyces*. Importantly, changes in the taxonomic composition of tick microbiota were not associated with tick mortality (Thapa *et al.*, 2019). Considering the robustness of the functional profiles of the tick microbiome (Estrada-Peña *et al.*, 2020a,b), I hypothesize that the decrease in taxonomic diversity associated with changes in temperature may not dramatically impact the functional profiles of tick microbiomes (Fig. 7.1).

7.2 Future Predictions

Climate change impacts the existence of all trophic life forms. Major shifts in the composition and structure of microbial communities on Earth are expected as a consequence (Cavicchioli *et al.*, 2019). Climate change towards warmer temperatures will modify the microbial communities of the soil, water, plants, host microbiota and, accordingly, the tick microbiome. Environmental disruption of the tick microbiome may interfere with tick holobiont co-adaptation. The microbiome will adapt faster than the tick host to new

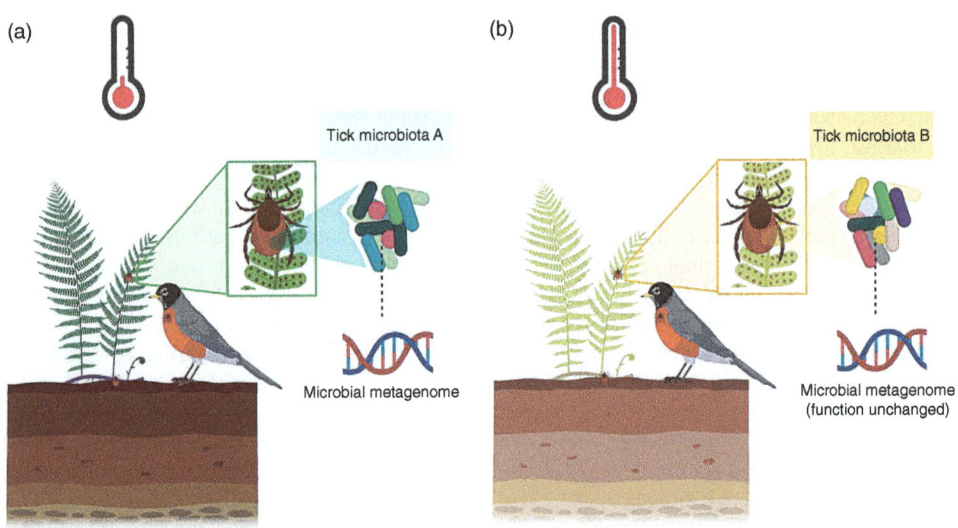

Fig. 7.1. Impact of temperature on the taxonomic and functional profiles of the tick microbiome. Long-term exposure to temperatures higher than the current average may modify the composition and diversity of tick microbiota from normal microbiota (a) to high-temperature-adapted microbiota (b). However, functionality of the microbial metagenomes associated with the tick microbiome may remain unaffected. (Author's own figure.)

stressful environments; however, this imbalance may diminish as the tick and its microbiome establish a new equilibrium. Experimental manipulation of the tick microbiota using antibiotics impacts the life cycle of ticks and tick-borne pathogens. The effect of tick microbiota disruption on the vector is variable and includes, for example, decreased fitness, increased engorgement and developmental anomalies, depending on the species and the antibiotic. Community disruption also impacts the pathogens by enhancing colonization or decreasing the microbial burden in tick tissues. As the tick holobiont interacts with tick-borne pathogens, the imbalance between the tick and its microbiome due to climate change may impact disease ecology. A new epidemiological landscape with dramatic changes in the distribution and incidence of tick-borne diseases may be a challenging scenario for preserving both human and animal health.

References

Abraham, N.M., Liu, L., Jutras, B.L., Yadav, A.K., Narasimhan, S., *et al.* (2017) Pathogen-mediated manipulation of arthropod microbiota to promote infection. *Proceedings of the National Academy of Sciences USA* 114, E781–E790.

Cavicchioli, R., Ripple, W.J., Timmis, K.N., Azam, F., Bakken, L.R., *et al.* (2019) Scientists' warning to humanity: microorganisms and climate change. *Nature Reviews Microbiology* 17, 569–586.

Díaz-Sánchez, S., Estrada-Peña, A., Cabezas-Cruz, A. and de la Fuente, J. (2019) Evolutionary insights into the tick hologenome. *Trends in Parasitology* 35, 725–737.

Duron, O., Morel, O., Noël, V., Buysse, M., Binetruy, F., *et al.* (2018) Tick–bacteria mutualism depends on B vitamin synthesis pathways. *Current Biology* 28, 1896–1902.

Estrada-Peña, A., Cabezas-Cruz, A. and Obregón, D. (2020a) Resistance of tick gut microbiome to anti-tick vaccines, pathogen infection and antimicrobial peptides. *Pathogens* 9, 309.

Estrada-Peña, A., Cabezas-Cruz, A. and Obregón, D. (2020b) Behind taxonomic variability: the functional redundancy in the tick microbiome. *Microorganisms* 8, 1829.

Narasimhan, S. and Fikrig, E. (2015) Tick microbiome: the force within. *Trends in Parasitology* 31, 315–323.

Narasimhan, S., Rajeevan, N., Liu, L., Zhao, Y.O., Heisig, J., *et al.* (2014) Gut microbiota of the tick vector *Ixodes scapularis* modulate colonization of the Lyme disease spirochete. *Cell Host & Microbe* 15, 58–71.

Obregón, D., Bard, E., Abrial, D., Estrada-Peña, A. and Cabezas-Cruz, A. (2019) Sex-specific linkages between taxonomic and functional profiles of tick gut microbiomes. *Frontiers in Cellular and Infection Microbiology* 9, 298.

Thapa, S., Zhang, Y. and Allen, M.S. (2019) Effects of temperature on bacterial microbiome composition in *Ixodes scapularis* ticks. *MicrobiologyOpen* 8, e719.

eo8 Climate Influence on Tick Neurobiology

Ladislav Šimo*
French National Research Institute for Agriculture, Food and Environment, Maisons-Alfort, France

8.1 Expert Opinion

8.1.1 Introduction

Both nervous and sensory systems play a key role in the interaction between animals and their environment. The abiotic factors associated with climate change can directly affect the functioning of these systems, which subsequently influences animal actions (Musolin and Saulich, 2012). The tick nervous and sensory systems direct responses to environmental stimuli and any changes in tick behaviour and physiology are reflected in these systems. In essence, this represents the phenotypic expression of the molecular changes.

Climate-mediated effects on performance or rapid evolution of tick neural and sensory systems is a currently neglected field and its improved understanding may contribute to a better appreciation of tick responses and adaptations to environmental pressure at both the individual and population levels. Based on current theories and knowledge acquired from various animal species, this expert opinion provides a short view of how climate change may impact the tick nervous system and its responses that enhance the tick's biological fitness. Herein, the aim is to primarily focus this work at the level of tick neural/sensory systems and discuss the potential effects of environmental variation on their performance and adaptation.

8.1.2 Tick nervous and sensory systems

The tick central nervous system – the synganglion – is represented by a fused nerve mass consisting of multiple lobes (also called ganglions) that supplies nerves to the visceral and external tissues (Šimo *et al.*, 2014). The tick synganglion functions via an array of neural and endocrine signalling pathways which contribute largely to tick physiological functions and behaviour (e.g. reproduction, growth, life cycle, metabolism, development, feeding, migration and host-seeking). The peripheral sensory system is comprised of a variety of epidermal sensilla whose shape is based on the information they gather from the external environment. These include chemosensilla, mechanosensilla, photosensilla and thermosensilla. While many are dispersed throughout the body and appendages, some are concentrated within specialized organs (e.g. the Haller's organ). Each sensillum consists of one or a few neurons detecting stimuli from both internal and external environments, whereas the tick synganglion serves as a central perception site for the transduced

*Email: ladislav.simo@vet-alfort.fr

DOI: 10.1079/9781789249637.0008

electric or chemical signals arriving from the sensory organs (Šimo *et al.*, 2014).

8.1.3 Impact of directional thermal stress on tick nervous system

The ability of ticks to recognize environmental stressors is central to understanding their adaptive responses to various environmental factors, at both individual and population levels. Ticks spend the majority of their life cycle off the host, where abiotic factors within the ambient environment directly affect their physiology and behaviour (Fig. 8.1).

Increasing temperatures caused by global warming is one of the major abiotic factors impacting animal central nervous systems. A recent study suggests that animal nervous systems respond to directional climate change (global warming) in one of three ways (O'Donnell, 2018): (i) impairment – here the species is constrained within its geographical range despite the environmental shift, but fitness and performance decline as a result of impaired neural and sensory systems; (ii) accommodation/adaptation – changes in nervous system function achieved via neural/sensory plasticity and/or genetic adaptation enable the species to adjust to the new thermal conditions; and (iii) active abiotic stasis – changing behaviour to pursue favourable thermal conditions similar to the original abiotic environment.

The immediate physiological and behavioural responses of ticks to variations in thermal conditions can be relatively easily monitored, e.g. by detecting shifts in their phenology or geographical distribution (eo9: Ronai). However, as in other animal species, the knowledge gained regarding genetic adaptation and sensory and neural plasticity caused by long-term environmental stress remains rather sparse (Kelley *et al.*, 2018). It is generally believed that tick populations

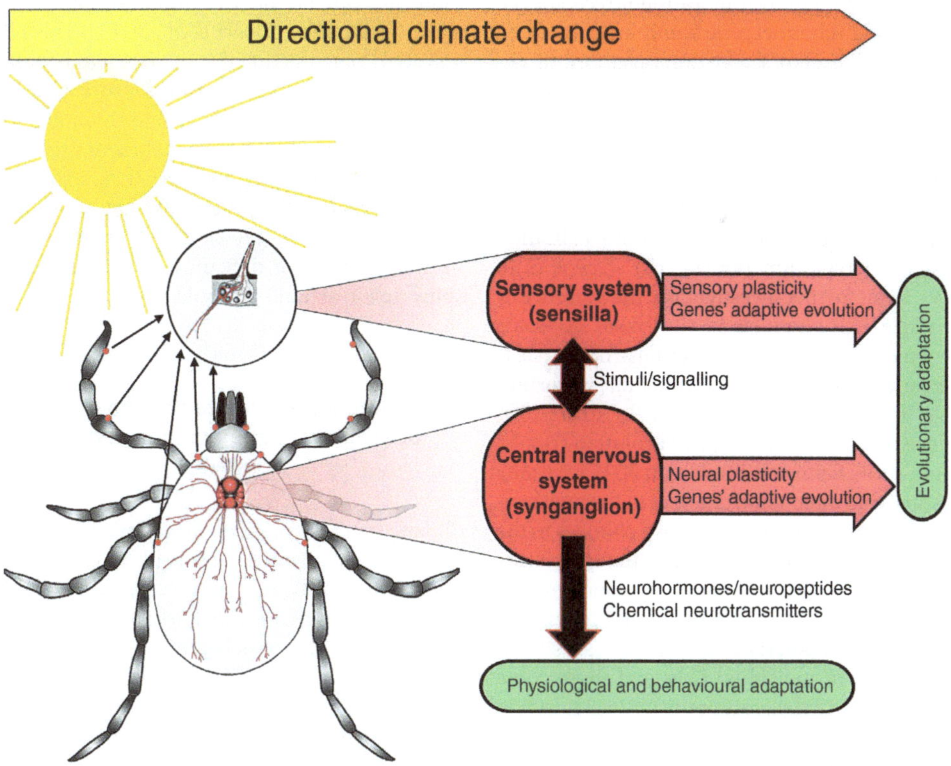

Fig. 8.1. A schema illustrating the responses of the tick nervous and sensory systems to directional climate change. (Author's own figure.)

are increasing in abundance (although this remains to be confirmed by scientific studies) and this phenomenon could be the result of altered neural and sensory performance, most likely caused by climate change.

It appears that tick nervous and sensory systems are highly adaptable to new thermal conditions both latitudinally and altitudinally that can favour tick population expansion. There is good reason to assume that in previously inhospitable regions, temperature was the main limiting factor for tick nervous and sensory system performance, thus rendering ticks unable to maintain homeostasis within the extreme ranges of the thermal environment. At this point, I am unable to conclude if any impairment of neural/sensory system has appeared in those individuals that remained in the original habitat with shifted thermal conditions. However, it is reasonable to assume that changes in gene expression and/or adaptive evolution of key elements of the tick thermal sensory system play a major role in the tick's acclimatization processes. Proteins of the transient receptor potential (TRP) channel gene family are considered key neural receptors, sensing a plethora of environmental stimuli including changes in ambient temperature. These ancient, evolutionarily conserved proteins are known to be expressed within the sensory nerve terminals and a variety of animal behaviours depend upon their activity (Matsuura *et al.*, 2009). While the localization and phenotypic function of TRP receptors in ticks await further investigation, the presence of multiple TRP channels in the *Ixodes scapularis* genome sequence (L. Šimo, Maisons-Alfort, 2021, personal observation) suggests an important role in sensory signalling. Due to strong evidence of *trp* genes in adaptive evolution (Dhaka *et al.*, 2006; Saito and Tominaga, 2017), we may assume that directional changes in thermal conditions trigger rapid evolutionary changes in the tick's nervous system, leading to the natural selection of positive mutations in their *trp* loci. Specifically, allelic variations of *trp* genes may correspond to different temperature sensitivity at the level of individual ticks and those could become the key elements in the natural selection of suitable genetic variations leading to thermal resilience of tick populations. In addition to positive mutation selection, we should also consider the possibility of adaptive evolution of gene expression, as these features were shown to be important factors in regulating multiple physiological processes including sensory perception in insects (Nourmohammad *et al.*, 2017). In either case, exploration of the tick *trp* gene family, in relation to directional environmental change, opens an important field for investigating the evolutionary dynamics of the tick nervous system.

One may argue that tick habitat expansion is predominantly linked to climate change-mediated migration of their vertebrate hosts. This scenario is partially correct, as ticks are more effective hitchhikers than long-distance walkers. However, when host-transferred to a new geographical region, their further development and survival largely depend on neural and sensory adaptation to the new thermal and humidity conditions they experience. Indeed, variation in host-seeking behaviour is a clear example illustrating the resilience of tick nervous systems to directional temperature change. It has been shown that the European tick, *Ixodes ricinus*, from cooler climates quests at lower temperatures than individuals of the same species from warmer climates (Gilbert *et al.*, 2014). This observation suggests that distinct *I. ricinus* populations adapt their host-seeking sensory abilities and subsequent behaviour to the thermal shift of their environment. Such adaptations are likely related to functionally appropriate adjustments of peripheral sensory neurons that transform temperature changes to electric or chemical signals, although it is difficult to predict whether synganglion signalling would also alter.

8.1.4 Impact of climate change-mediated desiccation stress on tick nervous systems

Climate change-triggered humidity imbalances along with temperature gradients represent another large hurdle for tick survival. Many ticks, especially species of the *I. ricinus* group, are highly susceptible to dehydration, and their entire physiology is heavily affected in unfavourable dry environments. Warmer and drier summers caused by climate change could definitively alter tick behaviour via their moisture sensing. The physiology of moisture-sensitive neurons in ticks is a relatively understudied field.

Here, I assume that the functional features of hygro-sensitive TRP channels (Montell, 2008; also see Section 8.1.3 above), expressed on the peripheral sensory neurons, enable ticks to adapt to environmental changes in humidity. Furthermore, the synganglion-controlled tick salivary glands are the key osmoregulatory organs regulating water sorption from humid atmospheres during the tick off-host period (Kim *et al.*, 2019; Mateos-Hernandéz *et al.*, 2020). Briefly, in desiccated conditions the salivary gland type II and III acini secrete hygroscopic saliva (Kim *et al.*, 2017) on to the tick's mouthparts to form a crystallized matrix. When environmental humidity increases, this salty deposit is liquidized by moisture in the air, allowing it to be ingested by the tick (Gaede and Knülle, 1997). Subsequently, most of the water and ions are absorbed via salivary gland type I acini thus maintaining tick hydration (Kim *et al.*, 2017). Synganglion cholinergic neurons regulate water absorption by type I acini, while production of hygroscopic saliva by type II and III acini is likely modulated by neuropeptidergic neurons (Šimo *et al.*, 2009b; Kim *et al.*, 2019; Mateos-Hernandéz *et al.*, 2020). Therefore, I speculate that during the off-host stage, climate-mediated humidity imbalance may alter the performance of these neuronal cells to boost rehydration effects. It is possible that here hygro-sensory TRP channel receptor neurons could act as upstream controls, where they perceive extrinsic stimuli and subsequently their converted signal alters the functions of salivary gland-innervating neurons within the synganglion. In addition, severely unfavourable humid conditions may impair the tick nervous system, causing declines in developmental and survival activities. To avoid such a state, the co-mediation of the sensory and neuronal system may alter tick behaviour to seek new favourable microenvironments and/or change their phenology.

8.1.5 Climate influence on the tick neuroendocrine system

The tick neuroendocrine system, along with signalling mediators such as neuropeptides and chemical neurotransmitters that bind to their specific transmembrane receptors at various body sites, regulate all physiological functions and maintain body balance under both regular and stressful situations (Lees and Bowman, 2007; Šimo *et al.*, 2009a, 2014). The majority of these signalling molecules act in both central and peripheral tick nervous systems but may also be released to the tick haemolymph as neurohormones or act locally via paracrine or autocrine signals in various tissues. In general, the specific functions of tick neuropeptides and neurotransmitters are not well understood, but it can be expected that changes in abiotic conditions challenge the tick's homeostasis and the complex cascade of the actions of these molecules mediates their physiological and behavioural adaptability to new environments. As a result of environmental challenge, the sensitivity of the tick neuroendocrine system may trigger quantitative differences in neuropeptide/neurotransmitter levels in particular organs, changes in their release and transport, as well as responsiveness of their target tissues due to altered affinities of their cognate receptors (Lubawy *et al.*, 2020). Due to these features, a plethora of physiological and behavioural events, e.g. metabolic regulation, storing body reserves, ion homeostasis, muscle contraction, feeding, mating, excretion, sensory perception and/or circadian rhythms, can be altered as a result of specific adaptation to new climate conditions. Furthermore, increasing evidence suggests that endocrine system regulatory activity shapes the animal's developmental plasticity under shifting environments (Lema, 2020). Thus, in addition to maintaining homeostasis during prolonged climate change, neuropeptides/neurohormones and possibly neurotransmitters may contribute to the generation of relevant phenotypic variants at both the level of the individual tick and populations.

8.1.6 Other human-induced environmental factors impacting tick neural functions

Sensory signalling input to the tick's synganglion is the primary process by which ticks respond to stimuli from the ambient environment. Human-induced environmental change may negatively affect these processes, which could have a major impact on the tick's nervous system.

For example, air pollutants may seriously affect ticks' sensing capabilities, which may lead to impairment of information detection and reception as well as affecting signal transmission to the synganglion, ultimately resulting in changing tick behaviour. Specifically, tick olfactory sensilla in the Haller's organ, located on the tarsus of the first pair of legs (Šimo *et al.*, 2014), can be directly blocked, thus disrupting the sensory signal. The malfunction of the tick olfactory system may, for example, seriously affect tick host-seeking behaviour or reproductive physiology by disrupting pheromone sensing. Furthermore, the increasing rate of greenhouse gas emissions may also impact tick behaviour and physiology. For example, the change in atmospheric composition, mediated by an increase in carbon dioxide and other greenhouse gases, may directly affect behaviours of tick populations. Moreover, recent studies have also shown that ticks are capable of sensing magnetic and electric fields, thus altering their movement behaviour (Vargová *et al.*, 2018). Although these experiments were carried out under laboratory conditions, the authors' data indicate that electromagnetic fields may affect tick distribution in the natural environment. Taken together, anthropogenic activities are important factors that can affect tick sensory and nervous systems, thus altering their ability to respond to ecosystem variation.

8.2 Future Predictions

Ticks have existed for hundreds of millions of years and their ability to survive the challenges of multiple episodes of dramatic climatic change firmly place them as highly adaptable and evolutionarily successful creatures. From the neurobiological perspective, I suggest that the adaptive processes of their nervous system have played crucial roles in shaping their entire physiology throughout the ages (Fig. 8.1). I predict that recent climate change is not likely to negatively affect their evolutionary path. In fact, the opposite is true, and there is every reason to believe that tick bites will become an escalating problem in the years to come. Thus, it is highly plausible that the remarkably plastic nervous system and its subsequent physiological activities represent the primary mechanism by which ticks respond to directional climate changes, enabling them to increase in abundance and expand their geographic ranges. With that in mind, ticks will surely pose even greater dangers to public health.

Furthermore, as a tick neurobiologist, I suggest that future predictions of tick behaviour at the population level should take account of the extreme ranges of tolerance and adaptability of their nervous system, which are likely the main limiting factors to adapting to new environmental conditions. Determining the breadth of these ranges, and their molecular determinants, may help predict the development and behaviour of different tick species populations around the globe. Therefore, the current climatic changes call for increased attention to be paid to this extremely important research field investigating the dynamics of neural and sensory adaptations of these medically important arthropods (Fig. 8.1).

Acknowledgement

UMR BIPAR is supported by the French Government's Investissement d'Avenir programme, Laboratoire d'Excellence 'Integrative Biology of Emerging Infectious Diseases' (grant no. ANR-10-LABX-62-IBEID).

References

Dhaka, A., Viswanath, V. and Patapoutian, A. (2006) Trp ion channels and temperature sensation. *Annual Review of Neuroscience* 29, 135–161.

Gaede, K. and Knülle, W. (1997) On the mechanism of water vapour sorption from unsaturated atmospheres by ticks. *Journal of Experimental Biology* 200, 1491–1498.

Gilbert, L., Aungier, J. and Tomkins, J.L. (2014) Climate of origin affects tick (*Ixodes ricinus*) host-seeking behavior in response to temperature: implications for resilience to climate change? *Ecology and Evolution* 4, 1186–1198.

Kelley, J.L., Chapuis, L., Davies, W.I.L. and Collin, S.P. (2018) Sensory system responses to human-induced environmental change. *Frontiers in Ecology and Evolution* 6, 95.

Kim, D., Maldonado-Ruiz, P., Zurek, L. and Park, Y. (2017) Water absorption through salivary gland type I acini in the blacklegged tick, *Ixodes scapularis*. *PeerJ* 5, e3984.

Kim, D., Šimo, L., Vancová, M., Urban, J. and Park, Y. (2019) Neural and endocrine regulation of osmoregulatory organs in tick: recent discoveries and implications. *General and Comparative Endocrinology* 278, 42–49.

Lees, K. and Bowman, A.S. (2007) Tick neurobiology: recent advances and the post-genomic era. *Invertebrate Neuroscience* 7, 183–198.

Lema, S.C. (2020) Hormones, developmental plasticity, and adaptive evolution: endocrine flexibility as a catalyst for 'plasticity-first' phenotypic divergence. *Molecular and Cellular Endocrinology* 502, 110678.

Lubawy, J., Urbański, A., Colinet, H., Pflüger, H.-J. and Marciniak, P. (2020) Role of the insect neuroendocrine system in the response to cold stress. *Frontiers in Physiology* 11, 376.

Mateos-Hernandéz, L., Defaye, B., Vancová, M., Hajdusek, O., Sima, R., *et al.* (2020) Cholinergic axons regulate type I acini in salivary glands of *Ixodes ricinus* and *Ixodes scapularis* ticks. *Scientific Reports* 10, 16054.

Matsuura, H., Sokabe, T., Kohno, K., Tominaga, M. and Kadowaki, T. (2009) Evolutionary conservation and changes in insect TRP channels. *BMC Evolutionary Biology* 9, 228.

Montell, C. (2008) TRP channels: it's not the heat, it's the humidity. *Current Biology* 18, R123–R126.

Musolin, D.L. and Saulich, A.Kh. (2012) Responses of insects to the current climate changes: from physiology and behavior to range shifts. *Entomological Review* 92, 715–740.

Nourmohammad, A., Rambeau, J., Held, T., Kovacova, V., Berg, J. and Lässig, M. (2017) Adaptive evolution of gene expression in *Drosophila*. *Cell Reports* 20, 1385–1395.

O'Donnell, S. (2018) The neurobiology of climate change. *The Science of Nature* 105, 11.

Saito, S. and Tominaga, M. (2017) Evolutionary tuning of TRPA1 and TRPV1 thermal and chemical sensitivity in vertebrates. *Temperature (Austin)* 4, 141–152.

Šimo, L., Slovák, M., Park, Y. and Žitňan, D. (2009a) Identification of a complex peptidergic neuroendocrine network in the hard tick, *Rhipicephalus appendiculatus*. *Cell and Tissue Research* 335, 639–655.

Šimo, L., Žitňan, D. and Park, Y. (2009b) Two novel neuropeptides in innervation of the salivary glands of the black-legged tick, *Ixodes scapularis*: myoinhibitory peptide and SIFamide. *The Journal of Comparative Neurology* 517, 551–563.

Šimo, L., Sonenshine, S.E., Park, Y. and Žitňan, D. (2014) The nervous and sensory systems: structure, function, proteomics and genomics. In: Sonenshine, D.E. and Roe, R.M. (eds) *Biology of Ticks*. Oxford University Press, New York, pp. 309–367.

Vargová, B., Majláth, I., Kurimský, J., Cimbala, R., Kosterec, M., *et al.* (2018) Electromagnetic radiation and behavioural response of ticks: an experimental test. *Experimental and Applied Acarology* 75, 85–95.

e09 The Impact of Climate Change on Tick Host-Seeking Behaviour

Isobel Ronai*
The University of Sydney, Sydney, Australia

9.1 Expert Opinion

9.1.1 Introduction

As obligate blood feeders, ticks require a blood meal from a host for survival and reproduction. Tick feeding is a complex process with multiple associated behaviours (Waladde and Rice, 1982). These behaviours determine which pathogens a tick receives from its host, the pathogens a tick can transmit to its next host and which pathogens a tick can transmit to its offspring (transovarial transmission). The behaviour of ticks is an underappreciated determinant of their vectorial capacity.

Ticks that are exophilic do not live with their host and so need to seek out a host for their blood meal, whereas endophilic (also referred to as nidicolous) ticks live in the shelter of their host (Estrada-Peña *et al.*, 2013). The majority of hard ticks (Ixodidae) are exophilic and the majority of soft ticks (Argasidae) are endophilic. The host-seeking behaviour of exophilic ticks can be divided into two strategies. While some exophilic ticks use an active host-seeking strategy (i.e. hunting strategy), the vast majority use a passive strategy (i.e. ambush strategy) where the tick positions itself to directly attach on to a passing potential host. This questing behaviour (Fig. 9.1) is defined by ticks extending their forelegs (where their key sensory structure, the Haller's organ, is located) and commonly 'waving' them while located on ground cover, such as vegetation (Lees, 1948; Leal *et al.*, 2020). Questing ticks are therefore highly exposed to environmental conditions.

To be in position to quest for a host, ticks ascend from and descend to the environmentally optimal ground layer. There are multiple behavioural components to questing, such as position on ground cover (i.e. height), initiation, orientation and termination. Questing inflicts a metabolic cost on the tick and, while questing, the tick is more vulnerable to desiccation and predation. The trade-off between nutrition and survival is different for each of the life stages of ticks because they differ in size (Clark, 1995; Leal *et al.*, 2020; Alasmari and Wall, 2021). For example, immature ticks quest at a lower height in vegetation (Mejlon and Jaenson, 1997).

Questing is rapidly triggered in response to environmental stimuli in the immediate vicinity of the tick. Stimuli include host odour, host vibrations, host body heat, ambient temperature, relative humidity, photoperiod and light intensity (Waladde and Rice, 1982; Carr and Salgado, 2019; Leal *et al.*, 2020). The abiotic stimuli that affect questing are being impacted by climate change (Easterling *et al.*, 2000) and are important for understanding the effects of climate change on ticks and tick-borne diseases.

*Email: isobel.ronai@sydney.edu.au

DOI: 10.1079/9781789249637.0009

Fig. 9.1. Questing behaviour of adult female *Ixodes ricinus*. (Photograph courtesy of Jan Erhart, Institute of Parasitology, České Budějovice, Czech Republic.)

9.1.2 Abiotic environmental stimuli of questing behaviour

Ambient temperature is the most important determinant of the initiation of questing. This stimulus also plays a role in the position on ground cover and termination of questing. In general terms, when the ambient temperature increases, there is a corresponding increase in questing activity. As ectotherms, ticks require a certain ambient temperature for metabolic activity (Alasmari and Wall, 2021). For example, when temperatures are below freezing, no *Ixodes ricinus* nymphs and adults are questing (Perret *et al.*, 2000). After the optimal ambient temperature for questing activity for a particular tick species and life stage is reached (e.g. 25°C for *Ixodes scapularis* nymphs in the laboratory) (Vail and Smith, 2002), questing activity then becomes negatively correlated with ambient temperature. At high temperatures, ticks are at risk of desiccation, but while they may respond to desiccation by changing their position on ground cover, they do not change overall questing activity (Thomas *et al.*, 2020). At these high temperatures the termination of questing occurs regardless of the moisture gradient (Portugal *et al.*, 2020). Further work is needed to understand how temperature affects questing. What is the role of minimum temperature, maximum temperature, average temperature, time period at a particular temperature, for the different behavioural components of questing?

Relative humidity is important for the position on ground cover and termination of questing. At very low relative humidities, ticks will not quest (Randolph and Storey, 1999) as they are unable to survive. For example, when the relative humidity is below 63%, *Rhipicephalus* larvae die (Davey *et al.*, 1991). If the relative humidity is below a tick's critical equilibrium activity (for larvae, 80–85% relative humidity; Leal *et al.*, 2020) but the tick is questing then, its water balance will deplete over time and it will eventually need to rehydrate to survive. The tick will therefore terminate questing and descend to the ground (Lees, 1948). At relative humidities below the critical equilibrium activity, a tick is continuously spending time rehydrating, so the total time spent questing is reduced. Notably, if a tick has been in a high relative humidity environment, such as the ground layer, it will actively seek out a lower relative humidity (i.e. the edges of ground cover) (Lees, 1948; Randolph and Storey, 1999). As the relative humidity increases, more questing occurs (Portugal *et al.*, 2020). Furthermore, the higher the relative humidity, the higher the height of questing (Vail and Smith, 2002; Portugal *et al.*, 2020). If the relative humidity is above a tick's critical equilibrium activity, then its water balance will not be depleted over time and modelling suggests that ticks could always be questing (McClure and Diuk-Wasser, 2019).

Two aspects of light, photoperiod (i.e. light duration) and light intensity, have an effect on questing activity and the termination of questing. Long light periods cause the commencement of questing activity and short light periods, the stoppage of questing activity. For example, *I. ricinus* nymphs and adults commence questing when there is 12 h of daylight in a day and stop questing when there is 9 h of daylight in a day (Daniel *et al.*, 2015). Whether ticks quest diurnally or nocturnally is dependent on the tick species and life stage (Thomas *et al.*, 2020). In addition, laboratory studies indicate *I. scapularis* adults change their response to light depending on the season (Madden and Madden, 2005; Perret *et al.*, 2003). An increase in light intensity does not cause the initiation of questing (Lees, 1948). However, under strong light intensity, ticks will terminate questing and seek shelter (Wilkinson, 1953). It is also notable that if light intensity is blocked for a short period of time, then the initiation of questing occurs (Lees, 1948). An interruption in light intensity is likely used by ticks as a good proxy for a shadow being cast from a potential host. Further work is needed to better understand how the response of questing to photoperiod and light intensity is dependent on the tick species and life stage.

9.2 Future Predictions

In a changing world, an organism's first reaction is likely to be behavioural (Wong and Candolin, 2015). As ticks are not capable of rapidly dispersing to a more favourable location (they rely on host movement), behavioural plasticity will likely be important for the resilience of ticks to climate change. Behavioural plasticity will enable ticks to adjust their questing behaviour to the new environmental conditions under climate change. For example, ticks could shift the timing of questing across the day or year. Laboratory experiments using the projected temperatures under climate change indicate *I. ricinus* will increase the length of its questing season (Gilbert *et al.*, 2014). However, there will be constraints on the behavioural plasticity of ticks and, if the environmental changes under climate change become too severe, it is unlikely any adjustment in questing behaviour would be able to counter the effects. For example, at extreme temperatures ticks will likely die. The role of behavioural plasticity in questing has only started to be explored (Gilbert *et al.*, 2014) and further experimental work is needed.

In the long term, behavioural adaptation of questing could be important for ticks coping with climate change. The speed of behavioural evolution for ticks is unknown. However, field studies indicate that ticks have the capacity to adapt their questing behaviour. For example, *I. scapularis* from the northern USA are more likely to quest higher above the ground cover than those from the southern USA; crossing studies indicate this change in questing behaviour has a genetic basis (Arsnoe *et al.*, 2015) and is possibly an adaptation to desiccation stress (Ginsberg *et al.*, 2017). In the future, a laboratory evolution experiment could be conducted to establish how questing behaviour would be impacted under the scenarios of predicted climate change. Understanding the evolutionary potential for ticks to behaviourally adapt to climate change would allow a better assessment of the potential impact of climate change on ticks.

As a result of questing behaviour changing in response to climate change, tick–host interactions will be affected. In particular, if the temporal and spatial occurrence of questing changes, it will affect the length of host exposure periods and frequency of interactions with hosts. Potentially, the overlap of ticks and particular host species will be disrupted. Understanding how questing behaviour will change in response to climate change is therefore crucial to understanding pathogen transmission dynamics. It is also worth noting that changes in questing behaviour, under climate change, should be considered separately from the changes in the geographic range of ticks as this is dependent on their development and survival requirements.

In conclusion, the host-seeking (questing) behaviour of ticks is highly dependent on the environment, so climate change is likely to have a large impact on this behaviour. For example, the proportion of ticks questing is likely to increase (Gilbert *et al.*, 2014). As a consequence, the risk of tick-associated diseases will be affected in unpredictable ways. Furthermore, tick control efforts to prevent tick-borne diseases by targeting tick behaviour may also be affected. This expert opinion has focused on the impact of climate change on only one tick behaviour, questing, but it is important to remember that other tick behaviours (such as behavioural diapause, otherwise known as overwintering) will also be impacted by climate change.

References

Alasmari, S. and Wall, R. (2021) Metabolic rate and resource depletion in the tick *Ixodes ricinus* in response to temperature. *Experimental and Applied Acarology* 83, 81–93.

Arsnoe, I.M., Hickling, G.J., Ginsberg, H.S., McElreath, R. and Tsao, J.I. (2015) Different populations of blacklegged tick nymphs exhibit differences in questing behavior that have implications for human Lyme disease risk. *PLoS ONE* 10, e0127450.

Carr, A.L. and Salgado, V.L. (2019) Ticks home in on body heat: a new understanding of Haller's organ and repellent action. *PLoS ONE* 14, e0221659.

Clark, D.D. (1995) Lower temperature limits for activity of several ixodid ticks (Acari: Ixodidae): effects of body size and rate of temperature change. *Journal of Medical Entomology* 32, 449–452.

Daniel, M., Maly, M., Danielová, V., Kríž, B. and Nuttall, P. (2015) Abiotic predictors and annual seasonal dynamics of *Ixodes ricinus*, the major disease vector of Central Europe. *Parasites & Vectors* 8, 478.

Davey, R.B., Cooksey, L.M. and Despins, J.L. (1991) Survival of larvae of *Boophilus annulatus*, *Boophilus microplus*, and *Boophilus* hybrids (Acari: Ixodidae) in different temperature and humidity regimes in the laboratory. *Veterinary Parasitology* 40, 305–313.

Easterling, D.R., Meehl, G.A., Parmesan, C., Changnon, S.A., Karl, T.R. and Mearns, L.O. (2000) Climate extremes: observations, modeling, and impacts. *Science* 289, 2068–2074.

Estrada-Peña, A., Farkas, R., Jaenson, T.G.T., Koenen, F., Madder, M., *et al.* (2013) *Ticks and Tick-borne Diseases: Geographical Distribution and Control Strategies in the Euro-Asia Region.* CAB International, Wallingford, UK.

Gilbert, L., Aungier, J. and Tomkins, J.L. (2014) Climate of origin affects tick (*Ixodes ricinus*) host-seeking behavior in response to temperature: implications for resilience to climate change? *Ecology and Evolution* 4, 1186–1198.

Ginsberg, H.S., Albert, M., Acevedo, L., Dyer, M.C., Arsnoe, I.M., *et al.* (2017) Environmental factors affecting survival of immature *Ixodes scapularis* and implications for geographical distribution of Lyme disease: the Climate/Behavior hypothesis. *PLoS ONE* 12, e0168723.

Leal, B., Zamora, E., Fuentes, A., Thomas, D.B. and Dearth, R.K. (2020) Questing by tick larvae (Acari: Ixodidae): a review of the influences that affect off-host survival. *Annals of the Entomological Society of America* 113, 425–438.

Lees, A. (1948) The sensory physiology of the sheep tick, *Ixodes ricinus* L. *Journal of Experimental Biology* 25, 145–207.

Madden, S.C. and Madden, R.C. (2005) Seasonality in diurnal locomotory patterns of adult blacklegged ticks (Acari: Ixodidae). *Journal of Medical Entomology* 42, 582–588.

McClure, M. and Diuk-Wasser, M.A. (2019) Climate impacts on blacklegged tick host-seeking behavior. *International Journal for Parasitology* 49, 37–47.

Mejlon, H.A. and Jaenson, T.G.T. (1997) Questing behaviour of *Ixodes ricinus* ticks (Acari: Ixodidae). *Experimental and Applied Acarology* 21, 747–754.

Perret, J.-L., Guigoz, E., Rais, O. and Gern, L. (2000) Influence of saturation deficit and temperature on *Ixodes ricinus* tick questing activity in a Lyme borreliosis-endemic area (Switzerland). *Parasitology Research* 86, 554–557.

Perret, J.-L., Guerin, P.M., Diehl, P.A., Vlimant, M. and Gern, L. (2003) Darkness induces mobility, and saturation deficit limits questing duration, in the tick *Ixodes ricinus*. *Journal of Experimental Biology* 206, 1809–1815.

Portugal, J.S., Wills, R. and Goddard, J. (2020) Laboratory studies of questing behavior in colonized nymphal *Amblyomma maculatum* ticks (Acari: Ixodidae). *Journal of Medical Entomology* 57, 1480–1487.

Randolph, S.E. and Storey, K. (1999) Impact of microclimate on immature tick–rodent host interactions (Acari: Ixodidae): implications for parasite transmission. *Journal of Medical Entomology* 36, 741–748.

Thomas, C.E., Burton, E.S. and Brunner, J.L. (2020) Environmental drivers of questing activity of juvenile black-legged ticks (Acari: Ixodidae): temperature, desiccation risk, and diel cycles. *Journal of Medical Entomology* 57, 8–16.

Vail, S.G. and Smith, G. (2002) Vertical movement and posture of blacklegged tick (Acari: Ixodidae) nymphs as a function of temperature and relative humidity in laboratory experiments. *Journal of Medical Entomology* 39, 842–846.

Waladde, S.M. and Rice, M.J. (1982) The sensory basis of tick feeding behaviour. In: Obenchain, F.D. and Galun, R. (eds) *Physiology of Ticks*. Pergamon Press, Oxford, pp. 71–118.
Wilkinson, P.R. (1953) Observations on the sensory physiology and behaviour of larvae of the cattle tick, *Boophilus microplus* (Can.) (Ixodidae). *Australian Journal of Zoology* 1, 345–356.
Wong, B.B.M. and Candolin, U. (2015) Behavioral responses to changing environments. *Behavioral Ecology* 26, 665–673.

eo10 Expected Transitions in Ticks and Their Heritable Endosymbionts Under Environmental Changes

Yuval Gottlieb[1]* and Olivier Duron[2,3]
[1]*The Hebrew University of Jerusalem, Rehovot, Israel;* [2]*Université de Montpellier, Montpellier, France;* [3]*Centre de Recherche en Écologie et Évolution de la Santé, Montpellier, France*

10.1 Expert Opinion

Environmental changes caused by climate change and land use influence the abundance and diversity of arthropods (Sánchez-Bayo and Wyckhuys, 2019). Arthropods are commonly engaged in close and long-term interactions with heritable microbial symbionts (often from mother to progeny) including bacteria, fungi, viruses and protozoans, which directly contribute to host fitness. Some of these symbioses are obligatory and are associated with the nutritional capabilities of the arthropod, while others are facultative and are associated with various phenotypes in arthropods (Russell and Oliver, 2020). Recently, a diversity of heritable bacterial endosymbionts has been emphasized as a source of evolutionary innovations, including protection against biotic and abiotic environmental stresses such as defence against natural enemies and plant exploitation, as well as amelioration of the detrimental effects of heat, desiccation and pollutants (Ferrari and Vavre, 2011; Sudakaran *et al.*, 2017; Lemoine *et al.*, 2020). Thus, alteration of arthropod phenotypes by their heritable endosymbionts can shape the adaptation of arthropods to their environments.

Within arthropods, ticks harbour among the highest diversity of heritable endosymbionts reported so far, encompassing at least ten different bacterial genera (Duron *et al.*, 2017). While the function of these heritable endosymbionts has not been fully determined yet, they have the potential to deeply influence the ecology of ticks impacted by climate change (Narasimhan and Fikrig, 2015; Bonnet *et al.*, 2017). Heritable endosymbionts belonging to genera such as *Coxiella* and *Francisella* enable ticks to specialize on vertebrate blood (Guizzo *et al.*, 2017; Duron *et al.*, 2018; Ben-Yosef *et al.*, 2020). Because blood is a rich but unbalanced diet, ticks have developed obligatory associations with bacteria capable of producing essential B vitamins to avoid nutritional deficiencies (Duron and Gottlieb, 2020). These associations are critical for tick growth, reproduction and survival, and as such are present in most individuals of a given tick species. However, in the majority of cases, heritable endosymbionts of ticks are facultative and exhibit variable infection frequencies across temporal and spatial gradients in tick populations (Duron *et al.*, 2017). Although widely present, these facultative endosymbionts have received much less attention than tick-borne pathogens

*Email: gottlieb.yuval@mail.huji.ac.il

DOI: 10.1079/9781789249637.0010

and little is known of the effect they may have on their tick hosts, specifically under changing environments (Bonnet *et al.*, 2017). Nevertheless, by using current knowledge of other arthropod symbiotic systems, we may predict potential outcomes in ticks.

The interaction stability between arthropods and their facultative endosymbionts depends on various selection pressures including external environmental conditions. For example, the pea aphid, *Acyrthosiphon pisum*, harbours a facultative endosymbiont, *Serratia symbiotica*, that confers heat tolerance. Aphid clones infected with this bacterium display substantially improved survival during heat stress (Chen *et al.*, 2000; Montllor *et al.*, 2002). As a result, *S. symbiotica* prevalence and distribution are most influenced by climate as its infection frequency increases significantly during warmer seasons in natural populations (Russell and Moran, 2006) and also reaches higher densities in arid regions than in temperate regions (Chen and Purcell, 1997; Henry *et al.*, 2013). Along with facultative endosymbionts, the obligatory endosymbiont of aphids, *Buchnera aphidicola*, which provides essential amino acids for the carbohydrate-based aphid diet, can also influence thermal tolerance via modulating its own abundance and its abilities to produce heat-shock proteins (Zhang *et al.*, 2019). Interestingly, aphid facultative endosymbionts can further aid increases in *Buchnera* titres under heat stress (Heyworth *et al.*, 2020). An additional example comes from the saw-toothed grain beetle, *Oryzaephilus surinamensis*, in which facultative endosymbionts affect cuticle thickness, melanization and hydrocarbon profile, thereby promoting desiccation resistance (Engl *et al.*, 2018). Therefore, the facultative association with an endosymbiont can facilitate the persistence of arthropod populations under a range of ecological stressors such as heat and aridity, either directly or via interaction with an obligatory endosymbiont.

Ticks show ambient thermal preference and tolerance limits depending on their species and geographic distribution, which are also reflected in their metabolic rate (Fieler *et al.*, 2019). Additionally, ticks are physiologically adapted for immediate temperature changes, as during feeding they usually encounter a different temperature compared with the ambient temperature in their environment (Benoit *et al.*, 2019). Although tick obligatory endosymbionts are essential during high metabolic demand (Duron *et al.*, 2018; Ben-Yosef *et al.*, 2020), multiple physiological adaptations (e.g. diet and stress tolerance) via endosymbionts, as demonstrated in aphids, have not been reported in ticks. We can postulate that these adaptations can be achieved directly by producing heat-shock proteins, or indirectly by contributing to tick fitness parameters such as higher reproductive success and protection against other environmental stressors. Heritable endosymbionts may also be involved in behaviour and locomotion to direct ticks to find better shelters or hosts. Changes in tick behaviour (locomotion and feeding) and resistance to environmental stressors as a result of infection with various pathogenic agents have already been demonstrated (Kagemann and Clay, 2013; Benelli, 2020).

While conferring advantages under some environmental or biotic stressors, and despite a high fidelity of maternal inheritance, facultative symbionts may be costly and can result in decreased fecundity or survival under normal settings (Ferrari and Vavre, 2011). These transitions are difficult to predict but can be expected as was shown in pea aphids, where dramatic increases and decreases in the proportion of facultative endosymbionts were detected in the presence or absence of environmental pressures, respectively (Oliver *et al.*, 2008). Indirect evidence for this probable cost comes from other arthropods, where the prevalence of facultative endosymbionts is reflected in specific geographic and climatic patterns (e.g. Liu *et al.*, 2006; Morag *et al.*, 2012). If similar patterns exist in ticks, those with a greater tolerance to environmental stress thanks to endosymbionts may also suffer decreased fecundity or survival if compared with ticks without endosymbionts in the absence of environmental pressure.

Additionally, the presence of these endosymbionts can also shape tick evolutionary dynamics by affecting mitochondrial DNA (mtDNA) population genetics, as commonly observed in other arthropods (Hurst and Jiggins, 2005). All cytoplasmic endosymbionts that are mostly maternally inherited will affect the population genetics of arthropod mtDNA because these are co-inherited. Thus, when a maternally inherited endosymbiont drives through the arthropod population because it confers thermal tolerance, it takes with it the mtDNA type it is initially associated with, leading to reduced global mtDNA diversity (Hurst and Jiggins, 2005). Thus, inferences from mtDNA variation in tick populations can be confounded by the presence of such endosymbionts.

10.2 Future Predictions

Climate change is expected to affect tick–endosymbiont interactions towards adaptations to novel environmental conditions. This can be achieved via three major mechanisms: (i) specific symbiont evolution to maintain genes that will support the expected environmental stress; (ii) replacement of current endosymbionts by endosymbionts that are preadapted to function under certain environments; and (iii) acquisition of novel endosymbionts with relevant traits.

Heritable endosymbionts are characterized by a highly reduced genome size (Bennett and Moran, 2015; McCutcheon *et al.*, 2019). While numerous genes coding for unnecessary or redundant functions are pseudogenized or lost, the endosymbionts are under constant selection to maintain genes essential for both the endosymbiont and host survival. It can be expected that under environmental stress conditions, a beneficial endosymbiont will thrive. This can include higher replication rate to enable high symbiont titre, thus providing better nutritional support and better maternal transmission to promote offspring survival.

Heritable endosymbionts are also characterized by small population size, which under stochastic selection may result in the elimination of essential genes and even the extinction of the endosymbiont (Bennett and Moran, 2015; McCutcheon *et al.*, 2019). A mechanism to overcome such loss is symbiont replacement, in which a pre-existing facultative endosymbiont takes over the function of the lost obligatory endosymbiont. This has been demonstrated in various insect symbiosis systems (Bennett and Moran, 2015; McCutcheon *et al.*, 2019) and also in ticks (Binetruy *et al.*, 2020a). As facultative symbionts may carry other beneficial traits, these can additionally support tick fitness under changing environments.

Finally, another mechanism for adaptation to novel environmental conditions is the acquisition of novel endosymbionts (Bennett and Moran, 2015; McCutcheon *et al.*, 2019). Since climate change is expected to result in changes in tick distribution, it will create novel opportunities for horizontal transfer of symbionts from other tick species and, possibly, from other arthropods. Tick species, previously found in distinct distribution areas, will enter into contact and these ecological connections will facilitate symbiont exchanges via co-feeding to create novel associations. Novel ecological connections will also be created with more distantly related arthropods and similarly favour symbiont exchanges. Indeed, a recent examination of *Spiroplasma ixodetis*, a common facultative endosymbiont of hard ticks, showed that exchanges within a diversity of arthropods are common and widespread (Binetruy *et al.*, 2019). The best examples are found in the *Ixodes* tick genus. The *S. ixodetis* strains found in *Ixodes ricinus* and *Ixodes uriae* are more closely related to strains found in aphids, leafhoppers and flies than the *S. ixodetis* strains found in other *Ixodes* species (such as *Ixodes pacificus* or *Ixodes arboricola*). In ticks, the effect of *S. ixodetis* is currently unknown (Van Oosten *et al.*, 2018; Binetruy *et al.*, 2019), but in aphids, it was shown to manipulate the sex ratio of aphid progeny (Simon *et al.*, 2011) and to induce changes in synthesis of proteins involved in antimicrobial activity, carbohydrate processing, amino acid synthesis and metabolism (Guidolin *et al.*, 2018). In this context, acquisition of such novel endosymbionts by ticks can serve as an immediate and powerful mechanism of rapid adaptation to a changing environment. The mutational source of adaptation – an endosymbiont rather than a mutation of existing genetic material – may actually change our understanding of tick evolution.

Considering the two latter possibilities, we predict that defensive symbionts that protect arthropods against attack by natural enemies will be the major change in tick endosymbiont populations under changing environments. Since under climate changes ticks are expected to show modification of their biogeographical ranges, they may encounter novel antagonistic interactions with natural enemies that were originally absent. Cases of defensive symbiosis in pea aphids include the facultative endosymbiont, *Hamiltonella defensa*, that provides resistance against parasitoid wasps by killing the developing wasp larvae or embryos in the aphid haemocoel (Oliver *et al.*, 2003).

Through reducing the mortality risk after wasp oviposition, the pea aphid thus relies heavily on *H. defensa* for its protection. Another defensive endosymbiont, *Rickettsiella*, increases the resistance of the pea aphid to its natural enemies in a few different ways: (i) infection alters the colour of the aphid body, reducing the host's attractiveness to predators and parasitoids (Tsuchida *et al.*, 2010); and (ii) protects the pea aphid against pathogenic fungi, strongly reducing individual mortality (Łukasik *et al.*, 2013). However, the ecology of *Rickettsiella* interactions remains poorly understood in most host systems. While *H. defensa* is not present in ticks, *Rickettsiella* is widespread and common in ticks (Duron *et al.*, 2015, 2017), as best shown in the polar seabird tick, *I. uriae*, with more than 16 distinct *Rickettsiella* genotypes (Duron *et al.*, 2016). However, the effects of *Rickettsiella* remain entirely unknown in ticks. As a hypothesis, such defensive endosymbionts could significantly enhance the capacity of ticks to persist in hostile habitats or spread in novel habitats filled with native enemies.

The likelihood of radical changes in endosymbiont populations is very high and can proceed extremely fast. Best examples are documented in populations of a whitefly, *Bemisia tabaci*: fast evolution of these populations was observed after the spread of a new facultative endosymbiont belonging to the *Rickettsia bellii* clade, that spread from less than 1% of individuals infected to 97% in only 6 years and dramatically increased whitefly performance (Himler *et al.*, 2011). *Rickettsia*-infected whiteflies produce offspring at nearly twice the rate of individuals lacking the infection, have a higher proportion of offspring that survive to adulthood and have a lower mortality rate due to *Pseudomonas syringae*, a common environmental bacterium, of which some strains are pathogenic to insects (Hendry *et al.*, 2014). The simultaneous expression of these distinct effects (e.g. enhancing fecundity and survival, defensive symbiosis) by *R. bellii* demonstrates the profound and potentially rapid effects of heritable endosymbionts in whiteflies. Since a diversity of *R. bellii* strains are commonly observed in ticks, especially in the Americas (Ogrzewalska *et al.*, 2008; Labruna, 2009; Binetruy *et al.*, 2020b), they could be increasingly recognized as major players in the ecology of ticks and will certainly be a key feature of tick studies in years to come.

References

Benelli, G. (2020) Pathogens manipulating tick behavior – through a glass, darkly. *Pathogens* 9, 664.

Bennett, G.M. and Moran, N.A. (2015) Heritable symbiosis: the advantages and perils of an evolutionary rabbit hole. *Proceedings of the National Academy of Sciences USA* 112, 10169–10176.

Benoit, J.B., Lazzari, C.R., Denlinger, D.L. and Lahondère, C. (2019) Thermoprotective adaptations are critical for arthropods feeding on warm-blooded hosts. *Current Opinion in Insect Science* 34, 7–11.

Ben-Yosef, M., Rot, A., Mahagna, M., Kapri, E., Behar, A. and Gottlieb, Y. (2020) *Coxiella*-like endosymbiont of *Rhipicephalus sanguineus* is required for physiological processes during ontogeny. *Frontiers in Microbiology* 11, 493.

Binetruy, F., Bailly, X., Chevillon, C., Martin, O.Y., Bernasconi, M.V. and Duron, O. (2019) Phylogenetics of the *Spiroplasma ixodetis* endosymbiont reveals past transfers between ticks and other arthropods. *Ticks and Tick-borne Diseases* 10, 575–584.

Binetruy, F., Buysse, M., Lejarre, Q., Barosi, R., Villa, M., *et al.* (2020a) Microbial community structure reveals instability of nutritional symbiosis during the evolutionary radiation of *Amblyomma* ticks. *Molecular Ecology* 29, 1016–1029.

Binetruy, F., Buysse, M., Barosi, R. and Duron, O. (2020b) Novel *Rickettsia* genotypes in ticks in French Guiana, South America. *Scientific Reports* 10, 2537.

Bonnet, S.I., Binetruy, F., Hernández-Jarguín, A.M. and Duron, O. (2017) The tick microbiome: why non-pathogenic microorganisms matter in tick biology and pathogen transmission. *Frontiers in Cellular and Infection Microbiology* 7, 236.

Chen, D.-Q. and Purcell, A.H. (1997) Occurrence and transmission of facultative endosymbionts in aphids. *Current Microbiology* 34, 220–225.

Chen, D.-Q., Montllor, C.B. and Purcell, A.H. (2000) Fitness effects of two facultative endosymbiotic bacteria on the pea aphid, *Acyrthosiphon pisum*, and the blue alfalfa aphid, *A. kondoi*. *Entomologia Experimentalis et Applicata* 95, 315–323.

Duron, O. and Gottlieb, Y. (2020) Convergence of nutritional symbioses in obligate blood feeders. *Trends in Parasitology* 36, 816–825.
Duron, O., Noël, V., McCoy, K.D., Bonazzi, M., Sidi-Boumedine, K., *et al.* (2015) The recent evolution of a maternally-inherited endosymbiont of ticks led to the emergence of the Q fever pathogen, *Coxiella burnetii*. *PLoS Pathogens* 11, e1004892.
Duron, O., Cremaschi, J. and McCoy, K.D. (2016) The high diversity and global distribution of the intracellular bacterium *Rickettsiella* in the polar seabird tick *Ixodes uriae*. *Microbial Ecology* 71, 761–770.
Duron, O., Binetruy, F., Noël, V., Cremaschi, J., McCoy, K.D., *et al.* (2017) Evolutionary changes in symbiont community structure in ticks. *Molecular Ecology* 26, 2905–2921.
Duron, O., Morel, O., Noël, V., Buysse, M., Binetruy, F., *et al.* (2018) Tick–bacteria mutualism depends on B vitamin synthesis pathways. *Current Biology* 28, 1896–1902.e5.
Engl, T., Eberl, N., Gorse, C., Krüger, T., Schmidt, T.H., *et al.* (2018) Ancient symbiosis confers desiccation resistance to stored grain pest beetles. *Molecular Ecology* 27, 2095–2108.
Ferrari, J. and Vavre, F. (2011) Bacterial symbionts in insects or the story of communities affecting communities. *Philosophical Transactions of the Royal Society B* 366, 1389–1400.
Fieler, A., Rosendale, A., Farrow, D., Dunlevy, M., Davies, B., *et al.* (2019) Larval thermal characteristics of multiple ixodid ticks underlie their range dynamics. *bioRxiv* 2019.12.11.873042. https://doi.org/10.1101/2019.12.11.873042
Guidolin, A.S., Cataldi, T.R., Labate, C.A., Francis, F. and Cônsoli, F.L. (2018) *Spiroplasma* affects host aphid proteomics feeding on two nutritional resources. *Scientific Reports* 8, 2466.
Guizzo, M.G., Parizi, L.F., Nunes, R.D., Schama, R., Albano, R.M., *et al.* (2017) A *Coxiella* mutualist symbiont is essential to the development of *Rhipicephalus microplus*. *Scientific Reports* 7, 17554.
Hendry, T.A., Hunter, M.S. and Baltrus, D.A. (2014) The facultative symbiont *Rickettsia* protects an invasive whitefly against entomopathogenic *Pseudomonas syringae* strains. *Applied and Environmental Microbiology* 80, 7161–7168.
Henry, L.M., Peccoud, J., Simon, J.-C., Hadfield, J.D., Maiden, M.J.C., *et al.* (2013) Horizontally transmitted symbionts and host colonization of ecological niches. *Current Biology* 23, 1713–1717.
Heyworth, E.R., Smee, M.R. and Ferrari, J. (2020) Aphid facultative symbionts aid recovery of their obligate symbiont and their host after heat stress. *Frontiers in Ecology and Evolution* 8, 56.
Himler, A.G., Adachi-Hagimori, T., Bergen, J.E., Kozuch, A., Kelly, S.E., *et al.* (2011) Rapid spread of a bacterial symbiont in an invasive whitefly is driven by fitness benefits and female bias. *Science* 332, 254–256.
Hurst, G.D.D. and Jiggins, F.M. (2005) Problems with mitochondrial DNA as a marker in population, phylogeographic and phylogenetic studies: the effects of inherited symbionts. *Proceedings of the Royal Society B* 272, 1525–1534.
Kagemann, J. and Clay, K. (2013) Effects of infection by *Arsenophonus* and *Rickettsia* bacteria on the locomotive ability of the ticks *Amblyomma americanum*, *Dermacentor variabilis*, and *Ixodes scapularis*. *Journal of Medical Entomology* 50, 155–162.
Labruna, M.B. (2009) Ecology of *Rickettsia* in South America. *Annals of the New York Academy of Sciences* 1166, 156–166.
Lemoine, M.M., Engl, T. and Kaltenpoth, M. (2020) Microbial symbionts expanding or constraining abiotic niche space in insects. *Current Opinion in Insect Science* 39, 14–20.
Liu, Y., Miao, H. and Hong, X.Y. (2006) Distribution of the endosymbiotic bacterium *Cardinium* in Chinese populations of the carmine spider mite *Tetranychus cinnabarinus* (Acari: Tetranychidae). *Journal of Applied Entomology* 130, 523–529.
Łukasik, P., van Asch, M., Guo, H., Ferrari, J. and Godfray, H.C.J. (2013) Unrelated facultative endosymbionts protect aphids against a fungal pathogen. *Ecology Letters* 16, 214–218.
McCutcheon, J.P., Boyd, B.M. and Dale, C. (2019) The life of an insect endosymbiont from the cradle to the grave. *Current Biology* 29, R485–R495.
Montllor, C.B., Maxmen, A. and Purcell, A.H. (2002) Facultative bacterial endosymbionts benefit pea aphids *Acyrthosiphon pisum* under heat stress. *Ecological Entomology* 27, 189–195.
Morag, N., Klement, E., Saroya, Y., Lensky, I. and Gottlieb, Y. (2012) Prevalence of the symbiont *Cardinium* in *Culicoides* (Diptera: Ceratopogonidae) vector species is associated with land surface temperature. *FASEB Journal* 26, 4025–4034.
Narasimhan, S. and Fikrig, E. (2015) Tick microbiome: the force within. *Trends in Parasitology* 31, 315–323.
Ogrzewalska, M., Pacheco, R.C., Uezu, A., Ferreira, F. and Labruna, M.B. (2008) Ticks (Acari: Ixodidae) infesting wild birds in an Atlantic forest area in the state of São Paulo, Brazil, with isolation of *Rickettsia* from the tick *Amblyomma longirostre*. *Journal of Medical Entomology* 45, 770–774.

Oliver, K.M., Russell, J.A., Moran, N.A. and Hunter, M.S. (2003) Facultative bacterial symbionts in aphids confer resistance to parasitic wasps. *Proceedings of the National Academy of Sciences USA* 100, 1803–1807.

Oliver, K.M., Campos, J., Moran, N.A. and Hunter, M.S. (2008) Population dynamics of defensive symbionts in aphids. *Proceedings of the Royal Society B* 275, 293–299.

Russell, J.A. and Moran, N.A. (2006) Costs and benefits of symbiont infection in aphids: variation among symbionts and across temperatures. *Proceedings of the Royal Society B* 273, 603–610.

Russell, J.A. and Oliver, K.M. (2020) *Mechanisms Underlying Microbial Symbiosis. Advances in Insect Physiology*, vol. 58, 1st edn. Academic Press, London and New York.

Sánchez-Bayo, F. and Wyckhuys, K.A. (2019) Worldwide decline of the entomofauna: a review of its drivers. *Biological Conservation* 232, 8–27.

Simon, J.-C., Boutin, S., Tsuchida, T., Koga, R., Le Gallic, J.-F., *et al.* (2011) Facultative symbiont infections affect aphid reproduction. *PLoS ONE* 6, e21831.

Sudakaran, S., Kost, C. and Kaltenpoth, M. (2017) Symbiont acquisition and replacement as a source of ecological innovation. *Trends in Microbiology* 25, 375–390.

Tsuchida, T., Koga, R., Horikawa, M., Tsunoda, T., Maoka, T., *et al.* (2010) Symbiotic bacterium modifies aphid body color. *Science* 330, 1102–1104.

Van Oosten, A.R., Duron, O. and Heylen, D.J. (2018) Sex ratios of the tick *Ixodes arboricola* are strongly female-biased, but there are no indications of sex-distorting bacteria. *Ticks and Tick-borne Diseases*, 9, 307–313.

Zhang, B., Leonard, S.P., Li, Y. and Moran, N.A. (2019) Obligate bacterial endosymbionts limit thermal tolerance of insect host species. *Proceedings of the National Academy of Sciences USA* 116, 24712–24718.

eo11 Drought and Tick Dynamics During Climate Change

Joshua Benoit* and Kennan Oyen
University of Cincinnati, Cincinnati, Ohio, USA

11.1 Expert Opinion

11.1.1 Maintenance of water balance is critical to tick off-host survival

Dehydration is a major factor in the survival of terrestrial arthropods, and ticks are no exception (Needham and Teel, 1991; Herrmann and Gern, 2015). Preventing dehydration is critical to tick survival during their off-host periods (Table 11.1). Maintenance of water balance is particularly relevant for blood-feeding arthropods, which must eliminate water following large blood meals and suppress water loss while off a host and searching for their next meal (Benoit and Denlinger, 2010). In general, both hard (Ixodidae) and soft (Argasidae) ticks are more susceptible to dehydration in juvenile stages (Rodgers *et al.*, 2007; Brown *et al.*, 2014). Eggs are typically more resistant to dehydration than larvae but lack the ability to move to favourable microclimates (Yoder *et al.*, 2006, 2012; Benoit *et al.*, 2007). Female selection of stable and humid microhabitats to deposit eggs, and clustering of juvenile stages (eggs and larvae), are critical mechanisms to suppress water loss until the first feeding (Yoder and Knapp, 1999; Benoit *et al.*, 2007). Fed stages are significantly more resistant to dehydration, which is likely critical in preventing excessive water loss during moulting while transitioning from one life stage to the next. Soft ticks are generally more resistant to dehydration than hard ticks (Knülle and Rudolph, 1982; Dautel, 1999; Benoit and Denlinger, 2010). Questing (= host seeking) represents the greatest risk of dehydration to ticks due to direct exposure to extreme environmental conditions compared with the more stable conditions within substrates near the soil or other debris (Needham and Teel, 1991). Larvae and nymphs typically quest low to the ground which likely decreases exposure to dry conditions and wind, preventing excessive dehydration (Mejlon and Jaenson, 1997; Randolph and Storey, 1999; Vail and Smith, 2002). As soft ticks are predominantly found in closer association with their hosts (e.g. nest parasites), questing is likely not as prolonged and stressful, but extended periods when no hosts are available presents a unique challenge (Dautel, 1999).

One of the most interesting aspects of tick biology is that water replenishment is primarily accomplished by the process of water vapour uptake (or absorption). Glycerol, salts and other hydrophilic substances are secreted on to the mouthparts (Gaede and Knülle, 1997; Sonenshine and Roe, 2013; Rosendale *et al.*, 2016), capturing atmospheric water which the tick

*Email: joshua.benoit@uc.edu

DOI: 10.1079/9781789249637.0011

Table 11.1. Mechanisms used by ticks to prevent and respond to dehydration stress. (Authors' own table.)

	Mechanism		
	Behavioural	Physiological	Molecular/Biochemical
Factors that prevent or reduce tick dehydration	1. Females lay eggs in moist, stable microhabitats	1. Water vapour uptake from air	1. Increased expression of stress and antioxidant factors during dehydration stress
	2. Questing at night prevents excessive water loss (critical for juvenile stages)	2. Low respiration rates reduce water loss	2. Specific amino acids increased to prevent dehydration-induced damage
	3. Increased attraction to humidity	3. Thick, hydrocarbon-covered cuticle prevents water loss	3. Dehydration primes water vapour uptake, allowing immediate hydration in favourable conditions
	4. Long periods of quiescence	4. Egg clusters are resistant to water loss	4. Utilization of lipid reserves to accumulate glycerol and other metabolites to prevent cellular damage
	5. Clustering of juvenile stages to prevent water loss	5. Production of water from metabolism	

can drink. The critical equilibrium humidity (CEH) is where water gains from the humid air offset water loss, leading to a net gain of water and maintenance of water homeostasis (Wharton, 1985). In general, ixodid larvae have the lowest CEH (82–86% relative humidity (RH)), followed by nymphs (86–89% RH) and adults (89–96% RH) (Benoit *et al.*, 2007; Yoder *et al.*, 2012). These values are generally higher for soft ticks (Knülle and Rudolph, 1982; Dautel, 1999; Benoit and Denlinger, 2010). Drinking of free water has been noted in multiple hard tick species, which improves survival under dehydrating conditions, but does not represent the primary mode of hydration (Kahl and Alidousti, 1997; Yoder *et al.*, 2006; Maldonado-Ruiz *et al.*, 2020).

Each bout of dehydration, along with subsequent rehydration, increases carbon dioxide output and oxygen utilization (Rudolph and Knülle, 1974; Fielden and Lighton, 1996; Gaede and Knülle, 1997; Rosendale *et al.*, 2017), leading to discernible reductions in nutrient reserves and eventually earlier death (Rosendale *et al.*, 2017). Soft ticks increase metabolic rate in response to dehydration, most likely to produce metabolic water and prevent dehydration-induced mortality at the risk of exhausting energy reserves (Dautel, 1999). Because hard ticks feed only once at each life stage, nutrient reserves of off-host ticks are finite and therefore starvation, which can be exacerbated by increased metabolism during stress exposure, is a major factor regulating tick survival (Fielden *et al.*, 1999; Rosendale *et al.*, 2019). Soft ticks can feed multiple times as juveniles or adults, thus the dynamic between starvation and longevity is not as direct when compared with hard ticks. Because the water loss of soft ticks is often insignificant, discerning between death from either dehydration or starvation is difficult at moderate RH values (50–75%) (Knülle and Rudolph, 1982; Dautel, 1999). Ticks also use prolonged periods of quiescence to lower respiration (Fielden and Lighton, 1996; Fielden *et al.*, 1999), reducing water lost by breathing, and lowering energy utilization. Infection status also impacts dehydration, starvation and subsequent survival dynamics (Herrmann and Gern, 2015). For instance, lipid reserves are increased in *Borrelia*-infected ticks (Herrmann *et al.*, 2013) and increased longevity/starvation resistance has been noted in relation to infection status (Fielden *et al.*, 1999; Herrmann and Gern, 2015; Rosendale *et al.*, 2019). Little is known about the interactions between infection status and dehydration or starvation in soft ticks. Of interest is that fungal exposure associated with wet conditions may be reduced during xeric periods, representing a potential benefit to ticks.

Ticks vary substantially in dehydration resistance, with some species being susceptible to dehydration, e.g. *Ixodes uriae* and *Ixodes scapularis* (Benoit *et al.*, 2007), and others surviving much more effectively under dry conditions (Hafez *et al.*, 1970). Recent RNA-seq and metabolomic studies have examined the mechanisms underlying the response of ticks to dehydration (Rosendale *et al.*, 2016). Nearly 500 genes were differentially expressed during dehydration stress in the American dog tick, *Dermacentor variabilis*, including many stress proteins, and metabolites such as amino acids, glycerol and γ-aminobutyric acid (GABA) were increased (Rosendale *et al.*, 2016). Importantly, treatment of ticks with glycerol and GABA increased the rate of water vapour absorption, suggesting their critical role in preparation for water vapour uptake when dehydrated ticks can move into more humid microhabitats. Glycerol production during water vapour uptake and response to stress likely underlies the increased metabolic output and nutrient utilization during each dehydration and subsequent rehydration bout (Rosendale *et al.*, 2016, 2017).

Along with physiological changes, hard tick behaviour is altered by shorter dehydration bouts or periods of drought (Table 11.1). American dog ticks increase their activity when exposed to dehydrating conditions, specifically during dark periods (Rosendale *et al.*, 2016). This higher activity likely serves two roles: (i) increasing encounters with potential hosts; and (ii) locating humid microhabitats (Randolph and Storey, 1999; Rosendale *et al.*, 2016). For other ticks, such as *Ixodes ricinus*, questing is reduced by a saturation deficit (Perret *et al.*, 2000, 2003), but in *I. scapularis* there was no distinct association with dry conditions and questing (Thomas *et al.*, 2020). There has been a general trend to find fewer ticks questing in dry conditions (Diuk-Wasser *et al.*, 2010; Tagliapietra *et al.*, 2011; Berger *et al.*, 2014; Eisen *et al.*, 2016; Thomas *et al.*, 2020) and more when conditions are wet (Vail and Smith, 1998). To prevent dehydration, *I. ricinus* and other tick species will primarily quest throughout the night when desiccation risk is lower (Perret *et al.*, 2003; Rosendale *et al.*, 2016). RH also influences questing height as ticks tend to quest further off the ground when RH values are high compared with those far below the CEH (Vail and Smith, 2002). Because dry periods do not reduce questing in all tick species, shifts in seasonal quiescence, altered daily activity patterns or higher mortality could explain low counts when surveying for specific tick species during dry periods (Bertrand and Wilson, 1996; Diuk-Wasser *et al.*, 2010). Furthermore, behavioural phenotypes, including altered questing and total activity levels, which impact tick water balance, are influenced by infection status (Herrmann and Gern, 2015), suggesting potential for interactions between infestation status, ticks and biological changes during drought. Studies directly relating soft tick activity with dehydration are lacking, but dehydration and starvation are likely to increase host-seeking as a source of nourishment to restore water homeostasis, similar to hard ticks.

11.1.2 Drought, climate change and the impact on ticks

Climate change will shift local to worldwide weather patterns, leading to increased variability in precipitation and periods of drought (Dai, 2013). Extended dry periods are likely to diminish tick populations, specifically *Ixodes* species, as water balance will be difficult to maintain when soil moisture falls below physiologically relevant levels necessary to hydrate between questing bouts (Brown *et al.*, 2014). Several potential scenarios may characterize tick responses to humid conditions, cyclical dry spells or chronic drought (Fig. 11.1). Both the total population and the proportion of infected *Ixodes* ticks fall following periods of drought (Doby and Bigaignon, 1992; Jones and Kitron, 2000). Drought-induced changes in the abundance of host species, such as rodents, will also impact tick survival if ticks fail to locate hosts before succumbing to dehydration or starvation (Doby and Bigaignon, 1992; Ogden and Lindsay, 2016; Dhawan *et al.*, 2018). Along with increased drought, periods of heavy rainfalls are predicted to occur more frequently with climate change, which can reduce the overall activity of ticks and yield only

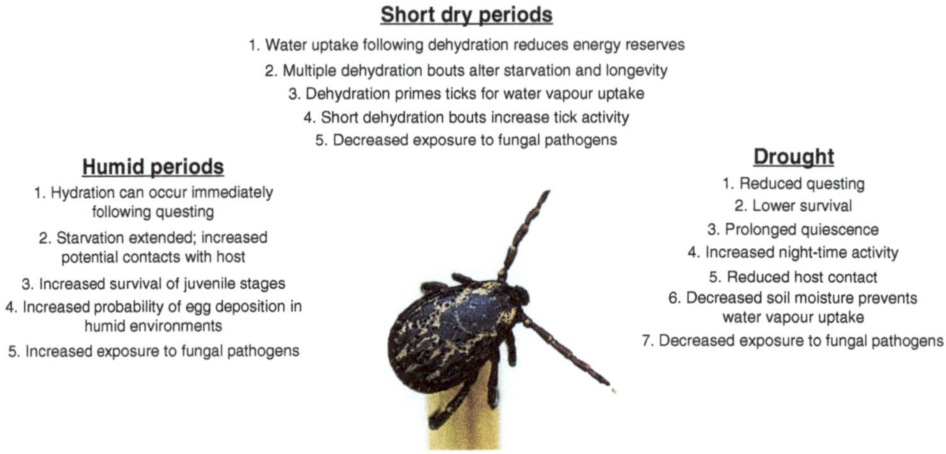

Fig. 11.1. Summary of specific factors related to tick physiology and behaviour that will be impacted by short dehydration exposure or when exposed to severe dehydration stress during drought. (Authors' own figure.)

short wet periods between longer xeric intervals (Dai, 2013; Ogden and Lindsay, 2016). These abbreviated periods of favourable conditions will allow for only short questing windows that might not directly match those when hosts are available.

11.2 Future Predictions

Based on these observations and predicted climate changes, how will ticks be impacted by periods of increased drought and more sporadic precipitation? Likely, species that require wet conditions to maintain water balance will experience rapid population declines during drought which can lead to local, and even regional, extinction if periods of drought are long or occur more frequently, as predicted (Dai, 2013). Importantly, these effects could be exacerbated by warmer temperatures, which drive water loss by increasing the water vapour pressure deficit, and drought-induced declines in host species (Dai, 2013; Dhawan *et al.*, 2018). The potential reduction in host species may be most impactful for soft ticks, which are restricted to small areas where their hosts reside, and inaccessibility to these hosts could lead to local population collapse. It is unlikely that the population of any specific tick species will increase with elevated drought frequency and duration within its native range. An alternative is that the cumulative effects of climate changes, including increased temperature variation and shifted host ranges, could lead to the redistribution of dry- and heat-adapted tick species, replacing those with higher water requirements. These redistributions among tick species are already occurring. For example, ticks that are more resistant to dry and hot periods common in the USA are expanding their ranges northward, e.g. the Gulf Coast tick, *Amblyomma maculatum* (Sonenshine, 2018). Increased dry periods could lead to local areas where tick abundances are higher (e.g. near rivers and lakes) relative to other, previously favourable habitats. These reprieves would be the result of increased overlap between ticks and hosts in favourable microclimates, potentially leading to very localized increases in disease transmission. Lastly, selection pressures associated with dehydration may alter tick behaviour, leading to questing within or closer to leaf litter along with more activity during the night (Ginsberg *et al.*, 2017; Ogden *et al.*, 2020). The changes are likely to impact host species preferences. The cumulative effects of drought will depend on the specific tick species, duration and severity of dry periods, thermal changes and how weather conditions impact host availability. Severe, prolonged periods of drought are likely to generate unsuitable conditions for the survival of most, if not all, ticks within their native ranges, leading to local extinction. Droughts over extended periods could prompt desertification, creating entire regions where few or no ticks are present, agricultural practices are impractical and conditions are inhospitable for prolonged human residence without significant resource allocation.

Acknowledgements

Both authors contributed to the idea development and the writing of this expert opinion. This work was funded, in part, by the National Science Foundation (DEB-1654417 to J.B.B.), the US Department of Agriculture National Institute of Food and Agriculture (2018-67013-28495 to J.B.B.), the National Institutes of Health (1R01AI148551-01A1 to J.B.B.) and the David H. Smith Conservation Research Fellowship (to K.J.O.). Due to reference limitations, we apologize for articles related to the enclosed content that were not cited.

References

Benoit, J.B. and Denlinger, D.L. (2010) Meeting the challenges of on-host and off-host water balance in blood-feeding arthropods. *Journal of Insect Physiology* 56, 1366–1376.

Benoit, J.B., Yoder, J.A., Lopez-Martinez, G., Elnitsky, M.A., Lee, R.E. Jr and Denlinger, D.L. (2007) Habitat requirements of the seabird tick, *Ixodes uriae* (Acari: Ixodidae), from the Antarctic Peninsula in relation to water balance characteristics of eggs, nonfed and engorged stages. *Journal of Comparative Physiology. B, Biochemical, Systemic, and Environmental Physiology* 177, 205–215.

Berger, K.A., Ginsberg, H.S., Gonzalez, L. and Mather, T.N. (2014) Relative humidity and activity patterns of *Ixodes scapularis* (Acari: Ixodidae). *Journal of Medical Entomology* 51, 769–776.

Bertrand, M.R. and Wilson, M.L. (1996) Microclimate-dependent survival of unfed adult *Ixodes scapularis* (Acari: Ixodidae) in nature: life cycle and study design implications. *Journal of Medical Entomology* 33, 619–627.

Brown, L., Medlock, J. and Murray, V. (2014) Impact of drought on vector-borne diseases – how does one manage the risk? *Public Health* 128, 29–37.

Dai, A. (2013) Increasing drought under global warming in observations and models. *Nature Climate Change* 3, 52–58.

Dautel, H. (1999) Water loss and metabolic water in starving *Argas reflexus* nymphs (Acari: Argasidae). *Journal of Insect Physiology* 45, 55–63.

Dhawan, R., Fischhoff, I.R. and Ostfeld, R.S. (2018) Effects of weather variability on population dynamics of white-footed mice (*Peromyscus leucopu*) and eastern chipmunks (*Tamias striatus*). *Journal of Mammalogy* 99, 1436–1443.

Diuk-Wasser, M.A., Vourc'h, G., Cislo, P., Hoem, A.G., Melton, F., *et al.* (2010) Field and climate-based model for predicting the density of host-seeking nymphal *Ixodes scapularis*, an important vector of tick-borne disease agents in the eastern United States. *Global Ecology and Biogeography* 19, 504–514.

Doby, J.M. and Bigaignon, G. (1992) Survey of the infestation level of the *Ixodes ricinus* tick by *Borrelia burgdorferi*. *Bulletin de la Societe de Pathologie Exotique* 85, 322–323.

Eisen, R.J., Eisen, L. and Beard, C.B. (2016) County-scale distribution of *Ixodes scapularis* and *Ixodes pacificus* (Acari: Ixodidae) in the Continental United States. *Journal of Medical Entomology* 53, 349–386.

Fielden, L.J. and Lighton, J.R.B. (1996) Effects of water stress and relative humidity on ventilation in the tick, *Dermacentor andersoni* (Acari: Ixodidae). *Physiological Zoology* 69, 599–617.

Fielden, L.J., Jones, R.M., Goldberg, M. and Rechav, Y. (1999) Feeding and respiratory gas exchange in the American dog tick, *Dermacentor variabilis*. *Journal of Insect Physiology* 45, 297–304.

Gaede, K. and Knülle, W. (1997) On the mechanism of water vapour sorption from unsaturated atmospheres by ticks. *Journal of Experimental Biology* 200, 1491–1498.

Ginsberg, H.S., Albert, M., Acevedo, L., Dyer, M.C., Arsnoe, I.M., *et al.* (2017) Environmental factors affecting survival of immature *Ixodes scapularis* and implications for geographical distribution of Lyme disease: the climate/behavior hypothesis. *PLoS ONE* 12, e0168723.

Hafez, M., El-Ziady, S. and Hefnawy T. (1970) Biochemical and physiological studies of certain ticks (Ixodoidea). Cuticular permeability of *Hyalomma (H.) dromedarii* Koch (Ixodidae) and *Ornithodoros (O.) savignyi* (Audouin) (Argasidae). *Journal of Parasitology* 1, 154–168.

Herrmann, C. and Gern, L. (2015) Search for blood or water is influenced by Borrelia burgdorferi in *Ixodes ricinus*. *Parasites & Vectors* 8, 6.

Herrmann, C., Voordouw, M.J. and Gern, L. (2013) *Ixodes ricinus* ticks infected with the causative agent of Lyme disease, *Borrelia burgdorferi* sensu lato, have higher energy reserves. *International Journal for Parasitology* 43, 477–483.

Jones, C.J. and Kitron, U.D. (2000) Populations of *Ixodes scapularis* (Acari: Ixodidae) are modulated by drought at a Lyme disease focus in Illinois. *Journal of Medical Entomology* 37, 408–415.

Kahl, O. and Alidousti, I. (1997) Bodies of liquid water as a source of water gain for *Ixodes ricinus* ticks (Acari: Ixodidae). *Experimental and Applied Acarology* 21, 731–746.

Knülle, W. and Rudolph, D. (1982) Humidity relationships and water balance of ticks. In: Obenchain, F.D. and Galun, R. (eds) *Physiology of Ticks*. Pergamon, Oxford, pp. 43–70.

Maldonado-Ruiz, L.P., Park, Y. and Zurek, L. (2020) Liquid water intake of the lone star tick, *Amblyomma americanum*: implications for tick survival and management. *Scientific Reports* 10, 6000.

Mejlon, H.A. and Jaenson, T.G.T. (1997) Questing behaviour of *Ixodes ricinus* ticks (Acari: Ixodidae). *Experimental and Applied Acarology* 21, 747–754.

Needham, G.R. and Teel, P.D. (1991) Off-host physiological ecology of ixodid ticks. *Annual Review of Entomology* 36, 659–681.

Ogden, N.H. and Lindsay, L.R. (2016) Effects of climate and climate change on vectors and vector-borne diseases: ticks are different. *Trends in Parasitology* 32, 646–656.

Ogden, N.H., Beard, C.B., Ginsberg, H.S. and Tsao, J.I. (2020) Possible effects of climate change on ixodid ticks and the pathogens they transmit: predictions and observations. *Journal of Medical Entomology* 2020, tjaa220.

Perret, J.-L., Guigoz, E., Rais, O. and Gern, L. (2000) Influence of saturation deficit and temperature on *Ixodes ricinus* tick questing activity in a Lyme borreliosis-endemic area (Switzerland). *Parasitology Research* 86, 554–557.

Perret, J.-L., Guerin, P.M., Diehl, P.A., Vlimant, M. and Gern, L. (2003) Darkness induces mobility, and saturation deficit limits questing duration, in the tick *Ixodes ricinus*. *Journal of Experimental Biology* 206, 1809–1815.

Randolph, S.E. and Storey, K. (1999) Impact of microclimate on immature tick–rodent host interactions (Acari: Ixodidae): implications for parasite transmission. *Journal of Medical Entomology* 36, 741–748.

Rodgers, S.E., Zolnik, C.P. and Mather, T.N. (2007) Duration of exposure to suboptimal atmospheric moisture affects nymphal blacklegged tick survival. *Journal of Medical Entomology* 44, 372–375.

Rosendale, A.J., Romick-Rosendale, L.E., Watanabe, M., Dunlevy, M.E. and Benoit, J.B. (2016) Mechanistic underpinnings of dehydration stress in the American dog tick revealed through RNA-Seq and metabolomics. *Journal of Experimental Biology* 219, 1808–1819.

Rosendale, A.J., Dunlevy, M.E., Fieler, A.M., Farrow, D.W., Davies, B. and Benoit, J.B. (2017) Dehydration and starvation yield energetic consequences that affect survival of the American dog tick. *Journal of Insect Physiology* 101, 39–46.

Rosendale, A.J., Dunlevy, M.E., McCue, M.D. and Benoit, J.B. (2019) Progressive behavioural, physiological and transcriptomic shifts over the course of prolonged starvation in ticks. *Molecular Ecology* 28, 49–65.

Rudolph, D. and Knülle, W. (1974) Site and mechanism of water vapour uptake from the atmosphere in ixodid ticks. *Nature* 249, 84–85.

Sonenshine, D.E. (2018) Range expansion of tick disease vectors in North America: implications for spread of tick-borne disease. *International Journal of Environmental Research and Public Health* 15, 148.

Sonenshine, D.E. and Roe, M.R. (2013) *Biology of Ticks*, 2nd edn. Oxford University Press, Oxford.

Tagliapietra, V., Rosà, R., Arnoldi, D., Cagnacci, F., Capelli, G., *et al.* (2011) Saturation deficit and deer density affect questing activity and local abundance of *Ixodes ricinus* (Acari, Ixodidae) in Italy. *Veterinary Parasitology* 183, 114–124.

Thomas, C.E., Burton, E.S. and Brunner, J.L. (2020) Environmental drivers of questing activity of juvenile black-legged ticks (Acari: Ixodidae): temperature, desiccation risk, and diel cycles. *Journal of Medical Entomology* 57, 8–16.

Vail, S.G. and Smith, G. (1998) Air temperature and relative humidity effects on behavioral activity of black-legged tick (Acari: Ixodidae) nymphs in New Jersey. *Journal of Medical Entomology* 35, 1025–1028.

Vail, S.G. and Smith, G. (2002) Vertical movement and posture of blacklegged tick (Acari: Ixodidae) nymphs as a function of temperature and relative humidity in laboratory experiments. *Journal of Medical Entomology* 39, 842–846.

Wharton, G.W. (1985) Water balance of insects. In: Kerkut, G.A. and Gilbert, L.I. (eds) *Regulation: Digestion, Nutrition, Excretion*. Pergamon, Oxford, pp. 565–601.

Yoder, J.A. and Knapp, D.C. (1999) Cluster-promoted water conservation by larvae of the American dog tick, *Dermacentor variabilis* (Acari: Ixodidae). *International Journal of Acarology* 25, 55–57.

Yoder, J.A., Benoit, J.B., Rellinger, E.J. and Tank, J.L. (2006) Developmental profiles in tick water balance with a focus on the new Rocky Mountain spotted fever vector, *Rhipicephalus sanguineus*. *Medical and Veterinary Entomology* 20, 365–372.

Yoder, J.A., Hedges, B.Z. and Benoit, J.B. (2012) Water balance of the American dog tick, *Dermacentor variabilis*, throughout its development with comparative observations between field-collected and laboratory-reared ticks. *International Journal of Acarology* 38, 334–343.

eo12 Climate Influences on Reproduction and Immunity in the Soft Tick, *Ornithodoros moubata*

DeMar Taylor[1]* and Mari Ogihara[2]
[1]University of Tsukuba, Tsukuba, Japan; [2]Institute of Livestock and Grassland Science, Tsukuba, Japan

12.1 Expert Opinion

12.1.1 Introduction

Ticks must obtain a blood meal to acquire the necessary nutrients for growth, development and reproduction. Pathogens enter the tick vector when they feed on infected vertebrate hosts and are then transmitted to non-infected hosts during subsequent feedings. The tick vector is not normally affected by the pathogens, but the blood meal is necessary for them to survive and reproduce. Understanding physiological mechanisms of ticks is necessary to determine the vector competence of ticks and identify targets for effective control strategies. In particular, knowledge about the regulation of reproduction and immune responses is essential. This expert opinion considers the physiological processes regulating reproduction and immune responses in the soft tick, *Ornithodoros moubata*, and the potential impact of climate change.

12.1.2 Life cycle

The life cycles of hard and soft ticks differ greatly, with soft ticks normally requiring multiple nymphal stages and hosts to complete their growth and development into reproducing adults. Most hard ticks have four life stages: egg, larva, nymph and adult. Soft ticks have a similar life cycle except for the number of nymphal stages that varies with the species and environmental conditions (Horigane and Taylor, 2009). *O. moubata* requires four nymphal stages to become adult males and five nymphal stages to become adult females, under conditions of 30 ± 1°C and 75 ± 5% relative humidity (RH). Of course, some females may emerge after four nymphal moults and some males after five moults, but all nymphs become adults after five nymphal moults. The only stage that does not require a blood meal is the larval stage. The amount of egg protein, vitellin (Vn), stored in the egg is sufficient to sustain the larval stage until it moults into a first instar nymph. All nymphal stages require a blood meal before they can moult. Adult females must also obtain a blood meal to produce viable eggs, but a small percentage, approximately 10%, are able to lay small batches of eggs without a blood meal. These females have likely sufficient blood meal reserves from the nymphal stages stored in their midgut to produce the eggs, but the reason only a few females lay eggs remains unknown. The lengths of time needed for feeding have also evolved differently in the hard

*Email: taylor.de.mar.ge@u.tsukuba.ac.jp

DOI: 10.1079/9781789249637.0012

and soft ticks. Female hard ticks usually require 7 to 10 days of attachment to the host to complete the feeding processes. On the other hand, female soft ticks can complete engorgement within a few hours. These differences are important for the reproductive cycles. In addition to blood feeding, ticks must mate in order to produce viable eggs, except for a few parthenogenic strains and species (Oliver, 1971; Kiszewski *et al.*, 2001).

Most hard tick species must mate while feeding on a host. Normally, males will feed for a short time and females require mating before they can complete the feeding process (Apanaskevich and Oliver, 2014). The females will feed slowly until mated and then enter a rapid feeding stage during which most of the blood meal is obtained. After becoming completely engorged they will detach, drop from the host, lay their eggs and die, completing only one reproductive cycle during their life. On the other hand, soft ticks can mate before or after feeding and lay eggs multiple times. *O. moubata* males will take a blood meal but it is not necessary for reproduction. Females can store the endospermatophores indefinitely in their seminal receptacle and lay eggs after each blood meal with or without subsequent mating (Ogihara and Taylor, 2014). Also, a female can store sperm from multiple males, increasing the genetic diversity in its offspring. As a result, soft ticks have much longer life cycles than hard ticks and are able to survive long periods of starvation. *Ornithodoros turicata* and *Ornithodoros lahorensis* adults can survive for 7 and 18 years, respectively, in the laboratory (Sonenshine, 1993). We have also observed females surviving and subsequently feeding after being forgotten in the drawer of a laboratory for 7 years. These ticks were exposed to various humidities and temperatures during the 7 years because the laboratory was not maintained under the same conditions during the temperate nights of Japan.

The environmental conditions also differ greatly within hard and soft ticks (Vial, 2009; Gray *et al.*, 2014). Hard ticks are known to occur and quest for hosts in fields, shrub areas and forests throughout the tropics to the subarctic areas of the world. Soft tick distributions are similar except they are normally found in the nests and burrows of reptiles, birds and mammals. These nest environments provide protection from harsh conditions. In addition, these nesting conditions appear to have allowed them to develop the ability to withstand long periods of starvation and live much longer than hard ticks. Soft ticks appear to show little host specificity except when in ecological isolation such as in bat caves (Gray *et al.*, 2014). Although nidicolous ticks do not often come into contact with various species, they also appear to be able to easily adapt to new hosts so there is a greater potential to create reservoirs of pathogens. They also tend to remain in colonies of mixed stages in the same nest, which may have allowed them to develop the ability to reproduce with mating before or after feeding. The ability of *O. moubata* to survive long periods of starvation, blood feed in multiple stages on various hosts, engorge within a short time and mate before or after feeding has allowed them to evolve into an important reservoir and vector of numerous pathogens.

12.1.3 Regulation of reproduction

Blood feeding and mating are both necessary for reproduction in ticks. The soft tick, *O. moubata*, is an excellent model species to study factors regulating reproduction in ticks because it can easily be maintained under laboratory situations. As discussed above, *O. moubata* can fully engorge within a few hours and can mate before or after feeding, so blood feeding and mating can be easily regulated and separated to study the physiological processes of engorgement and mating. In addition, the effects of different factors on these processes can be carefully evaluated. Our research has focused on the elucidation of factors regulating the synthesis of egg proteins in order to understand the processes that regulate reproduction in ticks (Taylor *et al.*, 1997). Early studies focused on elucidation of egg proteins in *O. moubata* and showed that the egg protein vitellogenin (Vg) is produced as a precursor to the mature egg protein Vn (Chinzei and Taylor, 1994; Taylor and Chinzei, 2002). Vg is synthesized in the midgut and fat body after which it is carried by the haemolymph to the ovary where it is absorbed by the oocytes. Egg protein synthesis is regulated by humoral factors: a fat body-stimulating factor (FSF) and a vitellogenin-inducing factor (VIF) (Chinzei and Taylor, 1990). FSF appears to be a peptide hormone synthesized in the synganglion

(Chinzei *et al.*, 1992). Steroid hormones, ecdysteroids, have been shown to act as the second hormone (VIF) to upregulate the production of Vg (Horigane *et al.*, 2007, 2008; Ogihara *et al.*, 2007). The synthesis of ecdysteroids in the Acari remains somewhat controversial but the ovary appears to play a major role in this process (Ogihara *et al.*, 2019). The precursor of active ecdysteroids, ecdysone, and its active form, 20-hydroxyecdysone, have been detected in ticks (Ogihara *et al.*, 2019). Early studies indicated the synthesis of ecdysteroids takes place in the epidermis and fat body of ticks (Connat *et al.*, 1986; Zhu *et al.*, 1991; Dotson *et al.*, 1993). However, Ogihara *et al.* (2015) showed that the ecdysteroidogenic enzyme CYP307A1 (encoded by *spook* gene) of *O. moubata* females functions in the synthesis of ecdysteroids in the ovary and RNAi knockdown of *spook* causes an absence of both the precursor ecdysone and the final product, 20-hydroxyecdysone.

Without mating, ecdysteroid titres in the haemolymph do not increase and Vg synthesis is not upregulated (Ogihara *et al.*, 2007; Horigane *et al.*, 2010). Vg titres in the haemolymph of virgin females remain persistently low and oocyte maturation and egg laying do not occur. On the other hand, in mated females, *Vg* gene expression persists at a high level (Horigane *et al.*, 2010) and Vg titres in the haemolymph increase to a peak at day 6 after blood feeding (Ogihara *et al.*, 2007). Subsequently, the Vg titres decrease to levels similar to those found in virgin females as Vg is incorporated into the oocytes. Egg laying begins on approximately day 10 after blood feeding and continues for 20 to 30 days only in mated females (Ogihara *et al.*, 2007). Therefore, mating must take place for the production of ecdysteroids to stimulate the upregulation of egg protein (Vg) synthesis and development of mature oocytes to produce viable offspring.

Blood feeding is also essential and nutrient signalling pathways appear to be important for regulating early Vg synthesis and ensure sufficient levels are present in the haemolymph (Horigane *et al.*, 2010). Injection of rapamycin, an inhibitor of the target of rapamycin (TOR) kinase, decreases *Vg* gene expression and Vg titres in the haemolymph as well as delays oocyte development and egg laying, indicating that the TOR nutrient signalling pathway is essential for the regulation of reproduction in *O. moubata* (M. Behri, H. Teshima, K. Kutsuwada, S. Nakatake, M.H. Ogihara and D. Taylor, 2021, unpublished results). Horigane *et al.* (2010) showed that virgin females can quickly (within 3 days) produce mature eggs if allowed to mate even 22 days after feeding. Stimulation of Vg synthesis by feeding appears to be an adaptation for soft ticks to store large amounts of Vg in the haemolymph and oocytes, so that when mating stimulation of the ovary by ecdysteroids occurs, this Vg can quickly be incorporated into oocytes to produce offspring. Although feeding and mating processes are very different between hard and soft ticks, the regulation of egg protein synthesis appears to be similar (Boldbaatar *et al.*, 2010; Umemiya-Shirafuji *et al.*, 2012; Ogihara *et al.*, 2019; Waltero *et al.*, 2019). Based on the above studies of *O. moubata*, we can conclude that Vg synthesis is regulated by mating-dependent and nutrient-dependent factors (Fig. 12.1). Mating-dependent regulation is necessary to produce mature oocytes and occurs only in mated females with upregulation of Vg synthesis in enlarged fat body tissues stimulated by increases in ecdysteroids synthesized in the ovary. Previously, Pound and Oliver (1979) reported that juvenile hormone (JH) regulated reproduction in ticks, but later Chinzei *et al.* (1991) showed that JH does not function in *O. moubata* and Neese *et al.* (2000) subsequently showed that JH and its precursor, methyl farnesoate, are not present in the Acari. Nutrient-dependent regulation results in Vg synthesis occurring in both virgin and mated females, but at much lower levels in virgins. Common Vg synthesis is regulated by the TOR nutrient signalling pathway with synthesis primarily in the midgut tissues. Therefore, Vg is likely an important storage protein as virgin females can incorporate it into and reabsorb it from the oocytes, as needed, but cannot produce mature oocytes in the absence of hormonal stimulation. Further studies are underway to clarify the mechanisms for these processes by characterizing the Vg receptor and its functions. Understanding reproduction is a key area of physiology that may provide effective targets for control of ticks and is essential for determining the potential of ticks to spread and become established in new areas as climate changes occur.

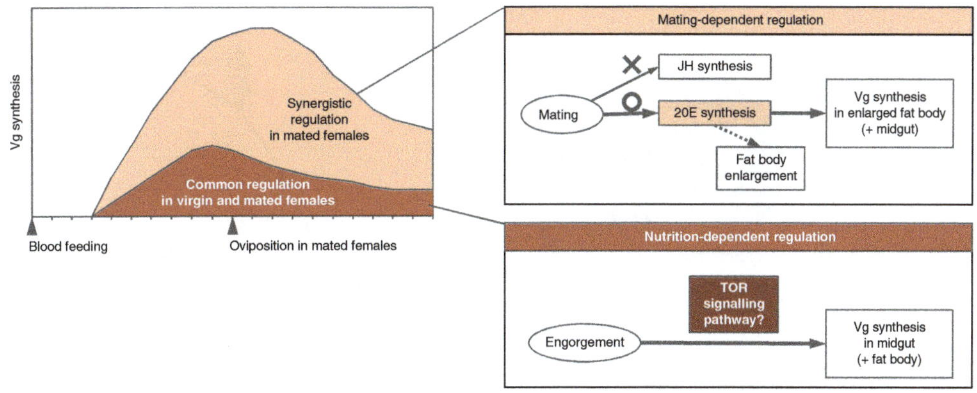

Fig. 12.1. Model of mating-dependent and nutrition-dependent regulation of vitellogenin (Vg) synthesis in the soft tick *Ornithodoros moubata*. See Section 12.1.3 for detailed explanations. (Authors' own figure.)

12.1.4 Influences of climate on reproduction

Early studies clarified optimal temperature and humidity for soft tick survival (reviewed in Vial, 2009). Walton (1957) observed that *O. moubata* ticks were most numerous in cool, wet habitats above 4000 ft, and abundant in foothills and plains, but absent from the hot dry areas of East Africa. Other early studies report that all stages survive longer at high humidities, and eggs can develop into first instar nymphs, which show the highest resistance to desiccation (Brett, 1939). Similar effects of temperature and humidity were also seen in laboratory studies on two other species of *Ornithodoros*. El Shoura (1987) showed the life cycle of *Ornithodoros erraticus* became shorter with rising temperatures and lower humidities, with 28°C and 75% RH as the optimum conditions.

Temperature and humidity of microenvironments in the habitat remarkably influence reproduction in *Ornithodoros* species. Cunliffe (1921) observed that increases in temperature above 22°C resulted in shortening of the period required for increases in the rate of oviposition, decreases in the viability of eggs and reduction in the longevity of the female tick under laboratory conditions. Phillips and Adeyeye (1996) concluded the optimum reproductive conditions for *O. turicata* were 24°C and 90–95% RH. They also speculated that ticks may survive longer by moving to lower temperatures when not feeding. Females may also be able to delay oviposition for an optimal microenvironment in the burrows, but this ability appears limited by the amount of blood imbibed, as ticks with lower engorgement weights appear to delay oviposition longer. Studies to predict distribution changes in *Ornithodoros* species by climate change have not been conducted in Africa. However, modelling of soft tick habitats in North America suggests possible habitat changes to higher altitudes (Sage *et al.*, 2017). Future climate change (i.e. global increase in mean temperature) may limit distribution of *Ornithodoros* species and/or result in movements to higher elevations and latitudes along with their hosts (Fig. 12.2). In contrast, climate change may also enhance seasonal blood feeding and reproduction of some soft tick species that show quiescence at low temperatures (Fig. 12.2) (Oleaga-Pérez *et al.*, 1990).

In the context of recent climate change, increases in the global mean temperature may also induce extreme climate events, such as extreme heat, flooding and deep freezing. Our recent studies on the effects of low temperature on Vg synthesis, egg laying and egg hatching of *O. moubata* revealed interesting effects of cold stress on reproduction (T. Takada and D. Taylor, 2021, unpublished results). Unfed mated females that were exposed to cold stress at −10, −5 or 0°C for 12 h showed significant increases in Vg protein concentrations from 10 to 20 days after exposure, whereas Vg concentrations in females exposed to either 5 or 25°C remained the same

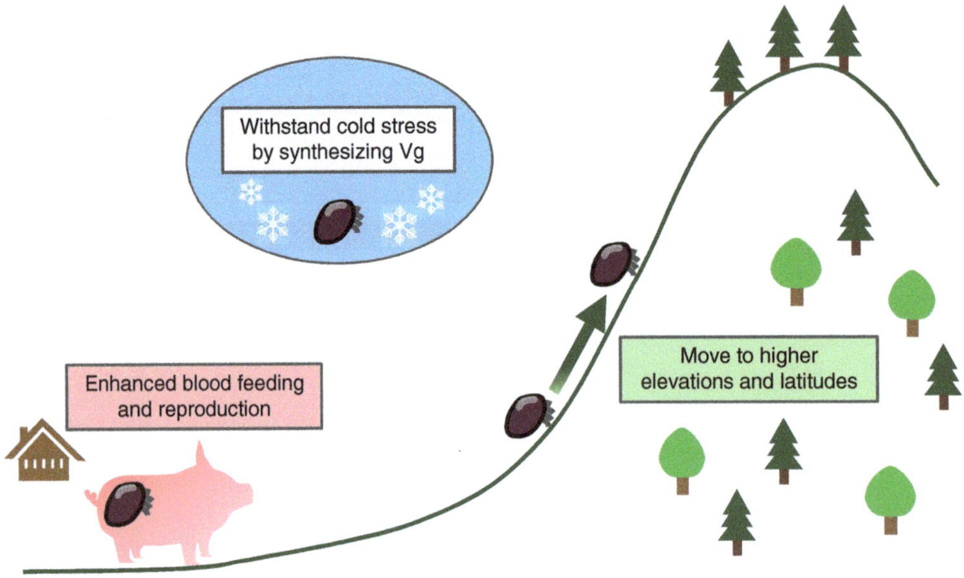

Fig. 12.2. Expected influences of climate change on soft ticks. See Section 12.1.4 for detailed explanations. Vg, vitellogenin. (Authors' own figure.)

as in unfed females normally kept at 30 ± 1°C and 75 ± 5% RH. Similar results were seen in females exposed to the same temperatures for 96 h, except that the increase in titres was delayed until approximately 15 days after exposure. Females exposed to 2°C for 2 weeks also showed similar increases in Vg concentrations 15 to 20 days after return to 30°C. Several reports on insects demonstrate Vg proteins are involved in cold tolerance (Poikela *et al.*, 2020), act as an antioxidant (Seehuus *et al.*, 2006; Havukainen *et al.*, 2013) and are upregulated by cold stress (Ramirez *et al.*, 2017; Qin *et al.*, 2019). Soft ticks may also produce Vg for cold tolerance (Fig. 12.2). In addition, it was shown that Vg synthesized in cold-exposed females was incorporated in eggs and enhanced egg maturation (T. Takada and D. Taylor, 2021, unpublished results). The percentage of unfed ticks laying eggs increased from greater than 15% in the controls to less than 80% in females exposed to –5°C for both 12 and 96 h. The percentage of eggs laid by unfed females that hatched was between 50 and 60%; this percentage decreased at all temperatures with exposure for 96 h. Surprisingly, the eggs (35%) from females kept at –10°C for 96 h showed higher hatching rates than eggs from females treated at warmer temperatures. Only 10% of unfed mated females exposed to 40°C laid eggs, but greater than 90% of these eggs were viable. These findings indicate that *O. moubata* is resilient to short exposures to very low temperatures and such exposures trigger a response to use their nutrient reserves from previous feedings during the nymphal stages to produce offspring that can survive harsh environments. Studies on the regulation by hormones and nutrition described in Section 12.1.2 have set the stage for further studies on this resiliency and the physiological mechanisms that trigger cold tolerance and increased production of offspring by unfed females when exposed to cold temperatures.

12.1.5 Immune responses

Soft ticks are vectors of numerous pathogens, such as tick relapsing fever bacteria and African swine fever (ASF) virus (Cutler, 2006; Burrage, 2013). Tick-borne relapsing fever is attributed to over 17 species of *Borrelia* and primarily transmitted by *Ornithodoros* ticks throughout East, West and North Africa (Talagrand-Reboul *et al.*, 2018). In most cases of tick-borne relapsing fever, humans are accidental intermediate hosts. However, *Borrelia duttonii* transmitted by *O. moubata* appears to be a strictly human infection with no known animal reservoir (Trape *et al.*, 2013). *O. moubata* can be both a reservoir and vector and is well-adapted to human dwellings and domestic animal shelters. ASF virus causes minimal disease in wild hogs, warthogs and wild pigs, but high mortality in domestic swine. ASF can be transmitted through *Ornithodoros* ticks, live or dead pigs, contaminated feed as well as by humans from their shoes or clothing. This disease has spread worldwide, causing significant problems in swine of Spain, Central Europe, Africa and recently China (Oleaga-Pérez *et al.*, 1990; Pietschmann *et al.*, 2016; Ma *et al.*, 2019). An additional important concern is that the ASF virus can be transmitted vertically by transovarial transmission in *Ornithorodos* species (Rennie *et al.*, 2001). Therefore, molecular and physiological analyses of soft tick reproduction can contribute to the understanding and possible control of transovarial pathogen transmission. In addition, molecular studies on immune responses and other physiological processes can lead to the identification of targets for the development of better control strategies (Blisnick *et al.*, 2017; de la Fuente *et al.*, 2017; Ali *et al.*, 2020). Several authors have reviewed what is known about the immune system of ticks (Taylor, 2006; Hynes, 2014; Sonenshine and Macaluso, 2017), but there is still much that must be done to fully understand how ticks can transmit vertebrate pathogens while having a robust immune system to protect them in environments with numerous potential tick pathogens.

In collaboration with Dr Minoru Yamakawa, formerly of the Innate Immunity Laboratory, National Institute of Agrobiological Sciences, Tsukuba, Japan, we have undertaken studies on the immune responses of *O. moubata* by identifying and characterizing antimicrobial peptides and analysing the regulatory mechanisms underlying this immunity. Nakajima *et al.* (2001, 2002a,b, 2003a,b,c, 2005) identified four defensin peptides and a haemoglobin fragment that provide upregulated antibacterial defence against Gram-positive bacteria in the midgut. Nakajima (2003) also identified a nuclear factor-κB (NF-κB) transcription factor, *OmRel* gene, included in the Toll regulatory pathway of innate immunity. Tsubokawa (2012) showed the OmRel protein can bind to the promoter region of the defensin genes to enhance transcription; when ticks were treated with *OmRel* RNAi, expression of all four defensin genes was decreased significantly. Tsubokawa (2012) also identified another NF-κB transcription factor included in the Imd regulatory pathway, *OmRelish* gene, that also appears to be important in the regulation of defensin gene expression. These results indicate that the regulation of antibacterial peptides in the immune response of *O. moubata* is similar to that seen in *Drosophila* and other insects, with both the Toll and Imd regulatory pathways involved (Lemaitre and Hoffmann, 2007). Recently, in collaboration with Dr Ai Takano of the Laboratory of Epidemiology, Yamaguchi University, Yamaguchi, Japan, we are investigating the importance of the immune regulatory factors and defensins in the vector competence of *O. moubata* to transmit relapsing fever, reptile-associated and Lyme *Borrelia* spp. (Takano *et al.*, 2013). Investigating immune mechanisms in the tick as well as the host is essential to achieve a full understanding of disease transmission mechanisms and development of methods for controlling pathogen transmission.

12.1.6 Influences of climate on immune responses

Climate plays extremely important roles in the spread, seasonality and abundance patterns of ticks. Furthermore, tick stress responses at the molecular level may counter the effects of heat shock and pathogen transmission to increase tick survival (Estrada-Peña *et al.*, 2012). Effects of temperature on the vector competence of mosquitoes for arboviruses appear to vary depending on the virus and the mosquito species. Vector competence is enhanced with lower temperatures for some diseases, whereas the opposite is apparent for others (Samuel *et al.*, 2016). This is likely the result of different effects on virus replication and/or the immune response of the vector. Kramer *et al.* (1983) demonstrated a western equine encephalitis virus infection was better modulated by the mosquito at higher temperatures and concluded that infection rates increased because the vector was not able to modulate the virus at lower temperatures. Herrmann and Gern (2010) showed that *Ixodes ricinus* infected with *Borrelia burgdorferi* survived challenging temperatures and humidities better than uninfected ticks. Numerous studies have looked at interactions between vectors and pathogens, but almost none of these studies looked at how climate may affect these interactions (de la Fuente *et al.*, 2017; Sonenshine and Macaluso, 2017). A recent review by Ogden *et al.* (2020) begins to address these problems and iterates the importance of including these effects in modelling studies. It is difficult to predict what effects temperature will have on a specific pathogen and the ability of a vector to transmit that pathogen, but clearly physiology of the tick vector also plays an important role in vector capacity and climatic factors may also affect the immune responses of the tick vector. Therefore, it is essential to include such factors in future models to predict the distribution of ticks and tick-borne diseases in the wake of global climate change.

12.2 Future Predictions

As discussed above in Section 12.1.2, the nidicolous lifestyle of soft ticks has given them the ability to survive longer and go through multiple gonotrophic cycles, making them potentially excellent reservoirs of pathogens. This is evidenced by *O. moubata* acting as both the vector and reservoir of *B. duttonii*. They also appear to be able to transmit pathogens transovarially and transstadially as with the ASF virus, raising other concerns. In addition, soft ticks appear to lack the host specificities seen in many hard ticks but are only limited to feeding on specific hosts when ecologically isolated. This indicates a greater potential for transmitting pathogens to a larger variety of hosts. However, the nidicolous lifestyle may play an important role in preventing these potentials from being fully exploited. As climate changes expand the areas of hosts, it is possible that these potentials will create greater problems with the spread of soft ticks, especially ticks that feed on birds and bats. Therefore, the roles and extents of these potentials need to be investigated extensively in soft ticks, a group often ignored in most areas of research.

Understanding the physiological processes of ticks is essential for identifying targets to control tick populations and including vector characteristics in models to effectively predict the spread of tick-borne diseases. Reproduction is a key process that can be exploited for effective control of tick populations. As outlined in Section 12.1.3, we have characterized and elucidated key factors for the regulation of egg production in *O. moubata*, providing a model for further research to better understand the effects of climate on reproduction and to develop novel targets for tick-control methods. In addition, understanding the immune responses in the tick vector is also essential for predicting vector competence and capacity as well as understanding the effects of climate on the tick vector and pathogen interactions. As explained in Section 12.1.5, we have also identified key antibacterial proteins and started to understand the processes regulating the immune system of *O. moubata*. Further research in this area can also provide a better understanding of the tick vector for its inclusion into models to predict the spread of diseases. *O. moubata* provides an excellent model for further research on the physiological processes of ticks because this species can be easily reared and manipulated in the laboratory, especially in terms of reproduction, because feeding and mating can be easily separated. They can also be easily used in immune studies because they can be infected with various pathogens through injection or artificial feeding.

At the same time, it is important to understand the mode of their life in the field for the prevention of tick-borne disease. Compared with hard tick research, accumulation of basic knowledge is lacking in soft ticks. However, recent efforts using in-depth analysis of earlier research and modelling provide new implications of soft tick habitats. Vial *et al.* (2018) predicted possible habitats of *Ornithodoros* species in the Palaearctic region with ecological criteria for soft tick survival, i.e. above 20°C in a 3-month summer, and precipitation between 60 and 750 mm per year. Furthermore, Sage *et al.* (2017) predicted possible change of soft tick habitats in North America along with future climate change in North America, showing soft tick habitats increasing in higher elevations. Research based on physiological analyses, field surveillance and mathematical modelling may help to establish effective solutions for the prevention of soft tick-borne as well as hard tick-borne diseases.

References

Ali, A., Mulenga, A. and Vaz, I.S. Jr (2020) Tick and tick-borne pathogens: molecular and immune targets for control strategies (editorial). *Frontiers in Physiology* 11, 744.

Apanaskevich, D.A. and Oliver, J.H. Jr (2014) Life cycles and natural history of ticks. In: Sonenshine, D.E. and Roe, R.M. (eds) *Biology of Ticks*, vol. 1, 2nd edn. Oxford University Press, Madison, New York, pp. 59–73.

Blisnick, A.A., Foulon, T. and Bonnet, S.I. (2017) Serine protease inhibitors in ticks: an overview of their role in tick biology and tick-borne pathogen transmission. *Frontiers in Cellular and Infection Microbiology* 7, 199.

Boldbaatar, D., Umemiya-Shirafuji, R., Liao, M., Tanaka, T., Xuan, X. and Fujisaki, K. (2010) Multiple vitellogenins from the *Haemaphysalis longicornis* tick are crucial for ovarian development. *Journal of Insect Physiology* 56, 1587–1598.

Brett, G.A. (1939) On the period of survival of the egg, larva, and first nymph stages of the argasid tick *Ornithodorus moubata*, Murray, at different relative humidities. *Bulletin of Entomological Research* 30, 247–253.

Burrage, T.G. (2013) African swine fever virus infection in *Ornithodoros* ticks. *Virus Research* 173, 131–139.

Chinzei, Y. and Taylor, D. (1990) Regulation of vitellogenesis induction by engorgement in the soft tick, *Ornithodoros moubata*. In: Hoshi, M. and Yamashita, O. (eds) *Advances in Invertebrate Reproduction 5*. Elsevier, Amsterdam, pp. 565–570.

Chinzei, Y. and Taylor, D. (1994) Hormonal regulation of vitellogenin synthesis in ticks. In: Harris, K.F. (ed.) *Advances in Disease Vector Research*, vol. 10. Springer, New York, pp. 1–22.

Chinzei, Y., Taylor, D. and Ando, K. (1991) Effects of juvenile hormone and its analogs on vitellogenin synthesis and ovarian development in *Ornithodoros moubata* (Acari: Argasidae). *Journal of Medical Entomology* 28, 506–513.

Chinzei, Y., Taylor, D., Miura, K. and Ando, K. (1992) Vitellogenesis induction by a synganglion factor in adult female ticks, *Ornithodoros moubata* (Acari: Argasidae). *Journal of the Acarological Society of Japan* 1, 15–26.

Connat, J.L., Ducommun, J., Diehl, P.A. and Aeschlimann, A. (1986) Some aspects of the control of the gonotrophic cycle in the tick, *Ornithodoros moubata* (Ixodoidea, Argasidae). In: Sauer, J.R. and Hair, J.A. (eds) *Morphology, Physiology, and Behavioral Biology of Ticks*. Ellis Horwood, Chichester, UK, pp. 194–216.

Cunliffe, N. (1921) Some observations on the biology and structure of *Ornithodorus moubata*, Murray. Together with a note on the external characters which serve to differentiate the sexes. *Parasitology* 13, 327–347.

Cutler, S.J. (2006) Possibilities for relapsing fever reemergence. *Emerging Infectious Diseases* 12, 369–374.

de la Fuente, J., Antunes, S., Bonnet, S., Cabezas-Cruz, A., Domingos, A.G., *et al.* (2017) Tick–pathogen interactions and vector competence: identification of molecular drivers for tick-borne diseases. *Frontiers in Cellular and Infection Microbiology* 7, 114.

Dotson, E.M., Connat, J.-L. and Diehl, P.A. (1993) Metabolism of [^{3}H]-ecdysone in embryos and larvae of the tick *Ornithodoros moubata*. *Archives of Insect Biochemistry and Physiology* 23, 67–78.

El Shoura, S.M. (1987) Effect of temperature and relative humidity on the life cycle of *Ornithodoros (Pavlovskyella) erraticus* (Ixodoidea: Argasidae). *Journal of Parasitology* 73, 1102–1108.

Estrada-Peña, A., Ayllón, N. and de la Fuente, J. (2012) Impact of climate trends on tick-borne pathogen transmission. *Frontiers in Physiology* 3, 64.

Gray, J.S., Estrada-Peña, A. and Vial, L. (2014) Ecology of nidicolous ticks. In: Sonenshine, D.E. and Roe, R.M. (eds) *Biology of Ticks*, vol. 2, 2nd edn. Oxford University Press, Madison, New York, pp. 59–73.

Havukainen, H., Munch, D., Baumann, A., Zhong, S., Halskau, O., *et al.* (2013) Vitellogenin recognizes cell damage through membrane binding and shields living cells from reactive oxygen species. *Journal of Biological Chemistry* 288, 28369–28381.

Herrmann, C. and Gern, L. (2010) Survival of *Ixodes ricinus* (Acari: Ixodidae) under challenging conditions of temperature and humidity is influenced by *Borrelia burgdorferi* sensu lato infection. *Journal of Medical Entomology* 47, 1196–1204.

Horigane, M. and Taylor, D. (2009) Genetic studies on tick ecdysteroid regulation for development of sustainable tick control strategies. *Journal of Developments in Sustainable Agriculture* 4, 73–78.

Horigane, M., Ogihara, K., Nakajima, Y., Shinoda, T. and Taylor, D. (2007) Cloning and expression of the ecdysteroid receptor during ecdysis and reproduction in females of the soft tick, *Ornithodoros moubata* (Acari: Argasidae). *Insect Molecular Biology* 16, 601–612.

Horigane, M., Ogihara, K., Nakajima, Y. and Taylor, D. (2008) Isolation and expression of the retinoid X receptor from last instar nymphs and adult females of the soft tick *Ornithodoros moubata* (Acari: Argasidae). *General and Comparative Endocrinology* 156, 298–311.

Horigane, M., Shinoda, T., Honda, H. and Taylor, D. (2010) Characterization of a vitellogenin gene reveals two phase regulation of vitellogenesis by engorgement and mating in the soft tick *Ornithodoros moubata* (Acari: Argasidae). *Insect Molecular Biology* 19, 501–515.

Hynes, W.L. (2014) How ticks control microbes. In: Sonenshine, D.E. and Roe, R.M. (eds) *Biology of Ticks*, vol. 1, 2nd edn. Oxford University Press, Madison, New York, pp. 129–146.

Kiszewski, A.E., Matuschka, F.-R. and Spielman, A. (2001) Mating strategies and spermiogenesis in Ixodid ticks. *Annual Review of Entomology* 46, 167–182.

Kramer, L.D., Hardy, J.L. and Presser, S.B. (1983) Effect of temperature of extrinsic incubation on the vector competence of *Culex tarsalis* for western equine encephalomyelitis virus. *American Journal of Tropical Medicine and Hygiene* 32, 1120–1139.

Lemaitre, B. and Hoffmann, J. (2007) The host defense of *Drosophila melanogaster*. *Annual Review of Immunology* 25, 697–743.

Ma, J., Chen, H., Gao, X., Xiao, J. and Wang, H. (2019) African swine fever emerging in China: distribution characteristic and high-risk areas. *Preventive Veterinary Medicine* 175, 104861.

Nakajima, Y. (2003) Identification, characterization and gene regulation of antibacterial defense peptides in the soft tick *Ornithodoros moubata* (Acari: Argasidae). Doctoral thesis, The University of Tsukuba, Tsukuba, Japan.

Nakajima, Y., van der Goes van Naters-Yasui, A., Taylor, D. and Yamakawa, M. (2001) Two isoforms of a member of the arthropod defensin family from the soft tick, *Ornithodoros moubata* (Acari: Argasidae). *Insect Biochemistry and Molecular Biology* 31, 747–751.

Nakajima, Y., van der Goes van Naters-Yasui, A., Taylor, D. and Yamakawa, M. (2002a) Antibacterial peptide defensin is involved in midgut immunity of the soft tick, *Ornithodoros moubata* (Acari: Argasidae). *Insect Molecular Biology* 11, 611–618.

Nakajima, Y., Taylor, D. and Yamakawa, M. (2002b) Involvement of antibacterial peptide defensin in tick midgut defense. *Experimental and Applied Acarology* 28, 135–140.

Nakajima, Y., Ogihara, K., Taylor, D. and Yamakawa, M. (2003a) Antibacterial hemoglobin fragments from the midgut of the soft tick, *Ornithodoros moubata* (Acari: Argasidae). *Journal of Medical Entomology* 40, 78–81.

Nakajima, Y., Ishibashi, J., Yukuhiro, F., Asaoka, A., Taylor, D. and Yamakawa, M. (2003b) Antibacterial activity and mechanism of action of tick defensin against Gram-positive bacteria. *Biochimica et Biophysica Acta* 1624, 125–130.

Nakajima, Y., Saido-Sakanaka, H., Taylor, D. and Yamakawa, M. (2003c) Up-regulated humoral immune response in the soft tick, *Ornithodoros moubata* (Acari: Argasidae). *Parasitology Research* 91, 476–481.

Nakajima, Y., Saido-Sakanaka, H., Ogihara, K., Taylor, D. and Yamakawa, M. (2005) Antibacterial peptides are secreted into the midgut lumen to provide antibacterial midgut defense in the soft tick, *Ornithodoros moubata* (Acari: Argasidae). *Applied Entomology and Zoology* 40, 391–397.

Neese, P.A., Sonenshine, D.E., Kallapur, V.L., Apperson, C.S. and Roe, R.M. (2000) Absence of insect juvenile hormones in the American dog tick, *Dermacentor variabilis* (Say) (Acari: Ixodoidea: Ixodidae), and in *Ornithodoros parkeri* Cooley (Acari: Argasidae). *Journal of Insect Physiology* 46, 477–490.

Ogden, H.H., Beard, C.B., Ginsberg, H.S. and Tsao, J.I. (2020) Possible effects of climate change on Ixodid ticks and the pathogens they transmit: predictions and observations. *Journal of Medical Entomology* 2020, tjaa220.

Ogihara, K., Horigane, M., Nakajima, Y., Moribayashi, A. and Taylor, D. (2007) Ecdysteroid hormone titer and its relationship to vitellogenesis in the soft tick, *Ornithodoros moubata* (Acari: Argasidae). *General and Comparative Endocrinology* 150, 371–380.

Ogihara, M.H. and Taylor, D. (2014) Female reproductive system: anatomy, physiology and molecular biology. In: Sonenshine, D.E. and Roe, R.M. (eds) *Biology of Ticks*, vol. 1, 2nd edn. Oxford University Press, Madison, New York, pp. 449–483.

Ogihara, M.H., Hikiba, J., Suzuki, Y., Taylor, D. and Kataoka, H. (2015) Ovarian ecdysteroids in both immature and mature stages of an Acari, *Ornithodoros moubata*. *PLoS ONE* 10, e0124953.

Ogihara, M.H., Taylor, D. and Kataoka, H. (2019) Steroid hormones in Acari, their functions and synthesis. *Applied Entomology and Zoology* 54, 323–338.

Oleaga-Pérez, A., Pérez-Sánchez, R. and Encinas-Grandes, A. (1990) Distribution and biology of *Ornithodoros erraticus* in parts of Spain affected by African swine fever. *Veterinary Record* 126, 32–37.

Oliver, J.H. Jr (1971) Parthenogenesis in mites and ticks (Arachnida: Acari). *American Zoologist* 11, 283–299.

Phillips, J.S. and Adeyeye, O.A. (1996) Reproductive bionomics of the soft tick, *Ornithodoros turicata* (Acari: Argasidae). *Experimental and Applied Acarology* 20, 369–380.
Pietschmann, J., Mur, L., Blome, S., Beer, M., Pérez-Sánchez, R., *et al.* (2016) African swine fever virus transmission cycles in Central Europe: evaluation of wild boar–soft tick contacts through detection of antibodies against *Ornithodoros erraticus* saliva antigen. *BMC Veterinary Research* 12, 1.
Poikela, N., Tyukmaeva, V., Hoikkala, A. and Kankare, M. (2020) Multiple paths to cold tolerance: the role of environmental cues, morphological traits and the circadian clock gene *vrille*. *bioRxiv* 2020.04.20.050351. https://doi.org/10.1101/2020.04.20.050351
Pound, J.J. and Oliver, J.H. Jr (1979) Juvenile hormone; evidence of its role in the reproduction of ticks. *Science* 206, 355–357.
Qin, M., Wang, H., Liu, Z., Wang, Y., Zhang, W. and Xu, B. (2019) Changes in cold tolerance during the overwintering period in *Apis mellifera ligustica*. *Journal of Apicultural Research* 58, 702–709.
Ramirez, L., Negri, P., Sturla, L., Guida, L., Vigliarolo, T., *et al.* (2017) Abscisic acid enhances cold tolerance in honeybee larvae. *Proceedings of the Royal Society B* 284, 20162140.
Rennie, L., Wilkinson, P.J. and Mellor, P.S. (2001) Transovarial transmission of African swine fever virus in the argasid tick *Ornithodoros moubata*. *Medical and Veterinary Entomology* 15, 140–146.
Sage, K.M., Johnson, T.L., Teglas, M.B., Nieto, N.C. and Schwan, T.G. (2017) Ecological niche modeling and distribution of *Ornithodoros hermsi* associated with tick-borne relapsing fever in western North America. *PLoS Neglected Tropical Diseases* 11, e0006047.
Samuel, G.H., Adelman, Z.N. and Myles, K.M. (2016) Temperature-dependent effects on the replication and transmission of arthropod-borne viruses in their insect hosts. *Current Opinion in Insect Science* 16, 108–113.
Seehuus, S.-C., Norberg, K., Gimsa, U., Krekling, T. and Amdam, G.V. (2006) Reproductive protein protects functionally sterile honey bee workers from oxidative stress. *Proceedings of the National Academy of Sciences USA* 103, 962–967.
Sonenshine, D.E. (1993) *Biology of Ticks*, vol. 2. Oxford University Press, Oxford.
Sonenshine, D.E. and Macaluso, K.R. (2017) Microbial invasion vs tick immune regulation. *Frontiers in Cellular and Infection Microbiology* 7, 390.
Takano, A., Tsubokawa, S., Taylor, D. and Kawabata, H. (2013) Transmission of *Borrelia* spp. via tick vector. *Medical Entomology and Zoology* 64, 11–15.
Talagrand-Reboul, E., Boyer, P.H., Bergstrom, S., Vial, L. and Boulanger, N. (2018) Relapsing fevers: neglected tick-borne diseases. *Frontiers in Cellular and Infection Microbiology* 8, 98.
Taylor, D. (2006) Tick innate immunity: a review. *Journal of the Acarological Society of Japan* 15, 109–127.
Taylor, D. and Chinzei, Y. (2002) Vitellogenesis in ticks. In: Raikhel, A.S. and Sappington, T.W. (eds) *Recent Progress in Vitellogenesis. Reproductive Biology of Invertebrates*, vol. XII. Science Publishers, Inc., Enfield, New Hampshire, pp. 175–199.
Taylor, D., Moribayashi, A., Shono, T. and Chinzei, Y. (1997) Hormonal regulation of vitellogenesis in the soft tick, *Ornithodoros moubata*. In: Kawashima, S. and Kikuyama, S. (eds) *Advances in Comparative Endocrinology*. Monduzzi Editore, Bologna, Italy, pp. 213–220.
Trape, J.-F., Diatta, G., Arnathau, C., Bitam, I., Sarih, M., *et al.* (2013) The epidemiology and geographic distribution of relapsing fever borreliosis in West and North Africa, with a review of the *Ornithodoros erraticus* complex (Acari: Ixodida). *PLoS ONE* 8, e78473.
Tsubokawa, S. (2012) Roles of *Ornithodoros moubata* Rel/NF-κB in the regulation of defensin genes. MSc thesis, The University of Tsukuba, Tsukuba, Japan.
Umemiya-Shirafuji, R., Boldbaatar, D., Liao, M., Battur, B., Rahman, Md M., *et al.* (2012) Target of rapamycin controls vitellogenesis via activation of S6 kinase in the fat body of the tick *Haemaphysalis longicornis*. *International Journal for Parasitology* 42, 991–998.
Vial, L. (2009) Biological and ecological characteristics of soft ticks (Ixodida: Argasidae) and their impact for predicting tick and associated disease distribution. *Parasite* 16, 191–202.
Vial, L., Ducheyne, E., Filatov, S., Gerilovych, A., McVey, D.S., *et al.* (2018) Spatial multi-criteria decision analysis for modelling suitable habitats of *Ornithodoros* soft ticks in the Western Palearctic region. *Veterinary Parasitology* 249, 2–16.
Waltero, C.F., Abreu, L.A.D., Alonso, T., Fonseca, R.N.D., Vaz, I.D.S. Jr and Logullo, C. (2019) TOR as a regulatory target in *Rhipicephalus microplus* embryogenesis. *Frontiers in Physiology* 10, 965.
Walton, G.A. (1957) Observations on biological variation in *Ornithodoros moubata* (Murr.) (Argasidae) in East Africa. *Bulletin of Entomological Research* 48, 669–710.
Zhu, X.X., Oliver, J.H. Jr and Dotson, E.M. (1991) Epidermis as the source of ecdysone in an argasid tick. *Proceedings of the National Academy of Sciences USA* 88, 3744–3747.

eo13 Climate Change and Ticks: Measuring Impacts

Richard Wall*[1,2] and Saeed Alasmari[1,2]
[1]*University of Bristol, Bristol, UK;* [2]*Najran University, Najran, Saudi Arabia*

13.1 Expert Opinion

13.1.1 Introduction

Changes in the abundance and distribution of ticks and increases in the incidence of tick-borne disease in Europe, that result from climate change, are issues of very real and immediate concern (Rose *et al.*, 2016). Climate projections suggest that Europe will experience an increase in high-temperature extremes, droughts and heavy precipitation events, but with considerable variation across the continent: the majority of published assessments suggest that temperatures may increase by 1 to 4°C per century (IPCC, 2013). Such changes would be expected to have direct effects on ticks, acting on tick development and survival, mediated by the physiological adaptations of the tick species in question, and indirect effects, acting through host availability and the presence of suitable habitat. However, any changes in tick population dynamics or tick-borne disease incidence will inevitably depend on the interactions between a wide range of factors and outcomes are likely to be complex and non-linear. Hence, the specific consequences of climate change on components of the system may be hard to isolate (Gray *et al.*, 2009). Nevertheless, substantial changes have been predicted and, it is argued, may already be evident (Beugnet and Marié, 2009).

In terms of direct effects, warming might be expected to increase development rates and extend the seasonal periods during which conditions are favourable for survival and pathogen transmission. However, this will be strongly dependent on precipitation: drier, hotter conditions in summer will reduce survival and activity for the most abundant tick species found in Europe, such as *Ixodes ricinus*, while the opposite will be the case in winter in parts of North-Western Europe that experience warmer wetter winters. Analysis of the effects of hotter summer temperatures indicated that they resulted in a transfer of *I. ricinus* ticks between development cohorts, resulting in increased questing activity and therefore increased disease transmission in late autumn and early spring (Gray, 2008). Similarly, changing rainfall patterns may affect the feeding patterns. If unusually dry conditions occur early in the season when ticks are questing (April–June), nymphal ticks may be more likely to feed on small rodents nearer the ground and increase the rodent reservoir of pathogens such as *Borrelia burgdorferi* sensu lato and tick-borne encephalitis (TBE) virus (Randolph and Storey 1999), whereas if dry conditions occur later in the season (June–August) then the ticks

*Email: richard.wall@bristol.ac.uk

DOI: 10.1079/9781789249637.0013

may starve for want of sufficient hosts prior to the winter when questing usually stops (Randolph *et al.*, 2002; Randolph, 2004). Climate changes may also facilitate range expansion and it has been suggested that the impacts of climate change are likely to be greatest at the edge of the range of environmental tolerances for particular species (Lindgren *et al.*, 2000; Gray *et al.*, 2009). In addition, it is more likely that tropical and subtropical species will establish at higher latitudes and altitudes. For example, predicted changes could result in the permanent establishment of the more desiccation-tolerant species, *Rhipicephalus sanguineus*, which is presently found only in heated or sheltered buildings such as kennels, in many locations where it is currently unable to establish (Gray *et al.*, 2009). Changes in phenology will vary greatly across climatic zones; in some regions sharp changes between seasons may serve to synchronize tick populations. Alternatively, in parts of Northern and North-Western Europe, warmer, wetter winter conditions might make populations more heterogeneous, removing distinct peaks in the tick season, increasing the proportion of the tick population questing at any one time, but increasing the rates of starvation (Gilbert *et al.*, 2014).

In terms of indirect effects, climate change will have a range of particularly important impacts on the availability of habitat suitable for ticks or their hosts; these effects may be natural or anthropogenic. In parts of Europe there are large-scale plans for wetland creation to mitigate the effects of climate change on biodiversity loss. Initiatives are also being undertaken to promote biodiversity enhancement, to increase woodland management for biodiversity and to link fragments of existing biodiverse-rich habitat to increase habitat connectivity (Bailey, 2007). This has led to increases in wildlife populations and expansion of their distributions, particularly deer (Stigum *et al.*, 2019), which are important reproduction hosts, and rodents, which are important reservoirs of a range of disease pathogens such as for Lyme borreliosis. For example, over the last 20–30 years populations of roe deer, *Capreolus capreolus*, have expanded in areas of increased tick abundance (Tälleklint and Jaenson, 1998; Scharlemann *et al.*, 2008) and their incursion into peri-urban areas is becoming more common (Jameson and Medlock, 2011), increasing the risk of human exposure to tick bites in these areas. The compounded effects of climate on tick abundance and changes in phenology will also affect the transmission of pathogens and the increase in temperature and extreme weather events predicted under most climate change scenarios are likely to affect the incidence of tick-borne disease in many parts of Europe (Gray *et al.*, 2009).

13.1.2 Evidence and its limitations

Despite extensive discussion, unambiguous links between climate change and changes in tick abundance, phenology and disease incidence are difficult to demonstrate. This is, in part, because of the complexity of the interacting drivers, as discussed above. In addition, good empirical studies of the effects of climate on tick populations are rare, partly due to the long timescales that are required to demonstrate long-term changes that map to the expected rates of climate change. Hence, many existing studies suffer from the fact that the duration of the time series is too short, or the frequency and quality of sampling effort may not have been consistent throughout the decades required. As a result, conclusions drawn from many of these studies need to be interpreted with some caution.

It is not the intention to present an assessment of the literature on this subject here, but two studies in particular highlight the rigour required to provide good evidence. Between 1977 and 2011, questing *I. ricinus* were monitored in the Tula region of western Russia by Korotkov *et al.* (2015). Three sites were situated in coniferous–deciduous forests and a fourth site was in the forest–steppe zone. In these zones, 20 to 40 transects of 1 km were set up and questing *I. ricinus* ticks were collected from vegetation every 10 m along the transects. For the initial 14 years, the recorded tick abundance was at 0.1 to 0.9 specimens per 1-km transect – the lower limit of the sensitivity of the methods used. However, during the following 21 years there was a substantial increase in tick abundance; in some floodplain areas tick abundance reached about 18 specimens per 1-km transect and about 5 specimens per 1-km transect in the xerophilic woods. The abundance of small and large wild mammals was monitored; fluctuations in these potential hosts were small and their numbers remained relatively constant. Despite this, an overall increase in tick abundance was observed over the 35-year span of the study. Correlation analysis showed a significant association between climate variables and the abundance of questing *I. ricinus*. As a result, it was concluded that the increase in tick abundance was mainly a result of both climate and anthropogenic changes in the environment rather than changes in host abundance.

Similarly, Žygutienė (2009) conducted a study in Lithuania into the changing patterns of *I. ricinus* between 1991 and 2008 and sampled many areas of Lithuanian woodland. In the Klaipeda region from 1995 to 2008, the density of ticks during the spring peak of activity increased threefold from 19 ticks per 1 km to 57 ticks per 1 km. In addition, the duration of the tick biting season was extended: by 2007 the last ticks were collected in mid-December and the first of the 2008 season were found in mid-March. In association there was a significant rise in morbidity due to TBE between 1997 and 2008 in Lithuania. It was concluded that the abundance of *I. ricinus* numbers is increasing, their seasonal period of activity is extending, and this is leading to the increase in the prevalence of tick-borne disease (Žygutienė, 2009). However, while this latter study highlights the extensive nature of the surveillance needed to demonstrate changes, interestingly this author considered that changes in wild host abundance had a greater influence on tick numbers and disease risk than climate, but that weather conditions may influence human risk from tick bites through increased exposure.

As highlighted by Žygutienė (2009), when considering tick-borne disease incidence in particular, it is important to appreciate that underlying changes in vector abundance may be only one possible driver which must be distinguished from changes in behaviours that increase host exposure and disease incidence, greater awareness among public and medical professions compounded by changes in social and economic conditions (Godfrey and Randolph, 2011), improved surveillance, improved communications among the infectious disease community and better diagnosis. These may all lead to increases (or decreases) in tick-borne disease incidence without any necessary change in vector abundance. Climate-mediated changes may also affect land-use patterns and population movement and this in turn allows for increased transmission of vector-borne diseases (Godfrey and Randolph, 2011). A change in the perception of disease risk may lead to changes in approach to intervention, with perhaps a greater willingness to treat prophylactically or to intervene with treatment earlier. This is likely to further complicate the identification of long-term associations between climate and reported tick-borne disease incidence.

13.1.3 Novel tools: metabolic rate measurement for detecting the impact of climate on ticks

Given the difficulty of inferring the effects of climate on ticks through measurement of seasonal patterns of abundance alone, novel approaches that are able to detect climate-mediated impacts directly will be of value. One approach is through the analysis of metabolic rate.

Tick nutrition is derived from the blood meal and sufficient must be obtained to allow larvae and nymphs to undergo development to the next life-cycle stage and eventually initiate repeated questing until another host is located. For adults, accumulated metabolic reserves must also provide the resources for reproduction (Obenchain and Oliver, 1973). The longer ticks can survive between blood meals, the higher the chance of encountering a new host. However, the rate of metabolite resource depletion is strongly affected by temperature and precise measurement of the rates of depletion may allow insights into the effects of climate on tick activity as well as the more exact determination of feeding patterns.

The study of metabolite-use patterns has been greatly facilitated by the spectrophotometric techniques developed by van Handel (1985a,b) and recent studies have demonstrated that these methods can be used also to determine the entire energy budget of individual ticks (Alasmari and Wall, 2020). The primary immediate energy source for ticks is glucose, which is stored in a hydrated polymeric form, glycogen, but this is rapidly depleted (Moraes *et al.*, 2007). Carbohydrate can be used to replenish glycogen reserves. Lipids also play key roles in tick metabolism, both as an energy source and structurally. Lipid is stored primarily in the form of triglycerides in adipocytes, the main fat body cell, and additionally as cytoplasmic lipid droplets. Lipid analysis alone has been used to infer feeding patterns in field-derived cohorts of *I. ricinus* (Randolph *et al.*, 2002; Abdullah *et al.*, 2018) but when measured in conjunction with other metabolites, a more comprehensive picture can be formed. Proteins are essential for muscle development and the synthesis of cuticle, hormones and enzymes and egg yolk, as well as the production of sperm and gonadal proteins (Kongsuwan *et al.*, 2010). Proteins may also be metabolized as a long-term energy reserve (Williams *et al.*, 1986). Among ticks collected from the field, clear differences in metabolite concentrations between life-cycle stages are apparent – largely associated with differences in the relative amounts of protein compared with glycogen (Fig. 13.1). For ticks collected in late spring and early summer in the south-west of England, in nymphs, glycogen was found to represent about 40% of the total mass of metabolites, followed by carbohydrate and protein (at about 25% each). In contrast, in females, glycogen comprised only 19% and in males 12% of the mass of metabolites, whereas protein represented 53% of the mass in females and 61% in males (Alasmari and Wall, 2020). These differences were considered to reflect differences in the feeding history of the preceding life-cycle stage the previous year, with the nymphs collected at that time of year having more substantive reserves compared with the adults.

Further analysis of the rates of metabolite loss suggested that complete resource depletion would occur at between 45 and 70 days at temperatures of 25–30°C and 200 days at 5°C, assuming humidity was sufficiently high not to result in desiccation. The lower thresholds for metabolic activity were estimated to be between –10 and –5°C, which probably reflects the lower lethal temperature for *I. ricinus*. The thermal sensitivity of metabolic rate, often described by an organism's Q_{10}, is the magnitude of change in metabolic rate for a 10°C change in temperature. In insects Q_{10} values range from 1.5 to 3, with a mode of 2.5 (Woods *et al.*, 2003). In *I. ricinus*, Q_{10} values of 1.50 for nymphs, 1.71 for males and 1.63 for females have been recorded (Alasmari and Wall, 2021). These relatively low values reflect the low metabolic rate that helps to increase tick survival during the extended inter-feed intervals. It has previously been suggested that ixodid ticks have a metabolic rate which is typically 13% below that of most arthropods (Lighton and Fieldon, 1995). Hence, a more detailed understanding of the pattern of metabolite use and depletion, particularly in response to temperature, may provide tools that allow the likely impacts of climatic change on ticks to be assessed directly, allowing subtle changes in seasonal activity and feeding patterns to be detected (Alasmari and Wall, 2021). However, in the future, this approach may need to take account of the possibility that such relationships are likely to be affected by saturation deficit and daylength (Belozerov, 2009).

13.2 Future Predictions

Ticks and tick-borne disease are important worldwide, in terms of both mortality/morbidity

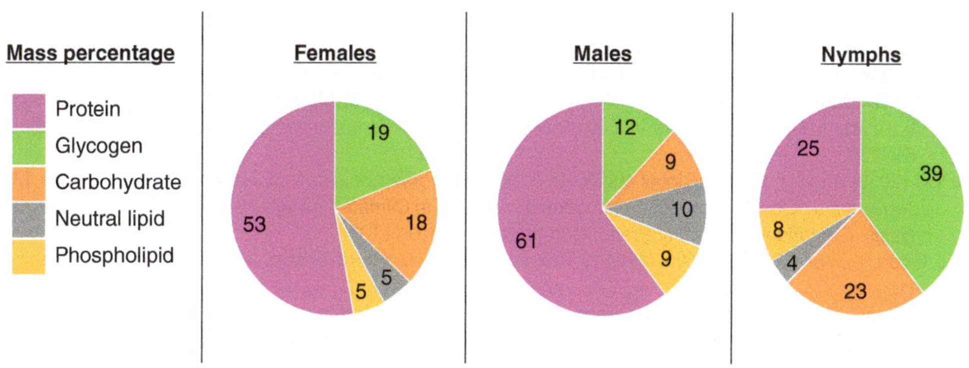

Fig. 13.1. The percentage mass for different categories of metabolite in individual *Ixodes ricinus* nymphs, males or females collected from the field in 2017 between March and May by blanket dragging. (Redrawn from Alasmari and Wall, 2020. Authors' own figure.)

and animal welfare – such impacts are likely to become more problematic in the future. Because of their lengthy off-host life-cycle stages, ticks are highly sensitive to climate change as their life-history processes such as development and survival are temperature- and moisture-dependent. It has been suggested that changes in the phenology and distribution of tick species have been observed in recent years, but relatively few robust studies have undertaken data collection over sufficiently long time periods to demonstrate changes unequivocally; such studies are essential. The complex interactions between tick development and mortality, environmental change and host behaviour can make accurate predictions of future changes difficult. Clearer data and more exact predictions will emerge in the future as new tools to allow a more exact understanding of the effects of the environment on ticks emerge. Future changes in exposure and risk must be considered more comprehensively in climate impact assessments since the nature and timing of these interactions may change with altered climate. However, given the complexity of the systems in question, in the short term, preparation for the unexpected and unpredicted outcomes is the essential take-home message from this expert opinion. Finally, and critically, although there is potential for climate change to result in an increase in tick abundance and tick-borne disease incidence, strategic changes to exposure also provide an opportunity to mitigate impacts of climate change (Rose *et al.*, 2016).

References

Abdullah, S., Davies, S. and Wall, R. (2018) Spectrophotometric analysis of lipid used to examine the phenology of the tick *Ixodes ricinus*. *Parasites & Vectors* 11, 523.

Alasmari, S. and Wall, R. (2020) Determining the total energy budget of the tick *Ixodes ricinus*. *Experimental and Applied Acarology* 80, 531–541.

Alasmari, S. and Wall, R. (2021) Metabolic rate and resource depletion in the tick *Ixodes ricinus* in response to temperature. *Experimental and Applied Acarology* 83, 81–93.

Bailey, S. (2007) Increasing connectivity in fragmented landscapes: an investigation of evidence for biodiversity gain in woodlands. *Forest Ecology and Management* 238, 7–23.

Belozerov, V.N. (2009) Diapause and quiescence as two main kinds of dormancy and their significance in life cycles of mites and ticks (Chelicerata: Arachnida: Acari) Part 2. Parasitiformes. *Acarina* 17, 3–32.

Beugnet, F. and Marié, J.L. (2009) Emerging arthropod-borne diseases of companion animals in Europe. *Veterinary Parasitology* 163, 298–305.

Gilbert, L., Aungier, J. and Tomkins, J.L. (2014) Climate of origin affects tick (*Ixodes ricinus*) host-seeking behaviour in response to temperature: implications for resilience to climate change? *Ecology and Evolution* 4, 1186–1198.

Godfrey, E.R. and Randolph, S.E. (2011) Economic downturn results in tick-borne disease upsurge. *Parasites & Vectors* 4, 35.

Gray, J.S. (2008) *Ixodes ricinus* seasonal activity: implications of global warming indicated by revisiting tick and weather data. *International Journal of Medical Microbiology* 298, 19–24.

Gray, J.S., Dautel, H., Estrada-Peña, A., Kahl, O. and Lindgren, E. (2009) Effects of climate change on ticks and tick-borne diseases in Europe. *Interdisciplinary Perspectives on Infectious Diseases* 2009, 593232.

IPCC (2013) Annex I: Atlas of Global and Regional Climate Projections. In: van Oldenborgh, G.J., Collins, M., Arblaster, J., Christensen, J.H., Marotzke, J., *et al.* (eds) *Climate Change 2013 – The Physical Science Basis*. Cambridge University Press, Cambridge, pp. 1311–1394.

Jameson, L.J. and Medlock, J.M. (2011) Tick surveillance in Great Britain. *Vector Borne and Zoonotic Diseases* 11, 403–412.

Kongsuwan, K., Josh, P., Zhu, Y., Pearson, R., Gough, J. and Colgrave, M.L. (2010) Exploring the midgut proteome of partially fed female cattle tick (*Rhipicephalus* (*Boophilus*) *microplus*). *Journal of Insect Physiology* 56, 212–226.

Korotkov, Y.U., Kozlova, T. and Kozlovskaya, L. (2015) Observations on changes in abundance of questing *Ixodes ricinus*, castor bean tick, over a 35-year period in the eastern part of its range (Russia, Tula region). *Medical and Veterinary Entomology* 29, 129–136.

Lighton, J.R. and Fielden, L.J. (1995) Mass scaling of standard metabolism in ticks: a valid case of low metabolic rates in sit-and-wait strategists. *Physiological Zoology* 68, 43–62.
Lindgren, E., Tälleklint, L. and Polfeldt, T. (2000) Impact of climatic change on the northern latitude limit and population density of the disease-transmitting European tick *Ixodes ricinus*. *Environmental Health Perspectives* 108, 119–123.
Moraes, J., Galina, A., Alvarenga, P.H., Rezende, G.L., Masuda, A., *et al.* (2007) Glucose metabolism during embryogenesis of the hard tick *Boophilus microplus*. *Comparative Biochemistry and Physiology Part A: Molecular and Integrative Physiology* 146, 528–533.
Obenchain, F.D. and Oliver, J.H. (1973) A qualitative analysis of the form, function and interrelationships of fat body and associated tissues in adult ticks (Acari–Ixodoidea). *Journal of Experimental Zoology* 186, 217–235.
Randolph, S.E. (2004) Evidence that climate change has caused 'emergence' of tick-borne diseases in Europe? *International Journal of Medical Microbiology Supplements* 293, 5–15.
Randolph, S.E. and Storey, K. (1999) Impact of microclimate on immature tick–rodent host interactions (Acari: Ixodidae): implications for parasite transmission. *Journal of Medical Entomology* 36, 741–748.
Randolph, S., Green, R., Hoodless, A. and Peacey, M. (2002) An empirical quantitative framework for the seasonal population dynamics of the tick *Ixodes ricinus*. *International Journal for Parasitology* 32, 979–989.
Rose, H., Ellse, L. and Wall, R. (2016) Climate change and arthropod ectoparasites and vectors of veterinary importance. In: Johnson, S.N. and Jones, T.H. (eds) *Global Climate Change and Terrestrial Invertebrates*. Wiley, Chichester, UK, pp. 111–125.
Scharlemann, J.P.W., Johnson, P.J., Smith, A.A., Macdonald, D.W. and Randolph, S.E. (2008) Trends in ixodid tick abundance and distribution in Great Britain. *Medical and Veterinary Entomology* 22, 238–247.
Stigum, V.M., Jaarsma, R.I., Sprong, H., Rolandsen, C.M. and Mysterud, A. (2019) Infection prevalence and ecotypes of *Anaplasma phagocytophilum* in moose *Alces alces*, red deer *Cervus elaphus*, roe deer *Capreolus capreolus* and *Ixodes ricinus* ticks from Norway. *Parasites & Vectors* 12, 1.
Tälleklint, L. and Jaenson, T.G.T. (1998) Increasing geographical distribution and density of *Ixodes ricinus* (Acari: Ixodidae) in Central and Northern Sweden. *Journal of Medical Entomology* 35, 521–526.
Van Handel, E. (1985a) Rapid determination of glycogen and sugars in mosquitoes. *Journal of the American Mosquito Control Association* 1, 299–301.
Van Handel, E. (1985b) Rapid determination of total lipids in mosquitoes. *Journal of the American Mosquito Control Association* 1, 302–304.
Williams, J.P., Sauer, J.R., McNew, R.W. and Hair, J.A. (1986) Physiological and biochemical changes in unfed lone star ticks, *Amblyomma americanum* (Acari: Ixodidae) with increasing age. *Journal of Medical Entomology* 23, 230–235.
Woods, H.A., Makino, W., Cotner, J.B., Hobbie, S.E., Harrison, J.F., *et al.* (2003) Temperature and the chemical composition of poikilothermic organisms. *Functional Ecology* 17, 237–245.
Žygutienė, M. (2009) Tick-borne pathogens and spread of *Ixodes ricinus* in Lithuania. *EpiNorth* 10, 63–71.

eo14 Scandinavia and Ticks in a Changing Climate

Lene Jung Kjær* and René Bødker
University of Copenhagen, Copenhagen, Denmark

14.1 Expert Opinion

14.1.1 Introduction

Climate change is expected to affect areas near the poles relatively more than areas near the equator. Scandinavia can therefore expect more dramatic changes than the rest of Europe. There is fear that a rise in temperature will cause an expansion in vector distribution at the vector distribution limits, or that a temperature rise may prolong the vegetation growing season in northern climates, thus facilitating a prolonged season for vector activity. Scandinavia has seen an increase in the incidence of diseases such as Lyme borreliosis and tick-borne encephalitis that pose a real risk to both human and animal health (Lindgren and Gustafson, 2001; Paulsen *et al.*, 2015; Soleng *et al.*, 2018) and new tick vectors and their exotic pathogens are emerging in both southern and northern Scandinavia.

14.1.2 Present trends in distribution of endemic *Ixodes ricinus* ticks in Scandinavia

Scandinavia constitutes the northernmost range of *Ixodes ricinus*, with large areas still free of this tick. *I. ricinus* is the most common tick in the region. In the last decades, Scandinavia has seen a latitudinal range expansion with the northernmost limit being approximately 66.2°N in Norway (Hvidsten *et al.*, 2020) and at 66°N in Sweden (Jaenson *et al.*, 2012). Additionally, an increase in abundance in already endemic areas has been reported in Scandinavia (Jensen and Jespersen, 2005; Jaenson *et al.*, 2012). Increasing tick abundance is of concern as it can lead to increases in tick–human encounters and thus increase the risk of infections with tick-associated pathogens.

14.1.3 Present trends in introduction of exotic ticks in Scandinavia

Exotic tick species have long been known to occur sporadically in Scandinavia, brought along by migrating wild birds and transport of domestic mammals. However, until recently, it was believed that the Scandinavian climate was too harsh for these tick species to survive long enough to feed or establish within the region. Now, milder winters and warmer summers appear to be driving an increasing risk of exotic ticks and their exotic pathogens.

The meadow tick, *Dermacentor reticulatus*, has previously been absent from Scandinavia although it is endemic to Central Europe. However,

*Email: lenju@sund.ku.dk

DOI: 10.1079/9781789249637.0014

D. reticulatus has recently spread northwards and westwards into northern Germany and the Netherlands. This tick carries several pathogens; among the most concerning are the malaria-like blood parasite, *Babesia canis*, and the bacterium, *Rickettsia raoultii* – pathogens that have previously been absent in Scandinavia. *B. canis* causes canine babesiosis, a disease with a high mortality and morbidity in dogs (Solano-Gallego *et al.*, 2016), whereas *R. raoultii* can cause mild disease in humans (Klitgaard *et al.*, 2017). The earliest records of *D. reticulatus* in Scandinavia are from 2003–2005, where it was found on migrating birds in Norway (Hasle *et al.*, 2009). Furthermore, in 2009, a dog from the capital region of Oslo was diagnosed with *B. canis* (Øines *et al.*, 2010). As *I. ricinus* is not a vector of *B. canis*, and as the dog had never been outside the Oslo area, this occurrence suggested that *D. reticulatus* had been present in the area. In Sweden, the first report of *D. reticulatus* was from Skåne in 2010, and the tick was also found in the same region in 2017 (Rudander, 2018). In Denmark, *D. reticulatus* in the wild was first found on a golden jackal (*Canis aureus*) that migrated from eastern Germany to northern Denmark in 2017 as part of a recent wave of long-migrating golden jackals and grey wolves entering Denmark from the south (Klitgaard *et al.*, 2017). The jackal carried several *D. reticulatus*, most of which were infected with *R. raoultii*. During 2018–2020, several dogs on the southern isles of Lolland and Falster in Denmark were diagnosed with canine babesiosis and have even been found with *D. reticulatus*, suggesting that *D. reticulatus* is now present in southern Denmark; however, so far, we have not been able to find *D. reticulatus* by flagging on the islands.

A highly unexpected and direct result of a warmer climate is the recent emergence of adult *Hyalomma* ticks in Scandinavia. The hard tick, *Hyalomma marginatum*, is mainly found in Southern Europe, Northern Africa and parts of Asia as it requires a warm climate to complete its life cycle (Grandi *et al.*, 2020). *H. marginatum* along with the closely related *Hyalomma rufipes* (endemically distributed in northern parts of Africa and around the Red Sea) have been sporadically reported from Northern Europe. Both ticks are transported by migrating birds in the larval and nymphal stages. After feeding, the larvae remain on the migrating bird and take a second blood meal as nymphs before detaching. This allows the tick to stay on a bird while the bird disperses over very long distances. Previously, only one single adult host-seeking *Hyalomma* tick had been recorded in Scandinavia. However, between 2018 and 2019, 41 adult stages of both *H. marginatum* and *H. rufipes* were found on animals and one human in Sweden (Grandi *et al.*, 2020). During the same period, adult stages of *H. marginatum* were also found in Denmark on a human and horses (Bødker *et al.*, 2019). The finding of adult stages of these tick species suggest they were likely introduced as larvae/nymphs by migrating birds in the spring of 2018. The exceptionally warm summer of 2018 in Northern Europe then allowed their maturation to the adult stage in autumn 2018 and spring 2019 when they started appearing on large mammalian hosts. Ticks belonging to *Hyalomma* spp. are of concern, as they are the vector of Crimean–Congo haemorrhagic fever virus, causing viral haemorrhagic fever in humans, and the vector of Alkhumra haemorrhagic fever virus that may infect cattle and subsequently spread to humans in contact with cattle, especially at slaughterhouses. Both diseases show a high case-fatality rate in humans and may spread to hospital staff not expecting a haemorrhagic virus in a Scandinavian patient without a recent travel history. No virus was found in these bird-introduced ticks, but the exotic zoonotic bacterium, *Rickettsia aeschlimannii*, was isolated from adult *Hyalomma* ticks in Sweden in 2018.

The taiga tick, *Ixodes persulcatus*, is closely related to *I. ricinus*. It originates from Eastern Europe and Northern Asia but has expanded its range within the last 15 years, both in Eastern Europe as well as towards Scandinavia, potentially due to increases in mean annual temperatures (Jääskeläinen *et al.*, 2006; Baeten *et al.*, 2007; Tokarevich *et al.*, 2011; Bugmyrin *et al.*, 2013; Jaenson *et al.*, 2016). *I. persulcatus* may carry the Siberian and Far Eastern subtypes of tick-borne encephalitis virus (TBEV) along with a range of other pathogens (Jääskeläinen *et al.*, 2016; Jaenson *et al.*, 2016; Süss, 2011). These subtypes of TBEV are perceived to cause more severe symptoms than the European subtype (Lindquist and Vapalahti, 2008; Heyman *et al.*, 2010; Jääskeläinen *et al.*, 2016). *I. persulcatus* has previously been absent from Scandinavia, but was found in 2015 in northern Sweden, presumably spreading from Finland by land north of the Bothnian Bay, where it was reported in 2004 and 2008 (Jääskeläinen *et al.*, 2006, 2010).

14.2 Future Predictions

The Scandinavian climate is changing, and along with it the environment in which ticks are living and the populations of wild animals on which ticks are feeding. A change in the distribution and abundance of disease-causing tick vectors is therefore inevitable and is currently already happening. However, predicting exactly how this change will develop in the future is complex.

14.2.1 Future distribution of endemic *Ixodes ricinus* ticks in Scandinavia

Empirically based models have shown that climatic variables (especially day- and night-time land surface temperatures) are strong predictors of the current distribution and abundance of *I. ricinus* in Scandinavia (Kjær *et al.*, 2019a,b). This suggests that *I. ricinus* will spread north into areas presently free of ticks. However, whereas these models have proved highly accurate for distribution modelling, modelling the tick abundance with environmental and climatic drivers alone has shown less accuracy and predictive power (Kjær *et al.*, 2019b), suggesting that other drivers, such as distribution and abundance of host species, also play an important part in determining the abundance of *I. ricinus* in Scandinavia. Climatic warming will most likely expand the present distribution of *I. ricinus* further north, while the impact on tick abundance in already endemic areas is less clear. In Denmark, spread of *I. ricinus* to previous tick-free areas along the sandy north-west coast appears to be driven by a fivefold increase in the roe deer population since the 1980s and not by a warmer climate. There are concerns that an additional latitudinal range expansion of *I. ricinus*, increasing the tick-endemic area in Scandinavia, may proportionally increase the incidence of tick-borne diseases in humans in Scandinavia. However, this concern appears to be exaggerated. In a recent joint Scandinavian study, we modelled the distribution of *I. ricinus* in Scandinavia and related the predicted distribution to human population density maps (Kjær *et al.*, 2019a). The study found that 70–90% of the population within Denmark, Norway and Sweden are already clustered within 5 km of the present habitats where ticks are likely to occur (Fig. 14.1). The northern regions of Norway and Sweden presently free of ticks are sparsely populated and thus a range expansion into these regions will affect only a relatively small proportion of the human population in Scandinavia. This suggests that any efforts proposed to control or prevent ticks and tick-borne diseases should be targeting areas where *I. ricinus* is already established but where climate change may affect tick abundance.

14.2.2 Future risk of exotic ticks in Scandinavia

The increasing evidence of *D. reticulatus* presence in southern Scandinavia and the emergence of canine babesiosis suggest that *D. reticulatus* may well be on the way to becoming established in southern Scandinavia, particularly in Denmark. As *D. reticulatus* prefers more open areas, as opposed to the forested habitats of *I. ricinus*, Denmark could potentially provide optimal habitats for this tick species. While the fast introduction of infected meadow ticks by large migrating carnivores is a result of improved wildlife protection in the European Union, the successful establishment of the introduced ticks in new areas is more likely the result of the warmer climate. Random introductions are likely to have occurred with migrating birds for centuries without establishing the tick in the cooler Northern Europe. Because migrating birds are introducing *Hyalomma* ticks every spring, adult host-seeking *Hyalomma* ticks will have to be considered a zoonotic risk in future warmer summers even though nothing indicates *Hyalomma* species will be able to establish in a warmer Scandinavia within the next 50 years. This emerging risk is a highly unexpected result of climate change as the warmer summers permit the maturation of the introduced nymphs to the adult stage.

I. persulcatus seems to be able to establish in areas where *I. ricinus* cannot, potentially due to its ability to live in areas with a short vegetation

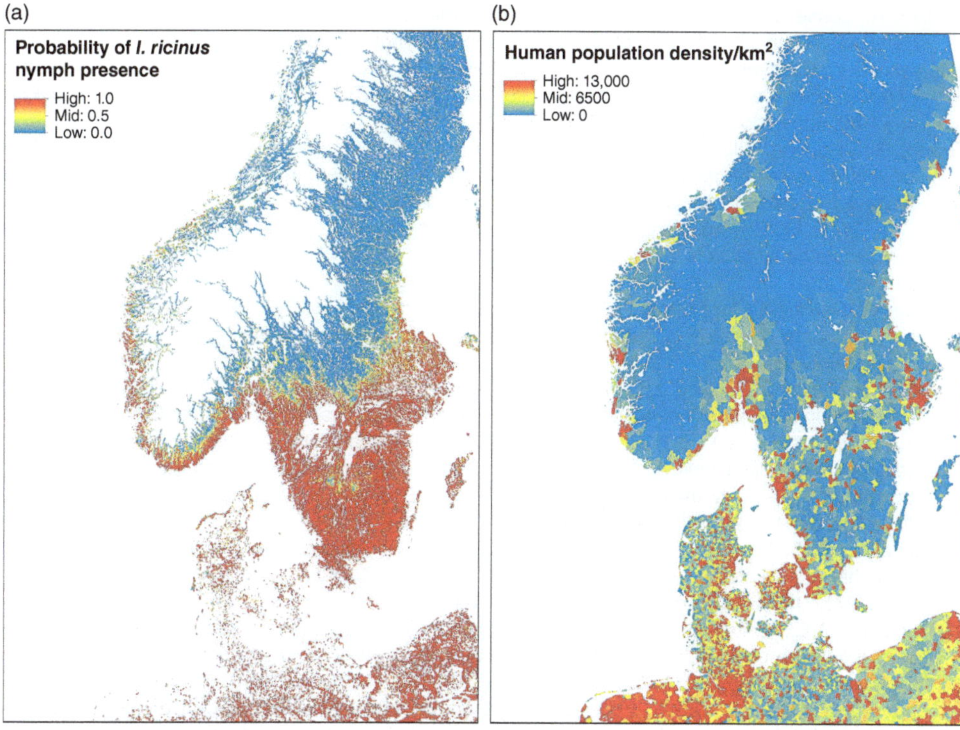

Fig. 14.1. Predicted probability of presence of nymphal *Ixodes ricinus* in Scandinavia (redrawn from Kjær *et al.*, 2019a) (a) and a human population density map of Scandinavia from the Gridded Population of the World data set (Socioeconomic Data and Applications Center, NASA, 2015) (b). White areas within the prediction map are altitudes above 450 m, or lakes, rivers and streams, or habitats other than sampled forest and meadow habitats (not predicted). Large areas of Scandinavia are presently free from *I. ricinus* ticks and a climate-driven expansion of ticks into these areas will have a limited effect on public health as the majority of the Scandinavian population is already living close to tick-endemic locations. (Authors' own figure.)

growing period. Thus, this seemingly resilient tick species may not only increase vector-borne disease incidence in the most northern Scandinavian regions but also – after now having crossed to Sweden north of the Bothnian Bay from Finland – expand its range towards more southern regions in Sweden, potentially bringing along the Siberian and Far Eastern subtypes of TBEV.

Because the abundance of endemic ticks is partly driven by non-climatic factors, Scandinavia may be able to develop climate adaptation strategies by targeting these other non-climatic drivers of tick abundance (e.g. the abundance of large mammals near densely populated areas) and thus prevent a net increase in abundance of ticks in a warmer climate. However, Scandinavia needs to be vigilant and prepare for the incoming exotic tick species and their associated pathogens. The emerging risk posed by the exotic ticks may be very difficult to counteract.

References

Baeten, L.A., Powers, B.E., Jewell, J.E., Spraker, T.R. and Miller, M.W. (2007) A natural case of chronic wasting disease in a free-ranging moose (*Alces alces shirasi*). *Journal of Wildlife Diseases* 43, 309–314. https://doi.org/10.7589/0090-3558-43.2.309

Bødker, R., Lohse, L. and Kjær, L.J. (2019) Vector-borne zoonoses. In: *Annual Report on Zoonoses in Denmark*. Technical University of Denmark, Kongens Lyngby, Denmark, pp. 17–19.

Bugmyrin, S.V., Bespyatova, L.A., Korotkov, Y.S., Burenkova, L.A., Belova, O.A., *et al.* (2013) Distribution of *Ixodes ricinus* and *I. persulcatus* ticks in southern Karelia (Russia). *Ticks and Tick-borne Diseases* 4, 57–62. https://doi.org/10.1016/J.TTBDIS.2012.07.004

Grandi, G., Chitimia-Dobler, L., Choklikitumnuey, P., Strube, C., Springer, A., *et al.* (2020) First records of adult *Hyalomma marginatum* and *H. rufipes* ticks (Acari: Ixodidae) in Sweden. *Ticks and Tick-borne Diseases* 11, 101403. https://doi.org/10.1016/j.ttbdis.2020.101403

Hasle, G., Bjune, G., Edvardsen, E., Jakobsen, C., Linnehol, B., *et al.* (2009) Transport of ticks by migratory passerine birds to Norway. *Journal of Parasitology* 95, 1342–1351. https://doi.org/10.1645/GE-2146.1

Heyman, P., Cochez, C., Hofhuis, A., van der Giessen, J., Sprong, H., *et al.* (2010) A clear and present danger: tick-borne diseases in Europe. *Expert Review of Anti-infective Therapy* 8, 33–50. https://doi.org/10.1586/eri.09.118

Hvidsten, D., Frafjord, K., Gray, J.S., Henningsson, A.J., Jenkins, A., *et al.* (2020) The distribution limit of the common tick, *Ixodes ricinus*, and some associated pathogens in north-western Europe. *Ticks and Tick-borne Diseases* 11, 101388. https://doi.org/10.1016/j.ttbdis.2020.101388

Jääskeläinen, A.E., Tikkakoski, T., Uzcategui, N.Y., Alekseev, A.N., Vaheri, A. and Vapalahti, O. (2006) Siberian subtype tickborne encephalitis virus, Finland. *Emerging Infectious Diseases* 12, 1568–1571. https://doi.org/10.3201/eid1210.060320

Jääskeläinen, A.E., Sironen, T., Murueva, G.B., Subbotina, N., Alekseev, A.N., *et al.* (2010) Tick-borne encephalitis virus in ticks in Finland, Russian Karelia and Buryatia. *Journal of General Virology* 91, 2706–2712. https://doi.org/10.1099/vir.0.023663-0

Jääskeläinen, A.E., Tonteri, E., Pieninkeroinen, I., Sironen, T., Voutilainen, L., *et al.* (2016) Siberian subtype tick-borne encephalitis virus in *Ixodes ricinus* in a newly emerged focus, Finland. *Ticks and Tick-borne Diseases* 7, 216–223. https://doi.org/10.1016/J.TTBDIS.2015.10.013

Jaenson, T.G.T., Jaenson, D.G.E., Eisen, L., Petersson, E. and Lindgren, E. (2012) Changes in the geographical distribution and abundance of the tick *Ixodes ricinus* during the past 30 years in Sweden. *Parasites & Vectors* 5, 8. https://doi.org/10.1186/1756-3305-5-8

Jaenson, T.G.T., Värv, K., Fröjdman, I., Jääskeläinen, A., Rundgren, K., *et al.* (2016) First evidence of established populations of the taiga tick *Ixodes persulcatus* (Acari: Ixodidae) in Sweden. *Parasites & Vectors* 9, 377. https://doi.org/10.1186/s13071-016-1658-3

Jensen, P.M. and Jespersen, J.B. (2005) Five decades of tick–man interaction in Denmark – an analysis. *Experimental and Applied Acarology* 35, 131–146. https://doi.org/10.1007/s10493-004-1991-7

Kjær, L.J., Soleng, A., Edgar, K.S., Lindstedt, H.E.H., Paulsen, K.M., *et al.* (2019a) Predicting and mapping human risk of exposure to *Ixodes ricinus* nymphs using climatic and environmental data, Denmark, Norway and Sweden, 2016. *Eurosurveillance* 24, 1800101. https://doi.org/10.2807/1560-7917.es.2019.24.9.1800101

Kjær, L.J., Soleng, A., Edgar, K.S., Lindstedt, H.E.H., Paulsen, K.M., *et al.* (2019b) Predicting the spatial abundance of *Ixodes ricinus* ticks in southern Scandinavia using environmental and climatic data. *Scientific Reports* 9, 18144. https://doi.org/10.1038/s41598-019-54496-1

Klitgaard, K., Chriél, M., Isbrand, A., Jensen, T.K. and Bødker, R. (2017) Identification of *Dermacentor reticulatus* ticks carrying *Rickettsia raoultii* on migrating jackal, Denmark. *Emerging Infectious Diseases* 23, 2072–2074. https://doi.org/10.3201/eid2312.170919

Lindgren, E. and Gustafson, R. (2001) Tick-borne encephalitis in Sweden and climate change. *The Lancet* 358, 16–18. https://doi.org/10.1016/S0140-6736(00)05250-8

Lindquist, L. and Vapalahti, O. (2008) Tick-borne encephalitis. *The Lancet* 371, 1861–1871. https://doi.org/10.1016/S0140-6736(08)60800-4

Øines, Ø., Storli, K. and Brun-Hansen, H. (2010) First case of babesiosis caused by *Babesia canis canis* in a dog from Norway. *Veterinary Parasitology* 171, 350–353. https://doi.org/10.1016/J.VETPAR.2010.03.024

Paulsen, K.M., Pedersen, B.N., Soleng, A., Okbaldet, Y.B., Pettersson, J.H.-O., *et al.* (2015) Prevalence of tick-borne encephalitis virus in *Ixodes ricinus* ticks from three islands in north-western Norway. *APMIS* 123, 759–764. https://doi.org/10.1111/apm.12412

Rudander, S.H. (2018) *Dermacentor reticulatus* risk att etablera sig i Sverige [The risk of *Dermacentor reticulatus* establishing itself in Sweden]. Fakulteten för Veterinärmedicin och, *Husdjursvetenskap*, Institutionen för Biomedicin och Veterinär Folkhälsovetenskap, Uppsala, Sverige.

Socioeconomic Data and Applications Center, NASA (2015) Gridded Population of the World, v4. Available at: http://sedac.ciesin.columbia.edu/data/collection/gpw-v4 (accessed 18 February 2019).

Solano-Gallego, L., Sainz, Á., Roura, X., Estrada-Peña, A. and Miró, G. (2016) A review of canine babesiosis: the European perspective. *Parasites & Vectors* 9, 336. https://doi.org/10.1186/s13071-016-1596-0

Soleng, A., Edgar, K.S., Paulsen, K.M., Pedersen, B.N., Okbaldet, Y.B., *et al.* (2018) Distribution of *Ixodes ricinus* ticks and prevalence of tick-borne encephalitis virus among questing ticks in the Arctic Circle region of northern Norway. *Ticks and Tick-borne Diseases* 9, 97–103. https://doi.org/10.1016/J.TTBDIS.2017.10.002

Süss, J. (2011) Tick-borne encephalitis 2010: epidemiology, risk areas, and virus strains in Europe and Asia – an overview. *Ticks and Tick-borne Diseases* 2, 2–15. https://doi.org/10.1016/J.TTBDIS.2010.10.007

Tokarevich, N.K., Tronin, A.A., Blinova, O.V., Buzinov, R.V., Boltenkov, V.P., *et al.* (2011) The impact of climate change on the expansion of *Ixodes persulcatus* habitat and the incidence of tick-borne encephalitis in the north of European Russia. *Global Health Action* 4, 8448. https://doi.org/10.3402/gha.v4i0.8448

eo15 Birds, Ticks and Climate Change

Tom Kelly*, John Healy and Neil Coughlan
University College Cork, Republic of Ireland

15.1 Expert Opinion

15.1.1 Introduction

Birds are ideal carriers of ticks. In particular, most songbirds migrate at night and for this and other reasons their survival is notably higher than that of mammals of the same size. Powered flight enables birds to move over short, medium or long distances, and in some conditions longer movements can occur very rapidly. Due to the flexible attachment times of some ixodid ticks they can be carried with their bird hosts wherever the latter go. It would appear that billions of birds migrate from southern latitudes to the northern hemisphere every spring. If even a tiny fraction, e.g. 1%, carry ticks, then millions must be carried over at least part of the distances the birds migrate every year. The key question, therefore, is what are the chances of the transported ticks surviving and establishing themselves in the receiving area or environment? For thermophilic ticks, such as Afrotropical species, it would appear that climate warming at higher latitudes is enabling some species such as *Hyalomma marginatum* and *Hyalomma rufipes* to establish – albeit temporarily – viable populations. In the cool temperate zones of the Nearctic and Western Palaearctic it is clear that *Ixodes scapularis* and *Ixodes ricinus*, respectively, are edging their way northwards as well as into higher altitudes. At least part of these distributional changes is explained by songbird-assisted transport and climate change.

15.1.2 Birds as vectors

Derived from the dinosaur Diapsid/Archosaurian lineage, birds (Avialae/Aves) originated in the mid-Jurassic some 150 million years ago (Brusette *et al.*, 2015; de la Fuente *et al.*, 2015). Modern birds, the Neornithes, appear to have radiated after the Cretaceous mass extinctions *c.*66 million years ago and are considered the most speciose of the terrestrial tetrapods, numbering some 20,000 extant and recently extinct taxa (species and subspecies) (Gill *et al.*, 2020). Feathers are a key feature of birds and cover the integument which is without sweat glands. Inter alia, the feathers 'and the air trapped between them' (Jenni and Winkler, 2020) reduce heat conductance, and in many species the maintenance of the plumage is assisted by preening including 'oiling', i.e. grooming with the secretions from the uropygial gland (Jenni and Winkler, 2020), which is situated dorsally just in front of the tail.

In general, birds are smaller sized than mammals. Both groups are endothermic, but birds –assisted in part by the plumage – maintain

*Email: t.kelly@ucc.ie

DOI: 10.1079/9781789249637.0015

a higher core body temperature (≥42°C) than mammals (e.g. Clarke and Rothery, 2007). Avian red blood cells, unlike those of mammals, are nucleated.

Most birds have an exceptionally high basal metabolic rate and exhibit very rapid growth; both these attributes are associated with the almost universal capacity for powered flight. This remarkable adaptation, which birds share with bats (Chiroptera), explains in part their relatively high life expectancy; birds have up to four times greater life expectancy than similar-sized mammals (e.g. Healy *et al.*, 2014; Nabi *et al.*, 2021). Powered flight allows birds to escape many predators but, in addition, although having the smallest genome of any vertebrate group, birds have a sophisticated immune system with well-developed innate and adaptive immune responses (e.g. Kaiser, 2010; Owen *et al.*, 2010; Nabi *et al.*, 2021). Birds lack lymph nodes but possess the bursa of Fabricius where B cells receive their identity. Since birds lack lymph nodes it is not known for certain where antigen presentation occurs (e.g. by dendritic cells) (Kaiser, 2010). In addition, maternal antibodies are passed from the female bird to her offspring (Ramos *et al.*, 2014; Boulinier and Staszewski, 2018) and recently interest has developed in transgenerational immune priming (Roth *et al.*, 2018). Nevertheless, birds are host to many micro- and macroparasites including ticks (Thomas *et al.*, 2007; de la Fuente *et al.*, 2015; Nicholson *et al.*, 2019).

In terms of their global biomass, birds are (despite their greater life expectancies) less abundant than mammals (mammals, 0.007 Gt C; birds, 0.002 Gt C; 1 Gt C = 10^{15} g carbon) (Bar-On *et al.*, 2018). In addition, humans (0.06 Gt C) and livestock (0.1 Gt C), together with the 'negligible' biomasses of Amphibia and Reptilia (Bar-On *et al.*, 2018), provide the resources – mainly blood – upon which the global tick fauna (and other haematophagous arthropods) depend. Notably, birds are considered major agents in the dispersal of plants, animals, fungi and microbes (Coughlan *et al.*, 2017). Birds are parasitized by both ixodid and argasid ticks. Broadly, bird ticks may be classified as being either exophilic or endophilic. Endophilic ticks – also known as nidicoles – more or less permanently inhabit the nest burrows, roosts and nest aggregations of their hosts, i.e. colonies of seabirds, herons and swallows. Consequently, many of these species are either highly host-specific (monoxenous), such as the sand martin (*Riparia riparia*) tick, *Ixodes lividus*; or oligoxenous, e.g. *Ixodes arboricola*, an inhabitant of tree hole-nesting birds, which appears to parasitize a wider range of species successfully (e.g. Estrada-Peña *et al.*, 2017). Exophilic ticks are opportunistic parasites and 'quest' (see Nicholson *et al.*, 2019) in the habitat of their potential hosts, to which they attach and engorge before dropping off elsewhere. In the Western Palaearctic, *I. ricinus* is a highly successful exophilic tick, being one of the most widely distributed and numerous species within the region; it does not show evidence of host specificity (polyxenous) and is a notorious vector of numerous micropathogens including tick-borne encephalitis virus and Lyme disease *Borrelia burgdorferi* sensu lato (s.l.). Many bird species are parasitized by *I. ricinus*, particularly the small perching birds (passerines and near passerines), but it appears that very few adult female ticks are able to complete their blood meals on these hosts (see Heylen and Matthysen, 2010; Buczek *et al.*, 2020; and references therein). The same is true of, for example, *I. scapularis* in the Nearctic (e.g. Leighton *et al.*, 2012).

15.1.3 Climate change impacts on birds and bird migration

The phenomenon of climate change is widely accepted in the scientific community and there is now a voluminous and growing literature. The ecological consequences have received detailed attention and birds in particular have been the focus of much intensive research (e.g. Møller *et al.*, 2006). Changes in phenology and distribution are the observed (and predicted) responses to global warming – together with increased mortality – that appear to occur among populations of different bird species (Both *et al.*, 2006; Chan and Thomas, 2011; Hewson *et al.*, 2016; Zurrell *et al.*, 2018; Conradie *et al.*, 2019). However, the debate about the responses to climate change continues (Knudsen *et al.*, 2011). Bairlein (2016) describes the declines in long-distance migrant species, i.e. those that fly from sub-Saharan Africa to Europe, and among those wintering in 'Central and South America'. However, while habitat change resulting from intensification in agriculture leading to the large-scale removal of natural vegetation and the illegal killing (especially along the northern shores of the Mediterranean) of millions of birds are a partial explanation, climate change – leading, inter alia, to a mismatch between the food resources on the breeding grounds and the timing of arrival of the long-distance migrants – is also a major cause of reproductive failure and population decline. Interestingly, at least in Europe, short-distance migrants (i.e. those wintering north of the Sahara) are not declining at the same rate as long-distance migrants (Bairlein, 2016).

'At their simplest the annual cycles of birds are made up of 2–3 processes: breeding, moult and (in migratory species only) migration' (Newton, 2011). Newton (2011) also points out that migration, along with the other 'processes', are the 'most energy-demanding events in the yearly cycles of birds'. Bird migration is one of the greatest behavioural and ecological events on Earth and it has been intensively studied. The result is voluminous literature which, however, is well reviewed (e.g. Alerstam, 1990; Newton, 2008). Nevertheless, it should be noted that migration is but one of the several types of movement shown by birds (e.g. Paradis *et al.*, 1998; Wernham *et al.*, 2002; Hertel *et al.*, 2020). Migration is defined as 'the return journeys that many individuals make between geographically separated breeding and wintering areas' (Newton, 2011). It is widely recognized that migratory behaviour may be highly labile (e.g. Alerstam *et al.*, 2003). Long-distance migrants include those species such as the swift (*Apus apus*) which moves from its high-latitude breeding quarters in the Western Palaearctic to its wintering grounds in southern Africa (Wernham *et al.*, 2002) and the wheatear (*Oenanthe oenanthe*) which undertakes a remarkable journey from eastern Canada, across the Atlantic, to western sub-Saharan Africa (Bairlein *et al.*, 2012).

In the Western Palaearctic, short-distance migrants, i.e. those wintering north of the Sahara, include the blackbird (*Turdus merula*), song thrush (*Turdus philomelos*) and robin (*Erithacus rubecula*) (Lack, 1954). However, many populations in the western and southern parts of their breeding ranges appear to be sedentary (Wernham *et al.*, 2002). For example, the adult male blackbird may be effectively sedentary, with the female being a partial short-distance migrant, while the immature blackbird is a habitual migrant – albeit over relatively short distances (Lack, 1954).

Bird migration is thought to have evolved to cope with the necessary optimal exploitation of strongly seasonal environments in the aftermath of the Last Glacial Maximum, which ended about 11,000 to 10,000 years BP (Zink, 2011). But despite the seasonality associated with the present interstadial (i.e. the interval between successive glaciations), perhaps less than 30% or even fewer of bird species are migratory (e.g. Alerstam, 1990; Newton, 2008). Moreover, the number and proportion of migratory species vary considerably with geography; for example, there is a steep increase in the number of long-distance migrants from Ireland and Britain compared with Central Europe (Fuller *et al.*, 2007). To a large extent, Ireland has an almost non-seasonal climate, whereas in Central Europe relatively severe winters alternate with warm summers. Recently, the impact of the Last Glacial Maximum on the evolution of 'modern'-day migration has been challenged by Someville *et al.* (2020), whose study suggests that migratory behaviour has changed relatively little in the Old World but more profoundly in the Americas. The models suggest that the migratory routes in the Palaearctic–African system have been in place for ≥50,000 years with presumably corresponding implications for the transfer of ticks between e.g. Europe and sub-Saharan Africa. In North America, the impact of the Last Glacial Maximum appears to have been more profound and post-glacial migratory journeys may have increased by up to 40% compared with those during the last Ice Age (Someville *et al.*, 2020). While traditional methods of bird ringing to study migration continue (e.g. Wernham *et al.*, 2002; Newton, 2008), new technical developments are profoundly influencing our understanding of these mass movements of birds and these in turn should allow a more robust testing of hypotheses regarding the long-distance dispersal of ticks and the potential impacts of climate change.

Most migration is nocturnal, especially in passerine and near passerine birds – a fact understood since the late 19th century (e.g. Barrington, 1900). The discovery of radar during the Second World War gave rise to a science known as radar ornithology which is still practised today even though it is difficult to identify the species involved in the huge movements of birds that are detected (e.g. Maniola *et al.*, 2020). Radar is highly informative in relating weather conditions to observable migration (e.g. Shamoun-Baranes *et al.*, 2017). Even more informative but, perforce, on a much smaller scale (in terms of the number of individual birds being studied) is the use of biotelemetry – especially satellite tags (e.g. Cooke *et al.*, 2004; Kays *et al.*, 2015; López-López, 2016). These devices have revealed extraordinary detail about the journeys taken by e.g. the cuckoo (*Cuculus canorus*) from Britain and continental Europe to and from its wintering grounds in sub-Saharan Africa (Hewson *et al.*, 2016; Vega *et al.*, 2016). The insights gained by biotelemetry studies of migratory birds will be highly relevant in testing hypotheses concerning the long-distance dispersal of ticks between the southern and northern hemispheres.

Moreau (1972) suggested that five billion birds migrate annually from the Western Palaearctic to Africa. A more recent estimate of the numbers migrating in the 'World's biggest migratory system' suggests that in autumn an average of 2.1 (upper bound 2.91; lower bound 1.52) billion birds, 95% of which are passerines, move south across the Mediterranean on their way to sub-Saharan Africa (Hahn *et al.*, 2009; see also Wallmenius *et al.*, 2014). There does not appear to be any comparable figure for the numbers of birds returning the following spring to the Western Palaearctic but, assuming for the sake of argument a 60% survival of those overwintering in Africa, an average of 1.26 billion birds (upper bound 1.746 billion; lower bound 912 million) journey north into Europe to breed. The most abundant passerine species making these journeys every year include the willow warbler (*Phylloscopus trochilus*), tree pipit (*Anthus trivialis*), barn swallow (*Hirundo rustica*) and garden warbler (*Sylvia borin*), whereas the 'wetland'-inhabiting warblers, e.g. the sedge warbler (*Acrocephalus schoenobaenus*) and Eurasian reed warbler (*Acrocephalus scirpaceus*), appear to be less abundant (Hahn *et al.*, 2009).

15.1.4 Bird migration, ticks and climate change

Many authors have addressed the issue of climate change and ticks (e.g. Gray *et al.*, 2009; Levi *et al.*, 2015; Semenza and Suk, 2018). However, while global warming is occurring, the picture of its relationship with tick populations is partly confounded by parallel changes inter alia in host populations, especially deer (Cervidae), and socio-economic factors which on the one hand are associated with more leisure time in the countryside among the wealthy and on the other the need to gather food in woodland and comparable habitats by those with lower economic resources (e.g. Randolph, 2004, 2010; Medlock *et al.*, 2013; Ostfeld and Brunner, 2015). Nevertheless, there is clear evidence of range extensions in both *I. ricinus* in Europe and *I. scapularis* in the USA and Canada (e.g. Lindgren *et al.*, 2000; Zeman and Benes, 2004; Scott *et al.*, 2018) and many suggest that migratory birds are involved (see Reed *et al.*, 2003; Scott *et al.*, 2018; Buczek *et al.*, 2020; and references therein).

It has long been postulated that exotic ticks can be transported over great distances by migratory birds (e.g. Hoogstraal *et al.*, 1961, 1963; Hoogstraal, 1979). More recently, it has been suggested that climate change has enhanced the probability of subtropical and tropical ticks being successfully transported into the northern hemisphere (e.g. de la Fuente *et al.*, 2015; Levi *et al.*, 2015). Also, ticks can be transported over great distances. For example, Britain's first record of a grey-cheeked thrush (*Hylocichla minima*, now known as *Catharus minima*) was mist-netted on Fair Isle in Scotland, on 5 October 1953. A larval tick, subsequently identified as the three-host North American rabbit tick, *Haemaphysalis leporispalustris*, was removed from its chin. Therefore, this individual tick attached somewhere in North America and survived a trans-Atlantic journey of perhaps several thousand miles (see Williamson, 1954). Obviously, such long-distance transport from sub-Saharan Africa to Europe is also a possibility (e.g. Hoogstraal *et al.*, 1961, 1963; de la Fuente *et al.*, 2015). How might thermophilic ticks be transported by birds from sub-Saharan Africa to Europe? We suggest a scenario in which there are three stages extending from the source to the receiving areas.

Stage 1: source area – sub-Sahara

Western Palaearctic migrants such as the tree pipit (*A. trivialis*) and the willow warbler (*P. trochilus*) may pick up ticks – probably *H. rufipes* (Hoogstraal *et al.*, 1961; Vatansever, 2017), a two-host tick (Nicholson *et al.*, 2019) – before embarking on their long journey north which includes crossing the Sahara Desert. The key variables here are: (i) the health and energetic status of the bird, i.e. is it free of 'disease' and has it accumulated sufficient fat to enable it to successfully complete the long journey? (e.g. Newton, 2008); and (ii) has it been infested by ticks and if so, what developmental stages are involved and how many? For example, larval ixodid ticks take less blood and drop off the host earlier than do the nymphs (e.g. Balashov, 1972). Therefore, given that migration is 'one of the most energy demanding processes' (Newton, 2011) experienced by birds, the fewer ticks that attach the better and preferably these would be larvae.

Because of the energy demands incurred by migration, migrants fuel up before departure and this may involve a period of sustained foraging during which they are likely to encounter questing ticks. Numerous studies have shown, however, that the distribution of ticks within a host population is generally overdispersed, i.e. the variance is significantly greater than the mean (e.g. Randolph *et al.*, 1996), and so a few hosts will carry many ticks while a majority of host individuals are uninfested. The negative binomial distribution is often fitted to these data; the clumping parameter k of this distribution is inversely proportional to the degree of overdispersion. Clearly, while the fewer the ticks attached to the bird the greater its chance of completing the journey, this will make it more difficult for the tick to establish itself in a new area, i.e. at the end of the bird's journey (see Stage 3).

Stage 2: the journey

When the migrant embarks on its journey north, it is in its interest to complete the journey as quickly as possible (e.g. Nilsson *et al.*, 2013). Several studies have shown that birds on the spring migration to the breeding grounds complete the journey more rapidly than do birds migrating south to the winter grounds in the autumn (e.g. Nilsson *et al.*, 2013; Schmaljohann, 2018). So, in the 'journey phase', flight speed and the ability to complete the journey quickly is an important variable. Ellgren (1993) and Hedenstrom (2008) give details of the speed of some migrants: for example, wheatear, 110 km/day; tree pipit, 98 km/day; willow warbler, 84 km/day; and swift, 150 km/day. The duration of the migrant's journey has vital implications for ticks, because the latter are constrained by the number of days they remain attached to the bird. On the other hand, there appears to be surprising variation in the duration of attachment in perhaps a majority of tick species (e.g. Balashov, 1972; Hoogstraal *et al.*, 1961; Troughton and Levin, 2007). For example, Hoogstraal *et al.* (1961) stated that in the important bird-dispersed two-host tick, *Hyalomma marginatum rufipes* (now known as *H. rufipes*) (Estrada-Peña *et al.*, 2017): 'Larvae and nymphs remain on avian hosts for at least two weeks and sometimes for six weeks or more'. Likewise, Troughton and Levin (2007) showed that considerable variation occurs in the attachment times of North American tick species including *I. scapularis* and *H. leporispalustris*. In the latter species, the female remained attached for 9 to 23 days, and larva and nymph from 11 to 16 days. Similarly, Heylen and Matthysen (2010) showed that remarkable variation occurs in the detachment times of *I. arboricola* – but not *I. ricinus* – when fed on songbirds in the laboratory. It should be relatively simple to model the distance a tick can be carried by a bird as the speed in km/day multiplied by the duration of attachment would give an approximate estimate. There is some suggestion that migration temporarily weakens the immune response but delayed detachment in particular could ensure that the tick(s) would be carried 400 to 500 km further by, for example, the wheatear. Abiotic factors may also play a role in bird migration and the time it takes to complete a journey (e.g. Shamoun-Baranes *et al.*, 2017). Some, but not all individuals, interrupt their journey – the so-called stopover – and these appear to be more frequent and of longer duration in the autumn, when flying south. This, according to Schmaljohann (2018), is the main reason why the autumn migratory journey takes longer than that in the spring; Nilsson *et al.* (2013) add that the autumn movements contain a lot of juvenile birds (see also Hahn *et al.*, 2009) which are not as expert at foraging as the adults – perhaps they are also less able to deal with abiotic problems like headwinds and sandstorms (see Strandberg *et al.*, 2010 and references therein). There are two flyways across the Sahara, the so-called Western European– and Eastern European–African Flyways (see Bairlein, 2016). The desert is a major barrier which passerine species fly over at night whereas raptor species cross by day (Strandberg *et al.*, 2010). The western flyway crosses the West African Sahara which '5000 years ago was covered by extensive areas of Mediterranean scrub and dry woodland (Moreau, 1966)' (Strandberg *et al.*, 2010), although a recent survey by Brandt *et al.* (2020) has found an 'unexpectedly' high number of trees within the southern part of this region. The desertification of the Sahel in the 1960s led to a spectacular decline in the number of whitethroats breeding in Britain and Ireland (Winstanley *et al.*, 1974).

Stage 3: the receiving environment and climate change

The successful transport of Afrotropical ticks by birds across the Sahara (in terms of their successful colonization and establishment) and into Europe will depend on conditions in the receiving environment. Thus, when the tick drops off it must meet both suitable hosts (large animals – especially ungulates, which may be scarce in warmer parts of Europe) and a microclimate that enables it to complete all stages of the life cycle. Given the large numbers of migrants making the trans-Saharan journey (our estimate is greater than one billion birds based on Hahn *et al.*, 2009) in excess of one million ticks may detach in Europe each spring (based on 1% of these migrants being infested). Therefore, if climate change is creating microclimates which enable some of these ticks to survive, then detections of e.g. *H. rufipes* should begin to accumulate. This indeed has happened with records in, for example, Sweden, Hungary, Malta and the Ukraine (e.g. Hornok and Horvath, 2012; Vatansever, 2017). It would be interesting to know, therefore, if *H. rufipes* has permanently established itself north of the Western Sahara – from which short-distance migrants could continually transport it into Southern Europe and from there, potentially, as climate warming continues, into higher latitudes. However, the taxonomy of the *H. marginatum* species–subspecies complex is difficult and *H. marginatum* is itself well established on the Iberian, Ligurian and Balkan peninsulas, as well as in Turkey and, more recently, France (Santos-Silva and Vatansever, 2017), but some of the reports do not appear to consider *H. rufipes* as a possible colonizer. This is a serious difficulty. Interestingly, Estrada-Peña *et al.* (2011) modelled the likely distribution of *H. marginatum* and showed that while it is likely to be continually dispersed from the south by birds, it is highly improbable that it will become established further north (i.e. outside its existing distribution) (see Map in Estrada-Peña *et al.*, 2011) because available microclimates – particularly in relation to the requirements of the nymphal moult – do not exist there (see also Estrada-Peña *et al.*, 2015; Vial *et al.*, 2016).

There is extensive migration of birds from the Mediterranean to northern latitudes in Europe (e.g. Wernham *et al.*, 2002). In theory, as already mentioned, climate change could set up a relay of recently established thermophilic tick species to more temperate ecosystems further north. However, the main 'traffic' in the conveyor belt of bird-assisted tick dispersal is likely to involve exophilic species such as *I. ricinus* and ornithophilic forms such as *Ixodes frontalis*. True endophilic species such as *I. lividus* and, to a lesser extent, *I. arboricola*, seem to rarely be carried by migratory birds (see Buczek *et al.*, 2020 for a comprehensive review). Hoogstraal *et al.* (1961) is the only study that discovered argasid ticks being transported by migrant birds. However, these surveys were conducted in Egypt and involved only *Argas* larvae, which were not identified to species. Interestingly, argasid ticks do not appear to have been recorded on migratory birds in Europe save for one record of *Argas vulgaris* (now thought to be a synonym of *Argas polonicus*) in Italy (Vial, 2009; Hasle, 2013; Buczek *et al.*, 2020; Toma *et al.*, 2021). The dispersal of ticks by birds in Europe has been comprehensively and authoritatively reviewed (Buczek *et al.*, 2020). Earlier, Hasle (2013) summarized the situation in Norway. Numerous studies have been undertaken in a wide range of European countries; ground-feeding birds such as the song thrush, European blackbird, robin and tree pipit have been found to be the main carriers of ticks. In the Western European–African Flyway (*sensu* Bairlein, 2016) it is difficult sometimes to separate local movements of birds from those that might have originated in the North Mediterranean littoral. Consequently, the explicit influence of climate change is difficult to detect.

15.1.5 Climate change, birds and *Ixodes ricinus*

I. ricinus, a common European tick, has attracted increased research interest in recent years due to its role as a vector of *B. burgdorferi* s.l., the causative agent of Lyme borreliosis. A complexity of factors, both biotic and abiotic, underpin the considerable growth in tick population sizes throughout Northern Europe. Principal among biotic factors has been an increase in both state-managed and privately owned afforested land. This, together with a political and social tolerance, has allowed an expansion in population sizes and distributional ranges of the adult tick's critical hosts, i.e. roe deer (*Capreolus capreolus*) in continental Europe and red deer (*Cervus elaphus*), sika deer (*Cervus nippon*) and fallow deer (*Dama dama*) principally in the UK and Ireland (e.g. Holt *et al.*, 2011). Adding to the complexity, and potentially influencing the host balance, has been the so-called Bambi effect whereby increasing deer populations have been shown to correlate with declines in woodland birds due to damage to nesting and foraging habitat by deer (Chollet and Martin, 2013).

There is evidence that climate effects are facilitating the extension of the distributional range of *I. ricinus* to both higher latitudes and higher altitudes (Jaenson *et al.*, 2012; Hvidsten *et al.*, 2020). Tick occurrences have been reported as high as 66°N in Norway and Sweden and at inland locations in the north of Sweden. With respect to altitude, studies in the Czech Republic (Daniel *et al.*, 2009) have shown an increase of over 500 m in the altitudinal range of *I. ricinus* over a 50-year period and this increase correlates significantly with increasing mean temperature at the locations surveyed. Separately, even within established areas of tick distribution, increasing rainfall inevitably generates more high-humidity land area suitable for tick survival. Milder winters, earlier cessation of spring frosts and later commencement of autumn frosts extend the temporal distribution of *I. ricinus*, allowing for host-seeking and reproduction for an extended portion of the year. We examined eight recently slaughtered red deer carcases in the west of Ireland in January 2018 and found at least 20 adult female ticks per deer.

The dispersal of ticks by birds to higher altitudes and latitudes is entirely dependent on movement of host animals. Passerine birds serve as important hosts for the larval and nymphal tick instars. Ground-visiting and ground-foraging species are those most likely to become tick-infested and principal among these are the European blackbird, song thrush and European robin, as well as the starling (*Sturnus vulgaris*), dunnock (*Prunella modularis*) and great tit (*Parus major*) (Comstedt *et al.*, 2006). It is likely that short-distance movements by birds are at least partly responsible for the extended geographical range and altitudinal shift in *I. ricinus* distribution. Widening of the tick's temporal distribution away from the historic 'spring and autumn' activity peaks in Ireland and the UK, and the typical 'single peak' of activity seen in Northern Europe, may significantly alter the pattern and extent of infestation of passerine species and hence influence tick dispersal. This is particularly so when tick activity coincides with major influx or exodus of migratory species. There is considerable evidence that passerine spring migrations are occurring earlier than previously although no clear conclusions can be drawn with regard to timings of autumn departures (reviewed by Newton, 2010). Northward extension of *I. ricinus* is ultimately limited by winter temperatures inducing failure to reproduce and unsustainable tick mortality. Coristine and Kerr (2015) argue that additional northward expansion of the ranges of passerine species is unlikely to be incremented further by increasing ambient temperatures since, in any case, temperature is not a limiting factor during the breeding season. Accordingly, further poleward dispersal and population maintenance of *I. ricinus* is more likely to be mediated by mammalian hosts. Changing patterns of dispersal and distribution of *I. ricinus* brought about by climate change will, in turn, have implications for the distribution and prevalence of the tick-borne pathogens causing Lyme borreliosis and tick-borne encephalitis (TBE).

15.2 Future Predictions

The pattern of bird migration is responding to climate change (see Fiedler *et al.*, 2006; and references therein) and predictions about future changes in tick distributional ranges must take account of this. Fiedler *et al.* (2006) state that climate change-induced alterations to migratory behaviour are the result of four processes: (i) changes in population composition such as ratio of migrants to non-migrants; (ii) changes in the distance migrated; (iii) shifts in the direction of migration; and (iv) changes in the seasonal timing and duration of migration. Each of these processes has been shown to occur (e.g. Berthold *et al.*, 1992; Wernham *et al.*, 2002; Bearhop *et al.*, 2005; Fiedler *et al.*, 2006; Visser *et al.*, 2009; Delmore *et al.*, 2020). Overall, in the Western Palaearctic, this may mean at least a partial reduction in long-distance dispersal from the Afrotropical regions and an increase in the transportation of ticks from north of the Sahara Desert, i.e. the Mediterranean Basin and the surrounding littoral.

Apart from the possibility of reducing long-distance dispersal, a patchy tick distribution may develop as engorged ticks drop from avian hosts along a migration route, resulting in the development of a pattern similar to metapopulations (e.g. Hanski, 1998; Brooker *et al.*, 1999). Hence, the 'stepping stone' model (e.g. Saura *et al.*, 2014) may provide an appropriate way of viewing and monitoring changes in the ixodid tick fauna. This does not apply in the case of argasid ticks as these species do not appear to be dispersed by migratory birds in the Holarctic (e.g. Buczek *et al.*, 2020).

The dynamics of populations within patches, in relation to the processes of immigration and extinction, is well described (e.g. MacArthur and Wilson, 1967; MacArthur, 1972). Populations in small patches have a relatively higher risk of stochastic extinction whereas the opposite is the case, broadly speaking, in large patches. The establishment of a population in a new patch may depend on threshold effects and the sustained input of propagules (e.g. Gotelli, 1991; Kalachev *et al.*, 2011; Coughlan *et al.*, 2017; Kada *et al.*, 2017). Kada *et al.* (2017) mention Allee effects as being important in establishing ixodid tick populations. So, where tick populations are large, there will be intraspecific density-dependent competition which will lower the number of individuals within the population. However, if the colonizing tick numbers are so small that the probability of males and females meeting and mating is low, then this may inhibit the establishment of a sustainable population even in the face of repeated seasonal larval and nymphal input to the patch by migrating birds (however, see Healy and Bourke, 2004, 2008). This negative density dependence (the Allee effect), and its inclusion in tick population and dispersal models, is an important advance (Kada *et al.*, 2017). We see four possible future changes to tick distributional ranges mediated by birds.

1. *H. marginatum* to advance to higher latitudes. *H. marginatum* appears to be well established in Spain, Italy, Greece and Turkey, and in France along the Mediterranean littoral east of Montpellier (e.g. Vial *et al.*, 2016). We see this as an advancing front which will extend beyond the Pyrenees and the Italian Alps. The drivers will be climate change, bird-assisted dispersal of the larval stages and an increasing availability of large herbivorous hosts, especially cattle and horses (Eurostat, 2020) in France, Germany and the Low Countries. It will be important to monitor potential pockets of *H. marginatum* colonization by sampling areas with the necessary large-host capacity. In this regard, work on the distribution of species such as roe deer (Wint *et al.*, 2015) and red deer (Milner *et al.*, 2006) will be relevant.

2. Further extension to higher latitudes and higher altitudes of *I. ricinus* plus possible contraction of its southern range northwards. It has already been explained why *I. ricinus* is extending its range northwards and to higher altitudes, with changes in climatic conditions favouring its ascent to higher latitudes and altitudes (see Section 15.1.5). However, increasing temperatures and decreasing humidity are also likely to make its continued occupancy of Iberia, southern France and northern Italy more problematic, especially at lower altitudes. Hence its southernmost populations may be depleted and its range at lower latitudes in the Iberian and Ligurian peninsulas may contract northwards.

3. Extension of the current distributional front of the taiga tick, *Ixodes persulcatus*, westwards and southwards. The taiga tick is present in Finland and well established in Sweden, with records also from Estonia and Latvia (Jaenson *et al.*, 2016; Laaksonen *et al.*, 2017). Sirotkin and Korenberg (2018) provide a detailed description of the main biotic and abiotic variables which drive the life cycle of *I. persulcatus*. With recent changes in distribution, it seems possible that larvae and nymphs questing in early autumn can be picked up by birds migrating southwards from the east. Although *I. persulcatus* is more cold-adapted than the 'thermophilous' *I. ricinus* (e.g. Sirotkin and Korenberg, 2018) and is extending its range northwards in western Russia, it has also been discovered to be widely distributed in the Gulf of Bothnia along the lands abutting Finland's west coast and Sweden's east coast (see Jääskeläinen *et al.*, 2017 and references therein). Bird species from this region and further east as far as Siberia may undertake movements towards Central Europe in the autumn – a behaviour suggested to be due to 'reverse migration' (e.g. Rabøl, 1969; Gilroy and Lees, 2003; Thorup, 2004; Pfeifer *et al.*, 2007). Although the numbers are small, the possibility that the larger species like Asiatic thrushes – e.g. the black-throated thrush (*Turdus astrogularis*), eyebrowed thrush (*Turdus obscurus*), Naumann's thrush (*Turdus naumanni*), scaly thrush (*Zoothera daumi*), red-flanked bluetail (*Tarsiger cyanurus*) and Siberian thrush (*Geokichla sibirica*) – may carry ticks, including *I. persulcatus* and/or their micropathogens, would appear to be worthy of investigation.

4. The permanent establishment of *H. rufipes* in Europe and Turkey due to long-distance and short-distance dispersal by birds. Climate change will increase the probability of *H. rufipes* becoming established at certain points in Europe. This will be a major change to the ixodid tick fauna of the Western Palaearctic. Estrada-Peña *et al.* (2011) predicted that microclimatic conditions in Central Europe will prevent moulting of nymphs of *H. marginatum* and *H. rufipes* which have successfully dispersed from Africa. However, a number of recent reports from Sweden (Grandi *et al.*, 2020), Germany (Chitima-Dobler *et al.*, 2016), the Czech Republic (Hubálek *et al.*, 2020), Hungary (Hornok and Horvath, 2012) and the UK (McGinley *et al.*, 2021) show that this is not the case and clearly indicate that locations, albeit randomly distributed, can be colonized by *H. rufipes* (see review of Hubálek *et al.*, 2020 and references therein) with major implications for the tick-borne virome of temperate Europe. The pattern of occurrence of *H. rufipes* is not one of an advancing front; rather, it is a long-distance dispersal transportation phenomenon into effectively unpredictable locations. This might be followed by a 'stepping stone'-type short-distance dispersal process followed by the establishment of metapopulations as discussed above (Brooker *et al.*, 1999; Saura *et al.*, 2014). Tracking such a process by monitoring birds would probably have limited success due to the overdispersion of ticks on trapped birds and the consequent need to sample very large numbers (England *et al.*, 2016). The more critical information regarding the establishment of *H. rufipes* populations relates to the number of adult ticks. Since adult ticks do not usually attach to birds, systematic surveying of beef cattle herds, horses and wild deer populations should provide a better indicator of newly developing metapopulations.

References

Alerstam, T. (1990) *Bird Migration*. Cambridge University Press, Cambridge.

Alerstam, T., Hedenstrom, A. and Akesson, S. (2003) Long distance migration: evolution and determinants. *Oikos* 103, 247–260.

Bairlein, F. (2016) Migratory birds under threat. *Science* 354, 547–548.

Bairlein, F., Norris, D.R., Nagel, R., Bulte, M., Voight, C.C., *et al.* (2012) Cross-hemisphere migration of a 25 g songbird. *Biological Letters* 8, 505–507.

Balashov, Y.S. (1972) Blood sucking ticks (Ixodoidea) vectors of disease of man and domestic animals. *Miscellaneous Publications of the Entomological Society of America* 8, 163–376.

Bar-On, Y.M., Phillips, R. and Milo, R. (2018) The biomass distribution on Earth. *Proceedings of the National Academy of Sciences USA* 115, 6506–6511.

Barrington, R.M. (1900) *The Migration of Birds as Observed at Irish Lighthouses and Lightships*. Ponsonby and Porter, Dublin and London.

Bearhop, S., Fiedler, W., Furness, R.W., Votier, S.C., Waldron, S., *et al.* (2005) Assortative mating as a mechanism for rapid evolution of a migratory divide. *Science* 310, 502–504.

Berthold, P., Helbig, A.J., Mohr, G. and Querner, U. (1992) Rapid microevolution of migratory behaviour in a wild bird. *Nature* 360, 668–670.
Both, C., Brouwhuis, S., Lessells, C.M. and Visser, M.E. (2006) Climate change and population declines in a long-distance migratory bird. *Nature* 441, 81–85.
Boulinier, T. and Staszewski, V. (2018) Maternal transfer of antibodies: raising immuno-ecology issues. *Trends in Ecology & Evolution* 23, 282–288.
Brandt, M., Tucker, C.J., Keriryaa, A., Rasmussen, K., Abel, C., *et al.* (2020) An unexpectedly large amount of trees in the West African Sahara and Sahel. *Nature* 587, 78–82.
Brooker, L., Brooker, M. and Cale, P. (1999) Animal dispersal in fragmented landscapes: measuring habitat connectivity, corridor use and dispersal mortality. *Conservation Ecology* 3, 4.
Brusette, S.L., O'Connor, J.K. and Jarvis, E.D. (2015) The origin and diversification of birds. *Current Biology* 25, 888–898.
Buczek, A.M., Buczek, W., Buczek, A. and Bartosik, K. (2020) The potential role of migratory birds in the rapid spread of ticks and tick-borne pathogens in the changing climatic and environmental conditions of Europe. *International Journal of Environmental Research and Public Health* 17, 2117.
Chan, I.-C. and Thomas, C. (2011) Rapid range shifts of species associated with high levels of climate warming. *Science* 333, 1024–1026.
Chitima-Dobler, L., Navu, S., Besterhorn, M., Dobler, G. and Wölfil, S. (2016) First detection of *Hyalomma rufipes* in Germany. *Ticks and Tick-Borne Diseases* 7, 1135–1138.
Chollet, S. and Martin, J.-L. (2013) Declining woodland birds in North America – should we blame Bambi? *Diversity and Distributions* 19, 481–483.
Clarke, A. and Rothery, P. (2007) Scaling of body temperatures in mammals and birds. *Functional Ecology* 22, 58–67.
Comstedt, P., Bergstrom, S., Olsen, B., Garpmo, U., Marjavaara, E., *et al.* (2006) Migratory passerine birds as reservoirs of Lyme borreliosis in Europe. *Emerging Infectious Diseases* 12, 1087–1095.
Conradie, S.R., Woodborne, S.M., Cunningham, S.J. and McKechnie, A.E. (2019) Chronic sub-lethal effects of high temperatures will cause severe declines in southern African arid zones during the 21st century. *Proceedings of the National Academy of Sciences USA* 116, 14065–14070.
Cooke, S.J., Hinch, S.G., Wikelski, M., Andrews, R.D., Kuchel, L., *et al.* (2004) Biotelemetry: a mechanistic approach to ecology. *Trends in Ecology & Evolution* 19, 334–338.
Coristine, L. and Kerr, J. (2015) Temperature-related geographical shifts among passerines: contrasting processes along poleward and equatorward range margins. *Ecology and Evolution* 5, 5162–5176.
Coughlan, N.E., Kelly, T.C., Davenport, J. and Jansen, M.A.K. (2017) Up, up and away: bird mediated dispersal between aquatic environments. *Freshwater Biology* 62, 631–648.
Daniel, M., Materna, J., Hönig, V., Metelka, L., Danielová, V., *et al.* (2009) Vertical distribution of the tick *Ixodes ricinus* and tick-borne pathogens in the northern Moravian mountains correlated with climate warming (Jeseníky Mts., Czech Republic). *Central European Journal of Public Health* 17, 139–145.
de la Fuente, B., Estrada-Peña, A., Cabezas-Cruz, A. and Brey, A. (2015) Flying ticks: anciently evolved associations that constitute a risk of infectious disease spread. *Parasites & Vectors* 8, 538.
Delmore, K.E., van Doren, B., Convay, G., Curk, T., Garrido-Garduna, T., *et al.* (2020) Individual viability and versatility in an eco-evolutionary model of avian migration. *Proceedings of the Royal Society B* 278, 20201339.
Ellgren, H. (1993) Speed of migration and migratory flight lengths of passerine birds ringed during autumn migration in Sweden. *Ornis Scandinavica* 24, 220–228.
England, M.E., Phipps, P., Medlock, J.M., Atkinson, P.M., Atkinson, B., *et al.* (2016) *Hyalomma* ticks on northward migrating birds in southern Spain: implications for the risk of entry of Crimean–Congo haemorrhagic fever virus to Great Britain. *Journal of Vector Ecology* 41, 128–134.
Estrada-Peña, A., Martínez Avilés, M. and Muñoz Reoyo, M.J. (2011) A population model to describe the distribution and seasonal dynamics of the tick *Hyalomma marginatum* in the Mediterranean Basin. *Transboundary and Emerging Diseases* 58, 213–223.
Estrada-Peña, A., de la Fuente, J., Latapia, T. and Ortega, C. (2015) The impact of climate trends on a tick affecting public health: a retrospective modelling approach for *Hyalomma marginatum PLoS ONE* 10, e0125760.
Estrada-Peña, A., Mihalca, A.D. and Petney, T.N. (2017) *Ticks of Europe and North Africa*. Springer, Cham, Switzerland.

Eurostat (2020) Livestock population in numbers. Available at: https://ec.europa.eu/eurostat/web/products-eurostat-news/-/DDN-20200923-1?inheritRedirect=true&redirect=%2Feurostat%2Fnews%2Fwhats-new (accessed 18 June 2021).
Fiedler, W., Bairlein, F. and Köppen, U. (2006) Using large-scale data from ringed birds for the investigation of climate change on migrating birds: pitfalls and prospects. In: Moller, A.P., Fiedler, A. and Berthold, P (eds) *Birds and Climate Change*. Academic Press, London, pp. 49–67.
Fuller, R.J., Gaston, K.J. and Quine, C.P. (2007) Living on the edge: British and Irish woodland birds in a European context. *Ibis* 149, 56–63.
Gill, F., Donsker, D. and Rasmussen, P. (eds) (2020) *IOC World Bird List (v10.2)*. https://doi.org/10.14344/IOC.ML.10.2
Gilroy, J. and Lees, A.C. (2003) Vagrancy theories: are autumn vagrants really reverse migrants. *British Birds* 96, 427–438.
Gotelli, N.J. (1991) Metapopulation models: the rescue effect, propagule rain, and the core–satellite hypothesis. *The American Naturalist* 138, 768–776.
Grandi, G., Chitima-Dobler, L., Choklikitumnuey, P., Strube, A.S. and Albhin, A (2020) First records of adult *Hyalomma marginatum* and *H. rufipes* ticks (Acari: Ixodidae) in Sweden. *Ticks and Tick-borne Diseases* 11, 101403.
Gray, J.S., Dautel, H., Estrada-Peña, A., Kahl, O. and Lindgren, E. (2009) Effects of climate change on ticks and tick-borne diseases. *Interdisciplinary Perspectives on Infectious Diseases* 2009, 593232.
Hahn, S., Bauer, S. and Liechti, F. (2009) The natural link between Europe and Africa – 2.1 billion birds on migration. *Oikos* 118, 624–626.
Hanski, I. (1998) Metapopulation dynamics. *Nature* 395, 41–49.
Hasle, G. (2013) Transport of Ixodid ticks and tick-borne pathogens by migratory birds. *Frontiers in Cellular and Infection Microbiology* 3, 48.
Healy, J.A.E. and Bourke, P. (2004) Field evidence of aggregating behaviour in the tick *Ixodes ricinus* L. *Acarologia XLIV*, 3–13.
Healy, J.A.E. and Bourke, P. (2008) Aggregation of the tick *Ixodes ricinus* (Acari: Ixodidae): use and reuse of questing points. *Journal of Medical Entomology* 45, 222–228.
Healy, K., Guillerme, T., Finlay, S., Kane, A., Kelly, S.B.A., *et al.* (2014) Ecology and mode-of-life explains life span variation in birds and mammals. *Proceedings of the Royal Society B* 281, 2140298.
Hedenstrom, A. (2008) Adaptations to migration in birds: behavioural strategies, morphology and scaling effects. *Proceedings of the Royal Society B* 363, 287–294.
Hertel, A.G., Niemelä, P.T., Dingemanse, N.S. and Mueller, T. (2020) A guide for studying among individual behavioural variation from movement data in the wild. *Movement Ecology* 8, 30.
Hewson, C.M., Thorup, K., Pearce-Higgins, J. and Atkinson, P.W. (2016) Population decline is linked to migration route in the common cuckoo. *Nature Communications* 17, 12296.
Heylen, D. and Matthysen, E. (2010) Contrasting detachment strategies in two congeneric ticks (Ixodidae) parasitizing the same songbird. *Parasitology* 137, 661–667.
Holt, C., Fuller, R.J. and Dolman, P.M. (2011) Breeding and post-breeding responses of woodland birds to modification of habitat structure by deer. *Biological Conservation* 144, 2151–2162.
Hoogstraal, H. (1979) The epidemiology of tick-borne Crimean-Congo haemorrhagic fever in Asia, Europe and Africa. *Journal of Medical Entomology* 15, 307–417.
Hoogstraal, H., Kaiser, M.N., Traylor, M.L., Gaber, S. and Guindy, E. (1961) Ticks (Ixodoidea) on birds migrating from Africa to Europe and Asia. *Bulletin of the World Health Organization* 24, 197-212.
Hoogstraal, H., Kaiser, M.N., Traylor, M.L., Gaber, S. and Guindy, E, (1963) Ticks (Ixodoidea) on birds migrating from Europe and Asia to Africa. *Bulletin of the World Health Organization* 28, 235–262.
Hornok, S. and Horvath, G. (2012) First report of adult *Hyalomma marginatum rufipes* (vector of Crimean-Congo haemorrhagic fever virus) on cattle under a continental climate in Hungary. *Parasites & Vectors* 5, 170.
Hubálek, Z., Sedláček, P., Estrada-Peña, A., Vojtíšek, J. and Rudolf, I. (2020) First record of *Hyalomma rufipes* in the Czech Republic, with a review of relevant cases in other parts of Europe. *Ticks and Tick-borne Diseases* 11, 101421.
Hvidsten, D., Frafjord, K., Gray, J.S., Henningsson, A.J., Jenkins, A., *et al.* (2020) The distribution limit of the common tick, *Ixodes ricinus*, and some associated pathogens in north-western Europe. *Ticks and Tick-borne Diseases* 11, 101388.
Jääskeläinen, A.E., Tonteri, E., Sironen, T., Pakarinen, L., Vaheri, A. and Vapalahti, O. (2017) European subtype tick-borne encephalitis virus in *Ixodes persulcatus* ticks. *Emerging Infectious Diseases* 17, 323–325.

Jaenson, T.G.T., Jaenson, D.G., Eisen, L., Petersson, E. and Lindgren, E. (2012) Changes in the geographical distribution and abundance of the tick *Ixodes ricinus* during the past 30 years in Sweden. *Parasites & Vectors* 5, 8.

Jaenson, T.G.T., Värv, K., Fröjdman, I. Jaaaselainen, R., Rundgren, K., *et al.* (2016) First evidence of established populations of the taiga tick *Ixodes persulcatus* (Acari: Ixodidae) in Sweden. *Parasites & Vectors* 9, 317.

Jenni, L. and Winkler, R. (2020) *The Biology of Moult in Birds*. Croom Helm, London.

Kada, S., McCoy, K. and Boulinier, T. (2017) Impact of stage-dependent dispersal on the colonization dynamics of host patches by ticks and tick-borne pathogens. *Parasites & Vectors* 10, 375.

Kaiser, P. (2010) Advances in avian immunology – prospects for disease control: a review. *Avian Pathology* 39, 309–324.

Kalachev, L., Kelly, T.C., O'Callaghan, M.J.A., Pokrovskii, A.V. and Pokrovksy, A.V. (2011) Analysis of threshold-type behaviour in mathematical models of the intrusion of a novel macroparasite in a host colony. *Mathematical Medicine and Biology* 28, 287–333.

Kays, R., Crofoot, M.C., Jetz, W. and Wikelski, M. (2015) Terrestrial animal tracking as an eye on life and the planet. *Science* 348, 1255642.

Knudsen, E., Lindén, A., Both, C., Jonzén, N., Pulido, F., *et al.* (2011) Challenging claims in the study of migratory birds and climate change. *Biological Reviews of the Cambridge Philosophical Society* 86, 928–946.

Laaksonen, M., Sajanti, E., Sormunen, J.J., Penttinen, R., Hänninen, J., *et al.* (2017) Crowdsourcing-based nationwide tick collection reveals the distribution of *Ixodes ricinus*, *I. persulcatus* and associated pathogens in Finland. *Emerging Microbes & Infections* 6, e31.

Lack, D.L. (1954) *The Natural Regulation of Animal Numbers*. Clarendon Press, Oxford.

Leighton, P.A., Koffi, J.K., Pelcat, Y., Lindsay, L.R. and Ogden, N.A (2012) Predicting the speed of tick invasion: an empirical model of range expansion for the Lyme disease vector *Ixodes scapularis* in Canada. *Journal of Applied Ecology* 49, 457–464.

Levi, T., Keesing, F., Oggenfuss, K. and Ostfeld, R.S. (2015) Accelerated phenology of blacklegged ticks under climate warming. *Transactions of the Royal Society B* 370, 20130556.

Lindgren, E., Tälleklint, L. and Polfeldt, T. (2000) Impact of climate change on the northern latitude limit and population density of the disease-transmitting European tick *Ixodes ricinus*. *Environmental Health Perspectives* 108, 119–123.

López-López, P. (2016) Individual-based tracking systems in ornithology: welcome to the era of big data. *Ardeola* 63, 103–136.

MacArthur, R. (1972) *Geographical Ecology: Patterns in the Distribution of Species*. Harper and Row, New York.

MacArthur, R.H. and Wilson, E.O. (1967) *The Theory of Island Biogeography. Monographs in Population Biology*, vol. 1. Princeton University Press, Princeton, New Jersey.

Maniola, I., Bradaric, N., Groenland, R., Fijn, R., Bouten, W. and Shamoun-Baranes, J. (2020) Associations of synoptic weather conditions with nocturnal bird migration over the North Sea. *Frontiers in Ecology and Evolution* 8, 542438.

McGinley, L., Hansford, K.M., Cull, B., Gillingham, E.L., Carter, D.P., *et al.* (2021) First report of exposure to *Hyalomma marginatum* in England: further evidence of a *Hyalomma* moulting event in north-western Europe? *Ticks and Tick-borne Diseases* 12, 10154.

Medlock, J.M., Hansford, K., Bormane, A., Derdakova, M., Estrada-Peña, A., *et al.* (2013) Driving forces for changes in geographical distribution of *Ixodes ricinus* ticks. *Parasites & Vectors* 6, 1.

Milner, J.M., Bonenfant, C., Mysterud, A., Gaillard, J.-M., Csanyi, S. and Stenseth, N.S. (2006) Temporal and spatial development of red deer harvesting in Europe: biological and cultural factors. *Journal of Applied Ecology* 43, 721–734.

Møller, A.P., Fiedler, A, and Berthold, P. (2006) *Birds and Climate Change*. Academic Press, London.

Moreau, R.E. (1966) *The Bird Faunas of Africa and Its Islands*. Academic Press, London.

Moreau, R.E. (1972) *The Palaearctic–African Bird Migration Systems*. Academic Press, London.

Nabi, G., Wang, Y., Lu, L., Jiang, C., Ahmad, S., *et al.* (2021) Birds and bats as viral reservoirs: a physiological and ecological perspective. *Science of the Total Environment* 754, 142323.

Newton, I. (2008) *The Migration Ecology of Birds*. Academic Press, London.

Newton, I. (2010) *Bird Migration*. HarperCollins, London.

Newton, I. (2011) Migration within the annual cycle: species, sex and age differences. *Journal of Ornithology* 152, 169–185.

Nicholson, W.L., Sonenshine, D., Noden, B.H. and Brown, R.N. (2019) Ticks (Ixodida). In: Mullen, G.R. and Durden, L.A. (eds) *Medical and Veterinary Entomology*. Academic Press/Elsevier, London, pp. 603–672.

Nilsson, C., Klassen, R.H. and Alerstam, T. (2013) Differences in speed and duration of bird migration between spring and autumn. *The American Naturalist* 181, 837–845.

Ostfeld, R.S. and Brunner, J.L. (2015) Climate change and Ixodes tick-borne diseases of humans. *Philosophical Transactions of the Royal Society B* 370, 20140051.

Owen, J.P., Nelson, A.C. and Clayton, D.H. (2010) Ecological immunology of bird–ectoparasite systems. *Trends in Parasitology* 26, 530–539.

Paradis, E., Baillie, S.R., Sutherland, W.J. and Gregory, R.D. (1998) Patterns of natal and breeding dispersal in birds. *Journal of Animal Ecology* 67, 518–536.

Pfeifer, R., Stadler, J. and Brandi, R. (2007) Birds from the Far East in Central Europe: a test of the reverse migration hypothesis. *Journal of Ornithology* 148, 379–385.

Rabøl, J. (1969) Reversed migration as the cause of westward vagrancy in four *Phylloscopus* warblers. *British Birds* 62, 89–92.

Ramos, R., Garnier, R., Gonalez-Solas, J. and Boulinier, T. (2014) Long antibody persistence and transgenerational transfer of immunity in a long-lived seabird. *The American Naturalist* 184, 764–776.

Randolph, S.E. (2004) Evidence that climate change has caused the 'emergence' of tick-borne diseases in Europe. *International Journal of Medical Microbiology* 293, 5–15.

Randolph, S.E. (2010) Human activities predominate in determining changing evidence of tick-borne zoonoses in Europe. *Eurosurveillance* 15, 19606.

Randolph, S.E., Gern, L. and Nuttall, P.A. (1996) Co-feeding ticks: epidemiological significance for tick-borne pathogen transmission. *Parasitology Today* 12, 472–479.

Reed, K.D., Meece, J.K., Henkel, J.S. and Shukla, S. (2003) Birds, migration, and emerging zoonoses: West Nile fever virus, Lyme disease, influenza A and enteropathogens. *Clinical Medicine & Research* 1, 5–12.

Roth, O., Beemelmanns, A., Barnbeau, S.M. and Sadd, B.M. (2018) Recent advances in vertebrate and invertebrate transgenerational immunity in the light of ecology and evolution. *Heredity* 121, 225–238.

Santos-Silva, M.M. and Vatansever, Z. (2017) *Hyalomma marginatum* Koch 1844. In: Estrada-Peña, A., Mihalca, A.D. and Petney, T.N. (eds) *Ticks of Europe and North Africa*. Springer, Cham, Switzerland, pp. 349–354.

Saura, S., Bodin, O. and Fortin, M.-S. (2014) Stepping stones are crucial for species' long distance dispersal and range expansion through habitat networks. *Journal of Applied Ecology* 51, 171–182.

Schmaljohann, H. (2018) Proximate mechanisms affecting seasonal differences in the migration speed of avian species. *Scientific Reports* 8, 4016.

Scott, J.D., Clark, K.L., Foley, J.E., Bierman, B.C. and Durdon, L.A. (2018) Far-reaching dispersal of *Borrelia burgdorferi* sensu lato – infected blacklegged ticks by migratory songbirds in Canada. *Healthcare* 6, 89.

Semenza, J.C. and Suk, J.E. (2018) Vector-borne diseases and climate change: a European perspective. *FEMS Microbiology Letters* 365, fnx244.

Shamoun-Baranes, J., Liechti, F. and Vansteelant, W.M.G. (2017) Atmospheric conditions create freeways, detours and tailbacks for migrating birds. *Journal of Comparative Physiology* 203, 509–529.

Sirotkin, M.B. and Korenberg, E.I. (2018) Influences of abiotic factors on different developmental stages of the taiga tick *Ixodes persulcatus* and the sheep tick *Ixodes ricinus*. *Entomological Review* 98, 496–513.

Someville, M., Wikelski, M., Beyer, R.M., Rodrigues, A.S.L., Manica, A. and Jetz, W. (2020) Simulation-based reconstruction of global bird migration over the past 50,000 years. *Nature Communications* 11, 801.

Strandberg, R., Klassen, R.H.G., Hake, M. and Alerstam, T. (2010) How hazardous is the Sahara Desert crossing for migratory birds? Indications from satellite tracking of raptors. *Biology Letters* 6, 297–300.

Thomas, N.J., Hunter, D.B. and Atkinson, C.T. (2007) *Infectious Diseases of Wild Birds*. Blackwell, Oxford.

Thorup, K. (2004) Reverse migration as a cause of vagrancy. *Bird Study* 51, 228–238.

Toma, L., Mancuso, E., d'Alessio, S.G., Menegon, M., Spina, F., *et al.* (2021) Tick species from Africa by migratory birds: a 3-year study in Italy. *Experimental and Applied Acarology* 83, 147–164.

Troughton, D.E. and Levin, M.E. (2007) Life cycles of seven Ixodid tick species (Acari: Ixodidiae) under standardised laboratory conditions. *Journal of Medical Entomology* 44, 732–740.

Vatansever, Z. (2017) *Hyalomma rufipes* Koch 1844. In: Estrada-Peña, A., Mihalca, A.D. and Petney, T.N. (eds) *Ticks of Europe and North Africa*. Springer, Cham, Switzerland, pp. 355–359.

Vega, M.L., Willemoes, M., Thomson, R.L., Tolvanen, J., Rutila, J., *et al.* (2016) First time migration in juvenile common cuckoos documented by satellite tracking. *PLoS ONE* 11, e0168940.

Vial, L. (2009) Biological and ecological characteristics of soft ticks (Ixodida: Argasdae) and their impacts for predicting tick and associated disease distribution. *Parasite* 16, 191–202.

Vial, L., Stachurski, F., Leblond, A., Huber, K., Vourc'h, G., *et al.* (2016) Strong evidence of the tick *Hyalomma marginatum* Koch 1844 in southern continental France. *Ticks and Tick-borne Diseases* 7, 1162–1167.

Visser, M.E., Perdeck, A.C., van Balen, J.H. and Both, C. (2009) Climate change leads to decreasing bird migration distances. *Global Change Biology* 15, 1859–1865.

Wallmenius, K., Barboutis, C., Fransson, T., Jaenson, T.G.T., Lingren, P.-E., *et al.* (2014) Spotted fever *Rickettsia* species in *Hyalomma* and *Ixodes* ticks infesting migratory birds in the European Mediterranean area. *Parasites & Vectors* 7, 318.

Wernham, C., Toms, M., Marchant, J., Clark, J., Siriwardina, G. and Baillie, S. (2002) *The Migration Atlas: Movement of Birds of Britain and Ireland*. Poyser, London.

Williamson, K. (1954) Grey-cheeked thrush at Fair Isle: a new British bird. *British Birds* XLVII, 266–268.

Winstanley, D.R., Spencer, R.G. and Williamson, K. (1974) Where have all the whitethroats gone? *Bird Study* 24, 1–14.

Wint, W., Morley, D.W., Searle, K.R. and Alexander, N.S. (2015). A first attempt at modelling roe deer (*Capreolus capreolus*) distributions in Europe. *Journal of Open Public Health Data* 2, p.e2.

Zeman, P. and Benes, C. (2004) A tick-borne encephalitis ceiling in Central Europe has moved upwards during the last 30 years: possible impact of global warming? *International Journal of Medical Microbiology* 293, 48–54.

Zink, R.M. (2011) The evolution of avian migration. *Biological Journal of the Linnean Society* 104, 237–250.

Zurrell, D., Graham, C.H., Gallien, L., Thullier, W. and Zimmerman, E. (2018) Long-distance migratory birds threatened by multiple independent risks from global change. *Nature Climate Change* 8, 992–996.

eo16 How Tick Vectors are Coping with Global Warming

André Wilke[1], John Beier[1], Domenico Otranto[2,3] and Giovanni Benelli[4*]
[1]*University of Miami, Miami, Florida, USA;* [2]*Università degli Studi di Bari, Bari, Italy;* [3]*Bu-Ali Sina University, Hamedan, Iran;* [4]*University of Pisa, Pisa, Italy*

16.1 Expert Opinion

16.1.1 Introduction

Arthropod vectors transmit some of the most important infectious diseases that affect both humans and animals. Approximately three billion people presently live in endemic regions and are vulnerable to arthropod-borne pathogens, which are directly responsible for the deaths of millions of people each year, contributing to numerous long-term illnesses and lifelong sequelae (Kleber de Oliveira *et al.*, 2016; Lorenz *et al.*, 2017; Rosenberg *et al.*, 2018).

16.1.2 Ticks, tick-borne diseases

Ticks comprise an important group of vectors, being responsible for the transmission of many pathogenic agents (Bowman and Nuttall, 2008). They are obligatory ectoparasites (Di Giovanni *et al.*, 2021) and feed on the blood of a wide variety of animal species, including humans (Black and Piesman, 1994; Mans *et al.*, 2008; Mendoza-Roldan *et al.*, 2020a,b). Ticks are organized in three families, Argasidae, Ixodidae and Nuttalliellidae, and act as key vectors for many flaviviruses (Dantas-Torres *et al.*, 2012; Guglielmone *et al.*, 2014), as well as *Rickettsia rickettsii* and *Borrelia burgdorferi* sensu lato (s.l.), the aetiological agents of Rocky Mountain fever and Lyme disease, respectively (Black and Piesman, 1994; Jongejan and Uilenberg, 2005; Süss *et al.*, 2008; Eisen, 2020). For example, between the years 2000 and 2010, over 300,000 cases of Lyme disease have been reported in the USA and Europe (Rosenberg *et al.*, 2018; Semenza and Suk, 2018).

16.1.3 Epidemiology of tick-borne diseases

There is a paucity of knowledge on the epidemiological importance of several tick species and their impact on the health of humans and animals is still poorly understood (McMichael *et al.*, 2006; Dantas-Torres *et al.*, 2012; Benelli, 2020). The prevalence of ticks is positively associated with wildlife prevalence (Medlock *et al.*, 2013). The recent increase in the numbers of human interactions with wildlife, either in recreational activities or due to the expansion of urban areas into natural areas, has directly impacted on the incidence of tick-borne diseases (TBDs) (Randolph and Rogers, 2000; Süss *et al.*, 2008; Gilbert, 2010). Furthermore, global warming contributes to warmer winters and hastens the start of

*Email: giovanni.benelli@unipi.it

DOI: 10.1079/9781789249637.0016

spring, which may be a factor that increases the contact between some tick species (e.g. *Rhipicephalus sanguineus* s.l.) and animal hosts. In addition, the rise in temperatures also affects tick behaviour, making ticks more active in their search for hosts. There is a clear association between the rise in temperature and the increase in incidence of some TBDs, such as those vectored by *Ixodes ricinus* (Parola *et al.*, 2008; Porretta *et al.*, 2013).

16.1.4 Effects of global warming in the epidemiology of tick-borne diseases

Global warming can have unpredictable effects on tick species that are not anthropophilic. These species can be driven by the more conducive warmer temperatures, becoming more active, thus increasing their host-seeking efforts (Wilke *et al.*, 2019). That is the case of *Ixodes scapularis* nymphs. Under normal circumstances, *I. scapularis* rarely seeks human hosts but warmer and more humid conditions can shift their behaviour, increasing their host-seeking activity and, therefore, increasing their contact with humans (Brownstein *et al.*, 2005).

Global warming would make it possible for ticks to invade and colonize new areas that were previously inhospitable (Capelli *et al.*, 2012; Bett *et al.*, 2019; Wilke *et al.*, 2019); for example, the rise in tick-borne encephalitis (TBE) cases in the north European subarctic zone of Russia (Arkhangelsk Oblast federal subject), which has been related to the northward expansion of *Ixodes persulcatus*, the main TBE vector, in this region (Tokarevich *et al.*, 2011). Suitable habitats for ticks are expected to double in area in North America and Europe by 2080 (Molaei *et al.*, 2019). Canada and Northern Europe will inevitably be more conducive to TBD transmission in the near future and the current predictions indicate a substantial increase in the incidence of TBDs in those regions (Sagurova *et al.*, 2019; Grandi *et al.*, 2020).

In addition, the increase in external temperature may contribute to reducing the pathogen pre-activation processes inside the tick vector, such as in the case of *Rhipicephalus appendiculatus* transmitting *Theileria parva*, the agent of East Coast fever, under controlled conditions (Otranto, 2018). Moreover, the increase in conducive conditions and suitable areas for tick proliferation is expected to become more abundant in Southern Europe (Lindgren *et al.*, 2000; Estrada-Peña, 2001; Zeman and Beneš, 2004; Brownstein *et al.*, 2005; Hartelt *et al.*, 2008). Ticks may become more abundant at higher altitudes as the environmental conditions become more suitable (Gilbert, 2010; Lukan *et al.*, 2010).

16.1.5 Impact of global warming in the range of tick species

The increase in the range of ticks to new areas could contribute to major changes in the future patterns of TBD transmission (Bouchard *et al.*, 2019). However, predictive models indicate that global warming will have the opposite impact on areas located at lower latitudes or that are predicted to become drier. A decrease in suitable areas for tick development and proliferation is expected in the southern USA (Brownstein *et al.*, 2005; see also Wilke *et al.*, 2019). Furthermore, Estrada-Peña *et al.* (2003) predicted that increasing the temperature by 2°C will chiefly contribute to the destruction of habitats of four major tick vectors in South Africa, i.e. *Boophilus decoloratus*, *Amblyomma hebraeum*, *R. appendiculatus* and *Hyalomma truncatum*. On the other hand, an increase of 1°C was predicted as responsible for a small but positive response in *A. hebraeum* and *B. decoloratus* (Estrada-Peña *et al.*, 2003).

Asia is also plagued by ticks, with 117 species being found in China alone (Chen *et al.*, 2010). Recent studies are pointing out an increase in tick occurrence records and a subsequent increase in their distribution range (Zhang *et al.*, 2019). The expansion in the range of tick vectors in Asia is of great public health concern since they are vectors of many arboviruses in the region, including severe fever with thrombocytopaenia syndrome virus, potentially endangering billions of people (Luo *et al.*, 2015; Zhuang *et al.*, 2018).

16.2 Future Predictions

Overall, many challenges will need to be overcome in the near future to prevent the increase in prevalence and incidence of TBDs. The availability of new suitable habitats for tick development in areas that were previously inhospitable can represent a major challenge in controlling tick populations and in preventing TBDs. Furthermore, migratory birds may rapidly introduce ticks to these newly suitable areas by carrying them while migrating (Buczek *et al.*, 2020). Of note, the range of bird dispersion and migration has greatly expanded in North America due to global warming (Roberts *et al.*, 2019). Such a regime shift can significantly impact tick dispersal and TBD transmission.

Disregarding the uncertainties from the predictive models in pointing out whether and to what extent ticks will invade new areas, or become more abundant and aggressive toward human hosts, systematic surveillance will play a pivotal role in the development of control strategies. Under the above circumstances, integrated vector management (WHO, 2004) is the most effective and efficient way to deal with the major public health concern of TBD transmission on a warming planet.

References

Benelli, G. (2020) Pathogens manipulating tick behavior – through a glass, darkly. *Pathogens* 9, 664. https://doi.org/10.3390/pathogens9080664

Bett, B., Lindahl, J. and Delia, G. (2019) Climate change and infectious livestock diseases: the case of Rift Valley fever and tick-borne diseases. In: Rosenstock, T.S., Nowak, A. and Girvetz, E. (eds) *The Climate-Smart Agriculture Papers*. Springer, Cham, Switzerland, pp. 29–37. https://doi.org/10.1007/978-3-319-92798-5_3

Black, W.C. and Piesman, J. (1994) Phylogeny of hard- and soft-tick taxa (Acari: Ixodida) based on mitochondrial 16S rDNA sequences. *Proceedings of the National Academy of Sciences USA* 91, 10034–10038. https://doi.org/10.1073/pnas.91.21.10034

Bouchard, C., Dibernardo, A., Koffi, J., Wood, H., Leighton, P. and Lindsay, L. (2019) Increased risk of tick-borne diseases with climate and environmental changes. *Canada Communicable Disease Report* 45, 83–89. https://doi.org/10.14745/ccdr.v45i04a02

Bowman, A.S. and Nuttall, P.A. (2008) *Ticks: Biology, Disease and Control*. Cambridge University Press, Cambridge.

Brownstein, J., Holford, T. and Fish, D. (2005) Effect of climate change on Lyme disease risk in North America. *EcoHealth* 2, 38–46. https://doi.org/10.1007/s10393-004-0139-x

Buczek, A.M., Buczek, W., Buczek, A. and Bartosik, K. (2020) The potential role of migratory birds in the rapid spread of ticks and tick-borne pathogens in the changing climatic and environmental conditions in Europe. *International Journal of Environmental Research and Public Health* 17, 2117. https://doi.org/10.3390/ijerph17062117

Capelli, G., Ravagnan, S., Montarsi, F., Ciocchetta, S., Cazzin, S., *et al.* (2012) Occurrence and identification of risk areas of *Ixodes ricinus*-borne pathogens: a cost-effectiveness analysis in north-eastern Italy. *Parasites & Vectors* 27, 61. https://doi.org/10.1186/1756-3305-5-61

Chen, Z., Yang, X., Bu, F., Yang, X., Yang, X. and Liu, J. (2010) Ticks (Acari: Ixodoidea: Argasidae, Ixodidae) of China. *Experimental and Applied Acarology* 51, 393–404. https://doi.org/10.1007/s10493-010-9335-2

Dantas-Torres, F., Chomel, B.B. and Otranto, D. (2012) Ticks and tick-borne diseases: a One Health perspective. *Trends in Parasitology* 28, 437–446. https://doi.org/10.1016/j.pt.2012.07.003

Di Giovanni, F., Wilke, A.B.B., Beier, J.C., Pombi, M., Mendoza-Roldan, J.A., *et al.* (2021) Parasitic strategies of arthropods of medical and veterinary importance. *Entomologia Generalis* (in press). https://doi.org/10.1127/entomologia/2021/1155

Eisen, L. (2020) Vector competence studies with hard ticks and *Borrelia burgdorferi* sensu lato spirochetes: a review. *Ticks and Tick-borne Diseases* 11, 101359. https://doi.org/10.1016/j.ttbdis.2019.101359

Estrada-Peña, A. (2001) Climate warming and changes in habitat suitability for *Boophilus microplus* (Acari: Ixodidae) in Central America. *Journal of Parasitology* 87, 978–987. https://doi.org/10.1645/0022-3395(2001)087[0978:CWACIH]2.0.CO;2

Estrada-Peña, A., Boomker, A. and Frederik, J.D. (2003) Climate change decreases habitat suitability for some tick species (Acari: Ixodidae) in South Africa. *Onderstepoort Journal of Veterinary Research* 70, 79–93.

Gilbert, L. (2010) Altitudinal patterns of tick and host abundance: a potential role for climate change in regulating tick-borne diseases? *Oecologia* 162, 217–225. https://doi.org/10.1007/s00442-009-1430-x

Grandi, G., Chitimia-Dobler, L., Choklikitumnuey, P., Strube, C., Springer, A., *et al.* (2020) First records of adult *Hyalomma marginatum* and *H. rufipes* ticks (Acari: Ixodidae) in Sweden. *Ticks and Tick-borne Diseases* 11, 101403. https://doi.org/10.1016/j.ttbdis.2020.101403

Guglielmone, A.A., Robbins, R.G., Apanaskevich, D.A., Petney, T.N., Estrada-Peña, A. and Horak, I.G. (2014) *The Hard Ticks of the World*. Springer, New York.

Hartelt, K., Pluta, S., Oehme, R. and Kimmig, P. (2008) Spread of ticks and tick-borne diseases in Germany due to global warming. *Parasitology Research* 103, 109–116. https://doi.org/10.1007/s00436-008-1059-4

Jongejan, F. and Uilenberg, G. (2005) The global importance of ticks. *Parasitology* 129, S3–S14. https://doi.org/10.1017/S0031182004005967

Kleber de Oliveira, W., Cortez-Escalante, J., De Oliveira, W.T.G.H., do Carmo, G.M.I., Henriques, C.M.P., *et al.* (2016) Increase in reported prevalence of microcephaly in infants born to women living in areas with confirmed Zika virus transmission during the first trimester of pregnancy – Brazil, 2015. *Morbidity and Mortality Weekly Report* 65, 242–247. https://doi.org/10.15585/mmwr.mm6509e2

Lindgren, E., Tälleklint, L. and Polfeldt, T. (2000) Impact of climatic change on the northern latitude limit and population density of the disease-transmitting European tick *Ixodes ricinus*. *Environmental Health Perspectives* 108, 119–123. https://doi.org/10.2307/3454509

Lorenz, C., Azevedo, T.S., Virginio, F., Aguiar, B.S., Chiaravalloti-Neto, F. and Suesdek, L. (2017) Impact of environmental factors on neglected emerging arboviral diseases. *PLoS Neglected Tropical Diseases* 11, e0005959. https://doi.org/10.1371/journal.pntd.0005959

Lukan, M., Bullova, E. and Petko, B. (2010) Climate warming and tick-borne encephalitis, Slovakia. *Emergent Infectious Diseases* 16, 524–526. https://doi.org/10.3201/eid1603.081364

Luo, L.M., Zhao, L., Wen, H.L., Zhang, Z.T., Liu, J.W., *et al.* (2015) Haemaphysalis longicornis ticks as reservoir and vector of severe fever with thrombocytopenia syndrome virus in China. *Emerging Infectious Diseases* 21, 1770–1776. https://doi.org/10.3201/eid2110.150126

Mans, B.J., Andersen, J.F., Francischetti, I.M.B., Valenzuela, J.G., Schwan, T.G., *et al.* (2008) Comparative sialomics between hard and soft ticks: implications for the evolution of blood-feeding behavior. *Insect Biochemical Molecular Biology* 38, 42–58. https://doi.org/10.1016/j.ibmb.2007.09.003

McMichael, A.J., Woodruff, R.E. and Hales, S. (2006) Climate change and human health: present and future risks. *The Lancet* 367, 859–869. https://doi.org/10.1016/S0140-6736(06)68079-3

Medlock, J.M., Hansford, K.M., Bormane, A., Derdakova, M., Estrada-Peña, A., *et al.* (2013) Driving forces for changes in geographical distribution of *Ixodes ricinus* ticks in Europe. *Parasites & Vectors* 6, 1. https://doi.org/10.1186/1756-3305-6-1

Mendoza-Roldan, J., Modry, D. and Otranto, D. (2020a) Zoonotic parasites of reptiles: a crawling threat. *Trends in Parasitology* 36, 677–687. https://doi.org/10.1016/j.pt.2020.04.014

Mendoza-Roldan, J., Ribeiro, S.R., Castilho-Onofrio, V., Grazziotin, F.G., Rocha, B., *et al.* (2020b) Mites and ticks of reptiles and amphibians in Brazil. *Acta Tropica* 208, 105515. https://doi.org/10.1016/j.actatropica.2020.105515

Molaei, G., Little, E.A.H., Williams, S.C. and Stafford, K.C. (2019) Bracing for the worst – range expansion of the lone star tick in the Northeastern United States. *New England Journal of Medicine* 381, 2189–2192. https://doi.org/10.1056/NEJMp1911661

Otranto, D. (2018) Arthropod-borne pathogens of dogs and cats: from pathways and times of transmission to disease control. *Veterinary Parasitology* 251, 68–77. https://doi.org/10.1016/j.vetpar.2017.12.021

Parola, P., Socolovschi, C., Jeanjean, L., Bitam, I., Fournier, P.E., *et al.* (2008) Warmer weather linked to tick attack and emergence of severe rickettsioses. *PLoS Neglected Tropical Diseases* 2, e338. https://doi.org/10.1371/journal.pntd.0000338

Porretta, D., Mastrantonio, V., Amendolia, S., Gaiarsa, S., Epis, S., *et al.* (2013) Effects of global changes on the climatic niche of the tick *Ixodes ricinus* inferred by species distribution modelling. *Parasites & Vectors* 6, 271. https://doi.org/10.1186/1756-3305-6-271

Randolph, S.E. and Rogers, D.J. (2000) Fragile transmission cycles of tick-borne encephalitis virus may be disrupted by predicted climate change. *Proceedings of the Royal Society B* 267, 1741–1744. https://doi.org/10.1098/rspb.2000.1204

Roberts, C.P., Allen, C.R., Angeler, D.G. and Twidwell, D. (2019) Shifting avian spatial regimes in a changing climate. *Nature Climate Change* 9, 562–566. https://doi.org/10.1038/s41558-019-0517-6

Rosenberg, R., Lindsey, N.P., Fischer, M., Gregory, C.J., Hinckley, A.F., *et al.* (2018) Vital Signs: trends in reported vectorborne disease cases – United States and Territories, 2004–2016. *Morbidity and Mortality Weekly Report* 67, 496–501. https://doi.org/10.15585/mmwr.mm6717e1

Sagurova, I., Ludwig, A., Ogden, N.H., Pelcat, Y., Dueymes, G. and Gachon, P. (2019) Predicted northward expansion of the geographic range of the tick vector *Amblyomma americanum* in North America under future climate conditions. *Environmental Health Perspectives* 127, 107014. https://doi.org/10.1289/EHP5668

Semenza, J.C. and Suk, J.E. (2018) Vector-borne diseases and climate change: a European perspective. *FEMS Microbiology Letters* 365, fnx244. https://doi.org/10.1093/femsle/fnx244

Süss, J., Klaus, C., Gerstengarbe, F.W. and Werner, P.C. (2008) What makes ticks tick? Climate change, ticks, and tick-borne diseases. *Journal of Travel Medicine* 15, 39–45. https://doi.org/10.1111/j.1708-8305.2007.00176.x

Tokarevich, N.K., Tronin, A.A., Blinova, O.V., Buzinov, R.V., Boltenkov, V.P., *et al.* (2011) The impact of climate change on the expansion of *Ixodes persulcatus* habitat and the incidence of tick-borne encephalitis in the north of European Russia. *Global Health Action* 4, 8448. https://doi.org/10.3402/gha.v4i0.8448

WHO (2004) *Global Strategic Framework for Integrated Vector Management (No. WHO/CDS/CPE/PVC/2004.10)*. World Health Organization, Geneva, Switzerland.

Wilke, A.B., Beier, J.C. and Benelli, G. (2019) Complexity of the relationship between global warming and urbanization – an obscure future for predicting increases in vector-borne infectious diseases. *Current Opinion in Insect Science* 35, 1–9. https://doi.org/10.1016/j.cois.2019.06.002

Zeman, P. and Beneš, C. (2004) A tick-borne encephalitis ceiling in Central Europe has moved upwards during the last 30 years: possible impact of global warming? *International Journal of Medical Microbiology Supplements* 293, 48–54. https://doi.org/10.1016/S1433-1128(04)80008-1

Zhang, G., Zheng, D., Tian, Y. and Li, S. (2019) A dataset of distribution and diversity of ticks in China. *Science Data* 6, 105. https://doi.org/10.1038/s41597-019-0115-5

Zhuang, L., Sun, Y., Cui, X.M., Tang, F., Hu, J.G., *et al.* (2018) Transmission of severe fever with thrombocytopenia syndrome virus by *Haemaphysalis longicornis* ticks, China. *Emerging Infectious Diseases* 24, 868–871. https://doi.org/10.3201/eid2405.151435

eo17 Possible Direct and Human-Mediated Impact of Climate Change on Tick Populations in Turkey

Sırrı Kar[1] and Aysen Gargili Keles[2*]
[1]*Tekirdağ Namık Kemal University, Tekirdağ, Turkey;* [2]*Marmara University, Istanbul, Turkey*

17.1 Expert Opinion

17.1.1 Introduction: basic geographical, ecological and climate characteristics of Turkey

Turkey (783.562 km^2) is geographically bordered by the Middle East, the Caucasus, the Black Sea, the Marmara Sea, the Aegean Sea and the Mediterranean Sea, and the climatic characteristics in the regions are mainly influenced by these neighbouring seas and lands. High and mostly continuous mountains run parallel to the coast along the Black Sea and Mediterranean coasts; as a result of this natural barrier, these two marine climates dominate the coasts, and the continental climate is predominant mainly in the interior. Turkey is located in the Mediterranean macroclimate zone, between temperate and subtropical zones. As a result of its geographical location and topography, it is possible to encounter a wide range of regional and seasonal variations throughout the country (https://www.csb.gov.tr/, https://www.mgm.gov.tr/, accessed 1 June 2021). Three main climate types (18% B/dry, 43% C/temperate and 39% D/continental) and at least ten different sub-climate types are seen according to the Köppen–Geiger climate classification (Fig. 17.1a). Average temperatures of B, C and D zones are 12.4, 14.2 and 9.5°C, respectively. Average annual maximum temperatures range from 10 to 26°C, and the values drop markedly from the south-west to the north-east direction (Fig. 17.1b) (Aydin *et al.*, 2019). Average annual total precipitation changes are between 412 and 575 mm in the three main climate zones (Ozturk *et al.*, 2017) and mostly between 283 and 1252 mm in sub-climate areas (Fig. 17.1c).

Average summer temperature is 30–35°C on the Mediterranean coastline and in south-east Anatolia, while the coastline in the south and west generally hovers around 10–15°C in winter. Average temperature in east Anatolia is about −9 and 18°C in winter and summer, respectively (https://www.mgm.gov.tr/, accessed 1 June 2021). Sub-climatic types of different character can be observed in areas quite close to each other. For example, even in Thrace with a relatively small surface area (approximately 24 km^2), four climatic subtypes can be observed (Peel *et al.*, 2007).

*Email: agargili@yahoo.com

DOI: 10.1079/9781789249637.0017

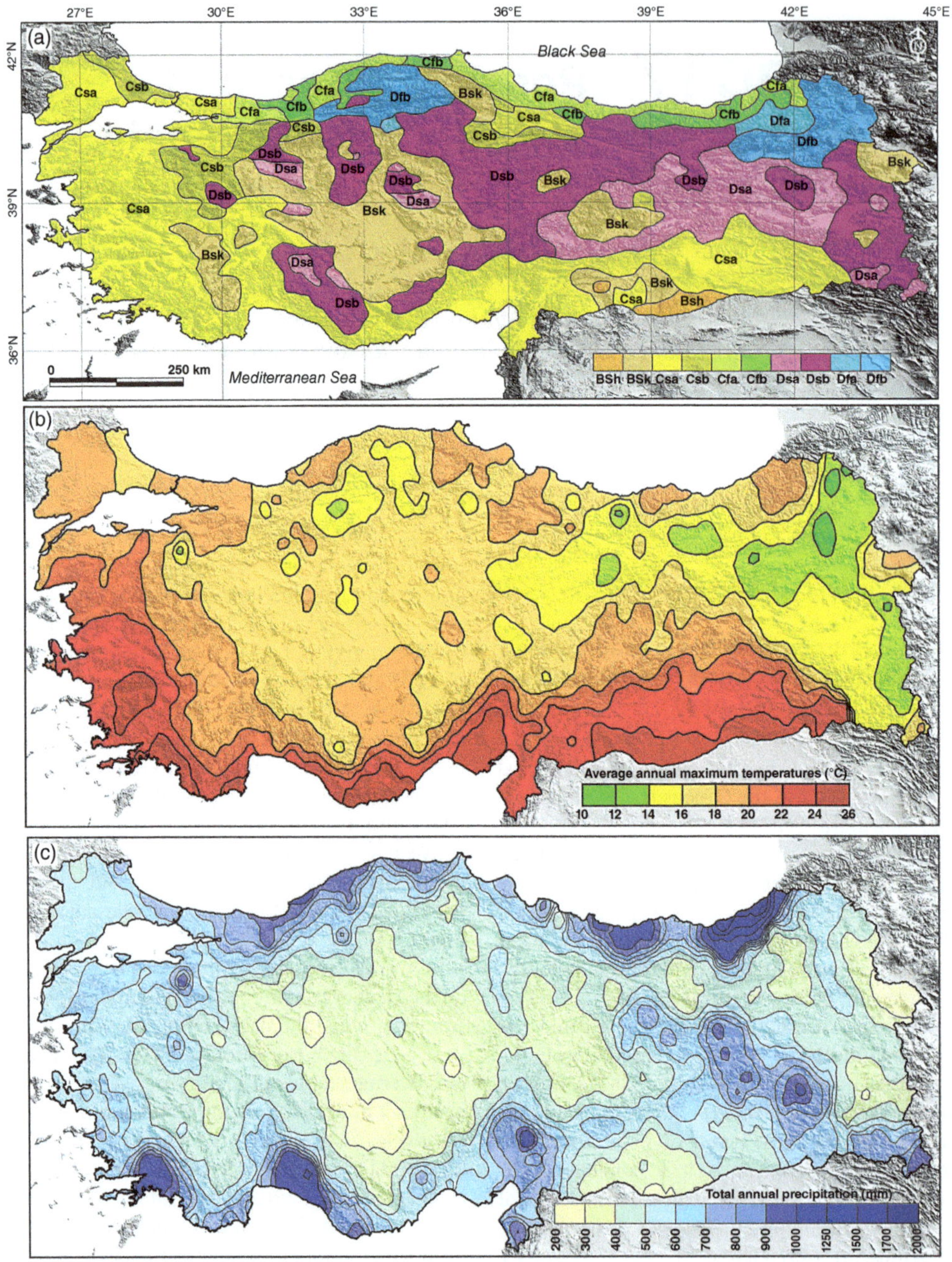

Fig. 17.1. Current climate characteristics of Turkey: (a) Köppen–Geiger sub-climate types (from Ozturk *et al*., 2017); (b) spatial distribution of long-term average annual maximum temperatures; (c) total rainfall (from Aydin *et al*., 2019). BSh, hot arid steppe climate; BSk, cold arid steppe climate; Csa, hot dry summer climate; Csb, warm dry summer climate; Cfa, hot without dry summer climate; Cfb, warm without dry summer climate; Dsa, hot dry summer continental climate; Dsb, warm dry summer continental climate. (Figures used with permission from cited authors.)

17.1.2 General information on tick fauna in Turkey

Dozens of tick species have been reported in Turkey to date (Aydin and Bakirci, 2007), with about 30 of them standing out in terms of transmission of human and animal pathogens. The main reasons for the richness and diversity of the tick fauna are the characteristics of climatic conditions, geographical location, relatively high number of total and grazing livestock, high number of stray dogs (estimated as >100,000 in Istanbul alone) and cats, and natural habitat and wildlife that still exist to varying degrees, especially in mountainous areas. Turkey, with its central location, is one of the most critical terrestrial transit routes between Europe, Asia and Africa. Furthermore, it is on the Black Sea Flyway of wild birds, which runs along South Africa, Madagascar, the eastern part of sub-Saharan Africa, the Rift Valley, the Nile Delta, the Middle East, Anatolia, Europe and the northern Black Sea. This flyway is used by millions of birds of different species (Kirwan *et al.*, 2014; Najdenski *et al.*, 2018). They carry several tick species, such as *Hyalomma marginatum*, *Hyalomma aegyptium*, *Hyalomma rufipes*, *Haemaphysalis punctata* and *Ixodes ricinus*, in both directions (Hoogstraal *et al.*, 1961, 1963).

Turkey, of course, receives its share of the potential consequences of global climate change, such as increase in average total temperature, decrease in total precipitation and humidity, and increase in the magnitude and frequency of extreme weather events. However, the impacts will not be the same in every region of the world, e.g. some dry/hot areas are estimated to get higher precipitation (Bett *et al.*, 2019). In the Eastern World, including Turkey, temperature values and some other characteristics of the climate are predicted to shift in a northerly direction and result in a mostly drier subtropical southern hemisphere (Pontes *et al.*, 2020). The severity, or at least the rate of occurrence, of this impact seems to vary from region to region in Turkey, because of its mosaic-like features in terms of ecology, topography and climate. In any case, most Turkish ticks adapted to subtropical conditions will not be positively affected by changes in climate, and the expected permanency and speed of climate change impacts will make effective physiological and/or genetic adaptation difficult. On the other hand, shifting of the climate character northwards may give invasion opportunities to new tick genotypes or species (Bett *et al.*, 2019) carried from Africa and the Middle East via the Black Sea Flyway of wild birds. However, it is difficult to predict how these ticks will be affected by the drivers other than temperature, such as extreme meteorological events and local host demography, and how interspecific or intergenotypic their interaction with existing ticks will be. Furthermore, possible changes in the migration phenology of wild birds due to changes in the ecosystem, whether or not causally related to climate (Calvert *et al.*, 2009), may result in differences in the features of predicted tick invasion in Turkey.

In addition to meteorological factors, the condition of a tick population in a region depends on many other factors such as human behaviour, vegetation and presence of attainable hosts; there is mostly a causal relationship among these drivers. Due to this multifactorial intricate relationship, making a widely overarching prediction about the future of a general tick population is difficult. Furthermore, the response of ticks to these parameters can vary significantly according to the species (Dantas-Torres, 2015; Bett *et al.*, 2019). The resistance of some ticks to certain meteorological changes can be high. However, predicted extreme meteorological events such as heavy rainfall and prolonged drought will of course be profound for almost all exophilic tick species in the subtropics and in many other regions. Human-mediated impacts, such as deforestation, qualitative and quantitative losses in natural vegetation and wildlife, changes in management of domestic animals and declines in grazing livestock rate, will affect the tick population in Turkey at least as much as climate change. They have already limited host availability for some ticks that feed on livestock and this restriction seems to be fuelled by climate change. On the other hand, the chaos in some foci caused by ecological and climatic changes can be favourable for some species (albeit temporarily in the short term) as a result, for example, of a decrease in the population of other possible tick competitors.

Overall, predicted climatic, environmental and ecological changes appear to adversely affect some resident tick species in particular, resulting in possible reductions in prevalence, density and species diversity of the total tick population in Turkey. However, these reductions do not necessarily mean that there will be a decrease in the general significance of tick-borne diseases. Long-term coexistence of certain hosts, ticks and disease agents in a given area can be expected to result in perpetual exposure of ticks to the agents, and of the hosts to the ticks and agents. Therefore, persistent immunity in the hosts to the agents and ticks, and possibly in the ticks to the agents, can be induced. Furthermore, in such an endemic area, possible cross-immunity and interspecific competitions among the ticks and agents for hosts and vectors are expected. As a result of all this, it can be expected that many tick-borne agents remain latent and the prevalence and incidence of possible diseases are low in such endemically stable communities (Tønnesen *et al.*, 2006). Climatic and environmental changes may lead to species changing their geographical dispersion or newly establishing in the country, differentiation in the proportional distribution of the species in favour of some species and isolation of some species in some areas, which may cause some deterioration in these endemically stable regions such as decrease in the host, tick and/or agent species' diversity or alteration. It is possible to predict that all these may increase the risk of some tick-borne diseases, in different ways and degrees.

17.2 Future Predictions

Here we consider the future of tick species having unique features and represented in Turkey, taking into account the following parameters: (i) direct effect of changes in temperature, precipitation and humidity on basic biology and general phenology; (ii) human-mediated impact on ecosystems due to increasing industrialization and population (e.g. deforestation; hunting and other pressures on wildlife; numerical, breed or species change in livestock; change in the general management and grazing regime of livestock); and (iii) impact of climate change on the ecosystem through vegetation, wildlife and human behaviour.

17.2.1 *Hyalomma marginatum*

For the immature stages of this two-host, exophilic tick, hares, hedgehogs, partridges and some other ground-feeding birds are the hosts, while cattle are the main hosts for the adults in Turkey. The prevalence and density of adults on some animals, such as sheep, goats, dogs and wild boar, are exceptionally low, even when they share the same pasture as cattle. Adults infest animals most of the year, but prevalence and density are higher in warm and hot months. This species, found in almost all regions of the country, can be encountered from the seacoast with an altitude of 0 m, where the Mediterranean climate prevails, to the eastern regions with an altitude of >1000 m, where the relatively harsh continental climate prevails. Overlap of cattle-grazing areas and the habitats of immature stage hosts, such as natural shrublands and fragmented forests, is the main determinant of the distribution patterns across the country, while altitude, climate and sub-climate types in Turkey have negligible influence.

The observations presented above suggest that changes in climatic parameters alone, up to certain levels (at least as much as the differences between regions within Turkey), may not dramatically affect this tick. In fact, possible resistance to some predicted levels of change in temperature, precipitation and humidity are not surprising for this species, which is a typical member of the *Hyalomma* genus, known as the arid environment ticks. Increasing temperatures and prolonged droughts can cause changes in phenology, extending its annual active period and increasing the likelihood of simultaneous feeding with some other species that have different seasonal dynamics, such as the one-host winter tick, *Hyalomma scupense*. It is difficult to predict how this possible co-feeding will affect this and other tick species. Although the degradation of nature will negatively affect the tick in the future, relative flexibility of immature stages in host selectivity will enable the tick to adapt to the changing ecosystem and wildlife demography, to some extent. However, gradual changes in grazing cattle population, due to changes in management procedures and the trend towards intensive farming, will be the most important pressure on the *H. marginatum* population in

Turkey. Furthermore, this decrease in cattle grazing is predicted to accelerate due to both the direct impact of climate change and its effect on vegetation (Fortuny *et al.*, 2020). Sheep, goats and wild ruminants are likely to maintain the species, but it is obvious that their possible contribution at this point will remain quite low compared with cattle. A decrease in grazing sheep and goat populations is also likely over time. The wild ruminant populations are already in decline (Sekercioglu *et al.*, 2011) and although some conservation strategies will provide partial recovery in their populations, an effective increment is not predicted for large wild animals that need relatively large natural space. Moreover, wild animals may be more resistant to many parasites, including ticks, compared with cattle and other domestic animals. Under all these circumstances, our prediction is that *H. marginatum* will be one of the losers from climate change and human impact in the future, albeit in the medium to long term.

17.2.2 *Hyalomma rufipes*

This species is similar to *H. marginatum* in many respects such as biology, seasonality, host and mostly habitat preference, except that it occurs in drier regions of Africa (Hoogstraal, 1956). It arrives in Turkey with bird migrations and is rarely reported on hosts; no established populations have yet been documented. This failure to establish, despite mating individuals being found on animals in some foci (Kar *et al.*, 2009), is associated with low winter temperatures in Europe (Hasle, 2013). Predicted temperature increases and decreases in humidity will likely favour establishment. However, the risks valid for *H. marginatum* are likely to determine the long-term future of *H. rufipes* as well.

17.2.3 *Hyalomma aegyptium*

Tortoises are primary hosts for all stages of this three-host exophilic tick species. Although immature stages can also successfully use many other animals as hosts such as rodents, lagomorphs, hedgehogs, lizards and ground-feeding birds, the adult is highly specific to tortoises and is rarely encountered on other animals. Feeding occurs in warm and hot months. *H. aegyptium* occurs in almost every region of Turkey and the distribution patterns are causally related to the tortoise (*Testudo* spp.) population (Aysul *et al.*, 2010; Kar *et al.*, 2011). For the near and medium term, the direct impact of climate change predicted for this tick species appears to be low. However, increases in temperature can extend the annual activation period for the tick and for tortoises that normally hibernate during cold months in Turkey. Prolonged drought, change in vegetation and elevated temperature in the hot season can cause an extended aestivation period in tortoises, which may affect the phenology or some other features of the tick. A possible decrease in the tortoise population due to human influence and climate-related ecological changes seem to be the most critical factors driving a decrease in the population of *H. aegyptium* in the future.

17.2.4 *Hyalomma excavatum*

For the immature stages of this three-host exophilic tick species, many kinds of small and medium-sized animals can be hosts such as burrow-dwelling small mammals, rodents, lagomorphs and birds. Similarly, the adult stages have a wide host range and can effectively feed on cattle, sheep, goats, horses, donkeys, medium- and large-sized wild animals, and some other large animals. Found in many regions of Turkey, *H. excavatum* is more common in large, lowland areas. The adults are encountered on animals in almost all months of the year, but prevalence and density are higher in warm and hot months (Aydin and Bakirci, 2007). As with other exophilic *Hyalomma* spp. possible increases in temperature and drought seem to extend its annual effective activation period. For reasons such as being an arid environment tick compatible with high temperature and low humidity, having no absolute need for natural woodlands, having an adaptation to deforested steppes and having a wide host spectrum, this species will have a higher resistance to the direct and indirect effects of expected climate change compared with other *Hyalomma* spp. in Turkey.

17.2.5 *Hyalomma anatolicum*

For this typical monotropic, endophilic, three-host tick species, the primary hosts are cattle in Turkey. However, the tick can infest goats, sheep, horses and some other large animals, albeit at low levels. The adults feed in the summer months and the immatures are seen on hosts in the summer and autumn. Endophilic off-host stages in and around barns can be encountered in crevices and cracks mostly located at different heights of the wall (Aydin and Bakirci, 2007). Climate types or subtypes in Turkey are not an effective driver for this tick. Unexpectedly, for a tick species that is endophilic (and is therefore expected to be less affected by environmental factors compared with exophilic ticks), some environmental parameters such as geography, ecology and vegetation are important in the distribution pattern of the species. *H. anatolicum*, which was common in many parts of the country until a few decades ago, is now quite rarely encountered in barns that are found in flat plains or river valleys (M. Aktas, Elazığ, 2019, personal communication; Z. Vatansever, Kars, 2019, personal communication). These barns, with mostly a traditional structure, are either apart from natural woodlands and shrublands, such as the habitat of *H. marginatum*, and/or the cattle housed in the barns are not grazed (Fig. 17.2). Modernization of barn structure and management systems, and utilization of more effective antiparasitic applications in recent decades in the country, seem to be responsible for the apparent decline in this tick population. Currently, increased awareness and widespread antiparasitic use on cattle (hosts for all stages of *H. anatolicum*), due to the recent outbreak of Crimean–Congo haemorrhagic fever, may have contributed to this decrease (S. Bakirci, Aydın, 2020, personal communication). Although expected climate change may not have much direct impact on this endophilic tick species, the population decline will continue due to human influence in Turkey.

17.2.6 *Hyalomma scupense*

This species consists of two phenotypes: one-host (syn. *H. scupense*) and two-host (syn. *Hyalomma detritum*). The endophilic two-host biological form is remarkably similar to *H. anatolicum* in several respects, such as basic features of host association, host spectrum, off-host life strategy, distribution pattern and seasonality (as an exception, the immatures can be more common on hosts in the winter period). Furthermore, the future trend for the population in Turkey is similar to that of *H. anatolicum*, with similar drivers of a decline in distribution, prevalence and density.

One-host biological forms of *H. scupense* have been found recently in Turkish Thrace, in areas where *H. marginatum* is common in warm and hot months (Kar *et al.*, 2015). It is a typical cattle-specific winter tick and its adaptation to cold weather is well known (Pomerantzev, 1950). As the weather cools in the second half of autumn, the larvae become active and engorged females usually detach from hosts by mid-spring. It is endophilic, like the two-host biological form and *H. anatolicum*, but engorged females do not climb high and lay eggs on the ground close to cattle manure, which is in or around paddocks or

Fig. 17.2. Outside and inside a typical barn. (Authors' own photographs.)

just outside barns. Because of these characteristics, the one-host tick is generally more likely to be affected by meteorological events than other endophilic species. It is difficult to predict how this species, which has cold affinity, will respond to rising temperatures in the future, but at least some changes in seasonality pattern seem likely. Direct vulnerability to human influence will be a primary determinant in the future of this species, as with many other endophilic species.

17.2.7 *Boophilus (Rhipicephalus) annulatus*

This typical one-host tick species is highly specific to cattle and always exhibits an endophilic character in Turkey. The presence of cattle in barns is the primary determinant in its activity. It does not show a significant infestation seasonality and it is possible to encounter the tick on cattle both in winter and in hot seasons (Aydin and Bakirci, 2007; Yay *et al.*, 2004). Our field observations indicate that relatively elevated temperature and humidity in the microenvironment are important for activation of this tick. In addition, type and condition of barns are important criteria for the presence and activation phenology of the tick. Infestations, especially in winter, are mostly encountered in traditional barns with low ventilation and high humidity. Off-host endophilic dynamics are similar to those of most other endophilic species, but this species does not climb high and is almost always found in ground-level cracks and crevices within the barn. Due to its typical endophilic character, it can be seen in every region of Turkey, regardless of geographical, ecological and climatic features. As a result of these characteristics, it does not seem to be much affected by the meteorological parameters of predicted climate change. This species used to be common in Turkey, but similar to those of endophilic *Hyalomma* spp. and most likely for similar reasons, it is no longer prevalent and our prediction for its possible future in Turkey is also similar.

17.2.8 *Rhipicephalus turanicus*

For the immature stages of the three-host tick, *R. turanicus*, rodents, lagomorphs, birds and many other small and medium-sized animals are hosts, while several kinds of medium-sized and large animals support the adults, such as sheep, goats, cattle, horses, donkeys, dogs, cats and lots of wild animals. It is one of the most prevalent tick species in Turkey and can be encountered in gardens, parks, pastures, grasslands, bushlands and forests in the urbanized, rural or natural areas at different altitudes and under different sub-climate types. Adults are seen in the warm spring and summer months (Aydin and Bakirci, 2007; Orkun, 2018). It has relatively high humidity preference and therefore its prevalence and density on hosts decrease in hot and dry summer months and increase in the days following rainfall. Possible prolonged drought periods seem to be the most challenging condition for this species in the future. It can easily adapt to different environments and is likely to increase in some currently restricted areas that become more suitable in future in terms of water, humidity and vegetation, due to human influence or natural events. Such suitable areas will also attract many domestic and wild animals, and this may provide an advantage to the tick owing to its extremely low host selectivity. For all these reasons, we anticipate that this species will be one of the most successful tick species in Turkey under the possible climatic and ecological changes expected in the future.

17.2.9 *Rhipicephalus sanguineus*

Three-host, monotropic, endophilic tick species, *R. sanguineus*, is specific to dogs and is rarely encountered on other animals. It is one of the most common tick species in Turkey, being found in urban, peri-urban and rural areas. The distribution pattern is causally related to dogs and the climatic, ecological and geographical features of the regions are not effective drivers on this pattern. Off-host stages are essentially nidicolous, mostly seen in dog shelters, in the cracks and crevices on the walls or some other elevated points but can also be encountered in peridomestic areas. The species has higher affinity for increased temperatures and higher drought resistance than *R. turanicus*. Adults are first seen on hosts in spring, but all stages peak in summer. All these features indicate that the predicted climatic and ecological changes will not be so devastating, and its future seems to be largely related to the population demography of dogs in Turkey.

17.2.10 *Rhipicephalus bursa*

For all stages of this two-host species, goats, sheep and cattle are primary hosts, but the tick can be encountered on other large and medium-sized domestic and wild animals. It is a more drought-resistant species compared with other members of the *Rhipicephalus* genus in Turkey. Adults infest animals in the hot season, peaking in June, and immatures feed in the cold season. *R. bursa* is prevalent in Turkey and can be seen on animals in farms located in woodland, shrubland or even in open pasture areas. Some features of the species, such as its thermophilic character, drought resistance, broad geographical preference and relatively flexible host preference, seem to provide resistance to changing environmental parameters. The decrease in grazing livestock population is probably the most important threat to this species in the future. In fact, the natural dynamics of this tick species are poorly understood. Some data suggest that exophilic *R. bursa* may have developed a kind of peri-domestic off-host dynamic occurring in the vicinity of livestock farms located in totally or relatively natural environments.

17.2.11 *Ixodes ricinus*

The host spectrum of the three-host tick, *I. ricinus*, is broad. Immature stages usually feed on small mammals, birds and lizards. Adults parasitize many domestic and wild animals of different sizes, albeit with varying degrees of interest, ranging from mice to deer and cattle (Perkins *et al.*, 2006; Medlock *et al.*, 2013). The species is common in almost all parts of Turkey, especially on the coast of the Black Sea and in the Marmara Sea basin (Vatansever *et al.*, 2008; Gargili *et al.*, 2010, 2011). The distribution patterns in other regions are mostly in foci (Aydin and Bakirci, 2007). It can be very prevalent both in natural forests and in certain sized peri-urban woodlands surrounded by residential areas. Although the activity of the species is higher in the spring, it can continue almost year-round. Despite the sensitivity to high temperature and drought above certain values, *I. ricinus* is often encountered in summer and this is probably a result of the tick's ability to use favourable micro niches (Medlock *et al.*, 2013). In the conditions of Istanbul and Thrace, adults and even immatures can be active throughout the winter unless there is snow on the ground (Kar *et al.*, 2017). It is likely that they have a much more active winter dynamic in the southern regions of the country under the influence of a Mediterranean climate. In the future, factors such as temperatures above critical values, prolonged drought and deforestation will negatively affect the species. On the other hand, the ability to use a broad host spectrum and microenvironments indicates that the population can continue at certain levels, albeit in isolated or more restricted foci. In addition, increased temperatures can cause some changes in seasonal dynamics, e.g. more effective winter activity may be observed.

17.2.12 *Haemaphysalis* spp. and *Dermacentor* spp.

Prominent members of this group in Turkey are *Haemaphysalis parva*, *H. punctata*, *Haemaphysalis inermis*, *Haemaphysalis sulcata* and *Dermacentor marginatus* (Aydin and Bakirci, 2007). All these species have three-host and exophilic characteristics. Although there may be variations between species, immatures feed on a wide range of medium and small-sized wild mammals, ground-feeding birds and reptiles, and adults infest large and medium-sized wild and domestic animals. Adult ticks are active in the cold months, particularly late autumn and early spring, and immatures are active in the hot months of the year. Their humidity preferences are mostly very high (except *H. sulcata*), especially in *H. inermis*, and most probably direct sunlight tolerance is also lower in these species. They are common in most parts of Turkey and can be encountered in a wide variety of habitats, such as bushlands, oak forests, forest margins, pastures, and river or mountain valleys. Their prevalence is higher in relatively cooler and more humid areas or foci. *H. sulcata* is an exception, as it is more prevalent in central and eastern Anatolia where a continental climate with more challenging conditions prevails.

When these general characteristics are evaluated, the decreasing trend of grazing livestock population, differentiation of wildlife population, destruction of natural habitats and predicted meteorological alterations will most likely adversely affect these species to varying degrees. However, their broad host spectrum may alleviate the severity of the impact to some extent. They may regress to more humid and temperate regions and foci in the future (particularly *H. inermis*), their prevalence may decrease and there may be some changes in seasonal dynamics. Compared with other species, *H. punctata*, a relatively more thermophilic species, and *H. sulcata*, a species better adapted to arid and harsh areas, may be less affected by the predicted changes in environmental parameters.

References

Aydin, L. and Bakirci, S. (2007) Geographical distribution of ticks in Turkey. *Parasitology Research* 101, 163–166.

Aydin, S., Simsek, M., Cetinkaya, G. and Ozturk M.Z. (2019) Regime characteristics of Turkey's climatic regions determined using the Erinç precipitation efficiency index. In: Gönençgil, B., Ertek, T.A., Akova, I. and Elbasi, E. (eds) *1st Istanbul International Geography Congress Proceedings*. Istanbul University Press, Istanbul, pp. 752–760.

Aysul, N., Kar, S., Yilmazer, N., Alp, H.G. and Gargili, A. (2010) Trakya Yöresi'ndeki kaplumbağalarda (*Testudo graeca*) *Hyalomma aegyptium* (Lineaus, 1758) 'un yaygınlığı [Prevalence of *Hyalomma aegyptium* (Lineaus, 1758) on tortoises (*Testudo graeca*) in Thrace]. *Pendik Veteriner ve Mikrobiyoloji Dergisi* 37, 53–56 (in Turkish).

Bett, B., Lindahl, J. and Delia, G. (2019) Climate change and infectious livestock diseases: the case of Rift Valley fever and tick-borne diseases. In: Rosenstock, T., Nowak, A. and Girvetz, E. (eds) *The Climate-Smart Agriculture Papers*. Springer, Cham, Switzerland, pp. 29–37.

Calvert, A.M., Taylor, P.D. and Walde, S. (2009) Cross-scale environmental influences on migratory stopover behaviour. *Global Change Biology* 15, 744–759.

Dantas-Torres, F. (2015) Climate change, biodiversity, ticks and tick-borne diseases: the butterfly effect. *International Journal for Parasitology: Parasites and Wildlife* 4, 452–461.

Fortuny, X., Carcaillet, C. and Chauchard, S. (2020) Selective and taxon-dependent effects of semi-feral cattle grazing on tree regeneration in an old-growth Mediterranean mountain forest. *Forest Ecosystems* 7, 11.

Gargili, A., Kar, S., Yilmazer, N., Cerit, C., Sönmez, G., *et al.* (2010) Evaluation of ticks biting humans in Thrace Province, Turkey. *Journal of Veterinary Faculty Kafkas University* 16(Suppl. A), 141–146.

Gargili, A., Kar, S., Yilmazer, N., Ergonul, O. and Vatansever, Z. (2011) Different abundances of human-biting ticks in two neighboring provinces in Turkey. *Journal of Veterinary Faculty Kafkas University* 17, 93–97.

Hasle, G. (2013) Transport of ixodid ticks and tick-borne pathogens by migratory birds. *Frontiers in Cellular and Infection Microbiology* 3, 48.

Hoogstraal, H. (1956) *African Ixodoidea. I. Ticks of the Sudan (With Special Reference to Equatoria Province and With Preliminary Reviews of the Genera Boophilus, Margaropus, and Hyalomma).* Department of the Navy, Washington, DC.

Hoogstraal, H., Kaiser, M.N., Traylor, M.A., Gaber, S. and Guindy, E. (1961) Ticks (Ixodidae) on birds migrating from Africa to Europe and Asia. *Bulletin of the World Health Organization* 24, 197–212.

Hoogstraal, H., Kaiser, M.N., Traylor, M.A., Guindy, E. and Gaber, S. (1963) Ticks (Ixodidae) on birds migrating from Europe and Asia to Africa. *Bulletin of the World Health Organization* 28, 235–262.

Kar, S., Guven, E. and Vatansever, Z. (2009) Tick infestation and *Hyalomma rufipes* presence in a stud farm in the Marmara Region, Turkey. In: Ozcel, M. (ed.) *XVI National Parasitology Congress Abstracts*. Turkish Parasitology Society, Turkey, p. 254.

Kar, S., Yilmazer, N., Midilli, K. and Gargili, A. (2011) Presence of the zoonotic *Borrelia burgdorferi* sl. and *Rickettsia* spp. in the ticks from wild tortoises and hedgehogs. *MUSBED* 1, 166–170.

Kar, S., Akyildiz, G. and Vatansever, Z. (2015) Trakya'da ineklerde *Hyalomma scupense* enfestasyonu ve enfestasyonun mevsimsel karakteristiği [*Hyalomma scupense* infestation in cows in Thrace and seasonal characteristics of infestation]. In: Ozcel, M. (ed.) *XIX National Parasitology Congress Abstracts*. Turkish Parasitology Society, Turkey, p. 79.

Kar, S., Yilmazer, N., Akyildiz, G. and Gargili, A. (2017) The human infesting ticks in the city of Istanbul and its vicinity with reference to a new species for Turkey. *Systematic and Applied Acarology* 22, 2245–2255.

Kirwan, G.M., Ozen, M., Ertuhan, M. and Atahan, A. (2014) Turkey bird report 2007–2011. *Sandgrouse* 36, 146–175.

Medlock, J.M., Hansford, K.M., Bormane, A., Derdakova, M., Estrada-Peña, A., *et al.* (2013) Driving forces for changes in geographical distribution of *Ixodes ricinus* ticks in Europe. *Parasites & Vectors* 6, 1.

Najdenski, H., Dimova, T., Zaharieva, M.M., Nikolov, B., Petrova-Dinkova, G., *et al.* (2018) Migratory birds along the Mediterranean–Black Sea Flyway as carriers of zoonotic pathogens. *Canadian Journal of Microbiology* 64, 915–924.

Orkun, O. (2018) Molecular characterization based on 16S rDNA phylogeny of some ixodid ticks in Turkey. *Turkiye Parazitoloji Dergisi* 42, 122–129.

Ozturk, M.Z., Cetinkaya, G. and Aydin, S. (2017) Climate types of Turkey according to Köppen–Geiger climate classification. *Istanbul University Journal of Geography* 35, 17–27.

Peel, M.C., Finlayson, B.L. and McMahon, T.A. (2007) Updated world map of the Koppen–Geiger climate classification. *Hydrology and Earth System Sciences* 11, 1633–1644.

Perkins, S.E., Cattadori, I.M., Tagliapietra, V., Rizzoli, A.P. and Hudson, P.J. (2006) Localized deer absence leads to tick amplification. *Ecology* 87, 1981–1986.

Pomerantzev, B.I. (1950) *Ixodid ticks (Ixodidae). In: Arachnoidea*, vol. 4(2). *Fauna of the USSR, New Series*, vol. 41. Nauk, Leningrad, USSR, p. 224 (in Russian).

Pontes, G.M., Wainer, I., Taschetto, A.S., Gupta, A.S., Abe-Ouchi, A., *et al.* (2020) Drier tropical and subtropical Southern Hemisphere in the mid-Pliocene Warm Period. *Scientific Reports* 10, 13458.

Sekercioglu, C., Anderson, S., Akcay, E., Bilgin, R., Can, O.E., *et al.* (2011) Turkey's globally important biodiversity in crisis. *Biological Conservation* 144, 2752–2769.

Tønnesen, M.H., Penzhorn, B.L., Bryson, N.R., Stoltsz, W.H. and Masibigiri, T. (2006) Seroprevalence of *Babesia bovis* and *Babesia bigemina* in cattle in the Soutpansberg region, Limpopo Province, South Africa, associated with changes in vector-tick populations. *Journal of the South African Veterinary Association* 77, 61–65.

Vatansever, Z., Gargili, A., Aysul, N.S., Sengoz, G. and Peña, A.S. (2008) Ticks biting humans in the urban area of Istanbul. *Parasitology Research* 102, 551–553.

Yay, M., Yazar, S., Aydin, L. and Sahin, I. (2004) Investigation of tick species on sheep and cattle around of Kayseri. *Erciyes University Journal of Health Sciences* 13, 25–29.

eo18 Climate Change Alone Cannot Explain Altered Tick Distribution Across Europe: A Spotlight on Endemic and Invasive Tick Species

Frederic Stachurski[1,2], Nathalie Boulanger[3,4], Adrien Blisnick[5], Laurence Vial[1,2] and Sarah Bonnet[6*]

[1]CIRAD, Montpellier, France; [2]Université de Montpellier, Montpellier, France; [3]Université de Strasbourg, Strasbourg, France; [4]Centre National de Référence Borrelia, France; [5]Institut Pasteur, Paris, France; [6]INRAE, Nouzilly, France

18.1 Expert Opinion

18.1.1 Introduction

In 2014, the Intergovernmental Panel on Climate Change reported that due to global warming, temperature increases of at least 1.5 to 2°C in Northern Europe are expected in the coming decades (IPCC, 2014). Concomitantly, the geographical distribution of several tick species has changed due to both socio-economic developments as well as environmental and climatic changes (Gray *et al.*, 2009; Barre and Uilenberg, 2010; Medlock *et al.*, 2013; Ogden, 2017; Eisen and Eisen, 2018; Wikel, 2018). Tick density and distribution within a biogeographic zone depend on the biotope, which can be more or less favourable to tick development (ecological preferences), the availability of preferred hosts (trophic preferences), but also the climatic conditions. Thus, several interdependent factors can influence tick population distribution and abundance, and any human action affecting these various parameters, whether voluntary or not, such as deforestation, agriculture, animal husbandry and herding, urbanization, breeding, hunting, etc. will have consequences for local tick populations (Kilpatrick *et al.*, 2017).

Like all living creatures faced with environmental change, tick populations have two options: either adapt to the new conditions or become extinct. However, and although ticks cannot move by themselves over great distances, they can be dispersed through host movement, whether during domestic animal transport or during the natural movement of wildlife, particularly migratory birds. If ticks are introduced into regions with a favourable climate and biotope, they can then colonize these areas and become permanently established. These phenomena are perfectly illustrated by both the recent colonization of West Africa by *Rhipicephalus microplus* (Madder *et al.*, 2011), a tick which was already established on several continents, and by the historical colonization of the Caribbean by *Amblyomma variegatum* (Barre and Uilenberg, 2010).

The effect of climate on the evolution of tick populations remains difficult to disentangle

*Email: sarah.bonnet@inrae.fr

DOI: 10.1079/9781789249637.0018

from other possible causes and undoubtedly varies depending on the region concerned and local tick species. Large-scale, long-term monitoring is therefore necessary to accurately assess climatic impact on tick populations. Climate change can alter tick populations, either indirectly by affecting vertebrate host populations or directly by increasing or decreasing their numbers. These ectoparasites, and in particular hard ticks, spend almost their entire life cycle in the external environment, thus climatic conditions influence their activity, viability and distribution. This expert opinion aims to illustrate the impact of climate change, and its association with other variables, on the distribution and abundance of tick populations in Europe using *Ixodes ricinus* and *Hyalomma marginatum* as typical examples of endemic and invasive species, respectively.

18.1.2 Impact on endemic populations: the *Ixodes ricinus* example

I. ricinus is the most abundant and widely distributed tick species, capable of feeding on more than 300 different host species. It is undoubtedly the most important species affecting human health in Europe (Rizzoli *et al.*, 2014). *I. ricinus* is endemic from the Atlantic coast to Central Asia, and from Iceland to the North African Atlas Mountains (Estrada-Peña *et al.*, 2006). It is a triphasic, exophilic tick which generally spends more than 95% of its life off a host out in the open, where it digests its blood meal, metamorphoses, lays eggs (for females), undergoes diapause during unfavourable conditions or quests for hosts (Földvári, 2016). As a result, this tick species – which is particularly sensitive to desiccation – is utterly dependent on environmental conditions (Milne, 1950).

Due to its relatively long life cycle, generally 2–3 years but which can reach 6 years, one can assume that *I. ricinus* adapts less quickly to environmental changes than would a tick with a shorter cycle. Nevertheless, its vast geographical distribution – covering almost all of Europe – demonstrates that it can survive in varied environments as long as vegetation is abundant, rain is moderate to heavy, and its numerous potential hosts are present. It withstands large temperature fluctuations or a lack of humidity by taking refuge in humus or litter to survive and rehydrate (Milne, 1950; Rudolph and Knulle, 1974). The periods during which *I. ricinus* lies in wait for hosts are thus very clearly dictated by the prevailing weather conditions, both in the months preceding and during questing periods. In temperate regions, these periods of activity vary with the season, as demonstrated by several studies describing a more intense activity peak in late spring/early summer, followed by a smaller activity peak from late summer and throughout autumn (Vassalo *et al.*, 2000; Randolph *et al.*, 2002; Lauterbach *et al.*, 2013; Daniel *et al.*, 2015; Estrada-Peña *et al.*, 2015; Paul *et al.*, 2016; Borsan *et al.*, 2020). However, if the climate is mild, ticks can be active from early spring right through to late autumn, or even all year round.

During recent decades, *I. ricinus* has spread throughout Europe due to several identified factors (Leger *et al.*, 2013; Medlock *et al.*, 2013). Typical examples of its evolving distribution include its northward expansion in Nordic countries, as has been observed since the 1980s (Talleklint and Jaenson, 1998; Jore *et al.*, 2014; Alkishe *et al.*, 2017), or its presence at higher and higher altitudes, beyond 1500 m (Gilbert, 2010). Depending on the regions considered, this expansion has been attributed to changes in plant cover and soil use (Brownstein *et al.*, 2005), to climate changes (milder and shorter winters, prolonged spring and autumn seasons) (Lindgren *et al.*, 2000) or to increases in host populations (Lindgren *et al.*, 2000; Mysterud *et al.*, 2016). Indeed, since the 1980s, deer populations have increased in Europe not only due to milder winters and longer favourable vegetation periods, but also due to decreased hunting quotas in some regions (Carpio *et al.*, 2021). Because they are such important hosts for *I. ricinus*, this has undoubtedly enabled tick populations to expand in certain zones.

Climate change can also effectively directly influence *I. ricinus* populations by leading to the appearance or disappearance of areas suitable for tick development and activity, as is predicted by several modelling studies (Estrada-Peña *et al.*, 2006, 2015; Estrada-Peña and Venzal, 2007; Williams *et al.*, 2015; Alkishe *et al.*, 2017). Globally, and at European latitudes, warmer temperatures in autumn and winter could increase tick survival and development, thus increasing the density of tick stages present during the following spring and summer. Warmer and shorter winters could also favour an earlier start to tick activity in the spring, possibly lengthening this activity period. Nevertheless, milder autumn and winter temperatures can also enable larvae to undergo earlier metamorphosis, which may cause greater nymph mortality when winter sets in, resulting in fewer *I. ricinus* nymphs the following spring. For those species originating from temperate climates that are particularly sensitive to desiccation, such as *I. ricinus*, very hot and dry summers can be profoundly detrimental to their survival and activity.

As climate thus limits ticks to areas in which they can survive, then in regions suited to their activity, the direct impact of temperature and relative humidity fluctuations will be much more subtle or even zero, compared with effects caused by changes in host abundance or altered vegetation and landscape configuration. It is nevertheless conceivable that increased temperatures in Europe could be harmful to *I. ricinus* by increasing mortality and reducing activity periods. However, as previously mentioned, such an overall increase in temperature could also increase the speed of development, lengthen activity periods during autumn and spring, increase host availability and decrease winter mortality. It has been demonstrated that, for *I. ricinus*, harmful effects from the cold are cumulative, where it takes more than 30 days at −10°C for ticks to die (Knulle and Dautel, 1997). Very few long-term studies have assessed the impact of climate change on endemic *I. ricinus* populations in Europe. One such study carried out between 2008 and 2014 in the Sénart forest in the south-east of Paris, France, showed that although minimum temperatures during the months preceding the collection had the most significant effect on tick density, inter-annual fluctuations in active nymph density were more likely related to variations in vertebrate host populations, as well as fauna and flora management practices, than annual climate variations (Paul *et al.*, 2016). Moreover, monthly forest sampling data collected from 2000 to 2014 in Switzerland revealed decreases in *I. ricinus* populations (Hauser *et al.*, 2018). Although only partially explaining the decline, the authors cite a link with the air saturation deficit that occurs earlier and earlier each year. In parallel, another study showed a doubling of nymph populations from 2004 to 2018 in another Swiss site (Bregnard *et al.*, 2020). Thus, for endemic tick populations such as *I. ricinus*, local extreme climatic events are likely to affect their population dynamics more than global trends linked to climate change.

18.1.3 Impact on invasive populations: the *Hyalomma marginatum* example

Could current climate change also alter the introduction and/or permanent establishment of new tick species – and their attendant pathogens – in Europe? This question concerns the Mediterranean region in particular. Its proximity to Africa combined with a similar climate makes Southern Europe an enticing area for African ticks, mainly belonging to the *Rhipicephalus* or *Hyalomma* genus, to permanently settle once introduced.

The recent arrival of *H. marginatum* in France testifies that this is more than just a hypothesis. This two-host tick has an annual life cycle. Adults infest large ungulates (mainly horses, but also domestic and wild ruminants, wild boars, etc.) in spring. Then, a few weeks later, the larvae parasitize their hosts (lagomorphs, birds) on which the larval metamorphosis, then the nymphal meal, take place. This immature phase lasts about 3 weeks. The species is found in the Iberian Peninsula, but also from Italy to Turkey, around the Black Sea and in North-West Africa (ECDC and EFSA, 2018). It is known to be one of the main vectors of Crimean–Congo haemorrhagic fever virus (Bente *et al.*, 2013), of which some local cases caused by tick bite – some fatal – have been reported in Spain since 2016 (Negredo *et al.*, 2017). *H. marginatum* was commonly thought to be present throughout Corsica and this has since been confirmed (Grech-Angelini *et al.*, 2016a). Notwithstanding this validation, the rare mentions of its presence in mainland France that followed were unconvincing: whether due to uncertain identification or the observation of immature ticks accidentally introduced by migratory birds (Vial *et al.*, 2016). However, during a study in the Camargue, in the south of continental France from 2007 to 2010, low numbers were found in several horse herds (Chastagner *et al.*, 2013).

Since then, recent surveys on horses have confirmed the presence of *H. marginatum* in almost all French departments with a Mediterranean coastline, thus validating its permanent establishment in this country (Stachurski and Vial, 2018). It has also been detected northward, up to 150 km from the Mediterranean, but is not found westward, in regions where the Mediterranean climate is replaced by a modified oceanic climate (around 50 km from the coast). In some of the visited farms, even near the Mediterranean Sea, only a single adult *H. marginatum* tick has been found despite examining many horses, or it is not seen every year. It cannot then be asserted that the species is truly installed at these sites, as the presence of one adult could be due to a single nymph surviving after detaching from a bird, metamorphosing into an adult and then infesting a horse. This circumstance has also been observed in both Germany (Chitimia-Dobler *et al.*, 2019) and Sweden (Grandi *et al.*, 2020), where *H. marginatum* is unlikely to thrive and persist due to its requirements for a long dry summer with low humidity (ECDC and EFSA, 2018).

Birds are one of the major hosts of immature *H. marginatum*, particularly passerines which spend a lot of time on the ground such as blackbirds, nightingales, thrushes, etc. Therefore, it is tempting to hypothesize that these ticks were introduced into France by migratory birds, especially as they remain infested for approximately 3 weeks (see above). This leaves ample time for parasitized birds to travel from North Africa or Spain to France or other more northern countries (Kampen *et al.*, 2007; Jameson *et al.*, 2012). Thus, the introduction of *H. marginatum* nymphs into mainland France is probably not recent. These introduced nymphs have almost certainly been detaching on to sites with adult hosts for a very long time. But why have large established tick populations been detected in France only recently? Can the current climate change alone explain this observation? According to studies carried out in Russia, it seems that nymph metamorphosis occurs in the laboratory only when a certain number of degree-days (cumulative average daily temperatures above the minimum development temperature of 14–16°C) has been reached (Gray *et al.*, 2009). Perhaps these conditions have only recently been met, thus enabling on-site metamorphosis of engorged nymphs disseminated by migrating birds.

However, other explanations are also possible. As the species is present in Spain, it may have been introduced into France via the transport of animals infested by adult ticks such as cattle or horses, imported by equestrian facilities in the south of the country. Hunters have also occasionally introduced Spanish rabbits and hares to the south of France, which may have been infested by immature ticks. Alternatively, the tick may have been introduced to the French Pyrenees via infested wild boar or deer from across the Catalonian border. Moreover, adult hosts are more likely involved in the spread of *H. marginatum* from the southern to northern Balkans in Eastern Europe (Gray *et al.*, 2009).

It is therefore currently very difficult to determine whether the recently altered climatic conditions in the south of France facilitated *H. marginatum* colonization from engorged nymphs introduced by infested birds, or whether the increased transport of ungulates parasitized by adult ticks, and thus the intensification of human activities, is responsible for this establishment. The same uncertainty surrounds the origins of the recent establishment of another *Hyalomma* species, *Hyalomma scupense*, in Corsica (Grech-Angelini *et al.*, 2016b). How did this tick settle in Corsica? It is common to North-West Africa and naturally infests cattle during its three stages, thus rendering introduction by migratory birds extremely unlikely. This second example proves that new tick species are perfectly capable of moving into and colonizing regions hitherto free from their presence. Although ticks cannot travel long distances on their own, this is by no means an insurmountable obstacle in their conquest of new territories.

18.2 Future Predictions

Regarding European endemic species like *I. ricinus*, we predict that the current climate changes will not necessarily result in an increase in the species' range, but rather in a shift of those areas in which it can survive and thrive. For invasive species like those belonging to the *Hyalomma* genus, it is clear that current climate change can foster conditions that are increasingly conducive to their development in Southern Europe. In France, *H. marginatum* will probably soon become permanently established throughout the majority of the currently uninfested favourable sites on the French Mediterranean coast, due to permanent introduction of ticks by birds parasitized by larvae in the already infested and neighbouring sites or to movements of domestic ruminants or horses infested by the adult ticks. In other French regions or in northern countries, birds will also continue to introduce engorged nymphs, but this is unlikely to result in the tick becoming established as long as the climate remains unsuitable. In southern regions of Europe, other *Hyalomma* species like *Hyalomma rufipes* could become established in the near future, since it is also frequently carried by migratory birds (Negredo *et al.*, 2017; Chitimia-Dobler *et al.*, 2019; Grandi *et al.*, 2020). However that may be, it is not possible at present to say that climatic changes are the only reason for the establishment of *H. marginatum* in southern France and other Mediterranean regions of Europe.

Indeed, and although it is evident that climate change influences the distribution of tick populations in Europe, it is equally clear that climate change is not necessarily the sole and/or main cause of the current increase in tick-borne diseases. The dynamics of these diseases are extremely complex, rendering it difficult to differentiate between the direct and indirect impacts of climate change on all vector system components. Vertebrate host abundance, habitat and biotope changes, socio-economic modifications (including the risk of human and animal exposure to ticks), as well as public or veterinary health measures, are also critically important in explaining both variations in tick populations and disease occurrence (Kilpatrick *et al.*, 2017). When modelling tick and tick-borne pathogen distribution, it is essential to take into account that populations of the same tick species may respond differently to altered climatic conditions, in addition to changes in climate, vegetation and hosts. Furthermore, laboratory results on tick survival and activity thresholds as a function of temperature or humidity are very rarely applicable in nature. Thus, it is clear to date that the link between climate change and tick-associated risks is not uniform, varying according to the region and the tick species and population involved. It is therefore not enough only to carefully map the current and future distribution of tick populations; we must also continue to deepen our knowledge of vector systems by including the ecology of different tick species, tick–host–pathogen interactions and the impact of climate on these factors, in order to predict the potential risk of tick-borne disease emergence.

References

Alkishe, A.A., Peterson, A.T. and Samy, A.M. (2017) Climate change influences on the potential geographic distribution of the disease vector tick *Ixodes ricinus*. *PLoS ONE* 12, e0189092.

Barre, N. and Uilenberg, G. (2010) Spread of parasites transported with their hosts: case study of two species of cattle tick. *Revue scientifique et technique* 29, 149–160, 135–147.

Bente, D.A., Forrester, N.L., Watts, D.M., McAuley, A.J., Whitehouse, C.A. and Bray, M. (2013) Crimean–Congo hemorrhagic fever: history, epidemiology, pathogenesis, clinical syndrome and genetic diversity. *Antiviral Research* 100, 159–189.

Borsan, S.D., Toma-Naic, A., Peter, A., Sandor, A.D., Pestean, C., *et al.* (2020) Impact of abiotic factors, habitat type and urban wildlife on the ecology of hard ticks (Acari: Ixodidae) in urban and peri-urban habitats. *Parasites & Vectors* 13, 476.

Bregnard, C., Rais, O. and Voordouw, M.J. (2020) Climate and tree seed production predict the abundance of the European Lyme disease vector over a 15-year period. *Parasites & Vectors* 13, 408.

Brownstein, J.S., Skelly, D.K., Holford, T.R. and Fish, D. (2005) Forest fragmentation predicts local scale heterogeneity of Lyme disease risk. *Oecologia* 146, 469–475.

Carpio, A., Apollonio, M. and Acevedo, P. (2021) Wild ungulate overabundance in Europe: contexts, causes, monitoring and management recommendations. *Mammal Review* 51, 95–108.

Chastagner, A., Bailly, X., Leblond, A., Pradier, S. and Vourc'h, G. (2013) Single genotype of *Anaplasma phagocytophilum* identified from ticks, Camargue, France. *Emerging Infectious Diseases* 19, 825–827.

Chitimia-Dobler, L., Schaper, S., Riess, R., Bitterwolf, K., Frangoulidis, D., *et al.* (2019) Imported *Hyalomma* ticks in Germany in 2018. *Parasites & Vectors* 12, 134.

Daniel, M., Maly, M., Danielova, V., Kriz, B. and Nuttall, P. (2015) Abiotic predictors and annual seasonal dynamics of *Ixodes ricinus*, the major disease vector of Central Europe. *Parasites & Vectors* 8, 478.

ECDC and EFSA (2018) *Hyalomma marginatum* – current known distribution, January 2018. European Centre for Disease Prevention and Control, Stockholm and European Food Safety Authority, Parma, Italy. Available at: https://www.ecdc.europa.eu/en/publications-data/hyalomma-marginatum-current-known-distribution-january-2018 (accessed 2 June 2021).

Eisen, R.J. and Eisen, L. (2018) The blacklegged tick, *Ixodes scapularis*: an increasing public health concern. *Trends in Parasitology* 34, 295–309.

Estrada-Peña, A. and Venzal, J.M. (2007) Climate niches of tick species in the Mediterranean region: modeling of occurrence data, distributional constraints, and impact of climate change. *Journal of Medical Entomology* 44, 1130–1138.

Estrada-Peña, A., Venzal, J.M. and Sanchez Acedo, C. (2006) The tick *Ixodes ricinus*: distribution and climate preferences in the western Palaearctic. *Medical and Veterinary Entomology* 20, 189–197.

Estrada-Peña, A., Estrada-Sanchez, A. and Estrada-Sanchez, D. (2015) Methodological caveats in the environmental modelling and projections of climate niche for ticks, with examples for *Ixodes ricinus* (Ixodidae). *Veterinary Parasitology* 208, 14–25.

Földvári, G. (2016) Life cycle and ecology of *Ixodes ricinus*: the roots of public health importance. In: Braks, M., van Wieren, S., Takken, W. and Sprong, H. (eds) *Ecology and Prevention of Lyme Borreliosis*. Wageningen Academic Publishers, Wageningen, the Netherlands, pp. 31–40.

Gilbert, L. (2010) Altitudinal patterns of tick and host abundance: a potential role for climate change in regulating tick-borne diseases? *Oecologia* 162, 217–225.

Grandi, G., Chitimia-Dobler, L., Choklikitumnuey, P., Strube, C., Springer, A., *et al.* (2020) First records of adult *Hyalomma marginatum* and *H. rufipes* ticks (Acari: Ixodidae) in Sweden. *Ticks and Tick-borne Diseases* 11, 101403.

Gray, J.S., Dautel, H., Estrada-Peña, A., Kahl, O. and Lindgren, E. (2009) Effects of climate change on ticks and tick-borne diseases in Europe. *Interdisciplinary Perspectives on Infectious Diseases* 2009, 593232.

Grech-Angelini, S., Stachurski, F., Lancelot, R., Boissier, J., Allienne, J.F., *et al.* (2016a) Ticks (Acari: Ixodidae) infesting cattle and some other domestic and wild hosts on the French Mediterranean island of Corsica. *Parasites & Vectors* 9, 582.

Grech-Angelini, S., Stachurski, F., Lancelot, R., Boissier, J., Allienne, J.F., *et al.* (2016b) First report of the tick *Hyalomma scupense* (natural vector of bovine tropical theileriosis) on the French Mediterranean island of Corsica. *Veterinary Parasitology* 216, 33–37.

Hauser, G., Rais, O., Moran Cadenas, F., Gonseth, Y., Bouzelboudjen, M., *et al.* (2018) Influence of climatic factors on *Ixodes ricinus* nymph abundance and phenology over a long-term monthly observation in Switzerland (2000–2014). *Parasites & Vectors* 11, 289.

IPCC (2014) *Climate Change 2014: Synthesis Report. Contribution of Working Groups I, II and III to the Fifth Assessment Report of the Intergovernmental Panel on Climate Change*. Intergovernmental Panel on Climate Change, Geneva, Switzerland.

Jameson, L.J., Morgan, P.J., Medlock, J.M., Watola, G. and Vaux, A.G.C. (2012) Importation of *Hyalomma marginatum*, vector of Crimean–Congo haemorrhagic fever virus, into the United Kingdom by migratory birds. *Ticks and Tick-borne Diseases* 3, 95–99.

Jore, S., Vanwambeke, S.O., Viljugrein, H., Isaksen, K., Kristoffersen, A.B., *et al.* (2014) Climate and environmental change drives *Ixodes ricinus* geographical expansion at the northern range margin. *Parasites & Vectors* 7, 11.

Kampen, H., Poltz, W., Hartelt, K., Wölfel, R. and Faulde, M. (2007) Detection of a questing *Hyalomma marginatum marginatum* adult female (Acari, Ixodidae) in southern Germany. *Experimental and Applied Acarology* 43, 227–231.

Kilpatrick, A.M., Dobson, A.D.M., Levi, T., Salkeld, D.J., Swei, A., *et al.* (2017) Lyme disease ecology in a changing world: consensus, uncertainty and critical gaps for improving control. *Philosophical Transactions of the Royal Society B* 372, 20160117.

Knulle, W. and Dautel, H. (1997) Cold hardiness, supercooling ability and causes of low-temperature mortality in the soft tick, *Argas reflexus*, and the hard tick, *Ixodes ricinus* (Acari: Ixodoidea) from Central Europe. *Journal of Insect Physiology* 43, 843–854.

Lauterbach, R., Wells, K., O'Hara, R.B., Kalko, E.K. and Renner, S.C. (2013) Variable strength of forest stand attributes and weather conditions on the questing activity of *Ixodes ricinus* ticks over years in managed forests. *PLoS ONE* 8, e55365.

Leger, E., Vourc'h, G., Vial, L., Chevillon, C. and McCoy, K.D. (2013) Changing distributions of ticks: causes and consequences. *Experimental and Applied Acarology* 59, 219–244.

Lindgren, E., Talleklint, L. and Polfeldt, T. (2000) Impact of climatic change on the northern latitude limit and population density of the disease-transmitting European tick *Ixodes ricinus*. *Environmental Health Perspectives* 108, 119–123.

Madder, M., Thys, E., Achi, L., Toure, A. and De Deken, R. (2011) *Rhipicephalus* (*Boophilus*) *microplus*: a most successful invasive tick species in West-Africa. *Experimental and Applied Acarology* 53, 139–145.

Medlock, J.M., Hansford, K.M., Bormane, A., Derdakova, M., Estrada-Peña, A., *et al.* (2013) Driving forces for changes in geographical distribution of *Ixodes ricinus* ticks in Europe. *Parasites & Vectors* 6, 1.

Milne, A. (1950) The ecology of the sheep tick, *Ixodes ricinus* L.: microhabitat economy of the adult tick. *Parasitology* 40, 14–34.

Mysterud, A., Easterday, W.R., Stigum, V.M., Aas, A.B., Meisingset, E.L., *et al.* (2016) Contrasting emergence of Lyme disease across ecosystems. *Nature Communications* 7, 11882.

Negredo, A., de la Calle-Prieto, F., Palencia-Herrejón, E., Mora-Rillo, M., Astray-Mochales, J., *et al.* (2017) Autochthonous Crimean–Congo haemorrhagic fever in Spain. *New England Journal of Medicine* 377, 154–161.

Ogden, N.H. (2017) Climate change and vector-borne diseases of public health significance. *FEMS Microbiology Letters* 364, fnx186.

Paul, R.E., Cote, M., Le Naour, E. and Bonnet, S.I. (2016) Environmental factors influencing tick densities over seven years in a French suburban forest. *Parasites & Vectors* 9, 309.

Randolph, S.E., Green, R.M., Hoodless, A.N. and Peacey, M.F. (2002) An empirical quantitative framework for the seasonal population dynamics of the tick *Ixodes ricinus*. *International Journal for Parasitology* 32, 979–989.

Rizzoli, A., Silaghi, C., Obiegala, A., Rudolf, I., Hubalek, Z., *et al.* (2014) *Ixodes ricinus* and its transmitted pathogens in urban and peri-urban areas in Europe: new hazards and relevance for public health. *Frontiers in Public Health* 2, 251.

Rudolph, D. and Knulle, W. (1974) Site and mechanism of water vapour uptake from the atmosphere in ixodid ticks. *Nature* 249, 84–85.

Stachurski, F. and Vial, L. (2018) Installation de la tique *Hyalomma marginatum*, vectrice du virus de la fièvre hémorragique de Crimée–Congo, en France continentale. *Bulletin épidémiologique, santé animale et alimentation* 84, 1–5 (in French).

Talleklint, L. and Jaenson, T.G. (1998) Increasing geographical distribution and density of *Ixodes ricinus* (Acari: Ixodidae) in central and northern Sweden. *Journal of Medical Entomology* 35, 521–526.

Vassalo, M., Paul, R.E. and Perez-Eid, C. (2000) Temporal distribution of the annual nymphal stock of *Ixodes ricinus* ticks. *Experimental and Applied Acarology* 24, 941–949.

Vial, L., Stachurski, F., Leblond, A., Huber, K., Vourc'h, G., *et al.* (2016) Strong evidence for the presence of the tick *Hyalomma marginatum* Koch, 1844 in southern continental France. *Ticks and Tick-borne Diseases* 7, 1162–1167.

Wikel, S.K. (2018) Ticks and tick-borne infections: complex ecology, agents, and host interactions. *Veterinary Sciences* 5, 60.

Williams, H.W., Cross, D.E., Crump, H.L., Drost, C.J. and Thomas, C.J. (2015) Climate suitability for European ticks: assessing species distribution models against null models and projection under AR5 climate. *Parasites & Vectors* 8, 440.

eo19 Climate and Management Effects on Tick–Game Animal Dynamics

Andrew Hoodless* **and Rufus Sage**
Game & Wildlife Conservation Trust, Fordingbridge, UK

19.1 Expert Opinion

19.1.1 Introduction

The castor bean or sheep tick, *Ixodes ricinus*, is the most abundant and widespread tick species in Europe. It is a multicompetent vector of protozoa, bacteria and viruses to humans and animals and is important in the transmission of many zoonotic diseases, such as Lyme disease, babesiosis, anaplasmosis, erlichiosis and tick-borne encephalitis (TBE) (Dantas-Torres *et al.*, 2012). Documented cases of Lyme disease and TBE in humans are increasing and two reasons are likely to be increased recreational activity, and hence higher contact rates with ticks, coupled with better awareness and diagnosis among medical professionals. However, several interacting abiotic and biotic factors are likely to be at least as important (e.g. Jaenson *et al.*, 2012a). These include the effects of temperature and humidity on the rate of tick development and questing activity, the effect of seasonal shifts on tick and host habitats, and the effects of climate and management on host fecundity and abundance.

Several of the intermediate and large hosts for *I. ricinus*, including large mammals, lagomorphs and game birds, are classed as game species throughout Europe, with defined hunting seasons and often a long tradition of habitat management to boost local numbers. Those most widely hunted, and of greatest economic importance, are deer, wild boar (*Sus scrofa*) and game birds, which collectively can feed large numbers of all tick stages. In Central and Northern Europe, there tends to be a greater focus on hunting large mammals, whereas game bird shooting is more widely practised in Britain and Ireland, France and Iberia. The management of game birds for driven shooting typically involves intensive management of habitat and predators to produce high densities of wild-bred birds or the release of large numbers of artificially reared birds.

There is general agreement on the complexity of tick–host–disease dynamics and the difficulty in predicting the effects of a changing climate (e.g. Tabachnick, 2009). This expert opinion discusses the mechanisms by which climate and management could alter tick and game host densities and contact rates, as well as considering the likely relative importance of climate change and game management in affecting the distribution of ticks in Europe.

*Email: ahoodless@gwct.org.uk

DOI: 10.1079/9781789249637.0019

19.1.2 Evidence for climate effects on tick distribution and behaviour

I. ricinus lives in a wide range of environmental conditions throughout Europe. Temperature varies considerably across this geographic range and high humidity (*c.*85% relative humidity) in microhabitats at ground level is the abiotic prerequisite for tick population survival (Daniel *et al.*, 1998). In general, but not in all locations, tick questing is bimodal with a large spring peak and a smaller autumn peak (Gray, 1991). Perret *et al.* (2004) showed that questing activity of adult and nymphal *I. ricinus* was related to relative humidity, with declines in activity from April owing to periods of drought (mean saturation deficit >*c.*4.0 mmHg) during May–September.

Randolph *et al.* (2002) proposed a single-cohort, autumn-recruitment model, whereby an annual cohort of each tick stage starts in the autumn. Ticks either become active and feed immediately in early autumn or not until the following spring. Those which feed in the autumn undergo delayed development (diapause) and moult to the next stage in the following autumn, whereas those which feed in the spring or summer develop directly. Randolph *et al.* (2002) hypothesized that the size of the autumn peak in tick questing was dependent on the duration of development of spring-fed ticks, with warmer summers leading to more rapid development, and the data of Perret *et al.* (2004) support this.

Tick development rates and host-questing activity are dependent on soil temperature, resulting in geographic variation in the seasonal population dynamics of each life stage. Across the current *I. ricinus* core range, the timing of emergence of unfed ticks in spring will depend on local temperatures. Only in places with exceptionally warm temperatures throughout the spring and summer might ticks be expected to emerge sufficiently early to avoid diapause induction and develop directly to the next stage within the same calendar year. Warmer temperatures are likely to differentially affect the survival of different tick life stages: for instance, nymphs are likely to be more tolerant to hydric stress than larvae, but they quest higher in the vegetation and are potentially more exposed to water loss. Any influence of autumnal temperatures on the onset of overwinter quiescence in unfed ticks could be critical in determining the availability of questing ticks of each stage the following spring.

In the centre of the *I. ricinus* range, there is reasonable evidence for higher mean annual temperatures resulting in higher tick abundance. In northern Spain, Ruiz-Fons *et al.* (2012) found a positive effect of mean annual temperature on larval tick abundance and a positive effect of mean annual temperature and rainfall on nymphal abundance across nine sites. They concluded that for larvae, when soil moisture deficit is not a stressing factor, the abundance of ticks may be determined largely by the number of frost days during winter.

The distribution of *I. ricinus* is expanding within Europe and there is some evidence for climate effects at the edge of the range (Medlock *et al.*, 2013). The effect of raised temperature is more marked at lower temperatures at the northern or altitudinal limit due to the non-linear positive relationship between temperature and tick development rate, with evidence for the recent occurrence of ticks at high latitudes in Sweden providing one of the best examples (Jaenson *et al.*, 2012b).

19.1.3 Game species as tick hosts

Deer, particularly roe deer (*Capreolus capreolus*), are the most abundant wild hosts for adult ticks in Europe and commonly use deciduous woodlands that comprise suitable habitat for ticks. They feed large numbers of all stages of ticks: a study in Germany involving complete carcass examinations found that 50% were adult females, 27% nymphs and 10% larvae, with the remainder adult males (Muders, 2015). This study recorded an intensity of 172 ± 63 ticks in May and all roe deer were infested with ticks during spring and autumn, dropping to about 40% of animals in winter. Tick feeding intensity on roe deer has been shown to be related to vegetation and altitude (Carpi *et al.*, 2008; Vor *et al.*, 2010). However, no consistent influence of roe deer densities on tick abundance has been recorded across studies. Gray *et al.* (1992) showed that the availability of deer as hosts facilitates high tick densities, but even small numbers of deer can maintain very large tick populations (Wilson *et al.*, 1984; Robertson *et al.*, 2000; Hofmeester *et al.*, 2017).

Tick prevalence and burdens tend to be lower on wild boar than roe deer. For instance, mean intensities on wild boar and roe deer culled mainly during the winter were 0.13 ticks (range 0–10) and 6.4 ticks (0–261), respectively (Muders, 2015). Tick prevalence on boar reached 100% in spring but was less than 20% throughout the rest of the year. Other studies have recorded average intensities of 13–15 ticks on wild boar in spring (Ruiz-Fons *et al.*, 2006; Pacilly *et al.*, 2014).

The fact that they can occur at high densities, coupled with their role as hosts for adult ticks and their high reproductive potential, make deer and wild boar important hosts in tick life cycles. They also have greater capacity to spread ticks in the landscape than other tick hosts. Roe deer are thought to have been instrumental in the spread of TBE to northern Sweden (Jaenson *et al.*, 2012a). Roe deer home ranges are typically about 50–60 ha for resident animals and over 200 ha for migrant individuals (Mysterud, 1999; Carpi *et al.*, 2008). Although they feed fewer ticks than roe deer, wild boar move more widely throughout forest and agricultural habitats. Migration distance was <5 km for 88% of wild boar in a study in the Czech Republic, but migration of males during the rutting season of up to 40 km has been recorded (Havránek *et al.*, 2005 cited by Kriz *et al.*, 2014).

Pheasants (*Phasianus colchicus*) and red grouse (*Lagopus lagopus scoticus*) commonly feed larval and nymphal ticks. While the latter can occur at high densities because of intensive management for shooting in upland areas of the UK, the pheasant is more important in a European context. Since the intensification of farming in the 1960s and 1970s, pheasant shooting has grown in popularity, with shoots increasingly relying on annual releases of poults rather than wild-hatched birds. Pheasant releasing occurs in several European countries, but the practice is most common in the UK. Around 40 million pheasants are estimated to be released in late summer in the UK each year, at densities of about 8–19 birds/ha (Sage *et al.*, 2018; Aebischer, 2019). Pheasants are encouraged to make daily movements from release pens to use game crops planted on nearby farmland during the day. During the shooting season (October–January) they are driven from these crops over a line of guns and back to the release area.

Tick burdens on pheasants are highest in April, with medians of 7 larvae and 43 nymphs recorded on males, which feed about four times more ticks than females (Hoodless *et al.*, 1998). Pheasants can be competent amplifying hosts for *Borrelia burgdorferi* sensu lato (Kurtenbach *et al.*, 1998a). Released pheasants experience high mortality from predation and shooting between August and February, such that numbers are typically 90% lower during the spring and early summer than at the time of release (Sage *et al.*, 2018) (Fig. 19.1). They are important hosts for nymphal ticks during February–May and August–October, feeding similar numbers to grey squirrels (*Sciurus carolinensis*). They are quantitatively unimportant compared with wood mice (*Apodemus sylvaticus*) for larvae and with roe deer for adults (Hoodless *et al.*, 1998). Given the almost ubiquitous distribution of small mammals and roe deer in lowland woods, it seems likely that the addition of pheasants will maintain or increase the abundance of ticks in locations where they are already present and will shift the prevalence of *Borrelia* genospecies towards the bird-specific strains *Borrelia garinii* and *Borrelia valaisiana* (Kurtenbach *et al.*, 1998b). However, because pheasants typically travel less than 500 m and mainly use the part of the wood between the release pen and the game crop, they are unlikely to play a major role in the dispersal of ticks in the landscape.

19.1.4 Climate, habitat and management interactions

A temperature increase sufficient to advance the spring tick questing peak and accelerate the development rate of ticks is likely to affect the timing of vegetation growth in spring. Such an increase may result in earlier breeding and/or higher fecundity of some hosts if synchrony with food resources is maintained. It seems likely that a warming climate will result in more autumn questing ticks and that juvenile animals may be present to encounter ticks earlier in the summer. There is already good evidence for an advancement in the emergence and growth of woodland plants in spring because of a changing climate (Fitter and Fitter, 2002) and this undoubtedly affects the microclimate for ticks and seasonality of tick questing. However,

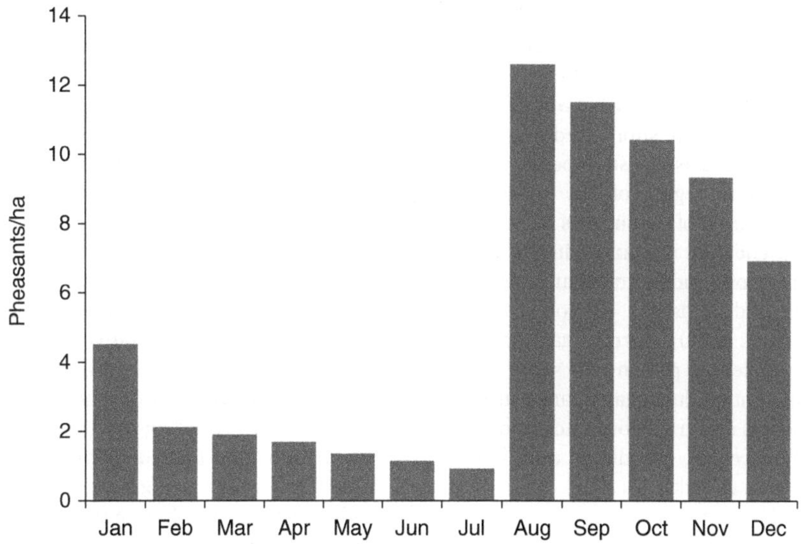

Fig. 19.1. Typical pheasant densities (number of birds per hectare of estate) on lowland English shoots at the start of each month. Mean values of estimated density at release and seasonal pheasant mortality have been taken from Sage *et al.* (2018). Pheasant poults are assumed to leave release pens in early August. (Authors' own figure.)

management of landscapes and habitats has the potential to outweigh the effects of changing climate, at least locally and in the short term.

In an extensive study of woodlands across Europe, Ehrmann *et al.* (2018) identified a range of habitat properties that promote tick hosts and ticks. Their analysis showed that, overall, local habitat factors had a fourfold stronger influence on *Borrelia* prevalence than landscape and macroclimate considerations, indicating more ticks and more interactions with hosts. Basic indicators of habitat attractiveness for potential tick hosts were the most important predictors.

Game management has had a widespread influence on the structure and configuration of woodland habitats throughout Europe. There is a long tradition of game parks for large mammals, particularly deer, which contain a high proportion of open space (Rackham, 1998; Benes *et al.*, 2006). They have wide rides that facilitate hunting and coppice stands that supply browsing and shelter for animals. This management not only supports higher densities of large game hosts, but also results in more open woodland with grassy areas suitable for ticks.

Management for game bird shooting also involves intensive habitat manipulation. Woods are widely planted and managed for pheasant releasing and shooting, particularly in England (Sage *et al.*, 2020). This increases habitat for tick hosts and ticks: for example, Perez *et al.* (2016) showed that increasing connectivity of woodlands in the landscape leads to more wood mice, which are a primary host for larval ticks. Management within woods involves reducing tree density inside and close to release pens, to create sunny areas for the pheasants and to increase shrub cover (Sage *et al.*, 2020). Wide rides are created and maintained to facilitate shooting within the wood, and grain is typically fed to released pheasants between September and February–April. At locations where pheasants congregate at high densities, such as the release site and feeding points, they affect the ground vegetation, often reducing it through mechanical damage and possibly changing its species composition through dunging and enrichment (Sage *et al.*, 2005). The management practices and pheasants themselves therefore affect habitats and microclimates for ticks throughout the wood, with some effects likely to reduce suitability, e.g. tree thinning and trampling of vegetation, and others likely to have positive effects on tick numbers, e.g. ride management and provision of grain through feeding small mammals.

Movement of game animals for reintroduction could result in the introduction of ticks and pathogens to new locations. There are several historic examples of this, with one of the best documented being the short-lived and localized circulation of Far-Eastern subtypes of TBE virus in the European region of the former Soviet Union, thousands of kilometres west of their recognized endemic region, owing to repeated introductions of hundreds of thousands of game mammals and birds for economic purposes (Kovalev *et al.*, 2010). Current game bird practices are unlikely to present the same issue because pheasant poults at game farms are reared on short grass and transported to shoots before they are placed in woodland and have the chance to pick up ticks.

19.2 Future Predictions

Modelling has predicted a climatically suitable area for ticks in Europe approximately two times larger than at present by 2050, under scenarios with continued human population growth and increased greenhouse gas emissions and with lower human population growth and more sustainable greenhouse gas emissions (Porretta *et al.*, 2013). This prediction includes latitudinal and longitudinal increases encompassing northern Scandinavia and Finland across to central Russia. In endemic tick zones, warmer summer temperatures will lead to quicker development of spring-fed ticks and a higher proportion moulting to the next stage in the same calendar year. Higher autumn and winter temperatures will lead to a longer questing season. This assumes that relative humidity is not appreciably reduced in microhabitats because of increased temperature, which could affect tick survival or reduce habitat space for ticks in some localities.

Increased host abundance leading to enhanced tick reproduction in endemic zones, and dispersal of roe deer at higher altitude and latitude, are key drivers for change in the geographical distribution of *I. ricinus* ticks in Europe (Medlock *et al.*, 2013). In economically sustainable game bird shooting systems that have the resources to undertake substantial management work aimed at improving habitats and maximizing bird densities, including those for red-legged partridges (*Alectoris rufa*) on farmland and grasslands in the UK and Spain, and for red grouse on UK moorlands, we suggest that management effects are likely to exceed the influence of climate change processes on local tick densities. Nevertheless, released pheasants are likely to feed more ticks (especially larvae) in future owing to greater seasonal synchrony in autumn and will affect the local circulation of *Borrelia* genospecies. Deer and wild boar will play a fundamental role in the spread of *I. ricinus* into areas of changing climatic suitability. Higher tick biting rates in autumn and early winter would coincide with the dispersal of rutting males and juveniles and increase the distribution of ticks in the landscape.

References

Aebischer, N.J. (2019) Fifty-year trends in hunting bags of birds and mammals and calibrated estimation of national bag size using GWCT's National Gamebag Census. *European Journal of Wildlife Research* 65, 64–76.

Benes, J., Cizek, O., Dovala, J. and Konvicka, M. (2006) Intensive game keeping, coppicing and butterflies: the story of Milovicky Wood, Czech Republic. *Forest Ecology and Management* 237, 353–365.

Carpi, G., Cagnacci, F., Neteler, M. and Rizzoli, A. (2008) Tick infestation on roe deer in relation to geographic and remotely sensed climatic variables in a tick-borne encephalitis endemic area. *Epidemiology & Infection* 136, 1416–1424.

Daniel, M., Kolar, J., Zeman, P., Pavelka, K. and Sadlo, J. (1998) Predictive map of *Ixodes ricinus* high-incidence habitats and a tick-borne encephalitis risk assessment using satellite data. *Experimental and Applied Acarology* 22, 417–433.

Dantas-Torres, F., Chomel, B.B. and Otranto, D. (2012) Ticks and tick-borne diseases: a One Health perspective. *Trends in Parasitology* 28, 437–446.

Ehrmann, S., Ruyts, S.C., Scherer-Lorenzen, M., Bauhus, J., Brunet, J., *et al.* (2018) Habitat properties are key drivers of *Borrelia burdorferi* (s.l.) prevalence in *Ixodes ricinus* populations of deciduous forest fragments. *Parasites & Vectors* 11, 23.

Fitter, A.G. and Fitter, R.S.R. (2002) Rapid change in flowering time in British plants. *Science* 296, 1689–1691.

Gray, J.S. (1991) The development and seasonal activity of the tick *Ixodes ricinus*: a vector of Lyme borreliosis. *Review of Medical and Veterinary Entomology* 79, 323–333.

Gray, J.S., Kahl, O., Janetzki, C. and Stein, J. (1992) Studies on the ecology of Lyme disease in a deer forest in County Galway, Ireland. *Journal of Medical Entomology* 29, 915–920.

Havránek, F., Bukovjan, K., Urbanec, R. and Rehak, L. (2005) *Wild Boar in the Populated Areas*. Czech Ministry of Agriculture, Prague (in Czech).

Hofmeester, T.R., Sprong, H., Jansen, P.A., Prins, H.H.T. and Van Wieren, S.E. (2017) Deer presence rather than abundance determines the population density of the sheep tick, *Ixodes ricinus*, in Dutch forests. *Parasites & Vectors* 10, 433.

Hoodless, A.N., Kurtenbach, K., Nuttall, P.A. and Randolph, S.E. (1998) The role of pheasants as hosts for ticks and Lyme disease spirochaetes in southern England. *Game & Wildlife* 15, 477–489.

Jaenson, T.G.T., Hjertqvist, M., Bergström, T. and Lundkvist, A. (2012a) Why is tick-borne encephalitis increasing? A review of the key factors causing the increasing incidence of human TBE in Sweden. *Parasites & Vectors* 5, 184.

Jaenson, T.G.T., Jaenson, D.G.E., Eisen, L., Petersson, E. and Lindgren, E. (2012b) Changes in the geographical distribution and abundance of the tick *Ixodes ricinus* during the past 30 years in Sweden. *Parasites & Vectors* 5, 8.

Kovalev, S.Y., Kokorev, V.S. and Belyaeva, I.V. (2010) Distribution of Far-Eastern tick-borne encephalitis virus subtype strains in the former Soviet Union. *Journal of General Virology* 91, 2941–2946.

Kriz, B., Daniel, M., Benes, C. and Maly, M. (2014) The role of game (wild boar and roe deer) in the spread of tick-borne encephalitis in the Czech Republic. *Vector-Borne and Zoonotic Diseases* 14, 801–807.

Kurtenbach, K., Carey, D., Hoodless, A.N., Nuttall, P.A. and Randolph, S.E. (1998a) Competence of pheasants as reservoirs for Lyme disease spirochetes. *Journal of Medical Entomology* 35, 77–81.

Kurtenbach, K., Peacey, M., Rijpkema, S.G.T., Hoodless, A.N., Nuttall, P.A. and Randolph, S.E. (1998b) Differential transmission of the genospecies of *Borrelia burgdorferi* sensu lato by gamebirds and small rodents in England. *Applied and Environmental Microbiology* 64, 1169–1174.

Medlock, J.M., Hansford, K.M., Bormane, A., Derdakova, M., Estrada-Peña, A., *et al.* (2013) Driving forces for change in geographical distribution of *Ixodes ricinus* in Europe. *Parasites & Vectors* 6, 1.

Muders, S.V. (2015) *The role of European big game (Capreolus capreolus and Sus scrofa) as hosts for ticks and in the epidemiological life cycle of tick-borne diseases*. PhD thesis, Karlsruhe Institute of Technology, Karlsruhe, Germany.

Mysterud, A. (1999) Seasonal migration pattern and home range of roe deer (*Capreolus capreolus*) in an altitudinal gradient in southern Norway. *Journal of Zoology* 247, 479–486.

Pacilly, F.C.A., Benning, M.E., Jacobs, F., Leidekker, J., Sprong, H., *et al.* (2014) Blood feeding on large grazers affects the transmission of *Borrelia burgdorferi* sensu lato by *Ixodes ricinus*. *Ticks and Tick-borne Diseases* 5, 810–817.

Perez, G., Bastian, S., Agoulon, A., Bouju, A., Durand, A., *et al.* (2016) Effect of landscape features on the relationship between *Ixodes ricinus* ticks and their small mammal hosts. *Parasites & Vectors* 9, 20.

Perret, J., Rais, O. and Gern, L. (2004) Influence of climate on the proportion of *Ixodes ricinus* nymphs and adults questing on a tick population. *Journal of Medical Entomology* 41, 361–365.

Porretta, D., Mastrantonio, V., Amendolia, S., Gaiarsa, S., Epis, S., *et al.* (2013) Effects of global changes on the climatic niche of the tick *Ixodes ricinus* inferred by species distribution modelling. *Parasites & Vectors* 6, 271.

Rackham, O. (1998) Savanna in Europe. In: Kirby, K.J. and Watkins, C. (eds) *The Ecological History of European Forests*. CAB International, Wallingford, UK, pp. 1–24.

Randolph, S.E., Green, R.M., Hoodless, A.N. and Peacey, M.F. (2002) An empirical quantitative framework for the seasonal population dynamics of the tick *Ixodes ricinus*. *International Journal for Parasitology* 32, 979–989.

Robertson, J.N., Gray, J.S. and Stewart, P. (2000) Tick bite and Lyme borreliosis risk at a recreational site in England. *European Journal of Epidemiology* 16, 647–652.

Ruiz-Fons, F., Fernández-de-Mera, I.G., Acevedo, P., Höfle, U., Vicente, J., *et al.* (2006) Ixodid ticks parasitizing Iberian red deer (*Cervus elaphus hispanicus*) and European wild boar (*Sus scrofa*) from Spain: geographical and temporal distribution. *Veterinary Parasitology* 140, 133–142.

Ruiz-Fons, F., Fernández-de-Mera, I.G., Acevedo, P., Gortázar, C. and de la Fuente, J. (2012) Factors driving the abundance of *Ixodes ricinus* ticks and the prevalence of zoonotic *I. ricinus*-borne pathogens in natural foci. *Applied and Environmental Microbiology* 78, 2669–2676.

Sage, R.B., Ludolf, C. and Robertson, P.A. (2005) The ground flora of ancient semi-natural woodlands in pheasant release pens in England. *Biological Conservation* 122, 243–252.
Sage, R.B., Turner, C.V., Woodburn, M.I.A., Hoodless, A.N., Draycott, R.A.H. and Sotherton, N.W. (2018) Predation of released pheasants *Phasianus colchicus* on lowland farmland in the UK and the effect of predator control. *European Journal of Wildlife Research* 64, 14.
Sage, R.B., Hoodless, A.N., Woodburn, M.I.A., Draycott, R.A.H., Madden, J.R. and Sotherton, N.W. (2020) Summary review and synthesis: effects on habitats and wildlife of the release and management of pheasants and red-legged partridges on UK lowland shoots. *Wildlife Biology* 2020, wlb.00766.
Tabachnick, W.J. (2009) Challenges in predicting climate and environmental effects on vector-borne disease episystems in a changing world. *Journal of Experimental Biology* 213, 946–954.
Vor, T., Kiffner, C., Hagedorn, P., Niedrig, M. and Ruhe, F. (2010) Tick burden on European roe deer (*Capreolus capreolus*). *Experimental and Applied Acarology* 51, 405–417.
Wilson, M.L., Levine, J.F. and Spielman, A. (1984) Effect of deer reduction on abundance of the deer tick (*Ixodes dammini*). *Yale Journal of Biology and Medicine* 57, 697–705.

eo20 Climate-Driven Livestock Management Shifts and Tick Populations

Georgia Titcomb*
University of California, Santa Barbara, California, USA

20.1 Expert Opinion

20.1.1 Introduction

Climate change can affect all aspects of the tick life cycle, from altering host distributions to modulating tick development and survival. However, humans are further altering tick populations via livestock management practices as managers in turn respond to climate change. Given that livestock biomass is now nearly 15 times greater than all wild mammal biomass on Earth (Bar-On *et al.*, 2018), humans can greatly affect tick populations via livestock management. This is likely to occur through at least three pathways: (i) by altering host abundance and composition; (ii) via habitat modification due to grazing and trampling; and (iii) by altering tick mortality via tick-control methods. Thus, climate-driven changes to livestock management can profoundly alter tick populations, especially in regions such as eastern and southern Africa, where high livestock density and worsening climate changes are coupled with high tick diversity and abundance.

20.1.2 Climate change affects management decisions

Human agricultural systems will be severely affected by climate change, especially in many regions of Africa, necessitating both gradual and transformative adaptations (Rippke *et al.*, 2016). One such adaptation is a shift from crop production to livestock husbandry in warmer and drier environments (Jones and Thornton, 2009), which has the potential to increase large mammalian tick hosts in these areas. Indeed, in the most arid regions of Africa, total herbivore biomass is now greater than it was 1000 years ago due to livestock increases (Hempson *et al.*, 2017). These changes could lead to increases in populations of ticks that feed on livestock when animals are untreated, or reductions when tick-control measures are applied.

Climate change is not only influencing whether humans choose livestock over crop production, but also which species they choose to rear. For example, one analysis across ten African countries showed that farmers are more likely to select goats and sheep instead of cattle in warmer regions (Seo and Mendelsohn, 2007).

*Email: georgiatitcomb@gmail.com

DOI: 10.1079/9781789249637.0020

Amid increased drought risk, many livestock owners are also investing in camels and goats that can tolerate water variability more effectively than cattle, especially in the East Africa region (Kagunyu and Wanjohi, 2014). For example, between 1980 and 2015, cattle populations across Kenya declined by 25%, concurrent with 75% increases in sheep and goats and a 13% increase in camels (and a 68% drop for wildlife) (Ogutu *et al.*, 2016) (Fig. 20.1). Local-scale changes were more extreme: in Laikipia county, an area heralded for successful wildlife conservation and livestock ranching, wildlife trends varied by species (six increased, nine decreased), cattle remained relatively unchanged (+7%), sheep and goats increased by 260%, and camels increased by an astounding 18,000% (Ogutu *et al.*, 2016). These dramatic shifts towards drought-resistant species could substantially affect ticks and their pathogens across large swathes of land, especially given that livestock vastly outweigh wildlife in this region and globally. Thus, while direct climate effects on ticks and their pathogens are important, it is also critical to consider the effects on the many additional host-mediated pathways that affect ticks and their pathogens (Fig. 20.1).

20.1.3 Humans affect ticks and pathogens via livestock management

Altered host composition

Rising global wildlife losses have yielded secondary parasite losses, especially for parasites that specialize on few host species (Dunn *et al.*, 2009). For example, specialist ticks decline with final wildlife host reductions, often as a result of fragmentation and agricultural intensification (Ogrzewalska *et al.*, 2011; Esser *et al.*, 2019). Conversely, ticks like *Rhipicephalus* (*Boophilus*) *microplus*, which feed predominantly on cattle, have expanded globally with cattle movements (Nyangiwe *et al.*, 2018; Silatsa *et al.*, 2019). Thus, increased livestock husbandry has likely fuelled increases in tick species that prefer livestock and concurrent decreases in certain wildlife specialists.

Although generalist tick species are thought to be influenced more strongly by climatic variables than host ranges (Cumming, 1999), dramatic changes in host composition and/or biomass have notable effects on generalist tick communities as well, and this effect varies according to climatic context (Titcomb *et al.*,

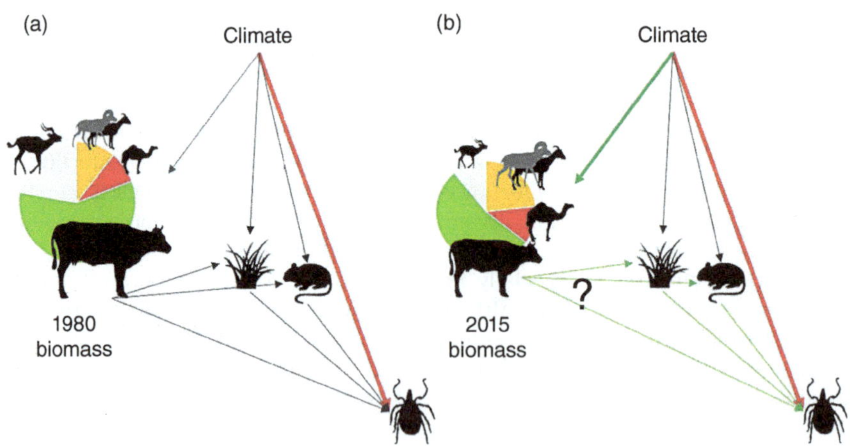

Fig. 20.1. Conceptual illustration of the intertwining effects of climate on hosts and ticks. (a) Direct climate effects on ticks (red arrow) form the primary basis of tick population projections, but indirect pathways (black arrows) are also important. (b) Recent drying trends in Kenya have substantially shifted large mammal composition over the past 35 years, yet these – and future – effects on ticks and pathogens (green arrows) are unknown. (Data for pie charts from Ogutu *et al.*, 2016. Author's own figure.)

2017). Furthermore, continent-level tick–host records suggest that different livestock species are parasitized by different arrays of ticks, although tick preferences will vary on local scales (McCoy *et al.*, 2013). For example, associations reported by Cumming (1998) demonstrate potential tick community differences among hosts, although replication across locations is needed (Fig. 20.2). If, on these local scales, livestock composition becomes increasingly dominated by sheep, goats and camels as opposed to cattle, ticks that thrive on these animals may increase (e.g. *Hyalomma dromedarii*; Fig. 20.2) while those that maximize their fitness on cattle may decrease (e.g. *Rhipicephalus zambeziensis*, *R. microplus*; Fig. 20.2). More generalist tick species (e.g. *Amblyomma gemma*, *Rhipicephalus pulchellus*; Fig. 20.2) are more likely to track overall host populations (in the absence of tick-control methods). However, more data on tick preferences, ability to adapt to alternative hosts and changes to the broader wildlife community across locations will be needed to make accurate predictions of tick compositional shifts in response to host compositional shifts.

Even when generalist tick populations are independent of host compositional changes, remaining host biology can affect the pathogens

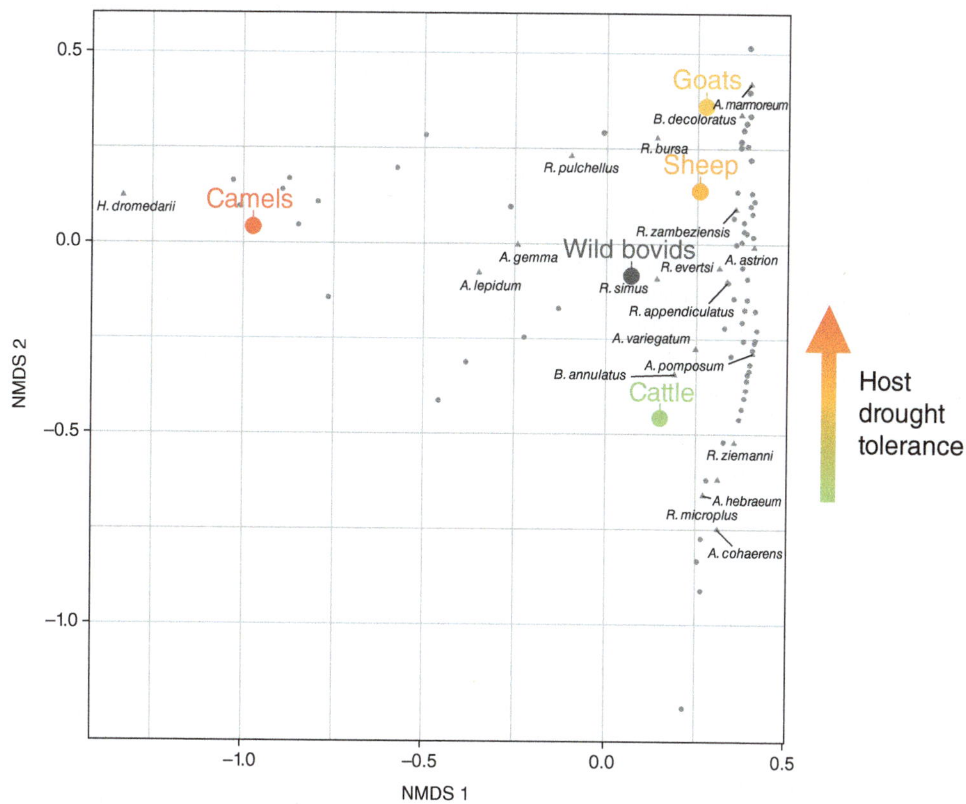

Fig. 20.2. Non-metric multidimensional scaling plot of host and tick species from records reported in the literature (data reported from Cumming, 1998; important ticks identified from de la Fuente *et al.*, 2008). To control for uneven sampling effort, host species records were scaled by the total number of records for each host. Tick communities on camels are more different, and therefore more distant, than communities parasitizing other herbivores, while more subtle differences in tick composition for cattle, wild bovids, and sheep and goats are reflected by their relative proximity on the plot. It is important to note that because records are collected on a continental scale, associations are also subject to geographic variation in host availability. Genera on the figure: *A.*, *Amblyomma*; *B.*, *Boophilus*; *H.*, *Hyalomma*; *R.*, *Rhipicephalus*. (Author's own figure.)

vectored. 'Dilution' and 'amplification' effect studies have shown that host community composition changes that shift overall pathogen competency can either reduce or increase pathogen risk. If different livestock species vary in pathogen competence, then pathogen distributions will shift when host composition changes. For example, in addition to parasitizing many wild mammals, *Rhipicephalus appendiculatus* can feed on cattle, sheep and goats (Cumming, 1998), but only cattle and buffalo are affected by and considered to be competent hosts for the vectored pathogen *Theileria parva*, which causes East Coast fever (Bishop *et al.*, 2004). Thus, replacing cattle with sheep and goats may reduce *T. parva* prevalence in these ticks (while potentially increasing prevalence of other pathogens).

Habitat modification

Two consequences of growing livestock biomass and shifting livestock species composition on tick life cycles are the secondary effects on vegetation and small mammal hosts (Fig. 20.1). Given that cattle, goats, sheep and camels occupy very different dietary niches (Rutagwenda *et al.*, 1990), it is possible that livestock composition shifts will impact vegetation cover, which has substantial effects on tick survival (e.g. Perry *et al.*, 1990). Furthermore, given that small mammals also respond to livestock management practices more strongly in arid contexts (Young *et al.*, 2015), shifts in livestock species composition and biomass may also alter the composition of these intermediate hosts for many tick species.

Altered tick mortality

Humans also disrupt tick survival via acaricide application. This practice has been shown to have landscape-scale effects on tick populations. For example, one large-scale experimental study in Kenya demonstrated three- to fourfold reductions in questing tick abundance in locations where treated cattle grazed (Keesing *et al.*, 2013). Similarly, other studies have found reductions in questing tick abundance after grazing acaricide-treated versus non-treated sheep. However, this was most effective when wildlife populations were low and will likely introduce toxins that impact other invertebrates and birds feeding on ticks (Keesing *et al.*, 2013; Van Wieren *et al.*, 2016).

Although acaricides are easily applied to a variety of livestock species, they are also costly to farmers who must balance tick-borne disease risk with chemical prices (Mugabi *et al.*, 2010). If acaricide application (or other control method) varies among livestock species, tick populations will likely reflect livestock compositional shifts. Furthermore, rising global concerns about growing acaricide resistance among certain tick species – notably the important cattle tick, *R. microplus* (Abbas *et al.*, 2014) – demonstrate the need for alternative tick-control strategies, avoidance of overapplication and better understanding of the effects across different livestock species and contexts.

When livestock act as competent hosts, management efforts such as vaccination or acaricide spraying can impact pathogens across a landscape. However, tick-control success will also depend on spatial overlaps with wildlife reservoirs. For example, buffalo are important wildlife reservoirs for several tick-borne diseases that strongly affect cattle, especially East Coast fever and corridor disease (de la Fuente *et al.*, 2008). However, if climate changes shift livestock populations toward animal species that either share fewer tick species with, or are less susceptible to, pathogens circulating among wildlife, then disease-control methods in these animals may be more easily achieved.

20.2 Future Projections

20.2.1 Key questions and considerations

Climate-driven alterations to underlying host community composition, population size and tick-removal strategies can substantially affect ticks and their pathogens across landscapes. However, this will likely vary by tick species, climatic region and economic factors. Making projections for future changes to tick populations and pathogen prevalence as a result of these shifts will require answering many questions, including:

1. To what extent do ticks select among different livestock species? How do livestock competencies vary for different tick-borne pathogens?
2. Do tick-control practices vary among livestock species, and to what extent do tick species vary in acaricide resistance?

3. How do livestock species alter vegetative structure and intermediate host communities that further affect tick survival?
4. To what extent do livestock species share ticks and pathogens with wildlife?

20.2.2 Climate futures and near-term projections

Tick population models based on climatic variables have shown range shifts for many economically important species across Africa (Olwoch *et al.*, 2007). However, agricultural responses to climate change likely have underappreciated secondary effects on ticks and their pathogens, especially in regions where livestock, wildlife and tick populations are high. Indeed, recent rainfall reductions in East Africa have already contributed to strategic livestock shifts (Kagunyu and Wanjohi, 2014) and projected increases in extreme weather events (Niang *et al.*, 2014) will continue to force farmers to make management decisions, such as transitioning to livestock from crops, or shifting from cattle to goats and sheep.

In East and Central Africa, long-term climate projections are at odds with recent drying trends; warmer and wetter climates are predicted for the latter part of this century (Niang *et al.*, 2014). Meanwhile, northern and southern regions of the continent are expected to experience significantly reduced rainfall (Niang *et al.*, 2014). Together these shifts will alter land suitability for different agricultural activities, potentially driving substantial livestock transitions in arid regions (Rippke *et al.*, 2016). While more integrative research is needed to better predict the specific effects of climate-induced agricultural shifts on tick and pathogen populations, increased livestock biomass in dry regions could have positive effects on generalist ticks by providing additional hosts, but negative effects by reducing vegetation cover. Shifts from cattle to other livestock species, paired with growing wildlife losses, will likely lead to corresponding shifts in tick communities and pathogen prevalence that reflect the new host assemblage. These effects will be context-dependent, subject to modification by tick-control strategies and unlikely to apply systematically across all tick species.

Future climate changes will continue to have direct effects on tick biology and pathogen transmission. However, given the enormous magnitude to which humans now control the movement and density of suitable tick hosts and vegetation via livestock management, it is critical to consider how human adaptations to climate change will further impact ticks and tick-borne diseases.

References

Abbas, R.Z., Zaman, M.A., Colwell, D.D., Gilleard, J. and Iqbal, Z. (2014) Acaricide resistance in cattle ticks and approaches to its management: the state of play. *Veterinary Parasitology* 203, 6–20.

Bar-On, Y.M., Phillips, R. and Milo, R. (2018) The biomass distribution on Earth. *Proceedings of the National Academy of Sciences USA* 115, 6506–6511.

Bishop, R., Musoke, A., Morzaria, S., Gardner, M. and Nene, V. (2004) Theileria: intracellular protozoan parasites of wild and domestic ruminants transmitted by ixodid ticks. *Parasitology* 129, S271–S283.

Cumming, G.S. (1998) Host preference in African ticks (Acari: Ixodida): a quantitative data set. *Bulletin of Entomological Research* 88, 379–406.

Cumming, G.S. (1999) Host distributions do not limit the species ranges of most African ticks (Acari: Ixodida). *Bulletin of Entomological Research* 89, 303–327.

de la Fuente, J., Estrada-Peña, A., Venzal, J.M., Kocan, K.M. and Sonenshine, D.E. (2008) Ticks as vectors of pathogens that cause disease in humans and animals. *Frontiers in Bioscience* 13, 6938–6946.

Dunn, R.R., Harris, N.C., Colwell, R.K., Koh, L.P. and Sodhi, N.S. (2009) The sixth mass coextinction: are most endangered species parasites and mutualists? *Proceedings of the Royal Society B* 276, 3037–3045.

Esser, H.J., Herre, E.A., Kays, R., Liefting, Y. and Jansen, P.A. (2019) Local host–tick coextinction in neotropical forest fragments. *International Journal for Parasitology* 49, 225–233.

Hempson, G.P., Archibald, S. and Bond, W.J. (2017) The consequences of replacing wildlife with livestock in Africa. *Scientific Reports* 7, 17196.

Jones, P.G. and Thornton, P.K. (2009) Croppers to livestock keepers: livelihood transitions to 2050 in Africa due to climate change. *Environmental Science and Policy* 12, 427–437.
Kagunyu, A.W. and Wanjohi, J. (2014) Camel rearing replacing cattle production among the Borana community in Isiolo County of Northern Kenya, as climate variability bites. *Pastoralism* 4, 13.
Keesing, F., Allan, B.F., Young, T.P. and Ostfeld, R.S. (2013) Effects of wildlife and cattle on tick abundance in central Kenya. *Ecological Applications* 23, 1410–1418.
McCoy, K.D., Léger, E. and Dietrich, M. (2013) Host specialization in ticks and transmission of tick-borne diseases: a review. *Frontiers in Cellular and Infection Microbiology* 3, 57.
Mugabi, K.N., Mugisha, A. and Ocaido, M. (2010) Socio-economic factors influencing the use of acaricides on livestock: a case study of the pastoralist communities of Nakasongola District, Central Uganda. *Tropical Animal Health and Production* 42, 131–136.
Niang, I., Ruppel, O.C., Abdrabo, M., Essel, A., Leonard, C., *et al.* (2014) Africa. In: Barros, V.R., Field, C.B., Dokken, D.J., Mastrandrea, M.D., Mach, K.J., *et al.* (eds) *Climate Change 2014: Impacts, Adaptation, and Vulnerability. Part B: Regional Aspects. Contribution of Working Group II to the Fifth Assessment Report of the Intergovernmental Panel on Climate Change*. Cambridge University Press, Cambridge and New York, pp. 1199–1265.
Nyangiwe, N., Yawa, M. and Muchenje, V. (2018) Driving forces for changes in geographic range of cattle ticks (Acari: Ixodidae) in Africa: a review. *South African Journal of Animal Science* 48, 829–841.
Ogrzewalska, M., Uezu, A., Jenkins, C.N. and Labruna, M.B. (2011) Effect of forest fragmentation on tick infestations of birds and tick infection rates by *Rickettsia* in the Atlantic Forest of Brazil. *EcoHealth* 8, 320–331.
Ogutu, J.O., Piepho, H.-P., Said, M.Y., Ojwang, G.O., Njino, L.W., *et al.* (2016) Extreme wildlife declines and concurrent increase in livestock numbers in Kenya: what are the causes? *PLoS ONE* 11, e0163249.
Olwoch, J.M., Van Jaarsveld, A.S., Scholtz, C.H. and Horak, I.G. (2007) Climate change and the genus *Rhipicephalus* (Acari: Ixodidae) in Africa. *Onderstepoort Journal of Veterinary Research* 74, 45–72.
Perry, B.D., Lessard, P., Norval, R.A.I., Kundert, K. and Kruska, R. (1990) Climate, vegetation and the distribution of *Rhipicephalus appendiculatus* in Africa. *Parasitology Today* 6, 100–104.
Rippke, U., Ramirez-Villegas, J., Jarvis, A., Vermeulen, S.J., Parker, L., *et al.* (2016) Timescales of transformational climate change adaptation in sub-Saharan African agriculture. *Nature Climate Change* 6, 605–609.
Rutagwenda, T., Lechner-Doll, M., Schwartz, H.J., Schultka, W. and von Engelhardt, W. (1990) Dietary preference and degradability of forage on a semiarid thornbush savannah by indigenous ruminants, camels and donkeys. *Animal Feed Science and Technology* 31, 179–192.
Seo, S.N. and Mendelsohn, R. (2007) *Climate Change Adaptation in Africa: A Microeconomic Analysis of Livestock Choice (Policy Research Working Paper No. 4277)*. The World Bank, Washington, DC.
Silatsa, B.A., Kuiate, J.R., Njiokou, F., Simo, G., Feussom, J.M.K., *et al.* (2019) A countrywide molecular survey leads to a seminal identification of the invasive cattle tick *Rhipicephalus* (*Boophilus*) *microplus* in Cameroon, a decade after it was reported in Cote d'Ivoire. *Ticks and Tick-borne Diseases* 10, 585–593.
Titcomb, G., Allan, B.F., Ainsworth, T., Henson, L., Hedlund, T., *et al.* (2017) Interacting effects of wildlife loss and climate on ticks and tick-borne disease. *Proceedings of the Royal Society B* 284, 20170475.
Van Wieren, S., Braks, M. and Lahr, J. (2016) Effectiveness and environmental hazards of acaricides applied to large mammals for tick control. In: Braks, M., Van Wieren, S., Takken, W. and Sprong, H. (eds) *Ecology and Prevention of Lyme Borreliosis. Ecology and Control of Vector-Borne Diseases*, vol. 4. Wageningen Academic Publishers, Wageningen, the Netherlands, pp. 265–278.
Young, H.S., McCauley, D.J., Dirzo, R., Goheen, J.R., Agwanda, B., *et al.* (2015) Context-dependent effects of large-wildlife declines on small-mammal communities in central Kenya. *Ecological Applications* 25, 348–360.

eo21 Potential Impacts of Climate Change on Medically Important Tick Species in North America

Geoffrey Lynn, Sukanya Narasimhan and Erol Fikrig*
Yale School of Medicine, New Haven, Connecticut, USA

21.1 Expert Opinion

21.1.1 Introduction

Climate change is currently underway, accelerated primarily by greenhouse gas-emitting human activities, and continued increases in global mean temperatures are expected, especially at higher latitudes (Pachauri *et al.*, 2014). In North America, tick-borne diseases are a significant public health concern in the eastern USA and increasingly in southern Canada. Models predict that these areas are likely to receive increased precipitation along with higher mean annual temperatures (Romero-Lankao *et al.*, 2014). Warming trends are likely to extend over all seasons, including winter, when higher minimum temperatures and less snowfall are expected, along with an increase in dry summers and extreme heat days. These changes have potentially broad-reaching impacts on ecosystems and biodiversity, including arthropod disease vectors.

Ticks are parasitic arthropods that are present in a diverse range of habitats and members of the family Ixodidae (hard ticks) can be found on all seven continents. As ectothermic organisms with limited capabilities to traverse large physical distances, physiological limitations require ticks to be well-adapted to local environmental conditions. The combined strong influences of abiotic factors including climate and soil composition, as well as biotic elements including host presence, vegetation and anthropogenic land usage, strongly suggest that climate change will have a definite impact on the distribution and abundance of ticks and tick-borne disease dynamics. Accordingly, because most tick-borne pathogens are very closely associated with specific tick species, which in turn are often spatially limited by habitat suitability, it can be expected that the occurrence of tick-borne diseases will be strongly influenced by climate-related ecological changes.

21.1.2 Tick biology

Nearly all ticks are constrained to certain geographical regions by a set of environmental factors, which are often interrelated. While they have few natural predators, many ticks are susceptible to desiccation and consequently air and soil humidity, as well as composition of local vegetation, are important determinants of distribution patterns for most species found in temperate climates. Upper and lower temperature maximums also impose some level of limitation on tick activity and distribution. Host-seeking (questing) ticks exposed to inhospitable elements have the ability to relocate short distances to

*Email: erol.fikrig@yale.edu

DOI: 10.1079/9781789249637.0021

protected microclimates with more favourable conditions, affording them some buffer to weather conditions, although this ability comes with energetic costs. In addition to short-term behavioural quiescence, ticks also may undergo photoperiod-induced diapause where developmental metamorphosis is arrested during seasons with unfavourable environmental conditions (Belozerov *et al.*, 2002). Similarly, maximum and minimum daily temperatures limit the periods when active questing for hosts can occur: many species are unable to survive freezing temperatures, while excessively hot temperatures can likewise result in accelerated water loss and increased mortality (Lindsay *et al.*, 1995; Vail and Smith, 1998; Ogden *et al.*, 2004). At the broader temporal scale, warmer temperatures accelerate interstadial development rates (Ogden *et al.*, 2006), potentially resulting in additional generations per calendar year. Furthermore, climate-induced alterations to tick phenology (seasonal activity patterns) are likely to influence composition of pathogens associated with specific ticks. Most tick-borne pathogens are transmitted horizontally, which requires that uninfected ticks acquire pathogens by feeding on an infected host. For species with asynchronous phenology, shorter winters may increase the temporal period between peak nymphal and larval activity, which extends the duration that an infection must persist within a host between transmission by nymphs and potential acquisition by uninfected larvae (Gatewood *et al.*, 2009). For blacklegged ticks (*Ixodes scapularis*), there is evidence that synchronized phenology may be necessary for enzootic maintenance of *Ehrlichia muris euclairensis*, an acute infection in rodent hosts (Stromdahl *et al.*, 2014; Lynn *et al.*, 2017). Likewise, cases of infection with *Anaplasma phagocytophilum*, also a short-lived infection when quantified in relation to *Borrelia burgdorferi* cases, are higher in cold-weather states with greater synchrony. Conversely, genotypes of *B. burgdorferi* that persist in hosts for longer periods are expected to increase as a result of climate-related changes in tick phenology (Gatewood *et al.* 2009; Levi *et al.*, 2015).

The local composition of fauna also has the potential to influence which tick species establish continuous populations in a given location. *I. scapularis*, *Dermacentor variabilis* (the American dog tick) and *Amblyomma maculatum* (the Gulf Coast tick) are moderate to strong generalists as parasites, feeding on small and medium-sized mammals, birds, lizards, as well as large animals to a lesser extent. In contrast, *Amblyomma americanum* (the lone star tick) is more of a host specialist, feeding primarily on larger animals including deer and canines throughout its life cycle. Likewise, it appears that *Haemaphysalis longicornis* (the Asian long-horned tick), a newly introduced species to the eastern USA, also prefers deer and other medium and large mammals, while largely ignoring bird and rodent hosts (Tufts *et al.*, 2021). Importantly, deer are key hosts for the adult stages of most medically important tick species in North America and heavy reduction or elimination of deer herds would likely limit the abundance and possibly long-term presence of certain tick species. However, white-tailed deer (*Odocoileus virginianus*) populations have been increasing for decades in much of the geographical USA and climate, specifically winter severity, is the most influential factor limiting the distribution of white-tailed deer in Canada. Unsurprisingly, increasingly moderate winters are expected to enable continued northward expansion of white-tailed deer range (Kennedy-Slaney *et al.*, 2018). Similarly, the white-footed mouse (*Peromyscus leucopus*) is a key host for immature stages of *I. scapularis* and *D. variabilis* and is present in much of eastern North America where it is frequently the most abundant mammal. The white-footed mouse is also a highly important reservoir for *B. burgdorferi*, with populations predicted to continue northward expansion in Quebec in part due to milder, shorter winters (Roy-Dufresne *et al.*, 2013).

21.1.3 Medically relevant hard tick species in the eastern USA and Canada

Much of western North America is expected to see increased droughts and greater soil aridity, which adds to the challenge of predicting climate change impact on ticks. However, in the Northeast, Southeast and Midwest regions, an increase in potentially favourable conditions for several tick species has prompted greater study than in the West. In the Midwest and eastern regions, *I. scapularis*, the primary vector of human tick-borne infections, has expanded its geographic distribution since at least the 1970s (Eisen *et al.*,

2016; Lantos *et al.*, 2017; Oliver *et al.*, 2017), coinciding with warmer winters in the areas where ticks were formerly absent (Sonenshine, 2018). Minimum and maximum temperatures and precipitation or humidity levels have been shown to be strong indicators of *I. scapularis* presence (Estrada-Peña, 2002; Diuk-Wasser *et al.*, 2010), which suggests that a warming climate will enable further expansion into higher latitudes. The northern limits of *I. scapularis* range are probably determined by insufficient degree-days for metamorphological development and extreme low temperatures, associated with increased mortality. Warmer and short winters appear to be driving extension of the northern edges of *I. scapularis* range, where these ticks are being identified with greater frequency in southern and eastern Canada (Ogden *et al.*, 2008; Clow *et al.*, 2017). *I. scapularis* is particularly sensitive to relative humidity and, with some exceptions, is typically found in shaded temperate deciduous forests, which may limit continued western expansion into semi-arid regions beyond North Dakota and the Great Plains (Russart *et al.*, 2014). However, a warming climate is likely to enable the transformation of large expanses of northern boreal forest, comprised of spruce and other coniferous softwood tree species, to mixed and mostly temperate deciduous forests. It is likely that a shift that leads to a greater concentration of hardwoods such as oaks could create improved conditions for *I. scapularis* and many mammalian hosts, including white-tailed deer (Frelich *et al.*, 2012). Although it is unclear how climate change may impact *I. scapularis* at the current southern limits of its range, lower survival at simulated southern conditions relative to northern conditions suggests that the southern limit of *I. scapularis* might retract north (Ginsberg *et al.*, 2014). This may also have important implications for disease transmission given that *I. scapularis* originating from southern locations exhibit different questing behaviours from northern counterparts, which are genetically and geographically isolated from the former. These differences account for dissimilar host associations and are likely to contribute to lower incidence of Lyme disease in southern locations (Arsnoe *et al.*, 2015).

D. variabilis ticks are present throughout much of the same range as *I. scapularis*, from the south-eastern USA into southern Canada, although the former is less sensitive to low humidity allowing it to exploit a greater range of habitat types. A northward expansion is currently underway, and it is expected that warmer temperatures will increase the range of *D. variabilis* further into Canada, as well as at higher elevations within the USA, while potentially decreasing suitability of southern range (James *et al.*, 2015; Minigan *et al.*, 2018). *D. variabilis* is the primary vector of *Rickettsia rickettsii* in the USA, where incidence of rickettsiosis is greatest in the Southeast region. It is therefore currently unclear how climate change might influence diseases associated with this species.

A. americanum is an important vector for multiple diseases in the eastern and south-eastern USA. Its current range expansion is progressing northward along the New England coast in moderate maritime climates, and west, beyond historical recorded range (Molaei *et al.*, 2019). Two recent models (Raghavan *et al.*, 2019b; Sagurova *et al.*, 2019) predict that climate change will enhance continued northerly and western expansion of populations. Individual specimens have been collected as far north and west as North Dakota (Russart *et al.*, 2014) and while there is no evidence of reproductive populations yet, this suggests bird dispersal of immature stages is occurring, which could seed endemic populations under more favourable environmental conditions. Low atmospheric relative humidity is another limiting factor that is expected to halt *A. americanum* establishment west of the Great Plains.

Similarly, it is possible that climate change will impact *A. maculatum*, a species most commonly found in the south-eastern and south-central USA and the primary vector for *Rickettsia parkeri* (Paddock and Goddard, 2015). Although this species was thought to prefer warm moist coastal climates, it has experienced a recent and dramatic expansion in its range, which includes isolated breeding populations further inland as far as Kansas and Arizona. *A. maculatum* is able to utilize a broad range of vertebrate hosts including birds, which suggests that temperature and/or humidity are the primary factors limiting distribution of this species. And while it is likely that the transport by animals to favourable microclimates explains the occurrence of isolated populations, a broader trend of increased presence in cooler and drier pastoral areas raises the possibility that physiological adaptation may contribute to survival (Sonenshine, 2018). Consequently, the extent that *A. maculatum* range will continue to expand is uncertain.

Owing to its recent introduction to the USA, the current geographic distribution of *H. longicornis* has not been definitively established. As of 2019, specimens have been identified in locations along the East Coast from Connecticut to South Carolina and inland as far as Arkansas. Models created using ecological inputs associated with *H. longicornis* distribution patterns in Asia have estimated that large areas of the South-east and East Coast, as well as smaller pockets along the West Coast and southern Canada, may currently be suitable habitat (Raghavan *et al.*, 2019a; Rochlin, 2019). Until information is known about its suitable climatic conditions and adaptability to native North American hosts, it is presently difficult to predict the extent that climate change may affect the distribution of this species and pathogens it may potentially interact with. However, *H. longicornis* is a vector of human and veterinary pathogens in Asia; in the USA, it was demonstrated to be a competent vector of *R. rickettsii* under laboratory conditions (Stanley *et al.*, 2020) and has been documented attacking humans.

The brown dog tick (*Rhipicephalus sanguineus*) is a globally distributed species that is also a vector of veterinary pathogens and, occasionally, human rickettsioses (Dantas-Torres, 2010). It is most closely associated with canines and, by extension, humans, and is found in urban and rural settings. *R. sanguineus* is both endophilic (adapted to indoor living) and capable of surviving outdoors in tropical, semi-tropical and some temperate climates where peri-domestic refuges are available. While *R. sanguineus* is resistant to desiccation and may not be greatly impacted by modest changes in relative humidity (Yoder *et al.*, 2006), it is likely that warmer temperatures allow for improved survival at higher latitudes (Gray *et al.*, 2013) and an increased number of annual generations. Additionally, warmer temperatures are associated with a greater propensity for this species to parasitize humans (Parola *et al.*, 2008).

21.1.4 Land use and anthropogenic factors

While climate is likely to be of primary significance, land use and anthropogenic disturbances should not be overlooked as a major influence of the future distribution of tick species and their associated pathogens. Historically, the arrival of Europeans in North America resulted in large-scale clearance of forests for agriculture and industrial purposes. When combined with hunting, formerly abundant deer populations were reduced or eliminated from much of the north-eastern USA and, likewise, black-legged ticks became scarce. Gradually, changes in land use resulted in widespread reforestation, which allowed deer and *I. scapularis* to repopulate these areas (Barbour and Fish, 1993). Land use has also been shown to impact yet another key host, white-footed mice, where abundance was positively associated with habitat fragmentation (Simon *et al.*, 2014). Additional factors that are difficult to account for, and yet may influence future distribution of ticks in North America, include importation of exotic species and intraspecies interactions, including potential competition between tick species with highly similar niches. Ultimately, however, future outcomes associated with climate change – such as rising sea level, altered growing conditions and water resources, and wildfires and flooding – suggest that land-use changes will also be driven in part by a changing climate and are likely to present similar challenges globally.

21.2 Future Predictions

The impact of climate change on ticks and tick-borne diseases is determined by a network of interactions between biotic and abiotic factors on tick biology and ecology. Climate change will also significantly influence human behaviour directly or indirectly (Evans, 2019), and this should be considered to the extent it is possible in order to accurately model potential outcomes of climate change on human disease. The development of computational algorithms that include multiple variables over time and space to accurately predict the impact of climate on tick-transmitted human diseases will be warranted. Molecular changes in the tick may precede successful adaptation to increasing temperatures and humidity and may also have a functional consequence on the vectorial capacity of the tick. It is also important to bear in mind that the infection status of the tick may influence tick survival under higher temperatures. For example, an earlier study by Neelakanta *et al.* (2010) showed that *A. phagocytophilum* infection increased the expression of an antifreeze glycoprotein, IAFGP, that provided a survival advantage to *I. scapularis* nymphs exposed to cold temperatures. If infection with a specific symbiont enhances the ability of the tick to survive at higher temperatures or lower relative humidity levels, it is conceivable that climate change may favour increased prevalence of certain microbes, including tick-borne pathogens. The kinetics of pathogen transmission may also be impacted by climate change. For example, gene expression in *B. burgdorferi* undergoes dramatic changes in response to changes in temperature and plays a critical role in transmission to the vertebrate host (Ojaimi *et al.*, 2003). In laboratory-reared ticks, spirochaete survival declined over time when ticks were incubated at temperatures in excess of 27°C (Shih *et al.*, 1995). Finally, impact on host immunity will be yet another significant determinant of how climate change will influence enzootic diseases. A recent study by Moriyama and Ichinohe (2019) reported that mice housed at increased temperatures were more susceptible to influenza virus infection due to dampened adaptive immune responses, whereas elevated temperatures induced stronger immune responses in mice infected with trypanosome parasites (Anderson and Kuhn, 1989). Therefore, it is critically important to improve collective understanding of the impact of climate change on tick–host–pathogen interactions at the molecular level, in order to make educated predictions for future tick-borne disease-related phenomena.

Acknowledgements

This work was supported by grants from the National Institutes of Health (AI126033, AI138949) and the Steven and Alexandra Cohen Foundation.

References

Anderson, K.J. and Kuhn, R.E. (1989) Elevated environmental temperature enhances immunity in experimental Chagas' disease. *Infection and Immunity* 57, 13–17. https://doi.org/10.1128/IAI.57.1.13-17.1989

Arsnoe, I.M., Hickling, G.J., Ginsberg, H.S., McElreath, R. and Tsao, J.I. (2015) Different populations of blacklegged tick nymphs exhibit differences in questing behavior that have implications for human Lyme disease risk. *PLoS ONE* 10, e0127450. https://doi.org/10.1371/journal.pone.0127450

Barbour, A.G. and Fish, D. (1993) The biological and social phenomenon of Lyme disease. *Science* 260, 1610–1616. https://doi.org/10.1126/science.8503006

Belozerov, V.N., Fourie, L.J. and Kok, D.J. (2002) Photoperiodic control of developmental diapause in nymphs of prostriate ixodid ticks (Acari: Ixodidae). *Experimental and Applied Acarology* 28, 163–168. https://doi.org/10.1023/A:1025377829119

Clow, K.M., Leighton, P.A., Ogden, N.H., Lindsay, L.R., Michel, P., *et al.* (2017) Northward range expansion of *Ixodes scapularis* evident over a short timescale in Ontario, Canada. *PLoS ONE* 12, e0189393. https://doi.org/10.1371/journal.pone.0189393

Dantas-Torres, F. (2010) Biology and ecology of the brown dog tick, *Rhipicephalus sanguineus*. *Parasites & Vectors* 3, 26. https://doi.org/10.1186/1756-3305-3-26

Diuk-Wasser, M.A., Vourc'h, G., Cislo, P., Hoen, A.G., Melton, F., *et al.* (2010) Field and climate-based model for predicting the density of host-seeking nymphal *Ixodes scapularis*, an important vector of tick-borne disease agents in the eastern United States. *Global Ecology and Biogeography* 19, 504–514. https://doi.org/10.1111/j.1466-8238.2010.00526.x

Eisen, R.J., Eisen, L. and Beard, C.B. (2016) County-scale distribution of *Ixodes scapularis* and *Ixodes pacificus* (Acari: Ixodidae) in the continental United States. *Journal of Medical Entomology* 53, 349–386. https://doi.org/10.1093/jme/tjv237

Estrada-Peña, A. (2002) Increasing habitat suitability in the United States for the tick that transmits Lyme disease: a remote sensing approach. *Environmental Health Perspectives* 110, 635–640. https://doi.org/10.1289/ehp.110-1240908

Evans, G.W. (2019) Projected behavioral impacts of global climate change. *Annual Review of Psychology* 70, 449–474. https://doi.org/10.1146/annurev-psych-010418-103023

Frelich, L.E., Peterson, R.O., Dovčiak, M., Reich, P.B., Vucetich, J.A. and Eisenhauer, N. (2012) Trophic cascades, invasive species and body-size hierarchies interactively modulate climate change responses of ecotonal temperate–boreal forest. *Philosophical Transactions of the Royal Society of London B* 367, 2955–2961. https://doi.org/10.1098/rstb.2012.0235

Gatewood, A.G., Liebman, K.A., Vourc'h, G., Bunikis, J., Hamer, S.A., *et al.* (2009) Climate and tick seasonality are predictors of *Borrelia burgdorferi* genotype distribution. *Applied and Environmental Microbiology* 75, 2476–2483. https://doi.org/10.1128/AEM.02633-08

Ginsberg, H.S., Rulison, E.L., Azevedo, A., Pang, G.C., Kuczaj, I.M., *et al.* (2014) Comparison of survival patterns of northern and southern genotypes of the North American tick *Ixodes scapularis* (Acari: Ixodidae) under northern and southern conditions. *Parasites & Vectors* 7, 394. https://doi.org/10.1186/1756-3305-7-394

Gray, J., Dantas-Torres, F., Estrada-Peña, A. and Levin, M. (2013) Systematics and ecology of the brown dog tick, *Rhipicephalus sanguineus*. *Ticks and Tick-borne Diseases* 4, 171–180. https://doi.org/10.1016/j.ttbdis.2012.12.003

James, A.M., Burdett, C., McCool, M.J., Fox, A. and Riggs, P. (2015) The geographic distribution and ecological preferences of the American dog tick, *Dermacentor variabilis* (Say), in the USA. *Medical and Veterinary Entomology* 29, 178–188. https://doi.org/10.1111/mve.12099

Kennedy-Slaney, L., Bowman, J., Walpole, A.A. and Pond, B.A. (2018) Northward bound: the distribution of white-tailed deer in Ontario under a changing climate. *Wildlife Research* 45, 220–228. https://doi.org/10.1071/WR17106

Lantos, P.M., Tsao, J., Nigrovic, L.E., Auwaerter, P.G., Fowler, V.G., *et al.* (2017) Geographic expansion of Lyme disease in Michigan, 2000–2014. *Open Forum Infectious Diseases* 4, ofw269. https://doi.org/10.1093/ofid/ofw269

Levi, T., Keesing, F., Oggenfuss, K. and Ostfeld, R.S. (2015) Accelerated phenology of blacklegged ticks under climate warming. *Philosophical Transactions of the Royal Society of London B* 370, 20130556. https://doi.org/10.1098/rstb.2013.0556

Lindsay, L.R., Barker, I.K., Surgeoner, G.A., McEwen, S.A., Gillespie, T.J. and Robinson, J.T. (1995) Survival and development of *Ixodes scapularis* (Acari: Ixodidae) under various climatic conditions in Ontario, Canada. *Journal of Medical Entomology* 32, 143–152. https://doi.org/10.1093/jmedent/32.2.143

Lynn, G.E., Oliver, J.D., Cornax, I., O'Sullivan, M.G. and Munderloh, U.G. (2017) Experimental evaluation of *Peromyscus leucopus* as a reservoir host of the *Ehrlichia muris*-like agent. *Parasites & Vectors* 10, 48. https://doi.org/10.1186/s13071-017-1980-4

Minigan, J.N., Hager, H.A., Peregrine, A.S. and Newman, J.A. (2018) Current and potential future distribution of the American dog tick (*Dermacentor variabilis*, Say) in North America. *Ticks and Tick-borne Diseases* 9, 354–362. https://doi.org/10.1016/j.ttbdis.2017.11.012

Molaei, G., Little, E., Williams, S.C. and Stafford, K.C. (2019) Bracing for the worst – range expansion of the lone star tick in the northeastern United States. *New England Journal of Medicine* 381, 2189–2192. https://doi.org/10.1056/NEJMp1911661

Moriyama, M. and Ichinohe, T. (2019) High ambient temperature dampens adaptive immune responses to influenza A virus infection. *Proceedings of the National Academy of Sciences USA* 116, 3118–3125. https://doi.org/10.1073/pnas.1815029116

Neelakanta, G., Sultana, H., Fish, D., Anderson, J.F. and Fikrig, E. (2010) *Anaplasma phagocytophilum* induces *Ixodes scapularis* ticks to express an antifreeze glycoprotein gene that enhances their survival in the cold. *Journal of Clinical Investigation* 120, 3179–3190. https://doi.org/10.1172/JCI42868

Ogden, N.H., Lindsay, L.R., Beauchamp, G., Charron, D., Maarouf, A., *et al.* (2004) Investigation of relationships between temperature and developmental rates of tick *Ixodes scapularis* (Acari: Ixodidae) in the laboratory and field. *Journal of Medical Entomology* 41, 622–633. https://doi.org/10.1603/0022-2585-41.4.622

Ogden, N.H., Maarouf, A., Barker, I.K., Bigras-Poulin, M., Lindsay, L.R., *et al.* (2006) Climate change and the potential for range expansion of the Lyme disease vector *Ixodes scapularis* in Canada. *International Journal for Parasitology* 36, 63–70. https://doi.org/10.1016/j.ijpara.2005.08.016

Ogden, N.H., St-Onge, L., Barker, I.K., Brazeau, S., Bigras-Poulin, M., *et al.* (2008) Risk maps for range expansion of the Lyme disease vector, *Ixodes scapularis*, in Canada now and with climate change. *International Journal of Health Geographics* 7, 24. https://doi.org/10.1186/1476-072X-7-24

Ojaimi, C., Brooks, C., Casjens, S., Rosa, P., Elias, A., *et al.* (2003) Profiling of temperature-induced changes in *Borrelia burgdorferi* gene expression by using whole genome arrays. *Infection and Immunity* 71, 1689–1705. https://doi.org/10.1128/iai.71.4.1689-1705.2003

Oliver, J.D., Bennett, S.W., Beati, L. and Bartholomay, L.C. (2017) Range expansion and increasing *Borrelia burgdorferi* infection of the tick *Ixodes scapularis* (Acari: Ixodidae) in Iowa, 1990–2013. *Journal of Medical Entomology* 54, 1727–1734. https://doi.org/10.1093/jme/tjx121

Pachauri, R.K., Allen, M.R., Barros, V.R., Broome, J., Cramer, W., *et al.* (2014) *Climate Change 2014: Synthesis Report. Contribution of Working Groups I, II and III to the Fifth Assessment Report of the Intergovernmental Panel on Climate Change*. Intergovernmental Panel on Climate Change, Geneva, Switzerland.

Paddock, C.D. and Goddard, J. (2015) The evolving medical and veterinary importance of the Gulf Coast tick (Acari: Ixodidae). *Journal of Medical Entomology* 52, 230–252. https://doi.org/10.1093/jme/tju022

Parola, P., Socolovschi, C., Jeanjean, L., Bitam, I., Fournier, P.E., *et al.* (2008) Warmer weather linked to tick attack and emergence of severe rickettsioses. *PLoS Neglected Tropical Diseases* 2, e338. https://doi.org/10.1371/journal.pntd.0000338

Raghavan, R.K., Barker, S.C., Cobos, M.E., Barker, D., Teo, E., *et al.* (2019a) Potential spatial distribution of the newly introduced long-horned tick, *Haemaphysalis longicornis* in North America. *Scientific Reports* 9, 498. https://doi.org/10.1038/s41598-018-37205-2

Raghavan, R.K., Peterson, A.T., Cobos, M.E., Ganta, R. and Foley, D. (2019b) Current and future distribution of the lone star tick, *Amblyomma americanum* (L.) (Acari: Ixodidae) in North America. *PLoS ONE* 14, e0209082. https://doi.org/10.1371/journal.pone.0209082

Rochlin, I. (2019) Modeling the Asian longhorned tick (Acari: Ixodidae) suitable habitat in North America. *Journal of Medical Entomology* 56, 384–391. https://doi.org/10.1093/jme/tjy210

Romero-Lankao, P., Smith, J.B., Davidson, D.J., Diffenbaugh, N.S., Kinney, P.L., *et al.* (2014) North America. In: Barros, V.R., Field, C.B., Dokken, D.J., Mastrandrea, M.D., Mach, K.J., *et al.* (eds) *Climate Change 2014: Impacts, Adaptation, and Vulnerability. Part B: Regional Aspects. Contribution of Working Group II to the Fifth Assessment Report of the Intergovernmental Panel on Climate Change*. Cambridge University Press, Cambridge and New York, pp. 1439–1498.

Roy-Dufresne, E., Logan, T., Simon, J.A., Chmura, G.L. and Millien, V. (2013) Poleward expansion of the white-footed mouse (*Peromyscus leucopus*) under climate change: implications for the spread of Lyme disease. *PLoS ONE* 8, e80724. https://doi.org/10.1371/journal.pone.0080724

Russart, N.M., Dougherty, M.W. and Vaughan, J.A. (2014) Survey of ticks (Acari: Ixodidae) and tick-borne pathogens in North Dakota. *Journal of Medical Entomology* 51, 1087–1090. https://doi.org/10.1603/me14053

Sagurova, I., Ludwig, A., Ogden, N.H., Pelcat, Y., Dueymes, G. and Gachon, P. (2019) Predicted northward expansion of the geographic range of the tick vector *Amblyomma americanum* in North America under future climate conditions. *Environmental Health Perspectives* 127, 107014. https://doi.org/10.1289/EHP5668

Shih, C.M., Telford, S.R. 3rd and Spielman, A. (1995) Effect of ambient temperature on competence of deer ticks as hosts for Lyme disease spirochetes. *Journal of Clinical Microbiology* 33, 958–961. https://doi.org/10.1128/JCM.33.4.958-961.1995

Simon, J.A., Marrotte, R.R., Desrosiers, N., Fiset, J., Gaitan, J., *et al.* (2014) Climate change and habitat fragmentation drive the occurrence of *Borrelia burgdorferi*, the agent of Lyme disease, at the northeastern limit of its distribution. *Evolutionary Applications* 7, 750–764. https://doi.org/10.1111/eva.12165

Sonenshine, D.E. (2018) Range expansion of tick disease vectors in North America: implications for spread of tick-borne disease. *International Journal of Environmental Research and Public Health* 15, 478. https://doi.org/10.3390/ijerph15030478

Stanley, H.M., Ford, S.L., Snellgrove, A.N., Hartzer, K., Smith, E.B., *et al.* (2020) The ability of the invasive Asian longhorned tick *Haemaphysalis longicornis* (Acari: Ixodidae) to acquire and transmit *Rickettsia rickettsii* (Rickettsiales: Rickettsiaceae), the agent of Rocky Mountain spotted fever, under laboratory conditions. *Journal of Medical Entomology* 57, 1635–1639. https://doi.org/10.1093/jme/tjaa076

Stromdahl, E., Hamer, S., Jenkins, S., Sloan, L., Williamson, P., *et al.* (2014) Comparison of phenology and pathogen prevalence, including infection with the *Ehrlichia muris*-like (EML) agent, of *Ixodes scapularis* removed from soldiers in the midwestern and the northeastern United States over a 15 year period (1997–2012). *Parasites & Vectors* 7, 553. https://doi.org/10.1186/s13071-014-0553-z

Tufts, D.M., Goodman, L.B., Benedict, M.C., Davis, A.D., VanAcker, M.C. and Diuk-Wasser, M. (2021) Association of the invasive *Haemaphysalis longicornis* tick with vertebrate hosts, other native tick vectors, and tick-borne pathogens in New York City, USA. *International Journal for Parasitology* 51, 149–157. https://doi.org/10.1016/j.ijpara.2020.08.008

Vail, S.G. and Smith, G. (1998) Air temperature and relative humidity effects on behavioral activity of blacklegged tick (Acari: Ixodidae) nymphs in New Jersey. *Journal of Medical Entomology* 35, 1025–1028. https://doi.org/10.1093/jmedent/35.6.1025

Yoder, J.A., Benoit, J.B., Rellinger, E.J. and Tank, J.L. (2006) Developmental profiles in tick water balance with a focus on the new Rocky Mountain spotted fever vector, *Rhipicephalus sanguineus*. *Medical and Veterinary Entomology* 20, 365–372. https://doi.org/10.1111/j.1365-2915.2006.00642.x

eo22 Climate Change and Tick Evolution: Lessons from the Past

Ben Mans[1,2,3*]
[1]*Onderstepoort Veterinary Research, Pretoria, South Africa;* [2]*University of Pretoria, Pretoria, South Africa;* [3]*University of South Africa, Florida, South Africa*

22.1 Expert Opinion

22.1.1 Introduction

Extant ticks have defined biogeographic distributions that are determined by their evolutionary history (geographic origin of speciation), but also by ecological preferences that include habitat, host and environmental factors such as temperature and vegetation (Estrada-Peña, 2008). By extension, the geographic distribution of tick-borne diseases is dependent on and integrally linked with tick distribution (Estrada-Peña *et al.*, 2012, 2013). If climate change leads to changes in environmental factors, habitat, host distribution, land cover and agricultural practices, it would certainly affect the biogeographic distribution of ticks as well as tick-borne disease prevalence. This may be addressed by predictions of future climate change informed by current knowledge of habitat preferences of tick species (Estrada-Peña *et al.*, 2012). A more interesting question is whether climate change would impact on tick evolution, both in the short and long term. Past climate change events and the impact on tick evolution may offer some general insights into the evolution of ticks and how future climate changes could affect this.

22.1.2 Tick evolution and origins

The origin of ticks, both geographic and temporal, determines how much we may deduce regarding the impact of climate change on their evolution. As such, molecular clock analysis based on mitochondrial genes suggests that ticks originated ~273 million years ago (mya) in the Permian, while soft and hard tick lineages diverged ~260 mya (Mans *et al.*, 2011, 2012, 2019). Major lineages such as the Ixodidae and Argasidae originated ~225 mya (Mans *et al.*, 2019). The Argasinae and Ornithodorinae originated ~160–180 mya, the *Carios* genus (Klompen and Oliver, 1993) ~125 mya, Prostriates ~220 mya, Metastriates ~180 mya, Amblyomminae ~175 mya, Haemaphysalinae ~170 mya and Rhipicephalinae ~150 mya (Mans *et al.*, 2019).

Alternatively, molecular clock analysis using the nuclear 18S rRNA gene placed the origin of ticks ~190 mya in the Jurassic and divergence of hard and soft ticks ~178 mya (Beati and Klompen, 2019). Major lineages such as the Ixodidae and Argasidae originated ~130–160 mya (Beati and Klompen, 2019). The Argasinae and Ornithodorinae originated ~95–100 mya, the *Carios* genus (Klompen and Oliver, 1993) ~46 mya, Prostriates ~112 mya, Metastriates ~97 mya,

*Email: manssb@arc.agric.za

DOI: 10.1079/9781789249637.0022

Amblyomminae ~71 mya, Haemaphysalinae ~33 mya and Rhipicephalinae ~36 mya (Beati and Klompen, 2019).

The difference in time of origin between these hypotheses is quite extensive and spans a number of geological periods. Since fossils for at least *Amblyomma* and *Haemaphysalis* found in Myanmar amber dated to ~100 mya (Chitimia-Dobler *et al.*, 2017, 2018), and for *Carios* (Klompen and Oliver, 1993) in New Jersey amber dated to ~90–94 mya (Klompen and Grimaldi, 2001), it would seem as if the nuclear dating is at the lower end of the scale. Given that amber fossils represent narrow slots in time and as such only minimum times of origin, the origins of ticks are likely much higher than the 90–100 mya amber fossil limit, since it would seem extremely unlikely that the origin of ticks and amber fossils would coincide. For the purpose of this discussion, we would consider a possible origin of ticks in the Late Permian/Early Triassic as postulated by Mans *et al.* (2019).

22.1.3 Climate change of major evolutionary periods

Sedimentary fossil records define the geological strata due to the clear transitionary records of the fossils found in various periods. This transition in fossil species is generally linked with major extinction events, the majority associated with climate change (MacLeod, 2015; Bond and Grasby, 2017). In ticks, the relatively long times of divergence between the basal lineages and extant genera have been interpreted as possible extinction events, with loss of lineages that may be linked to mass extinction events (Mans *et al.*, 2016).

The Permian mass extinction (~252 mya) was associated with a global warming event that may have triggered the extinction itself (Penn *et al.*, 2018). This mass extinction event resulted in the decimation of 75–90% of all marine and terrestrial species (Botha and Smith, 2006). In this regard, it has been proposed that the origin of ticks just before this mass extinction event may have resulted in the paucity of species in the monotypic Nuttalliellidae, leading to this species being termed a 'dead clade walking' and a living fossil (Mans *et al.*, 2011). A 'dead clade walking' refers to lineages that survive mass extinctions but never recover with regard to species diversity (Jablonski, 2002). It may be asked whether the same happened for the Deinocrotonidae, a sister family to the Nuttalliellidae, for which only a single species has been found in Myanmar amber (Peñalver *et al.*, 2017) and which seems to have become extinct, possibly during the Cretaceous–Tertiary (K–T) extinction event. It should also be considered whether the same happened to the lineages leading up to hard and soft tick lineages since these diverged around the Permian mass extinction and only expanded as families almost 30–40 million years later (Mans *et al.*, 2019). This divergence in tick lineages coincided with divergence of synapsids/diapsids, dinosaurs and mammaliaform stem clades, whose recovery and diversification were affected by the Permian extinction (Botha and Smith, 2006; Sahney and Benton, 2008).

The next mass extinction event was the end-Triassic mass extinction (~201 mya). This extinction event led to decimation of ~80% of all marine and terrestrial species and 50% of all tetrapods (Beerling and Berner, 2002). Climate change has also been associated with this extinction event (Beerling, 2002; Harris *et al.*, 2017; Soh *et al.*, 2017). Again, of interest is the occurrence of this extinction event after the divergence of the Prostriates/Metastriates and Argasinae/Ornithodorinae (~225 mya) but before divergence of the argasid families and the Metastriates (~160–180 mya) (Mans *et al.*, 2019). Similarly, there seems to be little diversification of tick lineages over this period, suggesting that a number of tick lineages may have gone extinct at that time. This may very well have been due to decimation of host lineages. In this regard, while dinosaurs and mammals survived the end-Triassic mass extinction relatively well, most of the archosauromorphs went extinct (Foth *et al.*, 2016), suggesting an association of ticks with these animals before subsequent host switching to dinosaurs and mammals.

The K–T extinction event occurred ~65 mya leading to decimation of the dinosaurs (Benton, 1989). This period of seeming stability of ~135 million years seemed to have been beneficial for the diversification of tick lineages since the majority of all extant genera originated within the Cretaceous (Mans *et al.*, 2019). While the K–T extinction event decimated the dinosaurs, modern tetrapod groups such as birds and mammals were not affected, but rather showed rapid radiation and occupation of new environmental niches (Benton, 1989). This linked with the now diverse array of tick genera might have allowed ticks to survive and recover from this mass extinction event better than the two previous events. Even so, evidence exists that a number of lineages went extinct over this period, notably the *Compluriscutula*, *Cornupalpatum* and the Deinocrotonidae (Poinar and Brown, 2003; Poinar and Buckley, 2008; Peñalver *et al.*, 2017). That many other species and lineages may also have become extinct is supported by the species-poor *Africaniella*, *Archaecroton*, *Bothriocroton* and *Robertsicus*, all genera proposed to have originated in the Cretaceous (Mans *et al.*, 2019).

As may be expected, the majority of expansions within extant genera occurred during the Tertiary (Mans *et al.*, 2019). This would also have been the time of greatest expansion of geographic ranges, associated no doubt with expansion of the geographic ranges of birds, mammals, ungulates and modern ecological habitats such as grasslands and woody savannah (Janis, 2003; Jardine *et al.*, 2012; Prum *et al.*, 2015; Guglielmone and Nava, 2017). Tectonics and climate changes had a significant impact on vertebrate evolution over this period (Figueirido *et al.*, 2012). This is echoed in the impact that tectonics and large-scale climatic changes had on the evolutionary history of ticks, notably *Hyalomma* (Sands *et al.*, 2017). These same factors probably also influenced the evolutionary histories of other tick genera.

22.1.4 The Anthropocene and the sixth mass extinction event

Recently, the Anthropocene concept has been introduced, for which no conclusive date has yet been defined (Corlett, 2015). The Anthropocene tries to define the impact that humans have on biodiversity, climate and the environment. With regard to ticks, the Agricultural Revolution linked with colonization of the continents (~10,000–7000 years ago) may be a better estimate, since the movement of people, their companion animals and livestock, as well as livestock expansion and husbandry, may have had a major impact on the geographic distribution of ticks of companion and commercial animals. As such, the extensive distribution of *Amblyomma hebraeum*, *Amblyomma variegatum*, *Rhipicephalus annulatus*, *Rhipicephalus appendiculatus*, *Rhipicephalus decoloratus*, *Rhipicephalus evertsi*, *Rhipicephalus microplus*, *Rhipicephalus sanguineus*, *Haemaphysalis longicornis*, *Haemaphysalis punctata*, *Hyalomma anatolicum*, *Hyalomma rufipes* and *Hyalomma truncatum* is certainly in part due to their ability to parasitize companion animals or livestock that allowed human-mediated dispersion throughout the geographic ranges where habitats are suitable for these tick species (Burrow *et al.*, 2019).

Also associated with the Anthropocene is the sixth mass extinction, which proposes that human-mediated climate change has a direct and tangible impact on biodiversity and extinction, with 515 species considered on the brink of extinction (Ceballos *et al.*, 2020). Human-mediated climate change has been proposed as one of the factors associated with this extinction event (Sage, 2020). In regard to ticks, a number of host species are in danger of extinction and some of these are preferred hosts for a number of tick species (Mihalca *et al.*, 2011). The latter study proposed that 63 tick species are co-endangered due to their hosts being on the Red List and that at least one species has gone extinct due to its inability to adapt to new hosts. The latter is *Ixodes nitens*, a parasite of the rat, *Rattus macleari*, last recorded in 1903 on Christmas Island. Ticks most likely to face extinction are host specialists with narrow host ranges that are unable to adapt to new hosts once their host of preference disappears. In contrast, host generalist species are more likely to survive mass extinction events and impacts of climate change. In this regard, it was proposed that the reason for the survival of *Nuttalliella namaqua* might be its ability to parasitize reptilian and mammalian hosts over the course of its evolutionary history (Mans *et al.*, 2014). Generalists also commonly have larger geographic ranges (Matthee, 2020), increasing the chances that local host extinction or perturbations in climate do not affect the entire species population. However, co-extinction has been identified as a significant threat to parasites in general since interdependent species are most affected by extinction events (Dunn *et al.*, 2009). In the case where species networks become extinct, generalist and specialist parasites may be equally affected.

22.2 Future Predictions

The above discussion on the possible impact of climate change and mass extinctions on tick biodiversity relies on an accurate molecular clock analysis. Even so, given that the oldest tick fossils are dated at 100 mya, it is very likely that ticks survived at least two major mass extinction events. It should be clear from the above that major climate changes, whether human-mediated or not, and their impact on associated host lineages, can lead to extinction of tick lineages, but also provide opportunities for lineage expansion if new hosts and/or environmental niches appear. While lineage extinction may be rapid, it is expected that lineage expansion associated with speciation will be a long-term outcome over several millennia.

On a more short-term basis, local changes in climate can influence geographic distribution of ticks. Estimates of this are based on geographic distribution modelling linked with habitat suitability mapping. Numerous studies have attempted to predict future changes in tick distribution using such habitat suitability mapping approaches; the references listed are not exhaustive (Estrada-Peña, 2003, 2008; Estrada-Peña and Venzal, 2007; Olwoch *et al.*, 2008; Gray *et al.*, 2009; Spickett, 2013; Estrada-Peña *et al.*, 2015). Estrada-Peña (2003) predicted that habitat suitability will decrease for the major economically important ticks in South Africa: *A. hebraeum*, *H. truncatum*, *R. appendiculatus* and *R. decoloratus*, from 1983 to 2015. A similar contraction in distribution was predicted for *R. appendiculatus* in the 'future' with an undetermined forecast range (Olwoch *et al.*, 2008). Conversely, it was predicted that the interior of South Africa will progressively become drier with expansion of wooded savannah, but linked with wetter, moister summers. Predictions based on temperature and rainfall that incorporate this drier interior scenario indicate that the habitat suitability of *R. appendiculatus* will probably expand westwards into central Mpumalanga, the eastern Free State and north-central North West provinces over the next 20–40 years (Spickett, 2013). The same holds true for *A. hebraeum*. Concomitantly, habitat suitability is predicted to decrease in the northern and eastern borders of KwaZulu-Natal, Limpopo and Mpumalanga due to wetter, moister summers, but increase for one-host tick species in these regions. To date, these decreases or increases in geographic distribution (Estrada-Peña, 2003; Olwoch *et al.*, 2008; Spickett, 2013) have not been recorded or confirmed (Horak *et al.*, 2018). This highlights several problems with habitat suitability modelling, notably that increased habitat suitability may not necessarily result in immediate range contraction or expansion, while excessive and prolonged countrywide surveillance is necessary to monitor changes in tick densities and distribution (Estrada-Peña, 2001) that would allow testing of habitat suitability predictions. These factors are likely to bias any predictive power that habitat suitability predictions may have. However, accurate future predictions of tick distributions may become feasible as the sophistication of modelling algorithms improves (Estrada-Peña *et al.*, 2016). Even so, the above predictions are micro scenarios for similar changes that could be expected on a global level. As such, geographic distributions of ticks are likely to show local changes over the next century with expansion and contraction of their ranges that will be dependent on local climate changes. In this regard, the expansion of *R. microplus* into habitats formerly considered outside its preferred habitat, with displacement of other more indigenous tick species such as *R. decoloratus*, should be noted (Burrow *et al.*, 2019). This has been reported in the Ivory Coast and Cameroon (Madder *et al.*, 2011; Mamoudou *et al.*, 2016; Boka *et al.*, 2017), South and southern Africa (Tønnesen *et al.*, 2004; Nyangiwe *et al.* 2011, 2013) and Tanzania and eastern Africa (Lynen *et al.*, 2008; Horak *et al.*, 2009). Expansion of *R. microplus* distribution has been linked with higher-temperature tolerance that may indicate adaptation in response to climate change and new environmental niches (Lynen *et al.*, 2008). Climate change-linked expansions and contractions in geographic ranges will likely lead to destabilization of endemic stability of diseases in regions of marginal habitat suitability, that may lead to increased disease outbreaks in these regions (Estrada-Peña, 2001). Conversely, integrated tick-control management linked with livestock-movement control and efficient monitoring of tick spread may slow or prevent the spread of ticks into new regions.

References

Beati, L. and Klompen, H. (2019) Phylogeography of ticks (Acari: Ixodida). *Annual Review of Entomology* 64, 379–397.

Beerling, D. (2002) CO_2 and the end-Triassic mass extinction. *Nature* 415, 386–387.

Beerling, D.J. and Berner, R.A. (2002) Biogeochemical constraints on the Triassic–Jurassic boundary carbon cycle event. *Global Biochemical Cycles* 16, 1036.

Benton, M.J. (1989) Mass extinctions among tetrapods and the quality of the fossil record. *Philosophical Transactions of the Royal Society of London B: Biological Sciences* 325, 369–385.

Boka, O.M., Achi, L., Adakal, H., Azokou, A., Yao, P., *et al.* (2017) Review of cattle ticks (Acari, Ixodida) in Ivory Coast and geographic distribution of *Rhipicephalus* (*Boophilus*) *microplus*, an emerging tick in West Africa. *Experimental and Applied Acarology* 71, 355–369.

Bond, D.P.G. and Grasby, S.E. (2017) On the causes of mass extinctions. *Palaeogeography, Palaeoclimatology, Palaeoecology* 478, 3–29.

Botha, J. and Smith, R.M.H. (2006) Rapid vertebrate recuperation in the Karoo Basin of South Africa following the end-Permian extinction. *Journal of African Earth Sciences* 45, 502–514.

Burrow, H.M., Mans, B.J., Cardoso, F.F., Birkett, M.A., Kotze, A.C., *et al.* (2019) Towards a new phenotype for tick resistance in beef and dairy cattle: a review. *Animal Production Science* 59, 1401–1427.

Ceballos, G., Ehrlich, P.R. and Raven, P.H. (2020) Vertebrates on the brink as indicators of biological annihilation and the sixth mass extinction. *Proceedings of the National Academy of Sciences USA* 117, 13596–13602.

Chitimia-Dobler, L., De Araujo, B.C., Ruthensteiner, B., Pfeffer, T. and Dunlop, J.A. (2017) *Amblyomma birmitum* a new species of hard tick in Burmese amber. *Parasitology* 144, 1441–1448.

Chitimia-Dobler, L., Pfeffer, T. and Dunlop, J.A. (2018) *Haemaphysalis cretacea* a nymph of a new species of hard tick in Burmese amber. *Parasitology* 145, 1440–1451.

Corlett, R.T. (2015) The Anthropocene concept in ecology and conservation. *Trends in Ecology and Evolution* 30, 36–41.

Dunn, R.R., Harris, N.C., Colwell, R.K., Koh, L.P. and Sodhi, N.S. (2009) The sixth mass coextinction: are most endangered species parasites and mutualists? *Proceedings of the Royal Society B* 276, 3037–3045.

Estrada-Peña, A. (2001) Climate warming and changes in habitat suitability for *Boophilus microplus* (Acari: Ixodidae) in Central America. *Journal of Parasitology* 87, 978–987.

Estrada-Peña, A. (2003) Climate change decreases habitat suitability for some tick species (Acari: Ixodidae) in South Africa. *Onderstepoort Journal of Veterinary Research* 70, 79–93.

Estrada-Peña, A. (2008) Climate, niche, ticks, and models: what they are and how we should interpret them. *Parasitology Research* 103(Suppl. 1), S87–S95.

Estrada-Peña, A. and Venzal, J.M. (2007) Climate niches of tick species in the Mediterranean region: modeling of occurrence data, distributional constraints, and impact of climate change. *Journal of Medical Entomology* 44, 1130–1138.

Estrada-Peña, A., Ayllón, N. and de la Fuente, J. (2012) Impact of climate trends on tick-borne pathogen transmission. *Frontiers in Physiology* 3, 64.

Estrada-Peña, A., Gray, J.S., Kahl, O., Lane, R.S. and Nijhof, A.M. (2013) Research on the ecology of ticks and tick-borne pathogens – methodological principles and caveats. *Frontiers in Cellular and Infectious Microbiology* 3, 29.

Estrada-Peña, A., de la Fuente, J., Latapia, T. and Ortega, C. (2015) The impact of climate trends on a tick affecting public health: a retrospective modeling approach for *Hyalomma marginatum* (Ixodidae). *PLoS ONE* 10, e0125760.

Estrada-Peña, A., Alexander, N. and Wint, G.R. (2016) Perspectives on modelling the distribution of ticks for large areas: so far so good? *Parasites & Vectors* 9, 179.

Figueirido, B., Janis, C.M., Pérez-Claros, J.A., De Renzi, M. and Palmqvist, P. (2012) Cenozoic climate change influences mammalian evolutionary dynamics. *Proceedings of the National Academy of Sciences USA* 109, 722–727.

Foth, C., Ezcurra, M.D., Sookias, R.B., Brusatte, S.L. and Butler, R.J. (2016) Unappreciated diversification of stem archosaurs during the Middle Triassic predated the dominance of dinosaurs. *BMC Evolutionary Biology* 16, 188.

Gray, J.S., Dautel, H., Estrada-Peña, A., Kahl, O. and Lindgren, E. (2009) Effects of climate change on ticks and tick-borne diseases in Europe. *Interdisciplinary Perspectives on Infectious Diseases* 2009, 593232.

Guglielmone, A.A. and Nava, S. (2017) Birds and hard ticks (Acari: Ixodidae), with discussions about hypotheses on tick evolution. *Revista FAVE – Sección Ciencias Veterinarias* 16, 13–29.

Harris, R., McCall, R., Randall, O., Tawang, M.H.B., Williams, R., *et al.* (2017) Climate change during the Triassic and Jurassic. *Geology Today* 33, 210–215.

Horak, I.G., Nyangiwe, N., De Matos, C. and Neves, L. (2009) Species composition and geographic distribution of ticks infesting cattle, goats and dogs in a temperate and in a subtropical region of south-east Africa. *Onderstepoort Journal of Veterinary Research* 76, 263–276.

Horak, I.G., Heyne, H., Williams, R., Gallivan, G.J., Spickett, A.M., *et al.* (2018) *The Ixodid Ticks (Acari: Ixodidae) of Southern Africa*. Springer, Cham, Switzerland.

Jablonski, D. (2002) Survival without recovery after mass extinctions. *Proceedings of the National Academy of Sciences USA* 99, 8139–8144.

Janis, C.M. (2003) Tertiary mammal evolution in the context of changing climates, vegetation, and tectonic events. *Annual Review of Ecology and Systematics* 24, 467–500.

Jardine, P.E., Janis, C.M., Sahney, S. and Benton, M.J. (2012) Grit not grass: concordant patterns of early origin of hypsodonty in Great Plains ungulates and Glires. *Palaeogeography, Palaeoclimatology, Palaeoecology* 365–366, 1–10.

Klompen, H. and Grimaldi, D. (2001) First Mesozoic record of a parasitiform mite: a larval argasid tick in Cretaceous amber (Acari: Ixodida: Argasidae). *Annals of the Entomological Society of America* 94, 10–15.

Klompen, J.S.H. and Oliver, J.H. (1993) Systematic relationships in the soft ticks (Acari: Ixodida: Argasidae). *Systematic Entomology* 18, 313–331.

Lynen, G., Zeman, P., Bakuname, C., Di Giulio, G., Mtui, P., *et al.* (2008) Shifts in the distributional ranges of *Boophilus* ticks in Tanzania: evidence that a parapatric boundary between *Boophilus microplus* and *B. decoloratus* follows climate gradients. *Experimental and Applied Acarology* 44, 147–164.

MacLeod, N. (2015) *The Great Extinctions. What Causes Them and How They Shape Life*. Firefly Books, London.

Madder, M., Thys, E., Achi, L., Touré, A. and De Deken, R. (2011) *Rhipicephalus* (*Boophilus*) *microplus*: a most successful invasive tick species in West-Africa. *Experimental and Applied Acarology* 53, 139–145.

Mamoudou, A., Nguetoum, N.C., Zoli, P.A. and Sevidzem, S.L. (2016) Identification and infestation of ticks on cattle in the peri-urban area of Ngaoundere, Cameroon. *Journal of Veterinary Science and Medical Diagnostics* 5, 4.

Mans, B.J., de Klerk, D., Pienaar, R. and Latif, A.A. (2011) *Nuttalliella namaqua*: a living fossil and closest relative to the ancestral tick lineage: implications for the evolution of blood-feeding in ticks. *PLoS ONE* 6, e23675.

Mans, B.J., de Klerk, D., Pienaar, R., de Castro, M.H. and Latif, A.A. (2012) The mitochondrial genomes of *Nuttalliella namaqua* (Ixodoidea: Nuttalliellidae) and *Argas africolumbae* (Ixodoidea: Argasidae): estimation of divergence dates for the major tick lineages and reconstruction of ancestral blood-feeding characters. *PLoS ONE* 7, e49461.

Mans, B.J., de Klerk, D.G., Pienaar, R. and Latif, A.A. (2014) The host preferences of *Nuttalliella namaqua* (Ixodoidea: Nuttalliellidae): a generalist approach to surviving multiple host-switches. *Experimental and Applied Acarology* 62, 233–240.

Mans, B.J., de Castro, M.H., Pienaar, R., de Klerk, D., Gaven, P., *et al.* (2016) Ancestral reconstruction of tick lineages. *Ticks and Tick-borne Diseases* 7, 509–535.

Mans, B.J., Featherston, J., Kvas, M., Pillay, K.A., de Klerk, D.G., *et al.* (2019) Argasid and ixodid systematics: implications for soft tick evolution and systematics, with a new argasid species list. *Ticks and Tick-borne Diseases* 10, 219–240.

Matthee, C.A. (2020) The influence of host dispersal on the gene flow and genetic diversity of generalist and specialist ectoparasites. *African Zoology* 55, 119–126.

Mihalca, A.D., Gherman, C.M. and Cozma, V. (2011) Coendangered hard-ticks: threatened or threatening? *Parasites & Vectors* 4, 71.

Nyangiwe, N., Goni, S., Hervé-Claude, L.P., Ruddat, I. and Horak, I.G. (2011) Ticks on pastures and on two breeds of cattle in the Eastern Cape Province, South Africa. *Onderstepoort Journal of Veterinary Research* 78, 1–9.

Nyangiwe, N., Harrison, A. and Horak, I.G. (2013) Displacement of *Rhipicephalus decoloratus* by *Rhipicephalus microplus* (Acari: Ixodidae) in the Eastern Cape Province, South Africa. *Experimental and Applied Acarology* 61, 371–382.

Olwoch, J.M., Reyers, B., Engelbrecht, F.A. and Erasmus, B.F.N. (2008) Climate change and the tick-borne disease, theileriosis (East Coast fever) in sub-Saharan Africa. *Journal of Arid Environments* 72, 108–120.

Peñalver, E., Arillo, A., Delclòs, X., Peris, D., Grimaldi, D.A., *et al.* (2017) Parasitised feathered dinosaurs as revealed by Cretaceous amber assemblages. *Nature Communications* 8, 1924.

Penn, J.L., Deutsch, C., Payne, J.L. and Sperling, E.A. (2018) Temperature-dependent hypoxia explains biogeography and severity of end-Permian marine mass extinction. *Science* 362, eaat1327.

Poinar, G. Jr and Brown, A.E. (2003) A new genus of hard ticks in Cretaceous Burmese amber (Acari: Ixodida: Ixodidae). *Systematic Parasitology* 54, 199–205.

Poinar, G.O. and Buckley, R. (2008) *Compluriscutula vetulum* (Acari: Ixodida: Ixodidae), a new genus and species of hard tick from lower Cretaceous Burmese amber. *Proceedings of the Entomological Society of Washington* 110, 445–450.

Prum, R.O., Berv, J.S., Dornburg, A., Field, D.J., Townsend, J.P., *et al.* (2015) A comprehensive phylogeny of birds (Aves) using targeted next-generation DNA sequencing. *Nature* 526, 569–573.

Sage, R.F. (2020) Global change biology: a primer. *Global Change Biology* 26, 3–30.

Sahney, S. and Benton, M.J. (2008) Recovery from the most profound mass extinction of all time. *Proceedings of the Royal Society B* 275, 759–765.

Sands, A.F., Apanaskevich, D.A., Matthee, S., Horak, I.G., Harrison, A., *et al.* (2017) Effects of tectonics and large scale climatic changes on the evolutionary history of *Hyalomma* ticks. *Molecular Phylogenetic and Evolution* 114, 153–165.

Soh, W.K., Wright, I.J., Bacon, K.L., Lenz, T.I., Steinthorsdottir, M., *et al.* (2017) Palaeo leaf economics reveal a shift in ecosystem function associated with the end-Triassic mass extinction event. *Nature Plants* 3, 17104.

Spickett, A.M. (2013) *Ixodid Ticks of Major Economic Importance and Their Distribution in South Africa. Monograph 1: Ticks and Tick-borne Diseases*, 1st edn. AgriConnect (Pty) Ltd, Pretoria, South Africa.

Tønnesen, M.H., Penzhorn, B., Bryson, N.R. and Masibigiri, T. (2004) Displacement of *Boophilus decoloratus* by *Boophilus microplus* in the Soutpansberg region, Limpopo Province, South Africa. *Experimental and Applied Acarology* 32, 199–208.

eo23 *Amblyomma* Ticks and Future Climates

Stefan Vilges de Oliveira*
University of Uberlândia, Uberlândia, Brazil

23.1 Expert Opinion

23.1.1 Introduction

Ticks are responsible for transmitting a variety of pathogenic microorganisms, including protozoa, rickettsia, spirochaetes and viruses. These blood-sucking arthropods are among the most important vectors of several diseases affecting livestock, humans and pets. Their ability to transmit pathogens, combined with promiscuous feeding behaviour and wide geographical distribution, represent a health threat (Jongejan and Uilenberg, 2004).

Ticks are notorious for causing economic losses in livestock production, not only for the depreciation of the leather from infested animals, but also for the decrease in animal production due to the spread of pathogens through parasitism, thus requiring high costs associated with their control (Barros-Battesti *et al.*, 2006; Ghosh *et al.*, 2007). Babesiosis, theileriosis and anaplasmosis are the major parasitic diseases transmitted by ticks that impose serious constraints on the productivity of cattle throughout the world (Demessie and Derso, 2015), being especially relevant in different countries of Asia, Africa and Latin America (Jongejan and Uilenberg, 2004).

Ticks of the genus *Amblyomma* are important parasites in several parts of the world (Guglielmone and Robbins, 2018). At least 63 species of *Amblyomma* have been recorded as human parasites. They are the main vectors of the zoonotic bacterium *Rickettsia rickettsii*, the aetiological agent of spotted fever in parts of South and Central America (Labruna, 2009). In the USA, *Amblyomma americanum* transmits *Ehrlichia chaffeensis* and *Ehrlichia ewingii* (which cause human ehrlichiosis), *Francisella tularensis* (tularaemia), Heartland virus (Heartland virus disease), Bourbon virus (Bourbon virus disease) and southern tick-associated rash illness (STARI). *Amblyomma maculatum* is the transmitter of the bacterium *Rickettsia parkeri* (*R. parkeri* rickettsiosis), which causes a form of spotted fever (CDC, 2020). In Africa, *Amblyomma hebraeum* and *Amblyomma variegatum* are the vectors of *Rickettsia africae*, the causative agent of African tick-bite fever. These ticks are common in western, central and southern Africa (Kelly *et al.*, 1996). In Australia, *Amblyomma triguttatum* is the vector of Q fever and some rickettsial infections (Dehhaghi *et al.*, 2019).

Scientific evidence indicates that variations in weather patterns, such as temperature and precipitation changes, have direct effects on tick survival rates (Sutherst, 2004; Dobson, 2009;

*Email: stefanbio@yahoo.com.br

DOI: 10.1079/9781789249637.0023

Lafferty, 2009; Randolph, 2009). However, in the vast majority of studies, only parts of a complex mechanism were evaluated that includes not only ticks, but also the distribution of hosts, bioagents and all ecological processes that are pivotal for the survival and establishment of a species in a certain area (Estrada-Peña *et al.*, 2013). This expert opinion discusses the evidence of how the climate can influence the distribution of ticks of the *Amblyomma* genus in climate change scenarios.

23.1.2 Climate change, ticks and tick-borne diseases

The Intergovernmental Panel on Climate Change (IPCC) defines climate changes as statistically significant variations in average climate parameters (including its natural variability) which persist over an extended period (typically decades or longer). Climate change can be caused by natural processes, but the recent alterations observed have been accelerated by human activities (IPCC, 2014).

The climate is recognized as one of the main determinants of the distribution of infectious disease vectors (Peterson, 2006) and climatic data have been used to predict the geographic distribution of tick species (Estrada-Peña, 2008). Studies reveal that variations in climate and environment may be related to the increased incidence of tick-borne diseases (Lindgren and Gustafson, 2001; Gray *et al.*, 2009). The term 'increased density' has been used to define one of the results of global warming for tick populations. The increase in density is an expected consequence owing to seasonal variations expressed in less cold, shorter winters and the rise in average annual temperature, hence allowing better development of tick populations (Estrada-Peña and Venzal, 2007). Climate change can also influence the risk of emergence of tick-borne diseases, due to the ability to alter the patterns of land use and occupation by human populations (e.g. agriculture, tourism, etc.). These oscillatory temperature patterns influence human behaviour in the short term (such as picnics, fishing, camping, human migration), thus promoting closer contact with natural areas that are tick vector habitats (Gray *et al.*, 2009; Estrada-Peña *et al.*, 2014; Oliveira *et al.*, 2017b). Since some of these conditions are in turn influenced by climate change, a complex chain of processes is triggered, so other factors become more important than climate change itself for the occurrence of diseases (Estrada-Peña *et al.*, 2014). In this sense, some authors suggest that climate has a secondary influence as the fragmentation of the landscape and other effects of human occupation have a dominant influence on the abundance and environmental suitability for niche occupation by ticks (Estrada-Peña *et al.*, 2012).

The above discussion provides an overview of the various factors involved in the distribution of tick populations. Notwithstanding, the fact is that climate change trends are certainly involved in the expansion and retraction of the historic tick distribution range, and several studies support this (Dantas-Torres, 2015).

23.1.3 Distribution of ticks in climate change scenarios

Although several studies have sought to generate models of geographical distribution for ticks under current and future scenarios, numerous factors are involved in delimiting the ecological niche of these vectors, thus promoting great complexity in the exercise of creating representative models (Estrada-Peña *et al.*, 2015).

Within this perspective, studies should include the variables that operate at different stages in the life cycle of this vector. The abundance of hosts, seasonal variations in the distribution of vectors and anthropic activities (such as the use and occupation of the soil) are determinant in the distribution pattern of ticks and, consequently, in the distribution of niches (Peterson, 2011; Estrada-Peña *et al.*, 2015). Another aspect that must be considered is that species with wide geographical distribution can vary considerably within their occurrence area, which can result in differences in species densities between regions depending on ecological variations (Gray *et al.*, 2009). Given the complexity of considering all these aspects, the vast majority of studies use bioclimatic variables in adequacy models that are generated from correlations of the distribution of these variables with the presence or absence of species (Peterson, 2011; Estrada-Peña *et al.*, 2015).

It has been argued that predictive models of the geographical distribution of ticks using future climatic conditions may be biased, whether due to the use of inappropriate data or due to the extrapolation of habitat suitability for each species (Crisp *et al.*, 2009). Another fact already reported by some authors is related to the adaptability of diverse species of ticks. In Europe, *Ixodes ricinus* is extremely flexible to climatic fluctuations and can exhibit quite different seasonal activity, which would promote a differentiated and dynamic geographical scope (Gray *et al.*, 2009). However, some ecologists believe that the vast majority of species cannot adapt to sudden climatic variations, including those that are projected for the coming decades. Thus, they believe conservation of the niche restricts areas of climatic suitability and may even lead to the extinction of species more sensitive to global climate variations (Oliveira *et al.*, 2017b).

Many efforts have been focused on assessing the distribution of ticks. Remarkably, it is evident that analyses need to take account of the dynamics of bioecological processes instead of simply the probability of occurrence or the climatic suitability for a given species (Oliveira *et al.*, 2017a,b).

23.1.4 Evidence against the impacts of climate change

Some studies have estimated the expansion of tick distribution due to climate changes (Estrada-Peña and Venzal, 2007; Jaenson *et al.*, 2012; Descamps, 2013; Porretta *et al.*, 2013). However, these investigations did not incorporate the uncertainty implicit in the forecasts (Beale and Lennon, 2012; Oliveira *et al.*, 2017a). Additionally, models that aim to predict species distribution in future climates should consider the theory of niche conservation, which has been demonstrated at local, regional and global scales, and shows that several species have a very limited ability to evolve into new environmental conditions (Peterson, 2011). Recent efforts to explore the effect of climate change on the occurrence of ticks more objectively suggest that, by using controlled and objective methods, climate change will result in unexpected reductions in the distribution of vector-borne diseases (Peterson and Samy, 2016; Oliveira *et al.*, 2017a). Previous evaluations have shown that not considering niche conservation can result in predictions of the occurrence of tropical vectors in freezing and boiling temperatures (Owens *et al.*, 2013), thereby generating models of ecological niches with a lack of biological realism.

An alternative for these situations is to avoid extrapolating the model under new climatic conditions (Anderson, 2013). Notwithstanding, current practices of vector distribution modelling in future climatic conditions do not consider the impacts of model extrapolation in new climates (Fischer *et al.*, 2013). Previous models have also assumed a high potential for tick dispersion in the short term (e.g. decades). Comparisons of current climate with climate change future scenarios indicate that new climates (i.e. substantially different from current conditions), commonly referred to as non-analogous, might evolve in the future and give rise to new ecosystems. Good practice for vector modelling in future climates should include estimates of the species' dispersion potential in order to avoid extrapolation, the identification of non-analogous environmental conditions for more solid conclusions (Soberón and Peterson, 2011) and avoid making predictions for ecological niches in which the species will not survive.

23.2 Future Predictions

The lone star tick is increasingly recognized as a vector species of significant public health importance in North America. Predictions about future distributional changes of *A. americanum* highlight areas where public health information campaigns could be initiated proactively and where field studies of tick ecology (e.g. investigating physiology at range limits, invasion dynamics) might be conducted. The results of these present-day habitat suitability models and future ensemble predictions provide a foundation upon which more comprehensive studies of *A. americanum* biogeography (e.g. incorporating information about other aspects of tick ecology including vegetation associations, host distributions and/or dispersal) can be based (Springer *et al.*, 2015). A later study expanded the area of recognized distribution of *A. americanum* and highlighted that the models do not necessarily represent the population densities of ticks and that these can be extremely heterogeneous, resulting in different levels of risk for humans and other hosts (Raghavan *et al.*, 2019). The authors emphasize that quantifying specific abundances of regions and their spatial heterogeneity will be an important next step from the point of view of public health in efforts to characterize the risk of contracting diseases transmitted by this tick.

The distribution of *A. hebraeum* and *A. variegatum* has been studied in Zimbabwe. Precipitation and temperature data were used for modelling the ecological niche of these species. The models predicted that *A. hebraeum* may occupy areas in the south and south-east of the country, while *A. variegatum* is distributed throughout all the country. Progressive rise in temperature seems to be forcing the dispersion of *A. variegatum* to areas outside the zones with prolonged drought period. By contrast, *A. hebraeum* is compelled to spread to the north, following areas that favour rainfall patterns. Despite trends in climate patterns providing insights on the suitable habitat for *A. hebraeum* and *A. variegatum* in the future, it is risky to speculate the details of their geographical distribution, as they are of a dynamic nature (Estrada-Peña *et al.*, 2008).

In the current scenario and in future projections of climate change, species of the *Amblyomma cajennense* complex (which includes *Amblyomma sculptum*) in Brazil show different patterns of climate suitability, inferring that climate change may influence the potential distribution of these vectors (Figs. 23.1 and 23.2). The analysis used climatic variables of recent decades (1950–2000) and projected for the years 2050 and 2070, obtained from the WorldClim project. They were analysed under two greenhouse gas emission scenarios or Representative Concentration Pathways (RCPs) (RCP2.6 and RCP8.5) associated with different patterns of economic development. Ecological

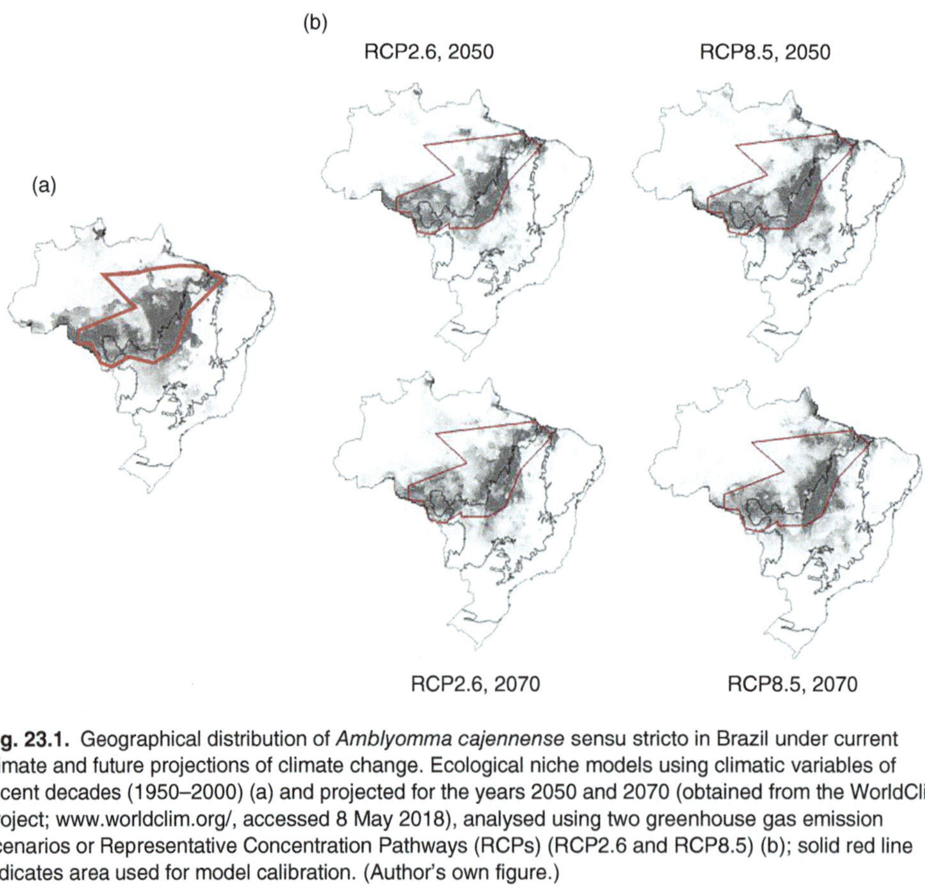

Fig. 23.1. Geographical distribution of *Amblyomma cajennense* sensu stricto in Brazil under current climate and future projections of climate change. Ecological niche models using climatic variables of recent decades (1950–2000) (a) and projected for the years 2050 and 2070 (obtained from the WorldClim project; www.worldclim.org/, accessed 8 May 2018), analysed using two greenhouse gas emission scenarios or Representative Concentration Pathways (RCPs) (RCP2.6 and RCP8.5) (b); solid red line indicates area used for model calibration. (Author's own figure.)

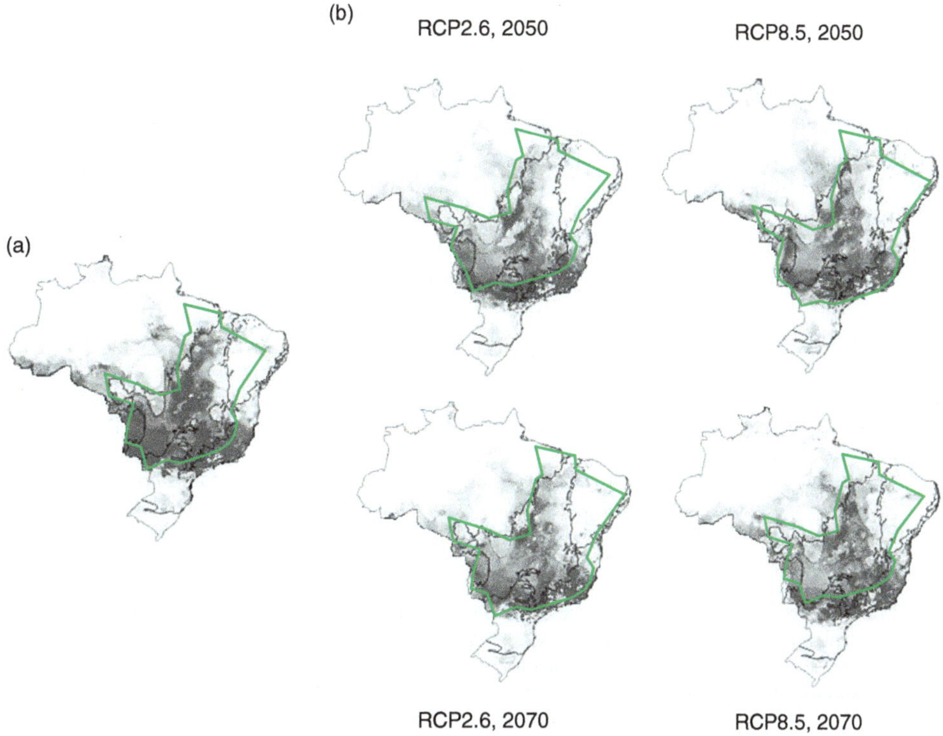

Fig. 23.2. Geographical distribution of *Amblyomma sculptum* in Brazil under current climate and future projections of climate change. Ecological niche models using climatic variables of recent decades (1950–2000) (a) and projected for the years 2050 and 2070 (obtained from the WorldClim project; www.worldclim.org/, accessed 8 May 2018), analysed using two greenhouse gas emission scenarios or Representative Concentration Pathways (RCPs) (RCP2.6 and RCP8.5) (b); solid green line indicates area used for model calibration. (Author's own figure.)

niche models revealed that these species have an impressive spatial heterogeneity in the uncertainties, both in suitable and inappropriate areas of occurrence (Oliveira *et al.*, 2017a). Uncertainty forecast estimates provide information to identify areas with greater precision for decision making. Here, uncertainties were apparently associated with regions with a low number of tick records included in the ecological niche model. The development of models for future climate, restricted only to climatically analogous areas and excluding new climates, estimated a reduction in suitable areas for *A. cajennense* sensu stricto and *A. sculptum*. This process, carried out during ecological niche modelling, ensures that the future climate projection finds only climates suitable for maintenance of the species. If the models include new climates from extrapolation, the species are predicted to expand their distribution limits in new regions of Brazil.

It is expected that the rise in temperature projected for the coming years (Estrada-Peña, 2008), associated with intense precipitation events, increases in most regions (IPCC, 2014). These climate changes can affect the ecology and distribution of *Amblyomma* ticks. Future predictions of the geographic distribution of ticks generally show expansion in conditions of increasing temperatures and decreasing rainfall (Estrada-Peña, 2008). However, changes in precipitation are more difficult to predict due to uncertainties in hydroclimate projections (Ljungqvist *et al.*, 2016).

The need to report the uncertainty of the model in a spatial context is a recent practice in ecological niche modelling (Peterson and Samy, 2016) and should be adopted in tick distribution modelling protocols. Health authorities, as well as reviewers of scientific journals, could assist in the task of making uncertainty maps an essential condition for current and future maps of the potential distribution of these vectors. More efforts are needed to understand the potential effects of climate change on the distribution of *Amblyomma* ticks. Informing decision makers of uncertain predictions in vector distribution may result in inadequate prioritization of areas for interventions and unnecessary resource expenditure (Oliveira *et al.*, 2017b).

References

Anderson, R.P. (2013) A framework for using niche models to estimate impacts of climate change on species distributions. *Annals of the New York Academy of Sciences* 1297, 8–28.

Barros-Battesti, D.M., Arzua, M. and Bechara, G.H. (2006) *Carrapatos de Importância Médico-Veterinária da Região Neotropical: Um Guia Ilustrado para Identificação de Espécies*. Vox/ICTTD-3/Butantan, São Paulo, Brazil.

Beale, C.M. and Lennon, J.J. (2012) Incorporating uncertainty in predictive species distribution modelling. *Philosophical Transactions of the Royal Society B: Biological Sciences* 367, 247–258.

CDC (2020) *Tickborne Diseases of the United States*. Centers for Disease Control and Prevention, Atlanta, Georgia. Available at: https://www.cdc.gov/ticks/tickbornediseases/index.html (accessed 19 November 2020).

Crisp, M.D., Arroyo, M.T., Cook, L.G., Gandolfo, M.A., Jordan, G.J., *et al.* (2009) Phylogenetic biome conservatism on a global scale. *Nature* 458, 754–756.

Dantas-Torres, F. (2015) Climate change, biodiversity, ticks and tick-borne diseases: the butterfly effect. *International Journal for Parasitology: Parasites and Wildlife* 4, 452–461.

Descamps, S. (2013) Winter temperature affects the prevalence of ticks in an Arctic seabird. *PLoS ONE* 8, e65374.

Demessie, Y. and Derso, S. (2015) Tick borne hemoparasitic diseases of ruminants: a review. *Advances in Biological Research*, 9, 210–224.

Dehhaghi, M., Panahi, H.K.S., Holmes, E.C., Hudson, B.J., Schloeffel, R., *et al.* (2019) Human tick-borne diseases in Australia. *Frontiers in Cellular and Infection Microbiology* 9, 3.

Dobson, A. (2009) Climate variability, global change, immunity, and the dynamics of infectious diseases. *Ecology* 90, 920–927.

Estrada-Peña, A. (2008) Climate, niche, ticks, and models: what they are and how we should interpret them. *Parasitology Research* 103, 87–95.

Estrada-Peña, A. and Venzal, J.M. (2007) Climate niches of tick species in the Mediterranean region: modeling of occurrence data, distributional constraints, and impact of climate change. *Journal of Medical Entomology* 44, 1130–1138.

Estrada-Peña, A., Horak, I.G. and Petney, T. (2008) Climate changes and suitability for the ticks *Amblyomma hebraeum* and *Amblyomma variegatum* (Ixodidae) in Zimbabwe (1974–1999). *Veterinary Parasitology* 151, 256–267.

Estrada-Peña, A., Ayllón, N. and de la Fuente, J. (2012) Impact of climate trends on tick-borne pathogen transmission. *Frontiers in Physiology* 3, 64.

Estrada-Peña, A., Gray, J.S., Kahl, O., Lane, R.S. and Nijhof, A.M. (2013) Research on the ecology of ticks and tick-borne pathogens – methodological principles and caveats. *Frontiers in Cellular and Infection Microbiology* 3, 29.

Estrada-Peña, A., Ostfeld, R.S., Peterson, A.T., Poulin, R. and de la Fuente, J. (2014) Effects of environmental change on zoonotic disease risk: an ecological primer. *Trends in Parasitology* 30, 205–214.

Estrada-Peña, A., Estrada-Sánchez, A. and Estrada-Sánchez, D. (2015) Methodological caveats in the environmental modelling and projections of climate niche for ticks, with examples for *Ixodes ricinus* (Ixodidae). *Veterinary Parasitology* 208, 14–25.

Fischer, D., Thomas, S.M., Suk, J.E., Sudre, B., Hess, A., *et al.* (2013) Climate change effects on Chikungunya transmission in Europe: geospatial analysis of vector's climatic suitability and virus' temperature requirements. *International Journal of Health Geographics* 12, 51.

Ghosh, S., Azhahianambi, P. and Yadav, M.P. (2007) Upcoming and future strategies of tick control: a review. *Journal of Vector Borne Diseases* 44, 79.

Gray, J.S., Dautel, H., Estrada-Peña, A., Kahl, O. and Lindgren, E. (2009) Effects of climate change on ticks and tick-borne diseases in Europe. *Interdisciplinary Perspectives on Infectious Diseases* 2009, 593232.

Guglielmone, A.A. and Robbins, R.G. (2018) *Hard Ticks (Acari: Ixodida: Ixodidae) Parasitizing Humans.* Springer, Cham, Switzerland.

IPCC (2014) *Climate Change 2014: Synthesis Report. Contribution of Working Groups I, II and III to the Fifth Assessment Report of the Intergovernmental Panel on Climate Change* [Core Writing Team, Pachauri, R.K. and Meyer, L.A. (eds)]. Intergovernmental Panel on Climate Change, Geneva, Switzerland.

Jaenson, T.G., Jaenson, D.G., Eisen, L., Petersson, E. and Lindgren, E. (2012) Changes in the geographical distribution and abundance of the tick *Ixodes ricinus* during the past 30 years in Sweden. *Parasites & Vectors* 5, 8.

Jongejan, F. and Uilenberg, G. (2004) The global importance of ticks. *Parasitology* 129(Suppl.), S3–S14.

Kelly, P.J., Beati, L., Mason, P.R., Matthewman, L.A., Roux, V., *et al.* (1996) *Rickettsia africae* sp. nov., the etiological agent of African tick bite fever. *International Journal of Systematic and Evolutionary Microbiology* 46, 611–614.

Labruna, M.B. (2009) Ecology of rickettsia in South America. *Annals of the New York Academy of Sciences* 1166, 156–166.

Lafferty, K.D. (2009) The ecology of climate change and infectious diseases. *Ecology* 90, 888–900.

Lindgren, E. and Gustafson, R. (2001) Tick-borne encephalitis in Sweden and climate change. *The Lancet* 358, 16–18.

Ljungqvist, F.C., Krusic, P.J., Sundqvist, H.S., Zorita, E., Brattström, G., *et al.* (2016) Northern hemisphere hydroclimate variability over the past twelve centuries. *Nature* 532, 94–98.

Oliveira, S.V. d., Romero-Alvarez, D., Martins, T.F., dos Santos, J.P., Labruna, M.B., *et al.* (2017a) *Amblyomma* ticks and future climate: range contraction due to climate warming. *Acta Tropica* 176, 340–348.

Oliveira, S.V. d., Gazeta, G.S. and Gurgel-Gonçalves, R. (2017b) Climate, ticks and tick-borne diseases: mini review. *Vector Biology Journal* 2, 1.

Owens, H.L., Campbell, L.P., Dornak, L.L., Saupe, E.E., Barve, N., *et al.* (2013) Constraints on interpretation of ecological niche models by limited environmental ranges on calibration areas. *Ecological Modelling* 263, 10–18.

Peterson, A.T. (2006) Ecologic niche modeling and spatial patterns of disease transmission. *Emerging Infectious Diseases* 12, 1822–1826.

Peterson, A.T. (2011) Ecological niche conservatism: a time-structured review of evidence. *Journal of Biogeography* 38, 817–827.

Peterson, A.T. and Samy, A.M. (2016) Geographic potential of disease caused by Ebola and Marburg viruses in Africa. *Acta Tropica* 162, 114–124.

Porretta, D., Mastrantonio, V., Amendolia, S., Gaiarsa, S., Epis, S., *et al.* (2013) Effects of global changes on the climatic niche of the tick *Ixodes ricinus* inferred by species distribution modelling. *Parasites & Vectors* 6, 271.

Raghavan, R.K., Peterson, A.T., Cobos, M.E., Ganta, R. and Foley, D. (2019) Current and future distribution of the lone star tick, *Amblyomma americanum* (L.) (Acari: Ixodidae) in North America. *PLoS ONE* 14, e0209082.

Randolph, S.E. (2009) Perspectives on climate change impacts on infectious diseases. *Ecology* 90, 927–931.

Soberón, J. and Peterson, A.T. (2011) Ecological niche shifts and environmental space anisotropy: a cautionary note. *Revista Mexicana de Biodiversidad* 82, 1348–1355.

Springer, Y.P., Jarnevich, C.S., Barnett, D.T., Monaghan, A.J. and Eisen, R.J. (2015) Modeling the present and future geographic distribution of the lone star tick, *Amblyomma americanum* (Ixodida: Ixodidae), in the continental United States. *American Journal of Tropical Medicine and Hygiene* 93, 875–890.

Sutherst, R.W. (2004) Global change and human vulnerability to vector-borne diseases. *Clinical Microbiology Reviews* 17, 136–173.

eo24 Climate Impacts on *Dermacentor reticulatus* Tick Population Dynamics and Range

Zbigniew Zając*
Medical University of Lublin, Lublin, Poland

24.1 Expert Opinion

24.1.1 Introduction

Dermacentor reticulatus (Fabricius, 1794), known as the ornate cow or meadow tick, is one of the most widespread tick species in Europe (Fig. 24.1). Due to the biology and ecology of the species, these ticks play a role as one of the most important reservoirs and vectors of infectious animal and human diseases on the continent (Földvári *et al.*, 2016; Rubel *et al.*, 2020).

To date, approximately 40 species of pathogens have been detected in *D. reticulatus* ticks. From the point of view of epidemiology and veterinary medicine, the ability of the tick species to transmit *Babesia canis* protozoa, the aetiological agent of canine babesiosis, is the most important issue. A growing problem, observed in recent years in many European countries, is the increasing incidence of this disease in dogs, which in most cases leads to death of the animal if left untreated (Matijatko *et al.*, 2012). *D. reticulatus* ticks have also been collected from livestock, companion animals and wild animals (Dautel *et al.*, 2006; Akimov and Nebogatkin, 2011; Mierzejewska *et al.*, 2015; Opalińska *et al.*, 2018; Karbowiak *et al.*, 2019; Zając *et al.*, 2020a). Infestation of dairy cows by this species poses a risk of transmission of tick-borne encephalitis virus to cattle and subsequently to humans through the consumption of unpasteurized milk and dairy products (Štefančíková *et al.*, 2002; Cisak *et al.*, 2010). Moreover, Omsk haemorrhagic fever virus, *Borrelia burgdorferi*, *Rickettsia raoultii*, *Rickettsia slovaca*, *Anaplasma marginale*, *Coxiella burnetii*, *Francisella tularensis*, *Bartonella* spp., *Theileria equi* and *Hepatozoon canis* have been detected in *D. reticulatus* (Földvári *et al.*, 2016).

There are documented cases of *D. reticulatus* feeding on humans, although the species bites humans less frequently than some other tick species (Estrada-Peña and Jongejan, 1999; Bartosik *et al.*, 2011a). The direct negative effects of feeding of these ticks include local and systemic symptoms in the host, as well as diseases such as *Dermacentor*-borne necrosis erythema and lymphadenopathy (DEBONEL) and scalp eschar and neck lymphadenopathy (SENLAT), which are diagnosed with increasing frequency (Oteo and Portillo, 2012; Földvári *et al.*, 2013).

24.1.2 Occurrence of *Dermacentor reticulatus* ticks in Europe versus progressive climate change

The current geographical range of *D. reticulatus* ticks in Europe is limited mainly to areas located in the temperate climate zone. Only a few isolated

*Email: zbigniew.zajac@umlub.pl

DOI: 10.1079/9781789249637.0024

populations of this species have been found in the Mediterranean zone (Fig. 24.2).

In recent years, there has been an increase in the abundance of local *D. reticulatus* populations (Buczek *et al.*, 2013) and progressive extension of the occurrence range of this species in Europe (Rubel *et al.*, 2016). Some authors refer to the process as expansion (Mierzejewska *et al.*, 2016). It is believed that this phenomenon is a result of the influence of progressive climate change, but the increase in populations of wild host animals is equally important. One of the determinants of the spectacular extension of the occurrence range of *D. reticulatus* is probably the interaction of these ticks with other tick species present in their habitats. The phenomenon of co-feeding of adult *D. reticulatus* and *Ixodes ricinus* ticks on the same host, which is frequently observed in natural conditions, has a positive effect on host-seeking and on embryonic development and survival of *D. reticulatus* larvae (Bartosik, 2006). This may explain the coexistence of both species in habitats that were previously dominated by one species, which is observed more frequently.

Fig. 24.1. Adult *Dermacentor reticulatus* ticks questing for a host (the top specimen is a male; the other ticks are females). (Author's own photograph.)

At the end of the 20th century, investigators of the biology and ecology of *D. reticulatus* agreed that there were two geographically separated populations of this species in Europe. They distinguished the Western European population

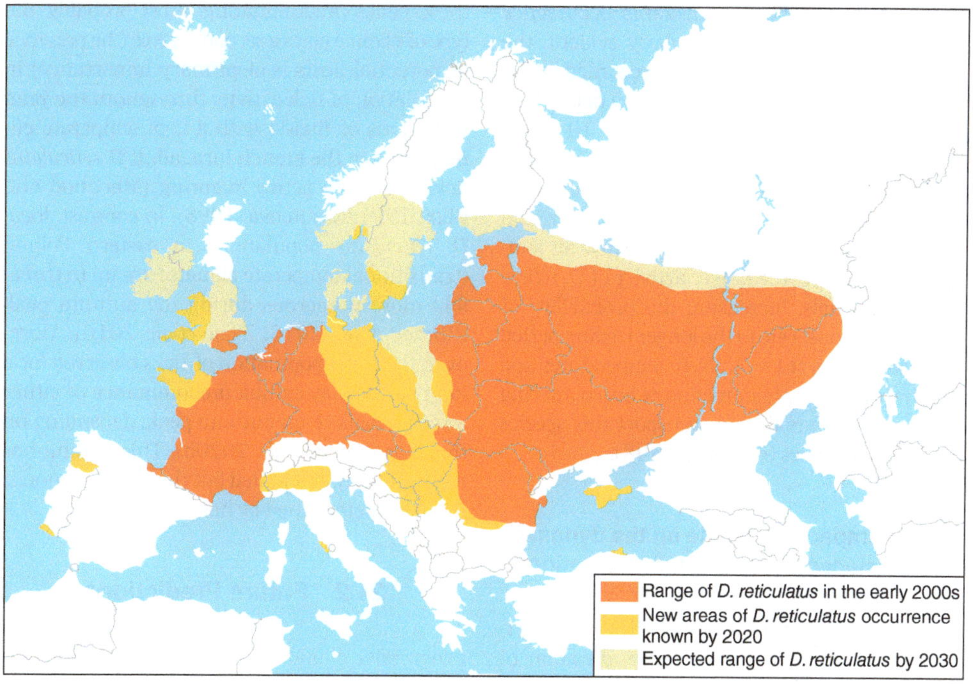

Fig. 24.2. *Dermacentor reticulatus* occurrence range in Europe. (Based on Karbowiak, 2014; Mierzejewska *et al.*, 2016; Rubel *et al.*, 2016; Kjær *et al.*, 2019; and own observations. Author's own figure.)

present in France, the Benelux countries and western regions of Germany, and the Eastern European population occurring east of the Vistula River in Poland (Karbowiak, 2014). In recent years, the occurrence range of the Eastern and Western *D. reticulatus* populations has extended into Central Europe, thus reducing the distance between these two groups (Mierzejewska *et al.*, 2016) (Fig. 24.2). There are also reports of *D. reticulatus* collected from vegetation in southern Sweden. Larvae and nymphs of the meadow tick have been found on migratory birds in Scandinavia. In Norway, adult stages have been removed from dogs, including animals that had never travelled abroad (Hasle *et al.*, 2009; Øines *et al.*, 2010; Rudander, 2018; Kjær *et al.*, 2019).

The expansion of *D. reticulatus*, especially northwards, is determined not only by global warming but also, to some extent, by changes in the range of tick hosts. Specimens of this species are characterized by high plasticity in their adaptation to prevailing weather conditions (Buczek *et al.*, 2014) and, as shown by data from Russia (Razumova, 1998), their populations are able to survive and develop in climatic conditions similar to those in central Scandinavia. In Central Europe, where extension in occurrence of the meadow tick is particularly evident, the average annual air temperature (recorded by the Tutiempo weather service; Tutiempo.net, 2020) has increased by 1.0°C over the last 10 years. Such conditions promote the survival of small mammals, e.g. rodents, which are the main hosts of juvenile stages of *D. reticulatus*. Milder and snowless winters with persistent positive air temperatures support the longer activity of *D. reticulatus* individuals, increasing the probability of host infestations. Due to the longer feeding period of meadow ticks at low ambient temperatures and the characteristic phenomenon of overwintering on the host, animals can transport this species over long distances (Izdebska, 1998).

24.1.3 Impact of climate on the dynamics of *Dermacentor reticulatus* populations

Ambient temperature is one of the main determinants of the occurrence range, duration of the life cycle and activity of *D. reticulatus* ticks. Additionally, photoperiod has an impact on the seasonal activity rhythms and diapause in ticks (Bartosik *et al.*, 2011b). Furthermore, temperature influences questing-related locomotor activity of these arthropods, the course of feeding and the transmission of tick-borne pathogens in the tick–host system. Temperature also stimulates the moulting process in larvae and nymphs (Zahler and Gothe, 1995; Buczek *et al.*, 2017).

D. reticulatus ticks are active in a wide temperature range. Adult specimens were collected from vegetation at a temperature exceeding 30°C and at approximately 3°C (Zając *et al.*, 2016, 2020b). They are also active during the winter months at a positive ambient temperature, although the number of active individuals in this period is substantially lower (Kiewra *et al.*, 2016; Kohn *et al.*, 2019). In eastern Poland, active *D. reticulatus* specimens have been collected from plots partially covered by snow patches (Z. Zając, 2019, unpublished results). In autumn, 60% of males and 68% of females that are active are able to survive winter in the natural conditions of the temperate climate zone (Zając *et al.*, 2020c).

Depending on the geographic region, meadow ticks exhibit two peaks of seasonal activity: in spring and in autumn, with one of these peaks predominating. Most probably, the type of climate plays a secondary role (the presence of potential hosts is of primary importance) in modulation of tick activity throughout the year. In regions of Russia with a cool temperate climate, and in the French Jura, adult *D. reticulatus* ticks are more active in spring (Martinod and Gilot, 1991; Razumova, 1998). In contrast, local *D. reticulatus* populations in eastern Poland (transitional temperate climate) are up to threefold more numerous during the autumn peak (Buczek *et al.*, 2013; Zając *et al.*, 2016). Moreover, the same population of ticks observed for a long period may exhibit predominance of either a spring peak or an autumn peak, depending on the year (Zając *et al.*, 2020b). This finding has not been fully elucidated but the phenomenon is probably determined by host availability.

24.2 Future Predictions

Progressive climate changes resulting in an increase in the average annual air temperature in Europe will indirectly contribute to the extension of the occurrence range of *D. reticulatus* ticks on the continent. Northward expansion is most likely. In the next 10 years, *D. reticulatus* ticks will be common in south Scandinavia and Great Britain. The so-called Eastern European and Western European populations of this species are likely to merge. A rise in the average annual temperature will also contribute to an increase in the size of local *D. reticulatus* populations, as this phenomenon is already beginning to be observed. This may result in enhanced risk of exposure to attacks of both animals and humans by this species. Given the role of *D. reticulatus* as a reservoir and vector of pathogens, the proportion of infected ticks is likely to increase.

References

Akimov, I. and Nebogatkin, I. (2011) Distribution of ticks of the genus *Dermacentor* (Acari, Ixodidae) in Ukraine. *Vestnik Zoologii* 45, 35–40.

Bartosik, K. (2006) Mutual interactions between Prostriata and Metastriata ticks (Acari: Ixodida: Ixodidae) in parasitic and nonparasitic phase of the life cycle. PhD thesis, Medical University of Lublin, Lublin, Poland (in Polish).

Bartosik, K., Sitarz, M., Szymańska, J. and Buczek, A. (2011a) Tick bites on humans in the agricultural and recreational areas in south-eastern Poland. *Annals of Agricultural and Environmental Medicine* 18, 151–157.

Bartosik, K., Wiśniowski, L. and Buczek, A. (2011b) Abundance and seasonal activity of adult *Dermacentor reticulatus* (Acari: Amblyommidae) in eastern Poland in relation to meteorological conditions and the photoperiod. *Annals of Agricultural and Environmental Medicine* 18, 340–344.

Buczek, A., Bartosik, K., Wiśniowski, Ł. and Tomasiewicz, K. (2013) Changes in population abundance of adult *Dermacentor reticulatus* (Acari: Amblyommidae) in long-term investigations in eastern Poland. *Annals of Agricultural and Environmental Medicine* 20, 269–272.

Buczek, A., Bartosik, K. and Zając, Z. (2014) Changes in the activity of adult stages of *Dermacentor reticulatus* (Ixodida: Amblyommidae) induced by weather factors in eastern Poland. *Parasites & Vectors* 7, 245.

Buczek, A., Zając, Z., Woźniak, A., Kulina, D. and Bartosik, K. (2017) Locomotor activity of adult *Dermacentor reticulatus* ticks (Ixodida: Ixodidae) in natural conditions. *Annals of Agricultural and Environmental Medicine* 24, 271–275.

Cisak, E., Wójcik-Fatla, A., Zając, V., Sroka, J., Buczek, A. and Dutkiewicz, J. (2010) Prevalence of tick-borne encephalitis virus (TBEV) in samples of raw milk taken randomly from cows, goats and sheep in eastern Poland. *Annals of Agricultural and Environmental Medicine* 17, 283–286.

Dautel, H., Dippel, C., Oehme, R., Hartelt, K. and Schettler, E. (2006) Evidence for an increased geographical distribution of *Dermacentor reticulatus* in Germany and detection of *Rickettsia* sp. RpA4. *International Journal of Medical Microbiology* 296, 149–156.

Estrada-Peña, A. and Jongejan, F. (1999) Ticks feeding on humans: a review of records on human-biting Ixodoidea with special reference to pathogen transmission. *Experimental and Applied Acarology* 23, 685–715.

Földvári, G., Rigó, K. and Lakos, A. (2013) Transmission of *Rickettsia slovaca* and *Rickettsia raoultii* by male *Dermacentor marginatus* and *Dermacentor reticulatus* ticks to humans. *Diagnostic Microbiology and Infectious Disease* 76, 387–389.

Földvári, G., Široký, P., Szekeres, S., Majoros, G. and Sprong, H. (2016) *Dermacentor reticulatus*: a vector on the rise. *Parasites & Vectors* 9, 314.

Hasle, G., Bjune, G., Edvardsen, E., Jakobsen, C., Linnehol, B., *et al.* (2009) Transport of ticks by migratory passerine birds to Norway. *Journal of Parasitology* 95, 1342–1351.

Izdebska, J. (1998) The occurrence of *Dermacentor reticulatus* (Fabricius, 1794) (Acari, Ixodidae) on the bison (*Bison bonasus*) from the Białowieża Primaeval Forest. *Przegląd Zoologiczny* 42, 219–221.

Karbowiak, G. (2014) The occurrence of the *Dermacentor reticulatus* tick – its expansion to new areas and possible causes. *Annals of Parasitology* 60, 37–47.

Karbowiak, G., Miklisová, D., Stanko, M., Werszko, J., Hajdul-Marwicz, M., *et al.* (2019) The competition between immatures of *Ixodes ricinus* and *Dermacentor reticulatus* (Ixodida: Ixodidae) ticks for rodent hosts. *Journal of Medical Entomology* 56, 448–452.

Kiewra, D., Czułowska, A. and Lonc, E. (2016) Winter activity of *Dermacentor reticulatus* (Fabricius, 1794) in the newly emerging population of Lower Silesia, south-west Poland. *Ticks and Tick-borne Diseases* 7, 1124–1127.

Kjær, L.J., Soleng, A., Edgar, K.S., Lindstedt, H.E.H., Paulsen, K.M., *et al.* (2019) A large-scale screening for the taiga tick, *Ixodes persulcatus*, and the meadow tick, *Dermacentor reticulatus*, in southern Scandinavia, 2016. *Parasites & Vectors* 12, 338.

Kohn, M., Krücken, J., McKay-Demeler, J., Pachnicke, S., Krieger, K. and von Samson-Himmelstjerna, G. (2019) *Dermacentor reticulatus* in Berlin/Brandenburg (Germany): activity patterns and associated pathogens. *Ticks and Tick-borne Diseases* 10, 191–206.

Martinod, S. and Gilot, B. (1991) Epidemiology of canine babesiosis in relation to the activity of *Dermacentor reticulatus* in southern Jura (France). *Experimental and Applied Acarology* 11, 215–222.

Matijatko, V., Torti, M. and Schetters, T.P. (2012) Canine babesiosis in Europe: how many diseases? *Trends in Parasitology* 28, 99–105.

Mierzejewska, E.J., Welc-Faleciak, R., Karbowiak, G., Kowalec, M., Behnke, J.M. and Bajer, A. (2015) Dominance of *Dermacentor reticulatus* over *Ixodes ricinus* (Ixodidae) on livestock, companion animals and wild ruminants in eastern and central Poland. *Experimental and Applied Acarology* 66, 83–101.

Mierzejewska, E.J., Estrada-Peña, A., Alsarraf, M., Kowalec, M. and Bajer, A. (2016) Mapping of *Dermacentor reticulatus* expansion in Poland in 2012–2014. *Ticks and Tick-borne Diseases* 7, 94–106.

Øines, Ø., Storli, K. and Brun-Hansen, H. (2010) First case of babesiosis caused by *Babesia canis canis* in a dog from Norway. *Veterinary Parasitology* 171, 350–353.

Opalińska, P., Wierzbicka, A., Asman, M. and Żarski, K. (2018) The potential role of roe deer (*Capreolus capreolus*) and meadow tick (*Dermacentor reticulatus*) as a reservoir of *Toxoplasma gondii*, *Borrelia burgdorferi* sensu lato, *Anaplasma phagocytophilum* and *Babesia microti* in western Poland. *Acta Scientiarum Polonorum Silvarum Colendarum Ratio et Industria Lignaria* 17, 317–323.

Oteo, J.A. and Portillo, A. (2012) Tick-borne rickettsioses in Europe. *Ticks and Tick-borne Diseases* 3, 271–278.

Razumova, I.V. (1998) The activity of *Dermacentor reticulatus* Fabr. (Ixodidae) ticks in nature. *Meditsinskaia Parazitologiia i Parazitarnye Bolezni* 4, 8–14.

Rubel, F., Brugger, K., Pfeffer, M., Chitimia-Dobler, L., Didyk, Y.M., *et al.* (2016) Geographical distribution of *Dermacentor marginatus* and *Dermacentor reticulatus* in Europe. *Ticks and Tick-borne Diseases* 7, 224–233.

Rubel, F., Brugger, K., Belova, O.A., Kholodilov, I.S., Didyk, Y.M., *et al.* (2020) Vectors of disease at the northern distribution limit of the genus *Dermacentor* in Eurasia: *D. reticulatus* and *D. silvarum*. *Experimental and Applied Acarology* 82, 95–123.

Rudander, S.H. (2018) The risk of *Dermacentor reticulatus* establishing in Sweden. MSc thesis, Swedish University of Agricultural Sciences, Uppsala, Sweden (in Swedish).

Štefančíková, A., Štěpánová, G., Derdáková, M., Pet'ko, B., Kysel'ová, J., *et al.* (2002) Serological evidence for *Borrelia burgdorferi* infection associated with clinical signs in dairy cattle in Slovakia. *Veterinary Research Communications* 26, 601–611.

Tutiempo.net (2020) Weather in Europe. Available at: https://en.tutiempo.net/europe.html (accessed 2 November 2020).

Zahler, M. and Gothe, R. (1995) Effect of temperature and humidity on egg hatch, moulting and longevity of larvae and nymphs of *Dermacentor reticulatus* (Ixodidae). *Applied Parasitology* 36, 53–65.

Zając, Z., Bartosik, K. and Buczek, A. (2016) Factors influencing the distribution and activity of *Dermacentor reticulatus* (F.) ticks in an anthropopressure-unaffected area in central-eastern Poland. *Annals of Agricultural and Environmental Medicine* 23, 270–275.

Zając, Z., Woźniak, A. and Kulisz, J. (2020a) Infestation of dairy cows by ticks *Dermacentor reticulatus* (Fabricius, 1794) and *Ixodes ricinus* (Linnaeus, 1758) in eastern Poland. *Annals of Parasitology* 66, 87–96.

Zając, Z., Woźniak, A. and Kulisz, J. (2020b) Density of *Dermacentor reticulatus* ticks in eastern Poland. *International Journal of Environmental Research and Public Health* 17, 2814.

Zając, Z., Bartosik, K., Kulisz, J. and Woźniak, A. (2020c) Ability of adult *Dermacentor reticulatus* ticks to overwinter in the temperate climate zone. *Biology* 9, 145.

eo25 Changes Expected in *Ixodes ricinus* Temporal and Spatial Distribution in Europe

Annapaola Rizzoli
Fondazione Edmund Mach, San Michele all'Adige, Italy

25.1 Expert Opinion

25.1.1 Introduction

The tick, *Ixodes ricinus*, is among the arthropod vectors of greatest public health relevance in Europe. This species, with its three genetic, spatially structured clusters (Poli *et al.*, 2020), has enlarged its distribution range during the last decades and currently is reported to occur across a vast area of the Western Palaearctic ecozone (ECDC, 2020). *I. ricinus* is recognized as a competent vector for an increasing number of zoonotic pathogens; recent exploration of its microbiome and virome has revealed that this tick can harbour a much broader range of microorganisms than expected (Carpi *et al.*, 2011; Pettersson *et al.*, 2017). Therefore, shifting the paradigm from single pathogen to pathobiome will enable us to better understand its ability to adapt and transmit zoonotic pathogens under a global change scenario, where temperature can affect pathogen transmission (Vayssier-Taussat *et al.*, 2015; Daniel *et al.*, 2018).

The most frequently reported human diseases associated with *I. ricinus* in Europe are Lyme borreliosis (LB) and tick-borne encephalitis (TBE), which are responsible for the largest disease burden of any vector-borne disease in the European Union. The average incidence of LB in Western Europe is 22 cases per 100,000 person-years as estimated in a recent literature review (Sykes and Makiello, 2017). However, the heterogeneity of surveillance systems across European countries limits our ability to accurately assess the actual disease burden. This limitation would be partially improved by the collection of comparable data on Lyme neuroborreliosis, which is notifiable since 2018 (van den Wijngaard *et al.*, 2017).

TBE is notifiable since 2012, with an incidence of 0.6 cases per 100,000 person-years (ECDC, 2018). There has been a nearly 400% increase of reported cases in European endemic areas over the past 30 years, despite increased efforts in terms of raising awareness and the availability of an effective vaccine. The overall increase in LB and TBE cases could be related not only to enhanced surveillance and diagnosis, but also to a number of factors including the direct and indirect consequences of climate change on the vector and its interaction with the abiotic and biotic components of its ecological niche (Semenza and Suk, 2018).

Changes in *I. ricinus* geographical distribution and prolongation in the duration of activity across various European countries are considered among the explanatory factors for the

*Email: annapaola.rizzoli@fmach.it

DOI: 10.1079/9781789249637.0025

enhanced risk for *I. ricinus*-borne diseases in Europe, and consequently for the increase in observed human disease cases. However, much less information is available regarding the changes which may have occurred at the same time in the tick to host ratio and consequently on other parameters such as tick abundance and infection rate that ultimately determine the hazard, often expressed in terms of density of infected nymphs (DIN), which are the developmental stage mostly involved in disease transmission to humans. Unfortunately, despite significant advances in understanding tick–host interactions at cellular and molecular levels, the paucity of experimental and long-term field research aimed at monitoring changes in tick–host interactions limits our ability to fully explain the changes observed in the risk. Early warning models based on field monitoring of tick–host interactions appear quite effective in predicting changes in the risk and clearly demonstrate the need for further investment in long-term ecological studies on tick–host interactions influenced by climate change (Rosà *et al.*, 2019; Bregnard *et al.*, 2020).

Nevertheless, it must be stressed that also significant spatial and temporal changes in human exposure across different countries to infected ticks has occurred during the last decades, especially after the Second World War, as a consequence of changes in socio-economic conditions, including attitude towards nature and outdoor activities, and these changes can explain part of the variation in incidence (Randolph, 2010). The latest example of an increased number of tick-borne diseases reported because of sudden changes in human behaviour (with increased exposure to infected ticks) is provided by the increase in outdoor activities (likely with reduced awareness of tick borne-disease risk) in natural areas as a consequence of the COVID-19 lockdown and post-lockdown restrictions in Europe, which have been associated with an increase in human cases in certain countries (Steffen *et al.*, 2020).

25.1.2 Evidence of climate change impact on *Ixodes ricinus* spatial and temporal distribution

Climate change is proven to affect *I. ricinus* both directly and indirectly, following a wide number of pathways which are strongly interlinked although often difficult to identify and parameterize individually (Gray *et al.*, 2009; Medlock *et al.*, 2013).

Changes in macroclimate are considered among the critical driving factors determining *I. ricinus* geographic limit and its occurrence at latitude and altitude, although variation in biotic factors can exert a greater effect in shaping long-term dynamics. In fact, for *I. ricinus* to complete its life cycle, each active stage requires a blood meal from a vertebrate host, for which this tick displays a certain degree of host selection despite the broad range of hosts reported (Kempf *et al.*, 2011). However, despite its dependency on hosts, this tick species spends more than 90% of its life cycle in the leaf litter. Changes in temperature, precipitation and snow cover are proven to dramatically affect tick occurrence but also survival, longevity, reproduction rate, host-seeking behaviour and even walking distance, and thus the possibility of finding a host, which in turn is strongly affected by climatic conditions. In particular, sensitivity is high in relation to relative humidity (which must be above 70–80%) and especially to saturation deficit, a measure of the drying power of the atmosphere (Hauser *et al.*, 2018).

A retrospective assessment of temperature trend in Northern Europe reveals a deep impact of changes in climate on the life cycle of *I. ricinus* (Estrada-Peña and Fernández-Ruiz, 2020). The temperature in 1950–2018 clearly increased especially within Northern European countries. Evidence of faster development rates of all developmental stages, faster oviposition and incubation rates, and faster moulting rates was observed in central Sweden, the Baltic countries and parts of Finland and Norway. The temperature trend for 1950–2018 shows a consistent inflection point around 1990, demonstrating that the increased annual accumulated temperature had a deeper impact on the life cycle of *I. ricinus* since approximately 1990.

Evidence of geographical range shift, especially in terms of altitudinal and latitudinal range shift, is frequently reported in Europe. Further invasions into new regions are repeatedly observed, as seen in Finland and Norway, where the upslope shift of *I. ricinus* was even greater than documented previously (Jore *et al.*, 2014; Cayol *et al.*, 2017). However, invasion of *I. ricinus* into new regions does not necessarily allow the tick to establish, as observed in Iceland, where this species has been reported since 1967 but its establishment has not been confirmed since no questing larvae have yet been detected, although it could possibly be established locally in low abundance. For establishment, *I. ricinus* requires favourable habitats with a required density of suitable hosts for all tick stages and a climate favourable for surviving the winter, a condition observed in Iceland since 1997. Historical records indicate that *I. ricinus* is acquired locally in Iceland and the number of records has increased over recent years (Alfredsson *et al.*, 2017).

Expansion of the distributional range of *I. ricinus* has also occurred at altitude as a consequence of variation in climatic conditions (Gilbert, 2010; Medlock *et al.*, 2013). New evidence of a shift to higher altitudes than those reported in the past decades has been recently reported. This was confirmed indirectly by an upsurge in new disease foci (especially for TBE) at higher altitudinal sites than previously reported, reflecting a threshold in tick density required to sustain TBE virus transmission through co-feeding on rodent hosts (Daniel *et al.*, 2016, 2018; Alfano *et al.*, 2020).

Establishment of *I. ricinus* has also occurred within highly anthropogenic habitat, such as urban and peri-urban areas, where microclimatic and biotic conditions have become more favourable during the last decades. The reporting of tick prevalence (but also of infection rates with a number of pathogens) comparable to or even higher than in rural or natural habitats demonstrates the increased risk of multiple infections in urban and peri-urban habitats (Rizzoli *et al.*, 2014; Rosà *et al.*, 2018; Vogelgesang *et al.*, 2020).

Changes have also occurred in the onset and duration of questing activity, more often reported for the nymphal stage. Seasonal variation in tick host-questing activity shows a significant relationship with microclimatic conditions (Daniel *et al.*, 2016; Hauser *et al.*, 2018). Prolongation of tick activity also has been reported by the observation of feeding ticks on wild birds and domestic dogs even during traditionally unfavourable conditions, as in winter (Furness and Furness, 2018; Zanet *et al.*, 2020).

Additional evidence that *I. ricinus* can adapt and therefore has the potential to be resilient to climate change has been reported. In a study by Gilbert *et al.* (2014), *I. ricinus* specimens were collected from a cline of climates using a latitudinal gradient, from Scotland to central France, and observed under laboratory conditions with variable temperatures. The theoretical potential to quest was then estimated for each population over the year for future climate change projections. As predicted, more ticks from cooler climates quested at lower temperatures than did ticks from warmer climates. The projections, based on temperature alone, suggested that populations could advance their activity season by a month under climate change, which has implications for exposure periods of hosts to tick-borne pathogens.

25.2 Future Predictions

Predicted climate change is expected to further affect *I. ricinus* spatial distribution and its seasonal dynamics in the next decades. For example, a rising temperature with following milder winters and warmer springs and autumns may allow ticks to quest for a prolonged period, but extreme events, such as heatwaves with droughts, would interrupt their questing and increase mortality.

Recent projections used to assess the potential future distribution of *I. ricinus* under current and future climate conditions indicate that predicted changes in macroclimatic conditions will exert a significant impact on the future geographical distribution. A recent study, based on a combination of niche modelling and projections based on climate data from 17 general circulation models (GCMs) under two Representative Concentration Pathway emissions scenarios (RCPs), for the years 2050 and 2070, indicated that the future potential distribution of *I. ricinus* will overlap across most of Western and Central Europe, and in more narrow zones in Eastern and Northern Europe and North Africa. Potential further expansion is predicted in Northern and Eastern Europe. However, in some important regions such as the Mediterranean, future distribution is still unclear owing to high uncertainty in model predictions (Alkishe *et al.*, 2017).

Furthermore, tick–host interaction is expected to be modified with consequences on the transmission cycles of pathogens, which in some cases could be totally disrupted. For example, a recent model aimed at assessing the possibility of permanent foci of *Borrelia afzelii* and *Borrelia garinii* for 2030 predicted an increase in latitude, mainly in the circulation of *B. garinii*, with large areas of expected permanent contact between vector and reservoirs in Nordic countries and Central Europe (Fernández-Ruiz and Estrada-Peña, 2020). However, in climate projections for the year 2050, this model predicted an unexpected scenario of contact disruption. Although large areas in Europe would be suitable for circulation of the pathogens, the predicted lack of niche overlap among ticks and reservoirs could promote a decrease in permanent foci.

The risk of TBE at country level, as in Hungary, is expected to increase along with the increase of temperature in 2021–2050 and 2071–2100. This increase in risk will be facilitated by the expected extension in tick questing season and the increase in ticks co-feeding on the same hosts, leading to the compounded increase of infections through non-systemic transmission (Nah *et al.*, 2020).

Despite continuous efforts to improve the power of mathematical models aimed at predicting future scenarios for *I. ricinus* and its transmitted pathogens (Estrada-Peña *et al.*, 2016), limitations are imposed by insufficient data and parameter estimations, and for this reason the uncertainty of future projections remains high.

References

Alfano, N., Tagliapietra, V., Rosso, F., Ziegler, U., Arnoldi, D. and Rizzoli, A. (2020) Tick-borne encephalitis foci in northeast Italy revealed by combined virus detection in ticks, serosurvey on goats and human cases. *Emerging Microbes & Infections* 9, 474–484.

Alfredsson, M., Olafsson, E., Eydal, M., Unnsteinsdottir, E.R., Hansford, K., *et al.* (2017) Surveillance of *Ixodes ricinus* ticks (Acari: Ixodidae) in Iceland. *Parasites & Vectors* 10, 466.

Alkishe, A.A., Peterson, A.T. and Samy, A.M. (2017) Climate change influences on the potential geographic distribution of the disease vector tick *Ixodes ricinus*. *PLoS ONE* 5, e0189092.

Bregnard, C., Rais, O. and Voordouw, M.J. (2020) Climate and tree seed production predict the abundance of the European Lyme disease vector over a 15-year period. *Parasites & Vectors* 13, 408.

Carpi, G., Cagnacci, F., Wittekindt, N.E., Zhao, F., Qi, J., *et al.* (2011) Metagenomic profile of the bacterial communities associated with *Ixodes ricinus* ticks. *PLoS ONE* 6, e25604.

Cayol, C., Koskela, E., Mappes, T., Siukkola, A. and Kallio, E.R. (2017) Temporal dynamics of the tick *Ixodes ricinus* in northern Europe: epidemiological implications. *Parasites & Vectors* 10, 166.

Daniel, M., Danielová, V., Kříž, B., Růžek, D., Fialová, A., *et al.* (2016) The occurrence of *Ixodes ricinus* ticks and important tick-borne pathogens in areas with high tick-borne encephalitis prevalence in different altitudinal levels of the Czech Republic, Part I. *Ixodes ricinus* ticks and tick-borne encephalitis virus. *Epidemiologie, Mikrobiologie, Imunologie* 65, 118–128.

Daniel, M., Danielová, V., Fialová, A., Malý, M., Kříž, B. and Nuttall, P.A. (2018) Increased relative risk of tick-borne encephalitis in warmer weather. *Frontiers in Cellular and Infection Microbiology* 8, 90.

ECDC (2018) *Tick Borne Encephalitis. Annual Epidemiological Report for 2018*. European Centre for Disease Prevention and Control, Stockholm.

ECDC (2020) Tick maps. European Centre for Disease Prevention and Control, Stockholm. Available at: https://www.ecdc.europa.eu/sites/default/files/images/Ixodes-ricinu-2020-09.png (accessed 3 June 2021).

Estrada-Peña, A. and Fernández-Ruiz, N. (2020) A retrospective assessment of temperature trends in northern Europe reveals a deep impact on the life cycle of *Ixodes ricinus* (Acari: Ixodidae). *Pathogens* 9, 345.

Estrada-Peña, A., Alexander, N. and Wint, G.R. (2016) Perspectives on modelling the distribution of ticks for large areas: so far so good? *Parasites & Vectors* 9, 179.

Fernández-Ruiz, N. and Estrada-Peña, A. (2020) Could climate trends disrupt the contact rates between *Ixodes ricinus* (Acari, Ixodidae) and the reservoirs of *Borrelia burgdorferi* s.l.? *PLoS ONE* 15, e0233771.

Furness, R.W. and Furness, E.N. (2018) *Ixodes ricinus* parasitism of birds increases at higher winter temperatures. *Journal of Vector Ecology* 43, 59–62.

Gilbert, L. (2010) Altitudinal patterns of tick and host abundance: a potential role for climate change in regulating tick-borne diseases? *Oecologia* 162, 217–225.

Gilbert, L., Aungier, J. and Tomkins, J.L. (2014) Climate of origin affects tick (*Ixodes ricinus*) host-seeking behavior in response to temperature: implications for resilience to climate change? *Ecology and Evolution* 4, 1186–1198.

Gray, J.S., Dautel, H., Estrada-Peña, A., Kahl, O. and Lindgren, E. (2009) Effects of climate change on ticks and tick-borne diseases in Europe. *Interdisciplinary Perspectives on Infectious Diseases* 2009, 593232.

Hauser, G., Rais, O., Morán Cadenas, F., Gonseth, Y., Bouzelboudjen, M. and Gern, L. (2018) Influence of climatic factors on *Ixodes ricinus* nymph abundance and phenology over a long-term monthly observation in Switzerland (2000–2014). *Parasites & Vectors* 11, 289.

Jore, S., Vanwambeke, S.O., Viljugrein, H., Isaksen, K., Kristoffersen, A.B., *et al.* (2014) Climate and environmental change drives *Ixodes ricinus* geographical expansion at the northern range margin. *Parasites & Vectors* 7, 11.

Kempf, F., De Meeûs, T., Vaumourin, E., Noel, V., Taragel'ová, V., *et al.* (2011) Host races in *Ixodes ricinus*, the European vector of Lyme borreliosis. *Infection, Genetics and Evolution* 11, 2043–2048.

Medlock, J.M., Hansford, K.M., Bormane, A., Derdakova, M., Estrada-Peña, A., *et al.* (2013) Driving forces for changes in geographical distribution of *Ixodes ricinus* ticks in Europe. *Parasites & Vectors* 6, 1.

Nah, K., Bede-Fazekas, Á., Trájer, A.J. and Wu, J. (2020) The potential impact of climate change on the transmission risk of tick-borne encephalitis in Hungary. *BMC Infectious Diseases* 20, 34.

Pettersson, J.H., Shi, M., Bohlin, J., Eldholm, V., Brynildsrud, O.B., *et al.* (2017) Characterizing the virome of *Ixodes ricinus* ticks from northern Europe. *Scientific Reports* 7, 10870.

Poli, P., Lenoir, J., Plantard, O., Ehrmann, S., Røed, K.H., *et al.* (2020) Strong genetic structure among populations of the tick *Ixodes ricinus* across its range. *Ticks and Tick-borne Diseases* 11, 101509.

Randolph, S.E. (2010) To what extent has climate change contributed to the recent epidemiology of tick-borne diseases? *Veterinary Parasitology* 167, 92–94.

Rizzoli, A., Silaghi, C., Obiegala, A., Rudolf, I., Hubálek, Z., *et al.* (2014) *Ixodes ricinus* and its transmitted pathogens in urban and peri-urban areas in Europe: new hazards and relevance for public health. *Frontiers in Public Health* 2, 251.

Rosà, R., Andreo, V., Tagliapietra, V., Baráková, I., Arnoldi, D., *et al.* (2018) Effect of climate and land use on the spatio-temporal variability of tick-borne bacteria in Europe. *International Journal of Environmental Research and Public Health* 15, 732.

Rosà, R., Tagliapietra, V., Manica, M., Arnoldi, D., Hauffe, H.C., *et al.* (2019) Changes in host densities and co-feeding pattern efficiently predict tick-borne encephalitis hazard in an endemic focus in northern Italy. *International Journal for Parasitology* 49, 779–787.

Semenza, J.C. and Suk, J.E. (2018) Vector-borne diseases and climate change: a European perspective. *FEMS Microbiology Letters* 365, fnx244.

Steffen, R., Lautenschlager, S. and Fehr, J. (2020) Travel restrictions and lockdown during the COVID-19 pandemic – impact on notified infectious diseases in Switzerland. *Journal of Travel Medicine* 29, taaa180.

Sykes, R.A. and Makiello, P. (2017) An estimate of Lyme borreliosis incidence in Western Europe. *Journal of Public Health* 39, 74–81.

van den Wijngaard, C.C., Hofhuis, A., Simões, M., Rood, E., van Pelt, W., *et al.* (2017) Surveillance perspective on Lyme borreliosis across the European Union and European Economic Area. *Eurosurveillance* 22, 27.

Vayssier-Taussat, M., Kazimirova, M., Hubalek, Z., Hornok, S., Farkas, R., *et al.* (2015) Emerging horizons for tick-borne pathogens: from the 'one pathogen–one disease' vision to the pathobiome paradigm. *Future Microbiology* 10, 2033–2043.

Vogelgesang, J.R., Walter, M., Kahl, O., Rubel, F. and Brugger, K. (2020) Long-term monitoring of the seasonal density of questing ixodid ticks in Vienna (Austria): setup and first results. *Experimental and Applied Acarology* 81, 409–420.

Zanet, S., Battisti, E., Pepe, P., Ciuca, L., Colombo, L., *et al.* (2020) Tick-borne pathogens in Ixodidae ticks collected from privately-owned dogs in Italy: a country-wide molecular survey. *BMC Veterinary Research* 16, 46.

eo26 Range Expansion of *Ixodes scapularis* in the USA

Durland Fish*
Yale School of Public Health, New Haven, Connecticut, USA

26.1 Expert Opinion

26.1.1 What is *Ixodes scapularis*?

Any discussion about the biology of *Ixodes scapularis* must recognize that there are two forms of this species. *I. scapularis* was originally described in 1821 (Say, 1821). The description was based solely upon gross morphology and the geographic origin of the specimen described is unknown. The distribution of *I. scapularis* in the USA ranges from Maine to Florida in the east and from Minnesota to Texas in the west (Eisen *et al.*, 2016). In 1979, a northern form of *I. scapularis* was described as a new species, *Ixodes dammini*, based upon minor morphological differences (Spielman *et al.*, 1979). At that time, *I. dammini* was only known to occur in coastal New England and northern Wisconsin. Because *I. dammini* was considered to be the only known vector for the agent of Lyme disease, Oliver *et al.* (1993) designated *I. dammini* as a junior synonym of *I. scapularis* so that Lyme disease cases could be diagnosed in the southern USA with *I. scapularis* being the vector. Considerable controversy resulted from this synonymy, which is still unresolved (Telford, 1998). Genetic studies on the northern and southern forms of *I. scapularis* have produced mixed results depending upon the genes investigated (Wesson *et al.*, 1993; Norris *et al.*, 1996), but most recent studies provide strong evidence supporting the existence of two genetically distinct forms or clades (Kelly *et al.*, 2014; Xu *et al.*, 2020). There is also an important difference in host-seeking behaviour of nymphal-stage ticks between the northern and southern forms that have epidemiological significance. Nymphal-stage ticks of the southern form are rarely collected with drag cloths and reports of human bites by nymphs are exceedingly rare (Diuk-Wasser *et al.*, 2006; Arsnoe *et al.*, 2019). Moreover, there are no reports of nymphs infected with the agent of Lyme disease in states south of coastal North Carolina. The nymphal stage is the primary vector of the agent of Lyme disease in the northern states, where it is easily collected in great numbers by drag cloth and human tick bites are common (Falco and Fish, 1988; Falco *et al.*, 1999; Daniels *et al.*, 2000). The average infection rate for the agent of Lyme disease in northern nymphal ticks is 21% (Diuk-Wasser *et al.*, 2012) and an estimated one-quarter of the human population is bitten by nymphs each year in hyperendemic areas (Campbell *et al.*, 1998). Consequently, 95% of the reported cases of Lyme disease in the USA are acquired from the northern form of *I. scapularis* (Pepin *et al.*, 2012).

*Email: durland.fish@yale.edu

DOI: 10.1079/9781789249637.0026

26.1.2 Relict populations

There is little evidence that *I. scapularis* is expanding its range significantly in the southern states. Reports of new locations are more likely due to increased surveillance. But the northern form has experienced a well-documented dramatic range expansion since the 1970s. These ticks were rare on the mainland Northeast during the 1960s, but common on Long Island and offshore islands of Massachusetts and Rhode Island. The first significant mainland population was discovered when cases of Lyme disease were first investigated in the coastal town of Lyme, Connecticut, in 1977 (Wallis *et al.*, 1978), just 6 miles (9.6 km) from Long Island. However, in the Midwest, the first report of *I. scapularis* occurred earlier in Wisconsin and concurrent with the first recognized case of Lyme disease in 1970 (Jackson and Defoliart, 1970; Scrimenti, 1970). From these isolated relict populations, the northern form of *I. scapularis* has expanded its range throughout the northern states from Maine to Minnesota and south to Virginia and Illinois (Diuk-Wasser *et al.*, 2006; Eisen *et al.*, 2016). This unprecedented rapid range expansion of ticks can be explained by major environmental change events that have occurred in the USA over the past 400 years.

26.1.3 Forest history in the USA

Deforestation and subsequent restoration of deciduous hardwood forests in the USA are the greatest environmental changes to have occurred in the USA in historic times. Early explorers during the 17th and 18th centuries reported a continuous old growth hardwood forest from the Northeast to the Midwest with sporadic clearings made for agriculture and hunting by the indigenous Amerindian population (MacCleery, 2011). By the end of the 19th century, nearly all of this forest was cut for agriculture, fuel, construction and export to England. The forests were usually clear-cut leaving only small patches of virgin forests remaining. As agriculture moved from the Northeast to the Midwest during the 19th century, abandoned agricultural fields in the Northeast went through natural succession and the forests began to regenerate (MacCleery, 2011). Forest cover increased steadily during the first half of the 20th century.

26.1.4 History of white-tailed deer

White-tailed deer (*Odocoileus virginianus*) is the definitive host of *I. scapularis* where the adult stage mates and females feed to repletion, resulting in oviposition of 2000 offspring (Fish, 1993). Populations of *I. scapularis* cannot be maintained in the absence of deer as evidenced by deer exclusion (Daniels and Fish, 1995) and elimination experiments (Rand *et al.*, 2004), absence of populations on islands where deer are absent (Wilson *et al.*, 1985) and absence in geographic areas where deer are absent in the Midwest. Deer are dependent upon deciduous forests for cover and food, and the history of forests is tightly linked to the history of deer populations.

Deer were common during precolonial times but were heavily hunted by Amerindians for food and clothing (Stanton, 1963). Clearings were maintained by burning to provide forest edge, the preferred habitat of deer. As colonization progressed, deer hunting with firearms and forest exploitation rapidly depleted deer, resulting in their total elimination from most of their former range by the late 19th century. The last known deer was shot in Indiana in 1893, when populations in the Northeast were already exterminated (DNR-Indiana, 2020). The only documented continuous population of deer in the coastal Northeast since colonial times occurred on Long Island, New York (Severinghaus and Brown, 1956). As the forests regenerated during the early 20th century, deer were imported from the Upper Midwest and released in the Northeast to restore populations for hunting (Witmer, 1991). Early successional forests provided excellent forage for deer. In the beginning, deer hunting was severely restricted and included only males to maximize population growth. As forests regenerated, deer populations increased, as did hunting. Deer were first introduced in Illinois in 1930 and by 1940 the state-wide herd was estimated to be 400 (DNR-Illinois, 2020). By 2005, over 200,000 deer were harvested by hunters. Populations continued to increase when suburbanization began to invade nearby forests, providing increased edge habitat and ornamental plantings for food (Sterba, 2013). Deer hunting was restricted or prohibited within suburban developments for safety reasons and opposition by residents. Populations exploded in these areas with deer densities greater than 50 per km^2 (Kilpatrick *et al.*, 2014). These historically high populations of deer set the stage for the expansion of *I. scapularis* from relict populations on Long Island. A similar sequence of events occurred in the Upper Midwest, resulting in a southward expansion from the northern Wisconsin relict population.

26.1.5 History of *Ixodes scapularis*

I. scapularis was a rather obscure species prior to 1969 when it was discovered to be the vector of human babesiosis on Nantucket Island, Massachusetts (Spielman *et al.*, 1979). Shortly afterwards, human babesiosis was found to occur on Long Island where *I. scapularis* was also known to occur (Spielman *et al.*, 1985). With the discovery of Lyme disease in 1975, *I. scapularis* (as *I. dammini*) became a focus of tick studies and its range expansion was tightly linked to reports of Lyme disease. It was well established east of the Connecticut River in the area of Lyme by 1977 (Wallis *et al.*, 1978), but the river seemed to be a temporary barrier for its westward expansion. A separate population became established in Westchester County, New York, on the northern border of New York City, in 1981 (D. Fish, 1981, personal observation), and a population was discovered north of Boston, Massachusetts, also in 1981 (Lastavica *et al.*, 1989). A fourth mainland population was first observed in coastal New Jersey in 1979 (Schulze *et al.*, 1984). The Westchester County population rapidly expanded northward, east of the Hudson River, and later, west of the river, again implicating rivers as temporary barriers to expansion (White *et al.*, 1991). Eventually, the Westchester population expanded eastward into Connecticut, while the Lyme population expanded westward until there was a continuous population throughout the state (Xue *et al.*, 2015). By 1989, *I. scapularis* had spread from eastern Massachusetts northward to coastal Maine (Rand *et al.*, 2007), from Westchester County northward to Albany, New York (White *et al.*, 1991), and from coastal New Jersey southward to Maryland (Amerasinghe *et al.*, 1992) and westward to Pennsylvania (Pak *et al.*, 2019). In the Midwest, the Wisconsin population spread southward to Illinois (Bouseman *et al.*, 1990) and eastward to Indiana and Michigan (Pinger *et al.*, 1991; Hamer *et al.*, 2010). Located between Indiana and Pennsylvania, Ohio reported its first established population in 2004 (Wang *et al.*, 2014), indicating that *I. scapularis* will soon regain its former distribution of a continuous population ranging from the east coast to the Midwest (Fig. 26.1). There may have been other unrecognized refugia in intermediate locations.

26.1.6 Mode of range expansion

The mechanism of spread by *I. scapularis* is most likely by deer movement. Deer can host hundreds of female *I. scapularis* and have a greater range of movement than any other mammalian host. Birds also serve as hosts, but only for immature stages and in much smaller, often single, numbers (Battaly and Fish, 1993). Immature stages experience 99% mortality and also must find a deer hosting a mate (Fish, 1993). Immature ticks imported from birds, and that successfully moult, can seed new areas in sufficient numbers to occasionally be found by drag sampling or on mammalian hosts (Ogden *et al.*, 2008), but are not enough to establish a reproductive population on deer. Such seeding from established populations may be predictive of expansion. Sampling of deer for adult ticks in nearby non-established areas usually results in low numbers for a few years, followed by a sudden increase in prevalence, indicating an established reproductive population from migrant deer. The rate of expansion is also consistent with deer as a mechanism of spread (Fig. 26.1).

26.1.7 Establishment of *I. scapularis*-borne pathogens

The expansion of *I. scapularis*-borne pathogens involves other mechanisms. The agent of Lyme disease, *Borrelia burgdorferi*, can be transported by reservoir-competent birds (Scott *et al.*, 2001). Thus, new populations of *I. scapularis* can quickly acquire infection when infected nymphs from

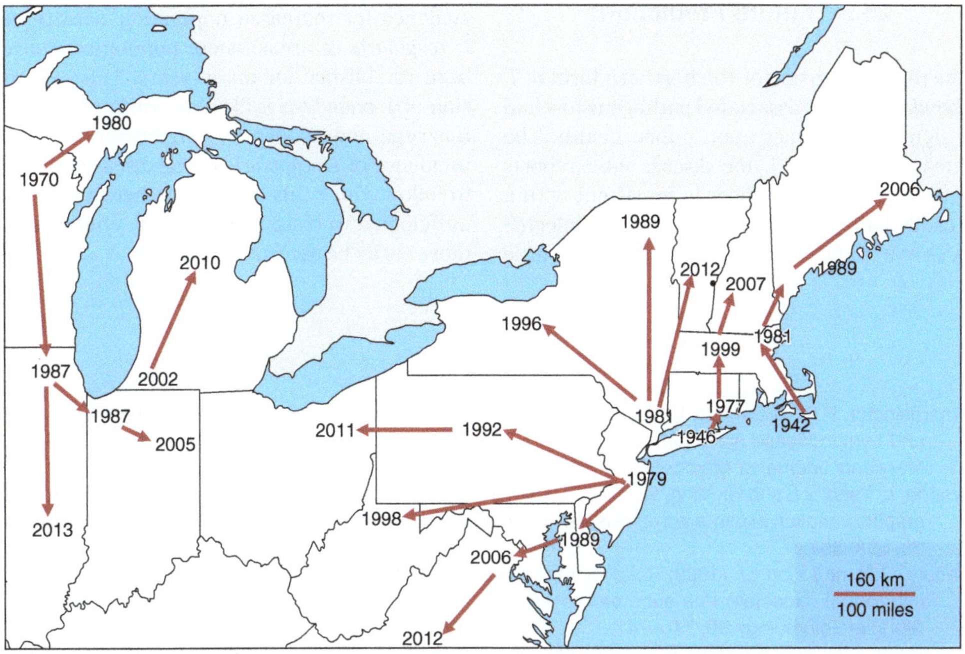

Fig. 26.1. Major pathways of *Ixodes scapularis* expansion in the eastern USA with first published records of occurrence. (Author's own figure.)

infected bird-derived larvae feed upon reservoir-competent mammals. A similar mechanism also applies to the agent of human anaplasmosis, *Anaplasma phyagocytophilum* (Ogden *et al.*, 2008). The agent of human babesiosis, *Babesia microti*, does not infect birds, and rodents are the dispersal agent. This is evident in the much slower movement of *B. microti* within established populations of *I. scapularis* (Diuk-Wasser *et al.*, 2014). Powassan encephalitis virus also does not infect birds and expansion of this virus may be dependent upon movement of reservoir-competent hosts, which include mice and mustelids (weasels, skunks, fishers, etc.). However, unlike other *I. scapularis*-borne pathogens, Powassan virus can also be transmitted vertically from female to offspring (Costero and Grayson, 1996), as is *Borrelia miyamotoi* (Scoles *et al.*, 2001), and these pathogens may appear concurrently with *I. scapularis* colonization.

26.1.8 Alternative theories

Several theories have been proposed to explain the expansion of *I. scapularis* and associated pathogens. These include recent absence of predation on hosts of immatures (Levi *et al.*, 2012) and warmer temperatures at higher latitudes due to climate change (Ogden *et al.*, 2006). Evidence against these alternative theories is lack of supportive data demonstrating that *I. scapularis* existed in areas prior to loss of predators and the multidirectional expansion of *I. scapularis*. Although climate change may be a cause of the northern expansion, it does not explain expansion in other directions, which have been equally rapid (Fig. 26.1). In the long term, climate change may permit a more northerly expansion (Brownstein *et al.*, 2005), but genetic adaptation independent of temperature would have to occur. Seasonality of host-seeking behaviour is a genetically inherited behavioural attribute in response to photoperiod (Belozerov and Naumov, 2002; Belozerov *et al.*, 2002; Gray *et al.*, 2016). Warming temperatures alone will not affect photoperiodic response at higher latitudes. Synchrony of host-seeking behaviour among stages is essential for population maintenance and it requires genetic adaptation. With a 2-year life cycle, *I. scapularis* is not likely to experience such genetic adaptation for many decades.

26.2 Future Predictions

The rapid expansion of the northern form of *I. scapularis* and its associated pathogens has had a significant impact upon public health. The linear increase in Lyme disease case reports over the past four decades is consistent with a steadily increasing human exposure to infected ticks caused by range expansion. There is little evidence for increased population density of *I. scapularis* in areas where populations have been established for many years. This expansion of *I. scapularis* is likely to continue until it has regained its former distribution and the incidence of *I. scapularis*-borne disease is likely to follow. Once this occurs, further expansion anticipated to result from climate change can more easily be ascertained.

References

Amerasinghe, F.P., Breisch, N.L., Azad, A.F., Gimpel, W.F., Greco, M., *et al.* (1992) Distribution, density, and Lyme disease spirochete infection in *Ixodes dammini* (Acari: Ixodidae) on white-tailed deer in Maryland. *Journal of Medical Entomology* 29, 54–61.

Arsnoe, I., Tsao, J.I. and Hickling, G.J. (2019) Nymphal *Ixodes scapularis* questing behavior explains geographic variation in Lyme borreliosis risk in the eastern United States. *Ticks and Tick-borne Diseases* 10, 553–563.

Battaly, G.R. and Fish, D. (1993) Relative importance of bird species as hosts for immature *Ixodes dammini* (Acari: Ixodidae) in a suburban residential landscape of southern New York State. *Journal of Medical Entomology* 30, 740–747.

Belozerov, V.N. and Naumov, R.L. (2002) Nymphal diapause and its photoperiodic control in the tick *Ixodes scapularis* (Acari: Ixodidae). *Folia Parasitologica* 49, 314–318.

Belozerov, V.N., Fourie, L.J. and Kok, D.J. (2002) Photoperiodic control of developmental diapause in nymphs of prostriate ixodid ticks (Acari: Ixodidae). *Experimental and Applied Acarology* 28, 163–168.

Bouseman, J.K., Kitron, U., Kirkpatrick, C.E., Siegel, J. and Todd, K.S. Jr (1990) Status of *Ixodes dammini* (Acari: Ixodidae) in Illinois. *Journal of Medical Entomology* 27, 556–560.

Brownstein, J.S., Holford, T.R. and Fish, D. (2005) Effect of climate change on Lyme disease risk in North America. *EcoHealth* 2, 38–46.

Campbell, G.L., Fritz, C.L., Fish, D., Nowakowski, J., Nadelman, R.B. and Wormser, G.P. (1998) Estimation of the incidence of Lyme disease. *American Journal of Epidemiology* 148, 1018–1026.

Costero, A. and Grayson, M.A. (1996) Experimental transmission of Powassan virus (Flaviviridae) by *Ixodes scapularis* ticks (Acari: Ixodidae). *American Journal of Tropical Medicine and Hygiene* 55, 536–546.

Daniels, T.J. and Fish, D. (1995) Effect of deer exclusion on the abundance of immature *Ixodes scapularis* (Acari: Ixodidae) parasitizing small and medium-sized mammals. *Journal of Medical Entomology* 32, 5–11.

Daniels, T.J., Falco, R.C. and Fish, D. (2000) Estimating population size and drag sampling efficiency for the blacklegged tick (Acari: Ixodidae). *Journal of Medical Entomology* 37, 357–363.

Diuk-Wasser, M.A., Gatewood, A.G., Cortinas, M.R., Yaremych-Hamer, S., Tsao, J., *et al.* (2006) Spatiotemporal patterns of host-seeking *Ixodes scapularis* nymphs (Acari: Ixodidae) in the United States. *Journal of Medical Entomology* 43, 166–176.

Diuk-Wasser, M.A., Hoen, A.G., Cislo, P., Brinkerhoff, R., Hamer, S.A., *et al.* (2012) Human risk of infection with *Borrelia burgdorferi*, the Lyme disease agent, in eastern United States. *American Journal of Tropical Medicine and Hygiene* 86, 320–327.

Diuk-Wasser, M.A., Liu, Y., Steeves, T.K., Folsom-O'Keefe, C., Dardick, K.R., *et al.* (2014) Monitoring human babesiosis emergence through vector surveillance New England, USA. *Emerging Infectious Diseases* 20, 225–231.

DNR-Illinois (2020) White-tailed deer Illinois: Management. Illinois Department of Natural Resources, Springfield, Illinois. Available at: https://deer.wildlifeillinois.org/management/history-of-management/ (accessed 23 November 2020).

DNR-Indiana (2020) White-tailed deer biology. Indiana Department of Natural Resources, Indianapolis, Indiana. Available at: https://www.in.gov/dnr/fishwild/3359.htm (accessed 23 November 2020).

Eisen, R.J., Eisen, L. and Beard, C.B. (2016) County-scale distribution of *Ixodes scapularis* and *Ixodes pacificus* (Acari: Ixodidae) in the continental United States. *Journal of Medical Entomology* 53, 349–386.

Falco, R.C. and Fish, D. (1988) Ticks parasitizing humans in a Lyme disease endemic area of southern New York State. *American Journal of Epidemiology* 128, 1146–1152.

Falco, R.C., McKenna, D.F., Daniels, T.J., Nadelman, R.B., Nowakowski, J., *et al.* (1999) Temporal relation between *Ixodes scapularis* abundance and risk for Lyme disease associated with erythema migrans. *American Journal of Epidemiology* 149, 771–776.

Fish, D. (1993) Population ecology of *Ixodes dammini*. In: Ginsberg, H. (ed.) *Ecology and Environmental Management of Lyme Disease*. Rutgers University Press, New Brunswick, New Jersey, pp. 25–42.

Gray, J.S., Kahl, O., Lane, R.S., Levin, M.L. and Tsao, J.I. (2016) Diapause in ticks of the medically important *Ixodes ricinus* species complex. *Ticks and Tick-borne Diseases* 7, 992–1003.

Hamer, S.A., Tsao, J.I., Walker, E.D. and Hickling, G.J. (2010) Invasion of the Lyme disease vector *Ixodes scapularis*: implications for *Borrelia burgdorferi* endemicity. *EcoHealth* 7, 47–63.

Jackson, J.O. and Defoliart, G.R. (1970) *Ixodes scapularis* Say in northern Wisconsin. *Journal of Medical Entomology* 7, 124–150.

Kelly, R.R., Gaines, D., Gilliam, W.F. and Brinkerhoff, R.J. (2014) Population genetic structure of the Lyme disease vector *Ixodes scapularis* at an apparent spatial expansion front. *Infection and Genetics Evolution* 27, 543–550.

Kilpatrick, H.J., Labonte, A.M. and Stafford, K.C. (2014) The relationship between deer density, tick abundance, and human cases of Lyme disease in a residential community. *Journal of Medical Entomology* 51, 777–784.

Lastavica, C.C., Wilson, M.L., Berardi, V.P., Spielman, A. and Deblinger, R.D. (1989) Rapid emergence of a focal epidemic of Lyme disease in coastal Massachusetts. *New England Journal of Medicine* 320, 133–137.

Levi, T., Kilpatrick, A.M., Mangel, M. and Wilmers, C.C. (2012) Deer, predators, and the emergence of Lyme disease. *Proceedings of the National Academy of Sciences USA* 109, 10942–10947.

MacCleery, D.W. (2011) *American Forests, A History of Resiliency and Recovery*. Forest History Society, Durham, North Carolina.

Norris, D.E., Klompen, J.S.H., Keirans, J.E. and Black, W.C. (1996) Population genetics of *Ixodes scapularis* (Acari: Ixodidae) based on mitochondrial 16S and 12S genes. *Journal of Medical Entomology* 33, 78–89.

Ogden, N.H., Maarouf, A., Barker, I.K., Bigras-Poulin, M., Lindsay, L.R., *et al.* (2006) Climate change and the potential for range expansion of the Lyme disease vector *Ixodes scapularis* in Canada. *International Journal for Parasitology* 36, 63–70.

Ogden, N.H., Lindsay, L.R., Hanincova, K., Barker, I.K., Bigras-Poulin, M., *et al.* (2008) Role of migratory birds in introduction and range expansion of *Ixodes scapularis* ticks and of *Borrelia burgdorferi* and *Anaplasma phagocytophilum* in Canada. *Applied and Environmental Microbiology* 74, 1780–1790.

Oliver, J.H., Owsley, M.R., Hutcheson, H.J., James, A.M., Chen, C.S., *et al.* (1993) Conspecificity of the ticks *Ixodes scapularis* and *Ixodes dammini* (Acari, Ixodidae). *Journal of Medical Entomology* 30, 54–63.

Pak, D., Jacobs, S.B. and Sakamoto, J.M. (2019) A 117-year retrospective analysis of Pennsylvania tick community dynamics. *Parasites & Vectors* 12, 189.

Pepin, K.M., Eisen, R.J., Mead, P.S., Piesman, J., Fish, D., *et al.* (2012) Geographic variation in the relationship between human Lyme disease incidence and density of infected host-seeking *Ixodes scapularis* nymphs in the eastern United States. *American Journal of Tropical Medicine and Hygiene* 86, 1062–1071.

Pinger, R.R., Holycross, J., Ryder, J. and Mummert, M. (1991) Collections of adult *Ixodes dammini* in Indiana, 1987–1990, and the isolation of *Borrelia burgdorferi*. *Journal of Medical Entomology* 28, 745–749.

Rand, P.W., Lubelczyk, C., Holman, M.S., Lacombe, E.H. and Smith, R.P. (2004) Abundance of *Ixodes scapularis* (Acari: Ixodidae) after the complete removal of deer from an isolated offshore island, endemic for Lyme disease. *Journal of Medical Entomology* 41, 779–784.

Rand, P.W., Lacombe, E.H., Dearborn, R., Cahill, B., Elias, S., *et al.* (2007) Passive surveillance in Maine, an area emergent for tick-borne diseases. *Journal of Medical Entomology* 44, 1118–1129.

Schulze, T.L., Lakat, M.F., Bowen, G.S., Parkin, W.E. and Shisler, J.K. (1984) *Ixodes dammini* (Acari, Ixodidae) and other ixodid ticks collected from white-tailed deer in New Jersey, USA. 1. Geographical distribution and its relation to selected environmental and physical factors. *Journal of Medical Entomology* 21, 741–749.

Scoles, G.A., Papero, M., Beati, L. and Fish, D. (2001) A relapsing fever group spirochete transmitted by *Ixodes scapularis* ticks. *Vector-Borne and Zoonotic Diseases* 1, 21–34.

Scott, J.D., Fernando, K., Banerjee, S.N., Durden, L.A., Byrne, S.K., *et al.* (2001) Birds disperse ixodid (Acari: Ixodidae) and *Borrelia burgdorferi*-infected ticks in Canada. *Journal of Medical Entomology* 38, 493–500.

Scrimenti, R.J. (1970) Erythema chronicum migrans. *Archives of Dermatology* 102, 104–105.
Severinghaus, C.W. and Brown, C.P. (1956) History of white-tailed deer in New York. *New York Fish and Game Journal* 3, 129–167.
Spielman, A., Clifford, C.M., Piesman, J. and Corwin, M.D. (1979) Human babesiosis on Nantucket Island, USA – description of the vector, *Ixodes* (*Ixodes*) *dammini*, n. sp. (Acarina, Ixodidae). *Journal of Medical Entomology* 15, 218–234.
Spielman, A., Wilson, M.L., Levine, J.F. and Piesman, J. (1985) Ecology of *Ixodes dammini*-borne human babesiosis and Lyme disease. *Annual Review of Entomology* 30, 439–460.
Stanton, D.C. (1963) *A History of White-Tailed Deer in Maine*. Maine Department of Inland Fisheries and Game, Augusta, Maine.
Sterba, J. (2013) *Nature Wars: The Incredible Story of How Wildlife Comebacks Turned Backyards into Battlegrounds*. Broadway Books, New York.
Telford, S.R. (1998) The name *Ixodes dammini* epidemiologically justified. *Emerging Infectious Diseases* 4, 132–134.
Wallis, R.C., Brown, S.E., Kloter, K.O. and Main, A.J. (1978) Erythema chronicum migrans and Lyme arthritis: field-study of ticks. *American Journal of Epidemiology* 108, 322–327.
Wang, P.M., Glowacki, M.N., Hoet, A.E., Needham, G.R., Smith, K.A., *et al.* (2014) Emergence of *Ixodes scapularis* and *Borrelia burgdorferi*, the Lyme disease vector and agent, in Ohio. *Frontiers in Cellular and Infection Microbiology* 4, 70.
Wesson, D.M., McLain, D.K., Oliver, J.H., Piesman, J. and Collins, F.H. (1993) Investigation of the validity of species status of *Ixodes dammini* (Acari, Ixodidae) using rDNA. *Proceedings of the National Academy of Sciences USA* 90, 10221–10225.
White, D.J., Chang, H.G., Benach, J.L., Bosler, E.M., Meldrum, S.C., *et al.* (1991) The geographic spread and temporal increase of the Lyme-disease epidemic. *Journal of the American Medical Association* 266, 1230–1236.
Wilson, M.L., Adler, G.H. and Spielman, A. (1985) Correlation between abundance of deer and that of the deer tick, *Ixodes dammini* (Acari, Ixodidae). *Annals of the Entomological Society of America* 78, 172–176.
Witmer, G.W. (1991) The need and difficulty of bringing the Pennsylvania deer herd under control. *5 – Fifth Eastern Wildlife Damage Control Conference, University of Nebraska, Lincoln, Nebraska* 5, 130–137.
Xu, G., Wielstra, B. and Rich, S.M. (2020) Northern and southern blacklegged (deer) ticks are genetically distinct with different histories and Lyme spirochete infection rates. *Scientific Reports* 10, 2089.
Xue, L., Scoglio, C, McVey, D., Boone, R. and Cohnstaedt, L.W. (2015) Two introductions of Lyme disease into Connecticut: a geospatial analysis of human cases from 1984 to 2012. *Vector-Borne and Zoonotic Diseases* 15, 523–528.

eo27 Distribution, Seasonal Occurrence and Biological Characteristics of *Haemaphysalis longicornis*, a Vector of Bovine Piroplasmosis in Japan

Rika Umemiya-Shirafuji
Obihiro University of Agriculture and Veterinary Medicine, Obihiro, Japan

27.1 Expert Opinion

27.1.1 Introduction

There are at least 43 species of hard tick in Japan (Yamaguti *et al.*, 1971; Fujita and Takada, 2007). Among them, certain species are problematic as harmful vectors of tick-borne diseases in veterinary and medical fields. *Haemaphysalis longicornis*, an extremely important tick species in the livestock industry, is a major vector of bovine piroplasmosis caused by *Theileria* and *Babesia* parasites. Piroplasmosis in cattle has caused considerable issues because there are no available vaccines against the protozoan parasites. For many years, seasonal occurrence and survival periods of *H. longicornis* in grazing lands have been investigated in order to control the distribution of ticks and tick infestations on animals. These studies provide valuable information including understanding of changes in tick distribution areas due to climate change, but most of this literature is written in Japanese. This expert opinion focuses on the distribution and seasonal occurrence of *H. longicornis*, and protozoan parasites transmitted by *H. longicornis* in Japan, and describes expected changes in the distribution areas and biological characteristics of this tick, as a vector, due to climate change.

27.1.2 Distribution of *Haemaphysalis longicornis* in Japan

Japan has four seasons with a climate ranging from subarctic in the north to subtropical in the south. Conditions are different between the Pacific side and the Sea of Japan side (Japan Meteorological Agency; https://www.data.jma.go.jp/gmd/cpd/longfcst/en/tourist.html, accessed 4 January 2021). Hard tick species distributed in Japan are three-host ticks except for *Rhipicephalus microplus*. *H. longicornis* is found throughout Japan and is a unique tick species having both triploid parthenogenetic and diploid bisexual races. *H. longicornis* parasitizes many animals from small to large mammals as well as birds (Yamaguti *et al.*, 1971). Male *H. longicornis* are predominantly recorded in the southern part of Japan, with a lack of collection records in northern Japan, indicating that the bisexual race is more abundant in warmer areas. The geographical distribution of both bisexual and parthenogenetic races overlaps in some prefectures, making it

*Email: umemiya@obihiro.ac.jp

DOI: 10.1079/9781789249637.0027

difficult to determine a clear boundary (Kitaoka, 1961; Fujita *et al.*, 2013). The habitat of *H. longicornis* appears to be limited even in identical prefectures, probably due to climatically different environments. In Hokkaido, northernmost of the four main islands of Japan, female *H. longicornis* can be collected in southern and central Hokkaido, while the male ticks are not detected or are extremely rare in these parts. *H. longicornis* is rarely found in eastern and northern Hokkaido at present; nevertheless, there are collection records of *H. longicornis* ticks from animals in both Okhotsk and Tokachi regions of eastern Hokkaido in 1927–1967 (Yamaguti *et al.*, 1971). According to this literature, ticks were collected from vegetation in the Okhotsk region, indicating that the ticks were established in the region during that period. However, epidemiological data reported for the last 14 years show that *H. longicornis* was not detectable in the Tokachi region, Pacific side of Hokkaido (Inokuma *et al.*, 2007; Yokoyama *et al.*, 2012). Bovine piroplasmosis, caused by *Theileria orientalis*, is still reported in eastern Hokkaido. In the Tokachi region, *Haemaphysalis japonica*, *Haemaphysalis megaspinosa*, *Ixodes ovatus* and *Ixodes persulcatus* can currently be collected from animals and vegetation, indicating that *T. orientalis* is transmitted by these tick species (Yokoyama *et al.*, 2012). Further experiments using these tick species and *T. orientalis* based on established experimental models (Hayashida *et al.*, 2018) are needed to demonstrate their vector capacities. The causes of the discrepancies between previous collection records and recent records of *H. longicornis* in eastern Hokkaido remain unknown. The relationships between these changes in *H. longicornis* distribution and climate change also remain unclear.

The impacts of climate change on tick migration and subsequent establishment are points to consider based on the biological features of each tick species. Humidity and temperature of the environment have important influences on development in the life cycle of ticks. In bisexual *H. longicornis*, more than 84.5% relative humidity is necessary to complete a life cycle (Yano *et al.*, 1988). At temperatures between 15 and 35°C, pre-oviposition, oviposition, egg hatching and moulting periods are shortened as the temperature rises in the environment (Fujisaki and Kamio, 1988). Compared with bisexual *H. longicornis*, lower temperatures appear to be suitable for egg development in the parthenogenetic ticks (Kitaoka, 1961). Differences in their temperature sensitivities may be attributed to the fact that the parthenogenetic race propagates in the northern part and the bisexual race in the southern part of Japan (Yano *et al.*, 1987).

Due to the wide range of vector capacities and host animals of *H. longicornis*, areas outside Japan have serious concerns about the expansion of the distribution range of this tick species. Recently, *H. longicornis* ticks were detected in the USA and the mitochondrial DNA sequences analysed (Egizi *et al.*, 2020). The results indicate that companion animals may be the sources of exotic ticks entering the USA. Important factors for changes in the distribution of ticks are the locomotion of host animals carrying ticks and the adaptive and survival ability of ticks in new areas. Certain areas can become new habitats for ticks if climate variability creates a suitable environment for the development and reproduction of those ticks. Other than *H. longicornis*, studies in Japan on the temperature sensitivities of other important tick species in both veterinary and medical fields are required to predict the possibilities for establishment of new habitats for ticks.

27.1.3 Seasonal occurrence of *Haemaphysalis longicornis* in Japan

To control tick infestations of livestock animals, it is very important to know the seasonal occurrence of ticks that exist in environments in which they thrive. Seasonal occurrence of ticks differs among tick species. Generally, occurrence of adult *Haemaphysalis* peaks during April to August (summer type: *H. longicornis*) or during October to March (winter type: *H. megaspinosa*) in Japan (Tsunoda and Fujimagari, 1996). Larval *H. longicornis* are found from spring and peak during autumn. Nymphal *H. longicornis* can be found from spring to autumn but do not show notable peaks. All developmental stages (except for the eggs of *H. longicornis*) overwinter in the soil of pastures (Ito *et al.*, 1983). Unfed adults and nymphs can survive more than one season after overwintering. Active periods of *H. longicornis* are well synchronized with the grazing season of cattle. Climate change may indirectly contribute to seasonal occurrence of ticks in the pasture area by leading environmental changes such as in vegetation. For example, certain climatic conditions are favourable for growth of pasture plants, providing abundant plants for grazing cattle. While such plant growth gives a suitable waiting area for ticks to parasitize their hosts and protect the surface habitat where they hatch and moult, it also raises the possibility that ticks on vegetation are eaten by cattle. It is believed that grazing cattle can be a factor suppressing tick populations by disturbing tick habitats due to greater grazing intake (Otake *et al.*, 1989). Seasonal occurrence of ticks in the grazing areas is not directly affected by climate change (Ootake *et al.*, 1985), but can fluctuate annually due to multiple indirect effectors including changes of vegetation, density and behaviour of host animals, and pasture management.

27.1.4 Protozoan parasites transmitted by *Haemaphysalis longicornis* in Japan

There are interesting reports of the effects of temperature on *Theileria* parasites in *H. longicornis*. Protozoan masses of *T. orientalis* are detectable in the salivary glands of nymphal *H. longicornis* collected from the soil of pasture during the winter (Fujisaki *et al.*, 1985). This means that *T. orientalis* overwinters in *H. longicornis* nymphs that moult after engorgement on infected cattle during the previous autumn and hide in the soil of pasture during the winter. These results also indicate that *T. orientalis* in *H. longicornis* may develop gradually into the infective stages during the overwintering period. After overwintering, they may play an important role in the spring outbreaks of piroplasmosis immediately after the introduction of cattle to a pasture.

Under laboratory conditions, high infection levels were discernible in ticks incubated at 20 and 25°C, but no maturation of *T. orientalis* was observed in the salivary glands of ticks exposed to 35°C (Fujisaki and Kamio, 1988). It was considered that within a certain period post-moulting, *T. orientalis* may complete development in the salivary glands up to a specific state and may be unable to initiate further maturation without the stimulation of feeding. In general, unfed *H. longicornis* can survive longer at 10 to 18°C than at 25°C, and thus infected ticks appear to be able to remain a long time without loss of infectivity if they are incubated at lower than 18°C after establishment of infection of the salivary glands by *T. orientalis* at 25°C. Another report showed that the development of *T. orientalis* was induced by only heat stimulation at 37°C and *T. orientalis* reached the stage infective to cattle (Sasaki *et al.*, 1990). These findings suggest that changes in environmental temperature may influence the efficiency of *T. orientalis* infection in host animals. However, it is largely unknown how climate change affects the transmission efficiency of *T. orientalis* from ticks to host animals.

In addition, little is known about relationships between temperature and development of *Babesia ovata*, another causative agent of bovine piroplasmosis transmitted by *H. longicornis* in Japan. Compared with *Theileria* parasites, *Babesia* parasites are characterized by both transstadial and transovarial transmission. Several studies demonstrated that lower temperatures contribute to maintaining a low infection level of *Babesia* in eggs laid by *Babesia*-infected ticks, but higher temperatures are detrimental (Ouhelli *et al.*, 1987; Ouhelli and Schein, 1988). Regarding epidemiological aspects, engorged females infected and incubated at constant low temperatures are more of a hazard to livestock than those incubated at higher temperatures. Further studies using experimental models (Umemiya-Shirafuji *et al.*, 2017) need to be done for understanding the effects of temperature change on the development of *B. ovata* in ticks.

27.2 Future Predictions

There is concern about the expansion of the distribution range of *H. longicornis* within Japan as well as other countries. At present, few studies have been conducted on how climate variability and global warming affect the geographical distribution of ticks in Japan. Changes in vegetation and animal habitat due to climate change are thought to result in changes in tick habitat. Knowledge of ecology, life cycle and biological features of ticks is extremely important in controlling ticks, especially in pasture areas. It is necessary to integrate research in different fields such as medical, veterinary and environmental, and to relate the interactions between ticks, pathogens and host animals from various angles.

Disclaimer

The findings and conclusions of this expert opinion are those of the author and do not necessarily represent the views of the National Research Centre for Protozoan Diseases, Obihiro University of Agriculture and Veterinary Medicine.

References

Egizi, A., Bulaga-Seraphin, L., Alt, E., Bajwa, W.I., Bernick, J., *et al.* (2020) First glimpse into the origin and spread of the Asian longhorned tick, *Haemaphysalis longicornis*, in the United States. *Zoonoses Public Health* 67, 637–650.

Fujisaki, K. and Kamio, T. (1988) Effect of constant temperatures on *Theileria sergenti* infection in salivary glands of nymphal *Haemaphysalis longicornis*. *Japanese Journal of Veterinary Science* 50, 529–536.

Fujisaki, K., Ito, Y., Kamio, T. and Kitaoka, S. (1985) The presence of *Theileria sergenti* in *Haemaphysalis longicornis* overwintering in pasture in Japan. *Annals of Tropical Medicine and Parasitology* 79, 519–524.

Fujita, H. and Takada, N. (2007) Identification of immature ticks in Japan. In: Organizing Committee of SADI (ed.) *Acari and Emerging/Reemerging Infectious Diseases*. Zenkoku Noson Kyoiku Kyokai Publishing, Tokyo, pp. 53–68 (in Japanese).

Fujita, H., Yano, Y., Takada, N., Ando, S., Kawabata, H. and Fujita, N. (2013) Tick species and tick-borne rickettsiae confirmed by the year of 2012 in Fukushima Prefecture, Japan. *Medical Entomology and Zoology* 64, 37–41 (in Japanese).

Hayashida, K., Umemiya-Shirafuji, R., Sivakumar, T., Yamagishi, J., Suzuki, Y., *et al.* (2018) Establishment of a mouse–tick infection model for *Theileria orientalis* and analysis of its transcriptome. *International Journal for Parasitology* 48, 915–924.

Inokuma, H., Ohashi, M., Jilintai, Tanabe, S. and Miyahara, K. (2007) Prevalence of tick-borne *Rickettsia* and *Ehrlichia* in *Ixodes persulcatus* and *Ixodes ovatus* in Tokachi district, eastern Hokkaido, Japan. *Journal of Veterinary Medical Science* 69, 661–664.

Ito, Y., Watanabe, N. and Matsumura, T. (1983) Overwintering of *Haemaphysalis longicornis* (Acari: Ixodidae) at a pasture, Nasu Plateau, Japan. *Bulletin of the National Grassland Research Institute* 24, 110–118 (in Japanese).

Kitaoka, S. (1961) Physiological and ecological studies on some ticks: VII. Parthenogenetic and bisexual races of *Haemaphysalis bispinosa* in Japan and experimental crossing between them. *National Institute of Animal Health Quarterly* 1, 142–149.

Ootake, H., Ito, I. and Hayashi, K. (1985) Relationships between population density of the tick, *H. longicornis*, and environmental factors in the permanent pasture: I. Ixodid fauna and seasonal fluctuation of the tick, *H. longicornis*, in Kawatabi. *Journal of Japanese Society of Grassland Science* 30, 452–457 (in Japanese).

Otake, H., Sugawara, I. and Ito, K. (1989) Relationship between population density of tick and environmental factors in the permanent pasture: III. Relationships between population density of tick (*Haemaphysalis longicornis* Numann) and stocking rate. *Journal of Japanese Society of Grassland Science* 35, 241–246 (in Japanese).

Ouhelli, H. and Schein, E. (1988) Effect of temperature on transovarial transmission of *Babesia bigemina* (Smith and Kilborne, 1893) in *Boophilus annulatus* (Say, 1821). *Veterinary Parasitology* 26, 229–235.

Ouhelli, H., Pandey, V.S. and Aboughal, A. (1987) Effect of infection by *Babesia* spp. on the development and survival of free-living stages of *Boophilus annulatus*. *Veterinary Parasitology* 23, 147–154.

Sasaki, N., Takahashi, K., Kawamoto, S., Kurosawa, T., Sonoda, M. and Kawai, S. (1990) Maturation of *Theileria sergenti* in salivary glands of *Haemaphysalis longicornis* induced by temperature stimulation. *Japanese Journal of Veterinary Science* 52, 495–501.

Tsunoda, T. and Fujimagari, M. (1996) Ecology of ticks in Chiba Prefecture. *Chiba Eiken Hokoku* 20, 1–3 (in Japanese).

Umemiya-Shirafuji, R., Hatta, T., Okubo, K., Sato, M., Maeda, H., *et al.* (2017) Transovarial persistence of *Babesia ovata* DNA in a hard tick, *Haemaphysalis longicornis*, in a semi-artificial mouse skin membrane feeding system. *Acta Parasitologica* 62, 836–841.

Yamaguti, N., Tipton, V.J., Keegan, H.L. and Toshioka, S. (1971) Ticks of Japan, Korea, and the Ryukyu Islands. *Brigham Young University Science Bulletin. Biological Series* 15, 1–226.

Yano, Y., Shiraishi, S. and Uchida, T.A. (1987) Effects of temperature on development and growth in the tick, *Haemaphysalis longicornis*. *Experimental and Applied Acarology* 3, 73–78.
Yano, Y., Shiraishi, S. and Uchida, T. (1988) Effects of humidity on development and growth in the tick, *Haemaphysalis longicornis*. *Journal of the Faculty of Agriculture, Kyushu University* 32, 141–146.
Yokoyama, N., Sivakumar, T., Ota, N., Igarashi, I., Nakamura, Y., *et al.* (2012) Genetic diversity of *Theileria orientalis* in tick vectors detected in Hokkaido and Okinawa, Japan. *Infection, Genetics and Evolution* 12, 1669–1675.

eo28 Climate and Vector Potential of Medically Important North American Ticks

Lars Eisen* and Rebecca Eisen
Centers for Disease Control and Prevention, Fort Collins, Colorado, USA

28.1 Expert Opinion

28.1.1 Introduction

In North America, tick-borne disease cases are increasing in number and expanding in geographic range, and human vaccines to counter infections are lacking. The negative impact of ticks on human health and well-being will likely continue to escalate into the future unless safe and effective tick-bite and tick-borne disease prevention strategies become broadly accessible and widely used. One root cause for the increase in tick-borne diseases in North America is the ongoing geographic range expansion of two human-biting tick species: the black-legged tick, *Ixodes scapularis*, and the lone star tick, *Amblyomma americanum* (Sonenshine, 2018). These two tick species collectively transmit more than ten human pathogens, including causative agents of anaplasmosis, babesiosis, ehrlichiosis, Lyme disease and viral illnesses (Eisen and Paddock, 2020). To predict how human exposure to ticks and tick-borne pathogens may change in the future in North America, we must better understand human behaviours leading to contact with ticks as well as the various drivers impacting tick species range expansion, tick population build-up and the intensity of enzootic pathogen transmission. Key drivers impacting tick vector potential in nature include: (i) abiotic factors related to temperature and humidity that influence tick development and survival, shape the phenology of host-seeking for the different life stages and set limits for where ticks can establish populations; (ii) presence of suitable microhabitats for ticks to seek hosts, moult and overwinter; and (iii) presence and abundance of blood meal hosts for all life stages, including hosts that facilitate long-range tick dispersal as well as reservoir hosts for horizontally maintained tick-borne pathogens (Eisen and Paddock, 2021; Ogden *et al.*, 2021). This expert opinion focuses on climate-related factors but also considers how these interact with landscape and tick host/pathogen reservoir factors to impact the vector potential for the three most notable North American human-biting tick species: *I. scapularis*, *A. americanum* and the American dog tick, *Dermacentor variabilis*.

28.1.2 The black-legged tick, *Ixodes scapularis*

I. scapularis is recognized as a vector of seven human pathogens, including the causative agents of Lyme disease, *Borrelia miyamotoi* disease (relapsing fever), anaplasmosis, babesiosis and Powassan virus disease, and one of the causative agents of ehrlichiosis (Eisen and Eisen,

*Email: evp4@cdc.gov

DOI: 10.1079/9781789249637.0028

2018). The current geographic range of *I. scapularis* extends from north-eastern Mexico in the south through the eastern, forested portion of the USA (limited to the west by the Great Plains) and into south-eastern Canada in the north (Eisen *et al.*, 2016b; McPherson *et al.*, 2017). This distributional area falls primarily within three Köppen–Geiger climate classes: (i) *temperate without dry season and with hot summer* (Cfa); (ii) *cold without dry season and with hot summer* (Dfa); and (iii) *cold without dry season and with warm summer* (Dfb) (Beck *et al.*, 2018). Following the colonization by European settlers of the northern part of the eastern USA in past centuries, *I. scapularis* underwent dramatic range contraction in this region presumably due to deforestation to support subsistence farming and near complete removal of white-tailed deer (a key reproductive tick host) through hunting. As this situation reversed in the last half-century, with reforestation creating suitable tick habitat and a rebounding deer population, *I. scapularis* sprawled from local refugia in the Northeast and Upper Midwest of the USA, where it had persisted, and started the still ongoing process of reclaiming its historical geographic range (Eisen *et al.*, 2016b). This continuing range expansion and tick population build-up complicates assessments of how specifically climate-related factors impact *I. scapularis* presently, as well as projections for climate-driven change in the geographic range and population dynamics in the future. Nevertheless, there is strong support from field studies and modelling for temperature-related factors (e.g. mean, minimum and maximum monthly or seasonal temperatures, or mean diurnal temperature range) setting limits for the presence and abundance of *I. scapularis* at the northern edge of its range in the USA and Canada (Estrada-Peña, 2002; Brownstein *et al.*, 2005; Ogden *et al.*, 2006; Hahn *et al.*, 2016; McPherson *et al.*, 2017). A warming climate should reasonably allow *I. scapularis* to expand its geographic range northward into forested areas with sufficiently humid conditions and where hosts are available for all life stages, or to higher altitudes in areas where it is established at lower elevations. Expansion of *I. scapularis* into new areas may occur gradually via shorter-range host movement, e.g. via deer or resident birds, or across larger distances via immature ticks feeding on migratory birds. As ubiquitous rodent species serve as reservoirs for notable pathogens (including *Anaplasma phagocytophilum*, *Babesia microti* and *Borrelia burgdorferi* sensu stricto) transmitted by *I. scapularis*, local population establishment and introduction of infected ticks is likely to be followed by the emergence of enzootic pathogen transmission. Human-driven impacts on *I. scapularis* appear to have been more modest in the south-eastern USA, where the tick is thought to have maintained more stable populations through the European settlement period until present. Model projections do not indicate major climate-driven changes for *I. scapularis* in the southern part of its range in the USA.

28.1.3 The lone star tick, *Amblyomma americanum*

A. americanum is a vector of pathogens causing ehrlichiosis, Bourbon virus disease and Heartland virus disease, and has been implicated as the cause of a medical condition (southern tick-associated rash illness) for which the causative agent has not yet been identified, as well as alpha-gal allergy associated with consumption of red meat (Eisen and Paddock, 2021). The current geographic range of *A. americanum* extends from north-eastern Mexico in the south through the eastern, forested portion of the USA as far north as Iowa, Indiana and New York (Springer *et al.*, 2014; Molaei *et al.*, 2019). This distributional area falls primarily into the *temperate without dry season and with hot summer* (Cfa) Köppen–Geiger climate class (Beck *et al.*, 2018). The preference of *A. americanum* for forested habitat mirrors that of *I. scapularis*, with *A. americanum* being even more closely linked to deer as they serve as important hosts for all tick life stages. Consequently, there is ongoing range expansion of *A. americanum* in the northern part of the eastern USA following the previously mentioned reforestation and deer population recovery over the last half-century. In contrast to *I. scapularis*, which persisted in and spread from refugia in the Northeast and Upper Midwest, the range expansion of *A. americanum* appears to have originated from its core distributional area in the Southeast USA where the tick has maintained stable populations. Notably, this tick's range extended as far north as southern Michigan, New York and Connecticut in the early 1900s. By 1945, the northern limit contracted to Missouri and Virginia and along the Atlantic coastline to southern New Jersey; and thereafter the tick is again expanding northward (Eisen and Paddock, 2021). Although *A. americanum* generally is considered less tolerant of cold climate conditions compared with *I. scapularis* or *D. variabilis*, its ongoing range expansion or reclamation complicates assessments of how specifically climate-related factors impact the potential for tick establishment to the north. Nevertheless, there is support for temperature-related factors (e.g. cumulative annual degree-days above certain temperature thresholds or number of growing degree-days for specific months) setting limits for the presence and abundance of *A. americanum* and climate change-based model projections agree on strong potential for northward spread across the eastern USA and potentially into south-eastern Canada in the future (Springer *et al.*, 2015; Raghavan *et al.*, 2019; Sagurova *et al.*, 2019). Introduction of *A. americanum* to new areas may occur via birds for the immature stages or via deer for all life stages. Establishment of local enzootic transmission cycles following tick establishment requires the presence of reservoir hosts such as deer for *Ehrlichia chaffeensis* and *Ehrlichia ewingii*, whereas the tick itself may serve as a reservoir for viral pathogens such as Heartland virus. How climate warming may impact *A. americanum* at the southern edge of its geographic range remains unclear (Springer *et al.*, 2015).

28.1.4 The American dog tick, *Dermacentor variabilis*

D. variabilis is a vector of pathogens causing Rocky Mountain spotted fever and tularaemia. The current geographic range of *D. variabilis* is similar to that of *I. scapularis* in eastern North America, extending from north-eastern Mexico through the USA to south-eastern Canada, but this tick species also extends into the Great Plains of the USA and south-central Canada, and has a disjunct, smaller distributional area in the Pacific Coast states separated from the main distributional area by the Intermountain West region (James *et al.*, 2015; Minigan *et al.*, 2018; Sonenshine, 2018; Boorgula *et al.*, 2020; Lehane *et al.*, 2020). This extensive distributional area falls primarily into the following Köppen–Geiger climate classes: (i) *temperate without dry season and with hot summer* (Cfa), (ii) *cold without dry season and with hot summer* (Dfa) and (iii) *cold without dry season and with warm summer* (Dfb) in eastern North America; and (iv) *arid cold steppe* (Bsk) and (v) *temperate with dry and warm summer* (Csb) in western North America (Beck *et al.*, 2018). Compared with *I. scapularis*, the habitat tolerance of *D. variabilis* is wider, including forested areas and brushland as well as adjacent grassland, and the adult ticks feed primarily on medium-sized mammals rather than deer. Consequently, *D. variabilis* did not undergo the same human-influenced dramatic geographic range contraction as *I. scapularis* in the northern part of the eastern USA and the current distributional area of *D. variabilis* should more closely reflect the climatic tolerances limiting tick establishment. There is model agreement that climate warming would allow *D. variabilis* to expand its geographic range (given presence of suitable habitat and hosts for all life stages) northward in eastern and central Canada, as well as to higher altitudes in areas where it is already established in the northern USA and Canada (James *et al.*, 2015; Minigan *et al.*, 2018; Boorgula *et al.*, 2020). However, the model projections disagree on how a warming climate may impact *D. variabilis* at the southern limit of its geographic range in Mexico.

28.2 Future Predictions

Future predictions for changes in the geographic ranges and local abundance of medically important tick species in North America need to consider a changing climate as well as availability of suitable tick habitat and blood meal hosts. Should temperatures increase and the period of the year allowing for tick development and activity become longer, then ticks and their associated pathogens are likely to continue expanding northward, or in some areas to higher altitudes, in temperate areas of North America. It remains unclear whether corresponding range contractions will occur at the southern limits of the tick species distributions under such a climate change scenario.

Disclaimer

The findings and conclusions of this expert opinion are by the authors and do not necessarily represent the views of the Centers for Disease Control and Prevention.

References

Beck, H.E., Zimmermann, N.E., McVicar, T.R., Vergopolan, N., Berg, A. and Wood, F. (2018) Present and future Köppen–Geiger climate classification maps at 1-km resolution. *Scientific Data* 5, 180214.

Boorgula, G.D.Y., Peterson, A.T., Foley, D.H., Ganta, R.R. and Raghavan, R.K. (2020) Assessing the current and future potential geographic distribution of the American dog tick, *Dermacentor variabilis* (Say) (Acari: Ixodidae) in North America. *PLoS ONE* 15, e0237191.

Brownstein, J.S., Holford, T.R. and Fish, D. (2005) Effect of climate change on Lyme disease risk in North America. *EcoHealth* 2, 38–46.

Eisen, R.J. and Eisen, L. (2018) The blacklegged tick, *Ixodes scapularis*: an increasing public health concern. *Trends in Parasitology* 34, 295–309.

Eisen, R.J. and Paddock, C.D. (2021) Tick and tickborne pathogen surveillance as a public health tool in the United States. *Journal of Medical Entomology* 58, 1490–1502.

Eisen, R.J., Eisen, L. Ogden, N.H. and Beard, C.B. (2016a) Linkages of weather and climate with *Ixodes scapularis* and *Ixodes pacificus* (Acari: Ixodidae), enzootic transmission of *Borrelia burgdorferi*, and Lyme disease in North America. *Journal of Medical Entomology* 53, 250–261.

Eisen, R.J., Eisen, L. and Beard, C.B. (2016b) County-scale distribution of *Ixodes scapularis* and *Ixodes pacificus* (Acari: Ixodidae) in the continental United States. *Journal of Medical Entomology* 53, 349–386.

Estrada-Peña, A. (2002) Increasing habitat suitability in the United States for the tick that transmits Lyme disease: a remote sensing approach. *Environmental Health Perspectives* 110, 635–640.

Hahn, M.B., Jarnevich, C.S., Monaghan, A.J. and Eisen, R.J. (2016) Modeling the geographic distribution of *Ixodes scapularis* and *Ixodes pacificus* (Acari: Ixodidae) in the contiguous United States. *Journal of Medical Entomology* 53, 1176–1191.

James, A.M., Burdett, C., McCool, M.J., Fox, A. and Riggs, P. (2015) The geographic distribution and ecological preferences of the American dog tick, *Dermacentor variabilis* (Say), in the USA. *Medical and Veterinary Entomology* 29, 178–188.

Lehane, A., Parise, C., Evans, C., Beati, L., Nicholson, W.L. and Eisen, R.J. (2020) Reported county-level distribution of the American dog tick (Acari: Ixodidae) in the contiguous United States. *Journal of Medical Entomology* 57, 131–155.

McPherson, M., García-García, A., Cuesta-Valero, F.J., Beltrami, H., Hansen-Ketchum, P., *et al.* (2017) Expansion of the Lyme disease vector *Ixodes scapularis* in Canada inferred from CMIP5 climate projections. *Environmental Health Perspectives* 125, 057008.

Minigan, J.N., Hager, H.A., Peregrine, A.S. and Newman, J.A. (2018) Current and potential future distribution of the American dog tick (*Dermacentor variabilis*, Say) in North America. *Ticks and Tick-borne Diseases* 9, 354–362.

Molaei, G., Little, E.A., Williams, S.C. and Stafford, K.C. III (2019) Bracing for the worst – range expansion of the lone star tick in the northeastern United States. *New England Journal of Medicine* 381, 2189–2192.

Ogden, N.H., Maarouf, A., Barker, I.K., Bigras-Poulin, M., Lindsay, L.R., *et al.* (2006) Climate change and the potential for range expansion of the Lyme disease vector *Ixodes scapularis* in Canada. *International Journal for Parasitology* 36, 63–70.

Ogden, N.H., Beard, C.B., Ginsberg, H.S. and Tsao, J.I. (2021) Possible effects of climate change on ixodid ticks and the pathogens they transmit: predictions and observations. *Journal of Medical Entomology* 58, 1536–1545.

Raghavan, R.K., Peterson, A.T., Cobos, M.E., Ganta, R.R. and Foley, D.H. (2019) Current and future distribution of the lone star tick, *Amblyomma americanum* (L.) (Acari: Ixodidae) in North America. *PLoS ONE* 14, e0209082.

Sagurova, I., Ludwig, A., Ogden, N.H., Pelcat, Y., Dueymes, G. and Gachon, P. (2019) Predicted northward expansion of the geographic range of the tick vector *Amblyomma americanum* in North America under future climate conditions. *Environmental Health Perspectives* 127, 107014.
Sonenshine, D.E. (2018) Range expansion of tick disease vectors in North America: implications for spread of tick-borne disease. *International Journal of Environmental Research and Public Health* 15, 478.
Springer, Y.P., Eisen, L., Beati, L., James, A.M. and Eisen, R.J. (2014) Spatial distribution of counties in the continental United States with records of occurrence of *Amblyomma americanum* (Ixodida: Ixodidae). *Journal of Medical Entomology* 51, 342–351.
Springer, Y.P., Jarnevich, C.S., Barnett, D.T., Monaghan, A.J. and Eisen, R.J. (2015) Modeling the present and future geographic distribution of the lone star tick, *Amblyomma americanum* (Ixodida: Ixodidae), in the continental United States. *American Journal of Tropical Medicine and Hygiene* 93, 875–890.

eo29 The Impact of Climate Change on the Biology of the Cattle Tick, *Rhipicephalus microplus*: Current Knowledge and Gaps to be Filled

Isabel Santos[1]*, Matias Pablo Juan Szabó[2], Alexandre Rodrigues Caetano[3] and Maristela Martins de Camargo[4]

[1]*Universidade de São Paulo, Ribeirão Preto, Brazil;* [2]*Universidade Federal de Uberlândia, Uberlândia, Brazil;* [3]*Embrapa Recursos Geneticos e Biotecnologia, Brasília, Brazil;* [4]*Universidade de São Paulo, São Paulo, Brazil*

29.1 Expert Opinion

29.1.1 Introduction

Rhipicephalus microplus, the cattle tick, is an important pest of livestock that causes huge losses for producers of beef, dairy and leather. It is presently difficult to control by any means, chemical or other. Adding to this problem is the lack of knowledge concerning the impact of climate change on the on- and off-host ecosystems where the tick maintains the parasitic and free-living stages of its life cycle, respectively. If ambient temperatures become excessively high, the on-host ecosystem of *R. microplus* may be affected because heat stress modifies and even impairs numerous components of immunity that are known to participate in the systemic and local reactions of hosts to ticks. Changes in ambient temperatures also impact the off-host ecosystem of this tick: temperatures modify chances for desiccation, predation and/or infection of female ticks and their eggs and larvae, as well as affect the host-seeking capacities of larvae and the vectorial capacity of the species, thus affecting even the zoonotic stability for tick-borne diseases. Effective management of ticks under novel climatic conditions, whether on or off a host, will first demand detailed observations made under the guidance of what is already known about tick ecology. Phenomena described in detail will support better hypotheses and, ultimately, more precise predictions and valuable strategies to confront ticks and justify specific measures to mitigate the effect of climate change on the biology of *R. microplus*. New research venues need to address this increasingly important issue.

We recently reviewed the factors affecting the on- and off-host ecology of *R. microplus* under current biogeographical conditions (De Miranda Santos *et al.*, 2018). This expert opinion revisits these factors in the context of climate change. The negative effects of heat stress caused by climate change on animal welfare, productivity and health of cattle will likely become more of an issue in the future if the Earth's climate continues to warm as predicted.

*Email: imsantos@fmrp.usp.br

DOI: 10.1079/9781789249637.0029

The current global distribution of *R. microplus* gives a clue about the potential for this parasite to adapt to increasing temperatures. *R. microplus* was introduced from the Indian subcontinent to many tropical and subtropical areas of the world (Estrada-Peña *et al.*, 2006). Its current distribution comprises the Americas, Africa and Southeast Asia (Estrada-Peña *et al.*, 2012), where it thrives on pasture grasslands. Besides the grasslands favoured by *R. microplus*, dry, anthropized-forested areas (Nava *et al.*, 2013) and desert-like areas (Nyangiwe *et al.*, 2013) can also harbour this parasite. These latter introductions to drier areas are unusual, because studies show that temperature and humidity strongly favour this species (Corson *et al.*, 2001), thus underscoring the ability of *R. microplus* to adapt to climate change.

Besides climate, the distribution of *R. microplus* depends on the movements of its hosts. Loss of grasslands or introduction of new grazing areas will therefore also affect the geographical range of *R. microplus*. The Brazilian Legal Amazon, the entire Pantanal in South America and the tick eradication quarantine area (TEQA) in the USA are hotspots for investigating the dynamics of populations of *R. microplus* in grasslands with climatic vulnerabilities that are constantly changing, as will be discussed below.

The Brazilian Legal Amazon includes the Savannah, Pantanal and Amazon Forest biomes, the latter an important regulator of rainfall (Salati *et al.*, 1979; Lovejoy and Nobre, 2018). It is currently an issue of contention between authorities, ecologists, climatologists and livestock producers regarding the role in climate change of widespread deforestation to provide new grasslands for cattle (Amigo, 2020). Significantly, 41.6% of the Brazilian herd of cattle (80 million head) is located in the Legal Amazon region; production of beef cattle has been the main economic activity of the region since the 1950s (Tooge, 2020). This activity has been necessarily accompanied by deforestation, with hitherto unknown consequences for the ecology of *R. microplus*. To the best of our knowledge, no studies formally address the problem of *R. microplus* in cattle raised in the Legal Amazon; there is a single report on finding one individual in this area (Gianizella *et al.*, 2018). Other reports of *R. microplus* occurring on cattle in the Brazilian Legal Amazon are anecdotal: the parasite is currently considered to be less problematic there than in the Southeast, Centre-West and South of Brazil according to producers who raise cattle in both areas; many producers introducing cattle to the Legal Amazon region quarantine them for ticks and also consider that tick loads on cattle are lower there because they are more susceptible to desiccation in the hotter climate (M. Coelho, Minas Gerais, 2020, personal communication). Together, these factors have become essential in predicting the distribution of this species, as well as the spread of the tick's resistance to acaricides (Chevillon *et al.*, 2013).

Biogeography and ecology of the *R. microplus* environment explain, in part, the specificity of its tick–host association, i.e. a tick with a one-host (monoxenous) life cycle because this increases its chances of encounters with originally migratory herbivorous mammals that roam grasslands (Balashov, 1972; Hoogstraal and Kim, 1985). The off-host environment is also of major importance for survival of tick populations and of any biological control agent to be used to control infestations in pastures. This is underscored by the simple fact that, even though ubiquitous throughout several parts of the globe, *R. microplus* is not present in all areas of cattle-raising, but rather at specific environments under certain climates. This raises questions regarding the climate changes that can increase populations of *R. microplus* where it already occurs or spread it to areas where it presently does not occur.

The situation of *R. microplus* in the USA is emblematic of what may come and should be closely observed. By the 1960s, an eradication programme that lasted approximately 50 years finally eliminated *R. microplus* from most of the territory of the USA. The exception is a narrow stretch of land on that country's border with Mexico, known as the TEQA. However, since 2004, *R. microplus* has been detected outside the TEQA with increasing frequency (Giles *et al.*, 2014). Modelled climate parameters associated with areas with persistence of *R. microplus* predict that, with increased temperatures, new favourable niches will appear that will drastically increase the chances for reintroducing *R. microplus* in major cattle-producing areas of the USA (Giles *et al.*, 2014).

In Brazil, which is home to approximately a tenth of the potential bovine hosts for *R. microplus* present on the globe, a recent study performed by the meteorological station at the São Paulo State University in Jaboticabal county reported that the local average temperatures have risen by 3.30 and 3.78°C over the last 20 and 30 years, respectively (cited in Labruna and Faccini, 2020). These data have been used to explain the increase in the number of generations for another important species of tick, *Dermacentor nitens* (Labruna and Faccini, 2020). The number of generations per year of a parasitic species is directly related to its success and capacity to evolve under selective pressures, how quickly changes will take place and, ultimately, its impact on its hosts. The Pantanal, the world's largest wetland, is home to approximately one-tenth of Brazil's 220 million head of cattle that are regularly infested with *R. microplus*. The Pantanal is very important for the beef cattle industry and is strongly negatively affected by both severe flooding and severe drought. In any event, the flooding and dry seasons of the Pantanal, each year, determine the movement of large numbers of cattle grazing there and, consequently, the movement of *R. microplus*; severe flooding can reduce movement of cattle by approximately 5% (Araujo *et al.*, 2018), but the impact of these changes, both flooding and number of cattle, on the ecology of *R. microplus* is not yet known. Obviously, the number of available hosts determines the size of the populations of *R. microplus* and flooding, besides causing movement, can kill or benefit predators of this tick, and ticks can be washed to new areas. A recent study showed that despite the harsh environment for this tick species, the cattle tick survives in the Pantanal, albeit in low numbers (Ramos *et al.*, 2016). In this regard, authors expressed their concern that any alteration in the Pantanal that accidentally favours this tick species could result in a greater infestation level of cattle. A recent analysis of the Pantanal's spatial and temporal dynamics of rainfall modelled together with global sea-surface temperature (SST) data indicates that extreme precipitation events will become more frequent in the highlands of the upper Paraguay river basin, while severe and prolonged droughts triggered by warming SSTs in the northern hemisphere oceans will strongly affect the Pantanal (Thielen *et al.*, 2020). In summary, being an ectotherm, all stages of the tick life cycle rely on finding the optimal temperature in their host(s) and environment. Hotter and wetter benefits *R. microplus* up to maximal limits.

29.1.2 Climate change and off-host ecology of *Rhipicephalus microplus*

The off-host, non-parasitic life cycle of *R. microplus* can be affected by climate changes at many levels. The non-parasitic life cycle involves the drop-off of engorged females followed by their search for an adequate microenvironment for a pre-oviposition period, oviposition, egg incubation, hatching and host-seeking of larvae. Based on the following data it is safe to predict that climate change can affect all of these stages. First, the drop-off rhythm of *R. microplus* is known to be influenced by season and period of the day, among other ambient factors. In Brazil, female ticks present distinct detachment periods according to the season: around 09.00 hours in humid summers and around 06.00 hours and then 15.00 hours in dry winters (De Paula and Furlong, 2002); the hypothesis to explain this phenomenon is that females detach most when a favourable off-host environment will be met. Indeed, females search for shelter on the soil for mean distances of 7.2 and 12.2 cm in summer and winter, respectively (Brovini *et al.*, 2003); mortality of female ticks seems to be higher in summer for those kept on the surface, whereas it is higher in winter for those kept underneath it. In addition, mean daily egg output of engorged females is significantly less in the hot summer if females are kept on the surface of grasslands (Oliveira *et al.*, 1974). Thus, climate change can affect survival of females after drop-off. Climate change that increases temperatures and decreases summer rainfall and dries grasslands may increase drop-off rates, but also increase desiccation of females. It goes without saying that the reproductive success of *R. microplus* is a driving force in its evolution.

Laid eggs and larvae are also affected by ambient temperatures and humidity. They are additionally vulnerable to infections with soil microbes and to predators, which, in turn, are also affected by ambient temperature and humidity. These features add to the complexity of climate change and the biology of *R. microplus*. The hatching rate of eggs is reduced in dry environments and at high ground temperatures (>30°C)

(Garris *et al.*, 1990); on the other hand, mean monthly temperatures below 20°C also negatively affect egg hatching (Canevari *et al.*, 2017). Host-seeking larvae, the next stage, must climb on vegetation and wait for encounters with hosts. At this very delicate stage, ticks are exposed to open spaces and more extreme environmental variations. Behavioural responses entail clumping together to retain moisture, hiding from the sun underneath leaves when necessary, or moving back to the ground to recover water content. These responses, however, consume energy and thus shorten the lifespan of larvae.

In studies on climate, the survival limits of *R. microplus* in the environment during its non-parasitic cycle are as follows. Temperatures >25°C reduce periods of oviposition and egg incubation, but >30°C causes mortality of females, eggs and larvae. Temperatures <25°C extend periods of oviposition and egg incubation, but at <15°C there is mortality of females and eggs in the environment. Relative humidity (RH) of over 80% favours tick survival without altering the length of the life cycle; however, below 80% RH, death of ticks may occur by desiccation (Davey *et al.*, 1991; Campos Pereira and Labruna, 2008). These are useful parameters with which to monitor tick populations in the field to model the effect of climate change on populations of *R. microplus*.

Finally, the duration of its life cycle is also a major issue for *R. microplus* and it depends greatly on the climate. While the parasitic, on-host phase duration is relatively constant (21–23 days, notwithstanding heat stress of bovine hosts due to climate change), the non-parasitic phase varies greatly. In Brazil it lasts generally 40–50 days in spring and summer, and 70–120 days in autumn and winter (Labruna, 2008). Thus, off-host conditions will determine survival or death for *R. microplus*, but also the length of its life cycle and, therefore, the number of generations per year, as well as its abundance. Under higher mean temperatures and favourable humidity, it may achieve as many as four or even five generations per year (Cerný and De la Cruz, 1971; Cruz *et al.*, 2020).

As stated above, the number of generations per year of a parasite is directly related to its capacity to evolve under diverse selective pressures and how quickly changes will take place. With the possible exception of resistance to acaricides, to the best of our knowledge, little is known about how the number of generations of haematophagous ectoparasites impacts on their adaptability. Therefore, to gauge this point, it is worth pointing out the example of pigeon lice: during four years of experimental evolution comprising 60 generations, lice transferred to different coloured rock pigeons were able to evolve heritable differences in the colour of their chitin exoskeletons, a complex trait, due to the selective pressure of the feather colour of their hosts (Bush *et al.*, 2019). Resistance of *R. microplus* to most acaricides took almost eight decades to become widespread globally, perhaps because the products used and the practices in using them were quite variable regionally and thus selective pressures also differed. Increases in ambient temperatures will place a more uniform pressure on all populations of this parasite at a global level. Thus, it is reasonable to predict that *R. microplus* will adapt more rapidly to this change than it did to acaricides.

Transcriptomes carried on different developmental stages of the butterfly *Bicyclus anynana*, bred at different temperatures, revealed temperature-regulated genes that modulate several traits in the resulting adult animal, including lifespan, frequency and output of reproductive cycles, and wing camouflage patterns and colours (reviewed by Rodrigues and Beldade, 2020). It is interesting to note that the changes observed in the adults occurred even when an early developmental stage of the animal was subjected to temperatures higher or lower than optimal. As opposed to vertebrates, the adaptations of which sometimes require several generations for a relevant change in phenotype, considerable and stable changes can be achieved at much faster rates in invertebrates. In response to increased environmental temperatures, it has been reported that several invertebrate species move to locations where the current temperatures are in their optimal range, changing their gene expression to better respond to the new condition, including expression of temperature-regulated genes, or, what appears to be rarer, decline or become extinct.

Ticks become more resistant to desiccation or increase their fat storage and expression of heat-shock proteins, when infected by *Borrelia* and *Anaplasma*, respectively. All three aspects are advantageous for the tick in the face of hotter environmental conditions, suggesting that carrying those pathogens might become a selective pressure for the survival of the ticks. If on one hand, a hotter climate is disadvantageous for ticks in general, on the other hand, the pathogen-loaded tick will be positively selected by it (Benelli, 2020).

29.1.3 Climate and management systems of cattle that affect the free-living stage of *Rhipicephalus microplus*, including biological control

Shade is a crucial factor for the survival of *R. microplus* larvae; in turn, shade for larvae is strongly affected by the type of pasture. Survival of 50% of *R. microplus* larvae in Zimbabwe ranged from 40 to 58 days in long grass and from 14 to 36 days in short grass (Short *et al.*, 1989), while larvae in Puerto Rico can survive for up to 20 days longer in shaded or wooded areas than open grass areas (Garris *et al.*, 1990). Heavy precipitation also has a negative impact on *R. microplus* larvae (Cerný and De la Cruz 1971; Urdaz-Rodríguez *et al.*, 2012). Cattle-raising with high stocking rates, a current and popular strategy to reduce grazing areas, reduces mobility of hosts and thus increases chances of larvae–host encounters. Agroforestry systems are also becoming popular because they effectively and significantly reduce ambient temperatures for cattle (Magalhães *et al.*, 2020). However, they may constitute a double-edged sword if they favour survival of dropped females, eggs and questing larvae. This situation raises the question about the need for agroforestry systems to adopt species of grass that do not favour *R. microplus*. On the other hand, as will be discussed in Section 29.1.4, heat stress may hamper host innate and acquired immune responses to the tick and, therefore, thermal comfort of the host may be more important for controlling *R. microplus* than promoting hot and dry environments that are unfavourable for the free-living stage.

How climate will affect pastures and producers' choices of grass is very important because, at any given time, the majority of a population of *R. microplus* ticks are in the pasture and only 5% are feeding upon their hosts (Campos Pereira and Labruna, 2008). Pasture for grazing animals can affect several aspects of the biology of *R. microplus* by affecting survival of females, eggs and larvae by providing more or less shade, as well as by affecting host-seeking behaviour with volatile botanical repellents or physical barriers to reach tips of leaves or reach the host. Indeed, some grasses protect ticks, while some immobilize, kill and/or repel ticks (De Miranda Santos *et al.*, 2018). The use of anti-tick grasses as part of an integrated management system for controlling cattle ticks was proposed many years ago by Thompson *et al.* (1978). The options for pasture grasses must now be considered in the context of climate change because, for example, *Melinis minutiflora* (molasses grass), which repels ticks (Prates *et al.*, 1998; Fernandez-Ruvalcaba *et al.*, 2004), is not compatible with fire and high stocking rates, like other grasses.

Physiological changes in *Brachiaria brizantha* (bread grass), including resistance to *Collaria oleosa*, a hemipteran pest of this pasture grass, were examined under current and projected levels of carbon dioxide (CO_2). High levels of CO_2 (700 ppm) triggered changes in both species, but in such a way that current resistance of the *B. brizantha* species was predicted to be maintained under future climate scenarios (Silva *et al.*, 2019). While individual species of grass have been studied to answer various questions regarding climate, animal nutrition, agronomical issues such as crop rotation and overall effects on ticks, there has been no systematic evaluation of the most popular species of pastures in terms of all these aspects. Drought, fire and precipitation are three aspects of climate change that undoubtedly will affect pastures. The abundance of different species of invasive pasture grasses is directly related to their resilience to fire: compared with *Urochloa brizantha* (syn. *B. brizantha*; bread grass), which favours the free-living stage of *R. microplus* (Fig. 29.1), the tick-unfavourable molasses grass is the least resilient species when these two grasses are subject to experimental fires (Damasceno and Fidelis, 2020). These aspects merit evaluation since changes in plant communities can benefit ticks (Estrada-Peña and Salman, 2013). At the very least, a meta-analysis of previous studies is warranted.

The role of entomopathogens and predators on populations of free-living *R. microplus* is relatively poorly studied; consequently, the impact

Fig. 29.1. Host-questing *Rhipicephalus microplus* larvae in a heavily infested pasture of *Brachiaria brizantha*, Uberlândia, Minas Gerais, Brazil. (Authors' own photographs.)

of climate change on these factors is completely unknown at this moment. Biological control for pests is generally not used due to the exquisite dependency of the agents on environmental factors, as well as the potential for harming non-target invertebrates (Domingos *et al.*, 2013). In spite of reported beneficial effects, producers are not adopting products based on fungi for biological control. Among possible reasons for this are requirements for high humidity for germination and sporulation (Chandler *et al.*, 2000; Samish *et al.*, 2004). If temperatures increase and rainfall declines, what will be the effect on certain strategies for biological control?

29.1.4 Climate change and on-host ecology of *Rhipicephalus microplus*

According to latest statistics published by the Food and Agriculture Organization of the United Nations for 2019, more than 1.5 billion cattle live on the globe (FAO, 2019). The proportion of so-called tropical breeds in this population presently constitutes over half of this total and is steadily increasing (Barendse, 2017). Tropical cattle, which include all indicine (zebu) cattle (*Bos taurus indicus*), are naturally more resistant to tick infestations than taurine cattle (*Bos taurus taurus*) (Utech *et al.*, 1978; Utech and Wharton, 1982; Wambura *et al.*, 1998; Silva *et al.*, 2007; Ibelli *et al.*, 2012), except for those taurine breeds descended from *Bos taurus* that entered Africa thousands of years ago and adapted to the tropical and epizootiological conditions in that continent. Accordingly, tick loads in the tropical and temperate breeds differ significantly and are highly heritable. Breeds that carry lower loads of *R. microplus* are also more tolerant of the environmental conditions of the tropics and present lower rectal and surface (skin) temperatures than tick-susceptible breeds of cattle (O'Kelly and Spiers, 1983). Their productivity is also less affected by higher ambient temperatures, which reflects many physiological differences between these breeds and temperate breeds, the characterization of which is an active area of research. The thermoneutral zone (TNZ) for dairy cattle lies in the range of 4 to 24°C, within which the animal maintains a body temperature of 38.4–39.1°C for normal physiological functions (Abbas *et al.*, 2020). Temperatures above or below the TNZ threshold can result in heat or cold stress, respectively. Abundant genetic variation has been reported in the DNA sequence of the *hsp70* gene and its promoter, and both are linked with distinct transcriptional profiles for the *hsp70* gene, as well as with the thermal tolerance capabilities of the animal (reviewed in Abbas *et al.*, 2020). Such genetic variants are potential candidates for use as biomarkers of thermal resistance.

These features result from the subspecies' evolutionary histories. Their domestication began about 10,000 years ago (Bradley *et al.*, 1998; Chen *et al.*, 2010), but in very different ecosystems. Domestication of indicine cattle occurred in the tropical climatic conditions of the Indian subcontinent (Harris, 1966; Hanotte *et al.*, 2002; Hansen, 2004), where *R. microplus* also was present. On the other hand, European taurine breeds were introduced relatively recently to tropical habitats during colonization, which also spread *R. microplus* around the globe. These distinct outcomes of adaptations to biogeography are reflected in the expression profiles of tick genes encoding secreted proteins: they differ significantly between salivary glands of cattle ticks feeding on genetically resistant and susceptible bovine hosts (Garcia *et al.*, 2020). The reproductive success of female ticks fed on hosts of different genetic composition is also dramatically different (Biegelmeyer *et al.*, 2015).

Any discussion of the on-host ecology of *R. microplus* in the context of climate change must include factors that generate the observed greater or lesser specificity of this parasite for their temperate or tropical bovine hosts, as well as the qualitatively distinct innate and acquired immune reactions that the hosts generate against the attachment and feeding processes of this parasite (Carvalho *et al.*, 2008, 2010a,b; Piper *et al.*, 2010; Franzin *et al.*, 2017; Garcia *et al.*, 2017). Although the genetic composition of bovine hosts modulates all factors that collectively result in contrasting phenotypes for tick loads, importantly, all are affected by ambient temperatures (see Section 29.1.5). Besides preferring certain cattle breeds according to their adaptability to local climatic conditions, producers also choose breeds based on their productive abilities, whether dairy or beef. This explains the increasing popularity of the Holstein Friesian breed among dairy producers in Brazil, for example, in spite of the good performance of the popular Girolando cross made between Holstein Friesians and the indicine Gir breed, which also has good dairy ability (Milkpoint, 2020), but greater tolerance to heat stress. Like the sizes of tick loads, tolerance to heat stress is heritable and introgression of genes involved in tolerance to heat stress has been proposed as a strategy to maintain animal productivity (Hansen, 2020) under climate change with increased temperatures. It will be interesting to see if and how this will affect numbers of *R. microplus* on temperate breeds of cattle bred for tolerance to heat stress.

29.1.5 Climate change and factors that affect host specificity

Tick–host specificity can be generated by the number of opportunities for encounters within a suitable ecosystem for the tick (Klompen *et al.*, 1996; De Meêus *et al.*, 2010), by skin semiochemicals that guide host-seeking (Poldy, 2020) and by the ability to mount immune reactions to tick bites that will more or less vigorously impair attachment and blood feeding of the parasite. Encounter means physical contact between tick and host. For such encounters to occur, there must be a temporo-spatial match determined by both host and tick behaviour in a suitable ecosystem/microenvironment. Regarding these opportunities for encounters, in certain biogeographical areas *R. microplus* can infest cervids (Kistner and Hayes, 1970), notably after habitat loss (Szabó *et al.*, 2003). However, it is noteworthy that other species of ticks change host preferences when exposed to higher ambient temperatures. For example, outbreaks of rickettsial spotted fever in humans occurred during record high ambient temperatures in France and this phenomenon was linked to the high temperatures that seemingly drove *Rhipicephalus sanguineus*, normally a species with high specificity for dogs, to bite humans (Parola *et al.*, 2008). Elevated ambient temperatures can also drive *R. sanguineus* to feed on rabbits (Socolovschi *et al.*, 2009). The mechanisms behind these changes with which high temperatures are correlated are not yet known, but they might involve skin semiochemistry and/or alterations in the behaviour of ticks that increase host-seeking. Regarding this component of the tick–host interface, it is not yet known if, nor to what degree, temperature and humidity modulate the 'semiochemical' conditions that govern host preferences. At present very little is known about the semiochemistry of bovine skin that affects the behaviour of *R. microplus* (Franzin *et al.*, 2017), let alone the role of temperature and humidity in these phenomena. In any event, the physical and chemical properties of volatiles that participate in the composition of skin semiochemistry, and that are known to affect behaviour of other haematophagous arthropods (Takken and Verhulst, 2013), suggest that the concentrations and evaporation rates of these compounds will probably be affected by ambient temperatures. In addition, it is possible that the state of hydration of *R. microplus*, which depends on ambient humidity, may promote greater or lesser attraction to its bovine hosts, as has been demonstrated for *Ixodes ricinus* ticks (Crooks and Randolph, 2006).

It must also be pointed out that, according to the fermentation hypothesis (Albone *et al.*, 1974), skin chemistry is profoundly affected by the composition of skin microbiota, which acts upon host chemical outputs, for example sweat and sebaceous glands, and generates semiochemicals that affect the behaviour of arthropod pests and parasites (Davis *et al.*, 2013). Importantly, activity of sweat glands, which regulate heat loss, is associated with tick loads (Allen, 1962; Schleger and O'Kelly, 1986). Furthermore, the interactions between host immunity and skin microbiota are mutually regulated, adding to the complexity of predicting the impact of climate change on the biology of *R. microplus*. Adding even more complexity is the fact that this tick abundantly expresses a large diversity of antimicrobial peptides, and the expression profiles are significantly differentially affected by host immunity (Garcia *et al.*, 2020) generated in a tropical and a temperate breed of cattle. The number of interleukin (IL)-17-producing γδ lymphocytes in mouse skin varies according to species of microbial commensals (Ridaura *et al.*, 2018), being more strongly affected by *Corynebacterium*. Noteworthy is the fact that this genus of bacterium is a notorious producer of volatiles from skin outputs (Leyden *et al.*, 1981). Innate lymphoid cells strongly regulate the activity of sebaceous glands surrounding hair follicles, which, in turn, produce different antimicrobial lipid compounds that differentially affect the microbial composition of skin (Kobayashi *et al.*, 2019). In all, lymphoid tissue associated with skin appendages participates in many effector and regulatory mechanisms against the numerous microbial, parasitic, chemical and physical insults that this largest of organs is subject to (Kabashima *et al.*, 2019). There is insufficient knowledge about the activities of these semiochemical, microbial and immune components in tropical and temperate breeds of cattle infested with *R. microplus*, and nothing is known about how they respond collectively to heat stress.

Fur is believed to protect ticks from desiccation. Indeed, genetic studies indicate that coat characteristics (Ibelli *et al.*, 2012), and, as stated above, anatomy and activity of sweat glands (Allen, 1962; Schleger and O'Kelly, 1986), are associated with tick loads. The role of the host's coat characteristics in resistance to infestations with *R. microplus* is reinforced by some of the following facts: short-haired cattle have fewer ticks than long-haired cattle (Veríssimo *et al.*, 2002; Fraga *et al.*, 2003; Marufu *et al.*, 2011; Hüe *et al.*, 2014); and bovines that carry the SLICK haplotype that results in shorter, slicker coats, besides being more heat tolerant (Dikmen *et al.*, 2014; Davis *et al.*, 2017), carry lower tick loads (Hüe *et al.*, 2014). Short hair likely generates an unfavourable environment for ticks. A difference of 40% fewer ticks was found between a shorn and unshorn side in cattle; the shorter the hair, the lower the tick loads (Veríssimo *et al.*, 2020). The range of optimal temperatures of the bovine host's skin for *R. microplus* attachment is between 31 and 38°C (Doube and Kemp, 1979). RH is a determining factor in survival of larvae only if they are not attached to host skin; values below 45% RH result in death of larvae by desiccation (Doube and Kemp, 1979). At the same time, coat characteristics are strongly involved in tolerance to high temperatures. Most, if not all, European breeds of cattle have thicker, woollier coats and consequently are highly sensitive to temperatures in the tropics (Ibelli *et al.*, 2012).

Ambient temperatures where cattle are raised is a matter that receives much attention because many productive aspects of cattle and, therefore, profits of producers depend on the thermal comfort of their animals. Recent reviews now list many immunophysiological aspects of cattle that are affected by high ambient temperatures (Baumgard *et al.*, 2015; Bagath *et al.*, 2019; Dahl *et al.*, 2020). Most reports describe correlations between ambient temperatures or artificial heat challenges and levels of immune components such as cytokines. The analysis of a transcriptome made with blood from male Holstein calves subjected to experimental heat stress and controls indicated several differentially expressed genes and involvement of several pathways (Srikanth *et al.*, 2017) that were also found by us to be altered in calves vaccinated against ticks and challenged with *R. microplus* (Maruyama *et al.*, 2019), such as antigen processing and platelet activation. This raises the question of how responses to ticks will be impacted by higher ambient temperatures. This question is not difficult to answer experimentally or even through directed observations in the field.

The small intestine is one of the first tissues to upregulate heat-shock proteins during heat stress (Flanagan *et al.*, 1995) and is highly sensitive to heat damage (Kregel, 2002). During heat stress, blood flow is diverted from internal organs to the skin in an attempt to dissipate heat (Lambert *et al.*, 2002); if extreme, this process leads to hypoxia of gut epithelial cells (Hall *et al.*, 1999). Lactating dairy cattle subject to experimental heat stress show compromised immunity: a faulty intestinal barrier ('leaky gut') of the mucosa and submucosa of the jejunum, which becomes infiltrated with activated macrophages, accompanied by upregulated expression of $CD163^{+}$, and permissive to bacteria and bacterial components such as lipopolysaccharide (Koch *et al.*, 2019). The levels of IL-1β, IL-6 and tumour necrosis factor-α in the gut increase and regulate the concentrations of the proteins claudin, occludin and zonula occludens 1 (Robinson *et al.*, 2015), important transmembrane proteins that form the gut's selective barrier. Interestingly, besides functioning as a sensor for bacteria, CD163 is the high-affinity scavenger receptor for the haemoglobin–haptoglobin (Hb–Hp) complex (Kristiansen *et al.*, 2001). We have shown that expression of claudin and CD163 in tick-infested bovine skin also differs significantly between a tropical and a temperate breed of cattle (Franzin *et al.*, 2017). We have also shown that levels of acute-phase proteins, including haptoglobin, also differ significantly between these breeds when infested with *R. microplus* (Carvalho *et al.*, 2008). Profiles of acute-phase proteins are considered to be important markers of inflammation and of animal well-being (Eckersall and Bell, 2010); besides indicating the mechanisms involved in reactions to heat stress, these proteins will be useful and convenient biomarkers to monitor the growing problem of heat stress. Engagement of the Hb–Hp complex by CD163 also mediates anti-inflammatory effects via the release of IL-10 and the induction of haem oxygenase-1 (Abraham and Drummond, 2006), which generates the immunosuppressive metabolites CO, Fe^{2+} and biliverdin from haem. Therefore, this population of macrophages might favour the tick in a haemoglobin-rich milieu, which is exactly the environment of the feeding pool. To the best of our knowledge, the effect of heat stress directly on the skin has only been examined in skin fibroblasts of sheep and cows: as expected, it upregulated expression of genes encoding heat-shock proteins among other candidate genes (Saadeldin *et al.*, 2020), which did not include those related to immune responses. How these components will affect responses to *R. microplus* in temperate and tropical breeds of cattle undergoing heat stress is not known, but it is reasonable to predict that heat stress may further favour the tick by enhancing this process through the mechanisms described above.

We have shown that a tropical and a temperate breed of cattle respond differently to tick saliva (Garcia *et al.*, 2017); however, variations in seasonal immunity to the saliva of *R. microplus* have not yet been examined. It would be reasonable to speculate that it can change based on what has been observed regarding fluctuations in immunity to saliva of sand flies, which is reduced when sand fly seasons are short in temperate regions. According to that study's authors, this may result in a diminished immune 'tone' of hosts to the vectors and the pathogens they transmit (Oliveira *et al.*, 2020). However, temperate breeds of cattle are already highly exposed to the saliva of *R. microplus* and, for some as yet unknown reason, tick saliva is poorly immunogenic for these breeds (Garcia *et al.*, 2017). Together, these findings make the point that not everything found for one species of vector can be extrapolated to other species, including where climate change is concerned.

29.1.6 Impact of climate change on the vectorial capacity of *Rhipicephalus microplus*

Finally, the impact of climate change on the vectorial capacity per se of *R. microplus* is unknown and must be examined. Based on its impact on other haematophagous vectors, it is reasonable to predict that it will be affected. For example, higher ambient temperatures decrease the vectorial capacity of *Aedes* spp. for Zika virus, especially so in *Aedes aegypti* relative to *Aedes albopictus* (Onyango *et al.*, 2020). Could a similar temperature-mediated effect explain the finding that Mogiana tick virus, a Jingmenvirus infecting *R. microplus*, was not detected in ticks collected to the east of the Paraná River in the Pantanal of Mato Grosso do Sul state, only in ticks collected from other regions in Brazil (Maruyama *et al.*, 2014).

When *Drosophila melanogaster* flies were kept at colder than optimal temperatures, a single clade (clade V) of *Wolbachia*, an endosymbiont intracellular α-proteobacterium of several insects, quickly dominated over other clades (Mazzucco *et al.*, 2020). At hotter than optimal temperatures, another clade predominated quickly but was then overturned by clade V (Versace *et al.*, 2014; Mazzucco *et al.*, 2020). These results suggest that extreme changes in environmental temperature might lead to considerable and dynamic changes in the balance of the microbiota of invertebrate hosts. It is unknown whether thermal adaptations of the flies contributed to the shifts in bacterial balance, or if clade V adaptation was due to acquired gene mutations (Mazzucco *et al.*, 2020). To the best of our knowledge, no similar study has been performed to date on any species of tick, including *R. microplus*, therefore nothing is known of the impacts that changes in temperature have on the prevalence of pathogens and composition of microbiota in these ectoparasites.

29.2 Future Predictions

Warmer temperatures are predicted to be one consequence of climate change. Breeding cattle for heat tolerance is an active area of research. However, physiological gains provided by heat tolerance in cattle must be accompanied by more effective responses against *R. microplus*. This is currently an open area of research and must receive attention. Development of high-density SNP (single nucleotide polymorphism) genotyping microarray chips was one of the key activities undertaken by the consortium that sequenced the bovine genome (Bovine HapMap Consortium, 2009). These tools allowed for the implementation of methods developed by Meuwissen *et al.* (2001) to perform genetic evaluations based on dense genome-wide molecular marker information. Results from the wide adoption of genomic selection technologies to accelerate genetic gains on production, productivity and product quality traits have been tremendous, with major improvements especially in low-heritability traits related to fertility, animal health and production longevity (Boichard *et al.* 2015; García-Ruiz *et al.*, 2016). In dairy cattle, these advances in animal breeding and genetic improvement technologies, coupled with associated increases in effective population sizes, changes in genetic selection differentials and shortening of generation intervals, provided gains in genetic progress of production traits so significant that negative trends in historically eroding traits were actually inverted, therefore providing major improvements in overall productivity and profitability (García-Ruiz *et al.*, 2016).

Methods to perform genomic selection for heat tolerance in cattle are effective when using the rate of decline in milk production under heat stress (Nguyen *et al.*, 2016) and internal body temperature measurements from heat challenges generated in climate-controlled chambers (Garner *et al.*, 2016) in dairy cattle in Australia. In addition, genomic selection for hair-shedding ability improves thermotolerance in beef cattle raised in temperate climates (Durbin *et al.*, 2020). Therefore, it can be concluded that selective breeding and genetic improvement could be used to reduce heat-induced stress in production dairy and beef cattle.

Likewise, methods to perform genomic selection for tick resistance in pure-bred and cross-bred cattle have been developed (Cardoso *et al.*, 2015) and are being fine-tuned (Mota *et al.*, 2016; Sollero *et al.*, 2017), aiming to increase accuracies of genomic predictions while reducing associated costs. Strategies combining genomic selection to heat-tolerance and tick-resistance, in addition to production and product quality traits, could provide reliable answers to challenges brought on by changes in global climate patterns.

Silvopastoral systems have not been evaluated specifically for their impact on the ecology of *R. microplus*; therefore this aspect is also an open area of research and must receive attention. The consequences of adopting silvopastoral systems on tick populations are currently unpredictable because the shade and the lower ambient temperatures that they provide favour both the host and the parasite. Many of the hotspots for climate change in Brazil mentioned here are also popular for implementing silvopastoral systems and would be suitable to make many useful, comparative observations about the tick–host–pasture interfaces under these conditions. Observations should include on- and off-host tick populations, types of pastures, correlations between tick populations and stocking densities, and profiles of host immunity under the different settings, among other parameters.

So-called temperate breeds of cattle can harbour hundreds of engorging female *R. microplus*. Under current climatic conditions, *R. microplus* may complete up to five generations a year and each engorged female may produce up to 4000 larvae. These numbers indicate that populations of the cattle tick have the potential to increase exponentially. These populations are relentlessly confronted by entomopathogens and predators of ticks, including entomofauna and amphibians, reptiles and birds, and all of these components that control tick populations are also predicted to be greatly affected by climate. Each component of the problem at hand is affected differently by ambient temperatures, humidity and rainfall, and without more detailed observations, the outcome of all of these vectors is currently not possible to calculate. The only safe prediction to make is that, since so many components of the free-living and parasitic stages of *R. microplus* are affected by climate, the dynamics of this parasite are surely changing as well.

References

Abbas, Z., Hu, L., Fang, H., Sammad, A., Kang, L., *et al.* (2020) Association analysis of polymorphisms in the 5′ flanking region of the HSP70 gene with blood biochemical parameters of lactating Holstein cows under heat and cold stress. *Animals* 10, e2016.

Abraham, N.G. and Drummond, G. (2006) CD163-mediated hemoglobin-heme uptake activates macrophage HO-1, providing an antiinflammatory function. *Circulation Research* 99, 911–914.

Albone, E.S., Eglinton, G., Walker, J.M. and Ware, G.C. (1974) The anal sac secretion of the red fox (*Vulpes vulpes*); its chemistry and microbiology. A comparison with the anal sac secretion of the lion (*Panthera leo*). *Life Sciences* 14, 387–400.

Allen, T.E. (1962) Responses of Zebu, Jersey and Zebu × Jersey cross-bred heifers to rising temperature with particular reference to sweating. *Australian Journal of Agricultural Research* 13, 165–179.

Amigo, I. (2020) When will the Amazon hit a tipping point? *Nature* 578, 505–507.

Araujo, A.G.J., Obregón, G.O., Sampaio, G., Monteiro, A.M.V., da Silva, L.T., *et al.* (2018) Relationships between variability in precipitation, river levels, and beef cattle production in the Brazilian Pantanal. *Wetlands Ecology and Management* 26, 829–848.

Bagath, M., Krishnan, G., Devaraj, C., Rashamol, V.P., Pragna, P., *et al.* (2019) The impact of heat stress on the immune system in dairy cattle: a review. *Research in Veterinary Science* 126, 94–102.

Balashov, Y.S. (1972) Bloodsucking ticks (Ixodoidea) – vectors of diseases of man and animals. *Miscellaneous Publications of the Entomological Society of America* 8, 161–376.

Barendse, W. (2017) Climate adaptation of tropical cattle. *Annual Review of Animal Bioscience* 5, 133–150.

Baumgard, L.H., Keating, K., Ross, J.W. and Rhoads, R.P. (2015) Effects of heat stress on the immune system, metabolism and nutrient partitioning: implications on reproductive success. *Revista Brasileira de Reprodução Animal* 39, 173–183.

Benelli, G. (2020) Pathogens manipulating tick behavior – through a glass, darkly. *Pathogens* 9, e664.

Biegelmeyer, P., Nizoli, L.Q., Da Silva, S.S., Dos Santos, T.R., Dionello, N.J., *et al.* (2015) Bovine genetic resistance effects on biological traits of *Rhipicephalus* (*Boophilus*) *microplus*. *Veterinary Parasitology* 208, 231–237.

Boichard, D., Ducrocq, V. and Fritz, S. (2015) Sustainable dairy cattle selection in the genomic era. *Journal of Animal Breeding and Genetics* 132, 135–143.

Bovine HapMap Consortium (2009) Genome-wide survey of SNP variation uncovers the genetic structure of cattle breeds. *Science* 324, 528–532.

Bradley, D.G., Loftus, R.T., Cunningham, P. and MacHugh, D.E. (1998) Genetics and domestic cattle origins. *Evolution and Anthropology Issues News Review* 6, 79–86.

Brovini, C.N., Furlong, J. and Chagas, A.C.S. (2003) Influence of climatic factors in the biology and behavior of *Boophilus microplus* engorged females in the field. *Bioscience Journal* 19, 71–76.

Bush, S.E., Villa, S.M., Altuna, J.C., Johnson, K.P., Shapiro, M.D., *et al.* (2019) Host defense triggers rapid adaptive radiation in experimentally evolving parasites. *Evolution Letters* 3, 120–128.

Campos Pereira, M. and Labruna, M.B. (2008) *Rhipicephalus* (*Boophilus*) *microplus*. In: Campos Pereira, M., Labruna, M.B., Szabó, M.P.J. and Klafke, G.M. (eds) *Rhipicephalus (Boophilus) microplus: Biologia, Controle e Resistência*. Medicina Veterinária, São Paulo, Brazil, pp. 15–55.

Canevari, J.T., Mangold, A.J., Guglielmone, A.A. and Nava, S. (2017) Population dynamics of the cattle tick *Rhipicephalus* (*Boophilus*) *microplus* in a subtropical sub-humid region of Argentina for use in the design of control strategies. *Medical and Veterinary Entomology* 31, 6–14.

Cardoso, F.F., Gomes, C.C., Sollero, B.P., Oliveira, M.M., Roso, V.M., *et al.* (2015) Genomic prediction for tick resistance in Braford and Hereford cattle. *Journal of Animal Science* 93, 2693–2705.

Carvalho, W.A., Bechara, G.H., Moré, D.D., Ferreira, B.R., Da Silva, J.S., *et al.* (2008) *Rhipicephalus* (*Boophilus*) *microplus*: distinct acute phase proteins vary during infestations according to the genetic composition of the bovine hosts, *Bos taurus* and *Bos indicus*. *Experimental Parasitology* 118, 587–591.

Carvalho, W.A., Franzin, A.M., Abatepaulo, A.R., De Oliveira, C.J., Moré, D.D., *et al.* (2010a) Modulation of cutaneous inflammation induced by ticks in contrasting phenotypes of infestation in bovines. *Veterinary Parasitology* 167, 260–273.

Carvalho, W.A., Maruyama, S.R., Franzin, A.M., Abatepaulo, A.R., Anderson, J.M., *et al.* (2010b) *Rhipicephalus* (*Boophilus*) *microplus*: clotting time in tick-infested skin varies according to local inflammation and gene expression patterns in tick salivary glands. *Experimental Parasitology* 124, 428–435.

Cerný, V. and De La Cruz, J. (1971) Development and survival of the tick *Boophilus microplus* (Can.) in laboratory and under natural conditions of Cuba. *Folia Parasitologica* 18, 73–78.

Chandler, D., Davidson, G., Pell, J.K., Ball, B.V., Shaw, K. and Sunderland, K.D. (2000) Fungal biocontrol of Acari. *Biocontrol Science & Technology* 10, 357–384.

Chen, S., Lin, B.Z., Baig, M., Mitra, B., Lopes, R.J., *et al.* (2010) Zebu cattle are an exclusive legacy of the South Asia Neolithic. *Molecular Biology and Evolution* 27, 1–6.

Chevillon, C., Ducornez, S., Meeüs, T., Koffi, B.B, Gaia, H., *et al.* (2007) Accumulation of acaricide resistance mechanisms in *Rhipicephalus* (*Boophilus*) *microplus* (Acari: Ixodidae) populations from New Caledonia Island. *Veterinary Parasitology* 147, 276–288.

Corson, M.S., Teel, P.D. and Grant, W.E. (2001) Influence of acaricide resistance on cattle-fever tick (*Boophilus* spp.) infestations in semi-arid thorn shrublands: a simulation approach. *Experimental and Applied Acarology* 25, 171–184.

Crooks, E. and Randolph, S.E. (2006) Walking by *Ixodes ricinus* ticks: intrinsic and extrinsic factors determine the attraction of moisture or host odour. *Journal of Experimental Biology* 209, 2138–2142.

Cruz, B.C., Mendes, A.F.L., Maciel, W.G., dos Santos, I.B., Gomes, L.V.C., *et al.* (2020) Arthropods and medical entomology – biological parameters for *Rhipicephalus microplus* in the field and laboratory and estimation of its annual number of generations in a tropical region. *Parasitology Research* 119, 2421–2430.

Dahl, G.E., Tao, S. and Laporta, J. (2020) Heat stress impacts immune status in cows across the life cycle. *Frontiers in Veterinary Science* 7, e116.

Damasceno, G. and Fidelis, A. (2020) Abundance of invasive grasses is dependent on fire regime and climatic conditions in tropical savannas. *Journal of Environmental Management* 271, e111016.

Davey, R.B., Cooksey, L.M. and Despins, J.L. (1991) Survival of larvae of *Boophilus annulatus*, *Boophilus microplus*, and *Boophilus* hybrids (Acari: Ixodidae) in different temperature and humidity regimes in the laboratory. *Veterinary Parasitology* 40, 305–313.

Davis, S.R., Spelman, R.J. and Littlejohn, M.D. (2017) Breeding heat tolerant dairy cattle: the case for introgression of the 'slick' prolactin receptor variant into dairy breeds. *Journal of Animal Science* 95, 1788–1800.

Davis, T.S., Crippen, T.L., Hofstetter, R.W. and Tomberlin, J.K. (2013) Microbial volatile emissions as insect semiochemicals. *Journal of Chemical Ecology* 39, 840–859.

De Meêus, T., Koffi, B.B., Barré, N., Garine-Wichatitsky, M. and Chevillon, C. (2010) Swift adaptation of a species of cattle tick to a new deer host in New Caledonia. *Infection, Genetics and Evolution* 10, 976–983.

De Miranda Santos, I.K.F., Garcia, G.R., Oliveira, P.S., Veríssimo, C.J., Katiki, L.M., *et al.* (2018) Acaricides: current status and sustainable alternatives for controlling the cattle tick, *Rhipicephalus microplus*, based on its ecology. In: Garros, C., Bouyer J., Takken, W. and Smallegange, R.C. (eds) *Pests and Vector-borne Diseases in the Livestock Industry*. Wageningen Academic Publishers, Wageningen, the Netherlands, pp. 91–134.

De Paula, P.H. and Furlong, J. (2002) The drop-off behavior of engorged females of the cattle tick, *Boophilus microplus*. *Ciência Rural* 32, 627–631.

Dikmen, S., Khan, F.A., Huson, H.J., Sonstegard, T.S., Moss, J.I., *et al.* (2014) The SLICK hair locus derived from Senepol cattle confers thermotolerance to intensively managed lactating Holstein cows. *Journal of Dairy Science* 97, 5508–5520.

Domingos, A., Antunes, S., Borges, L. and Estólio do Rosário, V. (2013) Approaches towards tick and tick-borne diseases control. *Revista da Sociedade Brasileira de Medicina Tropical* 46, 265–269.

Doube, B.M. and Kemp, D.H. (1979) The influence of temperature, relative humidity and host factors on the attachment and survival of *Boophilus microplus* (Canestrini) larvae to skin slices. *International Journal for Parasitology* 9, 449–454.

Durbin, H.J., Lu, D., Yampara-Iquise, H., Miller, S.P. and Decker, J.E. (2020) Development of a genetic evaluation for hair shedding in American Angus cattle to improve thermotolerance. *Genetics, Selection, Evolution* 52, 63.

Eckersall, P.D. and Bell, R. (2010) Acute phase proteins: biomarkers of infection and inflammation in veterinary medicine. *Veterinary Journal* 185, 23–27.

Estrada-Peña, A. and Salman, M. (2013) Current limitations in the control and spread of ticks that affect livestock: a review. *Agriculture* 3, 221–235.

Estrada-Peña, A., Bouattour, A., Camicas, J.L., Guglielmone, A., Horak, I., *et al.* (2006) The known distribution and ecological preferences of the tick subgenus *Boophilus* (Acari: Ixodidae) in Africa and Latin America. *Experimental and Applied Acarology* 38, 219–235.

Estrada-Peña, A., Venzal, J.M., Nava, S., Mangold, A., Guglielmone, A.A., *et al.* (2012) Reinstatement of *Rhipicephalus* (*Boophilus*) *australis* (Acari: Ixodidae) with redescription of the adult and larval stages. *Journal of Medical Entomology* 49, 794–802.

FAO (2019) *FAOSTAT: Data: Live Animals*. Food and Agriculture Organization of the United Nations, Rome. Available at: http://www.fao.org/faostat/en/#data/QA (accessed 4 June 2021).

Fernandez-Ruvalcaba, M., Torre, F.P., Cruz-Vasquez, C. and Garcia-Vasquez, Z. (2004) Anti-tick effects of *Melinis minutiflora* and *Andropogon gayanus* grasses on plots experimentally infested with *Boophilus microplus* larvae. *Experimental and Applied Acarology* 32, 293–299.

Flanagan, S.W., Ryan, A.J., Gisolfi, C.V. and Moseley, P.L. (1995) Tissue-specific HSP70 response in animals undergoing heat stress. *American Journal of Physiology* 268, 28–32.

Fraga, A.B., Alencar, M.M., Figueiredo, L.A., Razook, A.G. and Cyrillo, J.N.S.G. (2003) Análise de fatores genéticos e ambientais que afetam a infestação de fêmeas bovinos da raça Caracu por carrapatos (*Boophilus microplus*). *Revista Brasileira de Zootecnia* 32, 1578–1586.

Franzin, A.M., Maruyama, S.R., Garcia, G.R., Oliveira, R.P., Ribeiro, J.M., *et al.* (2017) Immune and biochemical responses in skin differ between bovine hosts genetically susceptible and resistant to the cattle tick *Rhipicephalus microplus*. *Parasites & Vectors* 10, e51.

Garcia, G.R., Maruyama, S.R., Nelson, K.T., Ribeiro, J.M., Gardinassi, L.G., *et al.* (2017) Immune recognition of salivary proteins from the cattle tick, *Rhipicephalus microplus*, differs according to the genotype of the bovine host. *Parasites & Vectors* 10, e144.

Garcia, G.R., Chaves Ribeiro, J.M., Maruyama, S.R., Gardinassi, L.G., Nelson, K., *et al.* (2020) A transcriptome and proteome of the tick *Rhipicephalus microplus* shaped by the genetic composition of its hosts and developmental stage. *Scientific Reports* 10, e12857.

García-Ruiz, A., Cole, J.B., VanRaden, P.M., Wiggans, G.R., Ruiz-López, F.J. and Van Tassell, C.P. (2016) Changes in genetic selection differentials and generation intervals in US Holstein dairy cattle as a result of genomic selection. *Proceedings of the National Academy of Sciences USA* 113, E3995–E4004.

Garner, J.B., Douglas, M.L., Williams, S.R., Wales, W.J., Marett, L.C., *et al.* (2016) Genomic selection improves heat tolerance in dairy cattle. *Scientific Reports* 6, 34114.

Garris, G.I., Popham, T.W. and Zimmerman, R.H. (1990) *Boophilus microplus* (Acari: Ixodidae): oviposition, egg viability, and larval longevity in grass and wooded environments of Puerto Rico. *Environmental Entomology* 19, 66–75.

Gianizella, S.L., Martins, T.F., Onofrio, V.C., Aguiar, N.O., Gravena, W., *et al.* (2018) Ticks (Acari: Ixodidae) of the state of Amazonas, Brazil. *Experimental and Applied Acarology* 74, 177–183.

Giles, J.R., Peterson, A.T., Busch, J.D., Olafson P.U., Scoles, G.A., *et al.* (2014) Invasive potential of cattle fever ticks in the southern United States. *Parasites & Vectors* 7, e189.

Hall, D.M., Baumgardner, K.R., Oberley, T.D. and Gisolfi, C.V. (1999) Splanchnic tissues undergo hypoxic stress during whole body hyperthermia. *American Journal of Physiology* 276, G1195–G1203.

Hanotte, O., Bradley, D.G., Ochieng, J.W., Verjee, Y., Hill, E.W. and Rege, J.E. (2002) African pastoralism: genetic imprints of origins and migrations. *Science* 296, 336–339.

Hansen, P.J. (2004) Physiological and cellular adaptations of zebu cattle to thermal stress. *Animal Reproduction Science* 82, 349–360.
Hansen, P.J. (2020) Prospects for gene introgression or gene editing as a strategy for reduction of the impact of heat stress on production and reproduction in cattle. *Theriogenology* 154, 190–202.
Harris, M. (1966) The cultural ecology of India's sacred cattle. *Current Anthropology* 7, 51–59.
Hoogstraal, H. and Kim, K.C. (1985) Tick and mammal's coevolution with emphasis on *Haemaphysalis*. In: Kim, K.C. (ed.) *Coevolution of Parasitic Arthropods and Mammals*. Wiley, New York, pp. 505–568.
Hüe, T., Horlin, J.C., Teurlai, M. and Naves, M. (2014) Comparison of tick resistance of crossbred Senepol × Limousin to purebred Limousin cattle. *Tropical Animal Health Production* 46, 447–453.
Ibelli, A.M.G., Ribeiro, A.R.B., Giglioti, R., Regitano, L.C.A., Alencar, M.M., *et al.* (2012) Resistance of cattle of various genetic groups to the tick *Rhipicephalus microplus* and the relationship with coat traits. *Veterinary Parasitology* 186, 425–430.
Kabashima, K., Honda, T., Ginhoux, F. and Egawa, G. (2019) The immunological anatomy of the skin. *Nature Reviews Immunology* 19, 19–30.
Kistner, T.P. and Hayes, F.A. (1970) White-tailed deer as hosts of cattle fever-ticks. *Journal of Wildlife Diseases* 6, 437–440.
Klompen, J.S.H., Black, W.C. IV, Keirans, J.E. and Oliver, J.H. Jr (1996) Evolution of ticks. *Annual Review of Entomology* 41, 141–161.
Kobayashi, T., Voisin, B., Kim, D.Y., Kennedy, E.A., Jo, J.H., *et al.* (2019) Homeostatic control of sebaceous glands by innate lymphoid cells regulates commensal bacteria equilibrium. *Cell* 176, 982–997.
Koch, F., Thom, U., Albrecht, E., Weikard, R., Nolte, W., *et al.* (2019) Heat stress directly impairs gut integrity and recruits distinct immune cell populations into the bovine intestine. *Proceedings of the National Academy of Sciences USA* 116, 10333–10338.
Kregel, K.C. (2002) Heat shock proteins: modifying factors in physiological stress responses and acquired thermotolerance. *Journal of Applied Physiology* 92, 2177–2186.
Kristiansen, M., Graversen, J.H., Jacobsen, C., Sonne, O., Hoffman, H.J., *et al.* (2001) Identification of the haemoglobin scavenger receptor. *Nature* 409, 198–201.
Labruna, M.B. (2008) As gerações anuais. In: Campos Pereira, M., Labruna, M.B., Szabó, M.P.J. and Klafke, G.M. (eds) *Rhipicephalus (Boophilus) microplus: Biologia, Controle e Resistência*. Medicina Veterinária, São Paulo, Brazil, pp. 57–63.
Labruna, M.B. and Faccini, J.L.H. (2020) The nonparasitic phase of *Dermacentor nitens* under field conditions in southeastern Brazil. *Revista Brasileira de Parasitologia Veterinária* 29, e008620.
Lambert, G.P., Gisolfi, C.V., Berg, D.J., Moseley, P.L., Oberley, L.W. and Kregel, K.C. (2002) Selected contribution: hyperthermia-induced intestinal permeability and the role of oxidative and nitrosative stress. *Journal of Applied Physiology* 92, 1750–1761.
Leyden, J.J., McGinley, K.J., Hölzle, E., Labows, J.N. and Kligman, A.M. (1981) The microbiology of the human axilla and its relationship to axillary odor. *Journal of Investigative Dermatology* 77, 413–416.
Lovejoy, T.E. and Nobre, C. (2018) Amazon tipping point. *Science Advances* 4, eaat2340.
Magalhães, C.A.S., Zolin, C.A., Lulu, J., Lopes, L.B., Furtini, I.V., *et al.* (2020) Improvement of thermal comfort indices in agroforestry systems in the southern Brazilian Amazon. *Journal of Thermal Biology* 91, e102636.
Marufu, M.C., Qokweni, L., Chimonyo, M. and Dzama, K. (2011) Relationships between tick counts and coat characteristics in Nguni and Bonsmara cattle reared on semiarid rangelands in South Africa. *Ticks and Tick-borne Diseases* 2, 172–177.
Maruyama, S.R., Castro-Jorge, L.A., Ribeiro, J.M., Gardinassi, L.G., *et al.* (2014) Characterisation of divergent flavivirus NS3 and NS5 protein sequences detected in *Rhipicephalus microplus* ticks from Brazil. *Memórias do Instituto Oswaldo Cruz* 109, 38–50.
Maruyama, S.R., Carvalho, B., González-Porta, M., Rung, J., Brazma, A., *et al.* (2019) Blood transcriptome profile induced by an efficacious vaccine formulated with salivary antigens from cattle ticks. *NPJ Vaccines* 4, e53.
Mazzucco, R., Nolte, V., Vijayan, T. and Schlötterer, C. (2020) Long-term dynamics among *Wolbachia* strains during thermal adaptation of their *Drosophila melanogaster* hosts. *Frontiers in Genetics* 11, e482.
Meuwissen, T.H.E., Hayes, B.J. and Goddard, M.E. (2001) Prediction of total genetic value using genome-wide dense marker maps. *Genetics* 157, 1819–1829.
Milkpoint (2020) Levantamento top 100 2020 os 100 maiores produtores de leite do Brasil. Available at: https://www.milkpoint.com.br/top100-2020-lp/ (accessed 4 June 2021).

Mota, R.R., Lopes, P.S., Tempelman, R.J., Silva, F.F., *et al.* (2016) Genome-enabled prediction for tick resistance in Hereford and Braford beef cattle via reaction norm models. *Journal of Animal Science* 94, 1834–1843.

Nava, S., Mastropaolo, M., Guglielmone, A.A. and Mangold, A.J. (2013) Effect of deforestation and introduction of exotic grasses as livestock forage on the population dynamics of the cattle tick *Rhipicephalus* (*Boophilus*) *microplus* (Acari: Ixodidae) in northern Argentina. *Research in Veterinary Science* 95, 1046–1054.

Nguyen, T.T.T., Bowman, P.J., Haile-Mariam, M., Pryce, J.E. and Hayes, B.J. (2016) Genomic selection for tolerance to heat stress in Australian dairy cattle. *Journal of Dairy Science* 99, 2849–2862.

Nyangiwe, N., Matthee, C., Horak, I. and Matthee, S. (2013) First record of the pantropical blue tick *Rhipicephalus microplus* in Namibia. *Experimental and Applied Acarology* 6, 503–507.

O'Kelly, J.C. and Spiers, W.G. (1983) Observations on body temperature of the host and resistance to the tick *Boophilus microplus* (Acari: Ixodidae). *Journal of Medical Entomology* 20, 498–505.

Oliveira, F., Giorgobiani, E., Guimarães-Costa, A.B., Abdeladhim, M., Oristian, J., *et al.* (2020) Immunity to vector saliva is compromised by short sand fly seasons in endemic regions with temperate climates. *Scientific Reports* 10, e7990.

Oliveira, G.P., Costa, R.P., De Mello, R.P. and Meneguelli, C.A. (1974) Estudo ecológico da fase não parasítica do *Boophilus microplus* (Canestrini, 1887) (Acarina, Ixodidae) no Estado do Rio de Janeiro. *Arquivos da Universidade Federal Rural* 4, 1–10.

Onyango, M.G., Bialosuknia, S.M., Payne, A.F., Mathias, N., Kuo, L., *et al.* (2020) Increased temperatures reduce the vectorial capacity of Aedes mosquitoes for Zika virus. *Emerging Microbes and Infections* 9, 67–77.

Parola, P., Socolovschi, C., Jeanjean, L., Bitam, I., Fournier, P.E., *et al.* (2008) Warmer weather linked to tick attack and emergence of severe rickettsioses. *PLoS Neglected Tropical Diseases* 2, e338.

Piper, E.K., Jackson, L.A., Bielefeldt-Ohmann, H., Gondro, C., Lew-Tabor, A.E., *et al.* (2010) Tick-susceptible *Bos taurus* cattle display an increased cellular response at the site of larval *Rhipicephalus* (*Boophilus*) *microplus* attachment, compared with tick-resistant *Bos indicus* cattle. *International Journal for Parasitology* 40, 431–441.

Poldy, J. (2020) Volatile cues influence host-choice in arthropod pests. *Animals* 10, e1984.

Prates, H.T., Leite, R.C., Craveiro, A.A. and Oliveira, A.B. (1998) Identification of some chemical components of the essential oil from molasses grass (*Melinis minutiflora* Beauv.) and their activity against cattle-tick (*Boophilus microplus*). *Journal of the Brazilian Chemical Society* 9, 193–197.

Ramos, V.N., Piovezan, U., Franco, A.H.A., Rodrigues, V.S., Nava, S. and Szabó, M.P.J. (2016) Nellore cattle (*Bos indicus*) and ticks within the Brazilian Pantanal: ecological relationships. *Experimental and Applied Acarology* 68, 227–240.

Ridaura, V.K., Bouladoux, N., Claesen, J., Chen, Y.E., Byrd, A.L., *et al.* (2018) Contextual control of skin immunity and inflammation by *Corynebacterium*. *Journal of Experimental Medicine* 215, 785–799.

Robinson, K., Deng, Z., Hou, Y. and Zhang, G. (2015) Regulation of the intestinal barrier function by host defense peptides. *Frontiers in Veterinary Science* 2, e57.

Rodrigues, Y.K. and Beldade, P. (2020) Thermal plasticity in insects' response to climate change and to multifactorial environments. *Frontiers in Ecology and Evolution* 18, e271.

Saadeldin, I.M., Swelum, A.A.A., Zakri, A.M., Tukur, H.A., *et al.* (2020) Effects of acute hyperthermia on the thermotolerance of cow and sheep skin-derived fibroblasts. *Animals* 10, e545.

Salati, E., Dall 'Ollio, A., Matsui, E. and Gat, J.R. (1979) Recycling of water in the Amazon, Brazil: an isotopic study. *Water Resource Research* 15, 1250–1258.

Samish, M., Ginsberg, H. and Glazer, I. (2004) Biological control of ticks. *Parasitology* 129(Suppl.), S389–S403.

Schleger, A.V. and O'Kelly, J.C. (1986) Esterase activity in the bovine sweat gland: genetic differences and the effect of temperature. *Acta Anatomica* 125, 229–232.

Short, N.J., Floyd, R.B., Norval, R.A. and Sutherst, R.W. (1989) Survival and behaviour of unfed stages of the ticks *Rhipicephalus appendiculatus*, *Boophilus decoloratus* and *Boophilus microplus* under field conditions in Zimbabwe. *Experimental and Applied Acarology* 6, 215–236.

Silva, A.M., Alencar, M.M., Regitano, L.C.A., Oliveira, M.C.S. and Barioni Júnior, W. (2007) Artificial infestation of *Boophilus microplus* in beef cattle heifers of four genetic groups. *Genetics and Molecular Biology* 30, 1150–1155.

Silva, D.M., Auad, A.M., Moraes, J.C. and Silva, S.E.B. (2019) How do *Collaria oleosa* and *Brachiaria* spp. respond to increase in carbon dioxide levels? *Neotropical Entomology* 48, 340–348.

Socolovschi, C., Raoult, D. and Parola, P. (2009) Influence of temperature on the attachment of *Rhipicephalus sanguineus* ticks on rabbits. *Clinical Microbiology and Infection* 15, 326–327.
Sollero, B.P., Junqueira, V.S., Gomes, C.C.G., Caetano, A.R. and Cardoso, F.F. (2017) Tag SNP selection for prediction of tick resistance in Brazilian Braford and Hereford cattle breeds using Bayesian methods. *Genetics, Selection, Evolution* 49, 49.
Srikanth, K., Lee, E., Kwan, A., Lim, Y., Lee, J., *et al.* (2008) Transcriptome analysis and identification of significantly differentially expressed genes in Holstein calves subjected to severe thermal stress. *International Journal of Biometeorology* 61, 1993–2008.
Szabó, M.P., Labruna, M.B., Pereira, M.C. and Duarte, J.M. (2003) Ticks (Acari: Ixodidae) on wild marsh-deer (*Blastocerus dichotomus*) from Southeast Brazil: infestations before and after habitat loss. *Journal of Medical Entomology* 40, 268–274.
Takken, W. and Verhulst, N.O. (2013) Host preferences of blood-feeding mosquitoes. *Annual Review of Entomology* 58, 433–453.
Thielen, D., Schuchmann, K.L., Ramoni-Perazzi, P., Marquez, M., Rojas, W., *et al.* (2020) Quo vadis Pantanal? Expected precipitation extremes and drought dynamics from changing sea surface temperature. *PLoS ONE* 15, e0227437.
Thompson, K.C., Roa, E.J. and Romero, N.R. (1978) Anti-tick grasses as the basis for developing practical tropical tick control packages. *Tropical Animal Health Production* 10, 179–182.
Tooge, R. (2020) *O que criadores de gado e frigoríficos fazem para evitar o desmatamento na Amazônia.* Grupo Globo, Rio de Janeiro, Brazil. Available at: https://g1.globo.com/economia/agronegocios/noticia/2020/10/25/o-que-criadores-de-gado-e-frigorificos-fazem-para-evitar-o-desmatamento-na-amazonia.ghtml (accessed 4 June 2021).
Urdaz-Rodríguez, J., Fosgate, F., Alleman, A.R., Rae, O., Donovan, A., *et al.* (2012) Association between ecological factors and the presence of *Rhipicephalus* (*Boophilus*) *microplus* larvae in Puerto Rico. *Experimental and Applied Acarology* 58, 145–157.
Utech, K.B. and Wharton, R.H. (1982) Breeding for resistance to *Boophilus microplus* in Australian Illawarra shorthorn and Brahman × Australian Illawarra shorthorn cattle. *Australian Veterinary Journal* 58, 41–46.
Utech, K.G.W., Wharton, R.H. and Kerr, J.D. (1978) Resistance to *Boophilus microplus* (Canestrini) in different breeds of cattle. *Australian Journal of Agricultural Research* 29, 885–895.
Veríssimo, C.J., Nicolau, V.L., Cardoso, V.L. and Pinheiro, M.G. (2002) Haircoat characteristics and tick infestation on Gyr (Zebu) and crossbred (Holstein × Gyr) cattle. *Archivos de Zootecnia* 51, 389–392.
Veríssimo, C.J., Giglioti, R., D'Agostino, S.M., Toledo, L.M., Katiki, L.M., *et al.* (2019) Cattle herd shearing can help to control *Rhipicephalus microplus* ticks. *Experimental and Applied Acarology* 79, 99–106.
Versace, E., Nolte, V., Pandey, R.V., Tobler, R. and Schlötterer, C. (2014) Experimental evolution reveals habitat-specific fitness dynamics among *Wolbachia* clades in *Drosophila melanogaster*. *Molecular Ecology* 23, 802–814.
Wambura, P.N., Gwakisa, P.S., Silayo, R.S. and Rugaimukamu, E.A. (1998) Breed-associated resistance to tick infestation in *Bos indicus* and their crosses with *Bos taurus*. *Veterinary Parasitology* 77, 63–70.

eo30 Climate Impacts on the Vector Capacity of Tropical and Temperate Populations of the Brown Dog Tick, *Rhipicephalus sanguineus* sensu lato

Gervasio Bechara* and Gustavo Sanches
Pontifical Catholic University of Paraná, Curitiba, Brazil

30.1 Expert Opinion

30.1.1 Introduction

Rhipicephalus sanguineus sensu lato (s.l.) is popularly known as the 'brown dog tick' since canids, especially domestic dogs, are its preferred hosts (Dantas-Torres, 2010). For over 200 years, *R. sanguineus* was considered a single taxon; however, at least two genetic lineages are currently accepted under the name '*Rhipicephalus sanguineus*', the tropical lineage and the temperate lineage (Moraes-Filho *et al.*, 2011; Nava *et al.*, 2018). Although morphologically similar (Oliveira *et al.*, 2005; Dantas-Torres *et al.*, 2013), these two lineages present strong differences in biology (Szabó *et al.*, 2005; Sanches *et al.*, 2016; Labruna *et al.*, 2017), in their geographical distribution (Moraes-Filho *et al.*, 2011; Zemtsova *et al.*, 2016) and in vector competence for *Ehrlichia canis* transmission (Cicuttin *et al.*, 2015; Moraes-Filho *et al.*, 2015).

The distribution and survival of a tick species are affected by several factors such as host availability, vegetation coverage, climatic conditions (humidity, photoperiod and temperature) and anthropic activities (Randolph, 2009). This expert opinion discusses the differences between both *R. sanguineus* lineages, how climate changes could affect their distribution and abundance, and consequently the incidence of the diseases transmitted by them to dogs, their preferred hosts.

30.1.2 Effect of climate changes on distribution of *Rhipicephalus sanguineus* lineages

R. sanguineus s.l. is one of the most widely distributed tick species globally (Walker *et al.*, 2005); however, the tropical and temperate lineages have more restricted distributions associated with typical climatic conditions, especially temperature and humidity. In a phylogeographical study performed with *R. sanguineus* s.l. collected from 23 countries worldwide, Zemtsova *et al.* (2016) demonstrated that ticks belonging to the tropical lineage were found between the latitudes 35°N and 20°S, while ticks belonging to the temperate lineage were found either north of the 30th parallel or south of the 43rd parallel. Moreover, both *R. sanguineus* lineages may coexist in some geographic regions, as reported in the Formosa

*Email: gervasio.bechara@pucpr.br

DOI: 10.1079/9781789249637.0030

Province in north-eastern Argentina between latitudes 24°S and 25°S (Nava *et al.*, 2012), in Texas (Jones *et al.*, 2017), in Chile between latitudes 20°S and 22°S (Díaz *et al.*, 2018), and for the first time both populations were reported inhabiting similar geoclimatic habitats sympatrically (Díaz *et al.*, 2018).

In terms of climatic conditions, the tropical and temperate lineages are found in areas with annual average near-surface temperatures between 20 and 30°C and between 10 and 20°C, respectively (Zemtsova *et al.*, 2016). Jones *et al.* (2017) in North America also confirmed this hypothesis; however, in Chile this threshold was not confirmed since *R. sanguineus* from the tropical lineage was detected in areas with annual average mean temperature of 16°C, partially due to a mild Pacific climate with low daily and seasonal temperature fluctuations (Díaz *et al.*, 2018).

Environmental pressures affect tick populations differently by selecting genotypes able to resist and to adapt to prevailing conditions (Esteve-Gassent *et al.*, 2016). Labruna *et al.* (2017) suggested that the absence of *R. sanguineus* tropical lineage in areas with cold weather could be associated with the fact that it does not enter diapause; thus, cold weather exerts a selection pressure caused by the lethal effects of temperature. In addition, according to Gray *et al.* (2009), *R. sanguineus* may undergo winter dormancy in cold regions, while in localities with warmer winters they are still active. In regions where there are no humidity restrictions for the development of ticks, and hosts are available, an increase of about 2–3°C in the average temperature in winter could result in the establishment of permanent populations where *R. sanguineus* is currently absent (Parola *et al.*, 2008; Gray *et al.*, 2009). This general trend of geographical expansion of a tick population due to an increase in temperature has already been observed or predicted for several species, such as *Ixodes ricinus*, *Ixodes scapularis*, *R. sanguineus* and *Rhipicephalus microplus* (Ogden *et al.*, 2006; Gray *et al.*, 2009; Porretta *et al.*, 2013; Hadgu *et al.*, 2019). In fact, a study showed that between 1960 and 2000, *R. sanguineus* s.l. extended its range by 66%, expanding up to 130,000 km^2 (Medlock and Leach, 2015).

30.1.3 Effect of climate changes on distribution and abundance of *Rhipicephalus sanguineus* lineages

Under natural conditions, *R. sanguineus* tropical lineage completes two to four generations per year, whereas *R. sanguineus* temperate lineage completes only one (Silveira *et al.*, 2009; Labruna *et al.*, 2017). An increase of temperature in winter favours *R. sanguineus* temperate lineage becoming active during this period, increasing questing behaviour. Moreover, it can produce adequate conditions for the development of immature stages, with reductions in moulting periods (Koch and Tuck, 1986; Tudela *et al.*, 2005). These new conditions could also contribute to the development of more generations per year of this population, thus increasing the abundance of the temperate lineage.

Studies of the influence of different temperatures and humidities on the life cycle of the *R. sanguineus* temperate lineage need be conducted to support this evidence, as well as to determine the temperature threshold tolerance. According to Ogden and Lindsay (2016), prolonged extreme values of temperature and low humidity adversely affect the tick's life cycle through the effect of temperature on metabolic processes, increasing mortality rate. For the tropical lineage of *R. sanguineus*, these temperature tolerance limits are better known: oviposition of engorged females is suppressed at <15°C, optimal at 30°C giving maximum egg yield which decreases at 37°C, while the pre-oviposition period decreases as the incubation temperature increases from 15 to 30°C (Jacobs *et al.*, 2004; Adejinmi and Akinboade, 2012). Moreover, high temperatures and low humidity tend to change the sex ratio, increasing the proportion of females as a biological strategy to try to compensate for low egg production of females under stressful climatic conditions (Dantas-Torres, 2010).

These findings suggest two scenarios for the abundance of both *R. sanguineus* lineages. Increases in temperature in areas where temperatures are already high, but still within the tolerance limit for the genotypes, could reduce the abundance of these *R. sanguineus* populations by increasing mortality due to the deleterious effect of extreme temperature on tick metabolism. Despite the fact this effect may occur in both

R. sanguineus lineages, it tends to be more noticeable in ticks of the *R. sanguineus* tropical lineage, since they already occupy areas of higher temperatures. To mitigate this effect, ticks would tend to refuge in woodland habitats or look for areas of optimum temperature. By contrast, *R. sanguineus* lineages located in tropical, subtropical and temperate areas with lower temperatures could become more abundant due to climatic conditions that favour their life cycle.

30.1.4 Effect of climate changes on vector competence of *Rhipicephalus sanguineus* lineages

The association between climate changes and vector-borne diseases is complex and remains poorly studied for some tick species, including those belonging to the *R. sanguineus* complex. Climate changes are predicted to increase the risk of tick-borne diseases by affecting the biology of ticks, hosts and pathogens (Ostfeld and Brunner, 2015; Ogden *et al.*, 2020). As mentioned above, a temperature rise within the *R. sanguineus* threshold tolerance with no humidity restrictions may promote tick abundance and expansion of geographical distribution. Another important factor is the host encounter rate by ticks (Ogden *et al.*, 2020). Studies show that when exposed to high temperatures, *R. sanguineus* ticks attach more rapidly on rabbits and humans (Parola *et al.*, 2008; Socolovschi *et al.*, 2009). According to Parola *et al.* (2008), the risk of parasitism may increase in warm areas, with a consequent increase in the risk of transmission of pathogens by *R. sanguineus* ticks to humans and other animals. Moreover, ticks exposed to light and high temperature become more active and excited, increasing their questing behaviour (Dantas-Torres *et al.*, 2013). Climate change also may increase the risk of tick-borne diseases by affecting the replication rates of different pathogens and other factors that influence vector competence (Ostfeld and Brunner, 2015). Environmental temperature crucially influences the course of infection of various vector-borne pathogens in different ways (Godsey *et al.*, 1987). For mosquito-borne arboviruses, some studies reported the association of lower temperatures with enhanced vector competence (Kay and Jennings, 2002; Westbrook *et al.*, 2010), while other studies showed greater vector competence at higher temperatures (Richards *et al.*, 2009; Anderson *et al.*, 2010). Considering the climate changes on tick–pathogen interactions, alterations in ambient temperature may affect the interactions of pathogens with ticks via the pathogen protein synthesis (Cluss and Boothby, 1990), pathogen developmental diapause (Lewengrub *et al.*, 1989) and stimulation of heat-shock proteins in tick vectors (Nakayama *et al.*, 1989; Villar *et al.*, 2010).

The vector competence of *R. sanguineus* tropical and the temperate lineages varies for different pathogens. Canine monocytic ehrlichiosis is a serious and potentially fatal canine vector-borne disease with high prevalence in animals from tropical, subtropical and partially temperate regions, coincidentally areas where *R. sanguineus* from the tropical lineage is predominant (Ewing, 1969; Irwin, 2001; Suksawat *et al.*, 2001). This fact, added to the importance of the disease, motivated studies on the vectorial competence of both strains of *R. sanguineus* for *E. canis* transmission. Moraes-Filho *et al.* (2015) experimentally demonstrated that only the *R. sanguineus* tropical lineage was a competent vector for *E. canis* to dogs, whereas *R. sanguineus* temperate lineage (populations of southern Brazil, Uruguay and Argentina) were not. Studies performed in Uruguay (Venzal *et al.*, 2007) and northern Argentina (Nava *et al.*, 2012; Cicuttin *et al.*, 2015) showed that *E. canis* was not detected in the temperate lineage of *R. sanguineus*. On the other hand, both *R. sanguineus* lineages seem to be competent vectors for *Anaplasma platys*, the aetiological agent of canine infectious cyclic thrombocytopaenia in dogs (Abarca *et al.*, 2007). In Argentina, for example, Cicuttin *et al.* (2015) detected the presence of *A. platys* DNA in *R. sanguineus* nymphs from the tropical lineage, while Oscherov *et al.* (2011) detected the presence of *A. platys* DNA in non-engorged *R. sanguineus* ticks from the temperate lineage collected in Corrientes Province, Argentina (Nava *et al.*, 2012). According to these authors, *A. platys* infection in *R. sanguineus* s.l. is a ubiquitous phenomenon worldwide, independent of the *R. sanguineus* lineage.

The effect of climate changes, especially the effect of ambient temperature on the distribution of both *R. sanguineus* lineages, their abundance, susceptibility to infection by pathogens and incidence of diseases associated with them, remains to be defined. As mentioned above, increased temperature may lead the *R. sanguineus* tropical lineage (mainly those populations located in areas near the tropics) to expand its geographical range to higher latitudes, occupying areas formerly inhabited only by the temperate lineage of *R. sanguineus*. Moreover, *R. sanguineus* populations located in areas with lower temperatures could become more abundant due to climatic conditions that favour their life cycle. Such changes may lead to the emergence of diseases associated only with the *R. sanguineus* tropical lineage, such as canine ehrlichiosis, in areas where previously the disease did not exist or was of low prevalence, and to an increase in incidence of diseases associated with both *R. sanguineus* lineages, such as canine anaplasmosis, due to the presence of more than one competent vector.

The effect of ambient temperature on the susceptibility of both *R. sanguineus* lineages to infection by different pathogens, as well as on their infectivity for vertebrate hosts, should be determined since each pathogen has an optimal growth temperature. Increased temperature within certain ranges may increase the multiplication of some pathogens in ticks, while excessive increases in temperature may limit *R. sanguineus* development due to deleterious effects on tick metabolism, as well as having a knock-on effect on vector competence by negatively affecting pathogen survival. In fact, Shih *et al.* (1995) demonstrated experimentally that temperatures >27°C limit the survival of *Borrelia burgdorferi* spirochaetes within *Ixodes dammini* nymphs, and temperatures that approach the surface temperature of an infected rodent host are lethal to the pathogen within 2 weeks after tick feeding. In addition, differences in spirochaete infection of deer ticks in response to various extrinsic incubation temperatures may induce variations in infection rates in nature (Spielman *et al.*, 1985; Telford *et al.*, 1992). Moreover, changes in ambient temperature affect the bacterial community composition in ticks. Thapa *et al.* (2019) showed that *I. scapularis* ticks incubated at 30 and 37°C exhibited significantly different bacterial diversity compared with the initial baseline microbiome. According to de la Fuente *et al.* (2017): 'Microbiome play[s] an essential role in various aspects of the arthropods life and may affect pathogen transmission either directly via nutrient competition or induced/reduced immunity, or indirectly by affecting tick populations (viability, reproduction) or fitness (affecting host-seeking success)'. In fact, Narasimhan *et al.* (2014) reported a perturbation of the *I. scapularis* microbiome that caused changes in the integrity of the peritrophic membrane, which may affect pathogen infection.

30.2 Future Predictions

Studies targeting the modelling of climate change impacts on the distribution of both *R. sanguineus* lineages are still non-existent, but essential to predict the future redistribution and the consequences of these changes on the incidence of the diseases transmitted by them. The increase in temperature, which is expected to occur in the near future, and likely to affect the ecology and geographical distribution of many organisms, could lead the brown dog tick *R. sanguineus* from the tropical lineage (mainly those populations located in areas near the tropics) to expand its geographical range to higher latitudes, occupying areas that previously were inhabited only by the temperate lineage of *R. sanguineus*.

References

Abarca, K., López, J., Perret, C., Guerrero, J., Godoy, P., *et al.* (2007) *Anaplasma platys* in dogs, Chile. *Emerging Infectious Diseases* 13, 1392–1395. https://doi.org/10.3201/eid1309.070021

Adejinmi, J.O. and Akinboade, O.A. (2012) Effect of photoperiodic regime and egg laying disturbance on the oviposition of the dog ticks: *Rhipicephalus sanguineus* and *Haemaphysalis leachi leachi* in Nigeria. *Israel Journal of Veterinary Medicine* 67, 96–101.

Anderson, S.L., Richards, S.L., Tabachnick, W.J. and Smartt, C.T. (2010) Effects of West Nile virus dose and extrinsic incubation temperature on temporal progression of vector competence in *Culex pipiens quinquefasciatus*. *Journal of the American Mosquito Control Association* 26, 103–107. https://doi.org/10.2987/09-5926.1

Cicuttin, G.L., Tarragona, E.L., De Salvo, M.N., Mangold, A.J. and Nava, S. (2015) Infection with *Ehrlichia canis* and *Anaplasma platys* (Rickettsiales: Anaplasmataceae) in two lineages of *Rhipicephalus sanguineus* sensu lato (Acari: Ixodidae) from Argentina. *Ticks and Tick-borne Diseases* 6, 724–729. https://doi.org/10.1016/j.ttbdis.2015.06.006

Cluss, R.G. and Boothby, J.T. (1990) Thermoregulation of protein synthesis in *Borrelia burgdorferi*. *Infection and Immunity* 58, 1038–1042. https://doi.org/10.1128/iai.58.4.1038-1042.1990

Dantas-Torres, F. (2010) Biology and ecology of the brown dog tick, *Rhipicephalus sanguineus*. *Parasites & Vectors* 3, 26. https://doi.org/10.1186/1756-3305-3-26

Dantas-Torres, F., Latrofa, M.S., Annoscia, G., Giannelli, A., Parisi, A. and Otranto, D. (2013) Morphological and genetic diversity of *Rhipicephalus sanguineus* sensu lato from the New and Old Worlds. *Parasites & Vectors* 6, 213. https://doi.org/10.1186/1756-3305-6-213

de la Fuente, J., Antunes, S., Bonnet, S., Cabezas-Cruz, A., Domingos, A.G., *et al.* (2017) Tick–pathogen interactions and vector competence: identification of molecular drivers for tick-borne diseases. *Frontiers in Cellular and Infection Microbiology* 7, 114. https://doi.org/10.3389/fcimb.2017.00114

Díaz, F.E., Martínez-Valdebenito, C., López, J., Weitzel, T. and Abarca, K. (2018) Geographical distribution and phylogenetic analysis of *Rhipicephalus sanguineus* sensu lato in northern and central Chile. *Ticks and Tick-borne Diseases* 9, 792–797. https://doi.org/10.1016/j.ttbdis.2018.03.004

Esteve-Gassent, M.D., Castro-Arellano, I., Feria-Arroyo, T.P., Patino, R., Li, A.Y., *et al.* (2016) Translating ecology, physiology, biochemistry, and population genetics research to meet the challenge of tick and tick-borne diseases in North America. *Archives of Insect Biochemistry and Physiology* 92, 38–64. https://doi.org/10.1002/arch.21327

Ewing, S.A. (1969) Canine ehrlichiosis. *Advances in Veterinary Sciences and Comparative Medicine* 13, 331–353.

Godsey, M.S. Jr, Amundson, T.E., Burgess, E.C., Schell, W., Davis, J.P., *et al.* (1987) Lyme disease ecology in Wisconsin: distribution and host preferences of *Ixodes dammini*, and prevalence of antibody to *Borrelia burgdorferi* in small mammals. *American Journal of Tropical Medicine and Hygiene* 37, 180–187. https://doi.org/10.4269/ajtmh.1987.37.180

Gray, J., Dautel, H., Estrada-Peña, A., Kahl, O. and Lindgren, E. (2009) Effects of climate change on ticks and tick-borne diseases in Europe. *Interdisciplinary Perspectives on Infectious Diseases* 2009, 593232. https://doi.org/10.1155/2009/593232

Hadgu, M., Menghistu, H.T., Girma, A., Abrha, H. and Hagos, H. (2019) Modeling the potential climate change induced impacts on future genus *Rhipicephalus* (Acari: Ixodidae) tick distribution in semi-arid areas of Raya Azebo district, Northern Ethiopia. *Journal of Ecology and Environment* 43, 43. https://doi.org/10.1186/s41610-019-0137-0

Irwin, P.J. (2001) The first report of canine ehrlichiosis in Australia. *Australian Veterinary Journal* 79, 552–553. https://doi.org/10.1111/j.1751-0813.2001.tb10746.x

Jacobs, P.A.H., Fourie, L.J. and Horak, I.G. (2004) A laboratory comparison of the life cycle of the dog ticks *Haemaphysalis leachi leachi* and *Rhipicephalus sanguineus*. *Onderstepoort Journal of Veterinary Research* 71, 15–28.

Jones, E.O., Gruntmeir, J.M., Hamer, S.A. and Little, S.E. (2017) Temperate and tropical lineages of brown dog ticks in North America. *Veterinary Parasitology: Regional Studies and Reports* 7, 58–61. https://doi.org/10.1016/j.vprsr.2017.01.002

Kay, B.H. and Jennings, C.D. (2002) Enhancement or modulation of the vector competence of *Ochlerotatus vigilax* (Diptera: Culicidae) for Ross River virus by temperature. *Journal of Medical Entomology* 39, 99–105. https://doi.org/10.1603/0022-2585-39.1.99

Koch, H.G. and Tuck, M.D. (1986) Molting and survival of the brown dog tick (Acari: Ixodidae) under different temperatures and humidities. *Annals of the Entomological Society of America* 79, 11–14. https://doi.org/10.1093/aesa/79.1.11

Labruna, M.B., Gerardi, M., Krawczak, F.S. and Moraes-Filho, J. (2017) Comparative biology of the tropical and temperate species of *Rhipicephalus sanguineus* sensu lato (Acari: Ixodidae) under different laboratory conditions. *Ticks and Tick-borne Diseases* 8, 146–156. https://doi.org/10.1016/j.ttbdis.2016.10.011

Lewengrub, S., Rudzinska, M.A., Piesman, J., Spielman, A. and Zung, J.L. (1989) The influence of heat on the development of *Babesia microti* in unfed nymphs of *Ixodes dammini*. *Canadian Journal of Zoology* 66, 1342–1345. https://doi.org/10.1139/z89-215

Medlock, J.M. and Leach, S.A. (2015) Effect of climate change on vector-borne disease risk in the UK. *The Lancet Infectious Diseases* 15, 721–30. https://doi.org/10.1016/S1473-3099(15)70091-5

Moraes-Filho, J., Marcili, A., Nieri-Bastos, F.A., Richtzenhain, L.J. and Labruna, M.B. (2011) Genetic analysis of ticks belonging to the *Rhipicephalus sanguineus* group in Latin America. *Acta Tropica* 117, 51–55. https://doi.org/10.1016/j.actatropica.2010.09.006

Moraes-Filho, J., Krawczak, F.S., Costa, F.B., Soares, J.F. and Labruna, M.B. (2015) Comparative evaluation of the vector competence of four South American populations of the *Rhipicephalus sanguineus* group for the bacterium *Ehrlichia canis*, the agent of canine monocytic ehrlichiosis. *PLoS ONE* 10, e139386. https://doi.org/10.1371/journal.pone.0139386

Nakayama, Y., Spielman, A. and James, A.A. (1989) Protein synthesis induced by heat in an *Ixodes* tick. *Insect Biochemistry* 19, 731–736. https://doi.org/10.1016/0020-1790(89)90053-X

Narasimhan, S., Rajeevan, N., Liu, L., Zhao, Y.O., Heisig, J., *et al.* (2014) Gut microbiota of the tick vector *Ixodes scapularis* modulate colonization of the Lyme disease spirochete. *Cell Host & Microbe* 15, 58–71. https://doi.org/10.1016/j.chom.2013.12.001

Nava, S., Mastropaolo, M., Venzal, J.M., Mangold, A.J. and Guglielmone, A.A. (2012) Mitochondrial DNA analysis of *Rhipicephalus sanguineus* sensu lato (Acari: Ixodidae) in the Southern Cone of South America. *Veterinary Parasitology* 190, 547–555. https://doi.org/10.1016/j.vetpar.2012.06.032

Nava, S., Beati, L., Venzal, J.M., Labruna, M.B., Szabó, M., *et al.* (2018) *Rhipicephalus sanguineus* (Latreille, 1806): neotype designation, morphological re-description of all parasitic stages and molecular characterization. *Ticks and Tick-borne Diseases* 9, 1573–1585. https://doi.org/10.1016/j.ttbdis.2018.08.001

Ogden, N.H. and Lindsay, L.R. (2016) Effects of climate and climate change on vectors and vector-borne diseases: ticks are different. *Trends in Parasitology* 32, 646–656. https://doi.org/10.1016/j.pt.2016.04.015

Ogden, N.H., Maarouf, A., Barker, I.K., Bigras-Poulin, M., Lindsay, L.R., *et al.* (2006) Climate change and the potential for range expansion of the Lyme disease vector *Ixodes scapularis* in Canada. *International Journal for Parasitology* 36, 63–70. https://doi.org/10.1016/j.ijpara.2005.08.016

Ogden, N.H., Beard, C.B., Ginsberg, H.S. and Tsao, J.I. (2020) Possible effects of climate change on ixodid ticks and the pathogens they transmit: predictions and observations. *Journal of Medical Entomology* 20, tjaa220. https://doi.org/10.1093/jme/tjaa220

Oliveira, P.R., Bechara, G.H., Denardi, S.E., Saito, K.C., Nunes, E.T., *et al.* (2005) Comparison of the external morphology of *Rhipicephalus sanguineus* (Latreille, 1806) (Acari: Ixodidae) ticks from Brazil and Argentina. *Veterinary Parasitology* 129, 139–147. https://doi.org/10.1016/j.vetpar.2005.01.001

Oscherov, E.B., Milano, A.M.F., Lobo, B., Anda, P. and Escudero, R. (2011) Detection of *Anaplasma platys* and other pathogens in ectoparasites from urban hosts in Northeast Argentine. *Revista Ibero-Latinoamericana de Parasitologia* 70, 42–48.

Ostfeld, R.S. and Brunner, J.L. (2015) Climate change and *Ixodes* tick-borne diseases of humans. *Philosophical Transactions of the Royal Society B* 370, 20140051. https://doi.org/10.1098/rstb.2014.0051

Parola, P., Socolovschi, C., Jeanjean, L., Bitam, I., Fournier, P.E., *et al.* (2008) Warmer weather linked to tick attack and emergence of severe rickettsioses. *PLoS Neglected Tropical Diseases* 2, e338. https://doi.org/10.1371/journal.pntd.0000338

Porretta, D., Mastrantonio, V., Amendolia, S., Gaiarsa, S., Epis, S., *et al.* (2013) Effects of global changes on the climatic niche of the tick *Ixodes ricinus* inferred by species distribution modelling. *Parasites & Vectors* 6, 271. https://doi.org/10.1186/1756-3305-6-271

Randolph, S. (2009) Epidemiological consequences of the ecological physiology of ticks. *Advances in Insect Physiology* 37, 297–339. https://doi.org/10.1016/S0065-2806(09)37006-X

Richards, S.L., Lord, C.C., Pesko, K. and Tabachnick, W.J. (2009) Environmental and biological factors influencing *Culex pipiens quinquefasciatus* Say (Diptera: Culicidae) vector competence for Saint Louis encephalitis virus. *American Journal of Tropical Medicine and Hygiene* 81, 264–272. https://doi.org/10.4269/ajtmh.2009.81.264

Sanches, G.S., Évora, P.M., Mangold, A.J., Jittapalapong, S., Rodriguez-Mallon, A., *et al.* (2016) Molecular, biological, and morphometric comparisons between different geographical populations of *Rhipicephalus sanguineus* sensu lato (Acari: Ixodidae). *Veterinary Parasitology* 215, 78–87. https://doi.org/10.1016/j.vetpar.2015.11.007

Shih, C.M., Telford, S.R. and Spielman, A. (1995) Effect of ambient temperature on competence of deer ticks as hosts for Lyme disease spirochetes. *Journal of Clinical Microbiology* 33, 958–961. https://doi.org/10.1128/jcm.33.4.958-961.1995

Silveira, J.A., Passos, L.M. and Ribeiro, M.F. (2009) Population dynamics of *Rhipicephalus sanguineus* (Latrielle, 1806) in Belo Horizonte, Minas Gerais state, Brazil. *Veterinary Parasitology* 161, 270–275. https://doi.org/10.1016/j.vetpar.2009.01.028

Socolovschi, C., Raoult, D. and Parola, P. (2009) Influence of temperature on the attachment of *Rhipicephalus sanguineus* ticks on rabbits. *Clinical Microbiology and Infection* 15, 326–327. https://doi.org/10.1111/j.1469-0691.2008.02260.x

Spielman, A., Wilson, M.L., Levine, J.F. and Piesman, J. (1985) Ecology of *Ixodes dammini*-borne human babesiosis and Lyme disease. *Annual Review of Entomology* 30, 439–460. https://doi.org/10.1146/annurev.en.30.010185.002255

Suksawat, J., Pitulle, C., Arraga-Alvarado, C., Madrigal, K., Hancock, S.I. and Breitschwerdt, E.B. (2001) Coinfection with three *Ehrlichia* species in dogs from Thailand and Venezuela with emphasis on consideration of 16S ribosomal DNA secondary structure. *Journal of Clinical Microbiology* 39, 90–93. https://doi.org/10.1128/JCM.39.1.90-93.2001

Szabó, M.P.J., Mangold, A.J., João, C.F., Bechara, G.H. and Guglielmone, A.A. (2005) Biological and DNA evidence of two dissimilar populations of the *Rhipicephalus sanguineus* tick group (Acai: Ixodidae) in South America. *Veterinary Parasitology* 130, 131–140. https://doi.org/10.1016/j.vetpar.2005.03.008

Telford, S.R., Urioste, S.S. and Spielman, A. (1992) Clustering of host-seeking nymphal deer ticks (*Ixodes dammini*) infected by Lyme disease spirochetes (*Borrelia burgdorferi*). *American Journal of Tropical Medicine and Hygiene* 47, 55–60. https://doi.org/10.4269/ajtmh.1992.47.55

Thapa, S., Zhang, Y. and Allen, M.S. (2019) Effects of temperature on bacterial microbiome composition in *Ixodes scapularis* ticks. *Microbiology Open* 8, e00719. https://doi.org/10.1002/mbo3.719

Tudela, M.I.M., Ruiz Mantecón, A. and Estrada-Peña, A. (2005) Impactos sobre el sector Agrario. In: *Evaluación Preliminar de los Impactos en España por Efecto del Cambio Climático*. Ministerio de Medio Ambiente, Madrid, pp. 437–468.

Venzal, J.M., Estrada-Peña, A., Castro, O., De Souza, C.G., Portillo, A. and Oteo, J.A. (2007) Study on seasonal activity in dogs and ehrlichial infection in *Rhipicephalus sanguineus* (Latreille, 1806) (Acari: Ixodidae) from southern Uruguay. *Parasitología Latinoamericana* 62, 23–26. https://doi.org/10.4067/S0717-77122007000100004

Villar, M., Ayllón, N., Busby, A.T., Galindo, R.C., Blouin, E.F., *et al.* (2010) Expression of heat shock and other stress response proteins in ticks and cultured tick cells in response to *Anaplasma* spp. infection and heat shock. *International Journal of Proteomics* 2010, 657261. https://doi.org/10.1155/2010/657261

Walker, J.B., Keirans, J.E. and Horak, I.G. (2005) *The Genus Rhipicephalus (Acari, Ixodidae): A Guide to the Brown Ticks of the World*. Cambridge University Press, Cambridge.

Westbrook, C.J., Reiskind, M.H., Pesko, K.N., Greene, K.E. and Lounibos, L.P. (2010) Larval environmental temperature and the susceptibility of *Aedes albopictus* Skuse (Diptera: Culicidae) to chikungunya virus. *Vector-Borne and Zoonotic Diseases* 10, 241–247. https://doi.org/10.1089/vbz.2009.0035

Zemtsova, G.E., Apanaskevich, D.A., Reeves, W.K., Hahn, M., Snellgrove, A. and Levin, M.L. (2016) Phylogeography of *Rhipicephalus sanguineus* sensu lato and its relationships with climatic factors. *Experimental and Applied Acarology* 69, 191–203. https://doi.org/10.1007/s10493-016-0035-4

eo31 Argasidae: Distribution and Vectorial Capacity in a Changing Global Environment

Serhii Filatov[1] and Ryan Rego[2,3]*
[1]*Institute of Experimental and Clinical Veterinary Medicine, Kharkiv, Ukraine;* [2]*Czech Academy of Sciences, České Budějovice, Czech Republic;* [3]*University of South Bohemia, České Budějovice, Czech Republic*

31.1 Expert Opinion

31.1.1 Introduction

Climate change is a pervasive topic in current biology. This is understandable because, at the fundamental level, every free-living organism is constrained by such extrinsic conditions as the environmental temperature and relative humidity, comprising their physiological limits. The Earth's biodiversity has been evolving concurrently with the climate systems and the current geographical distribution of species is a product of past climatic changes, as well as of their complex interactions with the abiotic environment and with each other. The current theory states that when biological species are challenged by changing environmental conditions, they must adapt *in situ* through genetic/phenotypic adjustments or migrate to a different habitat, tracing the environment more suitable for their survival and persistence (Chown *et al.*, 2010). For example, a rapid shift in the phenology of an organism can occur in the case of phenotypic adjustments, whereas a distributional range shift might take place when species migrate. Hence, we need to gain a better understanding of the physiological and ecological traits limiting species' ranges to anticipate their future distribution in the rapidly changing world. When it comes to ticks, such knowledge would also help to predict possible impacts of climate change on the distribution and risk of emergence of tick-borne pathogens, which is of considerable interest for a broad range of professionals in public and animal health. There is a large body of literature dedicated to the responses to climate in the hard tick family, Ixodidae. Indeed, it has been observed that some of the most notorious ixodid species expand their latitudinal and altitudinal ranges, as well as adjust the timing of seasonal activities (Gray *et al.*, 2009). In sharp contrast, our knowledge is merely exiguous in this respect as far as the soft tick family, Argasidae, is concerned.

All the possible variables that affect the stability of the vector in pathogen transmission come under the term 'vectorial capacity', which is partially dependent on temperature conditions. These include duration of infective life of the vector, vector abundance, competence and feeding rate, duration of the permissive season and host range. These variables can change easily with changes in climate over time (Rocklöv and Dubrow, 2020). As an example, Onyango *et al.* (2020) have shown that an increase in temperature in the Americas could significantly

*Email: ryanrego@paru.cas.cz

DOI: 10.1079/9781789249637.0031

impact mosquito competence, blood-feeding and longevity, and lead to a decrease in the overall vectorial capacity of *Aedes* mosquitoes in this region. At the same time, one has to be aware that most published data looking at vectorial capacity in different haematophagous arthropods place an exaggerated emphasis on changes in temperature and not enough consideration is given to other ecological complexities which include those that are influenced by rainfall, humidity and host exposure.

This expert opinion is shaped to discuss the available information on soft tick physiology and ecology, how they affect the ticks' current geographical distributions and how distributions might change in response to global climatic changes and anthropogenic transformation of the environment.

31.1.2 Lack of accessible baseline information on soft tick ecology and distributions. Is there more diversity than we currently know?

Argasidae is the second largest family within the Ixodida, with the current number of described species approximating 200, but more diversity is sure to be discovered as recent studies utilizing the integrative taxonomy approach suggest (Bakkes *et al.*, 2018; Mans *et al.*, 2019). Although, as a group, soft ticks have a nearly worldwide distribution, little effort has been made to assess their occurrence patterns at different scales and across biomes. Similarly, basic information about life histories and ecology is absent or incomplete for the prevailing majority of argasid species. On the other hand, it is quite often the case that even if such information exists it is poorly disseminated and scattered throughout obscure literature sources published in local languages, in museum collections or as fragmentary expert knowledge. Collation and analysis of such records is a formidable work requiring time, funding and expertise, while the data themselves are subject to various errors and uncertainties (Escribano *et al.*, 2016).

Gathering new data in field-based surveys of Argasidae, however, is an even more challenging task, which is mostly infeasible on large spatial and temporal scales. Albeit, without the primary data on distribution and ecology, many of the inferential approaches, such as predictive modelling, have only limited value for our understanding of species distributions in these ticks (Léger *et al.*, 2013). Similarly, it is difficult to make generalizations about soft tick physiology and its influence on the current and future tick geographical ranges, as only a few species have been studied in sufficient detail.

Commonly, all soft ticks are considered xerophiles occurring mainly in dry and hot climes but remarkable exceptions to this rule exist, such as in the lineages adapted to parasitize seabirds (i.e. *Argas macrostigmatus*, *Ornithodoros maritimus*, *Ornithodoros sawaii*) or rheophilic amphibians (*Ornithodoros faccinii*, *Ornithodoros saraivai*) that can be found in very humid habitats (Oliver, 1989; Muñoz-Leal *et al.*, 2017). Given the scarcity of baseline information, it is very likely that many more exciting lifestyles, host and habitat associations in argasids will be uncovered in the future, especially with more studies focused on under-sampled groupings of ticks and/or in under-represented regions of the globe.

31.1.3 Physiological limits as a basis for the mechanistic understanding of soft tick distributions

Because ticks spend most of their lifetime off a host, maintaining proper water balance is one of the main challenges imposed on them by the environment (Yoder *et al.*, 2012). Ticks hydrate mostly through active vapour absorption from the unsaturated ambient atmosphere and an important parameter of their water balance is the critical equilibrium humidity (CEH), which can be defined as a threshold humidity below which they cannot maintain normal water levels and continuously lose water. Interestingly, the CEH values are similar between all Ixodidae and Argasidae species, falling within the range of 75–95% relative humidity, independently of the ticks' habitat characteristics (Knülle and Rudolph, 1982). Therefore, the ability to conserve water when exposed to desiccating conditions is a critical requirement. Soft ticks are known to possess several physiological and behavioural adaptations that help them to cope with the challenge and thrive in arid landscapes. The cuticle in ticks has a specialized lipid layer conferring protection from dehydration below a certain temperature known as the critical transition temperature (CTT) (Knülle and Rudolph, 1982). Moreover, to avoid the loss of water, soft ticks can ventilate discontinuously and/or can respond to the changing humidity levels behaviourally, e.g. by forming aggregations or actively moving towards areas with preferred humidity levels within their microhabitat (George, 1981; Phillips *et al.*, 1995). Indeed, the water loss rates (WLRs) that have been measured in argasids are among the lowest reported in any arthropod species (Dautel, 1999). However, such important parameters cannot be directly related to climate constraints on tick distributions at large geographical scales. Rather it has been suggested that, since the distributions of tick species with different WLRs often overlap extensively, they interact with the environment at a much finer scale, making the availability of suitable microhabitats a more proximate factor limiting geographical ranges in ticks (Yoder *et al.*, 2012). Unfortunately, little is known about microclimate and other characteristics of microhabitats preferred by most Argasidae species (see Section 31.1.4).

Nevertheless, several macroclimatic variables have been proposed as factors limiting soft tick geographical distributions. Thus, for the Palaearctic Argasidae, the 15–20°C and 20–30°C mean daily July isotherms have been proposed as the northernmost range limits for *Argas* and *Ornithodoros* spp., respectively (Filippova, 1966). One of the possible explanations is that, although most temperate soft tick species can overwinter inside their protected habitats and avoid detrimental effects of low temperatures, the innate cold hardiness ability is much lower in the embryonic and some of the post-embryonic stages (i.e. larvae or early-stage nymphs); consequently, the sum of temperatures sufficient for the ticks' development acts as an environmental filter (Ershova and Vasil'eva, 1988; Dautel and Knülle, 1998). Such greater susceptibility of immature ticks to low temperatures would have a profound impact on the population dynamics of species, especially at the edges of their distribution; however, studies on demographic parameters or population genetics in soft ticks are insufficient to test the impact. Moreover, such dependence on a temperature threshold to predict distributions in ticks might not hold true because the relationship between tick developmental rates and temperature/relative humidity is non-linear and might vary depending on the geographical origin of populations (Dantas-Torres, 2015). Although data comparing life cycle parameters in different populations of the same species are virtually non-existent, studies of widespread species (such as *Argas persicus* and *Ornithodoros tartakovskyi*) showed some evidence of adaptations to local environmental conditions in geographically distinct populations (Melchakova, 1972; Dusbábek, 1985).

Another cited climatic variable with possible limiting influence on soft tick distribution is average annual precipitation (Vial *et al.*, 2018), although it is unclear how such a constraint could act from the physiological point of view. Indeed, most studies on water balance in ticks examined how these arachnids cope with desiccating conditions and none, to the best of our knowledge, sought to assess the effect of overhydration on tick performance, although it has been mentioned as a possible limiting factor for the arid-adapted hard tick, *Dermacentor andersoni* (Yoder *et al.*, 2007). Indeed, the use of precipitation data as a covariate for modelling habitat suitability in ticks has been justly criticized as a factor producing false estimates of good fit (Estrada-Peña *et al.*, 2013). It is difficult to fathom how the average rainfall data obtained from a few weather stations, and further interpolated to cover a larger territory, might relate to the actual relative humidity or saturation deficit within the ticks' microhabitat.

31.1.4 Characteristics of the unique ecological niches of soft ticks

Most argasids spend their entire lifespan within protected retreats (burrows, caves, crevices in cliff ledges, etc.) and, therefore, are branded nidicolous parasites of vertebrates (Vial, 2009). However convenient, this catch-all term is likely an oversimplification that does not capture the existing diversity of lifestyles in soft ticks or their true relationships with the environment. Soft ticks are not randomly distributed within a given territory although the understanding of mechanisms behind the spatial patterns of their distributions is currently missing. The microclimatic conditions important for the ticks' fitness can differ significantly even between distinct burrow/nest types of the same host species within a relatively small geographical area (Amat-Valero *et al.*, 2014; Espinaze *et al.*, 2020). Such differences would be especially important at the edges of a tick species' geographical distribution, where only a small proportion of microenvironments can sustain viable populations (Dautel *et al.*, 1991); but even within an area comprising the species' core range, this can have certain consequences for the ticks' relative abundances in habitat patches of varying quality (Vial *et al.*, 2018; Espinaze *et al.*, 2020; Veiga and Valera, 2020). Therefore, other factors such as local host communities, landscape topography and soil type are likely to drive fine-scale distributions in Argasidae but little effort has been made to clarify their influence on tick occurrence. Considering that soft ticks are largely indiscriminate feeders, a hypothesis that they represent habitat (or niche) specialists rather than specialize in a particular host species has been proposed repeatedly (Vasiljeva and Ershova, 1987; Klompen *et al.*, 1996), but has not received due attention in the literature. Among the most prominent features of the habitats commonly occupied by soft ticks are their remarkable persistence in time, the relative stability of abiotic conditions and the constant presence of a variety of potential blood meal hosts (Larson *et al.*, 2005; Vial, 2009; Coombes and Viles, 2015; Mukherjee *et al.*, 2019). These features might buffer possible harmful effects of changing climate on soft tick populations and would help them to survive the adversities *in situ* (Corlett and Tomlinson, 2020), which probably has already been the case during past climatic changes.

31.1.5 Global environmental change and anthropogenic influence on soft tick distributions

Modern human influences on natural ecosystems are so omnipresent that they go well beyond anthropologically induced climate change, directly impacting biotic communities through resource overexploitation resulting in habitat loss, degradation and fragmentation (Dantas-Torres, 2015; Jacobson *et al.*, 2019). Assuming the limited dispersal abilities in soft ticks (Vial, 2009; Kada *et al.*, 2017), it is likely that the ecological stability of landscapes is one of the most crucial factors to ensure the continued survival and resilience of populations in many argasid species. For example, an apparent reduction in numbers of habitats occupied by several *Ornithodoros* species has been reported in areas of Eurasia heavily impacted by human activities such as limestone mining, river damming, agricultural activities and urban development (Vial *et al.*, 2018). Similarly, the most recent limited soft tick surveys in the suburbs of Tbilisi, Georgia, conducted in 2018–2019, failed to detect populations of *Ornithodoros verrucosus* which had been reported from these localities in the first half of the 20th century (Filippova, 1966). At the same time, the tick species is still relatively common and abundant in certain natural areas of Georgia and Azerbaijan (S. Filatov, Kharkiv, 2021, personal communication). Apart from direct destruction of suitable habitats, widespread use of pesticides and general improvement in housing conditions might have contributed to the observed decline in species of medical or veterinary significance, such as *Ornithodoros papillipes* and *A. persicus* (Dusbabek and Rosicky, 1976; Vial *et al.*, 2018).

However, anthropogenic pressures described above are not uniform/do not apply to all soft tick species. There is growing evidence, from various parts of the world, that some argasid species might adapt to synanthropic conditions and increase in abundance or expand into new areas aided by such human activities as global travel and commerce (Dautel *et al.*, 1991; Keirans and Durden, 2001; Reck *et al.*, 2013; Bermúdez *et al.*, 2017; Bissett *et al.*, 2018). Nevertheless, the success of such events, as ecological theory predicts, will likely depend on how similar the newly explored habitats are to the original ecological conditions where a given soft tick species has evolved (Lundholm, 2011).

31.2 Future Predictions

It is difficult to generalize currently about global climatic change and its possible future effects on soft tick distributions or vectorial capacities as too few species have been studied in sufficient detail. Some soft tick species may decline in abundance and become extirpated in parts of their range because of anthropogenic pressures. Others may benefit from their ability to adapt to human-made habitats and expand to new territories outside their current natural range. Modelling efforts have started providing the first insights into distributions of soft ticks (Donaldson *et al.*, 2016; Sage *et al.*, 2017; Clifton *et al.*, 2018; Vial *et al.*, 2018), but more focused research efforts to reveal the basic biology and ecology of these understudied parasites are surely needed to better understand the effect of environmental change on their geographical distribution, as well as on the ecology of the pathogens they transmit.

References

Amat-Valero, M., Calero-Torralbo, M., Václav, R. and Valera, F. (2014) Cavity types and microclimate: implications for ecological, evolutionary, and conservation studies. *International Journal of Biometeorology* 58, 1983–1994.

Bakkes, D.K., De Klerk, D., Latif, A.A. and Mans, B.J. (2018) Integrative taxonomy of Afrotropical *Ornithodoros* (*Ornithodoros*) (Acari: Ixodida: Argasidae). *Ticks and Tick-borne Diseases* 9, 1006–1037.

Bermúdez, S.E., Castillo, E., Pohlenz, T.D., Kneubehl, A., Krishnavajhala, A., *et al.* (2017) New records of *Ornithodoros puertoricensis* Fox 1947 (Ixodida: Argasidae) parasitizing humans in rural and urban dwellings, Panama. *Ticks and Tick-borne Diseases* 8, 466–469.

Bissett, J., Ledet, S., Krishnavajhala, A., Armstrong, B., Klioueva, A., *et al.* (2018) Detection of tickborne relapsing fever spirochete, Austin, Texas, USA. *Emerging Infectious Diseases* 24, 2003–2009.

Chown, S.L., Hoffmann, A.A., Kristensen, T.N., Angilletta, M.J. Jr, Stenseth, N.C. and Pertoldi, C. (2010) Adapting to climate change: a perspective from evolutionary physiology. *Climate Research* 43, 3–15.

Clifton, S.M., Davis, C.L., Erwin, S., Hamerlinck, G., Veprauskas, A., *et al.* (2018) Modeling the argasid tick (*Ornithodoros moubata*) life cycle. In: Radunskaya, A., Segal, A. and Shtylla, B. (eds) *Understanding Complex Biological Systems with Mathematics*. Springer, Cham, Switzerland, pp. 63–87.

Coombes, M.A. and Viles, H.A. (2015) Population-level zoogeomorphology: the case of the Eurasian badger (*Meles meles* L.). *Physical Geography* 36, 215–238.

Corlett, R.T. and Tomlinson, K.W. (2020) Climate change and edaphic specialists: irresistible force meets immovable object? *Trends in Ecology & Evolution* 35, 367–376.

Dantas-Torres, F. (2015) Climate change, biodiversity, ticks and tick-borne diseases: the butterfly effect. *International Journal for Parasitology: Parasites and Wildlife* 4, 452–461.

Dautel, H. (1999) Water loss and metabolic water in starving *Argas reflexus* nymphs (Acari: Argasidae). *Journal of Insect Physiology* 45, 55–63.

Dautel, H. and Knülle, W. (1998) Seasonal oviposition and temperature requirements of eggs may limit northern distribution of European *Argas reflexus* (Acari: Argasidae). *Journal of Medical Entomology* 35, 26–37.

Dautel, H., Kahl, O. and Knülle, W. (1991) The soft tick *Argas reflexus* (F.) (Acari, Argasidae) in urban environments and its medical significance in Berlin (West). *Journal of Applied Entomology* 111, 380–390.

Donaldson, T.G., De León, A.A.P., Li, A.I., Castro-Arellano, I., Wozniak, E., *et al.* (2016) Assessment of the geographic distribution of *Ornithodoros turicata* (Argasidae): climate variation and host diversity. *PLoS Neglected Tropical Diseases* 10, e0004383.

Dusbábek, F. (1985) Biological comparison of different populations of *Argas* (*Persicargas*) *persicus* (Oken). *Folia Parasitologia* 32, 255–263.

Dusbabek, F. and Rosicky, R. (1976) Argasid ticks (Argasidae, Ixodoidea) of Czechoslovakia. *Acta Scientarium Naturalium* 10, 1–43.

Ershova, A. and Vasil'eva, I. (1988) Age composition of *Ornithodoros papillipes* nymphs (Argasidae) in intrapopulational groupings different in their number. *Parazitologiya* 22, 52–57.

Escribano, N., Ariño, A.H. and Galicia, D. (2016) Biodiversity data obsolescence and land uses changes. *PeerJ* 4, e2743.

Espinaze, M.P.A., Hui, C., Waller, L. and Matthee, S. (2020) Nest-type associated microclimatic conditions as potential drivers of ectoparasite infestations in African penguin nests. *Parasitology Research* 119, 3603–3616.

Estrada-Peña, A., Gray, J.S., Kahl, O., Lane, R.S. and Nijhof, A.M. (2013) Research on the ecology of ticks and tick-borne pathogens – methodological principles and caveats. *Frontiers in Cellular and Infection Microbiology* 3, 29.

Filippova, N. (1966) *Argasid Ticks (Argasidae). Fauna SSSR, Paukoobraznye 4*. Nauka, Moscow.

George, J.E. (1981) The influence of aggregation pheromones on the behavior of *Argas cooleyi* and *Ornithodoros concanensis* (Acari: Ixodoidea: Argasidae). *Journal of Medical Entomology* 18, 129–133.

Gray, J.S., Dautel, H., Estrada-Peña, A., Kahl, O. and Lindgren, E. (2009) Effects of climate change on ticks and tick-borne diseases in Europe. *Interdisciplinary Perspectives on Infectious Diseases* 2009, 593232.

Jacobson, A.P., Riggio, J., Tait, A.M. and Baillie, J.E.M. (2019) Global areas of low human impact ('Low Impact Areas') and fragmentation of the natural world. *Scientific Reports* 9, 14179.

Kada, S., McCoy, K.D. and Boulinier, T. (2017) Impact of life stage-dependent dispersal on the colonization dynamics of host patches by ticks and tick-borne infectious agents. *Parasites & Vectors* 10, 375.

Keirans, J.E. and Durden, L.A. (2001) Invasion: exotic ticks (Acari: Argasidae, Ixodidae) imported into the United States. A review and new records. *Journal of Medical Entomology* 38, 850–861.

Klompen, J., Black, W.T., Keirans, J. and Oliver, J.R. (1996) Evolution of ticks. *Annual Review of Entomology* 41, 141–161.

Knülle, W. and Rudolph, D. (1982) Humidity relationships and water balance of ticks. In: Obenchain, F.D. and Galun, R. (eds) *Physiology of Ticks*. Pergamon, Oxford, pp. 43–70.

Larson, D.W., Matthes, U. and Kelly, P.E. (2005) *Cliff Ecology: Pattern and Process in Cliff Ecosystems*. Cambridge University Press, Cambridge.

Léger, E., Vourc'h, G., Vial, L., Chevillon, C. and McCoy, K.D. (2013) Changing distributions of ticks: causes and consequences. *Experimental and Applied Acarology* 59, 219–244.

Lundholm, J. (2011) Urban cliffs. In: Douglas, I., Goode, D., Houck, M. and Wang, R. (eds) *The Routledge Handbook of Urban Ecology*. Routledge, New York, pp. 252–263.

Mans, B.J., Featherston, J., Kvas, M., Pillay, K.-A., De Klerk, D.G., *et al.* (2019) Argasid and ixodid systematics: implications for soft tick evolution and systematics, with a new argasid species list. *Ticks and Tick-borne Diseases* 10, 219–240.

Melchakova, E.D. (1972) Development of *Ornithodoros tartakovskyi* in different temperatures and relative humidity. *Trudy Instituta Zoologii Akademiia nauk Kazakhskoi SSR* 33, 123–129.

Mukherjee, A., Pal, A., Velankar, A.D., Kumara, H.N. and Bhupathy, S. (2019) Stay awhile in my burrow! Interspecific associations of vertebrates to Indian crested porcupine burrows. *Ethology Ecology & Evolution* 31, 313–328.

Muñoz-Leal, S., Toledo, L.F., Venzal, J.M., Marcili, A., Martins, T.F., *et al.* (2017) Description of a new soft tick species (Acari: Argasidae: *Ornithodoros*) associated with stream-breeding frogs (Anura: Cycloramphidae: *Cycloramphus*) in Brazil. *Ticks and Tick-borne Diseases* 8, 682–692.

Oliver, J.H. (1989) Biology and systematics of ticks (Acari: Ixodida). *Annual Review of Ecology and Systematics* 20, 397–430.

Onyango, M.G., Bialosuknia, S.M., Payne, A.F., Mathias, N., Kuo, L., *et al.* (2020) Increased temperatures reduce the vectorial capacity of *Aedes* mosquitoes for Zika virus. *Emerging Microbes & Infections* 9, 67–77.

Phillips, J.S., Adeyeye, O. and Bruni, D. (1995) Respiratory metabolism of the soft tick, *Ornithodoros turicata* (Dugès). *Experimental and Applied Acarology* 19, 103–115.

Reck, J., Marks, F.S., Guimarães, J.A., Termignoni, C. and Martins, J.R. (2013) Epidemiology of *Ornithodoros brasiliensis* (mouro tick) in the southern Brazilian highlands and the description of human and animal retrospective cases of tick parasitism. *Ticks and Tick-borne Diseases* 4, 101–109.

Rocklöv, J. and Dubrow, R. (2020) Climate change: an enduring challenge for vector-borne disease prevention and control. *Nature Immunology* 21, 479–483.

Sage, K.M., Johnson, T.L., Teglas, M.B., Nieto, N.C. and Schwan, T.G. (2017) Ecological niche modeling and distribution of *Ornithodoros hermsi* associated with tick-borne relapsing fever in western North America. *PLoS Neglected Tropical Diseases* 11, e0006047.

Vasiljeva, I. and Ershova, A. (1987) Sex and age structure of intrapopulational groupings of the tick *Ornithodoros papillipes*. *Parazitologiya* 21, 437–444.

Veiga, J. and Valera, F. (2020) Nest box location determines the exposure of the host to ectoparasites. *Avian Conservation and Ecology* 15, 11.

Vial, L. (2009) Biological and ecological characteristics of soft ticks (Ixodida: Argasidae) and their impact for predicting tick and associated disease distribution. *Parasite* 16, 191–202.

Vial, L., Ducheyne, E., Filatov, S., Gerilovych, A., McVey, D., *et al.* (2018) Spatial multi-criteria decision analysis for modelling suitable habitats of *Ornithodoros* soft ticks in the Western Palearctic region. *Veterinary Parasitology* 249, 2–16.
Yoder, J.A., Buchan, D.R., Ferrari, N.F. and Tank, J.L. (2007) Dehydration tolerance of the Rocky Mountain wood tick, *Dermacentor andersoni* stiles (Acari: Ixodidae), matches preference for a dry environment. *International Journal of Acarology* 33, 173–180.
Yoder, J.A., Hedges, B.Z. and Benoit, J.B. (2012) Water balance of the American dog tick, *Dermacentor variabilis*, throughout its development with comparative observations between field-collected and laboratory-reared ticks. *International Journal of Acarology* 38, 334–343.

eo32 Effects of Climate Change on Babesiosis Vectors

Jeremy Gray
University College Dublin, Dublin, Republic of Ireland

32.1 Expert Opinion

32.1.1 Introduction

Ixodid ticks are overwhelmingly the main means of transmission of babesiosis, with at least 22 tick species having been confirmed as vectors of 18 species of *Babesia* infecting domestic animals or humans (Gray *et al.*, 2019). The disease, characterized by infection of erythrocytes, is a major constraint on livestock production, particularly of cattle in the tropics and subtropics. Canine babesiosis has become increasingly prominent and zoonotic babesiosis, caused primarily by *Babesia microti* in the USA, continues to attract attention. The literature frequently contains statements that babesiosis is an emerging disease, implying that both its frequency and distribution are increasing, and climate change is often cited as a causal factor. However, the possible factors involved in these trends are complex and varied. No predictive models for the impact of climate change on *Babesia* spp. have been developed so far and direct observations of effects on babesiosis epidemiology appear to be scarce. Nevertheless, several studies have addressed climate change in relation to babesiosis vectors, and by implication transmission of *Babesia* spp., although endemic areas of babesiosis may not necessarily overlap with vector distribution.

32.1.2 Evidence for climate change impact on babesiosis vectors

Ixodes ricinus

Ixodes ricinus is the predominant tick species in Europe and the sole vector of bovine babesiosis caused by *Babesia divergens*. Whereas there is a general perception that tick-borne diseases in Europe are increasing, this particular disease appears to be in decline, mainly due to changes in the cattle industry. However, there is clear evidence that both the abundance and distribution of the vector have increased, and global warming appears to have a role in this, particularly at the northern edge of its distribution in Scandinavia. It could be expected that disease outbreaks would follow this expanding vector distribution; however, even in 1935, sporadic bovine babesiosis cases were reported many kilometres north of established tick populations (Jore *et al.*, 2011). It is probable such cases are caused by ticks developed from engorged individuals that were deposited by migratory birds or could possibly be recrudescences in cattle from endemic areas. While it is difficult to link isolated cases to global warming, an out-of-season outbreak in 20 beef cows in February 2019 in Devon, UK (Johnson *et al.*, 2020), was evidently due to unusually high temperatures (10°C more than the average), a probable manifestation of climate change.

*Email: jeremy.gray@ucd.ie

DOI: 10.1079/9781789249637.0032

Similar outbreaks have probably occurred elsewhere in Europe but are unlikely to be widely reported.

Dermacentor reticulatus

Dermacentor reticulatus is the vector of *Babesia canis*, the cause of canine babesiosis, a major health issue in European dogs, and of *Babesia caballi*, which causes equine babesiosis worldwide, although *D. reticulatus* is currently the main vector of equine babesiosis in Europe. This tick species is adapted to mild moist climates and occurs in a wide west–east band across Europe to Central Asia, approximately between the latitudes of 40° and 60°. It is absent from most of the Iberian Peninsula, the southern Balkans, Italy and Turkey. However, there are now many records of new sites of infestation and it is clear *D. reticulatus* is spreading in Central and Western Europe (Földvári *et al.*, 2016). These observations are supported by increasing reports of new *B. canis* foci but reports of *B. caballi* cases caused by this tick species in Europe are rare.

D. reticulatus is capable of completing its life cycle within a single season and rising temperatures would shorten development times and increase the proportion of ticks completing their life cycle in one year. Additionally, higher autumn and winter temperatures are likely to result in increased host acquisition. It is likely that these direct effects on *D. reticulatus* contribute to its spread; however, populations of deer and wild boar, the most important wild hosts of adult ticks, are also growing throughout Europe, boosting the *D. reticulatus* population and transporting the tick to new areas.

Haemaphysalis punctata

Haemaphysalis punctata is the vector of *Babesia major* to cattle and *Babesia motasi* to sheep, neither of which are regarded as major pathogens in Europe, although Asian strains of *B. motasi* are considered to be pathogenic. In recent years outbreaks of *H. punctata* have been reported in the south of England, resulting in some deaths of sheep. The contribution of specific pathogens to these deaths is uncertain, but it is possible that *B. motasi* is involved, especially when associated with tick pyaemia. *H. punctata* has a somewhat scattered distribution mainly occurring in the countries bordering the Mediterranean, with its range extending eastwards as far as Turkey. Small populations also occur in coastal regions of Northern Europe as far as Denmark. While there are no data indicating changes in global distribution, at the local scale there is evidence that it is spreading westward in southern England (Medlock *et al.*, 2018) and it has also been recorded in new foci on the coast of the Netherlands (Hofmeester *et al.*, 2016), probably as a result of higher summer and winter temperatures.

Hyalomma spp.

Hyalomma ticks are vectors of *B. caballi*, the cause of equine babesiosis, and of *Babesia occultans*, a mild pathogen of cattle. These ticks require a warmer climate for successful development than is generally available north of latitude 45°, but the immature stages are frequently found on migratory birds as far north as Scandinavia, particularly in coastal regions. Reports of unfed adults in Northern Europe, presumably moulted from nymphs brought in by birds, are increasing, possibly indicating an increased likelihood of tick population establishment. Detection in 20 cattle of the bovine piroplasm, *B. occultans*, has been claimed in eastern Poland, far north of the vector's distribution (Staniec *et al.*, 2018), but the true identity of this parasite is uncertain. There are no reports of the appearance of equine babesiosis in new areas in Europe or elsewhere.

32.1.3 Absence of evidence for climate change effects on babesiosis vectors

Rhipicephalus spp.

Several ixodid species have spread or changed their geographical distribution in the absence of clear evidence for a climate change effect, although it is impossible to rule out such an impact at some level in the complex series of processes that may be involved. For example, when introduced into new areas, *Rhipicephalus microplus*, which transmits both *Babesia bovis* and *Babesia bigemina* to cattle, gradually replaces competing ticks of the same genus and its recent spread has been observed in many parts of sub-Saharan Africa. Complex biological competitive advantages include a shorter life cycle, a tendency to induce resistance to indigenous *Rhipicephalus* species and a higher growth potential in areas of higher rainfall, but despite this last feature there is no evidence so far that climate change has a role in the spread of *R. microplus*. At present there are no clear examples of the other boophilid *Rhipicephalus* species, all of which transmit at least one species of *Babesia* to cattle, extending their ranges in Africa, Asia, Australasia or the Americas in response to climate change. However, Giles *et al.* (2014) observed that the burgeoning populations of *Rhipicephalus annulatus* and *R. microplus* south of the quarantine area on the Mexico/USA border are probably a result of climate change, presenting an increasing danger of outbreaks of babesiosis in the southern states of the USA.

Rhipicephalus sanguineus sensu lato (s.l.), the brown dog tick, is the vector of *Babesia vogeli*, one of the four *Babesia* species infecting dogs, and has a worldwide distribution occurring in warm climates between latitudes 38° and 44°. Judging from reported increases in rickettsioses and canine ehrlichiosis in Europe, rising temperatures may increase transmission of pathogens by this tick species and there are suspicions that tick populations are expanding northwards in France. However, there is no good evidence at present of the distribution of either *R. sanguineus* or *B. vogeli* infections being markedly affected by climate change. This tick species' nidicolous habit, which partly insulates it from climate change effects, may make it difficult to correlate distribution data with climate change.

Ixodes scapularis

Ixodes scapularis has shown a remarkable expansion of its range over the last few decades in the north-eastern USA and temperature rises have generally coincided with this northward advance, although deer movements are also important. *I. scapularis* does not transmit any *Babesia* spp. of veterinary significance but is the vector of a zoonotic strain of *B. microti*, the cause of the commonest form of human babesiosis. This pathogen lags far behind the spread of its vector, and its spread appears to depend on the rate and direction of travel of the small mammalian reservoir hosts. Currently, there is no direct evidence for climate change affecting either vector or pathogen, but it is probable that deer, which are vital for maintenance of the vector, and also the small mammal reservoirs of *B. microti*, have benefited from recent milder winter temperatures.

32.2 Future Predictions

Several ecological niche models, which take account of both the climatic and biotic requirements of the particular tick species, have been developed in attempts to predict the effects of climate change on tick abundance and geographical distribution. Such models suggest that in the next 30 years many new suitable areas will arise worldwide for *R. microplus* (Marques *et al.*, 2020). Conversely, some areas within the existing range of this and other *Rhipicephalus* boophilid species are likely to become unsuitable.

Despite the levels of uncertainty in predictive distribution models of *R. sanguineus* s.l. (Alkishe *et al.*, 2020), there seems little doubt that populations of this species complex will expand into cooler regions worldwide and indeed Beugnet *et al.* (2011) established that suitable habitat has been increasing in France since the 1960s.

The most recent study of *Hyalomma* spp. distribution suggested that *Hyalomma marginatum*, while apparently becoming more abundant in some locations, is unlikely to become established in Northern Europe in the near future (Estrada-Peña *et al.*, 2015). However, the increasing deposition by birds of engorged nymphal stages in Northern and Central Europe, and the consequent appearance of isolated adults far north of their usual European distribution, suggest that establishment may follow within the next 30 years.

While no recent studies have addressed the future distribution of either *D. reticulatus* or *H. punctata*, there are already indications of climate change affecting these species and it seems likely they will be found in more locations in the future and that *B. canis* infections will continue to spread.

I. scapularis has been predicted by Ogden *et al.* (2006) to occur many miles north of its current northern limit in southern Ontario by the 2050s. *B. microti*, the cause of human babesiosis, is not likely to adhere to the same schedule because of its dependence on migration of small mammalian reservoir hosts. Although *B. microti*-infected *I. scapularis* has been detected as far north as Canada, these parasite strains are evidently endemic to the region and do not seem to give rise to human cases, despite one suspected case in a medically compromised child who had received blood transfusions in infancy (Bullard *et al.*, 2014).

I. ricinus has been the subject of several studies seeking to predict future distribution in a climate change scenario, with two of the most recent (Porretta *et al.*, 2013; Alkishe *et al.*, 2017) coming to the same general conclusion that *I. ricinus* distribution will extend further north, occupying much of coastal Scandinavia, and further east, even beyond the Ural Mountains. Such models are mainly temperature-driven, as is the length of the vegetation growing season (VGS). Rising temperatures not only extend the VGS but also shorten *I. ricinus* development times, enabling tick populations to establish further north, as demonstrated by Hvidsten *et al.* (2020) who correlated the VGS with established tick populations, demonstrating a distinct movement of *I. ricinus* to the north of Norway over the last few decades. Indeed, Jaenson and Lindgren (2011) used the VGS and more indirect tick data (sampling and questionnaires) to produce forecasts that in many respects resemble those obtained with the ecosystem niche models mentioned above. In the southern part of the tick's range, where lack of humidity rather than low temperature predominantly determines the success of *I. ricinus* establishment, ecosystem niche models have a much higher level of uncertainty (Alkishe *et al.*, 2017).

References

Alkishe, A.A., Peterson, A.T. and Samy, A.M. (2017) Climate change influences on the potential geographic distribution of the disease vector tick *Ixodes ricinus*. *PLoS ONE* 12, e0189092.

Alkishe, A.A., Cobos, M.E., Peterson, A.T. and Samy, A.M. (2020) Recognizing sources of uncertainty in disease vector ecological niche models: an example with the tick *Rhipicephalus sanguineus* sensu lato. *Perspectives in Ecology and Conservation* 18, 91–102.

Beugnet, F., Kolasinski, M., Michelangeli, P.A., Vienne, J. and Loukos, H. (2011) Mathematical modelling of the impact of climatic conditions in France on tick activity and density since 1960. *Geospatial Health* 5, 255–263.

Bullard, J.M., Ahsanuddin, A.N., Perry, A.M., Lindsay, L.R., Iranpour, M., *et al.* (2014) The first case of locally acquired tick-borne *Babesia microti* infection in Canada. *Canadian Journal of Infectious Diseases and Medical Microbiology* 25, e87–e89.

Estrada-Peña, A., de la Fuente, J., Latapia, T. and Ortega, C. (2015) The impact of climate trends on a tick affecting public health: a retrospective modeling approach for *Hyalomma marginatum* (Ixodidae). *PLoS ONE* 10, e0125760.

Földvári, G., Široký, P., Szekeres, S., Majoros, G. and Sprong, H. (2016) *Dermacentor reticulatus*: a vector on the rise. *Parasites & Vectors* 9, 314.

Giles, J.R., Peterson, A.T., Busch, J.D., Olafson, P.U., Scoles, G.A., *et al.* (2014) Invasive potential of cattle fever ticks in the southern United States. *Parasites & Vectors* 7, 189.

Gray, J.S., Estrada-Peña, A. and Zintl, A. (2019) Vectors of babesiosis. *Annual Reviews of Entomology* 64, 149–165.

Hofmeester, T.R., van der Lei, P.B., Docters van Leeuwen, A., Sprong, H. and van Wieren, S.E. (2016) New foci of *Haemaphysalis punctata* and *Dermacentor reticulatus* in the Netherlands. *Ticks and Tick-borne Diseases* 7, 367–370.

Hvidsten, D., Frafjord, K., Gray, J.S., Henningsson, A.J., Jenkins, A., *et al.* (2020). The distribution limit of the common tick, *Ixodes ricinus*, and some associated pathogens in north-western Europe. *Ticks and Tick-borne Diseases* 11, 101388.

Jaenson, T.G. and Lindgren, E. (2011) The range of *Ixodes ricinus* and the risk of contracting Lyme borreliosis will increase northwards when the vegetation period becomes longer. *Ticks and Tick-Borne Diseases* 2, 44–49.

Johnson, N., Phipps P.L., McFadzean, H. and Barlow, A.M. (2020) An outbreak of bovine babesiosis in February, 2019, triggered by above average winter temperatures in southern England and co-infection with *Babesia divergens* and *Anaplasma phagocytophilum*. *Parasites & Vectors* 13, 305.

Jore, S., Viljugrein, H., Hofshagen, M., Brun-Hansen, H., Kristoffersen, A.B., *et al.* (2011) Multi-source analysis reveals latitudinal and altitudinal shifts in range of *Ixodes ricinus* at its northern distribution limit. *Parasites & Vectors* 4, 84.

Marques, R., Krüger, R.F., Peterson, A.T., de Melo, L.F., Vicenzi, N., *et al.* (2020) Climate change implications for the distribution of the babesiosis and anaplasmosis tick vector, *Rhipicephalus* (*Boophilus*) *microplus*. *Veterinary Research* 51, 81.

Medlock, J.M., Hansford, K.M., Vaux, A.G.C., Cull, B., Pietzsch, M.E., *et al.* (2018) Has the red sheep tick, *Haemaphysalis punctata*, recently expanded its range in England? *Medical Veterinary Entomology* 32, 473–480.

Ogden, N.H., Maarouf, A., Barker, I.K., Bigras-Poulin, M., Lindsay, L.R., *et al.* (2006) Climate change and the potential for range expansion of the Lyme disease vector *Ixodes scapularis* in Canada. *International Journal for Parasitology* 36, 63–70.

Porretta, D., Mastrantonio, V., Amendolia, S., Gaiarsa, S., Epis, S., *et al.* (2013) Effects of global changes on the climatic niche of the tick *Ixodes ricinus* inferred by species distribution modelling. *Parasites & Vectors* 6, 271.

Staniec, M., Adaszek, Ł., Buczek, K. and Winiarczyk, S. (2018) Molecular identification of *Babesia* spp. isolated from Polish cattle with asymptomatic protozoa infections. *Polish Journal of Veterinary Sciences* 21, 287–291.

Synopsis: Ticks

Pat Nuttall*
University of Oxford, Oxford, UK

The collection of expert opinions focused primarily on ticks considers whether or not changes in climate affect different aspects of the tick biosphere (from physiology to ecology). This synopsis gives a glimpse of what is covered.

Clearly, ticks have been through it all before. During >250 million years of evolution, the Ixodida (ticks) has experienced multiple episodes of extreme climate change. The survival of *Nuttalliella namaqua*, a 'living fossil' and only extant species of the tick family, Nuttalliellidae, is testament to their resilience (eo22: Mans). This resilience is reflected in modern-day species. For example, *Ixodes uriae* (the seabird tick) withstands a temperature range of −30 to 30°C (eo77: McCoy), while water loss rates in argasid ticks are among the lowest reported for any arthropod species (eo31: Filatov and Rego). Climate has surely left its imprint on tick evolution. In fact, it may be driving speciation today witnessed by different genotypes/phenotypes of *Rhipicephalus sanguineus* (tropical and temperate lineages) (eo30: Bechara and Sanches), *Haemaphysalis longicornis* (bisexual and parthenogenetic races) (eo27: Umemiya-Shirafuji), *Dermacentor reticulatus* (Eastern and Western European populations) (eo24: Zając), *Ixodes scapularis* (northern and southern clades) (eo9: Ronai; eo26: Fish) and *Ixodes ricinus* (different questing phenotypes) (eo25: Rizzoli).

Ticks are absolute parasites; they cannot survive without a blood meal. Nevertheless, even this dependence shows resilient extremes. Some argasid species can survive at least 7 years without a blood meal (eo12: Taylor and Ogihara). Although they depend on vertebrate blood, ticks spend up to 99% of their time off a host where their critical requirement for survival is water (eo11: Benoit and Oyen). Water availability is impacted by changes in climate; *ipso facto*, ticks are impacted by climate change. Because ticks spend much of their life cycle off a host, environmental conditions shape their physiology and behaviour. Recognizing environmental stressors via the synganglion, and its interconnected neural and sensory systems, is central to tick adaptive responses at individual and population levels. Changes in gene expression and/or adaptive evolution of tick sensory systems may play a major role in acclimatization (eo8: Šimo). A hint of this is shown by *I. ricinus* ticks from cooler climates, which quest at lower temperatures than those from warmer climates (eo25: Rizzoli). Likewise, there is some evidence of adaptations to local environmental conditions (e.g. differences in the developmental cycle) in geographically distinct populations of *Argas persicus* and *Ornithodoros tartakovskyi* (eo31: Filatov and Rego).

*Email: pat.nuttall@zoo.ox.ac.uk

DOI: 10.1079/9781789249637.syn2

Ixodid ticks replenish water loss by absorbing atmospheric water vapour via hygroscopic secretions from their salivary glands (eo11: Benoit and Oyen) controlled by their nervous system (eo8: Šimo). Argasid species may produce water by increasing their metabolic rate at the expense of food reserves (eo11: Benoit and Oyen). Ixodid ticks have mass-specific, basal metabolic rates typically 13% below those of other arthropods (eo13: Wall and Alasmari), which likely contributes to water conservation. Interestingly, metabolic rate measurement, e.g. using spectrophotometric techniques, may enable direct detection of climate effects on ticks (eo13: Wall and Alasmari). However, given that tick species with different water loss rates often inhabit the same geographical region, the availability of suitable microhabitats seems a more reliable indicator of climate constraints than general atmospheric humidity (eo31: Filatov and Rego). Indeed, overlooking the ability of ticks to occupy appropriate microhabitats even within harsh macrohabitats can result in predictions of future tick occurrence in freezing or 'boiling' temperatures (eo23: Oliveira).

Besides direct physiological responses to climate change (through sensing the environment via the nervous system and maintaining water balance), indirect responses may arise through the tick microbiota. Long-term exposure to raised temperatures eliminates certain groups of bacteria inhabiting ticks while favouring others. Such changes may not affect microbiota functionality as there is ample redundancy to compensate for any losses (eo7: Cabezas-Cruz). However, this may not be the case for endosymbionts comprising the tick microbiota. In arthropods, ticks harbour among the greatest diversity of heritable endosymbionts. Through their ability to alter tick phenotypes, endosymbionts can shape the adaptation of ticks to climate change. Gottlieb and Duron (eo10) predict that 'defensive' endosymbionts, which protect ticks against attack by natural enemies, will thrive as climate change induces ticks to seek new environments where they encounter new threats. The tick microbiota also includes tick-borne pathogens, some of which affect the tick phenotype. For example, *Anaplasma phagocytophilum* infection of *I. scapularis* ticks induces synthesis of tick heat-shock proteins, which promote tick survival by reducing the risk of desiccation, and 'antifreeze protein', a large glycoprotein that binds ice crystals, enhancing tolerance of freezing conditions (eo21: Lynn *et al.*).

Climate has additional direct effects on ticks, notably on survival and reproduction, but possibly also on tick immunity (although little is known). Temperature and humidity can impact the timing and rate of oviposition, egg viability and female longevity (eo12: Taylor and Ogihara). Both ixodid and argasid species produce the egg protein, vitellogenin, a precursor of vitellin found in mature eggs and seemingly an important storage protein that aids cold tolerance.

The dependence of ticks on blood means that any threat to their food supply threatens their existence. Thus, any impact of climate change on tick vertebrate hosts will impact ticks feeding on them. Fossil records provide evidence that major changes in climate can lead to extinction of tick lineages through lost associated host lineages, but also provide opportunities for tick lineage expansion if new hosts and/or environmental niches appear (eo22: Mans). Today, in the Anthropocene, we are witnessing the sixth mass extinction. At least one tick species has become extinct because of its inability to adapt to new hosts and 63 tick species have preferred hosts on the Red List of endangered species. The 'living fossil' (*N. namaqua*) may have survived six extinction events through its ability to parasitize both reptiles and mammals (eo22: Mans). Generalist feeders, like *N. namaqua*, are likely to be more resilient when a changing climate limits host availability. Nidicolous species such as many argasids, although limited in host choice (even though they are largely indiscriminate feeders), are protected from environmental extremes by their protected retreats (eo12: Taylor and Ogihara; eo31: Filatov and Rego). By contrast, for non-nidicolous (exophilic) species such as *I. ricinus*, questing represents the greatest risk (of dehydration) due to direct exposure to environmental conditions (eo11: Benoit and Oyen).

If today's climate is affecting ticks directly, we should expect to see changes in life-cycle development rates, phenology and seasonality (the timing and duration of tick host-seeking activity), as well as abundance and distribution of tick populations (eo9: Ronai; eo13: Wall and Alasmari; eo14: Kjær and Bødker; eo15: Kelly *et al.*; eo16: Wilke *et al.*; eo21: Lynn *et al.*; eo27: Umemiya-Shirafuji; eo28: Eisen and Eisen;

eo32: Gray). However, many factors impact tick abundance and distribution, besides climate. For example, the extensive geographical distribution of several *Amblyomma*, *Rhipicephalus*, *Haemaphysalis* and *Hyalomma* species that feed on domesticated animals can be linked to the development of agriculture and colonization of the continents by humans some 10,000 to 7000 years ago (eo22: Mans). Today, livestock biomass is almost 15 times greater than all wild mammal biomass; hence, livestock management (altering host abundance and distribution; modifying habitats; affecting tick mortality through tick control) has a huge impact on the abundance and distribution of ticks feeding on livestock. Such impacts may be climate-driven as, for example, warmer and drier environments cause a shift from crop production to livestock husbandry, including increased investment in camels and goats (eo20: Titcomb). While *I. scapularis* has expanded from relict populations in the north-eastern USA to almost fully reclaim its historical geographic range of pre-colonial times (eo26: Fish), *Amblyomma americanum* appears to have expanded from its core distribution in the south-eastern USA where stable populations historically were maintained (eo28: Eisen and Eisen). For both species, white-tailed deer (*Odocoileus virginianus*) are the 'reproductive host' of adult females and the source of nutrients for egg production. The coincident expansion of white-tailed deer populations and tick populations is one of several confounding factors that makes it difficult to tease out the influence of climate change. Likewise, in Europe, deer (notably *Capreolus capreolus*) populations have increased (since the 1980s) with concomitant increases in *I. ricinus* populations although not necessarily driven by climate change (eo18: Stachurski *et al.*; eo19: Hoodless and Sage).

Generating unambiguous evidence of direct climate change impacts on ticks ideally requires long-term, replicated studies (eo18: Stachurski *et al.*). Many studies of tick populations are relatively short, and the frequency and quality of sampling may be inadequate. But how long is long term? One study of *I. ricinus* in the Tula region of Russia (Eastern Europe) recorded an increased abundance of questing adult ticks over a 35-year period (numbers of mammalian hosts remained relatively constant over the period). The study claimed that increased duration of temperatures >5°C in late summer and autumn resulted in extension of the questing period, with climate-influenced succession of forest ecosystems also playing a role (eo13: Wall and Alasmari).

Another approach to pinpointing climate change impacts on ticks is to examine populations at the edge of their geographical distribution. In Scandinavia, for example, *I. ricinus* is at the limit of its extent no doubt in part because its lower threshold for metabolic activity is around −10 to −5°C (eo13: Wall and Alasmari). Photoperiod is also a range-limiting factor as it influences host-seeking behaviour (eo26: Fish). In Scandinavia, during the past 10 years, *I. ricinus* has expanded northwards to 66°N and increased in abundance, and reports of exotic species, such as *D. reticulatus*, *Hyalomma marginatum* and *Hyalomma rufipes*, are more common. In Denmark in particular, *D. reticulatus* appears now to be established, most likely as a result of a warmer climate (eo14: Kjær and Bødker; eo24: Zając). Likewise, in North America, *I. scapularis*, *A. americanum* and *Dermacentor variabilis* are pushing back the northern limits of their distribution (eo28: Eisen and Eisen). As yet, there is no clear evidence of retraction at the southern limits of these species. Altitude also limits tick distribution and here, too, changes are occurring with ticks climbing ever higher, driven directly and indirectly by climate change (eo12: Taylor and Ogihara; eo18: Stachurski *et al.*; eo19: Hoodless and Sage; eo25: Rizzoli).

While further tick expansion is predicted in northern latitudes, a decrease in suitable areas for tick development and reproduction is expected in more southerly regions such as the southern USA and South Africa (eo16: Wilke *et al.*). However, this depends on species. For example, expansion of *Rhipicephalus microplus* (often with displacement of other species such as *Rhipicephalus decoloratus*) has been linked with the relatively high temperature tolerance of this species, possibly reflecting adaptation to climate change and the emergence of new environmental niches (eo22: Mans; eo29: Santos *et al.*). Turkey is climatically diverse and has a rich tick fauna supplemented by ticks from migrating birds along the Black Sea Flyway. Decreasing cattle grazing in Turkey, driven in part by climate change and its effect on vegetation, is predicted to drive down populations of *H. marginatum* (eo17: Kar and Gargili Keles). However, in parts of France, *H. marginatum* feeds on horses and, having recently established there, appears to be on the increase encouraged by a warming, drier climate (eo18: Stachurski *et al.*).

Global warming can have startling effects on tick species that normally do not seek out humans but appear to do so as they become more active (eo16: Wilke *et al.*). However, it is always difficult to tease out cause and effect although sometimes the driver seems obvious, like the reported increase in human tick bites coincident with COVID-19 restrictions and more recreational activity outside (eo25: Rizzoli). Where increased temperatures differentially impact genotypes/phenotypes of a tick species, climate change may have consequences for vector capacity. For example, only the tropical lineage of *R. sanguineus* appears to be a competent vector of *Ehrlichia canis*, the aetiological agent of canine monocytic ehrlichiosis; increasing temperatures in Brazil are predicted to favour this vector-competent lineage (eo30: Bechara and Sanches).

Looking forward, it is critical to consider how human adaptations to climate change will further impact ticks and tick-borne diseases given the enormous magnitude to which livestock management controls the movement and density of suitable tick hosts and vegetation (eo20: Titcomb). This also applies to humankind's universal influence on natural ecosystems, including tick habitat creation and destruction, widespread use of pesticides and improvements in housing conditions (where once some argasid species thrived) (eo31: Filatov and Rego). When predictive models are used of future tick abundance, density and distribution, the uncertainty of the models must be explained. For example, ecological niche models of the potential future distribution of *Amblyomma* spp. in Brazil identified uncertainties apparently associated with regions where tick records were low (eo23: Oliveira). Modelling the potential impacts of climate change is particularly important for the many species that are important vectors of tick-borne infections of humans and domesticated animals. For example, predictive models of the impact of climate change on tick-borne *Babesia* have not so far been developed and direct observations of effects on babesiosis epidemiology appear scarce (eo32: Gray). Robust models of tick vectors can at least give some indication of what the future holds for the pathogens they transmit.

eo33 Conflict and Cooperation in Tick–Host–Pathogen Interactions Contribute to Increased Tick Fitness and Survival

José de la Fuente[1,2*] and Margarita Villar[1,3]
[1]*Instituto de Investigación en Recursos Cinegéticos, Ciudad Real, Spain;* [2]*Oklahoma State University, Stillwater, Oklahoma, USA;* [3]*University of Castilla-La Mancha, Ciudad Real, Spain*

33.1 Expert Opinion

33.1.1 A model for the study of tick–host–pathogen molecular interactions

The intracellular bacterium, *Anaplasma phagocytophilum*, is vectored primarily by *Ixodes* spp. ticks and is the causative agent of human and animal granulocytic anaplasmosis and tick-borne fever of ruminants (de la Fuente *et al.*, 2008). *A. phagocytophilum* primarily infects mammalian neutrophil granulocytes and tick gut and salivary gland cells. Therefore, human promyelocytic leukaemia HL60 cells and *Ixodes scapularis* ISE6 cells together with tick tissues are used for the study of tick–host–pathogen molecular interactions.

At the Genomics, Proteomics and Biotechnology of Health and Biotechnology (SaBio) group, we focus on the study of vector–host–pathogen molecular interactions using different omics technologies (genomics, transcriptomics, microRNAomics, proteomics, metabolomics, elementomics, interactomics, regulomics) and systems biology for data integration and analysis to identify targets for the development of control interventions for ticks and tick-borne diseases (Conflict and cooperation: the control of infectious diseases, 2018).

Using the *Ixodes*–HL60–*A. phagocytophilum* model, recent studies have shown the different biological processes involved in tick–host–pathogen interactions (de la Fuente *et al.*, 2005, 2016a,b, 2019, 2020a; Ayllón *et al.*, 2015; Villar *et al.*, 2015, 2016; Cabezas-Cruz *et al.*, 2016, 2019; Chmelař *et al.*, 2016; Gulia-Nuss *et al.*, 2016; Abraham *et al.*, 2017; Shaw *et al.*, 2017; Estrada-Peña *et al.*, 2018; Artigas-Jerónimo *et al.*, 2019a,b; Nelson *et al.*, 2020).

Ticks and tick-borne diseases are an increasing problem worldwide partly driven directly/indirectly by climate change. It has been proposed that climate change, distribution of tick hosts, and ecological- and anthropogenically induced changes contribute to the geographic expansion of ticks and tick-borne pathogens (Medlock *et al.*, 2013; de la Fuente *et al.*, 2015; Alkishe *et al.*, 2017; Andersen *et al.*, 2019; Rochlin and Toledo, 2020). In this context, the evolution of conflict and cooperation of tick–host–pathogen interactions is likely one of the factors supporting the geographic expansion of ticks and pathogens. These mechanisms may mediate the adaptation of tick species to climate change and hosts while guaranteeing completion of tick and pathogen life cycles. Our research focuses on dealing with how we can bring this increasing problem under control. These studies

*Email: jose_delafuente@yahoo.com

DOI: 10.1079/9781789249637.0033

advance our understanding of tick–host–pathogen interactions and suggest they evolved through conflict and cooperation, with possible implications in tick ecological adaptation and control of tick infestations and tick-borne diseases (Cabezas-Cruz *et al.*, 2017; Contreras *et al.*, 2017; Benelli, 2020).

33.1.2 Impact of *Anaplasma phagocytophilum* infection on tick and human cells

Sequencing of pathogen and tick genomes provides the basis to advance the characterization of tick–host–pathogen molecular interactions. Nevertheless, limited information is available on the genomes of *A. phagocytophilum* and *Ixodes* spp. ticks (Dugat *et al.*, 2014; Gulia-Nuss *et al.*, 2016; Jia *et al.*, 2020). Approximately 57% of the *I. scapularis* genome (2.1 Gb, ≈70% repeat content) is represented by assembled sequences (de la Fuente *et al.*, 2016c; Gulia-Nuss *et al.*, 2016).

Focusing on the *Ixodes*–HL60–*A. phagocytophilum* model, a comparative analysis of cell proteome and interactome showed that pathogen infection has a higher impact on molecular changes in tick than human cells, which probably reflects ancient ecological tick–pathogen associations with conflict and cooperation (de la Fuente *et al.*, 2017; Estrada-Peña *et al.*, 2018).

To further illustrate the impact of *A. phagocytophilum* infection on tick cells, we compiled data from Ayllón *et al.* (2015), Villar *et al.* (2015), Gulia-Nuss *et al.* (2016) and de la Fuente *et al.* (2016c). For example, transcriptomics analysis identified a total of 16,083 expressed genes in tick nymphs, representing 78% of the 20,486 predicted protein-coding genes. The impact of *A. phagocytophilum* infection on the transcriptome of *I. scapularis* nymphs affected 8516 genes, representing 53% of the expressed genes (de la Fuente *et al.*, 2016c). Proteomics analyses identified a total of 7418 proteins in *I. scapularis* ticks, representing 36% of the 20,486 predicted protein-coding genes. In *I. scapularis* ISE6 cells, a total of 14,234 proteins were identified, representing 69% of the 20,486 predicted protein-coding genes. The *A. phagocytophilum* infection resulted in differentially represented proteins equivalent to 2–9% of identified proteins in tick nymphs, adult female salivary glands and midguts, and ISE6 cells (Fig. 33.1).

These results support an effect of *A. phagocytophilum* pathogen infection on host and *I. scapularis* transcriptome and proteome, which is also reflected at the microRNAome, regulome, microbiome, elementome and host protein levels in the tick vector (Villar *et al.*, 2016; Abraham *et al.*, 2017; Artigas-Jerónimo *et al.*, 2019a,b; de la Fuente *et al.*, 2020a).

33.1.3 The relationship between conflict and cooperation of tick–host–pathogen interactions and climate change

Examples of conflict and cooperation in tick–host–pathogen interactions were described by de la Fuente *et al.* (2016b). The conflict is clearly reflected by the immune response mechanisms limiting tick feeding or pathogen infection, as well as diseases caused by tick-borne pathogens or tick-induced allergies such as the alpha-gal syndrome (AGS) (de la Fuente *et al.*, 2005, 2016a,b, 2019; Chmelař *et al.*, 2016). However, the cooperation is reflected by host factors and pathogen-induced gene expression that favour tick feeding and fitness (e.g. heat-shock proteins, subolesin, antifreeze glycoproteins), midgut microbiota composition (e.g. Brinkerhoff *et al.*, 2020) and changes in epigenetic regulatory mechanisms (e.g. miRNA isc-mir-79) that favour tick survival under extreme environmental conditions (Neelakanta *et al.*, 2010; Busby *et al.*, 2012; Cabezas-Cruz *et al.*, 2017, 2019; Artigas-Jerónimo *et al.*, 2019a; de la Fuente *et al.*, 2019; Benelli, 2020). Furthermore, *A. phagocytophilum* pathogen infection manipulates tick behaviour (Busby *et al.*, 2012; Benelli, 2020), midgut microbiota (Abraham *et al.*, 2017) and exoskeleton chemical composition resulting in increased biomineralization (de la Fuente *et al.*, 2020a), which may also contribute to increased ectoparasite fitness and survival.

Similar mechanisms have been described for interactions between other tick, host and pathogen species, thus highlighting that mechanisms of cooperation evolved during tick–host–pathogen interactions to increase tick survival and ecological adaptation under extreme environmental conditions (Belova *et al.*, 2012; Herrmann and Gern, 2015; Cabezas-Cruz *et al.*, 2017; de la Fuente *et al.*, 2017).

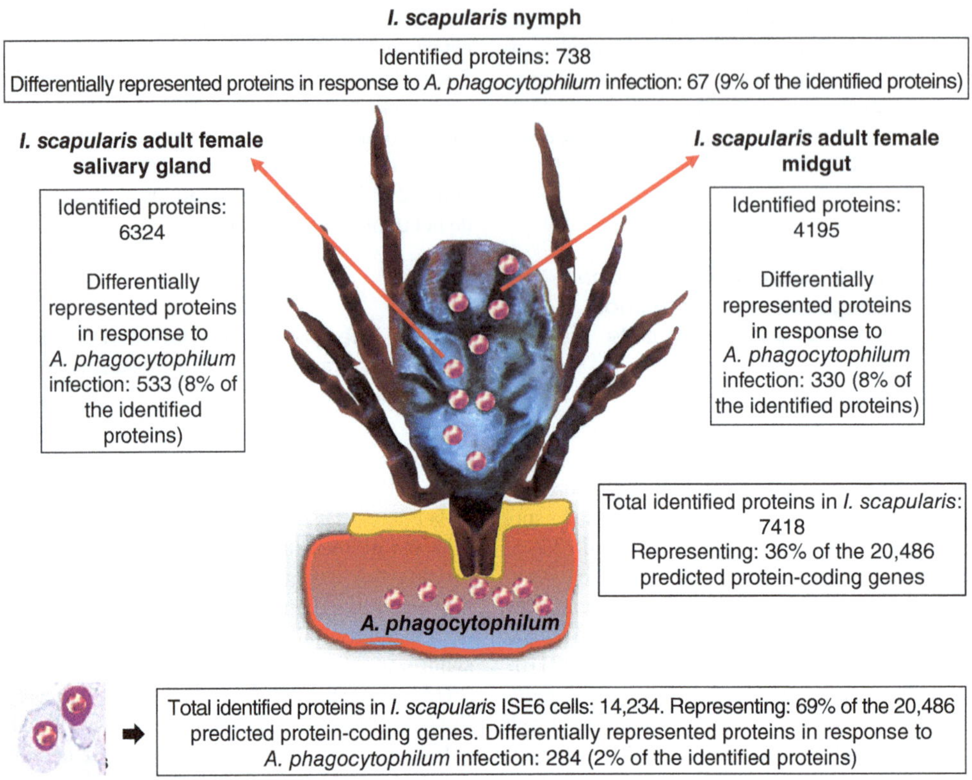

Fig. 33.1. The *Ixodes scapularis* proteome and tick–host–pathogen interactions. The impact in fed ticks of *A. phagocytophilum* infection on the tick proteome is represented in nymphs' and adult females' salivary glands and midguts by showing the total number of identified proteins and the differentially represented proteins in response to infection. (Data compiled from Ayllón *et al.*, 2015; Villar *et al.*, 2015; Gulia-Nuss *et al.*, 2016. Tick image courtesy of Katherine M. Kocan, Oklahoma State University, USA. Authors' own figure.)

Based on the possible contribution of climate change, distribution of tick hosts, and ecological- and anthropogenically-induced changes to the geographic expansion of ticks and tick-borne pathogens (Medlock *et al.*, 2013; de la Fuente *et al.*, 2015; Alkishe *et al.*, 2017; Andersen *et al.*, 2019; Rochlin and Toledo, 2020), the evolution of conflict and cooperation of tick–host–pathogen interactions is likely one of the factors supporting the geographic expansion of ticks and pathogens.

33.2 Future Predictions

Based on evidence that mechanisms of tick–host–pathogen interactions may mediate the adaptation of tick species to climate change and hosts while guaranteeing completion of tick and pathogen life cycles, it is likely that the prevalence of ticks and tick-borne diseases will continue to increase in the coming years. Therefore, the control of tick infestations and tick-borne pathogens constitutes a challenge and a priority that requires a multifactorial approach (de la Fuente, 2018) by targeting different mechanisms of conflict and cooperation in tick–host–pathogen interactions. To address this challenge, some of the aspects to be considered include:

1. Tick and pathogen genomic resources. To address this limitation, it is important to promote international collaborations to generate genomic databases and advance research on comparative, functional and evolutionary genomics of tick vectors and transmitted pathogens (de la

Fuente *et al.*, 2016c; Murgia *et al.*, 2019). This information is important to better understand tick–host–pathogen molecular interactions and select target pathways and proteins for the development of control interventions.

2. Tick midgut microbiome. Characterization of the tick hologenome and the functional role of the microbiota in tick–host–pathogen interactions is important to better understand their role in tick–host–pathogen interactions (Abraham *et al.*, 2017; Díaz-Sánchez *et al.*, 2019). These results will also guide the identification of key microorganisms for possible manipulations to reduce tick fitness and vector capacity.

3. Tick glycoproteins. Characterization of tick glycoproteins and their protective capacity may be relevant to advance control of tick-borne diseases (de la Fuente *et al.*, 2006; Cabezas-Cruz and de la Fuente, 2017; Cabezas-Cruz *et al.*, 2018). For example, the glycan Galα1–3Galβ1–(3)4GlcNAc–R (α-Gal) associated with AGS has been identified in tick glycoproteins, midgut microbiota and tick-borne pathogens, and induces in humans a protective antibody response against pathogens with this modification on their surface (Cabezas-Cruz and de la Fuente, 2017; Cabezas-Cruz *et al.*, 2018; Galili, 2019; de la Fuente *et al.*, 2019). Furthermore, understanding how tick saliva triggers different immune responses to α-Gal (immunoglobulin (Ig) E-mediated allergies or IgM/IgG, B-cell maturation, macrophage response, complement system and TLR2/nuclear factor κB (NF-κB) innate immune pathway-mediated protection to pathogen infection) in different individuals is important for the prevention and control of AGS and multiple infectious diseases (Cabezas-Cruz and de la Fuente, 2017; Chandrasekhar *et al.*, 2019; Contreras *et al.*, 2020; Pacheco *et al.*, 2020a; Urra *et al.*, 2020).

4. Tick epigenetic regulatory mechanisms. To expand research on tick epigenetic activation and repression mechanisms activated in response to feeding, and pathogen infection and multiplication, is a research area that requires special attention (Cabezas-Cruz *et al.*, 2017). The role of epigenesis in the synthesis and function of glycoproteins (Karamanos *et al.*, 2018) may also be key to better understanding of the role of these molecules in tick–host–pathogen interactions. Characterization of epigenetic regulatory pathways will contribute to the identification of candidate control targets to reduce tick infestations, fitness, and pathogen infection and transmission.

5. Tick exoskeleton chemical composition. Recent results have advanced the characterization of the chemical elements present in the tick exoskeleton and their putative function in tick–host–pathogen interactions (Suppan *et al.*, 2018; de la Fuente *et al.*, 2020a; Pacheco *et al.*, 2020b; Villar *et al.*, 2020). Further investigations in this area will contribute to better understanding of the origin and function of the tick elementome with the identification of possible targets for the development of control interventions.

6. Tick paratransgenesis. Paratransgenesis poses a challenge in ticks due to the limited methods available for their genetic manipulation (e.g. RNA interference; de la Fuente, 2018). However, paratransgenesis has been successfully implemented in mosquitoes to reduce vector populations and transmitted diseases (Hurwitz *et al.*, 2011; Adelman and Tu, 2016; Shane *et al.*, 2018; Macias *et al.*, 2020; Mancini *et al.*, 2020). Therefore, research to develop paratransgenic ticks using novel gene editing technologies (e.g. ReMOT Control; Macias *et al.*, 2020) and manipulation of ectoparasite midgut microbiota (e.g. Shane *et al.*, 2018; Mancini *et al.*, 2020) should be advanced for developing alternative control interventions for tick-borne diseases (de la Fuente, 2018).

7. Quantum vaccinomics. The combination of multiple tick-derived and pathogen-derived antigens is one of the challenges for the development of effective vaccines for the control of tick infestations and tick-borne diseases (de la Fuente and Contreras, 2015; de la Fuente, 2018; de la Fuente *et al.*, 2018; Ndawula and Tabor, 2020). Quantum vaccinomics is defined as the characterization of antigen protective epitopes as the immunological quantum (Artigas-Jerónimo *et al.*, 2020; de la Fuente *et al.*, 2020b). Our proposed pipeline for quantum vaccinomics consists of the characterization of vector–host–pathogen molecular interactions using omics technologies combined with multi-omics data integration, network analysis and systems biology. The role of the cell interactome and regulome in vector–host–pathogen interactions is then used in a vaccinomics and big data analytics approach for the identification of proteins playing a central role in the regulation of biological processes and with protective antigen capacity in vaccination trials. Key proteins involved in tick survival, fitness and ecological adaptation may be targeted in this platform for the identification of candidate protective antigens. The selected proteins are then used in quantum vaccinomics for the identification and characterization of the immunological quantum for the design and production of chimeric protective antigens. This approach will advance the development of effective vaccines based on a combination of multiple antigens in different formulations such as nanoparticles and virus-like particles.

8. Modelling tick vaccines. Even when effective formulations have been developed, modelling tick vaccines and other interventions is important to identify best implementation strategies for the control of ticks and tick-borne diseases (de la Fuente *et al.*, 2020c).

Altogether, results from these different research areas will advance our knowledge of the conflict and cooperation that evolved between ticks, hosts and pathogens to increase tick fitness and survival under extreme environmental conditions while highlighting targets and implementations for effective control strategies for ticks and tick-borne diseases.

References

Abraham, N.M., Liu, L., Jutras, B.L., Yadav, A.K., Narasimhan, S., *et al.* (2017) Pathogen-mediated manipulation of arthropod microbiota to promote infection. *Proceedings of the National Academy of Sciences USA* 114, E781–E790.

Adelman, Z.N. and Tu, Z. (2016) Control of mosquito-borne infectious diseases: sex and gene drive. *Trends in Parasitology* 32, 219–229.

Alkishe, A.A., Peterson, A.T. and Samy, A.M. (2017) Climate change influences on the potential geographic distribution of the disease vector tick *Ixodes ricinus*. *PLoS ONE* 12, e0189092.

Andersen, N.S., Larsen, S.L., Olesen, C.R., Stiasny, K., Kolmos, H.J., *et al.* (2019) Continued expansion of tick-borne pathogens: tick-borne encephalitis virus complex and *Anaplasma phagocytophilum* in Denmark. *Ticks and Tick-borne Diseases* 10, 115–123.

Artigas-Jerónimo, S., Alberdi, P., Villar Rayo, M., Cabezas-Cruz, A., Prados, P., *et al.* (2019a) *Anaplasma phagocytophilum* modifies tick cell microRNA expression and upregulates isc-mir-79 to facilitate infection by targeting the Roundabout protein 2 pathway. *Scientific Reports* 9, 9073.

Artigas-Jerónimo, S., Estrada-Peña, A., Cabezas-Cruz, A., Alberdi, P., Villar, M. and de la Fuente, J. (2019b) Modelling modulation of the tick regulome in response to *Anaplasma phagocytophilum* for the identification of new control targets. *Frontiers in Physiology* 10, 462.

Artigas-Jerónimo, S., Comín, J., Villar, M., Contreras, M., Alberdi, P., *et al.* (2020) A novel combined scientific and artistic approach for the advanced characterization of interactomes: the Akirin/Subolesin model. *Vaccines* 8, 77.

Ayllón, N., Villar, M., Galindo, R.C., Kocan, K.M., Šíma, R., *et al.* (2015) Systems biology of tissue-specific response to *Anaplasma phagocytophilum* reveals differentiated apoptosis in the tick vector *Ixodes scapularis*. *PLoS Genetics* 11, e1005120.

Belova, O.A., Burenkova, L.A. and Karganova, G.G. (2012) Different tick-borne encephalitis virus (TBEV) prevalences in unfed versus partially engorged ixodid ticks – evidence of virus replication and changes in tick behaviour. *Ticks and Tick-borne Diseases* 3, 240–246.

Benelli, G. (2020) Pathogens manipulating tick behaviour – through a glass, darkly. *Pathogens* 9, E664.

Brinkerhoff, R.J., Clark, C., Ocasio, K., Gauthier, D.T. and Hynes, W.L. (2020) Factors affecting the microbiome of *Ixodes scapularis* and *Amblyomma americanum*. *PLoS ONE* 15, e0232398.

Busby, A.T., Ayllón, N., Kocan, K.M., Blouin, E.F., de la Fuente, J., *et al.* (2012) Expression of heat shock proteins and subolesin affects stress responses, *Anaplasma phagocytophilum* infection and questing behaviour in the tick, *Ixodes scapularis*. *Medical and Veterinary Entomology* 26, 92–102.

Cabezas-Cruz, A. and de la Fuente, J. (2017) Immunity to α-Gal: toward a single-antigen pan-vaccine to control major infectious diseases. *ACS Central Science* 3, 1140–1142.

Cabezas-Cruz, A., Alberdi, P., Ayllón, N., Valdés, J.J., Pierce, R., *et al.* (2016) *Anaplasma phagocytophilum* increases the levels of histone modifying enzymes to inhibit cell apoptosis and facilitate pathogen infection in the tick vector *Ixodes scapularis*. *Epigenetics* 11, 303–319.

Cabezas-Cruz, A., Estrada-Peña, A., Rego, R.O. and de la Fuente, J. (2017) Tick–pathogen ensembles: do molecular interactions lead ecological innovation? *Frontiers in Cellular and Infection Microbiology* 7, 74.

Cabezas-Cruz, A., Espinosa, P.J., Alberdi, P., Šimo, L., Valdés, J.J., *et al.* (2018) Tick galactosyltransferases are involved in α-Gal synthesis and play a role during *Anaplasma phagocytophilum* infection and *Ixodes scapularis* tick vector development. *Scientific Reports* 8, 14224.

Cabezas-Cruz, A., Espinosa, P., Alberdi, P. and de la Fuente, J. (2019) Tick–pathogen interactions: the metabolic perspective. *Trends in Parasitology* 35, 316–328.

Chandrasekhar, J.L., Cox, K.M., Loo, W.M., Qiao, H., Tung, K.S. and Erickson, L.D. (2019) Cutaneous exposure to clinically relevant lone star ticks promotes IgE production and hypersensitivity through $CD4^+$ T cell- and MyD88-dependent pathways in mice. *Journal of Immunology* 203, 813–824.

Chmelař, J., Kotál, J., Kopecký, J., Pedra, J. and Kotsyfakis, M. (2016) All for one and one for all on the tick–host battlefield. *Trends in Parasitology* 32, 368–377.

Conflict and cooperation: the control of infectious diseases (2018) Genomics, Proteomics and Biotechnology at SaBio, IREC (CSIC-UCLM) mission: Control of infectious diseases. Available at: https://youtu.be/DhbBjQSuLYk (accessed 24 August 2020).

Contreras, M., Alberdi, P., Fernández De Mera, I.G., Krull, C., Nijhof, A., *et al.* (2017) Vaccinomics approach to the identification of candidate protective antigens for the control of tick vector infestations and *Anaplasma phagocytophilum* infection. *Frontiers in Cellular and Infection Microbiology* 7, 360.

Contreras, M., Pacheco, I., Alberdi, P., Díaz-Sánchez, S., Artigas-Jerónimo, S., *et al.* (2020) Allergic reactions and immunity in response to tick salivary biogenic substances and red meat consumption in the zebrafish model. *Frontiers in Cellular and Infection Microbiology* 10, 78.

de la Fuente, J. (2018) Controlling ticks and tick-borne diseases … looking forward. *Ticks and Tick-borne Diseases* 9, 1354–1357.

de la Fuente, J. and Contreras, M. (2015) Tick vaccines: current status and future directions. *Expert Review of Vaccines* 14, 1367–1376.

de la Fuente, J., Ayoubi, P., Blouin, E.F., Almazán, C., Naranjo, V. and Kocan, K.M. (2005) Gene expression profiling of human promyelocytic cells in response to infection with *Anaplasma phagocytophilum*. *Cellular Microbiology* 7, 549–559.

de la Fuente, J., Canales, M. and Kocan, K.M. (2006) The importance of protein glycosylation in development of novel tick vaccine strategies. *Parasite Immunology* 28, 687–688.

de la Fuente, J., Estrada-Peña, A., Venzal, J.M., Kocan, K.M. and Sonenshine, D.E. (2008) Overview: ticks as vectors of pathogens that cause disease in humans and animals. *Frontiers in Bioscience* 13, 6938–6946.

de la Fuente, J., Estrada-Peña, A., Cabezas-Cruz, A. and Brey, R. (2015) Flying ticks: anciently evolved associations that constitute a risk of infectious disease spread. *Parasites & Vectors* 8, 538.

de la Fuente, J., Estrada-Peña, A., Cabezas-Cruz, A. and Kocan, K.M. (2016a) *Anaplasma phagocytophilum* uses common strategies for infection of ticks and vertebrate hosts. *Trends in Microbiology* 24, 173–180.

de la Fuente, J., Villar, M., Cabezas-Cruz, A., Estrada-Peña, A., Ayllón, N. and Alberdi, P. (2016b) Tick–host–pathogen interactions: conflict and cooperation. *PLoS Pathogens* 12, e1005488.

de la Fuente, J., Waterhouse, R.M., Sonenshine, D.E., Roe, R.M., Ribeiro, J.M., *et al.* (2016c) Tick genome assembled: new opportunities for research on tick–host–pathogen interactions. *Frontiers in Cellular and Infection Microbiology* 6, 103.

de la Fuente, J., Antunes, S., Bonnet, S., Cabezas-Cruz, A., Domingos, A.G., *et al.* (2017) Tick–pathogen interactions and vector competence: identification of molecular drivers for tick-borne diseases. *Frontiers in Cellular and Infection Microbiology* 7, 114.

de la Fuente, J., Villar, M., Estrada-Peña, A. and Olivas, J.A. (2018) High throughput discovery and characterization of tick and pathogen vaccine protective antigens using vaccinomics with intelligent Big Data analytic techniques. *Expert Review of Vaccines* 17, 569–576.

de la Fuente, J., Pacheco, I., Villar, M. and Cabezas-Cruz, A. (2019) The alpha-gal syndrome: new insights into the tick–host conflict and cooperation. *Parasites & Vectors* 12, 154.

de la Fuente, J., Lima-Barbero, J.F., Prado, E., Pacheco, I., Alberdi, P. and Villar, M. (2020a) *Anaplasma* pathogen infection alters chemical composition of the exoskeleton of hard ticks (Acari: Ixodidae). *Computational and Structural Biotechnology Journal* 18, 253–257.

de la Fuente, J., Contreras, M., Artigas-Jerónimo, S. and Villar, M. (2020b) A quantum vaccinomics approach to vaccine development. *The Cuban Scientist* 1, 25–26.

de la Fuente, J., Estrada-Peña, A. and Contreras, M. (2020c) Modelling tick vaccines: a key tool to improve protection efficacy. *Expert Review of Vaccines* 19, 217–225.

Díaz-Sánchez, S., Estrada-Peña, A., Cabezas-Cruz, A. and de la Fuente, J. (2019) Evolutionary insights into the tick hologenome. *Trends in Parasitology* 35, 725–737.

Dugat, T., Loux, V., Marthey, S., Moroldo, M., Lagrée, A.C., *et al.* (2014) Comparative genomics of first available bovine *Anaplasma phagocytophilum* genome obtained with targeted sequence capture. *BMC Genomics* 15, 973.

Estrada-Peña, A., Villar, M., Artigas-Jerónimo, S., López, V., Alberdi, P., *et al.* (2018) Use of graph theory to characterize human and arthropod vector cell protein response to infection with *Anaplasma phagocytophilum*. *Frontiers in Cellular and Infection Microbiology* 8, 265.

Galili, U. (2019) Evolution in primates by 'catastrophic-selection' interplay between enveloped virus epidemics, mutated genes of enzymes synthesizing carbohydrate antigens, and natural anti-carbohydrate antibodies. *American Journal of Physical Anthropology* 168, 352–363.

Gulia-Nuss, M., Nuss, A.B., Meyer, J.M., Sonenshine, D.E., Roe, R.M., *et al.* (2016) Genomic insights into the *Ixodes scapularis* tick vector of Lyme disease. *Nature Communications* 7, 10507.

Herrmann, C. and Gern, L. (2015) Search for blood or water is influenced by *Borrelia burgdorferi* in *Ixodes ricinus*. *Parasites & Vectors* 8, 6.

Hurwitz, I., Fieck, A., Read, A., Hillesland, H., Klein, N., *et al.* (2011) Paratransgenic control of vector borne diseases. *International Journal of Biological Sciences* 7, 1334–1344.

Jia, N., Wang, J., Shi, W., Du, L., Sun, Y., *et al.* (2020) Large-scale comparative analyses of tick genomes elucidate their genetic diversity and vector capacities. *Cell* 182, 1328–1340.e13.

Karamanos, N.K., Piperigkou, Z., Theocharis, A.D., Watanabe, H., Franchi, M., *et al.* (2018) Proteoglycan chemical diversity drives multifunctional cell regulation and therapeutics. *Chemical Reviews* 118, 9152–9232.

Macias, V.M., McKeand, S., Chaverra-Rodriguez, D., Hughes, G.L., Fazekas, A., *et al.* (2020) Cas9-mediated gene-editing in the malaria mosquito *Anopheles stephensi* by ReMOT control. *G3: Genes, Genomes, Genetics* 10, 1353–1360.

Mancini, M.V., Damiani, C., Short, S.M., Cappelli, A., Ulissi, U., *et al.* (2020) Inhibition of *Asaia* in adult mosquitoes causes male-specific mortality and diverse transcriptome changes. *Pathogens* 9, 380.

Medlock, J.M., Hansford, K.M., Bormane, A., Derdakova, M., Estrada-Peña, A., *et al.* (2013) Driving forces for changes in geographical distribution of *Ixodes ricinus* ticks in Europe. *Parasites & Vectors* 6, 1.

Murgia, M.V., Bell-Sakyi, L., de la Fuente, J., Kurtti, T.J., Makepeace, B.L., *et al.* (2019) Meeting the challenge of tick-borne disease control: a proposal for 1000 *Ixodes* genomes. *Ticks and Tick-borne Diseases* 10, 213–218.

Ndawula, C. Jr and Tabor, A.E. (2020) Cocktail anti-tick vaccines: the unforeseen constraints and approaches toward enhanced efficacies. *Vaccines* 8, E457.

Neelakanta, G., Sultana, H., Fish, D., Anderson, J.F. and Fikrig, E. (2010) *Anaplasma phagocytophilum* induces *Ixodes scapularis* ticks to express an antifreeze glycoprotein gene that enhances their survival in the cold. *Journal of Clinical Investigation* 120, 3179–3190.

Nelson, C.M., Herron, M.J., Wang, X.R., Baldridge, G.D., Oliver, J.D. and Munderloh, U.G. (2020) Global transcription profiles of *Anaplasma phagocytophilum* at key stages of infection in tick and human cell lines and granulocytes. *Frontiers in Veterinary Science* 7, 111.

Pacheco, I., Contreras, M., Villar, M., Risalde, M.A., Alberdi, P., *et al.* (2020a) Vaccination with α-Gal protects against mycobacterial infection in the zebrafish model of tuberculosis. *Vaccines* 8, 195.

Pacheco, I., Acevedo, P., Prado, E., Mihalca, A.D. and de la Fuente, J. (2020b) Targeting the exoskeleton elementome to track tick geographic origins. *Frontiers in Physiology* 11, 572758.

Rochlin, I. and Toledo, A. (2020) Emerging tick-borne pathogens of public health importance: a mini-review. *Journal of Medical Microbiology* 69, 781–791.

Shane, J.L., Grogan, C.L., Cwalina, C. and Lampe, D.J. (2018) Blood meal-induced inhibition of vector-borne disease by transgenic microbiota. *Nature Communications* 9, 4127.

Shaw, D.K., Wang, X., Brown, L.J., Chávez, A.S., Reif, K.E., *et al.* (2017) Infection-derived lipids elicit an immune deficiency circuit in arthropods. *Nature Communications* 8, 14401.

Suppan, J., Engel, B., Marchetti-Deschmann, M. and Nürnberger, S. (2018) Tick attachment cement – reviewing the mysteries of a biological skin plug system. *Biological Reviews of the Cambridge Philosophical Society* 93, 1056–1076.

Urra, J.M., Ferreras-Colino, E., Contreras, M., Cabrera, C.M., Fernández de Mera, I.G., *et al.* (2021) The antibody response to the glycan α-Gal correlates with COVID-19 disease symptoms. *Journal of Medical Virology* 93, 2065–2075.

Villar, M., Ayllón, N., Alberdi, P., Moreno, A., Moreno, M., *et al.* (2015) Integrated metabolomics, transcriptomics and proteomics identifies metabolic pathways affected by *Anaplasma phagocytophilum* infection in tick cells. *Molecular & Cellular Proteomics* 14, 3154–3172.

Villar, M., López, V., Ayllón, N., Cabezas-Cruz, A., López, J.A., *et al.* (2016) The intracellular bacterium *Anaplasma phagocytophilum* selectively manipulates the levels of vertebrate host proteins in the tick vector *Ixodes scapularis*. *Parasites & Vectors* 9, 467.

Villar, M., Pacheco, I., Merino, O., Contreras, M., Mateos-Hernández, L., *et al.* (2020) Tick and host derived compounds detected in the cement complex substance. *Biomolecules* 10, 555.

eo34 Climate, Ticks and Pathogens: Gaps and Caveats

Agustín Estrada-Peña[1,2*], Natalia Fernández-Ruiz[1,2] and José de la Fuente[3,4]
[1]University of Zaragoza, Zaragoza, Spain; [2]Instituto Agroalimentario de Aragón, IA2, Zaragoza, Spain; [3]Instituto de Investigación en Recursos Cinegéticos, Ciudad Real, Spain; [4]Oklahoma State University, Stillwater, Oklahoma, USA

34.1 Expert Opinion

34.1.1 Background: exophilic and endophilic ticks

As with every ectothermic organism, the metabolic processes of ticks are deeply affected by climate. Ticks that live in the ground, between vegetation and decaying litter, are called exophilic. At the micro scale, vegetation modulates the effects of the weather and its impact on the development rates, seasonality and activity of exophilic ticks. Also, many tick species live in the shelter of their hosts, in a strategy aimed to cope with the effects of climate. These are called endophilic ticks and they have the ecological advantage of being more protected against unfavourable weather conditions. How the changing climate patterns are affecting the physiology of ticks is something still under debate, because many other factors may coincide and causes are sometimes difficult to separate.

34.1.2 The question: are tick-borne pathogens spreading?

To date, the range of several tick species has changed. Concomitantly, in the last two to three decades, tick-borne pathogens have been increasingly reported outside their 'historical' range. Researchers in both human and animal health are exploring the influence of climate change on the reported spread of pathogens. Initially, this commonly involved 'mechanistic' models aimed to capture the historical distribution of an organism and then project on to current climate conditions. However, because climate patterns impact on the many components involved in the ecology of ticks, such as vegetation canopy and phenology, density of hosts and abundance of pathogen reservoirs, all the partners involved in these interactions should be considered together and not separately, as is customary (but see Fernández-Ruiz and Estrada-Peña, 2020). Even if a spreading population of ticks can adapt to other hosts in its new range, we are lacking evidence of the adaptation of a tick-borne pathogen to a new vertebrate species that was not previously involved in the pathogen's circulation. The profusion of reports based on the detection of small fragments of DNA/RNA of pathogens in ticks adds confusion to the topic. Only a few studies have addressed the issue of viability of the pathogen in ticks spreading into new territories and/or into newly established communities of vertebrates.

Other confounding factors may blur our perception of the spread of tick-transmitted

*Email: aestrada@unizar.es

DOI: 10.1079/9781789249637.0034

pathogens, like increased awareness that drives larger active or passive surveillance programmes. New methods of pathogen detection, unavailable a few years ago, are also contributing to better tracking of the distribution of ticks and the pathogens they carry. Changes in land use (which may be driven by climate or derived from human actions, like farming), or the overprotection of some vertebrate species that are key for maintenance of the pathogen, may indirectly contribute to changes in the known distribution of tick-transmitted pathogens. It is impossible to disentangle these effects from the factors purely driven by climate traits: the effects of the host's and vegetal communities must be considered together with the impact of changes in the climate.

34.1.3 Evidence supporting the effects of climate on ticks

Empirical evidence indicates that several tick species have changed their historical range. There is a bias in records because most surveys are devoted to species with impact on either human health or prominent pests of livestock, like *Rhipicephalus microplus*. Consequently, empirical evidence exists for only a few species in proportionally small areas. The first study of the spread of *Ixodes ricinus* (a pivotal species in Europe for human health) towards northern latitudes was reported by Tälleklint and Jaenson (1998), who described an increase in range in the 1990s compared with previous surveys in the 1980s. Similar results were reported by Danielová *et al.* (2006) from data in the Šumava and Krkonoše Mountains in the Czech Republic, indicating a switch in altitude of *I. ricinus*. Through field surveys, the authors documented the occurrence of the tick and the pathogens it transmits at higher altitudes than previously recorded, suggesting that a warmer climate was also spreading the altitudinal range of reservoir vertebrates. Additional records were provided again for Sweden (Jaenson *et al.*, 2012), where surveys carried out in the 1990s were compared with data from the 2010s, demonstrating that *I. ricinus* had spread into large northern territories of the country, in harmony with a warmer climate; the authors also pointed to the increase in density of some key hosts for adult ticks as being responsible for the spread in some territories of Sweden. Reports from Norway (Hvidsten *et al.*, 2020) also indicate that *I. ricinus* has colonized large regions of coastal Norway where the tick was unknown before the 2000s.

There are notorious examples of spreading ticks in the northern USA and southern Canada. *Ixodes scapularis* is a prominent vector of pathogens to humans that colonizes large portions of the central and eastern USA. Since approximately the 2000s, a slow but sustained northern range increase of the species has been noticed in southern Canada (Clow *et al.*, 2017), where populations of the tick were scarce. Detailed studies exist about this sustained range increase and population models have correlated without question that a warmer climate is its promoter. In the USA, an expansion of the range of some tick species, namely *Dermacentor variabilis*, *Amblyomma americanum* and *Amblyomma maculatum*, has been reported (Sonenshine, 2018). In addition to host-dependent factors, *D. variabilis* is believed unable to survive over the winter in regions where temperatures fall and remain below the 0°C mean winter isotherm (December through February), giving a period of sustained freezing conditions. However, milder winters are now common in these regions, averaging above freezing temperatures during the winter months. Thus, if current trends in climate change continue, areas with suitable climate for *D. variabilis* could increase for the next several decades. However, it is not known whether hosts will also spread northward throughout most of Canada (Minigan *et al.*, 2017). *A. americanum* is increasing its density in the north-eastern USA and southern Canada (Springer *et al.*, 2015) and is spreading into new areas of the northern and mid-western USA. An interesting aspect of rapid and very extensive range expansion concerns genetic changes in these populations. Monzón *et al.* (2016) showed that *A. americanum* ticks from New York and Oklahoma are genetically distinct from historic range populations in North and South Carolina, suggesting adaptive evolution in these tick populations with implications for their transmission of disease-causing agents.

It is interesting to note that the ticks reported to spread northward are still present at their southern limit and that insignificant change in their density has been noticed in the last years. This is an important observation, since a 'spread to the north' driven by a warmer climate should immediately be correlated with a restriction of tick habitat suitability at the southern limit because conditions are too dry or too warm. In any case, no detailed analyses seem to exist about the reasons why these ticks still colonize their southern range while extending northwards.

34.1.4 Confounding effects of the existing evidence

There is no question that climate is modifying the distribution of tick-borne pathogens. However, confounding factors may alter our perception that climate is not the only trait driving the spread. Probably one of the most compelling examples is the finding of adult ticks of the genus *Hyalomma* in Central and Northern Europe. For decades, it has been known that migratory birds introduce immature feeding ticks from Africa and the southern Mediterranean into northern latitudes. Records of single adults, resulting from the moult of immatures, have rarely been reported in the 20th century in these territories, which are considered too cold for survival or moulting of the tick. Since around the 2010s, reports increased of single adults in several European countries outside their historical range. No evidence of permanent populations (i.e. completing the life cycle) is yet available for these sites.

The effects of climate on ticks are not only direct (e.g. impacting tick development or mortality rates) but may also indirectly affect ticks by modulating the population density of the vertebrates that ticks use as a blood source. This could change the prevalence of pathogens in the tick population. If a tick spreads into new regions, it must find suitable hosts for feeding and suitable reservoirs for the circulation of pathogens. For example, *Hyalomma lusitanicum* is spreading in Spain because it frequently feeds on wild boar, which has a fast-expanding population in the country. Another example is the cattle tick, *R. microplus*, spreading back into the USA from Mexico after a successful eradication programme. In this case, climate might play a role. However, areas on the Texas–Mexico border are now used for hunting, introducing wild ungulates that freely move throughout the territory invading tick-infested areas in Mexico and moving back into the USA. Such human-derived action is helping to promote the recirculation of *R. microplus* and introduction of pathogens into these areas.

Two factors are commonly overlooked while considering the effects of climate on ticks. One is the change of seasonality of ticks and/or hosts, which may create a lack of overlap of both populations and therefore the lack of the necessary contact between ticks and reservoirs that help circulate a pathogen. The second is the need for a critical threshold of reservoirs, necessary to support the circulation of a pathogen. Furthermore, human actions may profoundly modify the habitat (e.g. because of agricultural practices), making some vertebrates more abundant by creating inadequate habitats for others. The prevailing hosts may be better reservoirs of pathogens, giving the false impression that the change of climate was the actual factor contributing to the increase in density of ticks and prevalence of pathogens. Last, but not least, human habits may markedly affect the perceived pattern of human infections by tick-borne pathogens. For example, greater contact with the natural environment (as houses are built near forested areas) may give rise to the misinterpretation of factors driving an increase in human infections. Since data about persistence of tick-borne pathogens are based on historical records of human cases, the persistence of small foci of pathogens that do not culminate in human infections remains unnoticed: environmental and/or biotic traits maintaining these foci remain unrecorded. In modelling terms, small-scale developments aiming to bring together causes and effects could be built, but a generalist framework, integrating every process, is difficult to build and evaluate.

34.1.5 The actual impact of climate: an example using *Ixodes ricinus*

We wanted to demonstrate the usefulness of physiological models by assessing the impact of climate on the life cycle of the tick, *I. ricinus*. It is known that the tick has spread in Scandinavia in the last decades (i.e. Jaenson *et al.*, 2012). Thus, we aimed to demonstrate that weather conditions in the last 30 years improved conditions for the development of the tick in the region and its spread. Figure 34.1 displays the anomalies of temperature at four time slices for the period November–January which, historically, is a cold period in the region. However, Fig. 34.1 shows a sustained pattern of warmer winters in the region, in contrast to other parts of Central and Southern Europe.

We chose the field surveys by Jaenson *et al.* (2012) in Sweden to demonstrate how the changes of temperature, theoretically, shortened the tick's life cycle at unprecedented rates. Algorithms from a previous study were used to define the tick life cycle (Estrada-Peña and Estrada-Sánchez, 2014). Daily temperature of the 1980s and 2010s at points matching the field surveys of *I. ricinus* were used. Figure 34.2 shows that almost every site in Sweden under

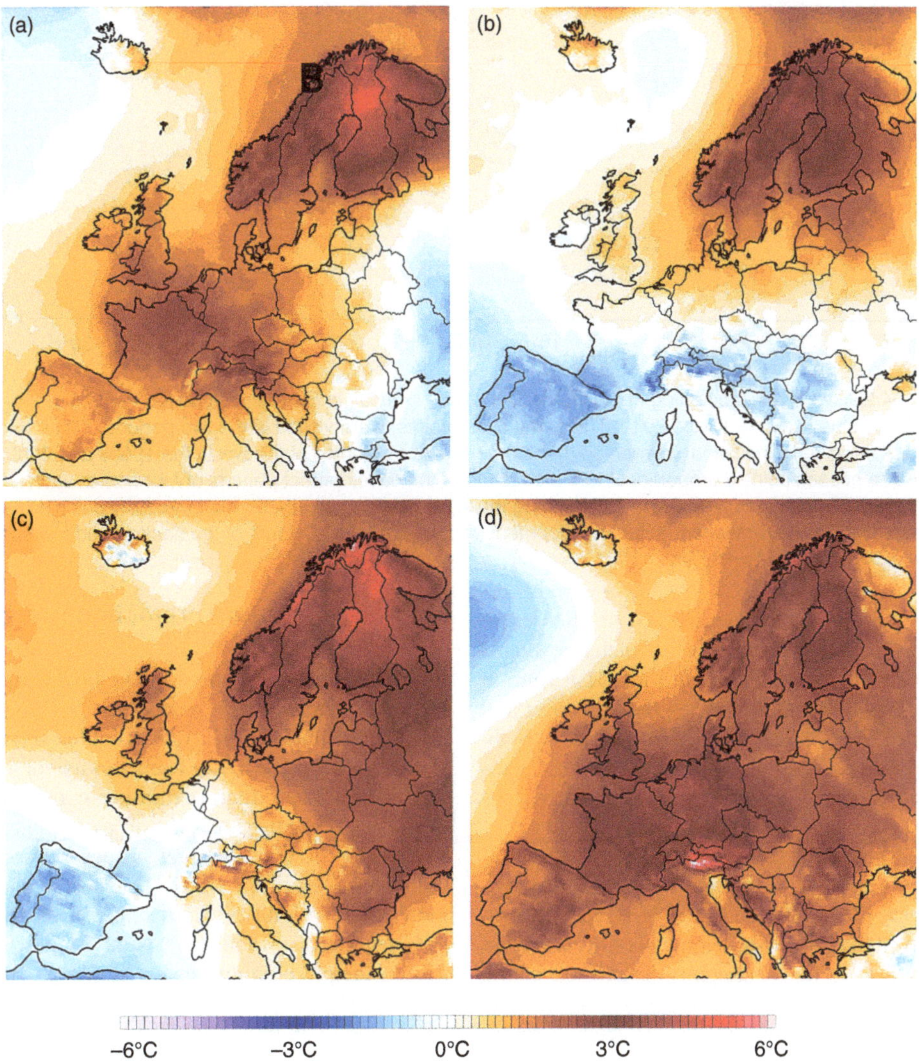

Fig. 34.1. Anomalies of temperature in Europe in the period November–January. Anomalies are calculated as the difference from the reference years, 1979–1990, and are displayed for the years 1995 (a), 2000 (b), 2005 (c) and 2016 (d). (Maps are derived from reanalysis of temperatures provided by https://climatereanalyzer.org, accessed 30 September 2021.)

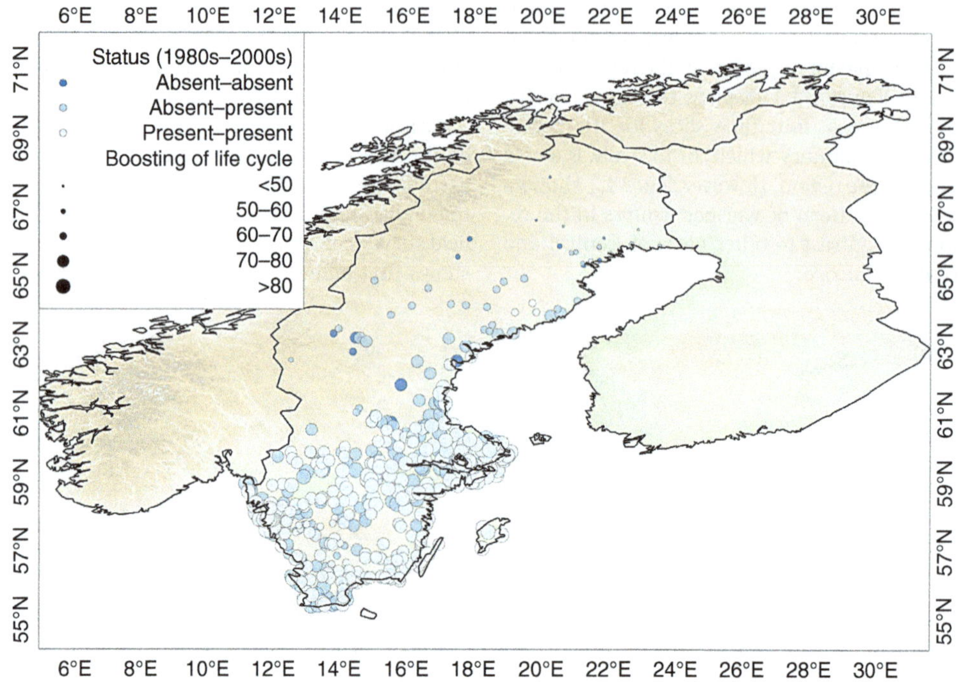

Fig. 34.2. Hypothetical boosting of the life cycle of *Ixodes ricinus* in Sweden calculated from field surveys in the years 1980 and 2010. The same sites (points) were visited in the two periods of time and the tick surveyed. The colour of each point means presence/absence of the tick in the 1980s and 2000s. The size of each point is proportional to the boosting of the development periods of *I. ricinus* at each surveyed point, based on algorithms describing these periods and obtained from laboratory colonies (i.e. a value of 100% means that the development period is halved). Colours of the background represent elevation. (Authors' own figure.)

the latitude 67°N experienced a theoretical boosting of the tick's life cycle, which is predicted to be twice faster (>90%) in the 2010s than in the 1980s. Many sites along the east coast and central parts of Sweden, where the tick was absent in the 1980s but present in the 2010s, experienced temperature ranges likely to reduce the length of the life cycle by up to 50–90%.

The climate trends in Northern Europe during the last decades appear to be one of the forces driving the northward spread of *I. ricinus*, shortening its life cycle. Warmer temperatures may not only lead to the geographical expansion of the tick but could also extend its period of annual activity and thereby impact the prevalence of pathogens in questing ticks. A warmer climate would probably reinforce the feedback of infections in questing ticks, through a longer cycle of interactions (infection–reinfection) between the tick vectors and the reservoirs of pathogens.

34.2 Future Predictions

Predictions by researchers in the field are incomplete and usually restricted to the expected distribution of the ticks. A different and innovative approach has been the use of the R_0 (the basic reproduction number, explaining the basic reproduction rate of a pathogen), evaluating how it changes under different scenarios (Nah *et al.*, 2020). However, a common framework for predictive projections has not yet been established; that framework should ideally combine every single partner involved in tick dynamics (climate, land use, vegetation, communities of vertebrates, tick physiology, stress, mortality, fecundity and activity). Available evidence points to the spread of ticks and the communities of vertebrates supporting tick-borne pathogen circulation into northern latitudes.

The link between weather, ticks and vertebrates is well established and long-term patterns should be captured and analysed. Modelling frameworks should aim for comparability (i.e. using the same set of explanatory variables describing the environmental niche of a tick) and provide ecological information. Multivariate statistics or interpolated climatology lack ecological information or have multiple statistical issues preventing their indiscriminate use. The limited number of weather stations, particularly in rural areas, means that spatial interpolation is required to create a continuous spatial coverage of weather data. Since temperature varies greatly both spatially and temporally, weather station data can be inadequate for epidemiological studies of the impact of climate change on ticks and pathogens and should be replaced with satellite-captured information (Sun *et al.*, 2014). In all likelihood, the inclusion of every trait affecting the circulation of a tick-borne pathogen is beyond current human capabilities. Other than maps displaying the probable presence of a tick species in an area, physiological models (not mechanistic correlations) could be the best approach to the problem. However, these approaches should also ideally include the densities of hosts and reservoirs, and how they interact with the ticks (i.e. host preferences, reservoir capacity).

The impact of global warming on tick-borne pathogens will be greatest at the coldest fringe of their current distribution, where sub-optimal temperatures are limiting the spread of vectors (and probably the reservoirs, too). Evaluation of the impact of climate on the spread of ticks and their hosts and pathogen reservoirs remains speculative until a dedicated network of active surveys can track the range and density of target species using harmonized collection procedures. Nevertheless, approaching an integrated framework addressing daily accumulated temperature at continental range, and its expected effect on the life cycle and activity of ticks and vertebrates, linked to actual tick surveys, looks promising for addressing the circulation of pathogens.

In conclusion, the climate trends are obvious and their effects on tick life cycles have been demonstrated at various sites and acting on several species. Aiming to mitigate the impact of tick-transmitted pathogens on health, we promote the active surveillance of ticks, together with the complete modelling of their life cycle and the effects of climate on the long-term abundance of the main hosts for ticks and pathogen reservoirs. These should be the mainstay of improving knowledge on the changing and complex relationships among ticks, vertebrates and pathogens, with a focus on future predictions. The mapping of long-term impacts of climate on the life cycle of ticks might guide health authorities in determining where to focus surveillance efforts and use preventive measures. Since adaptation is one of the main ways to prevent the effects of climate change on tick-borne pathogens, greater knowledge of the impact of climate on the tick life cycle will refine the available information.

References

Clow, K.M., Ogden, N.H., Lindsay, L.R., Michel, P., Pearl, D.L. and Jardine, C.M. (2017) The influence of abiotic and biotic factors on the invasion of *Ixodes scapularis* in Ontario, Canada. *Ticks and Tick-borne Diseases* 8, 554–563.

Danielová, V., Rudenko, N., Daniel, M., Holubová, J., Materna, J., *et al.* (2006) Extension of *Ixodes ricinus* ticks and agents of tick-borne diseases to mountain areas in the Czech Republic. *International Journal of Medical Microbiology* 296, 48–53.

Estrada-Peña, A. and Estrada-Sánchez, D. (2014) Deconstructing *Ixodes ricinus*: a partial matrix model allowing mapping of tick development, mortality and activity rates. *Medical and Veterinary Entomology* 28, 35–49.

Fernández-Ruiz, N. and Estrada-Peña, A. (2020) Could climate trends disrupt the contact rates between *Ixodes ricinus* (Acari, Ixodidae) and the reservoirs of *Borrelia burgdorferi* s.l.? *PLoS ONE* 15, e0233771.

Hvidsten, D., Frafjord, K., Gray, J.S., Henningsson, A.J., Jenkins, A., *et al.* (2020) The distribution limit of the common tick, *Ixodes ricinus*, and some associated pathogens in north-western Europe. *Ticks and Tick-borne Diseases* 11, 101–108.

Jaenson, T.G., Jaenson, D.G., Eisen, L., Petersson, E. and Lindgren, E. (2012) Changes in the geographical distribution and abundance of the tick *Ixodes ricinus* during the past 30 years in Sweden. *Parasites & Vectors* 5, 8.

Minigan, J.N., Hager, H.A., Peregrine, A.S. and Newman, J.A. (2017) Current and potential future distribution of the American dog tick (*Dermacentor variabilis*, Say) in North America. *Ticks and Tick-borne Diseases* 9, 354–362.

Monzón, J.D., Atkinson, E.G., Henn, B.M. and Benach, J.L. (2016) Population and evolutionary genomics of *Amblyomma americanum*, an expanding arthropod disease vector. *Genome Biology and Evolution* 8, 1351–1360.

Nah, K., Bede-Fazekas, Á., Trájer, A.J. and Wu, J. (2020) The potential impact of climate change on the transmission risk of tick-borne encephalitis in Hungary. *BMC Infectious Diseases* 20, 34.

Sonenshine, D.E. (2018) Range expansion of tick disease vectors in North America: implications for spread of tick-borne disease. *International Journal of Environmental Research and Public Health* 15, 478.

Springer, Y.P., Jarnevich, C.S., Barnett, D.T., Monaghan, A.J. and Eisen, R.J. (2015) Modelling the present and future geographic distribution of the lone star tick, *Amblyomma americanum* (Ixodida: Ixodidae), in the continental United States. *American Journal of Tropical Medicine and Hygiene* 93, 875–890.

Sun, H., Chen, Y., Gong, A., Zhao, X., Zhan, W. and Wang, M. (2014) Estimating mean air temperature using MODIS day and night land surface temperatures. *Theoretical and Applied Climatology* 118, 81–92.

Tälleklint, L. and Jaenson, T.G. (1998) Increasing geographical distribution and density of *Ixodes ricinus* (Acari: Ixodidae) in central and northern Sweden. *Journal of Medical Entomology* 35, 521–526.

eo35 Climate and Prediction of Tick-Borne Diseases Facing the Complexity of the Pathogen–Tick–Host Triad at Northern Latitudes

Atle Mysterud*
University of Oslo, Oslo, Norway

35.1 Expert Opinion

35.1.1 Introduction

Climate change has the potential to drastically alter the ecology of vector-borne zoonoses. Tick-borne diseases are among the most important group of emerging infectious diseases under global warming (Lindgren *et al.*, 2012). Climate change is stronger in northern areas compared with the global average, and this has already led to considerably shorter and milder winters. Climatic conditions are crucial for the emergence of tick-borne diseases, and the number of cases of tick-borne diseases is increasing in Europe and North America (Piesman and Gern, 2004). These are the kind of general statements often made when predicting the future of tick-borne diseases under global warming. But what is the actual evidence and are there any caveats?

35.1.2 Evidence for climate change impacts

In Northern Europe, tick-borne diseases are mainly linked to the sheep tick, *Ixodes ricinus*, a generalist species with a wide host range. Ticks are ectothermic organisms and therefore their development is extremely dependent on temperature. Ticks are most vulnerable to extreme weather during questing. The lower limit to tick questing activity is about 5°C (Qviller *et al.*, 2014). Questing activity of *I. ricinus* ticks from different latitudes is adapted to local climatic conditions (Gilbert *et al.*, 2014), but the effect is limited. Warming is therefore expected to have positive effects on tick abundance at northern latitudes. At the higher temperature range, ticks are prone to desiccation as seen in southern distribution ranges. In a northern area of Norway with humid coastal climate, we were unable to measure any impact of humidity on questing activity of *I. ricinus* (Qviller *et al.*, 2014). Humidity will possibly play a role in more inland areas of Scandinavia where summer drought is an issue of increasing concern. At northern latitudes and high elevation, warming comes in the form of earlier spring and later autumn, which will in general benefit ticks. There are some studies on how vector phenology is linked to annual climate variability, both in North America for *Ixodes scapularis* (Clow *et al.*, 2017) and in Europe for *I. ricinus* (Medlock *et al.*, 2013). *I. ricinus* is moving up in elevation in the Alps (Garcia-Vozmediano

*Email: atle.mysterud@ibv.uio.no

DOI: 10.1079/9781789249637.0035

et al., 2020), as well as towards higher elevation and more northern latitudes of Scandinavia (Jore *et al.*, 2011). These are patterns that are documented to some extent, although long-term data on tick abundances are limited and various sources are used for inference (Jore *et al.*, 2011).

Predicting the climate niche of the tick vector is therefore what is usually done when predicting the distribution for tick-borne diseases. An updated overview of development of climate indices and length of growing season for northern areas of Europe was recently presented (van Oort *et al.*, 2020). However, there are a number of other processes that may affect tick-borne disease load in humans and animals under climate change. Very rarely is the effect of host populations explicitly included in assessments of tick distribution (Fernández-Ruiz and Estrada-Peña, 2020), although host populations are often mentioned as important or assumed related to broad habitat classes.

35.1.3 Evidence for other drivers of change

The complex epidemiological cycle involving the pathogen–vector–host triad can be affected by multiple drivers. This gives the potential for limitations of disease that may counter or increase the effect of climate change on ticks. Host distribution in general does not limit tick distribution but affects the circulation of pathogens (Estrada-Peña and de la Fuente, 2017). Climate change has markedly altered the dynamics of some small mammalian host species with Europe-wide dampening of population cycles (Cornulier *et al.*, 2013), as well as bird (Both *et al.*, 2006) and deer (Rivrud *et al.*, 2019) migratory patterns. In general, the importance of indirect effects through host populations is strongly affected by land use and other limiting factors apart from climate. Deer are required for the reproduction of adult ticks and can limit tick numbers. The effect on tick populations is mainly from an absence to low population density of deer, while variation in deer densities at medium to higher population density levels has limited impact (Mysterud *et al.*, 2016). Deer numbers are strongly affected by human management and land use. Roe deer and red deer are at their northern distribution ranges and will likely experience more favourable habitat due to climate change, while habitat suitability for moose will decrease with warming. The effect of predation by red fox (*Vulpes vulpes*), lynx (*Lynx lynx*) and wolf (*Canis lupus*) on roe deer can be marked. Red fox in semi-urban habitat in the Netherlands lowered larval tick load on rodents (Hofmeester *et al.*, 2017). The large carnivore populations have increased in recent decades but are under fairly strict management regulations. My opinion is that expansion of carnivore populations, important to some tick hosts, is expected to have a fairly minor impact on prevalence of tick-borne disease.

35.1.4 Likely scenario

The sum of these factors is difficult to predict, but we have direct evidence from long-term tick-borne disease statistics for the last three to four decades. We see an emergence of tick-borne encephalitis (TBE) in Finland (Uusitalo *et al.*, 2020) and Norway (Andreassen *et al.*, 2012). Lyme disease is emerging in both Finland (Sajanti *et al.*, 2017) and Norway (Mysterud *et al.*, 2016). In Norway, we found a temporal increase in endemic areas and a spatial expansion in all four livestock and human tick-borne disease records: babesiosis in cattle, anaplasmosis in sheep and cattle, and Lyme disease in humans (Mysterud *et al.*, 2017). These long-term data provide strong evidence that all diseases were limited by the common tick vector, as the pathogens are linked to different transmission hosts, and predictions based on climate niches of *I. ricinus* are likely to give the bigger picture (Jaenson and Lindgren, 2011).

35.2 Future Predictions

The effect of warming on tick-borne diseases is complicated by the wide host range of ticks that differ across the life cycle (Fig. 35.1) and by the fact that hosts may also be affected by climate changes, affecting host abundances and overlap with ticks (Fig. 35.2). There are a number of critical questions for any future predictions of ecosystem function. Some obvious limitations

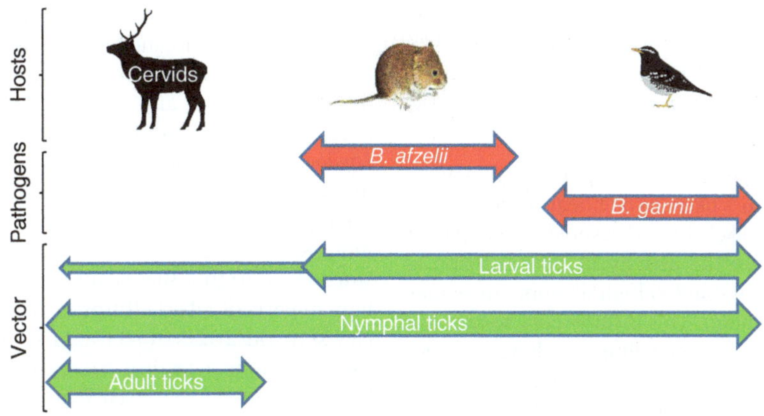

Fig. 35.1. An example of the complexity in regulation of tick-borne disease involving a pathogen–tick–host triad. The tick vector *Ixodes ricinus* has a wide biotic niche, but niche varies across life stages. The pathogens causing Lyme borreliosis have a more restricted niche; *Borrelia afzelii* has a reservoir in small mammals, while *Borrelia garinii* is bird-borne. This variation in host range across vector life stages and pathogens creates a complex host limitation of Lyme borreliosis, potentially causing indirect, host-mediated effects of climate and a role for other drivers. (Author's own figure.)

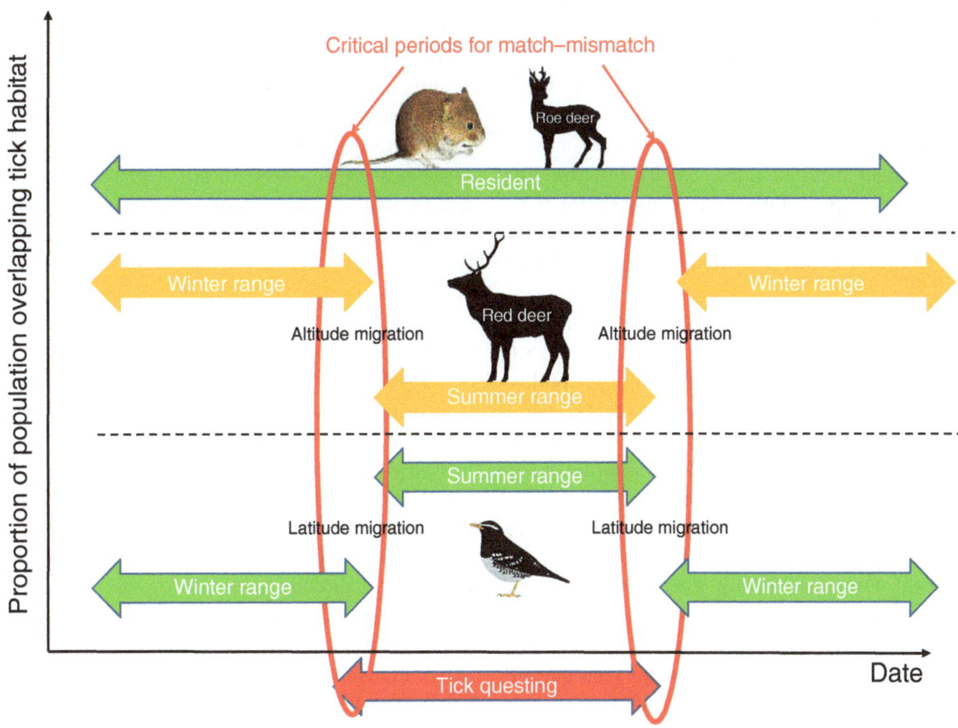

Fig. 35.2. Vertebrate hosts of pathogens and ticks are limited by a number of factors, including climate. Climate change will affect not only hosts' abundances, but also their seasonal overlap with tick questing activity. Deer are altitudinal migrants, while important bird transmission hosts are latitudinal migrants. This may cause increased ('match') or decreased ('mismatch') overlap with the questing activity of ticks, affecting the likelihood of successful feeding and transmission of pathogens. (Author's own figure.)

are: (i) uncertainties about the form of the link to drivers; (ii) uncertainties about the development of known drivers (to hosts); (iii) limiting factors of hosts may change abruptly; and (iv) new tick species may invade.

1. Non-linearities. Are responses to known limiting factors likely to be linear or non-linear, i.e. are there likely to be thresholds? Examples of thresholds are the activity of ticks starting at about 5°C and that too high a temperature may cause desiccation. This is fairly easy to account for in predictive modelling, as long as non-linearities are known and accounted for. Another point is the synchrony of (co-)feeding of larvae and nymphs that also may depend on temperature and has marked effects on the transmission of both *Borrelia burgdorferi* sensu lato (s.l.) and TBE virus (Randolph *et al.*, 1996).

2. Known host drivers. Our understanding of the role of hosts as drivers of tick-borne diseases has advanced in recent years. However, there are data limitations on all host populations that must be dealt with in explicit modelling. There are few studies providing quantitative predictions of future distributions of hosts linked explicitly to climate change scenarios. One study was undertaken recently for red deer (Rivrud *et al.*, 2019), but this has not been coupled with similar models for ticks and other hosts. Development of other drivers determining host distribution and abundance is uncertain.

3. Change of host limitations. A few hosts dominate as transmission hosts for *B. burgdorferi* s.l. (Hofmeester *et al.*, 2016) and most other pathogens. Disease outbreaks among transmission hosts appear the least predictable and potentially strongest factor potentially changing the trajectory of tick-borne disease incidences. For example, there is an ongoing emergence since 2001 of usutu virus among blackbirds in continental Europe, causing occasional massive die-offs (Ashraf *et al.*, 2015). This may lead to a (short-term?) absence of one of two or three important transmission hosts for *Borrelia garinii*. The bird-borne *B. garinii* is most often the cause of neuroborreliosis (Barstad *et al.*, 2018) and cases of neuroborreliosis dominate in the statistics of more severe, disseminated cases of Lyme disease (Mysterud *et al.*, 2019).

4. New tick species. The taiga tick *Ixodes persulcatus* has invaded Finland from the east (Pakanen *et al.*, 2020) and was recently discovered in Sweden (Jaenson and Wilhelmsson, 2019). Also, ticks with a southern distribution, like *Hyalomma marginatum*, may expand northwards in the longer term. Hybrids of the taiga tick and the Far Eastern tick (*Ixodes pavlovskyi*) are common (Rar *et al.*, 2019). Newspapers reported in summer 2020 that these caused a surge of tick bites in parts of Siberia (https://www.dtnext.in/News/World/2020/06/03003131/1233833/Russias-Siberia-region-hit-by-invasion-of-bloodsucking-.vpf, accessed 7 June 2021). New tick species or hybrids may introduce new pathogens and affect the circulation of endemic pathogens (Rar *et al.*, 2019).

All the listed uncertainties limit the value of detailed predictions based solely on climate effects on tick populations, but this is the best we have as of today. I consider patterns of exposure, vaccination and (early) treatment as outside the scope of this expert opinion, although it will certainly affect the disease statistics.

In the short term, using the previous decade to predict the next decade will give a fair approximation. The trend will be a further increase of all tick-borne diseases, but number of cases will be strongly dependent on annual variation in climate. Given the certainty of further warming, and likelihood of increases rather than decreases in main host populations, it is possible to make such a general, qualitative prediction with a fair level of confidence.

Predictions 50 years ahead are obviously very difficult due to the many possible changes in a suite of drivers. Denser populations of ticks in many regions will give further increases in Lyme disease. With increasing elevation of ticks and further expansion of red deer, anaplasmosis may become a more prominent problem for grazing sheep at somewhat higher elevations. The pathogens are differently affected by temperature. Similarly to many other viruses, the transmission of TBE virus is the most directly affected by temperature (Randolph *et al.*, 2000), beyond the effect of temperature on ticks. TBE is likely to increase more markedly compared with other tick-borne diseases.

References

Andreassen, A., Jore, S., Cuber, P., Dudman, S., Tengs, T., *et al.* (2012) Prevalence of tick-borne encephalitis virus in tick nymphs in relation to climatic factors on the southern coast of Norway. *Parasites & Vectors* 5, 177.

Ashraf, U., Ye, J., Ruan, X., Wan, S., Zhu, B. and Cao, S. (2015) Usutu virus: an emerging flavivirus in Europe. *Viruses* 7, 219–238.

Barstad, B., Quarsten, H., Tveitnes, D., Noraas, S., Ask, I.S., *et al.* (2018) Direct molecular detection and genotyping of *Borrelia burgdorferi* sensu lato in the cerebrospinal fluid of children with Lyme neuroborreliosis. *Journal of Clinical Microbiology* 56, e01868-17.

Both, C., Bouwhuis, S., Lessells, C.M. and Visser, M.E. (2006) Climate change and population declines in a long-distance migratory bird. *Nature* 441, 81–83.

Clow, K.M., Leighton, P.A., Ogden, N.H., Lindsay, L.R., Michel, P., *et al.* (2017) Northward range expansion of *Ixodes scapularis* evident over a short timescale in Ontario, Canada. *PLoS ONE* 12, e0189393.

Cornulier, T., Yoccoz, N.G., Bretagnolle, V., Brommer, J.E., Butet, A.E., *et al.* (2013) Europe-wide dampening of population cycles in keystone herbivores. *Science* 340, 63–66.

Estrada-Peña, A. and de la Fuente, J. (2017) Host distribution does not limit the range of the tick *Ixodes ricinus* but impacts the circulation of transmitted pathogens. *Frontiers in Cellular and Infection Microbiology* 7, 405.

Fernández-Ruiz, N. and Estrada-Peña, A. (2020) Could climate trends disrupt the contact rates between *Ixodes ricinus* (Acari, Ixodidae) and the reservoirs of *Borrelia burgdorferi* s.l.? *PLoS ONE* 15, e0233771.

Garcia-Vozmediano, A., Krawczyk, A.I., Sprong, H., Rossi, L., Ramassa, E. and Tomassone, L. (2020) Ticks climb the mountains: ixodid tick infestation and infection by tick-borne pathogens in the Western Alps. *Ticks and Tick-borne Diseases* 11, 101489.

Gilbert, L., Aungier, J. and Tomkins, J.L. (2014) Climate of origin affects tick (*Ixodes ricinus*) host-seeking behaviour in response to temperature: implications for resilience to climate change? *Ecology and Evolution* 4, 1186–1198.

Hofmeester, T.R., Coipan, E.C., Van Wieren, S.E., Prins, H.H.T., Takken, W. and Sprong, H. (2016) Few vertebrate species dominate the *Borrelia burgdorferi* s.l. life cycle. *Environmental Research Letters* 11, 043001.

Hofmeester, T.R., Jansen, P.A., Wijnen, H.J., Coipan, E.C., Fonville, M., *et al.* (2017) Cascading effects of predator activity on tick-borne disease risk. *Proceedings of the Royal Society B* 284, 20170453.

Jaenson, T.G.T. and Lindgren, E. (2011) The range of *Ixodes ricinus* and the risk of contracting Lyme borreliosis will increase northwards when the vegetation period becomes longer. *Ticks and Tick-borne Diseases* 2, 44–49.

Jaenson, T.G.T. and Wilhelmsson, P. (2019) First records of tick-borne pathogens in populations of the taiga tick *Ixodes persulcatus* in Sweden. *Parasites & Vectors* 12, 559.

Jore, S., Viljugrein, H., Hofshagen, M., Brun-Hansen, H., Kristoffersen, A., *et al.* (2011) Multi-source analysis reveals latitudinal and altitudinal shifts in range of *Ixodes ricinus* at its northern distribution limit. *Parasites & Vectors* 4, 84.

Lindgren, E., Andersson, Y., Suk, J.E., Sudre, B. and Semenza, J.C. (2012) Monitoring EU emerging infectious disease risk due to climate change. *Science* 336, 418–419.

Medlock, J.M., Hansford, K.M., Bormane, A., Derdakova, M., Estrada-Peña, A., *et al.* (2013) Driving forces for changes in geographical distribution of *Ixodes ricinus* ticks in Europe. *Parasites & Vectors* 6, 1.

Mysterud, A., Easterday, W.R., Stigum, V.M., Aas, A.B., Meisingset, E.L. and Viljugrein, H. (2016) Contrasting emergence of Lyme disease across ecosystems. *Nature Communications* 7, 11882.

Mysterud, A., Jore, S., Østerås, O. and Viljugrein, H. (2017) Emergence of tick-borne diseases at northern latitudes in Europe: a comparative approach. *Scientific Reports* 7, 16316.

Mysterud, A., Heylen, D.J.A., Matthysen, E., Garcia, A.L., Jore, S. and Viljugrein, H. (2019) Lyme neuroborreliosis and bird populations in northern Europe. *Proceedings of the Royal Society B* 286, 20190759.

Pakanen, V.M., Sormunen, J.J., Sippola, E., Blomqvist, D. and Kallio, E.R. (2020) Questing abundance of adult taiga ticks *Ixodes persulcatus* and their *Borrelia* prevalence at the north-western part of their distribution. *Parasites & Vectors* 13, 384.

Piesman, J. and Gern, L. (2004) Lyme borreliosis in Europe and North America. *Parasitology* 129, S191–S220.

Qviller, L., Grøva, L., Viljugrein, H., Klingen, I. and Mysterud, A. (2014) Temporal pattern of questing tick *Ixodes ricinus* density at differing elevations in the coastal region of western Norway. *Parasites & Vectors* 7, 179.

Randolph, S.E., Gern, L. and Nuttall, P.A. (1996) Co-feeding ticks: epidemiological significance for tick-borne pathogen transmission. *Parasitology Today* 12, 472–479.

Randolph, S.E., Green, R.M., Peacey, M.F. and Rodgers, D.J. (2000) Seasonal synchrony: the key to tick-borne encephalitis foci identified by satellite data. *Parasitology* 121, 15–23.

Rar, V., Livanova, N., Sabitova, Y., Igolkina, Y., Tkachev, S., *et al.* (2019) *Ixodes persulcatus/pavlovskyi* natural hybrids in Siberia: occurrence in sympatric areas and infection by a wide range of tick-transmitted agents. *Ticks and Tick-borne Diseases* 10, 101254.

Rivrud, I.M., Meisingset, E.L., Loe, L.E. and Mysterud, A. (2019) Future suitability of habitat in a migratory ungulate under climate change. *Proceedings of the Royal Society B* 286, 20190442.

Sajanti, E., Virtanen, M., Helve, O., Kuusi, M., Lyytikäinen, O., *et al.* (2017) Lyme borreliosis in Finland, 1995–2014. *Emerging Infectious Diseases* 23, 1282–1288.

Uusitalo, R., Siljander, M., Dub, T., Sane, J., Sormunen, J.J., *et al.* (2020) Modelling habitat suitability for occurrence of human tick-borne encephalitis (TBE) cases in Finland. *Ticks and Tick-borne Diseases* 11, 101457.

van Oort, B.E.H., Hovelsrud, G.K., Risvoll, C., Mohr, C.W. and Jore, S. (2020) A mini-review of *Ixodes* ticks climate sensitive infection dispersion risk in the Nordic region. *International Journal of Environmental Research and Public Health* 17, 5387.

eo36 Is the Clock 'Ticking' for Climate Change?

Ernest Gould* and Xavier de Lamballerie
University of Aix-Marseille, Marseille, France

36.1 Expert Opinion

36.1.1 Introduction

Yes, the clock is 'ticking' for climate change and there is little time to arrest and reverse the negative impact of global warming on the Earth's delicately balanced biotic and abiotic interactions. Here, we discuss the potential impact that climate change might have on the ticks that are the transmission vectors for the human and animal disease known as tick-borne encephalitis (TBE). This disease is caused by a group of related viruses collectively known as the tick-borne encephalitis virus (TBEV) serocomplex in the virus family *Flaviviridae*. In their natural habitat, forests and/or rough pastureland, these viruses infect humans, small and large domestic and wild animals, reptiles and birds. The hard-bodied ticks, *Ixodes ricinus* and *Ixodes persulcatus*, are the primary vectors of TBEV in Western Europe and Central Europe/Eurasia, respectively (Gritsun *et al.*, 2003). Other related tick-borne flaviviruses follow similar life cycles, transmission and evolutionary characteristics; in many cases they overlap geographically with the TBEV serocomplex viruses even though they may be primarily vectored by different ixodids.

Climate change is largely driven by the consequences of modern anthropological activities. These include industrialization, deforestation, the use of chlorofluorocarbons and other halons, and the exploitation of fossil fuels which generate heat-trapping emissions consequently accumulating excess carbon dioxide, nitrous oxide and other greenhouse gases. Their release into the stratosphere depletes the protective ozone layer, resulting in increased exposure of the Earth to ultraviolet-B (UVB) rays. In a worst-case scenario, the rate of melting of the ice at the Arctic and Antarctic poles will continue to accelerate, increasing the risk of flooding, increasing production and release of greenhouse gases, further depleting the protective ozone layer, and increasing the numbers of thermokarst lakes in Siberia and the Canadian tundra. Thermokarst lakes arise as the result of permafrost ice melting. The microorganisms in the relatively warmer underlying sediment ferment, producing methane gas which escapes to the atmosphere as a greenhouse gas, again contributing to further depletion of the ozone layer. These feedback processes and the release of greenhouse gases accelerate global warming and, in the opinion of some experts, may be reaching a point (the tipping point) where they could become irreversible.

*Email: eagvirology@gmail.com

DOI: 10.1079/9781789249637.0036

36.1.2 The protracted tick life cycle and its significance as a survival strategy

Figure 36.1 outlines the protracted life cycle of *I. ricinus* and *I. persulcatus* (family Ixodidae), the primary vectors of the viruses in the TBEV sero-complex, most of which circulate in these ticks in arboreal regions of the northern hemisphere in Europe and Eurasia (Zanotto *et al.*, 1995). Ticks use a sensory organ (Haller's organ) to identify when the environmental conditions are ideal to climb up the vegetation and quest for a blood meal by detecting the radiant heat from passing vertebrate hosts (Carr and Salgado, 2019). The life cycle starts when tick eggs are laid by adult females, usually in the litter on the humid soil mat, protected from dehydration by the natural covering of dense vegetation. On warm days, in spring, the eggs hatch releasing thousands of larvae that ascend the stalks of overlying vegetation and wait to ambush small forest animals from which they take a blood meal which may last up to 4 days. When replete, they drop off the animal and seek cover in the soil under the vegetation. Depending on the climate and time of the year, the larvae then moult to nymphs which subsequently quest, feed and moult before producing the next generation of adult ticks, thus completing the life cycle by feeding and then laying the subsequent generation of eggs. This life cycle may take from 2 to 5 years depending on the climatic conditions. During the blood-feeding process, larvae, nymphs or adults may become infected by a strain of TBEV that is infecting the animal host on which they are feeding. This virus replicates in the infected tick and may be transferred to a non-infected animal via a subsequent tick feeding stage. During its life cycle, the tick will spend endless hours maintaining an optimal level of hydration by extracting water from the humid (>80%) soil, underneath the vegetative cover. The ticks that are infected by virus may remain infected throughout the entire life cycle and in the case of adult females, the virus may even be transferred to a proportion of the eggs (vertical transmission) which, when they hatch as infectious larvae, can potentially initiate the next life cycle of the virus. Thus, this protracted life cycle has four significant elements that increase the likelihood of long-term tick and virus survival. First, the eggs can survive for months in the protective humid soil mat. Second, eggs may be infected by adult female ticks and a proportion of these may

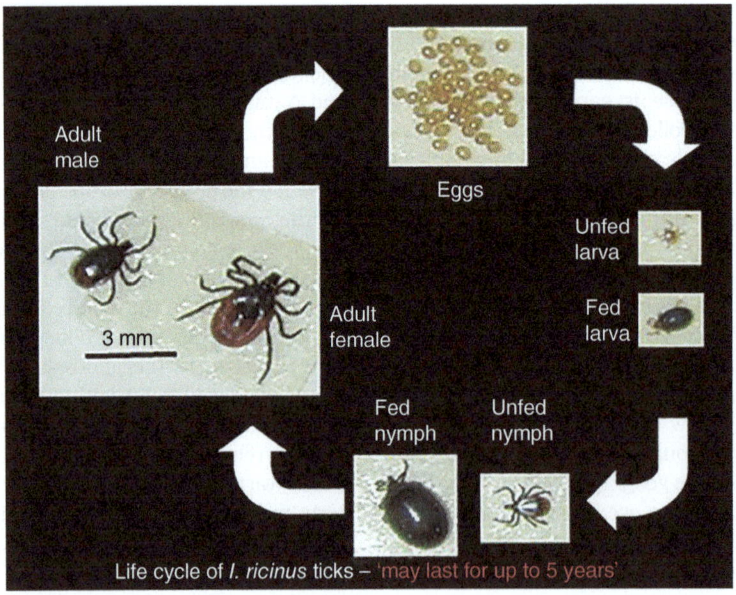

Fig. 36.1. Life cycle of *Ixodes ricinus* represented by the individual feeding and moulting stages through which the tick progresses from the egg to the engorged female adult tick. The white arrows indicate the moulting stages. (Authors' own figure.)

remain infected through each of the successive stages of the tick life cycle, aided by amplification of the virus during non-viraemic transmission (Nuttall *et al.*, 1994). Third, ticks can exploit the hydration process to retain viability for extended periods of time in the soil mat and finally, the relatively long period of 3 to 4 days while the ticks feed ensures a high likelihood of the infected tick being carried by the host either locally, within the forest, or to other climatic regions if the host is a bird (Carr and Salgado, 2019).

36.1.3 Virus evolution, dispersal and transmission characteristics

Figure 36.2 depicts a phylogenetic tree representing the TBEV serocomplex and related flaviviruses, overlaid on to an outline of the world, with the objective of displaying the biogeographic dispersal of the viruses from their presumed evolutionary origin. Kadam virus (KADV), the ancestral African lineage, and subsequent descendant lineages dispersed eastwards into Arabia and Asia before diverging and dispersing to northern and far-east Asia, westwards to Europe and northwards to Scandinavia. Gadgets Gully virus (GGYV) appears to have dispersed from Africa or Asia to Macquarie Island in the south-west Pacific Ocean, presumably carried in an infected tick(s) by a migratory bird(s). For the most part, this virus' dispersal and evolution pattern traces the arboreal forests and moorlands of Europe and Asia (Zanotto *et al.*, 1995), thus ensuring the essential vegetative cover, temperature and humidity required by the ticks to complete their life cycles (Gould *et al.*, 2003; Sumilo *et al.*, 2006). The viruses have evolved at least three strategies to ensure transmission, and therefore to increase their long-term survival probability, under a variety of environmental conditions. Viraemic transmission requires vertebrate hosts that are significantly susceptible to virus infection. The viraemia ensures a high level of virus in the blood meal taken by the feeding tick and thus a high likelihood of infection during feeding on the host. In contrast, non-viraemic transmission does not require vertebrate hosts to be highly susceptible to infection; they may, for example, be asymptomatic. When ticks co-feed, the virus may transfer directly between co-feeding ticks in the host (Jones *et al.*, 1987; Labuda *et al.*, 1993; Randolph *et al.*, 1999). This mode of infection increases the diversity of potential hosts available to the ticks and viruses. Third, eggs infected transovarially by the female adult tick hatch as infectious larvae. Under ideal conditions, infectious eggs may remain viable for months or even years, before hatching to produce larvae, potentially ensuring long-term survival of the virus.

36.2 Future Predictions

36.2.1 The next 50 years (2021–2071)

The global temperature has increased by an average rate of 0.18°C per decade since 1981 (NOAA/NCEI, 2020). If this rate continues for the next 50 years, the average global temperature will have increased by 0.9°C in 2071. Extreme weather events will also have increased significantly, including drought, hurricane-strength winds, further melting of the ice caps with associated flooding, and increasing release of carbon dioxide, nitrous oxide and methane. These predictions assume that the extent of global deforestation and exploitation of polluting industrial products will continue unabated unless positive progress is made internationally to focus on specific objectives that can alleviate the extreme weather events that have occurred during recent years.

36.2.2 What is the significance of tick-borne encephalitis virus dispersal in the context of climate change?

With the single exception of Powassan virus (POWV) which circulates in Russia, China, Canada and the USA, the TBEV serocomplex viruses are found only in the Old World (Fig. 36.2). Nevertheless, they have been isolated from Africa, Arabia, India, southern, northern and far-east Asia, the northern and southern mainland of Europe and Eurasia, Scandinavia and the British Isles (Fig. 36.2). Thus, these viruses have evolved and dispersed over a wide range of

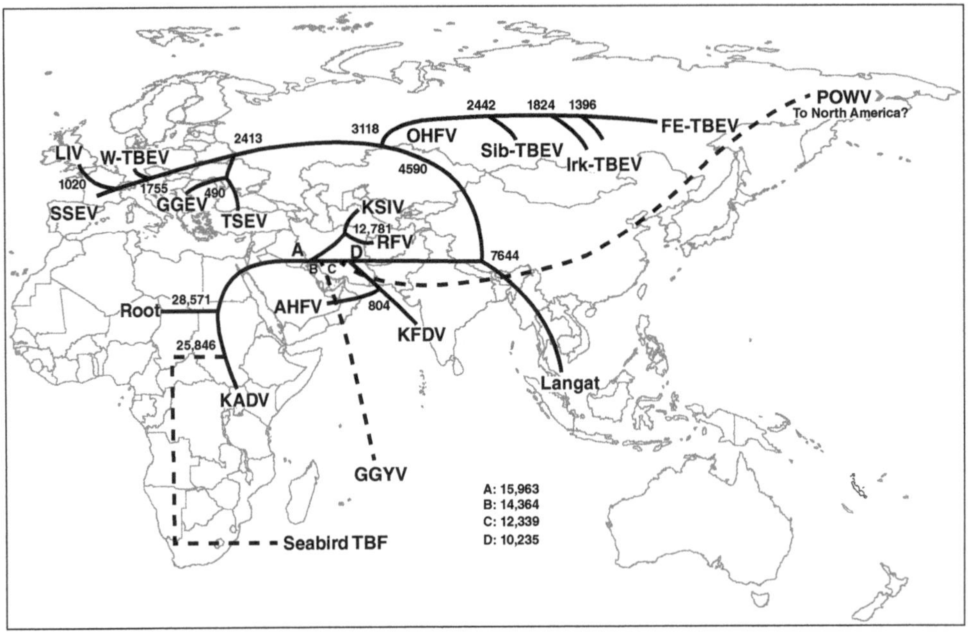

Fig. 36.2. Possible model for dispersal of tick-borne encephalitis virus (TBEV) serocomplex viruses. A maximum clade credibility tree was manually plotted on to a global map, showing the most parsimonious route of dispersal for viruses based on complete genomic sequences. Approximate times for the most recent common ancestor (tMRCA), in years, are shown at all major nodes in the tree. Viruses included in the TBEV-related group: louping ill virus (LIV), Western European TBEV (W-TBEV), Siberian TBEV (Sib-TBEV), Irkutsk TBEV (Irk-TBEV), Far East Asian TBEV (FE-TBEV), Greek goat encephalitis virus (GGEV), Turkish sheep encephalomyelitis virus (TSEV), Spanish sheep encephalomyelitis virus (SSEV), Omsk haemorrhagic fever virus (OHFV), Langat (LGTV), Karshi virus (KSIV), Royal Farm virus (RFV), Kyasanur forest disease virus (KFDV), Alkhurma haemorrhagic fever virus (AHFV), Powassan virus (POWV), Gadgets Gully virus, (GGYV), Kadam virus (KADV), seabird TBF (seabird-tick-borne flaviviruses). (Reproduced with permission from Heinze *et al.*, 2012.)

temperatures, from the warm tropics of Africa, India and southern Asia to the very cold permafrost of Siberia. In other words, this group of viruses has shown a high level of adaptability to ticks that have also adapted to very different ecological environments and climates. Moreover, ticks have survived extreme weather conditions and evolved for hundreds of millions of years since they first emerged. The scientific evidence also suggests that, at least for the next five decades, the predicted increased temperatures and shorter/milder winters will lead to increased tick population densities and associated TBE in more northerly regions of Europe and Eurasia and at higher altitudes (Daniel *et al.*, 2003; Tokarevich *et al.*, 2011; Frimmel *et al.*, 2014; Randolph, 2019). In further support of these observations, record levels of tick bites and human TBE at altitudes of up to 600 m have been recorded very recently in mountainous areas of southern Germany and neighbouring regions of Austria, Switzerland and France (reported on ProMed-mail; http://promedmail.org/post/20200907.7755195, accessed 7 June 2021).

36.2.3 What will happen beyond 50 years?

One can only speculate what will happen further into the future. However, factors additional to temperature will become more critical determinants. For example, as the ozone layer continues to be depleted by the impact of greenhouse gas accumulation, the rate of ice melting at the poles will accelerate, thus generating more greenhouse gases, particularly methane, via fermentation, depleting the ozone layer even further, and flooding of lowland areas will also increase. However, unless the rates of temperature increase and extreme events accelerate beyond current estimates, it is likely that under the protection of the Northern European/Eurasian forests, TBEV and related viruses would survive for at least 50 years, commencing from 2020. But, if methane release from the melting permafrost and the thermokarst lakes reaches sufficiently high levels, the Earth's atmosphere at or near the Arctic/Antarctic boundaries could become so polluted that all or most forms of life would not survive in those areas of the globe. Whether or not mankind can alleviate such potential catastrophes remains to be seen.

36.2.4 Can climate change be reversed?

In theory climate change could be reversed if appropriate action is taken in time, although there is a body of opinion that argues we may be too late. We already understand the basic causes of global warming: anthropological activities generate greenhouse gases; deforestation reduces the capacity of the planet to absorb carbon dioxide; industrialization, transportation and the use of fossil fuels generate greenhouse gases, which deplete ozone; the depleted ozone layer results in increased UVB penetrating the atmosphere, raising the temperature and melting the permafrost, which generates more greenhouse gases. This is known as a feedback process. Now, more than ever, the major industrial nations must increase their efforts to work together to continue to develop and put in place innovative and effective measures that will reverse these damaging processes. Measures must also be taken to reduce the risk of major disasters such as forest fires and flooding, both of which will continue to increase in severity and contribute to the accumulation of more greenhouse gases, if nothing is done to control climate change.

In brief, ticks have spent millions of years adapting strategies to survive extreme weather events and major global catastrophes. While many other life forms may be less-well prepared for climate change, it seems likely that at least for the next 50 to 100 years, TBEVs and their associated vectors and forest hosts will survive, probably in the northern regions of Europe and Eurasia. However, if global warming does accelerate significantly beyond its current rate of temperature increase, the estimate of 50–100 years for survival of ticks and TBEVs may need to be reduced.

Disclaimer

The findings and conclusions of this expert opinion are those of the authors and do not necessarily represent the views of the University of Aix-Marseille, France.

References

Carr, A.L. and Salgado, V.L. (2019) Ticks home in on body heat: a new understanding of Haller's organ and repellent action. *PLoS ONE* 14, e0221659.

Daniel, M., Danielová, V., Kríz, B., Jirsa, A. and Nozicka, J. (2003) Shift of the tick *Ixodes ricinus* and tick-borne encephalitis to higher altitudes in central Europe. *European Journal of Clinical Microbiology and Infectious Diseases* 22, 327–328.

Frimmel, S., Krienke, A., Riebold, D., Loebermann, M., Littmann, M., *et al.* (2014) Tick-borne encephalitis virus habitats in North East Germany: reemergence of TBEV in ticks after 15 years of inactivity. *BioMed Research International* 2014, 308371.

Gould, E.A., de Lamballerie, X., Zanotto, P.M. and Holmes, E.C. (2003) Origins, evolution, and vector/host coadaptations within the genus *Flavivirus*. *Advances in Virus Research* 59, 277–314.

Gritsun, T.S., Lashkevich, V.A. and Gould, E.A. (2003) Tick-borne encephalitis. *Antiviral Research* 57, 129–146.

Heinze, D.A., Gould, E.A. and Forrester, A.L. (2012) Revisiting the clinal concept of evolution and dispersal for the tickborne flaviviruses by using phylogenetic and biogeographic analyses. *Journal of Virology* 86, 8663–8671.

Jones, L.D., Davies, C.R., Steele, G.M. and Nuttall, P.A. (1987) A novel mode of arbovirus transmission involving a nonviremic host. *Science* 237, 775–777.

Labuda, M., Nuttall, P.A., Kozuch, O., Eleckova, E., Williams, T., *et al.* (1993) Non-viraemic transmission of tick-borne encephalitis virus: a mechanism for arbovirus survival in nature. *Experientia* 49, 802–805.

NOAA/NCEI (2020) State of the Climate: Global Climate Report for Annual 2019. National Oceanic and Atmospheric Administration/National Centers for Environmental Information, Asheville, North Carolina. Available at: https://www.ncdc.noaa.gov/sotc/global/201913 (accessed 7 June 2021).

Nuttall, P.A., Jones, L.D., Labuda, M. and Kaufman, W.R. (1994) Adaptations of arboviruses to ticks. *Journal of Medical Entomology* 31, 1–9.

Randolph, S.E. (2019) To what extent has climate change contributed to the recent epidemiology of tick-borne diseases? *Veterinary Parasitology* 167, 92–94.

Randolph, S.E., Miklisova, D., Lysy, J., Rogers, D.J. and Labuda, M. (1999) Incidence from coincidence: patterns of tick infestations on rodents facilitate transmission of tick-borne encephalitis virus. *Parasitology* 118, 177–186.

Sumilo, D., Bormane, A., Loreta, A., Lucenko, I., Vasilenko, V. and Randolph, S.E. (2006) Tick-borne encephalitis in the Baltic States: identifying risk factors in space and time. *International Journal of Medical Microbiology* 296, 76–79.

Tokarevich, N.K., Tronin, A.A., Blinova, O.V., Buzinov, R.V., Boltenkov, V.P., *et al.* (2011) The impact of climate change on the expansion of *Ixodes persulcatus* habitat and the incidence of tickborne encephalitis in the north of European Russia. *Global Health Action* 4, 8448–8458.

Zanotto, P.M., Gao, G.F., Gritsun, T., Marin, M.S., Jiang, W.R., *et al.* (1995) An arbovirus cline across the northern hemisphere. *Virology* 210, 152–159.

eo37 Climate Instability and Emerging Tick-Borne Disease

Janet Foley*
University of California, Davis, California, USA

37.1 Expert Opinion

Tick-borne diseases are emerging worldwide, but more importantly, their distributions are shifting. I like to think of multiple axes for a given emergence event: emergence (spreading) over space, increases in abundance, emergence to take advantage of new hosts and vectors, evolution of new and sometimes more virulent strains. Over the last several decades, some of the most important tick-borne diseases have emerged on some or all of these axes – expansion of Lyme disease and other diseases transmitted by *Ixodes ricinus*-group ticks into afforested areas and regions previously considered too cold (Klepeis *et al.*, 2013; Eisen *et al.*, 2016); expansion of exotic ticks that spread with agricultural products, wildlife trade and international travel (Mack, 2003; Rainey *et al.*, 2018; CDC, 2019); and spillover of animal pathogens into humans (Plowright *et al.*, 2017). What we see of the emergence event may be the 'tip of the iceberg', and details of emergence and underlying causes are often characterized poorly.

Underpinning each emergence event is change; for example, changing land use, ambient or microenvironmental temperature, trade routes or human–animal–vector interactions. Often, the scientific and lay literature tends to imply that emergence occurs linearly, in one direction. For example, with climate change, *Ixodes scapularis* can now occupy land further north and deeper into Canada in areas previously considered unsuitable (Hamer *et al.*, 2010; Clow *et al.*, 2017). Likewise, *I. ricinus* can now transmit disease at higher elevations and higher latitude (Dobson and Randolph, 2011). But climate scientists offer a particularly unsettling and important message: global temperatures will not just increase, rather climate change will induce globally unstable weather and impose societal disruption as a result. This means more hurricanes and severe weather events, more drought, drought and extreme precipitation and wind in new and varied places, hotter hot days, more hot days and sometimes colder cold days.

As tick-borne diseases respond to these changes, they will not necessarily emerge in one obvious direction and they certainly are unlikely to remain as they have been. Globally there already is a dearth of high-quality and well-analysed data on ticks and tick-borne disease. If we lack baseline data, and especially lack data on emerging tick-borne diseases, we will misunderstand the ecologies of tick-borne diseases; we will not have appropriate indices for suspicion of tick-borne disease for ill people and other animals; we will be unable to implement surveillance efficiently; we will fail to some degree in our ability to educate medical professionals; we will inform the

*Email: jefoley@ucdavis.edu

DOI: 10.1079/9781789249637.0037

public poorly; and we will fail to protect people and other animals from potentially lethal threats.

Multiple ecological and physiological characteristics impact how a tick might respond to global climate change: how tolerant a particular species and life stage is of moisture deficits, how able it is to survive and undertake necessary activities (e.g. questing for prey) at appropriate temperatures, whether it can tolerate or avoid freeze–thaw cycles, whether its hosts still have access to the specific niches it needs, and whether the off-host safe sites that ticks used during moulting and diapause are still able to support the tick (Ogden *et al.*, 2020). One of the greatest threats to a tick, besides desiccation, is entomopathogenic fungi which are likely increased in sites with increased moisture (Fernandes *et al.*, 2012). The subject of extreme cold is particularly interesting. Where it is particularly cold, the daytime temperature may be too low to allow a tick to seek a host and imbibe blood, and very cold regions may fail to support necessary tick hosts. However, even something as simple as cold may need to be further parsed: perhaps it is not the average coldness over the year that limits some ticks' distributions, but rather how long cold drags into the spring. Variable weather has been shown to have particularly deleterious impacts on ticks (Herrmann and Gern, 2013). And in between a zone of extreme cold on one end of the spectrum and warmer, suitable regions on the other is an intermediate zone in which freezing cold spells and thaws alternate, a condition most ticks find particularly challenging for survival.

Climate predictions are for rapid change, although the predictive models differ a bit in the details, in part depending on how much carbon is produced, whether mitigation measures are implemented, what part of the world is under consideration and whether or not the Earth is before or after 'tipping points' that mark a switch into an even more dire outcome that may resist intervention (UCAR Center for Science Education, 2020). Mean annual temperatures are predicted to rise and to do so to a greater extent in the north. Overall precipitation is also expected to increase, but, for example, in the New World, that increase will occur in the northern part of North America whereas southern North America and some of South America will experience drought. Unstable weather such as hurricanes, cyclones, heat events, drought and floods will increase in frequency and severity worldwide. At a more local level, more of the USA would have more months free of frost (Global Climate Change: Vital Signs of the Planet, 2020). Even though there will be drought in the south-west, what rain does occur would do so in heavy storms. These changes mirror predictions for much of Europe, with drought to the point of desertification, excessive heat, heavy rainstorms and floods (European Commission, 2020). Africa will be especially hard hit, experiencing more severe, frequent and longer droughts and elevated temperatures (UN Africa Renewal, 2019).

Coping with the impact of climate change on ticks and tick-borne disease will require an artful balance of data management, reliance on predictions and maintaining confidence in the approaches and the results. The current strategies for protecting people and other animals are inequitable, inconsistent and inadequate. Just in the USA, some jurisdictions maintain regular, active surveillance with scientifically valid methodology while other areas also at considerable risk rely on passive surveillance, or what data are available are never analysed. Even though Lyme disease is the most common vector-borne disease of people in North America (Rosenberg *et al.*, 2018), no disease seems to generate as much controversy as to how it produces symptoms, how to treat it and even what the epidemiological features are that lead to severe disease. It might seem that a new paradigm of ecological change as a result of climate change could yield hopeless gridlock at best and denial for the health risk at worst. Such scepticism can also taint attempts to manage trade, livestock health and the food supply, all of which can be disrupted by tick-borne disease. The way forward surely requires teamwork – of course including scientists, but also people who are adept at translating science as well as visionary leaders, intent on better policy.

37.2 Future Predictions

While much health policy in recent decades has been data-driven, there are notable exceptions. For example, extensive research supports that exercise, diet, microbiome, inheritance and other factors all influence body mass and health; yet there are anecdotes which circulate widely with insufficient data on their effects, such as the dangers of gluten, the benefits of turmeric and thousands of other examples. Policy around tick-borne disease will need to contend with public perception of data, scepticism of science, over-reliance on data (paralysis of policy where data are not available) and the fact that, due to climate change, the ecology of tick-borne disease will change rapidly. Protection of public health must be done in a place where people are wary of science and sometimes believe things that are not true, scientists are more comfortable with indisputable fact, and the situation is changing rapidly.

We lack knowledge of exactly what in the environment allows most medically important ticks to survive, mature and be active, both broadly (in a state, province or country) and at the very small spatial scale such as in forest leaf litter. For example, the brown dog tick (*Rhipicephalus sanguineus*) is an important vector of *Ehrlichia canis* and *Anaplasma platys* among dogs and *Rickettsia rickettsii*, the agent of Rocky Mountain spotted fever, in people. Publications document maximum and minimum temperatures that two important lineages of this tick (tropical and temperature) can tolerate (Dantas-Torres *et al.*, 2013). However, distributions of both lineages are changing (Villarreal *et al.*, 2018) and it seems clear that temperature probably was not the sole factor that limited previous distributions (Zemtsova *et al.*, 2016). In addition, this tick is peri-domestic and often found indoors, where it is less affected by outside environmental factors (Dantas-Torres, 2008).

Earlier studies on the ecology of Gulf Coast ticks (*Amblyomma maculatum*) were not aware that this is actually a complex of morphologically similar species including *A. maculatum* sensu stricto, *Amblyomma triste* and *Amblyomma tigrinum*. The presence of a population in Arizona indicates that the species can inhabit far more arid and hotter habitats than previously realized, implying that either the tick is capable of invading a broader variety of habitats subject to differing environmental conditions or that the Arizona population may be yet another new species within the complex. We may lack critical understanding even for the best-studied tick species such as *I. scapularis*. This tick is known to survive at temperatures as low as −10°C when protected by leaf litter and snow (Lindsay *et al.*, 1995; Linske *et al.*, 2019) and thus it has been theorized that the tick could not spread northward into regions where the temperature falls below this threshold. However, these hypothesized minimum temperatures do not appear to have limited northward expansion, for example in Ontario, Canada, where expansion is occurring much faster than temperatures are changing (Nelder *et al.*, 2014). Given such important gaps in data, it remains necessary to better document the true diversity of ticks and what specific features of their environment determine their distributions.

Overall tick data are truly inadequate to allow ecological synthesis: there have not been enough years of rapid global climate change yet; there are no long-term, multiple-site, international and methodologically comparable data sets of ticks and tick-borne disease; and there still are major gaps in basic understanding of what specifically limits the distribution and activity of many important ticks. As the pace of change accelerates, those data we do have become less directly applicable to the present and near future. For example, there are surprisingly few studies in California that document the phenology of feeding of the different stages of *Ixodes pacificus* and any data that are more than a decade old are out of date: adult ticks now emerge later in the year, because rain occurs later in the year, there is less of it and the weather often remains warm into autumn and winter. Our approach to the use of data will have to change; data will need to be used more efficiently and more predictively. For example, now that climatic conditions in many areas are different from those several decades ago and are continuing to change over the next several decades, we could extrapolate and predict tick species' expansion into newly suitable areas and contraction from those no longer suitable, provided we had good data on limiting environmental conditions. As importantly, with finer-scale environment data, we would be able to scale down from predictions of tick activity for larger units such as country, state and county to more local, and arguably more relevant, scales such as watershed, town and finer-scale geographical features.

People and organizations tasked with health management should approach the job with an adaptive management framework, a paradigm often incorporated into wildlife and wildlands management. According to the US Department of Interior, 'an adaptive approach involves exploring alternative ways to meet management objectives, predicting the outcomes of alternatives based on the current state of knowledge, implementing one or more of these alternatives, monitoring to learn about the impacts of management actions, and then using the results to update knowledge and adjust management actions', while the US Agency for International Development describes it as 'an intentional approach to making decisions and adjustments in response to new information and changes in context'. It is unrealistic that we can now acquire all the precise time- and space-specific data on ticks and tick-borne disease that are needed and yet managing changing risk remains crucial. Advances in machine learning, fuzzy logic, spatial statistics, Bayesian statistics and other fields can support efficient synthesis which must then be translated for policy makers and stakeholders to enact management strategies. At minimum, fine-scale geographical data on disease risk should be available to patients, people who work or recreate outdoors, medical professionals, and health and vector control agencies. Active and passive surveillance can incorporate citizen scientists and academic partners, as well as governmental health workers. Larger, inter-agency and even international surveillance programmes can evolve into programmes that acquire, troubleshoot and analyse data, feeding the data into adaptive, synthetic models that incorporate tick biology and environmental change.

Protecting people and other animals against tick-borne disease is by no means an easy task, especially as ecological conditions change. Even if we fully understood what limits each tick's distribution and activity, and could predict, for example, exactly where conditions would become more suitable as the Earth's climate changes, those ecological factors which limit ticks act synergistically: changing climate impacts not only temperature but also humidity, microenvironments for tick development, host abundance, etc. Changing climate impacts people and domestic animals as well, who can sometimes take new advantage of previously inhospitable regions and other times are forced to leave areas that are affected by drought, flooding or other impacts. The synergistic impacts on ticks make it especially difficult to predict tick-borne disease risk. However, the public has become increasingly aware of the risks of tick-borne disease and emerging infectious disease in general. Collaborative, predictive and synthetic data sharing and analysis, with adaptive management of risk, will best position public health authorities to respond to the needs of people everywhere on Earth where ticks pose a threat.

References

CDC (2019) What you need to know about Asian longhorned ticks – a new tick in the United States. Centers for Disease Control and Prevention, Atlanta, Georgia. Available at: https://www.cdc.gov/ticks/longhorned-tick/index.html (accessed 7 June 2021).

Clow, K.M., Leighton, P.A., Ogden, N.H., Lindsay, L.R., Michel, P., *et al.* (2017) Northward range expansion of *Ixodes scapularis* evident over a short timescale in Ontario, Canada. *PLoS ONE* 12, e0189393.

Dantas-Torres, F. (2008) The brown dog tick, *Rhipicephalus sanguineus* (Latreille, 1806) (Acari: Ixodidae): from taxonomy to control. *Veterinary Parasitology* 152, 173–185.

Dantas-Torres, F., Latrofa, M.S., Annoscia, G., Giannelli, A., Parisi, A. and Otranto, D. (2013) Morphological and genetic diversity of *Rhipicephalus sanguineus* sensu lato from the New and Old Worlds. *Parasites & Vectors* 6, 213.

Dobson, A.D. and Randolph, S.E. (2011) Modelling the effects of recent changes in climate, host density and acaricide treatments on population dynamics of *Ixodes ricinus* in the UK. *Journal of Applied Ecology* 48, 1029–1037.

Eisen, R.J., Eisen, L. and Beard, C.B. (2016) County-scale distribution of *Ixodes scapularis* and *Ixodes pacificus* (Acari: Ixodidae) in the continental United States. *Journal of Medical Entomology* 53, 349–386.

European Commission (2020) Adaptation to climate change. How will we be affected? European Commission, Brussels. Available at: https://ec.europa.eu/clima/policies/adaptation/how_en (accessed 16 December 2020).

Fernandes, É.K., Bittencourt, V.R. and Roberts, D.W. (2012) Perspectives on the potential of entomopathogenic fungi in biological control of ticks. *Experimental Parasitology* 130, 300–305.

Global Climate Change: Vital Signs of the Planet (2020) The effects of climate change. National Aeronautics and Space Administration, Washington, DC. Available at: https://climate.nasa.gov/effects/ (accessed 16 December 2020).

Hamer, S.A., Tsao, J.I., Walker, E.D. and Hickling, G.J. (2010) Invasion of the Lyme disease vector *Ixodes scapularis*: implications for *Borrelia burgdorferi* endemicity. *EcoHealth* 7, 47–63.

Herrmann, C. and Gern, L. (2013) Survival of *Ixodes ricinus* (Acari: Ixodidae) nymphs under cold conditions is negatively influenced by frequent temperature variations. *Ticks and Tick-borne Diseases* 4, 445–451.

Klepeis, P., Scull, P., LaLonde, T., Svajlenka, N. and Gill, N. (2013) Changing forest recovery dynamics in the northeastern United States. *Area* 45, 239–248.

Lindsay, L.R., Barker, I.K., Surgeoner, G.A., McEwen, S.A., Gillespie, T.J. and Robinson, J.T. (1995) Survival and development of *Ixodes scapularis* (Acari: Ixodidae) under various climatic conditions in Ontario, Canada. *Journal of Medical Entomology* 32, 143–152.

Linske, M.A., Stafford, K.C., Williams, S.C., Lubelczyk, C.B., Welch, M. and Henderson, E.F. (2019) Impacts of deciduous leaf litter and snow presence on nymphal *Ixodes scapularis* (Acari: Ixodidae) overwintering survival in coastal New England, USA. *Insects* 10, 227.

Mack, R.N. (2003) Phylogenetic constraint, absent life forms, and preadapted alien plants: a prescription for biological invasions. *International Journal of Plant Sciences* 164, S185–S196.

Nelder, M.P., Russell, C., Lindsay, L.R., Dhar, B., Patel, S.N., *et al.* (2014) Population-based passive tick surveillance and detection of expanding foci of blacklegged ticks *Ixodes scapularis* and the Lyme disease agent *Borrelia burgdorferi* in Ontario, Canada. *PLoS ONE* 9, e105358.

Ogden, N.H., Beard, C.B., Ginsberg, H.S. and Tsao, J.I. (2020) Possible effects of climate change on ixodid ticks and the pathogens they transmit: predictions and observations. *Journal of Medical Entomology* 57, tjaa220.

Plowright, R.K., Parrish, C.R., McCallum, H., Hudson, P.J., Ko, A.I., *et al.* (2017) Pathways to zoonotic spillover. *Nature Reviews Microbiology* 15, 502–510.

Rainey, T., Occi, J.L., Robbins, R.G. and Egizi, A. (2018) Discovery of *Haemaphysalis longicornis* (Ixodida: Ixodidae) parasitizing a sheep in New Jersey, United States. *Journal of Medical Entomology* 55, 757–759.

Rosenberg, R., Lindsey, N.P., Fischer, M., Gregory, C.J., Hinckley, A.F., *et al.* (2018) Vital signs: trends in reported vectorborne disease cases – United States and territories, 2004–2016. *Morbidity and Mortality Weekly Report* 67, 496–501.

UCAR Center for Science Education (2020) Predictions of global future climate. University Cooperation for Academic Research, Boulder, Colorado. Available at: https://scied.ucar.edu/learning-zone/impacts-climate-change/predictions-future-global-climate (accessed 16 December 2020).

UN Africa Renewal (2019) Global warming: severe consequences for Africa. *Africa Renewal*, December 2018–March 2019. United Nations, New York. Available at: https://www.un.org/africarenewal/magazine/december-2018-march-2019/global-warming-severe-consequences-africa (accessed 16 December 2020).

Villarreal, Z., Stephenson, N. and Foley, J. (2018) Possible northward introgression of a tropical lineage of *Rhipicephalus sanguineus* ticks at a site of emerging Rocky Mountain spotted fever. *Journal of Parasitology* 104, 240–245.

Zemtsova, G.E., Apanaskevich, D.A., Reeves, W.K., Hahn, M., Snellgrove, A. and Levin, ML. (2016) Phylogeography of *Rhipicephalus sanguineus* sensu lato and its relationships with climatic factors. *Experimental and Applied Acarology* 69, 191–203.

eo38 Co-infections of Ticks

Martin Pfeffer*, Nina Król and Anna Obiegala
University of Leipzig, Leipzig, Germany

38.1 Expert Opinion

38.1.1 Introduction

Hard ticks (Acari: Ixodida) play an increasingly important role in the transmission of zoonotic pathogens to humans and animals worldwide. In Europe, *Ixodes ricinus*, the castor bean tick, is of outstanding importance in this context, as it is the most common tick species in Central Europe (Claerebout *et al.*, 2013). Furthermore, it is known to transmit a variety of zoonotic pathogens such as *Borrelia burgdorferi* sensu lato (s.l.), which causes Lyme disease (LD), and tick-borne encephalitis virus (TBEV), the aetiological agent of tick-borne encephalitis (TBE) (Rizzoli *et al.*, 2014). *I. ricinus* is known to be host-opportunistic rather than host-specific. Thus, the possibility of having a wide range of different hosts increases the risk of ticks being infected with different pathogens simultaneously and thus their vector potential.

In Germany, humans and animals have the possibility of contracting a tick-borne zoonosis from species other than *I. ricinus*. There are more than 20 species of ticks known to occur in Germany and some of them are important vectors. *Dermacentor reticulatus*, known as the ornate cow tick, is the second most prevalent tick species in Germany and in Central Europe. It occurs focally in Germany, but it is increasingly localized in more and more places in Central Europe (Rubel *et al.*, 2016, 2020; Drehmann *et al.*, 2020). This tick species parasitizes mainly dogs, but humans as well as pigs, cats, goats and cattle can also be affected. *D. reticulatus* is primarily known as the main vector of *Babesia* that can lead to babesiosis in dogs (caused by *Babesia canis*), which in Germany was for a long time considered exclusively to be travel related, but it is autochthonous in many regions of Central Europe (Silaghi *et al.*, 2020). In Germany, *D. reticulatus* is considered also as the main vector of *Rickettsia raoultii*, a species of *Rickettsia* belonging to the spotted fever group of Rickettsiae. Recently, even TBEV has been described in *D. reticulatus* for the first time (Chitimia-Dobler *et al.*, 2019a).

Haemaphysalis concinna is the third most common tick species in Central Europe and is known as the relict tick. It is endemic in the warm temperate climate zone with a high degree of humidity in Europe and Asia where it is known to have a diverse host spectrum covering mammals, birds and reptiles (Rubel *et al.*, 2018). Further, this tick species likewise serves as a vector for the same spectrum of tick-borne pathogens (TBPs) as *I. ricinus* and *D. reticulatus*.

*Email: pfeffer@vetmed.uni-leipzig.de

DOI: 10.1079/9781789249637.0038

In addition, a new tick species was recently discovered in Central Europe. The lately described *Ixodes inopinatus* differs morphologically only slightly from the castor bean tick. This is why this tick species remained unrecognized for a long time. Its distribution range probably extends from Central Europe to Morocco (Estrada-Peña *et al.*, 2014). However, it is not yet clear what host preference and vector function this tick species has with regard to TBPs. Studies on TBPs in *I. inopinatus* reported similar prevalence rates regarding *Borrelia* spp., *Anaplasma* spp. and *Rickettsia* spp. compared with *I. ricinus* ticks (Hauck *et al.*, 2019). Thus, it is assumed that *I. inopinatus* as a potential vector covers a similar spectrum of TBPs as the closely related castor bean tick.

As a result of climate change, Germany can increasingly be expected to experience persistently higher temperatures. This may lead to the fact that tick species that are currently still endemic in warmer countries will spread into Germany. The increasing frequency of the tropical ticks, *Hyalomma marginatum* and *Hyalomma rufipes*, found in Germany and other temperate regions of Europe mirrors these changes (Chitimia-Dobler *et al.*, 2019b). These tick species are originally found in North Africa, Southern Europe and parts of Asia, and they are introduced via migratory birds. They are known carriers of Crimean–Congo haemorrhagic fever (CCHF) virus and *Rickettsia* spp.

The *Rhipicephalus sanguineus* complex comprises tick species that are grouped together and generally commonly named the so-called brown dog tick. The brown dog tick is a cosmopolite and is responsible for transmission of ehrlichiosis, hepatozoonosis and a mild form of babesiosis (caused by *Babesi vogeli*) in dogs. However, *R. sanguineus* also feeds sporadically on other mammals including humans. In Germany, this tick species is not considered a regular element of the fauna, but there are increasing numbers of records. Due to the cool temperatures in winter, the brown dog tick has not been able to reproduce outside buildings in Germany (Petney *et al.*, 2012). Nevertheless, it is believed that this tick species may soon be common in Germany due to climate change.

Regardless of the tick species, the prevalence rates of different pathogens may vary throughout Europe, e.g. with higher prevalence rates for *B. canis* or *Coxiella burnetii* in South-Western Europe compared with Northern Europe. This situation may change once the climatic conditions in temperate Europe equal the current environmental conditions in Southern Europe.

Although the vector competence for most pathogens is clear regarding which tick species is a vector, it is not yet clear how co-infections may influence tick survival and the transmission competence to their hosts. It is known, for example, that the presence of *B. burgdorferi* is boosting the presence of *Anaplasma phagocytophilum* in *I. ricinus* ticks (Diuk-Wasser *et al.*, 2016). Likewise, the concurrent infection of *A. phagocytophilum* and louping ill virus in *Ixodes* ticks and sheep is well recognized in Scotland (Reid *et al.*, 1986). Many pathogen prevalence studies that are aiming to detect a panel of pathogens report a certain frequency of co-infections in ticks. The extent of dual infections reaches from an average of >3% up to 22% and even triple infections are frequently reported (Galfsky *et al.*, 2019; Klitgaard *et al.*, 2019; Lernout *et al.*, 2019). Some of such reports found that particular co-infections occurred more often than simply by chance (Kjelland *et al.*, 2018; Ben and Lozynskyi, 2019). This phenomenon is not restricted to *I. ricinus*; it was likewise described for *Ixodes scapularis* (Sanchez-Vicente *et al.*, 2019) and *D. reticulatus* (Zając *et al.*, 2017), and likely will be found in any three-host tick species. We assume that this is more likely a matter of the detection technique applied than the real situation. As more whole-genome sequencing techniques are applied on ticks, we will gain more detailed information on the real pathogen burden of ticks. We are currently only scratching the surface as the pathogen prevalence data do not allow us to understand the full impact of co-infections in the tick with regard to either one another or the 'regular' tick microbiota. Another issue that needs to be explored is the influence of concurrent or subsequent co-infection. One obvious impact upon co-infections in ticks will be the effect of promiscuous compared with strict vertebrate host associations. Some particular hosts have been demonstrated to function as melting pots for tick-borne zoonotic agents, such as the European hedgehog (Jahfari *et al.*, 2017) and some rodent species (Galfsky *et al.*, 2019). The latter report screened rodents for the presence of seven tick-borne pathogens and found most rodents had a triple infection (30%) with up to quintuple infections in 6% of the investigated rodent hosts. As with the dynamic interplay within the tick itself, we currently cannot judge the impact of the hosts. This scenario will become even more complex when host communities change as a consequence of climate change. Thus, some researchers argue that within-host interactions are so strong that the dynamics of a single pathogen cannot be understood without the knowledge of other pathogens (Cattadori *et al.*, 2008; Telfer *et al.*, 2010; Cabezas-Cruz *et al.*, 2019).

There are some studies reporting that infection may impact the tick's fitness and behaviour. For example, it is known that *Borrelia* spp. and *Anaplasma* spp. may increase the desiccation resistance of *Ixodes* spp. by triggering physiological changes (*Borrelia*: higher fat reserves; *Anaplasma*: synthesis of heat-shock proteins) (Benelli, 2020). Further, *Bartonella henselae* can upregulate the *I. ricinus* serine protease inhibitor (IrSPI). Silencing this IrSPI leads to a weight reduction of feeding *I. ricinus* females (Liu *et al.*, 2014). From an observational standpoint, one may argue that the most frequently found co-infections are for bacteria which per se have an extraordinary genetic diversity/flexibility, e.g. *A. phagocytophilum* (Lagrée *et al.*, 2018; Silaghi *et al.*, 2018), *B. burgdorferi* s.l. (Herrmann *et al.*, 2013) and *Rickettsia* spp. Co-infections are still not the main focus in most studies of TBPs. Nevertheless, it is important to find out what mechanisms and combinations are beneficial for the transmission of pathogens and the survival of ticks as vectors.

38.1.2 Evidence for climate change impacts

As climate change is evolving and temperatures are rising, vectors in Southern European countries are expanding to the north. For example, *Aedes albopictus*, an important vector for e.g. chikungunya virus, is rapidly expanding northwards (Kraemer *et al.*, 2015). A clear expansion to the north has also been reported for the Mediterranean tick, *R. sanguineus*.

Furthermore, it is evident that climate has an impact on hosts and thus also on tick activity. *I. ricinus* normally has an annual activity rhythm from spring until autumn with the first activity peak in late spring/early summer and the second peak in September. Because of the hotter summers and the milder winters, annual tick activity is extended (Bouchard *et al.*, 2019). Additionally, mild winters with fewer days with temperatures below 0°C allow more generations of *Ixodes* and *Dermacentor* to develop during a year, whereas *R. sanguineus* is able to survive and establish at northern latitudes (Mahrenholz, 2008; Gray *et al.*, 2009). This may lead to more tick bites and increases in the incidence rates of diseases such as LD and TBE. Thus, one may assume that co-infections in ticks may likewise accumulate as the prevalence in hosts may be higher. As infection levels rise, higher co-infection rates are more likely. In contrast, warmer conditions could reduce host-seeking activity and lead to a stronger immune response in some tick species (Ostfeld and Brunner, 2015).

38.1.3 Evidence against climate change impacts

In a recent study by our group, tick-borne pathogens were detected in ticks and compared between the years within a decade. Prevalence rates for *Rickettsia* spp. and *Babesia* spp. were constant or even declining over the years rather than increasing (Galfsky *et al.*, 2019). Consequently, there are prevalence rates of pathogens in ticks that are not influenced by hosts or other external factors. *R. raoultii*, for example, is transmitted transovarially in *D. reticulatus* and the general high prevalence may be maintained without the existence of competent hosts.

38.2 Future Predictions

38.2.1 Next 10 years

Due to the temperature changes in Central Europe, it is likely that *H. marginatum*, *H. rufipes* and *R. sanguineus* will be more prevalent in this region. Further, tick-borne pathogens associated with these tick species may spread geographically to the north. Currently endemic tick species such as *I. ricinus* and *D. reticulatus* may take up these 'emerging' pathogens through feeding on infected hosts. This may lead to unpredictable co-infection combinations in the already established tick species populations. As climatic conditions in Central Europe increasingly favour the development of ticks, the density of ticks can rise, thus increasing the likelihood of humans being bitten by a tick. The emerging pathogens may have an impact on the combination of pathogens which may be transmitted to hosts simultaneously. These pathogens may have a boosting effect on other coexisting pathogens or a negative effect on their transmission. This may have an impact on the prevalence of pathogens and diseases in the hosts.

It is likely that the molecular biology methods will evolve in the future. Thus, the detection of co-infections, and their influencing factors in ticks and hosts, will be more precisely defined and less time-consuming to investigate. Diseases could be detected more often but also earlier, which leads to faster treatment with a better clinical outcome for the patient.

38.2.2 Next 20 years

In addition to the above (Section 38.2.1), the southern parts of Europe could be hotter and drier, therefore the distribution of *I. ricinus* and other tick species, such as *H. concinna*, which are dependent on high humidity and soil moisture, might diminish. Consequently, due to the lower diversity and the declining size of tick populations in Southern Europe, this might also result in lower pathogen prevalence rates and lower co-infection levels in hosts and ticks. A concentration towards tick species like *R. sanguineus* and the pathogens it transmits may be the consequence, which on the other hand could lead to less reported clinical cases in these southern areas.

38.2.3 Next 50 years

Many formerly exotic tick species as well as their pathogens are broadly established in formerly temperate Europe. New combinations of co-infections in tick species will lead to new double or triple infections in humans and animals which will challenge clinicians and veterinarians alike. On the other hand, new diagnostic techniques will be available allowing holistic tick analyses (i.e. detection of all putatively pathogenic organisms) by anyone in almost no time. Taxonomy will no longer be an issue due to a number of reasons; mainly because tick species classifications will no longer follow the traditional binomial Linnaean taxonomy as almost nobody will be able to identify ticks based on classical morphological criteria. Only epidemiologically and ecologically relevant factors of ticks will be used, i.e. where and when they are active, who they feed on and the level of pathogens they carry. As 'tick-site' diagnostics will be cheap and easy to perform, citizen science projects will allow accurate mapping of (multi-) pathogen-carrying vector ticks. This will be used for public health alerts in real time to anybody who is active in nature and exposed to tick bites. This will be an important step, as most ticks will show at least partial resistance to currently used acaricides. Luckily, most of the tick-borne bacteria will still be sensitive to antibiotic treatment, as selection pressure is low. For most tick-borne viruses, vaccines will become available and, due to the severity of the diseases (TBE, CCHF, severe fever with thrombocytopaenia syndrome, etc.), will be administered in the regions of their occurrence.

References

Ben, I. and Lozynskyi, I. (2019) Prevalence of *Anaplasma phagocytophilum* in *Ixodes ricinus* and *Dermacentor reticulatus* and coinfection with *Borrelia burgdorferi* and tick-borne encephalitis virus in Western Ukraine. *Vector-Borne and Zoonotic Diseases* 19, 793–801. https://doi.org/10.1089/vbz.2019.2450

Benelli, G. (2020) Pathogens manipulating tick behavior – through a glass, darkly. *Pathogens* 9, 664. https://doi.org/10.3390/pathogens9080664

Bouchard, C., Dibernardo, A., Koffi, J., Wood, H., Leighton, P.A. and Lindsay, L.R. (2019) Increased risk of tick-borne diseases with climate and environmental changes. *Canada Communicable Disease Report* 45, 83–89. https://doi.org/10.14745/ccdr.v45i04a02

Cabezas-Cruz, A., Espinosa, P., Alberdi, P. and de la Fuente, J. (2019) Tick–pathogen interactions: the metabolic perspective. *Trends in Parasitology* 35, 316–328. https://doi.org/10.1016/j.pt.2019.01.006
Cattadori, I.M., Boag, B. and Hudson, P.J. (2008) Parasite co-infection and interaction as drivers of host heterogeneity. *International Journal for Parasitology* 38, 371–380. https://doi.org/10.1016/j.ijpara.2007.08.004
Chitimia-Dobler, L., Lemhöfer, G., Król, N., Bestehorn, M., Dobler, G. and Pfeffer, M. (2019a) Repeated isolation of tick-borne encephalitis virus from adult *Dermacentor reticulatus* ticks in an endemic area in Germany. *Parasites & Vectors* 12, 90. https://doi.org/10.1186/s13071-019-3346-6
Chitimia-Dobler, L., Schaper, S., Riess, R., Bitterwolf, K., Frangoulidis, D., *et al.* (2019b) Imported *Hyalomma* ticks in Germany in 2018. *Parasites & Vectors* 12, 134. https://doi.org/10.1186/s13071-019-3380-4
Claerebout, E., Losson, B., Cochez, C., Casaert, S., Dalemans, A.-C., *et al.* (2013) Ticks and associated pathogens collected from dogs and cats in Belgium. *Parasites & Vectors* 6, 183. https://doi.org/10.1186/1756-3305-6-183
Diuk-Wasser, M.A., Vannier, E. and Krause, P.J. (2016) Coinfection by *Ixodes* tick-borne pathogens: ecological, epidemiological, and clinical consequences. *Trends in Parasitology* 32, 30–42. https://doi.org/10.1016/j.pt.2015.09.008
Drehmann, M., Springer, A., Lindau, A., Fachet, K., Mai, S., *et al.* (2020) The spatial distribution of *Dermacentor* ticks (Ixodidae) in Germany – evidence of a continuing spread of *Dermacentor reticulatus*. *Frontiers in Veterinary Science* 7, 578220. https://doi.org/10.3389/fvets.2020.578220
Estrada-Peña, A., Nava, S. and Petney, T. (2014) Description of all the stages of *Ixodes inopinatus* n. sp. (Acari: Ixodidae). *Ticks and Tick-borne Diseases* 5, 734–743. https://doi.org/10.1016/j.ttbdis.2014.05.003
Galfsky, D., Król, N., Pfeffer, M. and Obiegala, A. (2019) Long-term trends of tick-borne pathogens in regard to small mammal and tick populations from Saxony, Germany. *Parasites & Vectors* 12, 131. https://doi.org/10.1186/s13071-019-3382-2
Gray, J.S., Dautel, H., Estrada-Peña, A., Kahl, O. and Lindgren, E. (2009) Effects of climate change on ticks and tick-borne diseases in Europe. *Interdisciplinary Perspectives on Infectious Diseases* 2009, 593232. https://doi.org/10.1155/2009/593232
Hauck, D., Springer, A., Pachnicke, S., Schunack, B., Fingerle, V. and Strube, C. (2019) *Ixodes inopinatus* in northern Germany: occurrence and potential vector role for *Borrelia* spp., *Rickettsia* spp., and *Anaplasma phagocytophilum* in comparison with *Ixodes ricinus*. *Parasitology Research* 118, 3205–3216. https://doi.org/10.1007/s00436-019-06506-4
Herrmann, C., Gern, L. and Voordouw, M.J. (2013) Species co-occurrence patterns among Lyme borreliosis pathogens in the tick vector *Ixodes ricinus*. *Applied and Environmental Microbiology* 79, 7273–7280. https://doi.org/10.1128/AEM.02158-13
Jahfari, S., Ruyts, S.C., Frazer-Mendelewska, E., Jaarsma, R., Verheyen, K. and Sprong, H. (2017) Melting pot of tick-borne zoonoses: the European hedgehog contributes to the maintenance of various tick-borne diseases in natural cycles urban and suburban areas. *Parasites & Vectors* 10, 134. https://doi.org/10.1186/s13071-017-2065-0
Kjelland, V., Paulsen, K.M., Rollum, R., Jenkins, A., Stuen, S., *et al.* (2018) Tick-borne encephalitis virus, *Borrelia burgdorferi* sensu lato, *Borrelia miyamotoi*, *Anaplasma phagocytophilum* and *Candidatus* Neoehrlichia mikurensis in *Ixodes ricinus* ticks collected from recreational islands in southern Norway. *Ticks and Tick-borne Diseases* 9, 1098–1102. https://doi.org/10.1016/j.ttbdis.2018.04.005
Klitgaard, K., Kjær, L.J., Isbrand, A., Hansen, M.F. and Bødker, R. (2019) Multiple infections in questing nymphs and adult female *Ixodes ricinus* ticks collected in a recreational forest in Denmark. *Ticks and Tick-borne Diseases* 10, 1060–1065. https://doi.org/10.1016/j.ttbdis.2019.05.016
Kraemer, M.U.G., Sinka, M.E., Duda, K.A., Mylne, A.Q.N., Shearer, F.M., *et al.* (2015) The global distribution of the arbovirus vectors *Aedes aegypti* and *Ae. albopictus*. *eLife* 4, e08347. https://doi.org/10.7554/eLife.08347
Lagrée, A.-C., Rouxel, C., Maëllys, K., Dugat, T., Girault, G., *et al.* (2018) Co-circulation of different *A. phagocytophilum* variants within cattle herds and possible reservoir role for cattle. *Parasites & Vectors* 11, 163. https://doi.org/10.1186/s13071-018-2661-7
Lernout, T., de Regge, N., Tersago, K., Fonville, M., Suin, V. and Sprong, H. (2019) Prevalence of pathogens in ticks collected from humans through citizen science in Belgium. *Parasites & Vectors* 12, 550. https://doi.org/10.1186/s13071-019-3806-z
Liu, X.Y., de la Fuente, J., Cote, M., Galindo, R.C., Moutailler, S., *et al.* (2014) IrSPI, a tick serine protease inhibitor involved in tick feeding and *Bartonella henselae* infection. *PLoS Neglected Tropical Diseases* 8, e2993. https://doi.org/10.1371/journal.pntd.0002993

Mahrenholz, P. (2008) Climate change and adaptation needs. *Parasitology Research* 103(Suppl. 1), S139–S146. https://doi.org/10.1007/s00436-008-1058-5

Ostfeld, R.S. and Brunner, J.L. (2015) Climate change and *Ixodes* tick-borne diseases of humans. *Philosophical Transactions of the Royal Society B* 370, 1665. https://doi.org/10.1098/rstb.2014.0051

Petney, T.N., Pfäalle, M.P. and Skuballa, J.D. (2012) An annotated checklist of the ticks (Acari: Ixodida) of Germany. *Systematic and Applied Acarology* 17, 115–170. https://doi.org/10.11158/saa.17.2.2

Reid, H.W., Buxton, D., Pow, I., Brodie, T.A., Holmes, P.H. and Urquhart, G.M. (1986) Response of sheep to experimental concurrent infection with tick-borne fever (*Cytoecetes phagocytophila*) and louping-ill virus. *Research in Veterinary Science* 41, 56–62. https://doi.org/10.1016/S0034-5288(18)30572-1

Rizzoli, A., Silaghi, C., Obiegala, A., Rudolf, I., Hubálek, Z., *et al.* (2014) *Ixodes ricinus* and its transmitted pathogens in urban and peri-urban areas in Europe: new hazards and relevance for public health. *Frontiers in Public Health* 2, 251. https://doi.org/10.3389/fpubh.2014.00251

Rubel, F., Brugger, K., Pfeffer, M., Chitimia-Dobler, L., Didyk, Y.M., *et al.* (2016) Geographical distribution of *Dermacentor marginatus* and *Dermacentor reticulatus* in Europe. *Ticks and Tick-borne Diseases* 7, 224–233. https://doi.org/10.1016/j.ttbdis.2015.10.015

Rubel, F., Brugger, K., Walter, M., Vogelgesang, J.R., Didyk, Y.M., *et al.* (2018) Geographical distribution, climate adaptation and vector competence of the Eurasian hard tick *Haemaphysalis concinna*. *Ticks and Tick-borne Diseases* 9, 1080–1089. https://doi.org/10.1016/j.ttbdis.2018.04.002

Rubel, F., Brugger, K., Belova, O.A., Kholodilov, I.S., Didyk, Y.M., *et al.* (2020) Vectors of disease at the northern distribution limit of the genus *Dermacentor* in Eurasia: *D. reticulatus* and *D. silvarum*. *Experimental and Applied Acarology* 8, 95–123. https://doi.org/10.1007/s10493-020-00533-y

Sanchez-Vicente, S., Tagliafierro, T., Coleman, J.L., Benach, J.L. and Tokarz, R. (2019) Polymicrobial nature of tick-borne diseases. *mBio* 10, e02055-19. https://doi.org/10.1128/mBio.02055-19

Silaghi, C., Nieder, M., Sauter-Louis, C., Knubben-Schweizer, G., Pfister, K. and Pfeffer, M. (2018) Epidemiology, genetic variants and clinical course of natural infections with *Anaplasma phagocytophilum* in a dairy cattle herd. *Parasites & Vectors* 11, 20. https://doi.org/10.1186/s13071-017-2570-1

Silaghi, C., Weis, L. and Pfister, K. (2020) *Dermacentor reticulatus* and *Babesia canis* in Bavaria (Germany) – a georeferenced field study with digital habitat characterization. *Pathogens* 9, 541. https://doi.org/10.3390/pathogens9070541

Telfer, S., Lambin, X., Birtles, R., Beldomenico, P., Burthe, S., *et al.* (2010) Species interactions in a parasite community drive infection risk in a wildlife population. *Science* 330, 243–246. https://doi.org/10.1126/science.1190333

Zając, V., Wójcik-Fatla, A., Sawczyn, A., Cisak, E., Sroka, J., *et al.* (2017) Prevalence of infections and co-infections with 6 pathogens in *Dermacentor reticulatus* ticks collected in eastern Poland. *Annals of Agricultural and Environmental Medicine* 24, 26–32. https://doi.org/10.5604/12321966.1233893

eo39 Impact of Climate Change on Co-feeding Transmission

Jianhong Wu[1]* and Xue Zhang[2]
[1]*York University, Toronto, Canada;* [2]*Northeastern University, Shenyang, People's Republic of China*

39.1 Expert Opinion

39.1.1 Introduction

Tick-borne pathogen co-feeding transmission is a non-systemic transmission in the tick–host enzootic cycle, through which the host provides a bridge between co-feeding susceptible and infected ticks to facilitate pathogen transmission. Co-feeding transmission requires co-feeding of susceptible ticks in close (both spatially and temporally) proximity to other infected ticks on the bridging host; hence, the contribution of co-feeding transmission to tick-borne pathogen transmission in the tick–host enzootic cycle is highly affected by environmental conditions (Voordouw, 2015). Much has been done in modelling pathogen transmission dynamics through the systemic transmission route where a host, exposed to the pathogen, develops a widespread infection that acts as a reservoir from which susceptible vectors can acquire the pathogen. However, very little has been progressed for modelling and evaluating the impact of co-feeding transmission, although there have been increasing numbers of field and experimental studies reporting this non-systematic transmission for many vector-borne diseases in general, and for tick-borne diseases in particular (Jones *et al.*, 1987; Alekseev and Chunikhin, 1990; Labuda *et al.*, 1993; Hua *et al.*, 2003). For example, tick-borne encephalitis (TBE) virus is predominantly spread by infected *Ixodes* species, primarily *Ixodes ricinus* and *Ixodes persulcatus* (Taba *et al.*, 2017). Ticks responsible for TBE virus transmission have a complex life cycle with three post-egg stages: larva, nymph and adult. Three blood meals from respective hosts are needed for development from one stage to the next to complete the life cycle. The interstadial development periods are strongly influenced by the surrounding environmental conditions and host abundance affects tick questing behaviours. These tick vectors have host preferences, with larvae and nymphs preferring small to medium-sized mammals and adults tending to feed on large-sized mammals; at different stages, co-feeding on the same host can take place. The probability of co-feeding, and thus transstadial co-feeding transmission, is highly impacted by environmental conditions. Therefore, to model TBE virus transmission in the tick–host enzootic cycle and to quantify TBE virus transmission potential and risk, we need to stratify the ticks not only by their developmental stages, but also by their questing, feeding and engorging activities, and we need to describe the tick-on-host distribution during the infestation process. This usually requires a

*Email: wujh@yorku.ca

DOI: 10.1079/9781789249637.0039

multi-scale model framework as the timescale for infestation dynamics is comparatively short compared with the timescale of tick population dynamics and tick-borne disease transmission dynamics. The model parameterization and analysis are both significantly challenging and demand some interdisciplinary collaborative efforts (Wu and Zhang, 2020).

39.1.2 Impact of climate change on co-feeding transmission

To understand the impact of climate change patterns on TBE prevalence in the tick–host enzootic cycle in a specific region, Nah *et al.* (2019) developed a transmission dynamics model and integrated this with tick–human contacts during a surveillance interval. Key ecological and epidemiological parameters, region- and environment-specific, were estimated by fitting weekly human incidence data and incorporating local temperature data in Vas County, a TBE disease-endemic region of Hungary. Estimating these parameters is computationally challenging since the data we observe, i.e. the human incidence of TBE, is only the final outcome of the tick–host transmission and tick–human contact processes. The challenge also increases because the complex TBE virus transmission cycle involves the non-systemic route of transmission between co-feeding ticks. The approach of Nah *et al.* (2019) was based on the description of the hidden Markov transition process, using a novel TBE virus transmission–human case reporting cascade model that couples the susceptible-infected compartmental model describing the TBE virus transmission dynamics among ticks, animal hosts and humans, with the stochastic observation process of human TBE virus infection reporting. The study then used the parameterized cascade model to assess the transmission potential of TBE virus in the enzootic cycle with respect to climate change and to evaluate the contribution of non-systemic transmission. It was shown that TBE virus transmission potential in the enzootic cycle increased along with the increased temperature, although the TBE human incidence dropped since the 1990s. The study demonstrated that the co-feeding transmission pathway has been a significant factor in the transmission of TBE virus in Hungary, and it also concluded that the risk of TBE virus infection would be highly underestimated if the co-feeding transmission route is neglected in the risk assessment. These conclusions were also made in earlier studies using some mechanistic models (Rosà *et al.*, 2003; Norman *et al.*, 2004; Dunn *et al.*, 2013). In another study (Nah *et al.*, 2020), the cascade model developed in Nah *et al.* (2019) was used to project the transmission risk of TBE virus in the enzootic cycle with projected climate conditions derived from climate data in the study region (Vas County, Hungary) and during the study period: for 1961–1990, the CarpatClim-Hu database was used (Szalai *et al.*, 2013) and for future periods, the climate data were provided by the two regional climate models, ALADINClimate 4.5 (Csima and Horányi, 2008) and RegCM 3.1 (Torma, 2011; Torma *et al.*, 2011), driven by the A1B emission scenario of the Intergovernmental Panel on Climate Change's Special Report on Emissions Scenarios, which describes the radiative forcing of a carbon dioxide concentration of 850 ppm by 2100. In particular, using the established transmission model and the climate predictions of Vas County for 2021–2050 and 2071–2100, Nah *et al.* (2020) quantified the risk of TBE virus transmission by introducing and calculating a series of summative indices including the basic reproduction number, the duration of infestation, the stage-specific tick densities and the accumulated (tick) infections due to co-feeding transmission. The study also measured the significance of co-feeding transmission by observing the cumulative number of new transmissions through the non-systemic transmission route. It concluded that the transmission potential and the risk in the study site are expected to increase along with the increase in temperature for 2021–2050 and 2071–2100. This increase will be facilitated by the expected extension of the tick questing season, the increase in numbers of susceptible ticks (larval and nymphal) and the number of infected nymphal ticks co-feeding on the same hosts, leading to a compounded increase of infections through non-systemic transmission. An early study (Randolph and Rogers, 2000) provided ample evidence for the significance of TBE co-feeding transmission in Europe and showed that projected climate change would result in contraction of the range of TBE virus in some parts of Europe by lowering the degree of coincident feeding of larval and nymphal ticks. The study of Nah *et al.* (2020) showed that whether TBE prevalence increases or decreases in the tick–host enzootic cycle in a particular region depends highly on the patterns of the projected region-specific climate change.

39.1.3 Infestation dynamics and Allee effects

As co-feeding depends on local infection under the skin of the host rather than widespread distribution of the pathogen within the host, tick aggregation patterns are important for the efficacy of co-feeding transmission. How these patterns are formed from the interactions of vector attachment and host grooming behaviour, in the context of evolving vector and host populations due to birth, development and death, is an important question. The tick-on-host distribution is a consequence of multi-scale dynamical processes, including tick attaching and engorging as well as host grooming behaviours in a relatively shorter timescale (infestation); and the seasonal abundance of ticks in different post-egg stages and hosts in a relatively longer timescale (population dynamics). The multi-scale dynamical processes are influenced by many factors, including climate changes. As illustrated in Fig. 39.1, it is important to understand the infestation dynamics – the tick-on-host distribution dynamics – in order to describe pathogen co-feeding transmission. In particular, if we assume that the infection probability of a susceptible larval tick co-feeding with a single infected nymphal tick on the same host is c, then the infection probability of a susceptible larval tick co-feeding with n infected nymphal ticks on the same host is $1-(1-c)^n$. Therefore, the probability of the susceptible larval tick acquiring the infection from co-feeding transmission is $[1-(1-c)^8+1-(1-c)^2]/2$ and $1-(1-c)^5$, respectively, for the tick-on-host distribution patterns described in Fig. 39.1a and b, assuming the probability of the larval tick feeding on either host is the same. See also Zhang and Wu (2019) for further discussions.

Zhang and Wu (2020) developed a mathematical model for vector–host population dynamics that, in addition to the intrinsic demographic dynamics of both vector and host populations, incorporated the distribution dynamics of vector individuals on hosts governed by vector attachment and host grooming behaviours. In particular, a delay differential system was proposed that couples within-host (tick-on-host) dynamics and tick population dynamics by incorporating tick attachment and host grooming interactions into the dynamic change of membership of hosts indexed by their tick loads. The tick population is stratified according to a particular stage (nymph) of the tick population, for which the tick-on-host distribution is important for co-feeding pathogen transmission. The host population is further stratified according to

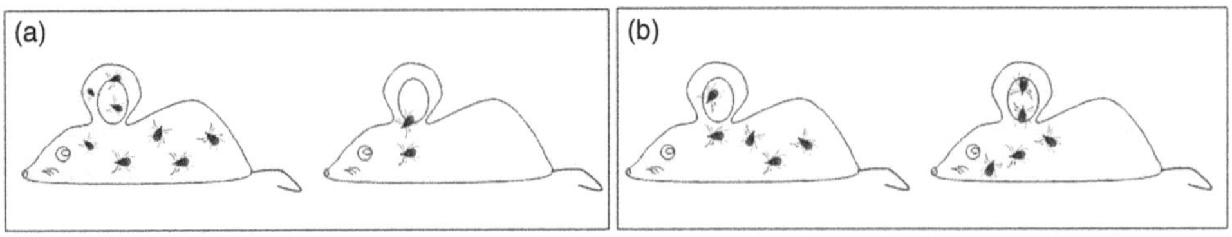

Fig. 39.1. Probability of a susceptible larval tick (not shown) acquiring an infection from co-feeding transmission. When a susceptible larval tick co-feeds in the proximity of n infected nymphal ticks, on the same host, the co-feeding transmission probability is $1-(1-c)^n$, assuming co-feeding transmission is an independent process and the co-feeding transmission probability between a susceptible larval tick and an infected nymphal tick is the constant, c. In the considered tick–host ecosystem with ten nymphal ticks, one larval tick and two hosts, the co-feeding probability of the susceptible larval tick acquiring infection transmitted by the infected nymphal ticks depends on the relative location of the larval tick and the distribution of nymphal ticks on the host. This co-feeding transmission probability is $1-(1-c)^8$ (respectively, $1-(1-c)^2$) in (a) when the larval tick feeds on the host on the left (respectively, right); and is $1-(1-c)^5$ in (b) regardless of which host the larval tick feeds on. In the figure, the larval tick is not shown, and it is assumed that this larval tick has equal probability of feeding on either host. (Authors' own figure.)

nymphal tick loads so the model system can be used to investigate how the membership dynamics of the host subpopulations (indexed by their tick loads) is governed by tick attachment preference and host grooming behaviour. It was shown that these individual tick and host behaviours have a strong influence on the long-term tick-on-host distribution and the tick population dynamics. Three special cases for different circumstances of tick attachment and host grooming behaviour were considered. In the case when the questing nymphal population is held at a constant value, the study observed a variety of tick-on-host distribution patterns. In particular, in the case when the tick attachment and host grooming rates are both constant, either the number of hosts is a decreasing function of the tick loads, or the number of hosts is decreasing first and then increasing when the tick loads increase. Which pattern can happen depends on whether the basic infestation number is less than or greater than unity. In the case when both attachment and grooming rates are dependent on tick load, a critical threshold was identified, with which the tick-on-host pattern can be fully characterized. Interesting patterns of tick-on-host distribution can be generated when the host grooms in response to tick bites, with these patterns being completely governed by tick attachment preference. In the special case where tick clusters on hosts sustain a certain level of tick load, a bimodal tick-on-host distribution was observed. These tick-on-host patterns, coupled with the developmental delay in tick populations, generate rich population dynamics, including extinction, logistic growth to a positive equilibrium, logistic growth to a positive periodic solution, and bi-stability where the coupled system has two positive equilibria and the limiting behaviour of the tick population and vector-on-host distribution both depend on the initial conditions. This seems to be the first work that reports this bi-stability (and Allee effects) of delay differential systems arising from the combination of individual tick attachment and host grooming behaviour. The tick aggregation pattern has been modelled using a negative binomial distribution; this binomial distribution was used by Rosà *et al.* (2003) in a study of the dynamics of tick-borne pathogen transmission through co-feeding. The study of Zhang and Wu (2020) takes a mechanistic approach to show that the negative binomial distribution pattern follows from density-dependent attachment and grooming behaviour; however, other distribution patterns are also possible. The particular tick attachment and host grooming behaviour combined influence the quality of the host species as a host (Keesing *et al.*, 2009).

39.2 Future Predictions

Future predictions of climate change impacts on co-feeding transmission dynamics require further refinement of the model formulation to account for diapause (see Zhang *et al.*, 2017 for some preliminary analyses). The role and importance of diapause have been examined in several modelling studies. The survey by Gray *et al.* (2016) identified gaps in current knowledge that require attention by researchers. Co-feeding depends on the co-occurrence of infected and susceptible ticks, perhaps in different stages, on the same host. This co-occurrence is highly seasonal. For example, in *I. ricinus*, larval ticks normally start their questing slightly later than nymphal and adult ticks and reach their highest activity peaks prior to the second questing peak of nymphal and adult ticks, creating opportunities for co-feeding transmission. Due to diapause, activity peaks of ticks in a particular stage may consist of individuals of different generational cohorts, participating in the same co-feeding process.

Disclaimer

The research was partially funded by the Natural Science Foundation of China (X.Z.) and by the GlaxoSmithKline's Investigator Sponsored Studies 'Estimating risk of tick-borne encephalitis with changes in climate, habitation and recreational activities', the Canada Research Chairs programme and the Natural Sciences and Engineering Research Council of Canada (J.W.). The opinions do not represent the views of the funding agencies.

References

Alekseev, A. and Chunikhin, S. (1990) Exchange of the tick-borne encephalitis virus between ixodid ticks feeding jointly on animals with a subthreshold level of viremia. *Medicinskaja Parazitologija I Parazitarnye Bolezni* 2, 48–50.

Csima, G. and Horányi, A. (2008) Validation of the ALADIN-climate regional climate model at the Hungarian Meteorological Service. *Időjárás* 112, 155–177.

Dunn, J., Davis, S., Stacey, A. and Diuk-Wasser, M. (2013) A simple model for the establishment of tick-borne pathogens of *Ixodes scapularis*: a global sensitivity analysis of R_0. *Journal of Theoretical Biology* 335, 213–221.

Gray, J.S., Kahl, O., Lane, R.S., Levin, M.L. and Tsao, J.I. (2016) Diapause in ticks of the medically important *Ixodes ricinus* species complex. *Ticks and Tick-borne Diseases* 7, 992–1003.

Hua, C., Cheminade, Y., Perret, J., Weynants, V., Lobet, V. and Gern, L. (2003) Early detection of *Borrelia burgdorferi* sensu lato infection in balb/c mice by co-feeding *Ixodes ricinus* ticks. *International Journal of Medical Microbiology* 293, 421–426.

Jones, L.D., Davies, C.R., Steele, G.M. and Nuttall, P.A. (1987) A novel mode of arbovirus transmission involving a nonviremic host. *Science* 237, 775–777.

Keesing, F., Brunner, J., Duerr, S., Killilea, M., LoGiudice, K., *et al.* (2009) Hosts as ecological traps for the vector of Lyme disease. *Proceedings of the Royal Society B* 276, 3911–3919.

Labuda, M., Jones, L.D., Williams, T., Danielova, V. and Nuttall, P.A. (1993) Efficient transmission of tick-borne encephalitis virus between cofeeding ticks. *Journal of Medical Entomology* 30, 295–299.

Nah, K., Magpantay, F.M.G., Bede-Fazekas, A., Rost, G., Trajer, A.J., *et al.* (2019) Assessing systemic and non-systemic transmission risk of tick-borne encephalitis virus in Hungary. *PLoS ONE* 14, e0217206.

Nah, K., Bede-Fazekas, Á., János Tráajer, A. and Wu, J. (2020) The potential impact of climate change on the transmission risk of tick-borne encephalitis in Hungary. *BMC Infectious Diseases* 20, 34.

Norman, R., Ross, D., Laurenson, M.K. and Hudson, P.J. (2004) The role of non-viraemic transmission on the persistence and dynamics of a tick-borne virus – louping ill in red grouse (*Lagopus lagopus scoticus*) and mountain hares (*Lepus timidus*). *Journal of Mathematical Biology* 48, 119–134.

Randolph, S.E. and Rogers, D.J. (2000) Fragile transmission cycles of tick-borne encephalitis virus may be disrupted by predicted climate change. *Proceedings of the Royal Society B* 267, 1741–1744.

Rosà, R., Pugliese, A., Norman, R. and Hudson, P.J. (2003) Thresholds for disease persistence in models for tick-borne infections including non-viraemic transmission, extended feeding and tick aggregation. *Journal of Theoretical Biology* 224, 359–376.

Szalai, S., Auer, I., Hiebl, J., Milkovich, J., Radim, T., *et al.* (2013) *Climate of the Greater Carpathian Region, Final Technical Report* (www.carpatclim-eu.org, CarpatClim datafiles 1961-01-01–2010-12-31). Joint Research Centre, European Commission and CARPATCLIM Consortium.

Taba, P., Schmutzhard, E., Forsberg, P., Lutsar, I., Ljøstad, U., *et al.* (2017) EAN consensus review on prevention, diagnosis and management of tick-borne encephalitis. *European Journal of Neurology* 24, 1214–1261.

Torma, C. (2011) Átlagos és szélsőséges hőmérsékleti és csapadék viszonyok modellezése a kárpát-medencére a xxi. századra a RegCM regionális klímamodell alkalmazásával [Modeling of average and extreme temperature and precipitation conditions for the Carpathian Basin in xxi. century using the RegCM regional climate model]. PhD thesis, Eötvös Loránd University, Budapest.

Torma, C., Coppola, E., Giorgi, F., Bartholy, J. and Pongrácz, R. (2011) Validation of a high-resolution version of the regional climate model RegCM3 over the Carpathian basin. *Journal of Hydrometeorology* 12, 84–100.

Voordouw, M.J. (2015) Co-feeding transmission in Lyme disease pathogens. *Parasitology* 142, 290–302.

Wu, J. and Zhang, X. (2020) *Transmission Dynamics of Tick-Borne Diseases with Co-Feeding, Developmental and Behavioural Diapause*. Springer, Cham, Switzerland.

Zhang, X. and Wu, J. (2019) Critical diapause portion for oscillations: parametric trigonometric functions and their applications for Hopf bifurcation analyses. *Mathematical Methods in the Applied Sciences* 42, 1363–1376.
Zhang, X. and Wu, J. (2020) Implications of vector attachment and host grooming behaviour for vector population dynamics and distribution of vectors on their hosts. *Applied Mathematical Modelling* 81, 1–15.
Zhang, X., Wu, X. and Wu, J. (2017) Critical contact rate for vector–host pathogen oscillation involving co-feeding and diapause. *Journal of Biological Systems* 2, 657–675.

eo40 Human Behaviour Trumps Entomological Risk

Sam Telford*
Tufts University, North Grafton, Massachusetts, USA

40.1 Expert Opinion

Although there is implicit recognition that the epidemiology of tick-borne infections is not necessarily represented by the ecology of the tick and its microbial guild, much research conflates the two. Studies and speculations about changes in prevalence or incidence of human disease that may be driven by climate are largely based in ecological terms centred on tick range expansion or local intensification of transmission cycles. I argue that human behaviour is as important a component to risk as is vectorial capacity. Speculation about climate change must incorporate human behavioural factors; future research agendas would be incomplete and likely very misleading if based solely on simplistic metrics like prevalence of infection in host-seeking ticks or range expansion anticipated from models of suitable tick habitat.

There is an apparent disconnect between the enzootic and the zoonotic transmission of tick-borne infections that is poorly recognized by research perspectives and the vast literature to date. Enzootic cycles (natural cycle, maintenance cycle) of tick-borne agents comprise the elements that are required to ensure that their basic reproduction number (R_0) remains >1 (≥1 is perpetuation, <1 is extinction). Enzootic cycles may be epidemiologically silent; that is, people are not exposed. Zoonotic transmission is a result of bridging from the enzootic cycle, usually due to: (i) tick anthropophily; (ii) spillover or overflow from the enzootic cycle (R_0 is vastly more than needed to ensure perpetuation); and/or (iii) increased human exposure. Human exposure is the *sine qua non* of the zoonotic condition, although there is a tautology here given that the definition of zoonosis is conditional on human infection and disease. Most of the contributions in this book focus on changes in the enzootic cycle that may result from climate change; without consideration of human exposure, such studies may have no epidemiological relevance.

One pertinent example of the research perspective disconnect may be seen in the epidemiology of Lyme disease and the other deer tick-transmitted infections in the eastern USA. There are literature reports ad nauseum regarding the prevalence of infection in host-seeking ticks, under the assumption that the entomological risk index (ERI) (Mather *et al.*, 1989) or density of infected nymphs (DIN) (Pepin *et al.*, 2012) provides a measure of human risk of acquiring these infections (Stafford *et al.*, 1998; Eisen and Eisen, 2016). ERI or DIN is also widely used as an index of the force of enzootic transmission (Levi *et al.*, 2016), but this requires

*Email: sam.telford@tufts.edu

DOI: 10.1079/9781789249637.0040

re-examination (Telford and Goethert, 2021). Generally, the prevalence of *Borrelia burgdorferi* sensu stricto (s.s.) infection for nymphal deer ticks (*Ixodes dammini*) in the north-eastern USA is 12–25%. For adult deer ticks, it is usually double that or more, 25–65%, because infection may be acquired by larval as well as nymphal feeding (Spielman *et al.*, 1985). However, >85% of all Lyme disease cases are reported during June and July when only nymphal deer ticks are seeking hosts (Fig. 40.1). Based solely on ERI, one would predict that more cases would be acquired in colder months, when adult deer ticks are active, if adult and nymphal tick abundance indices are equivalent. Although absolute numbers of sub-adults are vastly larger than those of adult ticks, when dragging or flagging, nymphal and adult *I. dammini* abundance indices within the same site tend to be similar (Ginsberg and Ewing, 1989; S. Telford, 1985–2020, personal observations).

Risk for Lyme disease is reduced during the adult deer tick season of activity in north-eastern states by four factors: (i) people are dressed for colder weather, which reduces the likelihood that ticks can get access to bare skin; (ii) people have fewer cumulative hours of outdoor activity (in part because daylength is shorter, but also maybe because the average duration of tolerable outdoor activity is shorter in colder weather than in warm) during the colder months when adult ticks are around; (iii) adult deer ticks are much larger (2–4 mm body length) than are nymphs (1 mm) and thus are found more readily and removed; and (iv) the duration of attachment required for effective transmission may differ (24 h for nymphs, 48 h for adults; Eisen, 2018), hence adult ticks are more likely to be removed before transmitting. These largely personally driven behavioural factors trump considerations of risk based on ERI.

In the case of the continued expansion of risk of Lyme disease and the other deer tick-transmitted zoonoses in the north-eastern USA, human factors may be more influential than are entomological factors for risk at the community or population level as well, particularly for temporal trends. The risk landscape is dynamic and based on changes in land use, human activity and fauna/flora, as well as human demography. The three environmental factors that led to the emergence of Lyme disease (and the other deer tick-transmitted infections) in the north-eastern USA are anthropogenic. The emergence of Lyme disease and the other deer tick-transmitted

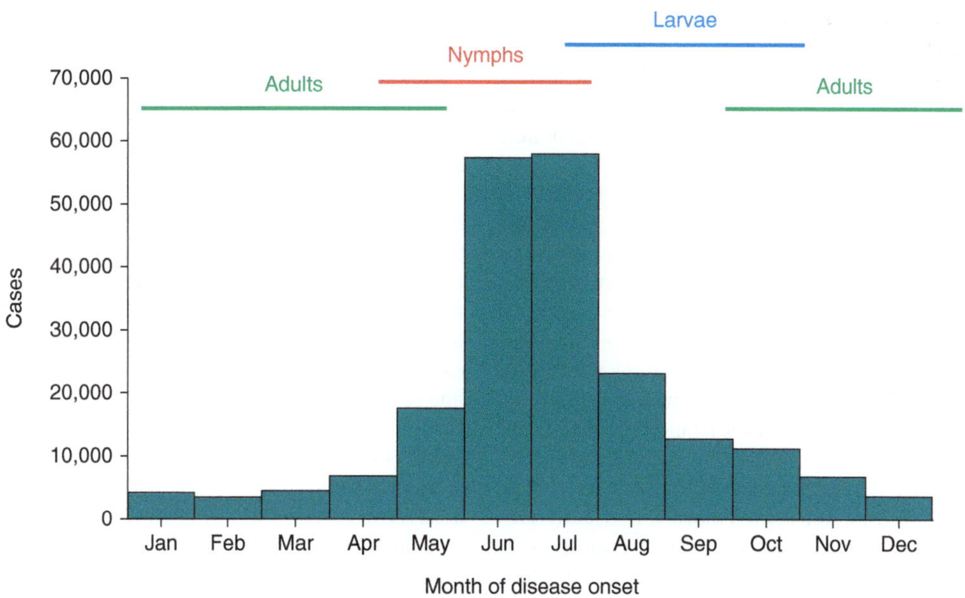

Fig. 40.1. Acute Lyme disease seasonality of onset, 2008–2018 (modified from https://www.cdc.gov/lyme/stats/graphs.html, accessed 18 October 2020), with phenology of deer ticks in the north-eastern USA. (Author's own figure.)

infections has been previously discussed (e.g. Spielman *et al.*, 1985) but the story is repeated briefly here, with an emphasis on human actions:

1. Forests were cut down during colonial times (Cronon, 1983) and the landscape at the turn of the 20th century was generally one of fields with scattered patches of shrubs and trees. With abandonment of farms and migration to cities, fields became reforested, yielding a dynamic successional landscape in which deer ticks, their hosts and microbial assemblages thrive. Microhabitats, fauna and flora change dynamically during succession, but even early successional forest can be extremely biodiverse (Swanson *et al.*, 2010). The deer tick is sensitive to microhabitats, with adults surviving longer in forested or edge sites than in fields (Bertrand and Wilson, 1996), implying that dense populations of ticks would be sustained in edge habitats characteristic of successional sites. In the context of a discussion on the relative importance of anthropogenic as opposed to biotic factors in promoting tick abundance, one might argue that habitat and microhabitat are biotic factors and can be analysed on their own; but these biotic factors would not exist without human action to generate and maintain the habitats.

2. Suburbanization and increased recreation promoted human exposure to the new forests. The process of suburbanization continues today. For example, in Massachusetts, 40 acres of land was cleared every day from 1985 to 1999 (Breunig, 2003) to make way for residential developments. Summer (often defined in the USA as starting the end of May, or Memorial Day weekend, and ending on Labour Day weekend, the first week of September) marks a time for vacation and leisure time. On Nantucket Island, the onset of good nymphal deer tick activity coincides with Memorial Day weekend (Piesman *et al.*, 1987). Nantucket Island, Massachusetts, perennially ranks in the top ten US sites for reported Lyme disease cases (Bacon *et al.*, 2008) and nearby Martha's Vineyard is usually very similar in incidence. Their year-round populations greatly expand during the summer (when nymphal deer ticks are most active): about 11,000 year-round residents swell to >50,000 people on Nantucket; and a year-round population of 17,000 swells to >200,000 on the Vineyard. Timing of vacations and summer recreational activity greatly promotes risk.

3. Game management, suburbanization, lack of natural predators and a decreasing population of hunters allowed deer to greatly expand their abundance and distribution from refugial sites (Severinghaus and Brown, 1956). The abundance of deer ticks is determined by that of their main reproductive host (Telford, 2017). Environmental factors are ecological and not really human behaviour per se, but human behaviour developed and promoted the environmental conditions and continues to do so. Any analysis of the potential effects of climate change on tick abundance and distribution, with implied consequences for human risk of tick-borne infection, must examine confounding concurrent trends such as suburbanization, new successional habitat and the proliferation of backyard wildlife as a result of animal protection and lack of hunting.

On top of the 'well known' environmental emergence factors, changes in the demography of the human populations at risk may serve as an important basis for the recent increase in prevalence of reported American cases of human granulocytic anaplasmosis, babesiosis and Powassan encephalitis (Rosenberg *et al.*, 2018). On Nantucket Island, the population in the early 1980s, when the first Lyme disease cases were reported, was 5087; in 2019, it was 10,172. A doubling in the population at risk, with no change in entomological risk and no change in tick preventive behaviour, would imply that there would be double the number of cases reported in more recent years compared with the early years. With all of the deer tick-transmitted infections other than Lyme disease, those older than 60 years are more likely to suffer overt disease and thus be seen by a healthcare provider and be reported to state surveillance programmes. Risk of hospitalization for babesiosis due to *Babesia microti*, for example, is greatest in those >50 years of age (Gray and Herwaldt, 2019). On Nantucket, a 12% increase in the >65-year-old population was seen from 2010 to 2019. Any changes in predicted risk for tick-borne infections must take human demography into consideration.

My own perspective on identifying predictors for risk has been greatly modified over the years, with my conclusion being that human behaviour trumps entomological factors in zoonotic risk for tick-borne infections, at least here in the north-eastern USA. Like all other medical entomologists, I was faithfully doing longitudinal measurements of tick abundance by dragging and mouse indices, along with 'infection rates', with the assumption that they meant something in measuring human risk. This assumption started to crumble for me when I found no association between entomological indices and case estimates based on acute serologies from Nantucket Cottage Hospital. This hospital is a unique defined source for case data given that it is the only healthcare facility on the island which is 30 km off the Massachusetts coast; all residents and long-term visitors must go there for their healthcare needs. Then, in 1993, I was working for a couple of weeks with Ed Korenberg and his team in his western Urals tick-borne encephalitis (TBE) study site (Mys, Chusavoi district, Russia), to look for *B. microti* and *Anaplasma phagocytophilum* transmission (Telford *et al.*, 2002). I asked Ed what predicted TBE cases. Ed: 'It is very simple. Look outside. What is happening?' Me: 'Ed, it is raining.' Ed: 'Yes, it is raining. Are ticks questing in the rain?' Me: 'No, Ed.' Ed: 'Are people outside in the rain?' Me: 'No, Ed.' Ed: 'Number of cases is inversely proportional to number of rainy days.' I was so gobsmacked by this common-sense logic that I remember that conversation, verbatim, to this day. I never think of the risk of tick-transmitted infections solely in terms of vectorial capacity anymore.

When standardized case report data became available through the Massachusetts Department of Public Health, I did a quick examination of Ed Korenberg's remarkable statement (Fig. 40.2). June rainfall was examined because the month of greatest nymphal deer tick activity on Nantucket is June. Although the correlation was not significant, nevertheless a trend is apparent, with those years in which June is marked by low rainfall tending to have more cases. The literature that examines the question of the influence of rainfall on Lyme disease incidence has generally looked for effects of an interaction with tick abundance or phenology and with time lags due to delayed effects from weather acting on larvae the previous year (Schauber *et al.*, 2005; Moore *et al.*, 2014; Eisen *et al.*, 2016). Atmospheric moisture tended to be related to nymphal tick abundance (with the inference that an absence of saturation deficit allowed for greater amounts of questing) (Subak, 2002; McCabe and Bunnell, 2004; Rodgers *et al.*, 2007). It may also be that the use of case reports from a larger area diminishes the capacity to identify local effects. Correlation of case reports with rainfall, of course, does not imply causation and there must be multifactorial associations working

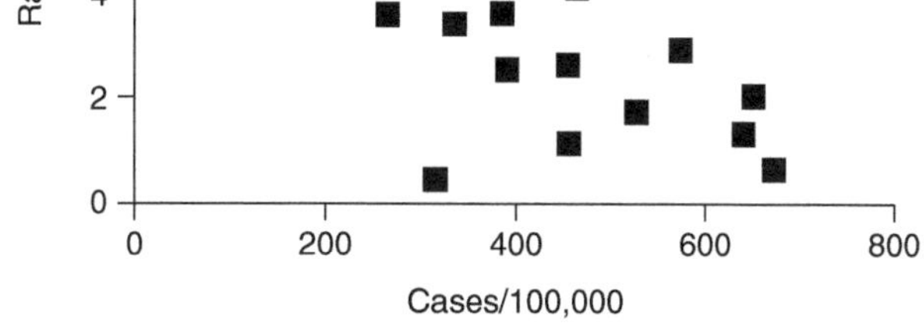

Fig. 40.2. Nantucket Island, 2000–2015: cases of Lyme disease reported to the Massachusetts Department of Public Health (expressed as number per 100,000) and their association with total June rainfall on Nantucket in inches (from https://www.ncdc.noaa.gov/cdo-web/datasets/GHCND/stations/GHCND:USW00014756/detail, accessed 18 October 2020) (Pearson $r = -0.44$, $P = 0.09$). (Author's own figure.)

with rainfall and other weather variables to better predict risk. At the very least, the Korenberg effect should prompt every one of us working on the epidemiology of tick-transmitted infections to remember that ultimately it is likely to be human behaviour that defines risk.

40.2 Future Predictions

Weather has a large influence on tick activity and survival via a well-established mechanism (saturation deficit). Dry weather and greater temperatures curtail tick activity (Rodgers *et al.*, 2007). Vector-borne agents in general may be affected by extrinsic incubation temperature although the upper and lower limits for influencing vectorial capacity differ between them and largely remain to be defined. Accordingly, there are ERI-modulating factors with good biological rationale that convince medical entomologists to focus on vector competence, longevity and abundance, in their pursuit of understanding possible changes in the ecology (distribution and abundance) and epidemiology of vector-borne infections under scenarios of climate change.

I argue that human aspects of risk are poorly studied and need to be considered in any discussion of the role of global climate change on risk for vector-borne infections, particularly for those transmitted by ticks. Many models exist that predict the future distribution of Lyme disease and other tick-borne infections in the USA and elsewhere, based on anticipated changes in weather. I note that everyone makes models, but few such models are empirically tested with ground truth data. Empirically, the warm extremes of the distribution of Lyme disease hold valuable information that might be used to anticipate the effects of a global climate change to a more variable, likely warmer environment. In the southern USA, Lyme disease is very rare even though enzootic cycles are stably present. The force of transmission of *B. burgdorferi* sensu lato (s.l.) in the southern USA is comparatively less than that in the north-east (Pepin *et al.*, 2012) although dogs are focally exposed (Little *et al.*, 2014). Human Lyme disease cases, interestingly, are rarely reported in southern states (Schwartz *et al.*, 2017). Vector behaviour, particularly nymphal anthropophily, certainly helps explain some of this, as I have argued in the context of my continued use of the junior subjective synonym *I. dammini* (*Ixodes scapularis*) (Telford, 1998). But adult deer ticks do bite people (Goddard, 2002) and can certainly transmit *B. burgdorferi*. There also appears to be a disconnect between the incidence of human ehrlichiosis (Heitman *et al.*, 2015) and the known abundance of lone star ticks in Florida and the Gulf states (Springer *et al.*, 2015). It may be that there are human behaviours specific to residents of the warmer southern states that modify risk of tick-borne infections.

I suggest the 'mad dogs and Englishmen' hypothesis. The weather during much of the day in more southern sites such as Florida is just not conducive to either ticks or people being out together. Ticks are likely to be more crepuscular or nocturnal as opposed to cathemeral. People other than madmen are going to be inside their air-conditioned or otherwise cooler buildings; they are less likely to have much activity except during the dawn, dusk or night-time hours. The powerful risk-modulating effect of air conditioning (and window screens) has been nicely demonstrated for dengue (Reiter *et al.*, 2003). A major flaw of speculation on the effect of 'global warming' on vector-borne disease risk is a failure to recognize the compensatory behaviours that people may develop as a result of temperatures. With warming will come increased use of air conditioners and mid-day siestas. I predict that there will be no net gain in global risk for tick-borne infections that is directly attributable to weather trends; yes, there may be focal increases in new sites, but such increases may be balanced by reduced zoonotic transmission and/or changes in human behaviour in many currently afflicted sites.

As Sarah Randolph and her associates persuasively argued in what I regard as a seminal paper, human factors are as important as enzootic parameters, if not more so, in TBE risk (Sumilo *et al.*, 2007); we should expect the same for any other tick-borne zoonosis. Medical entomologists need to spend more time studying humans and their interactions with the environment where ticks live than just studying ticks and the autecology of the microbes that they transmit.

Acknowledgements

I am supported by grants from the National Institutes of Health (R01 AI 137424; R01 AI 130105). Sarah Randolph would have written a much better commentary than I with this perspective of human behaviour and risk for tick-borne infection.

References

Bacon, R.M., Kugeler, K.J., Mead, P.S. and Centers for Disease Control and Prevention (2008) Surveillance for Lyme disease – United States, 1992–2006. *MMWR Surveillance Summaries* 57, 1–9.

Bertrand, M.R. and Wilson, M.L. (1996) Microclimate-dependent survival of unfed adult *Ixodes scapularis* (Acari: Ixodidae) in nature: life cycle and study design implications. *Journal of Medical Entomology* 33, 619–627. https://doi.org/10.1093/jmedent/33.4.619

Breunig, K. (2003) *Losing Ground: At What Cost? Changes in Land Use and Their Impact on Habitat, Biodiversity, and Ecosystem Services in Massachusetts*. Massachusetts Audubon Society, Lincoln, Massachusetts. Available at: https://www.massaudubon.org/content/download/8599/149714/file/LosingGround_All.pdf (accessed 8 June 2021).

Cronon, W. (1983) *Changes in the Land: Indians, Colonists, and the Ecology of New England*. Hill and Wang, New York.

Eisen, L. (2018) Pathogen transmission in relation to duration of attachment by *Ixodes scapularis* ticks. *Ticks and Tick-borne Diseases* 9, 535–542. https://doi.org/10.1016/j.ttbdis.2018.01.002

Eisen, L. and Eisen, R.J. (2016) Critical evaluation of the linkage between tick-based risk measures and the occurrence of Lyme disease cases. *Journal of Medical Entomology* 53, 1050–1062. https://doi.org/10.1093/jme/tjw092

Eisen, R.J., Eisen, L., Ogden, N.H. and Beard, C.B. (2016) Linkages of weather and climate with *Ixodes scapularis* and *Ixodes pacificus* (Acari: Ixodidae), enzootic transmission of *Borrelia burgdorferi*, and Lyme disease in North America. *Journal of Medical Entomology* 53, 250–261. https://doi.org/10.1093/jme/tjv199

Ginsberg, H.S. and Ewing, C.P. (1989) Comparison of flagging, walking, trapping, and collecting from hosts as sampling methods for northern deer ticks, *Ixodes dammini*, and lone-star ticks, *Amblyomma americanum* (Acari: Ixodidae). *Experimental and Applied Acarology* 7, 313–322. https://doi.org/10.1007/BF01197925

Goddard, J. (2002) A ten-year study of tick biting in Mississippi: implications for human disease transmission. *Journal of Agromedicine* 8, 25–32. https://doi.org/10.1300/J096v08n02_06

Gray, E.B. and Herwaldt, B.L. (2019) Babesiosis surveillance – United States, 2011–2015. *MMWR Surveillance Summaries* 68, 1–11. https://doi.org/10.15585/mmwr.ss6806a1

Heitman, K., Dahlgren, F., Drexler, N., Massung, R. and Behravesh, C.B. (2015) Increasing incidence of ehrlichiosis in the United States: a summary of national surveillance of *Ehrlichia chaffeensis* and *Ehrlichia ewingii* infections in the United States, 2008–2012. *American Journal of Tropical Medicine and Hygiene* 94, 52–60. https://doi.org/10.4269/ajtmh.15-0540

Levi, T., Keesing, F., Holt, R.D., Barfield, M. and Ostfeld, R.S. (2016) Quantifying dilution and amplification in a community of hosts for tick-borne pathogens. *Ecological Applications* 26, 484–498. https://doi.org/10.1890/15-0122

Little, S.E., Beall, M.J., Bowman, D.D., Chandrashekar, R. and Stamaris, J. (2014) Canine infection with *Dirofilaria immitis*, *Borrelia burgdorferi*, *Anaplasma* spp., and *Ehrlichia* spp. in the United States, 2010–2012. *Parasites & Vectors* 7, 257. https://doi.org/10.1186/1756-3305-7-257

Mather, T.N., Wilson, M.L., Moore, S.I., Ribeiro, J.M. and Spielman, A. (1989) Comparing the relative potential of rodents as reservoirs of the Lyme disease spirochete (*Borrelia burgdorferi*). *American Journal of Epidemiology* 130, 143–150. https://doi.org/10.1093/oxfordjournals.aje.a115306

McCabe, G.J. and Bunnell, J.E. (2004) Precipitation and the occurrence of Lyme disease in the northeastern United States. *Vector-Borne and Zoonotic Diseases* 4, 143–148. https://doi.org/10.1089/1530366041210765

Moore, S.M., Eisen, R.J., Monaghan, A. and Mead, P. (2014) Meteorological influences on the seasonality of Lyme disease in the United States. *American Journal of Tropical Medicine and Hygiene* 90, 486–496. https://doi.org/10.4269/ajtmh.13-0180

Pepin, K.M., Eisen, R.J., Mead, P.S., Piesman, J., Fish, D., *et al.* (2012) Geographic variation in the relationship between human Lyme disease incidence and density of infected host-seeking *Ixodes scapularis* nymphs in the eastern United States. *American Journal of Tropical Medicine and Hygiene* 86, 1062–1071. https://doi.org/10.4269/ajtmh.2012.11-0630

Piesman, J., Mather, T.N., Dammin, G.J., Telford, S.R. 3rd, Lastavica, C.C., *et al.* (1987) Seasonal variation of transmission risk of Lyme disease and human babesiosis. *American Journal of Epidemiology* 126, 1187–1189. https://doi.org/10.1093/oxfordjournals.aje.a114757

Reiter, P., Lathrop, S., Bunning, M., Biggerstaff, B., Singer, D., *et al.* (2003) Texas lifestyle limits transmission of dengue virus. *Emerging Infectious Diseases* 9, 86–89. https://doi.org/10.3201/eid0901.020220

Rodgers, S.E., Zolnik, C.P. and Mather, T.N. (2007) Duration of exposure to suboptimal atmospheric moisture affects nymphal blacklegged tick survival. *Journal of Medical Entomology* 44, 372–375. https://doi.org/10.1603/0022-2585(2007)44[372:doetsa]2.0.co;2

Rosenberg, R., Lindsey, N.P., Fischer, M., Gregory, C.J., Hinckley, A.F., *et al.* (2018) Vital signs: trends in reported vectorborne disease cases – United States and territories, 2004–2016. *MMWR Morbidity and Mortality Weekly Report* 67, 496–501. https://doi.org/10.15585/mmwr.mm6717e1

Schauber, E.M., Ostfeld, R.S. and Evans, A.S. Jr (2005) What is the best predictor of annual Lyme disease incidence: weather, mice, or acorns? *Ecological Applications* 15, 575–586. https://doi.org/10.1890/03-5370

Schwartz, A.M., Hinckley, A.F., Mead, P.S., Hook, S.A. and Kugeler, K.J. (2017) Surveillance for Lyme disease – United States, 2008–2015. *MMWR Surveillance Summaries* 66, 1–12. https://doi.org/10.15585/mmwr.ss6622a1

Severinghaus, C. and Brown, C. (1956) History of the white tailed deer in New York. *New York Fish and Game Journal* 3, 130–132.

Spielman, A., Wilson, M.L., Levine, J.F. and Piesman, J. (1985) Ecology of *Ixodes dammini*-borne zoonoses. *Annual Reviews of Entomology* 30, 439–460. https://doi.org/10.1146/annurev.en.30.010185.002255

Springer, Y.P., Jarnevich, C.S., Barnett, D.T., Monaghan, A.J. and Eisen, R.J. (2015) Modeling the present and future geographic distribution of the lone star tick, *Amblyomma americanum* (Ixodida: Ixodidae), in the continental United States. *American Journal of Tropical Medicine and Hygiene* 93, 875–890. https://doi.org/10.4269/ajtmh.15-0330

Stafford, K.C. 3rd, Cartter, M.L., Magnarelli, L.A., Ertel, S.H. and Mshar, P.A. (1998) Temporal correlations between tick abundance and prevalence of ticks infected with *Borrelia burgdorferi* and increasing incidence of Lyme disease. *Journal of Clinical Microbiology* 36, 1240–1244. https://doi.org/10.1128/JCM.36.5.1240-1244.1998

Subak, S. (2002) Analysis of weather effects on variability in Lyme disease incidence in the northeastern United States. *Experimental and Applied Acarology* 28, 249–256. https://doi.org/10.1023/a:1025367020459

Sumilo, D., Asokliene, L., Bormane, A., Vasilenko, V., Golovljova, I., *et al.* (2007) Climate change cannot explain the upsurge of tick-borne encephalitis in the Baltics. *PLoS ONE* 2, e500. https://doi.org/10.1371/journal.pone.0000500

Swanson, M.E., Franklin, J.E., Beschta, R.L., Crisafulli, C.M., Dominick, A., *et al.* (2010) The forgotten stage of forest succession: early successional ecosystems on forest sites. *Frontiers in Ecology and the Environment* 9, 117–125. https://doi.org/10.1890/090157

Telford, S.R. (1998) The name *Ixodes dammini* epidemiologically justified. *Emerging Infectious Diseases* 4, 132–134. https://doi.org/10.3201/eid0401.980126

Telford, S.R. (2017) Deer reduction is a cornerstone of integrated deer tick management. *Journal of Integrated Pest Management* 8, 25. https://doi.org/10.1093/jipm/pmx024

Telford, S.R. and Goethert, H.K. (2021) Perpetuation of Borreliae. In: Radolf, J. and Samuels, S. (eds) *Borrelia: Molecular Biology, Host Interaction, and Pathogenesis*, 2nd edn. Caister Academic Press, Norfolk, UK (in press).

Telford, S.R., Korenberg, E.I., Goethert, H.K., Kovalevskii, Iu.V., Gorelova, N.B., *et al.* (2002) Detection of natural foci of babesiosis and granulocytic ehrlichiosis in Russia. *Zhurnal Mikrobiologii, Epidemiologii i Immunobiologii* 6, 21–25 (in Russian).

eo41 It's All in the Timing: Effect of Tick Phenology on Pathogen Transmission Dynamics

Maria Diuk-Wasser*
Columbia University, New York, USA

41.1 Expert Opinion

41.1.1 Introduction

The maintenance of tick-borne zoonotic pathogens involves complex interactions between tick vectors, a wide range of vertebrate hosts and pathogenic microorganisms. This complexity can enhance pathogen persistence if the multiple hosts exploited by the tick provide functional redundancy for pathogen transmission (Estrada-Peña *et al.*, 2017). This appears to be the case for *Borrelia burgdorferi* sensu lato (s.l.), the agent of Lyme disease (or Lyme borreliosis), which infects the majority of its *Ixodes scapularis* tick vector populations in the USA (Diuk-Wasser *et al.*, 2012). Accordingly, it has been called 'one of the most robust and biodiverse zoonotic disease systems known' (Randolph and Dobson, 2012). On the other hand, the complexity may lead to 'fragile' transmission cycles (Randolph and Rogers, 2000), exemplified by the tick-borne encephalitis (TBE) virus, which requires stringent timing of immature ticks' seasonal activity to persist in restricted geographic foci (Randolph, 2001). Identifying the strengths and vulnerabilities in the enzootic transmission cycle of tick-borne pathogens is key to predicting their current and forecasting their future distribution and to develop effective public health interventions.

The 'fragility' of pathogen cycles in ixodid three-host ticks arises because they feed only once per life stage (as a larva, nymph and adult), which provides limited opportunities for pathogen transmission. Furthermore, feedings are spaced across a 2-year (or longer) life cycle, requiring horizontal transmission across overlapping generations for pathogen survival. Pathogen persistence depends on the 'infecting' (generally nymphal) stage infecting the vertebrate hosts and hosts remaining alive and infectious long enough to infect the 'receiving' (generally larval) stage (Ogden *et al.*, 2007). This delicate balance is determined by the interactions between tick phenology (the relative timing of different tick life stages), the duration of infection or infectivity in the host and the life expectancy of hosts. This expert opinion discusses how the prevalence and persistence of common *Ixodes* spp.-borne pathogens in the northern hemisphere depends on this delicate balance; how alternative pathways can enhance persistence; and how these climate-sensitive processes may drive tick and pathogen geographic distribution.

*Email: mad2256@columbia.edu

DOI: 10.1079/9781789249637.0041

41.1.2 The 'canonical' seasonal feeding inversion: robust persistence of *Borrelia burgdorferi*

Lyme disease is the most prevalent tick-borne disease in the northern hemisphere. Caused by *B. burgdorferi* s.l., it is transmitted by ticks of the *Ixodes ricinus* species complex, including *I. scapularis* (eastern USA), *Ixodes pacificus* (western USA), *I. ricinus* (Europe) and *Ixodes persulcatus* (Asia) (Diuk-Wasser *et al.*, 2016). After hatching, larval ticks take one blood meal during the late summer months, moult to nymphs, undergo diapause over the winter and take another blood meal during their second year in late spring and early summer (Wilson and Spielman, 1985; Yuval and Spielman, 1990). Because the seasonal activity peak of the first cohort nymphs (May–June) precedes the majority of activity of second cohort larvae (August) in the northeastern USA, infected nymphs can transmit the pathogen to a large proportion of hosts before larvae feed on these same hosts in late summer, maximizing transmission efficiency. This seasonal feeding inversion and asynchrony of tick immature activity has been postulated to explain the high prevalence of *B. burgdorferi* in questing *I. scapularis* nymphs and Lyme disease incidence in the USA, particularly in the Northeast region (Wilson and Spielman, 1985; Fish, 1993; Thompson *et al.*, 2001; Ogden *et al.*, 2007).

41.1.3 Synchrony in immature activity increases the fitness of pathogens and strains with short infectious periods: *Anaplasma phagocytophilum* and some *Borrelia burgdorferi* sensu stricto strains

The high efficiency of horizontal transmission of *B. burgdorferi* via the seasonal feeding inversion is dependent on the survival of the hosts and the persistence of a transmissible infection between the two feeding peaks (Gatewood *et al.*, 2009; Hamer *et al.*, 2012). While many *B. burgdorferi* strains produce lifelong infections in white-footed mice (the key reservoir host) (Donahue *et al.*, 1987), some *B. burgdorferi* strains, as well as other *I. scapularis*-borne pathogens (e.g. *Anaplasma phagocytophilum*), are rapidly cleared in their rodent hosts (Ogden *et al.*, 2007; Hanincova *et al.*, 2008; Dunn *et al.*, 2014), posing the question of how these strains and pathogens persist in nature. Dynamic modelling illustrates how these pathogens/strains require more synchronous tick phenology to persist (Ogden *et al.*, 2007). In the Northeast, *I. scapularis* nymphs are typically active several months before the larvae, imposing selective pressure on strains/pathogens for persistent infections (Fig. 41.1a). In contrast, nymphs and larvae more often feed more synchronously in the Upper Midwest USA, enabling survival of strains/pathogens with short infectious periods and/or transmitted via co-feeding (Fig. 41.1b).

Consistent with model predictions, persistent *B. burgdorferi* strains were found to be relatively more prevalent in the Northeast USA where immature tick activity is asynchronous, presumably because the larger temporal gap between nymphal and larval feeding confers a selective advantage to these strains in relation to strains with a shorter infectious period (Gatewood *et al.*, 2009). These genotypic differences in *B. burgdorferi* were found to be important predictors of human Lyme disease incidence, in combination with climatic factors and acarological indices (tick density) (Pepin *et al.*, 2012). In contrast, the more synchronous phenology in the Upper Midwest may increase the fitness of pathogens and strains with short infectious periods such as *A. phagocytophilum*. This may explain the higher ratio of human granulocytic anaplasmosis (HGA) relative to Lyme disease cases in Upper Midwest relative to Northeast states (CDC, 2020a,b). A newly discovered *Borrelia* spp., *Borrelia mayonii*, has only been detected in the Upper Midwest, although it remains to be determined if this is due to a shorter infectious period (Pritt *et al.*, 2016).

41.1.4 Stringent seasonal synchrony required for co-feeding transmission: tick-borne encephalitis virus in Europe, Powassan virus and *Ehrlichia muris eauclairensis* in the USA

Immature feeding synchrony has been found to be a prerequisite for the persistence of TBE virus, a flavivirus transmitted by *I. ricinus* ticks in Europe (Randolph *et al.*, 1999, 2000). Precise

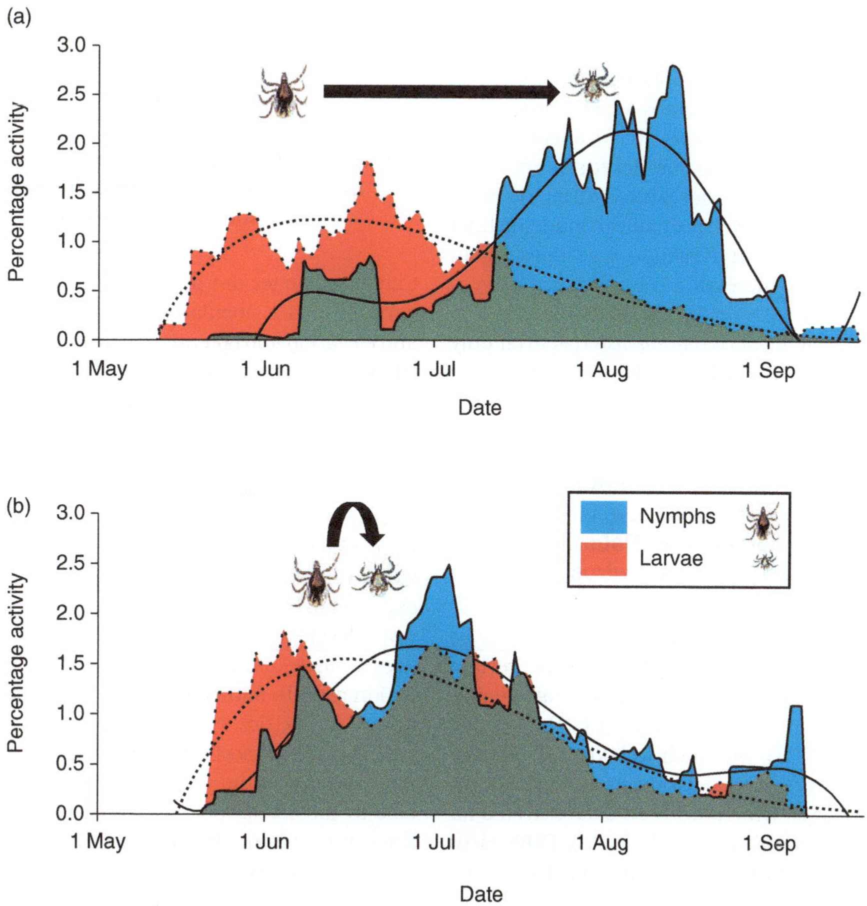

Fig. 41.1. Seasonal activity curves of immature *Ixodes scapularis* in the Northeast (a) and Upper Midwest (b) USA. Two-week moving averages are shown as solid lines with red shading for larvae, and as dotted lines with teal shading for nymphs. Fifth-order polynomial trend lines fitted for illustration purposes are shown in solid black (larvae), stippled black (nymphs). The arrows illustrate the approximate peak time between nymphal infection of the host and larval pathogen acquisition: approximately 2 months in the Northeast; under a month in the Upper Midwest (the curvature illustrating the possibility of co-feeding). (Adapted from Gatewood *et al*., 2009.)

synchrony is required because of the very short infectious period in rodent hosts and also for transmission between co-feeding ticks. Co-feeding or non-systemic transmission is a form of horizontal transmission between infected and uninfected ticks feeding in close proximity on hosts that develop non-systemic infection; it can even occur on immune hosts although at a reduced level (Labuda *et al.*, 1997; Ogden *et al.*, 1997; Randolph, 2011). Modelling studies postulate co-feeding has a greater contribution (>50%) to amplification of TBE virus than the systemic pathway (Randolph *et al.*, 1996) and empirical studies have found an association between frequency of co-feeding and TBE hazard (the occurrence of the pathogen in the enzootic cycle) over time (Rosà *et al.*, 2019) and over space (Randolph *et al.*, 2000). Co-feeding also plays an important role in the maintenance of the deer tick virus lineage of Powassan (POW) virus, a flavivirus transmitted by *I. scapularis* in the USA. Seasonal synchrony may thus explain the higher ratio of POW viral encephalitis relative to Lyme disease in the Upper Midwest compared with the Northeast USA (CDC, 2020b,c). Co-feeding transmission has also been demonstrated for non-viral pathogens, namely *B. burgdorferi* in Europe (Gern and Rais, 1996) and in the USA for short infectious strains (States *et al.*, 2017) as well as for *Ehrlichia muris eauclairensis* (Pritt *et al.*, 2011; Johnson *et al.*, 2015), a recently discovered *I. scapularis*-borne rickettsial pathogen with a short infectious period (Karpathy *et al.*, 2016). This pathogen is unique in inducing high mortality rates in *Peromyscus leucopus* (Lynn *et al.*, 2017), increasing the relevance of synchronous tick phenology for rapid transmission. Consistently, *E. muris eauclairensis* has so far only been detected in the Upper Midwest USA.

41.1.5 Alternative (non-horizontal) transmission pathways enhance pathogen persistence: *Babesia microti* and Powassan virus

Horizontal and co-feeding transmission may be insufficient for the maintenance of certain strains or pathogens, requiring alternative transmission pathways to persist (Nuttall and Labuda, 2003). One such well-known pathway is transovarial transmission from female adult ticks to eggs, which has been demonstrated for TBE virus (Danielová *et al.*, 2002), POW virus (Kemenesi and Bányai, 2018) and *Borrelia miyamotoi* (Han *et al.*, 2019).

In addition, vertical transmission from female rodents to offspring has been demonstrated for TBE virus (Bakhvalova *et al.*, 2009) and *Babesia microti* in three *Microtus* spp. in Poland (Tolkacz *et al.*, 2017) as well as for *B. microti* in *P. leucopus* and *Microtus pennsylvanicus* in the Northeast USA (Tufts and Diuk-Wasser, 2018, 2021). The role of vertical transmission of *B. microti* in *P. leucopus* depends on the host's survival over winter and maintaining a persistent infection, the latter of which has been demonstrated in the laboratory (Tufts and Diuk-Wasser, 2021).

The relative importance of different transmission pathways in nature remains to be determined. Modelling studies indicate that multiple pathways are likely to play a role in maintaining POW virus (Nonaka *et al.*, 2010), while the high incidence of babesiosis compared with Lyme disease in the Northeast versus the Midwest USA (CDC, 2020b,d) indicates that warmer temperatures may facilitate host overwinter survival enabling vertical transmission of persistent infections in mice.

41.1.6 Effects of climate and weather on tick phenology and pathogen transmission dynamics

Climate changes are predicted to drive the expansion of tick populations to higher latitudes and elevation and potentially constrain their distribution in warmer, lower-latitude areas (Brownstein *et al.*, 2005; Medlock *et al.*, 2013; Simon *et al.*, 2014; Eisen *et al.*, 2016; Jaenson *et al.*, 2018; Rosà *et al.*, 2018; Sonenshine, 2018; Ogden *et al.*, 2020). However, the factors driving the fitness and prevalence of pathogens transmitted by these vectors are less well understood. Climate (in particular temperature) is not expected to influence the duration of the pathogen's development in the arthropod vector, as it does for mosquito vectors (i.e. the extrinsic incubation period). Rather, pathogen transmission is limited by the tick's off-host interstadial period, i.e. the period of post-engorgement development and questing for the next host. This period is specific to the vector, not the pathogen, and lasts from 1 to 12 months, depending on the tick stage, the temperature-dependent rate of development and whether diapause occurs. Pathogen development in the endothermic vertebrate host is similarly mostly decoupled from environmental conditions.

Here we focus on the indirect effect of weather and climate on the occurrence and force of infection in pathogen transmission cycles as driven by tick phenology. The seasonal activity patterns of ticks depend on the combined effects of climate on development rates from one life stage to the next and weather conditions suitable for host-seeking activity (Ogden *et al.*, 2020). Interstadial development rates of ixodid ticks are determined in part by temperature, with rates generally increasing, usually nonlinearly, with increasing temperature (Randolph *et al.*, 2002; Ogden *et al.*, 2004). However, simple predictions of the seasonality of ticks from relationships between developmental rates and temperature are often confounded by diapause phenomena, i.e. delayed development (morphogenetic diapause) or delayed host-seeking activity (behavioural diapause) (Ogden *et al.*, 2007). Cues for the onset and termination of diapause may be complex and include changes in day-length and exposure to different maxima and minima of temperature (Belozerov and Naumov, 2002; Belozerov *et al.*, 2002; Randolph *et al.*, 2002), with potentially strong impacts on geographic variation in phenological patterns (Ogden *et al.*, 2018).

Long-term studies have detected earlier immature host-seeking activity with increased temperature. In a time series study between 1994 and 2012 in the Northeast USA, Levi *et al.* (2015) found that warmer years, through May and August, were associated with a nearly 3-week advance in host-seeking activity of nymphal and larval ticks relative to colder years. This advance is consistent with the observed advance in the onset of Lyme disease in the USA (Moore *et al.*, 2014; Monaghan *et al.*, 2015), indicating increased human exposure to these earlier host-seeking ticks. In contrast, the peak in nymphal activity of *I. ricinus* was not found to advance in a long-term study (2000–2014) in Switzerland (Hauser *et al.*, 2018).

Of more relevance for tick-borne pathogen cycles, climate changes may influence the feeding synchrony between larvae and nymphs. Levi *et al.* (2015) found that, as autumn temperatures increase in the Northeast USA, a smaller number and fraction of larvae fed the following spring, when they are much more likely to feed shortly after host infection by nymphs or to co-feed with nymphs. That is, feeding becomes increasingly more asynchronous. This contrasts with patterns in the Upper Midwest, where the faster rate of cooling causes unfed larvae to pass the winter in quiescence or behavioural diapause, from which they emerge and host-seek synchronously with nymphs in the spring. However, with increasingly warmer weather, the larval peak may advance so much that larvae moult into nymphs the same year, thus reversing the canonical feeding inversion and disrupting the transmission cycle (Ogden *et al.*, 2008). This represents one of the explanations for the low incidence of Lyme disease in the south of the USA (Ogden *et al.*, 2018), in addition to genetic differences in nymphs' host-seeking behaviour (Arsnoe *et al.*, 2019).

Spatial variation in rates of autumn cooling has been associated with tick feeding synchrony and TBE virus endemicity in Central Europe (Randolph *et al.*, 1999; Randolph and Rogers, 2000; Randolph, 2001), as well as with increased tick synchrony in the USA, as described above (Gatewood *et al.*, 2009). Based on temperature seasonality and other climate covariates, Randolph and Rogers (2000) developed a multivariate description of TBE virus areas of risk and predicted future distribution based on climate change scenarios. Based on a predicted summer rise in temperature and decrease in moisture, the model predicted TBE virus to spread into higher latitudes and elevations progressively through the 2020s, 2050s and 2080s, and to contract in the southern range. Twenty years later, while the prediction of a northerly expansion has held up, the southern contraction appears not to, presumably because of the complex interactions between changes in climate, host abundance and socio-economic determinants (Jaenson *et al.*, 2012, 2018; Medlock *et al.*, 2013; Rubel *et al.*, 2020; Walter *et al.*, 2020).

41.1.7 A call for integration: R_0 models

In order to capture the complexity involved in the persistence of tick-borne pathogens, R_0 models have been developed that incorporate both the complex life cycle of ticks and the multiple transmission routes that allow tick-borne pathogens to spread and persist (Hartemink *et al.*, 2008; Matser *et al.*, 2009; Davis and Bent, 2011; Dunn *et al.*, 2013). The stage-structured approach to R_0 was validated by Harrison *et al.* (2011) and extended further by Nonaka *et al.* (2010), Davis and Bent (2011) and Dunn *et al.* (2013) to explicitly include the phenologies of immature ticks and their interaction with pathogens of different infectious period lengths. Such models for R_0 have generated insights into how shifts in the seasonality of tick feeding might affect the persistence of tick-borne pathogens. For example, Davis and Bent (2011) found that the greater overlap of larvae and nymph feeding activity in the Upper Midwest of the USA substantially increased R_0, especially in those pathogens with shorter infectious periods, such as *B. microti*, *A. phagocytophilum* and POW virus. R_0 models have also been used to examine how co-feeding may increase strain diversity in *B. burgdorferi* (States *et al.*, 2017). Nonaka *et al.* (2010) demonstrated that the consistent prevalence of POW virus observed in tick populations could be maintained by a combination of low transovarial, intermediate co-feeding and high systemic transmission rates.

Because tick-borne pathogens are so often present as co-infections (Diuk-Wasser *et al.*, 2016), future modelling efforts should assess joint probabilities of emergence. Such a modelling approach would also be used to explicitly identify threshold targets to stop emergence. Finally, as our understanding of the key drivers of tick-borne pathogens' R_0 advances, the next step would be to develop spatially explicit R_0 maps to validate the models and develop risk maps to inform the public and predict emergence 'hotspots'.

41.2 Future Predictions

Climate change is predicted to contribute to the spread of tick vectors to higher latitudes and elevation; pathogens would generally follow. However, pathogen establishment will be further limited by the availability of competent hosts and the seasonality of the tick stages, with pathogens with short infectious periods displaying a lower R_0, i.e. a more fragile transmission cycle. There is evidence that the nymphal host-seeking season has been starting earlier with climate warming in some areas, which will affect human exposure depending on people's outdoor activity schedule. There is also some evidence that a warming climate will increase tick feeding synchrony, thus favouring pathogens with short infectious periods or those relying on co-feeding for transmission. However, the effect of climate on phenology depends on the extent to which the tick life cycle is controlled by temperature-independent diapause phenomena. Warming climates may also increase the fitness of pathogens transmitted transovarially in ticks or transplacentally in vertebrate hosts. Looking back to predictions such as the shift in the distribution of TBE, we note that, while the predicted northerly expansion held up, the southerly retraction did not, indicating that host abundance and human behavioural factors may be more important drivers. Thus, we need to understand more fully the relative importance of the factors increasing the system's resilience (e.g. host abundance) versus increasing its vulnerabilities (e.g. disruption of transmission by changes in tick phenology) if we hope to gain a foothold in controlling these emerging diseases.

References

Arsnoe, I., Tsao, J.I. and Hickling, G.J. (2019) Nymphal *Ixodes scapularis* questing behavior explains geographic variation in Lyme borreliosis risk in the eastern United States. *Ticks and Tick-borne Diseases* 10, 553–563. https://doi.org/10.1016/j.ttbdis.2019.01.001

Bakhvalova, V.N., Potapova, O.F., Panov, V.V. and Morozova, O.V. (2009) Vertical transmission of tick-borne encephalitis virus between generations of adapted reservoir small rodents. *Virus Research* 140, 172–178. https://doi.org/10.1016/j.virusres.2008.12.001

Belozerov, V.N. and Naumov, R.L. (2002) Nymphal diapause and its photoperiodic control in the tick *Ixodes scapularis* (Acari: Ixodidae). *Folia Parasitologica* 49, 314–318. https://doi.org/10.14411/fp.2002.058

Belozerov, V.N., Fourie, L.J. and Kok, D.J. (2002) Photoperiodic control of developmental diapause in nymphs of prostriate ixodid ticks (Acari: Ixodidae). *Experimental and Applied Acarology* 28, 163–168. https://doi.org/10.1023/a:1025377829119

Brownstein, J.S., Holford, T.R. and Fish, D. (2005) Effect of climate change on Lyme disease risk in North America. *EcoHealth* 2, 38–46. https://doi.org/10.1007/s10393-004-0139-x

CDC (2020a) Anaplasmosis statistics and epidemiology. Centers for Disease Control and Prevention, Atlanta, Georgia. Available at: http://www.cdc.gov/anaplasmosis/stats/index.html (accessed 15 December 2020).

CDC (2020b) Lyme disease data and statistics. Centers for Disease Control and Prevention, Atlanta, Georgia. Available at: http://www.cdc.gov/lyme/stats/index.html (accessed 15 December 2020).

CDC (2020c) Powassan virus statistics & maps. Centers for Disease Control and Prevention, Atlanta, Georgia. Available at: http://www.cdc.gov/powassan/statistics.html (accessed 15 December 2020).

CDC (2020d) Babesiosis data & maps. Centers for Disease Control and Prevention, Atlanta, Georgia. Available at: https://www.cdc.gov/parasites/babesiosis/data-statistics/index.html (accessed 15 December 2020).
Danielová, V., Holubová, J., Pejčoch, M. and Daniel, M. (2002) Potential significance of transovarial transmission in the circulation of tick-borne encephalitis virus. *Folia Parasitologica* 49, 323–325. https://doi.org/10.14411/fp.2002.060
Davis, S. and Bent, S.J. (2011) Loop analysis for pathogens: niche partitioning in the transmission graph for pathogens of the North American tick *Ixodes scapularis*. *Journal of Theoretical Biology* 269, 96–103. https://doi.org/10.1016/j.jtbi.2010.10.011
Diuk-Wasser, M.A., Hoen, A.G., Cislo, P., Brinkerhoff, R., Hamer, S.A., *et al.* (2012) Human risk of infection with *Borrelia burgdorferi*, the Lyme disease agent, in eastern United States. *American Journal of Tropical Medicine and Hygiene* 86, 320–327. https://doi.org/10.4269/ajtmh.2012.11-0395
Diuk-Wasser, M.A., Vannier, E. and Krause, P.J. (2016) Coinfection by *Ixodes* tick-borne pathogens: ecological, epidemiological, and clinical consequences. *Trends in Parasitology* 32, 30–42. https://doi.org/10.1016/j.pt.2015.09.008
Donahue, J.G., Piesman, J. and Spielman, A. (1987) Reservoir competence of white-footed mice for Lyme disease spirochetes. *American Journal of Tropical Medicine and Hygiene* 36, 92–96. https://doi.org/10.4269/ajtmh.1987.36.92
Dunn, J.M., Davis, S., Stacey, A. and Diuk-Wasser, M.A. (2013) A simple model for the establishment of tick-borne pathogens of *Ixodes scapularis*: a global sensitivity analysis of R_0. *Journal of Theoretical Biology* 335, 213–221. https://doi.org/10.1016/j.jtbi.2013.06.035
Dunn, J.M., Krause, P.J., Davis, S., Vannier, E.G., Fitzpatrick, M.C., *et al.* (2014) *Borrelia burgdorferi* promotes the establishment of *Babesia microti* in the northeastern United States. *PLoS ONE* 9, e115494. https://doi.org/10.1371/journal.pone.0115494
Eisen, R.J., Eisen, L., Ogden, N.H. and Beard, C.B. (2016) Linkages of weather and climate with *Ixodes scapularis* and *Ixodes pacificus* (Acari: Ixodidae), enzootic transmission of *Borrelia burgdorferi*, and Lyme disease in North America. *Journal of Medical Entomology* 53, 250–261. https://doi.org/10.1093/jme/tjv199
Estrada-Peña, A., de la Fuente, J. and Cabezas-Cruz, A. (2017) Functional redundancy and ecological innovation shape the circulation of tick-transmitted pathogens. *Frontiers in Cellular and Infection Microbiology* 7, 234. https://doi.org/10.3389/fcimb.2017.00234
Fish, D. (1993) Population ecology of *Ixodes dammini*. In: Ginsberg, H. (ed.) *Ecology and Environmental Management of Lyme Disease*. Rutgers University Press, New Brunswick, New Jersey, pp. 25–42.
Gatewood, A.G., Liebman, K.A., Vourc'h, G., Bunikis, J., Hamer, S.A., *et al.* (2009) Climate and tick seasonality are predictors of *Borrelia burgdorferi* genotype distribution. *Applied and Environmental Microbiology* 75, 2476–2483. https://doi.org/10.1128/AEM.02633-08
Gern, L. and Rais, O. (1996) Efficient transmission of *Borrelia burgdorferi* between cofeeding *Ixodes ricinus* ticks (Acari: Ixodidae). *Journal of Medical Entomology* 33, 189–192. https://doi.org/10.1093/jmedent/33.1.189
Hamer, S.A., Hickling, G.J., Sidge, J.L., Walker, E.D. and Tsao, J.I. (2012) Synchronous phenology of juvenile *Ixodes scapularis*, vertebrate host relationships, and associated patterns of *Borrelia burgdorferi* ribotypes in the midwestern United States. *Ticks and Tick-borne Diseases* 3, 65–74. https://doi.org/10.1016/j.ttbdis.2011.11.004
Han, S., Lubelczyk, C., Hickling, G.J., Belperron, A.A., Bockenstedt, L.K. and Tsao, J.I. (2019) Vertical transmission rates of *Borrelia miyamotoi* in *Ixodes scapularis* collected from white-tailed deer. *Ticks and Tick-borne Diseases* 10, 682–689. https://doi.org/10.1016/j.ttbdis.2019.02.014
Hanincova, K., Diuk-Wasser, M., Pappas, C.J., Iyer, R., Fish, D., *et al.* (2008) Fitness variation of *Borrelia burgdorferi* sensu stricto strains in mice. *Applied and Environmental Microbiology* 74, 153–157. https://doi.org/10.1128/AEM.01567-07
Harrison, A., Montgomery, W.I. and Bown, K.J. (2011) Investigating the persistence of tick-borne pathogens via the R_0 model. *Parasitology* 138, 896–905. https://doi.org/10.1017/S0031182011000400
Hartemink, N.A., Randolph, S.E., Davis, S.A. and Heesterbeek, J.A.P. (2008) The basic reproduction number for complex disease systems: defining R_0 for tick-borne infections. *The American Naturalist* 171, 743–754. https://doi.org/10.1086/587530
Hauser, G., Rais, O., Cadenas, F.M., Gonseth, Y., Bouzelboudjen, M. and Gern, L. (2018) Influence of climatic factors on *Ixodes ricinus* nymph abundance and phenology over a long-term monthly observation in Switzerland (2000–2014). *Parasites & Vectors* 11, 289. https://doi.org/10.1186/s13071-018-2876-7

Jaenson, T.G.T., Hjertqvist, M., Bergström, T. and Lundkvist, Å. (2012) Why is tick-borne encephalitis increasing? A review of the key factors causing the increasing incidence of human TBE in Sweden. *Parasites & Vectors* 5, 184. https://doi.org/10.1186/1756-3305-5-184
Jaenson, T.G.T., Petersson, E.H., Jaenson, D.G.E., Kindberg, J., Pettersson, H.-O., *et al.* (2018) The importance of wildlife in the ecology and epidemiology of the TBE virus in Sweden: incidence of human TBE correlates with abundance of deer and hares. *Parasites & Vectors* 11, 477. https://doi.org/10.1186/s13071-018-3057-4
Johnson, D.K.H., Schiffman, E.K., Davis, J.P., Neitzel, D.F., Sloan, L.M., *et al.* (2015) Human infection with *Ehrlichia muris*–like pathogen, United States, 2007–2013. *Emerging Infectious Diseases* 21, 1794–1799. https://doi.org/10.3201/eid2110.150143
Karpathy, S.E., Allerdice, M.E.J., Sheth, M., Dasch, G.A. and Levin, M.L. (2016) Co-feeding transmission of the *Ehrlichia muris*-like agent to mice (*Mus musculus*). *Vector-Borne and Zoonotic Diseases* 16, 145–150. https://doi.org/10.1089/vbz.2015.1878
Kemenesi, G. and Bányai, K. (2018) Tick-borne flaviviruses, with a focus on Powassan virus. *Clinical Microbiology Reviews* 32, e00106-17. https://doi.org/10.1128/CMR.00106-17
Labuda, M., Kozuch, O., Zuffová, E., Elecková, E., Hails, R.S. and Nuttall, P.A. (1997) Tick-borne encephalitis virus transmission between ticks cofeeding on specific immune natural rodent hosts. *Virology* 235, 138–143. https://doi.org/10.1006/viro.1997.8622
Levi, T., Keesing, F., Oggenfuss, K. and Ostfeld, R.S. (2015) Accelerated phenology of blacklegged ticks under climate warming. *Philosophical Transactions of the Royal Society B* 370, 20130556. https://doi.org/10.1098/rstb.2013.0556
Lynn, G.E., Oliver, J.D., Cornax, I., O'Sullivan, M.G. and Munderloh, U.G. (2017) Experimental evaluation of *Peromyscus leucopus* as a reservoir host of the *Ehrlichia muris*-like agent. *Parasites & Vectors* 10, 48. https://doi.org/10.1186/s13071-017-1980-4
Matser, A., Hartemink, N., Heesterbeek, H., Galvani, A. and Davis, S. (2009) Elasticity analysis in epidemiology: an application to tick-borne infections. *Ecology Letters* 12, 1298–1305. https://doi.org/10.1111/j.1461-0248.2009.01378.x
Medlock, J.M., Hansford, K.M., Bormane, A., Derdakova, M., Estrada-Peña, A., *et al.* (2013) Driving forces for changes in geographical distribution of *Ixodes ricinus* ticks in Europe. *Parasites & Vectors* 6, 1. https://doi.org/10.1186/1756-3305-6-1 z
Monaghan, A.J., Moore, S.M., Sampson, K.M., Beard, C.B. and Eisen, R.J. (2015) Climate change influences on the annual onset of Lyme disease in the United States. *Ticks and Tick-borne Diseases* 6, 615–622. https://doi.org/10.1016/j.ttbdis.2015.05.005
Moore, S.M., Eisen, R.J., Monaghan, A. and Mead, P. (2014) Meteorological influences on the seasonality of Lyme disease in the United States. *American Journal of Tropical Medicine and Hygiene* 90, 486–496. https://doi.org/10.4269/ajtmh.13-0180
Nonaka, E., Ebel, G.D. and Wearing, H.J. (2010) Persistence of pathogens with short infectious periods in seasonal tick populations: the relative importance of three transmission routes. *PLoS ONE* 5, e11745. https://doi.org/10.1371/journal.pone.0011745
Nuttall, P.A. and Labuda, M. (2003) Dynamics of infection in tick vectors and at the tick–host interface. *Advances in Virus Research* 61, 233–272. https://doi.org/10.1016/S0065-3527(03)60007-2
Ogden, N.H., Nuttall, P.A. and Randolph, S.E. (1997) Natural Lyme disease cycles maintained via sheep by co-feeding ticks. *Parasitology* 115, 591–599. https://doi.org/10.1017/s0031182097001868
Ogden, N.H., Lindsay, L.R., Beauchamp, G., Charron, D., Maarouf, A., *et al.* (2004) Investigation of relationships between temperature and developmental rates of tick *Ixodes scapularis* (Acari: Ixodidae) in the laboratory and field. *Journal of Medical Entomology* 41, 622–633. https://doi.org/10.1603/0022-2585-41.4.622
Ogden, N.H., Bigras-Poulin, M., O'Callaghan, C.J., Barker, I.K., Kurtenbach, K., *et al.* (2007) Vector seasonality, host infection dynamics and fitness of pathogens transmitted by the tick *Ixodes scapularis*. *Parasitology* 134, 209–227. https://doi.org/10.1017/S0031182006001417
Ogden, N.H., Bigras-Poulin, M., Hanincová, K., Maarouf, A., O'Callaghan, C.J. and Kurtenbach, K. (2008) Projected effects of climate change on tick phenology and fitness of pathogens transmitted by the North American tick *Ixodes scapularis*. *Journal of Theoretical Biology* 254, 621–632. https://doi.org/10.1016/j.jtbi.2008.06.020
Ogden, N.H., Pang, G., Ginsberg, H.S., Hickling, G.J., Burke, R.L., *et al.* (2018) Evidence for geographic variation in life-cycle processes affecting phenology of the Lyme disease vector *Ixodes scapularis* (Acari: Ixodidae) in the United States. *Journal of Medical Entomology* 55, 1386–1401. https://doi.org/10.1093/jme/tjy104

Ogden, N.H., Beard, C.B., Ginsberg, H.S. and Tsao, J.I. (2020) Possible effects of climate change on ixodid ticks and the pathogens they transmit: predictions and observations. *Journal of Medical Entomology* 2020, tjaa220. https://doi.org/10.1093/jme/tjaa220

Pepin, K.M., Eisen, R.J., Mead, P.S., Piesman, J., Fish, D., *et al.* (2012) Geographic variation in the relationship between human Lyme disease incidence and density of infected host-seeking *Ixodes scapularis* nymphs in the eastern United States. *American Journal of Tropical Medicine and Hygiene* 86, 1062–1071. https://doi.org/10.4269/ajtmh.2012.11-0630

Pritt, B.S., Sloan, L.M., Johnson, D.K.H., Munderloh, U.G., Paskewitz, S.M., *et al.* (2011) Emergence of a new pathogenic *Ehrlichia* species, Wisconsin and Minnesota, 2009. *New England Journal of Medicine* 365, 422–429. https://doi.org/10.1056/NEJMoa1010493

Pritt, B.S., Respicio-Kindry, L.B., Sloan, L.M., Schriefer, M.E., Replogle, A.J., *et al.* (2016) *Borrelia mayonii* sp. nov., a member of the *Borrelia burgdorferi* sensu lato complex, detected in patients and ticks in the upper midwestern United States. *International Journal of Systematic and Evolutionary Microbiology* 66, 4878–4880. https://doi.org/10.1099/ijsem.0.001445 z

Randolph, S.E. (2001) The shifting landscape of tick-borne zoonoses: tick-borne encephalitis and Lyme borreliosis in Europe. *Philosophical Transactions of the Royal Society B* 356, 1045–1056. https://doi.org/10.1098/rstb.2001.0893

Randolph, S.E. (2011) Transmission of tick-borne pathogens between co-feeding ticks: Milan Labuda's enduring paradigm. *Ticks and Tick-borne Diseases* 2, 179–182. https://doi.org/10.1016/j.ttbdis.2011.07.004

Randolph, S.E. and Dobson, A.D.M. (2012) Pangloss revisited: a critique of the dilution effect and the biodiversity–buffers–disease paradigm. *Parasitology* 139, 847–863. https://doi.org/10.1017/S0031182012000200

Randolph, S.E. and Rogers, D.J. (2000) Fragile transmission cycles of tick-borne encephalitis virus may be disrupted by predicted climate change. *Proceedings of the Royal Society B* 267, 1741–1744. https://doi.org/10.1098/rspb.2000.1204

Randolph, S.E., Gern, L. and Nuttall, P.A. (1996) Co-feeding ticks: epidemiological significance for tick-borne pathogen transmission. *Parasitology Today* 12, 472–479. https://doi.org/10.1016/s0169-4758(96)10072-7

Randolph, S.E., Miklisová, D., Lysy, J., Rogers, D.J. and Labuda, M. (1999) Incidence from coincidence: patterns of tick infestations on rodents facilitate transmission of tick-borne encephalitis virus. *Parasitology* 118, 177–186. https://doi.org/10.1017/s0031182098003643

Randolph, S.E., Green, R.M., Peacey, M.F. and Rogers, D.J. (2000) Seasonal synchrony: the key to tick-borne encephalitis foci identified by satellite data. *Parasitology* 121, 15–23. https://doi.org/10.1017/s0031182099006083

Randolph, S.E., Green, R.M., Hoodless, A.N. and Peacey, M.F. (2002) An empirical quantitative framework for the seasonal population dynamics of the tick *Ixodes ricinus*. *International Journal for Parasitology* 32, 979–989. https://doi.org/10.1016/s0020-7519(02)00030-9

Rosà, R., Andreo, V., Tagliapietra, V., Baráková, I., Arnoldi, D., *et al.* (2018) Effect of climate and land use on the spatio-temporal variability of tick-borne bacteria in Europe. *International Journal of Environmental Research and Public Health* 15, 732. https://doi.org/10.3390/ijerph15040732

Rosà, R., Tagliapietra, V., Manica, M., Arnoldi, D., Hauffe, H.C., *et al.* (2019) Changes in host densities and co-feeding pattern efficiently predict tick-borne encephalitis hazard in an endemic focus in northern Italy. *International Journal for Parasitology* 49, 779–787. https://doi.org/10.1016/j.ijpara.2019.05.006

Rubel, F., Walter, M., Vogelgesang, J.R. and Brugger, K. (2020) Tick-borne encephalitis (TBE) cases are not random: explaining trend, low- and high-frequency oscillations based on the Austrian TBE time series. *BMC Infectious Diseases* 20, 448. https://doi.org/10.1186/s12879-020-05156-7

Simon, J.A., Marrotte, R.R., Desrosiers, N., Fiset, J., Gaitan, J., *et al.* (2014) Climate change and habitat fragmentation drive the occurrence of *Borrelia burgdorferi*, the agent of Lyme disease, at the northeastern limit of its distribution. *Evolutionary Applications* 7, 750–764. https://doi.org/10.1111/eva.12165

Sonenshine, D.E. (2018) Range expansion of tick disease vectors in North America: implications for spread of tick-borne disease. *International Journal of Environmental Research and Public Health* 15, 478. https://doi.org/10.3390/ijerph15030478

States, S.L., Huang, C.I., Davis, S., Tufts, D.M. and Diuk-Wasser, M.A. (2017) Co-feeding transmission facilitates strain coexistence in *Borrelia burgdorferi*, the Lyme disease agent. *Epidemics* 19, 33–42. https://doi.org/10.1016/j.epidem.2016.12.002

Thompson, C., Spielman, A. and Krause, P.J. (2001) Coinfecting deer-associated zoonoses: Lyme disease, babesiosis, and ehrlichiosis. *Clinical Infectious Diseases* 33, 676–685. https://doi.org/10.1086/322681

Tolkacz, K., Bednarska, M., Alsarraf, M., Dwużnik, D., Grzybek, M., *et al.* (2017) Prevalence, genetic identity and vertical transmission of *Babesia microti* in three naturally infected species of vole, *Microtus* spp. (Cricetidae). *Parasites & Vectors* 10, 66. https://doi.org/10.1186/s13071-017-2007-x

Tufts, D.M. and Diuk-Wasser, M.A. (2018) Transplacental transmission of tick-borne *Babesia microti* in its natural host *Peromyscus leucopus*. *Parasites & Vectors* 11, 286. https://doi.org/10.1186/s13071-018-2875-8

Tufts, D.M. and Diuk-Wasser, M.A. (2021) Vertical transmission: a vector-independent transmission pathway of *Babesia microti* in the natural reservoir host *Peromyscus leucopus*. *Journal of Infectious Diseases* 223, 1787–1795. https://doi.org/10.1093/infdis/jiaa595

Walter, M., Vogelgesang, J.R., Rubel, F. and Brugger, K. (2020) Tick-borne encephalitis virus and its European distribution in ticks and endothermic mammals. *Microorganisms* 8, 1065. https://doi.org/10.3390/microorganisms8071065

Wilson, M.L. and Spielman, A. (1985) Seasonal activity of immature *Ixodes dammini* (Acari: Ixodidae). *Journal of Medical Entomology* 22, 408–414. https://doi.org/10.1093/jmedent/22.4.408

Yuval, B. and Spielman, A. (1990) Duration and regulation of the developmental cycle of *Ixodes dammini* (Acari: Ixodidae). *Journal of Medical Entomology* 27, 196–201. https://doi.org/10.1093/jmedent/27.2.196

eo42 *Anaplasma* Species' Novel Tick–Host–Pathogen Relationships and Effects of Climate Change

Daniel Sonenshine*
National Institute of Allergy and Infectious Diseases, Rockville, Maryland, USA

42.1 Expert Opinion

42.1.1 Introduction

Anaplasma species comprise a group of Gram-negative obligatory intracellular bacteria of the order Rickettsiales that colonize the midgut and salivary glands of ixodid ticks and also infect the blood of humans and other vertebrate animals (Rikihisa, 2010). *Anaplasma phagocytophilum*, transmitted mostly by *Ixodes scapularis* and *Ixodes pacificus* in North America, and *Ixodes ricinus* and *Ixodes persulcatus* in Europe and Northern Asia, is an important human pathogen, infecting neutrophils and occasionally other granulocytes in the blood where it develops within morulae, distinctive membrane-bound inclusions (Dumler *et al.*, 2005). In humans, infection with *A. phagocytophilum* causes anaplasmosis, also known as human granulocytic anaplasmosis (HGA), and often severe illness and even death (Bakken and Dumler, 2015). According to the US Centers for Disease Control and Prevention (CDC, 2020), cases of anaplasmosis in the USA have been increasing steadily over the past several decades to a peak of 5762 in 2017. However, mortality has remained low, less than 1%. Most cases of HGA occur in the north-eastern and north-central regions of the USA. In Europe and Northern Asia, the illness caused by this pathogen is known as tick-borne fever (TBF) and is widespread over most of the continent although at relatively low incidence (Cochez *et al.*, 2011; Matei *et al.*, 2019). In contrast to the USA, the pathogen in Europe and Northern Asia is transmitted by a wide range of tick vectors, including *Dermacentor marginatum*, *Dermacentor reticulatus*, *Rhipicephalus sanguineus* and others, although *I. ricinus* is generally regarded as the most important species transmitting the pathogen to humans (de la Fuente *et al.*, 2016). Transstadial transmission is important in maintaining *A. phagocytophilum* within its tick host. A wide range of rodents and other wildlife serve as reservoirs for *A. phagocytophilum* and other *Anaplasma* spp. (Nieto *et al.*, 2010). Similar illnesses have been reported to occur in pet animals and livestock (Cabezas-Cruz *et al.*, 2016). A strain of *A. phagocytophilum* transmitted by *I. ricinus* also infects sheep and cattle in the UK and is widespread throughout most of Europe (Lihou *et al.*, 2020). Other *Anaplasma* species are deadly pathogens of domestic livestock. Among the most important are *Anaplasma marginale* and *Anaplasma centrale*, which cause bovine anaplasmosis, a disease of cattle throughout most of the world.

A brief review of *A. phagocytophilum* growth in its tick host is needed to understand the extent to which climate change may affect persistence and spread. *A. phagocytophilum* is a novel pathogen because it uses unusual tactics to invade host cells,

*Email: daniel.sonenshine@nih.gov

DOI: 10.1079/9781789249637.0042

hijack the cell's genetic machinery and proliferate within protected enclosures (morulae). In ticks, this process begins when the bacteria initially invade the midgut epithelial cells and then migrate to the salivary glands for transmission to vertebrate hosts when host ticks feed again (de la Fuente *et al.*, 2017). Rather than following the invasive procedures common to many other rickettsia-like species, *Anaplasma* invade their cell hosts by caveolae-mediated endocytosis which enables them to bypass conventional phagolysosomes, avoid killing by reactive oxygen species and form specialized endosomes. Once internalized, these bacteria hijack the cell's cholesterol for membrane formation, ultimately sequestering themselves within a sheltered microenvironment, the morulae (Rikihisa, 2010). Another novel feature of *A. phagocytophilum* is the way it alters expression of certain genes to upregulate the tick's JAK/STAT pathway, thereby inhibiting cell apoptosis and facilitating multiplication in the midgut epithelial cells (Ayllón *et al.*, 2015). Following completion of the midgut cell phase (Fig. 42.1), the dense elementary bodies break out of the host cells, migrate through the tick's haemolymph to the salivary glands, bind to specific salivary gland proteins (SALP 16), invade the salivary gland and are transmitted with the tick's saliva to infect their mammalian hosts.

42.1.2 Evidence for climate change impacts

The geographic distribution of *A. phagocytophilum* is essentially tied to the distribution of its tick vectors, especially *I. scapularis* and *I. pacificus* in North America, and *I. ricinus* and *I. persulcatus* in Europe and Northern Asia. Since ticks spend most of their life cycle in the natural environment, away from their vertebrate hosts, they are especially reliant on favourable environmental and microenvironmental conditions for their long-term survival (Estrada-Peña *et al.*, 2012). Numerous studies have documented increases in geographic range expansion of these vector ticks, primarily northward but also into higher elevations in mountainous areas in the USA (Sonenshine, 2018) and in Europe, where *I. ricinus* has spread into much of northern Scandinavia and northern Russia (Jaenson *et al.*, 2012; Semenza and Suk, 2018). Other tick vectors of *A. phagocytophilum*, e.g. *I. persulcatus* and *Dermacentor nuttalli*, are even found in Mongolia in Northern Asia (Javkhlan *et al.*, 2014). In Canada, Bouchard *et al.* (2019) reported that the most important driver of range expansion and overwinter survival is increasing temperature. The increasingly benign temperature conditions in both Canada and the northern USA have fostered increased abundance of ticks where they were already established, as well as increased spread into new localities and increased tick seasonal activity, including longer periods of tick questing activity. *I. scapularis* is now established in large areas of southern Ontario, some areas of Quebec province, New Brunswick, Nova Scotia and even the prairie regions of southern Manitoba (Gasmi *et al.*, 2016; Clow *et al.*, 2017; Gabriele-Rivet *et al.*, 2017; McPherson *et al.*, 2017).

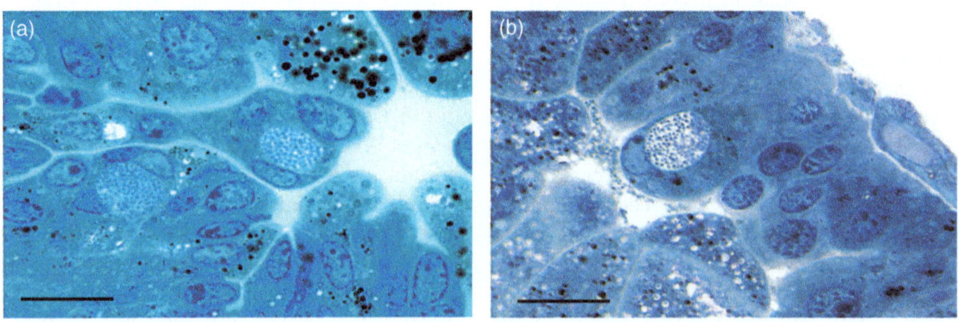

Fig. 42.1. Photomicrographs of histological sections of the midgut of *Dermacentor variabilis* infected with *Anaplasma phagocytophilum* showing colonies of the pathogen: (a) two midgut epithelial cells with membrane-bound vacuoles filled with early-stage elementary bodies; (b) a single midgut epithelial cell with a membrane-bound vacuole with (later) dense-stage elementary bodies. Bar = 10 μM. (Photograph courtesy of Dr Katherine M. Kocan, Oklahoma State University, Stillwater, Oklahoma, USA.)

A. phagocytophilum-positive ticks (*I. scapularis*) collected from wildlife were found in southern Ontario, as well as in questing ticks sampled in the Thousand Islands region of Ontario and even from a small percentage of ticks removed from humans, but the percentage infection with this pathogen was only 2.3% (Smith, 2019). However, further south, in the Hudson Valley region of New York State (USA), sampling of *I. scapularis* adults showed 18% infection with *A. phagocytophilum* (Aliota *et al.*, 2014). In Europe, there is substantial evidence of range expansion of the primary vector, *I. ricinus*, which has occurred during the past several decades, presumably due to warmer temperatures, especially milder winters, forest expansion and human-induced ecological changes (Materna *et al.*, 2005; Jore *et al.*, 2011; Jaenson *et al.*, 2012; Medlock *et al.*, 2013). Clear evidence of *A. phagocytophilum* spread into Scandinavia was shown by comparing surveillance for this pathogen in Finland, where *I. ricinus* ticks were found negative in 2000 but where 9.2% of adult and 3.1% of nymphal *I. ricinus* were found positive in specimens sampled in 2013 (Sormunen *et al.*, 2016), and in northern Norway, where 3.0% of *I. ricinus* were positive in samples collected north of the Arctic Circle (Henningsson *et al.*, 2015). In view of the overwhelming evidence of tick vector geographic spread, it has been widely assumed that this also applies to their tick-borne pathogens, including *A. phagocytophilum* (Bouchard *et al.*, 2019). However, it is not clear whether this is actually happening. To answer this question, a more in-depth examination of the evidence is warranted.

Many tick species, including *I. scapularis*, have been reported to tolerate sub-zero (Celsius) temperatures, the so-called cold hardiness phenomenon, although the degree and duration of tolerance of temperature extremes vary considerably among different tick species. Cold tolerance is largely dependent upon the supercooling point, i.e. the temperature at which ice crystals form in tissues. Ice formation disrupts membranes, killing cells and, unless reversed quickly, soon leads to death. Some species produce glycerol in sufficient amounts to act as an antifreeze at temperatures slightly above their supercooling point (Burks *et al.*, 1996). *I. scapularis* ticks express a protein, the so-called 'antifreeze protein' (IAFGP), a high-molecular-weight glycoprotein that provides protection against ice formation by binding ice crystals, allowing increased tolerance of temperatures as low as −10°C for ~2 h. Remarkably, infection of these ticks with *A. phagocytophilum* enhances expression of this glycoprotein. As a result, a tick harbouring this microbe in its tissues can survive extreme sub-zero temperatures more readily than uninfected individuals (Neelakanta *et al.*, 2010). Overall, it is likely that *I. scapularis* will expand its range substantially during the succeeding decades (Sonenshine, 2018) and thereby expand the prevalence of *A. phagocytophilum*-infected vectors further north as well as into localities at higher elevations. However, infection rates for *A. phagocytophilum* in *I. scapularis* are relatively low in North America, e.g. from 0.9% in Ontario, Canada (Kulkarni *et al.*, 2018), to as high as 14% in New York, USA (Sanchez-Vicente *et al.*, 2019), and similarly in Europe, e.g. 3.8% in Germany (Blazejak *et al.*, 2017). Hence, it is not clear how great an advantage IAFGP induction will be for further dispersal of this pathogen as the climate warms in future decades.

The effects of *A. phagocytophilum* on *I. scapularis* ticks' cold-hardy fitness is but one example of a much broader phenomenon, as described by Benelli (2020). Infection with *Borrelia* spp. and tick-borne encephalitis virus enhances questing and walking activity, while *A. phagocytophilum* infection of *I. scapularis* is known to induce the synthesis of heat-shock proteins, especially HSP20 and HSP70, which enhance tick survival by reducing the risk of desiccation. When invading the salivary glands, *A. phagocytophilum* inhibits the intrinsic apoptosis pathway in the salivary acini. In response, tick cells upregulate the extrinsic apoptosis pathway, thereby limiting pathogen multiplication and facilitating transmission by enhanced tick feeding and survival (summarized by Benelli, 2020).

Although species of the *I. ricinus* complex (*I. ricinus*, *I. scapularis*, *I. pacificus* and *I. persulcatus*) are generally considered to be the primary vectors for *A. phagocytophilum*, evidence has emerged implicating other tick species as vectors that also may expand the geographic range for this pathogen. Of particular interest are reports of infection in the European tick, *D. reticulatus*, a tick widespread in the continent including the far eastern region, e.g. Ukraine (Ben and

Lozynskyi, 2019) and far northern regions (Wirtgen *et al.*, 2011).

Habitat also plays an important role in tick survival and questing activity. As the climate warms, will vector ticks spreading north and into high elevations encounter habitats suitable for their survival? For most of its current and recent history, *I. scapularis* is and has been predominantly associated with deciduous forests in the eastern USA and Canada. Tick distribution, relative abundance and questing activity are greater in hardwood or mixed hard/pine forests than in predominantly pine forest habitats. Many more nymphal ticks are collected where leaf litter is deepest and where there is high shrub cover, features more characteristic of hardwood forests than in pine forests. Deciduous forests generally provide more humid environments and microenvironments and more mast (acorns, nuts, etc.) than pine forests, both conditions favourable to ticks as well as food for their small mammalian hosts (Schulze and Jordan, 2005). In a mountainous region of Switzerland, *I. ricinus* nymphal population density almost doubled over a 15-year period, largely in relation to periodic increases in mast production by beech trees which was followed 2 years later by a corresponding increase in nymphal tick density (Bregnard *et al.*, 2020). A similar phenomenon was reported for *I. scapularis* in relation to acorn production, where cyclical episodes of high acorn and other seed production from the deciduous forests in the north-eastern USA led to higher densities of small mammal populations, especially white-footed mice, and concomitant increases in nymphal tick densities, again with a 2-year time lag (Ostfeld *et al.*, 2006).

The question at issue here is whether increasing global warming will favour the northward and altitudinal spread of deciduous forests, gradually replacing the cold boreal forests, e.g. Laurentian–Acadian and spruce–fir forests characteristic of these regions that are currently unsuitable for *I. scapularis* survival. According to a recent study using satellite imagery of north-western Canada and Alaska over a 30-year period, a 14% decrease in the evergreen forest area of the boreal biome was replaced by a 15% increase in deciduous forest area (Wang *et al.*, 2020). This study also found a 'climate-driven' 7% expansion of herbaceous and shrub vegetation in the Arctic biome. The process is predicted to expand over the rest of the 21st century, with broadleaf deciduous forest communities increasing (e.g. to 58% of the Alaskan ecosystem net primary productivity by 2100) and corresponding declines in evergreen forests (Mekonnen *et al.*, 2019).

42.1.3 Evidence against climate change impacts

As ticks (especially species of the genus *Ixodes*) expand their range in response to climate change, they also encounter a variety of ecological challenges to which they may not be able to adapt. For ticks such as *I. ricinus* and *I. scapularis* and the pathogens they transmit, this may include increasing aridity, loss of snow cover and other factors different from their previous geographic range. In Northern Europe, studies comparing a variety of factors – e.g. numbers of days with snow cover, temperature fluctuations, relative humidity, intensity and frequency of precipitation – present a much more complex picture of how climate change is affecting *A. phagocytophilum* distribution. Large fluctuations in temperature, rather than steady annual increases, were positively associated with the presence of *Anaplasma*-infected ticks. As the climate warmed, coastal areas were the localities mostly impacted during winter, with less snow and increased temperatures in spring. *I. ricinus* ticks overwinter most successfully below the snow cover, where milder temperatures and atmospheric moisture saturation in the subnivean microenvironment provide favourable conditions for their survival. However, in recent decades, warmer winters with far fewer snow days, and warmer spring temperatures, have led to bare ground and greater loss of soil moisture and drying vegetation. Tick questing activity also is strongly dependent upon the duration of snow cover (Jore *et al.*, 2011). Thus, global warming, leading to reduced snowfall and less persistent snow cover, especially in the northern hemisphere, may actually reduce tick overwinter survival and reduce the spread of pathogens such as *A. phagocytophilum*, despite favouring increased questing activity of those few individuals that successfully survive the winter. In North America, *A. phagocytophilum* has not been found throughout all of the recently colonized regions of the northern USA and southern Canada. As an example, surveillance studies in Ontario indicate low tick infection rates ranging from 0.4 to 1.1% and mostly from the southern regions of the province (Nelder *et al.*, 2019).

In addition to these climate-related factors, changes in the variety and abundance of hosts and vegetation also play an important role in the survival of ticks, their spread into new geographic regions and the potential for pathogen transmission. Lizards serve as common hosts for *I. scapularis* in the south-eastern and south-central USA, animals that are refractory to *Borrelia burgdorferi* (Grigery *et al.*, 2005; Salkeld and Lane, 2010) and are reportedly poor reservoir hosts for *A. phagocytophilum* (Nieto *et al.*, 2010). Consequently, as lizard populations spread northward, it is likely that *A. phagocytophilum* infection also will decline. All *I. ricinus* found feeding on lizards in different regions of Europe were negative for this pathogen (Jahfari *et al.*, 2014).

42.2 Future Predictions

Predictions of future vector-borne disease spread must be treated with caution in view of the complex array of biotic, abiotic and human-associated factors that will determine its outcome. Despite the cautionary warning, my prediction, based on the evidence described above, is that *A. phagocytophilum* and its tick vectors (primarily *I. scapularis*, *I. pacificus*, *I. ricinus* and *I. persulcatus*) important for transmission to humans and livestock will indeed spread to new regions, especially in Canada and elsewhere in North America. Models developed by Minigan *et al.* (2018) for the American dog tick, *Dermacentor variabilis* (a known vector of *A. marginale*) show that the area of suitable climate for this species is projected to increase 30–42% beyond present conditions by year 2050, and possibly by as much as 30–53% by year 2070. It is likely that similar trends may apply to *I. scapularis* and its pathogens, including *A. phagocytophilum*. In the UK, most of Central and Northern Europe, and Northern Asia, *A. phagocytophilum* is already established (e.g. Norway, Sweden, Finland, Karelia region of Russia), but I predict that the tick vector populations that transmit it will increase in abundance. Increased tick densities will likely lead to higher percentage pathogen infection rates and, consequently, higher rates of transmission to livestock and humans.

References

Aliota, M.T., Dupuis, A.P., Wilczek, M.P., Peters, R.J., Ostfeld, R.S. and Kramer, L.D. (2014) The prevalence of zoonotic tick-borne pathogens in *Ixodes scapularis* collected in the Hudson Valley, New York State. *Vector-Borne and Zoonotic Diseases* 1, 245–250. https://doi.org/10.1089/vbz.2013.1475

Ayllón, N., Villar, M., Galindo, R.C., Kocan, K.M., Šíma, R., *et al.* (2015) Systems biology of tissue-specific response to *Anaplasma phagocytophilum* reveals differentiated apoptosis in the tick vector *Ixodes scapularis*. *PLoS Genetics* 11, e1005120. https://doi.org/10.1371/journal.pgen.1005120

Bakken, J.S. and Dumler, J.S. (2015) Human granulocytic anaplasmosis. *Infectious Disease Clinics of North America* 29, 341–355. https://doi.org/10.1016/j.idc.2015.02.007

Ben, I. and Lozynskyi, I. (2019) Prevalence of *Anaplasma phagocytophilum* in *Ixodes ricinus* and *Dermacentor reticulatus* and coinfection with *Borrelia burgdorferi* and tick-borne encephalitis virus in Western Ukraine. *Vector-Borne and Zoonotic Diseases* 19, 793–801. https://doi.org/10.1089/vbz.2019.2450

Benelli, G. (2020) Pathogens manipulating tick behavior – through a glass, darkly. *Pathogens* 9, 664. https://doi.org/10.3390/pathogens9080664

Blazejak, K., Janecek, E. and Strub, C. (2017) A 10-year surveillance of Rickettsiales (*Rickettsia* spp. and *Anaplasma phagocytophilum*) in the city of Hanover, Germany, reveals *Rickettsia* spp. as emerging pathogens in ticks. *Parasites & Vectors* 10, 588. https://doi.org/10.1186/s13071-017-2537-2

Bouchard, C., Dibernardo, A., Koffi, J., Wood, H., Leighton, P.R. and Lindsey, L.R. (2019) Increased risk of tick-borne diseases with climate and environmental changes. *Canada Communicable Disease Report* 45, 83–89. https://doi.org/10.14745/ccdr.v45i04a02

Bregnard, C., Rais, O. and Voordouw, M.J. (2020) Climate and tree seed production predict the abundance of the European Lyme disease vector over a 15-year period. *Parasites & Vectors* 13, 408. https://doi.org/10.1186/s13071-020-04291-z

Burks, C.S., Stewart, R.J., Needham, G.R. and Lee, R.E. (1996) The role of direct chilling injury and inoculative freezing in cold tolerance of *Amblyomma americanum*, *Dermacentor variabilis* and *Ixodes scapularis*. *Physiological Entomology* 21, 44–50. https://doi.org/10.1111/j.1365-3032.1996.tb00833.x

Cabezas-Cruz, A., Alberdib, P., Ayllón, N., Valdes, J.J., Piercea, R., *et al.* (2016) *Anaplasma phagocytophilum* increases the levels of histone modifying enzymes to inhibit cell apoptosis and facilitate pathogen infection in the tick vector *Ixodes scapularis*. *Epigenetics* 11, 303–319. https://doi.org/10.1080/15592294.2016.1163460

CDC (2020) Anaplasmosis: epidemiology and statistics. Centers for Disease Control and Prevention, Atlanta, Georgia. Available at: https://www.cdc.gov/anaplasmosis/stats/ (accessed 24 June 2021).

Clow, K.M., Leighton, P.A., Ogden, N.H., Lindsay, L.R., Miche, P., *et al.* (2017) Northward range expansion of *Ixodes scapularis* evident over a short timescale in Ontario, Canada. *PLoS ONE* 12, e0189393. https://doi.org/10.1371/journal.pone.0189393

Cochez, C., Ducoffre, C., Vandenvelde, C., Luyasu, V. and Heyman, P. (2011) Human anaplasmosis in Belgium: a 10-year seroepidemiological study. *Ticks and Tick-borne Diseases* 2, 156–159. https://doi.org/10.1016/j.ttbdis.2011.06.004

de la Fuente, J., Estrada-Peña, A., Cabezas-Cruz, A. and Kocan, K.M. (2016) *Anaplasma phagocytophilum* uses common strategies for infection of ticks and vertebrate hosts. *Trends in Microbiology* 24, 173–180. https://doi.org/10.1016/j.tim.2015.12.001

de la Fuente, J., Antunes, S., Bonnet, S., Cabezas-Cruz, A., Domingos, A.G., *et al.* (2017) Tick–pathogen interactions and vector competence: identification of molecular drivers for tick-borne diseases. *Frontiers in Cellular and Infection Microbiology* 7, 114. https://doi.org/10.3389/fcimb.2017.00114 z

Dumler, J.S., Choi, K.S., Garcia-Garcia, J.C., Barat, N.S., Scorpio, D.G., *et al.* (2005) Human granulocytic anaplasmosis and *Anaplasma phagocytophilum*. *Emerging Infectious Diseases* 11, 1828–1834. https://doi.org/10.3201/eid1112.050898

Estrada-Peña, A., Ayllón, N. and de la Fuente, J. (2012) Impact of climate trends on tick-borne pathogen transmission. *Frontiers in Physiology* 3, 64. https://doi.org/10.3389/fphys.2012.00064

Gabriele-Rivet, V., Koffi, J.K., Pelcat, Y., Arsenault, J., Cheng, A., *et al.* (2017) A risk model for the Lyme disease vector *Ixodes scapularis* (Acari: Ixodidae) in the prairie provinces of Canada. *Journal of Medical Entomology* 54, 862–868. https://doi.org/10.1093/jme/tjx036

Gasmi, S., Ogden, N.H., Leighton, P.A., Lindsay, L.R. and Thivierge, K. (2016) Analysis of the human population bitten by *Ixodes scapularis* ticks in Quebec, Canada: increasing risk of Lyme disease. *Ticks and Tick-borne Diseases* 7, 1075–1081. https://doi.org/10.1016/j.ttbdis.2016.09.006

Grigery, C.N., Moyer, P., Little, S.E. and Masters, E.J. (2005) Bacteriocidal activity of lizard and mouse serum for *Borrelia lonestari*, putative agent of a Lyme-like illness (aka STARI or Masters disease) in Missouri. *Missouri Medicine* 102, 442–446.

Henningsson, A.J., Hvidsten, D., Kristiansen, B.E., Matussek, A., Stuen, S. and Jenkins, A. (2015) Detection of *Anaplasma phagocytophilum* in *Ixodes ricinus* ticks from Norway using a real time PCR assay targeting the *Anaplasma* citrate synthase gene *gltA*. *BMC Microbiology* 15, 153. https://doi.org/10.1186/s12866-015-0486-5

Jaenson, T.G.T., Jaenson, D.G.E., Eisen, L., Petersson, E. and Lindgren, E. (2012) Changes in the geographical distribution and abundance of the tick *Ixodes ricinus* during the past 30 years in Sweden. *Parasites & Vectors* 5, 8. https://doi.org/10.1186/1756-3305-5-8

Jahfari, S., Coipans, E.C., Fonville, M., van Leeuwen, A.D. and Hengeveld, P. (2014) Circulation of four *Anaplasma phagocytophilum* ecotypes in Europe. *Parasites & Vectors* 7, 365. https://doi.org/10.1186/1756-3305-7-365

Javkhlan, G., Enkhtaivan, B., Baigal, B., Myagmarsuren, P., Battur, B. and Battsetseg, B. (2014) Natural *Anaplasma phagocytophilum* infection in ticks from a forest area of Selenge province, Mongolia. *Western Pacific Surveillance Response Journal* 5, 21–24. https://doi.org/10.5365/WPSAR.2013.4.3.001

Jore, S., Viljugrein, H., Hofshagen, M., Brun-Hansen, H., Kristoffersen, A.B., *et al.* (2011) Multi-source analysis reveals latitudinal and altitudinal shifts in range of *Ixodes ricinus* at its northern distribution limit. *Parasites & Vectors* 4, 84. https://doi.org/10.1186/1756-3305-4-84

Kulkarni, L.M., Kryuchkov, R., Statculescu, A., Thickstun, C., Dibernardo, A., *et al.* (2018) *Ixodes scapularis* tick distribution and infection rates in Ottawa, Ontario, 2017. *Canada Communicable Disease Report* 4, 237–242. https://doi.org/10.14745/ccdr.v44i10a02

Lihou, K., Vineer, H.R. and Wall, R. (2020) Distribution and prevalence of ticks and tick-borne disease on sheep and cattle farms in Great Britain. *Parasites & Vectors* 13, 406. https://doi.org/10.1186/s13071-020-04287-9

Matei, I.A., Estrada-Peña, A., Cutler, S.J., Vayssier-Taussat, M., Varela-Castro, L., *et al.* (2019) A review on the eco-epidemiology and clinical management of human granulocytic anaplasmosis and its agent in Europe. *Parasites & Vectors* 12, 599. https://doi.org/10.1186/s13071-019-3852-6

Materna, J., Daniel, M. and Danielová, V. (2005) Altitudinal distribution limit of the tick *Ixodes ricinus* shifted considerably towards higher altitudes in Central Europe: results of three years monitoring in the Krkonose Mts. (Czech Republic). *Central European Journal of Public Health* 13, 24–28.

McPherson, M., Garcia-Garcia, A., Cuesta-Valero, F.J.C., Beltrami, H., Ketchum, P., *et al.* (2017) Expansion of the Lyme disease vector *Ixodes scapularis* in Canada inferred from CMIP5 climate projections. *Environmental Health Perspectives* 125, 057008. https://doi.org/10.1289/EHP57

Medlock, J.M., Hansford, K.M., Bormane A., Derdakova, M., Estrada-Peña A., *et al.* (2013) Driving forces for changes in geographical distribution of *Ixodes ricinus* ticks in Europe. *Parasites & Vectors* 6, 1. https://doi.org/10.1186/1756-3305-6-1

Mekonnen, Z.A., Riley, W.J., Randerson, J.T., Grant, R.F. and Rogers, B.M. (2019) Expansion of high-latitude deciduous forests driven by interactions between climate warming and fire. *Nature Plants* 5, 952–958. https://doi.org/10.1038/s41477-019-0495-8

Minigan, J.N., Hager, H.A., Peregrine, A.S. and Newman, J.A. (2018) Current and potential future distribution of the American dog tick (*Dermacentor variabilis*, Say) in North America. *Ticks and Tick-borne Diseases* 9, 354–362. https://doi.org/10.1016/j.ttbdis.2017.11.012

Neelakanta, G., Sultana, H., Fish, D., Anderson, J.F. and Fikrig, E. (2010) *Anaplasma phagocytophilum* induces *Ixodes scapularis* ticks to express an antifreeze glycoprotein gene that enhances their survival in the cold. *Journal of Clinical Investigation* 120, 3179–3190. https://doi.org/10.1172/JCI42868

Nelder, M.P., Russell, C.B., Lindsay, L.R., Dibernardo, A., Brandon, N.C., *et al.* (2019) Recent emergence of *Anaplasma phagocytophilum* in Ontario, Canada: early serological and entomological indicators. *American Journal of Tropical Medicine and Hygiene* 101, 1249–1258. https://doi.org/10.4269/ajtmh.19-0166

Nieto, N.C., Madigan, J.E. and Foley, J.E. (2010) The dusky-footed woodrat (*Neotoma fuscipes*) is susceptible to infection by *Anaplasma phagocytophilum* originating from woodrats, horses, and dogs. *Journal of Wildlife Diseases* 46, 810–817. https://doi.org/10.7589/0090-3558-46.3.810

Ostfeld, R.S., Canham, C.D., Oggenfuss, K., Winchcombe, R.J. and Keesing, F. (2006) Climate, deer, rodents, and acorns as determinants of variation in Lyme-disease risk. *PLoS Biology* 4, e145. https://doi.org/10.1371/journal.pbio.0040145

Rikihisa, Y. (2010) *Anaplasma phagocytophilum* and *Ehrlichia chaffeensis*: subversive manipulators of host cells. *Nature Reviews Microbiology* 8, 328–339. https://doi.org/10.1038/nrmicro2318

Salkeld, D.J. and Lane, R.S. (2010) Community ecology and disease risk: lizards, squirrels, and the Lyme disease spirochete in California, USA. *Ecology* 91, 293–298. https://doi.org/10.1890/08-2106.1

Sanchez-Vicente, S., Tagliafierro, T., Coleman, J.L., Benach, J.L. and Tokarz, R. (2019) Polymicrobial nature of tick-borne disease. *mBio* 10, e02055-19. https://doi.org/10.1128/mBio.02055-19

Schulze, T.L. and Jordan, R.A. (2005) Influence of meso- and microscale habitat structure on focal distribution of sympatric *Ixodes scapularis* and *Amblyomma americanum* (Acari: Ixodidae). *Journal of Medical Entomology* 42, 285–294. https://doi.org/10.1093/jmedent/42.3.285

Semenza, J.C. and Suk, J.E. (2018) Vector-borne diseases and climate change: a European perspective. *FEMS Microbiology Letters* 365, fnx244. https://doi.org/10.1093/femsle/fnx244

Smith, K.A., Oesterle, P.T., Jardine, C.M., Dibernardo, A., Huynh, C., *et al.* (2019) Tick infestations of wildlife and companion animals in Ontario, Canada, with detection of human pathogens in *Ixodes scapularis* ticks. *Ticks and Tick-borne Diseases* 10, 72–76. https://doi.org/10.1016/j.ttbdis.2018.08.018

Sonenshine, D.E. (2018) Range expansion of tick disease vectors in North America: implications for spread of tick-borne disease. *International Journal of Environmental Research and Public Health* 15, 478. https://doi.org/10.3390/ijerph15030478

Sormunen, J.J., Penttinen, R., Klemola, T., Vesterinen, F.J. and Hanninen, J. (2016) *Anaplasma phagocytophilum* in questing *Ixodes ricinus* ticks in southwestern Finland. *Experimental and Applied Acarology* 70, 491–500. https://doi.org/10.1007/s10493-016-0093-7

Wang, J.A., Sulla-Menashe, D., Woodcock, C.E., Sonnentag, O., Keeling, R.F. and Friedl, M.A. (2020) Extensive land cover changes across Arctic–Boreal northwestern North America from disturbance and climate forcing. *Global Change Biology* 26, 807–822. https://doi.org/10.1111/gcb.14804

Wirtgen, A., Nahayo, A., Linden, A., Garigliany, M. and Desmecht, M. (2011) Detection of *Anaplasma phagocytophilum* in *Dermacentor reticulatus* ticks. *Veterinary Record* 168, 195. https://doi.org/10.1136/vr.d1053

eo43 Zoonotic Potential in the Genera *Anaplasma* and *Ehrlichia*

Snorre Stuen*
Norwegian University of Life Sciences, Sandnes, Norway

43.1 Expert Opinion

43.1.1 Introduction

Around 900 tick species have been described and several of them may transmit important pathogens, such as bacteria in the genera *Anaplasma* and *Ehrlichia* (Dumler *et al.*, 2001; Guglielmone *et al.*, 2014). These microorganisms are mainly associated with infections in animals, but several are recently known to have a zoonotic potential (Table 43.1).

Anaplasma capra was first described from humans (2015) in China (Li *et al.*, 2015). Little information is available concerning its distribution. *Ixodes* spp. and *Haemophysalis* spp. seem to be the main vectors involved. *A. capra* has been detected in ruminants in Asia but has also recently been reported from deer species in Europe (Jouglin *et al.*, 2019). Goats were initially supposed to be the main domestic reservoir, but the infections have been detected in dogs and several other ruminant species (Shi *et al.*, 2019). Clinical symptoms are mainly flu-like, but rash, eschar, lymphadenopathy and gastrointestinal manifestations have also been observed. Fewer than 30 human cases have so far been reported (Li *et al.*, 2015).

Anaplasma ovis was first described in sheep (1912), and has a worldwide distribution in small ruminant populations, especially in the Mediterranean countries and Central Europe (Renneker *et al.*, 2013). *A. ovis* is spread by a variety of tick species, particularly in the genera *Rhipicephalus* and *Dermacentor*. The main hosts are small ruminants, but the infection has also been detected in dogs and other species (Pereira *et al.*, 2016). The infection in small ruminants may cause haemolytic anaemia. Outbreaks of severe illness in sheep and goats are rare and seem to occur mainly under stressful conditions (Renneker *et al.*, 2013). Only one human case has so far been detected, in 2010 (Cyprus) (Chochlakis *et al.*, 2010).

Anaplasma phagocytophilum (formerly *Ehrlichia phagocytophila*) was first described in sheep (1932) and causes tick-borne fever (TBF) in ruminants (Gordon *et al.*, 1932). Besides ruminants, *A. phagocytophilum* may also cause disease in horses, dogs and cats (Stuen *et al.*, 2013). The infection has been identified worldwide. Different variants/strains of *A. phagocytophilum* exist which may be involved in different natural infection cycles and host species, although little is known about the natural reservoir hosts for the different strains/variants (Jahfari *et al.*, 2014). The infection is mainly associated with *Ixodes* spp., although other tick species may occasionally be involved as vectors. The infection can

*Email: snorre.stuen@nmbu.no

DOI: 10.1079/9781789249637.0043

Table 43.1. Vectors and hosts related to reported zoonotic infections in the genera *Anaplasma* and *Ehrlichia*. (Author's own table.)

Species	Vectors	Hosts	Human cases[a]
Anaplasma capra	*Ixodes* spp. especially *Ixodes persulcatus*, *Haemophysalis* spp. especially *Haemophysalis longicornis*	Mostly unknown – ruminants, dogs	<30
Anaplasma ovis	*Rhipicephalus* spp., *Dermacentor* spp.	Mainly small ruminants	1
Anaplasma phagocytophilum	Mainly *Ixodes* spp.	Several species – including small rodents, wild boar, cervids	>15,000
Anaplasma platys	*Rhipicephalus sanguineus* sensu lato (s.l.)	Mainly dogs	<10
Ehrlichia canis	*R. sanguineus* s.l.	Mainly dogs	>20
Ehrlichia chaffeensis	*Amblyomma americanum*, *Dermacentor variabilis*	White-tailed deer, several other species	>10,000
Ehrlichia ewingii	*A. americanum*	White-tailed deer, dogs	>100
Ehrlichia muris	*R. sanguineus*	Small rodents	>70
Ehrlichia ruminantium	*Amblyomma* spp. especially *Amblyomma variegatum* and *Amblyomma hebraeum*	Ruminants	<10
Panola Mountain *Ehrlichia*	*A. americanum*, *Amblyomma* spp.	Mostly unknown – white-tailed deer, goats	1

[a]Several cases may be under-reported.

cause severe economic and welfare challenges in the sheep industry (Brodie *et al.*, 1986). *A. phagocytophilum* is the most important zoonotic infection in the genus *Anaplasma* with most cases reported in the USA and China (Bakken and Dumler, 2015). The first human case was reported in the USA (1994). Since then, more than 15,000 human cases have been reported. Flu-like symptoms are common, but severe complications have been reported involving septic shock-like syndrome and acute respiratory distress symptoms. Hospitalization rate is estimated to be around 40%, while the mortality rate is around 1% (Bakken and Dumler, 2015).

Anaplasma platys was first reported from dogs (1978) in Florida, USA (French and Harvey, 1983). The infection seems to be widespread, however only scattered information is available. Dogs are the main host, but natural infection has been reported in other species (Ben Said *et al.*, 2017). *Rhipicephalus sanguineus* sensu lato (s.l.) seems to be the main vector involved. *A. platys* is known to cause canine cyclic thrombocytopaenia, which may vary from asymptomatic, mild to severe illness. Co-infection with other pathogens may increase the severity of an *A. platys* infection (French and Harvey, 1983). *A. platys* was first described as a human pathogen (2013) in Venezuela (Maggi *et al.*, 2013) and subsequently reported from the USA (Qurollo *et al.*, 2014). Clinical symptoms are flu-like and relatively mild (Maggi *et al.*, 2013; Arraga-Alvarado *et al.*, 2014). Fewer than ten cases have so far been verified (Maury de Tamí and Tamí-Maury, 2015).

Ehrlichia canis was first reported in 1953. The infection is a common infection in dogs worldwide and has occasionally been reported for other canine species. The main vector is *R. sanguineus* s.l., but *Rhipicephalus bursa* and *D. variabilis* may also be involved. Clinical signs include fever, anorexia and lymphadenopathy, which may develop into a severe disease known as tropical canine pancytopaenia (Ristic and Holland, 1993). Human infection was first reported (1996) from Venezuela (Perez *et al.*, 1996) and more recently from Mexico and Costa Rica (Bouza-Mora *et al.*, 2017). Clinical symptoms are in general mild, or even subclinical, and may involve flu-like symptoms (Perez *et al.*, 2006). Around 20 human cases have so far been reported, although the infection may be under-reported (Perez *et al.*, 2006; Bouza-Mora *et al.*, 2017).

Ehrlichia chaffeensis was first described as a human infection (1986) in the USA (Paddock and Childs, 2003). *E. chaffeensis* seems to have a worldwide distribution, although it is mainly reported from the USA. The infection is associated with *Amblyomma americanum* and *Dermacentor variabilis* but has also been detected in other tick species. Several species seem to serve as reservoir hosts for the bacterium, especially white-tailed deer (*Odocoileus virginianus*), although other species such as canines can be involved (Paddock and Childs, 2003). *E. chaffeensis* is the most important zoonotic infection among *Ehrlichia* species. Several thousand cases have so far been reported from the USA. Flu-like symptoms are common, but the illness can progress to severe multisystem disease with a case-fatality rate of around 3% (Paddock and Child, 2003; Walker *et al.*, 2004).

Ehrlichia ewingii was first described from dogs (1971) and causes fever, thrombocytopaenia and polyarthritis (Ewing *et al.*, 1971). Serological studies indicate that this pathogen is the most widespread *Ehrlichia* infection affecting dogs in the USA (Qurollo *et al.*, 2019), while the distribution in other countries is unknown. Canids and white-tailed deer are supposed to be the reservoir hosts. The main vector is *A. americanum*, although *D. variabilis* and *R. sanguineus* s.l. may also be involved (Saito and Walker, 2016). Human infection was first described in 1999 (Buller *et al.*, 1999) and includes flu-like symptoms and disease progression milder than for *E. chaffeensis* (Paddock and Childs, 2003). Fatalities have not been reported so far. More than 50 cases have been described, although human disease seems to be under-reported due to limited availability of molecular tests and serological cross-reactivity with *E. chaffeensis* (Harris *et al.*, 2016).

Ehrlichia muris was originally isolated from mice in 1983 (Wen *et al.*, 1995). Recent reclassification suggests two subspecies, *E. muris* subsp. *muris* and *E. muris* subsp. *eauclairensis*, both having zoonotic potential (Pritt *et al.*, 2017). *Ixodes* and *Haemophysalis* species are implicated as vectors. Reservoir hosts are small rodents, although cattle and small ruminants may also be involved (Pritt *et al.*, 2017). The first human case (2008) was reported in Russia (Nefedova *et al.*, 2008), followed by the USA and Eastern Asia (Pritt *et al.*, 2017). Clinical symptoms are flu-like with leukopaenia, lymphocytopaenia and thrombocytopaenia, and there are no reported fatalities. More than 70 human cases have so far been described (Johnson *et al.*, 2015).

Ehrlichia ruminantium was described as long ago as 1838 and causes hydropericardium ('heartwater') in ruminants (Uilenberg and Camus, 1993). This endemic disease in domestic and wild ruminants in sub-Saharan Africa also occurs in Madagascar and the Caribbean Islands (Maillard and Maillard, 1998). The infection is associated with different *Amblyomma* species, especially *Amblyomma variegatum* and *Amblyomma hebraeum*. Wild and domesticated ruminants seem to be the main reservoir hosts (Allsopp, 2009). *E. ruminantium* may cause high mortality, characterized by sudden onset of high fever, nervous signs and abdominal breathing. Although heartwater has been known for more than a century, it is still considered a major obstacle to the expansion and development of the livestock industry in southern Africa (Mahan, 2006). Human infection was first described (2005) in South Africa. Clinical symptoms were severe headache, sleepiness and unsteady gait. Three cases, all fatal due to severe vasculitis, have been reported so far (Allsopp *et al.*, 2005).

Panola Mountain *Ehrlichia* (PME) was first reported in a goat (2006) in the USA (Loftis *et al.*, 2006). The infection is associated with *Amblyomma* spp., especially *A. americanum*. White-tailed deer and goats seem to be involved as hosts (Loftis *et al.*, 2016). Serological studies indicate that the infection also occurs in the Caribbean Islands and Africa (Loftis *et al.*, 2016). However, little information is available concerning pathogenicity, potential hosts and geographical distribution. PME was recorded (2008) as a human infection in the USA, with mild clinical signs (Reeves *et al.*, 2008). Only one human case has so far been verified (Loftis *et al.*, 2016).

43.1.2 Distribution

Climate change, increased globalization, loss of natural habitats, improved diagnostics and an active surveillance will increase records of zoonotic infections caused by *Anaplasma*/*Ehrlichia* species.

Anaplasma/*Ehrlichia* species occur on all continents, although active and ongoing surveillance is necessary to obtain reliable prevalence maps. Their distribution changes continuously due to migration and transport of both tick vectors and vertebrate hosts. Movements of ticks and hosts that are driven directly or indirectly by climate change will have a huge effect on the dissemination and establishment of both pathogen and vector populations (Shope, 1991). Climate change may provide new habitats for infections and cause climatic-driven behavioural adaptation that may ease the spread of infections (Bett *et al.*, 2017). Successful colonization by ticks in new areas, however, depends on both available hosts and suitable habitats.

Climate-warming models predict that several tick species are likely to establish more northern permanent populations (Gray *et al.*, 2009). In addition, millions of ticks are annually spread by migrating birds, increasing the likelihood that ticks and pathogens become established in new geographical areas (Jaenson *et al.*, 2012). However, limited data are available concerning the effect of global warming on the distribution of different tick vector species and their infections. Since dogs are involved as hosts of several pathogens, they should be used more actively as sentinels for human infections (Bowser and Anderson, 2018).

Globalization, including trade and movements of domesticated animals, affects the transmission of pathogens between livestock and humans via tick bites. Contact between wildlife and domesticated animals is occurring more frequently due to an expanding human population, a demand for more pasture and farmland, and loss of natural wildlife habitats (Shope, 1991). Consequently, globalization will most probably increase the number of zoonotic infections due to an expanding tick–pathogen–host interface.

43.1.3 Diagnostics

Several of the aforementioned infections cause mild, non-specific clinical reactions. The numbers of human cases related to these infections are most likely under-reported due to flu-like symptoms and lack of molecular diagnostic tests (Maury de Tamí and Tamí-Maury, 2015; Bouza-Mora *et al.*, 2017). Species identification is important to assess the epidemiology of each of the different pathogens. Improved diagnostics are necessary for proper treatment and prevention, and to investigate the real zoonotic potential for these microbes, especially since limited data are available. In addition, lack of data concerning natural hosts hampers proper risk assessments (Ben Said *et al.*, 2017; Shi *et al.*, 2019). Zoonotic potential among several *Anaplasma*/*Ehrlichia* species has only recently been realized and its recognition may increase in the years to come as new diagnostic tests become available.

Improved diagnostics will also reveal new pathogens and/or strains/variants of known agents, such as *A. phagocytophilum*-like, *E. muris*-like and *A. platys*-like variants (Johnson *et al.*, 2015; Zobba *et al.*, 2015; Ben Said *et al.*, 2017). Unknown pathogen potential of these strains will become apparent as we may just have seen ‘the tip of the iceberg’ concerning zoonotic interactions. The occurrence of co-infections (Breitschwerdt *et al.*, 1998; Moniuszko *et al.*, 2014) also will be revealed by improved diagnostics. A correct diagnosis is necessary for proper treatment and prevention, especially since co-infection may aggravate clinical symptoms and increase disease severity (Moniuszko *et al.*, 2014). Vaccine development and other prophylactic measures also depend on proper species identification and will hopefully reduce the zoonotic occurrence of these microbes in the future.

43.2 Future Predictions

Several pathogens in the genera *Anaplasma* and *Ehrlichia* infect humans, of which *A. phagocytophilum* and *E. chaffeensis* have the greatest zoonotic potential, with a fatality rate of 1–3%. Several thousand human cases related to these two serious infections have so far been reported. Climate change will have a huge effect on the distribution and establishment of all tick-borne infections. An increase in zoonoses will most probably occur in the future related to increased tick–pathogen–host interactions, but also due to improved diagnostics and more active surveillance. In summary, an increasing number of people with non-specific flu-like symptoms in areas where ticks occur may be infected with a variety of *Anaplasma*/*Ehrlichia* pathogens.

References

Allsopp, B.A. (2009) Trend in control of heartwater. *Onderstepoort Journal of Veterinary Research* 76, 81–88.

Allsopp, M.T., Louw, M. and Meyer, E.C. (2005) *Ehrlichia ruminantium*: an emerging human pathogen? *Annals of the New York Academy of Sciences* 1063, 358–360.

Arraga-Alvarado, C.M., Qurollo, B.A., Parra, O.C., Berrueta, M.A., Hegarty, B.C., *et al.* (2014) Case report: molecular evidence of *Anaplasma platys* infection in two women from Venezuela. *American Journal of Tropical Medicine and Hygiene* 91, 1161–1165.

Bakken, J.S. and Dumler, J.S. (2015) Human granulocytic anaplasmosis. *Infectious Disease Clinics of North America* 29, 341–355.

Ben Said, M., Belkahia, H., El Mabrouk, N., Saidani, M., Alberti, A., *et al.* (2017) *Anaplasma platys*-like strains in ruminants from Tunisia. *Infection, Genetics and Evolution* 49, 226–233.

Bett, B., Kiunga, P., Gachohi, J., Sindato, C., Mbotha, D., *et al.* (2017) Effects of climate change on the occurrence and distribution of livestock diseases. *Preventive Veterinary Medicine* 137, 119–129.

Bouza-Mora, L., Dolz, G., Solórzano-Morales, A., Romero-Zuniga, J.J., Salazar-Sanchez, L., *et al.* (2017) Novel genotype of *Ehrlichia canis* detected in samples of human blood bank donors in Costa Rica. *Ticks and Tick-borne Diseases* 8, 36–40.

Bowser, N.H. and Anderson, N.E. (2018) Dogs (*Canis familiaris*) as sentinels for human infectious disease and application to Canadian populations: a systematic review. *Veterinary Sciences* 5, 83.

Breitschwerdt, E.B., Hegarty, B.C. and Hancock, S.I. (1998) Sequential evaluation of dogs naturally infected with *Ehrlichia canis*, *Ehrlichia chaffeensis*, *Ehrlichia equi*, *Ehrlichia ewingii*, or *Bartonella vinsonii*. *Journal of Clinical Microbiology* 36, 2645–2651.

Brodie, T.A., Holmes, P.H. and Urquhart, G.M. (1986) Some aspects of tick-borne diseases of British sheep. *Veterinary Record* 118, 415–418.

Buller, R.S., Arens, M., Hmiel, S.P., Paddock, C.D., Sumner, J.W., *et al.* (1999) *Ehrlichia ewingii*, a newly recognized agent of human ehrlichiosis. *New England Journal of Medicine* 341, 148–155.

Chochlakis, D., Ioannou, I., Tselentis, Y. and Psaroulaki, A. (2010) Human anaplasmosis and *Anaplasma ovis* variant. *Emerging Infectious Diseases* 16, 1031–1032.

Dumler, J.S., Barbet, A.F., Bekker, C.P., Dasch, G.A., Palmer, G.H., *et al.* (2001) Reorganization of genera in the families Rickettsiaceae and Anaplasmataceae in the order Rickettsiales: unification of some species of *Ehrlichia* with *Anaplasma*, *Cowdria* with *Ehrlichia* and *Ehrlichia* with *Neorickettsia*, descriptions of six new species combinations and designation of *Ehrlichia equi* and 'HGE agent' as subjective synonyms of *Ehrlichia phagocytophila*. *International Journal of Systematic and Evolutionary Microbiology* 51, 2145–2165.

Ewing, S.A., Roberson, W.R., Buckner, R.G. and Hayat, C.S. (1971) A new strain of *Ehrlichia canis*. *Journal of the American Veterinary Medical Association* 159, 1771–1774.

French, T.W. and Harvey, J.W. (1983) Serologic diagnosis of infectious cyclic thrombocytopenia in dogs using an indirect fluorescent antibody test. *American Journal of Veterinary Research* 44, 2407–2411.

Gordon, W.S., Brownlee, A., Wilson, D.R. and MacLeod, J. (1932) 'Tick-borne fever' (a hitherto undescribed disease of sheep). *Journal of Comparative Pathology* 45, 301–307.

Gray, J.S., Dautel, H., Estrada-Peña, A., Kahl, O. and Lindgren, E. (2009) Effects of climate change on ticks and tick-borne diseases in Europe. *Interdisciplinary Perspectives on Infectious Diseases* 2009, 593232.

Guglielmone, A.A., Robbins, R.G., Apanaskevich, D.A., Petney, T.N., Estrada-Peña, A. and Horak, I.G. (2014) *The Hard Ticks of the World (Acari: Ixodida: Ixodididae)*. Springer, Dordrecht, the Netherlands.

Harris, R.M., Couturier, B.A., Sample, S.C., Coulter, K.S., Casey, K.K., *et al.* (2016) Expanded geographic distribution and clinical characteristics of *Ehrlichia ewingii* infections, United States. *Emerging Infectious Diseases* 22, 862–865.

Jaenson, T.G., Jaenson, D.G., Eisen, L., Petersson, E. and Lindgren, E. (2012) Changes in the geographical distribution and abundance of the tick *Ixodes ricinus* during the last 30 years in Sweden. *Parasites & Vectors* 5, 8.

Jahfari, S., Coipan, E.C., Fonville, M., van Leeuwen, A.D., Hengeveld, P., *et al.* (2014) Circulation of four *Anaplasma phagocytophilum* ecotypes in Europe. *Parasites & Vectors* 7, 439.

Johnson, D.K., Schiffman, E.K., Davis, J.P., Neitzel, D.F., Sloan, L.M., *et al.* (2015) Human infection with *Ehrlichia muris*-like pathogen, United States, 2007–2013. *Emerging Infectious Diseases* 21, 1794–1799.

Jouglin, M., Blanc, B., de la Cotte, N., Bastian, S., Ortiz, K., *et al.* (2019) First detection and molecular identification of the zoonotic *Anaplasma capra* in deer in France. *PLoS ONE* 14, e0219184.
Li, H., Zheng, Y.C., Ma, L., Jia, N., Jiang, B.G., *et al.* (2015) Human infection with a novel tick-borne *Anaplasma* species in China: a surveillance study. *Lancet Infectious Diseases* 15, 663–670.
Loftis, A.D., Reeves, W.K., Spurlock, J.P., Mahan, S.M., Troughton, D.R., *et al.* (2006) Infection of a goat with a tick-transmitted *Ehrlichia* from Georgia, USA that is closely related to *Ehrlichia ruminantium*. *Journal of Vector Ecology* 31, 213–223.
Loftis, A.D., Kelly, P.J., Paddock, C.D., Blount, K., Johnson, J.W., *et al.* (2016) Panola Mountain *Ehrlichia* in *Amblyomma maculatum* from the United States and *Amblyomma variegatum* (Acari: Ixodidae) from the Caribbean and Africa. *Journal of Medical Entomology* 53, 696–698.
Maggi, R.G., Mascarelli, P.E., Havenga, L.N., Naidoo, V. and Breitschwerdt, E.B. (2013) Co-infection with *Anaplasma platys*, *Bartonella henselae* and *Candidatus* Mycoplasma haematoparvum in a veterinarian. *Parasites & Vectors* 6, 103.
Mahan, S.M. (2006) Diagnosis and control of heartwater, *Ehrlichia ruminantium* infection: an update. *CAB Reviews: Perspectives in Agriculture, Veterinary Science, Nutrition and Natural Resources* 1, 055.
Maillard, J.-C. and Maillard, N. (1998) Historique du peuplement bovin et de l'introduction de la tique *Amblyomma variegatum* dans les îles francaises des Antilles: synthèse bibliographique. *Ethnozootechnie* 61, 19–35 (in French).
Maury de Tamí, I. and Tamí-Maury, I. (2015) Brief comment on the article about human anaplasmosis by Arraga-Alvarado *et al.* *American Journal of Tropical Medicine and Hygiene* 93, 1124.
Moniuszko, A., Dunaj, J., Swięcicka, I., Zambrowski, G., Chmielewska-Badora, J., *et al.* (2014) Co-infections with *Borrelia* species, *Anaplasma phagocytophilum* and *Babesia* spp. in patients with tick-borne encephalitis. *European Journal of Clinical Microbiology and Infectious Diseases* 33, 1835–1841.
Nefedova, V.V., Korenberg, E.I., Kovalevskĭ, I.u.V., Gorelova, N.B. and Vorobeva, N.N. (2008) Microorganisms of the order Rickettsiales in taiga tick (*Ixodes persulcatus* Sch.) from the Pre-Ural region. *Vestnik Rossiiskoi Akademii Meditsinskikh Nauk* 7, 47–50 (in Russian).
Paddock, C.D. and Childs, J.E. (2003) *Ehrlichia chaffeensis*: a prototypical emerging pathogen. *Clinical Microbiology Reviews* 16, 37–64.
Pereira, A., Parreira, R., Nunes, M., Casadinho, A., Vieira, M.L., *et al.* (2016) Molecular detection of tick-borne bacteria and protozoa in cervids and wild boars from Portugal. *Parasites & Vectors* 9, 251.
Perez, M., Rikihisa, Y. and Wen, B. (1996) *Ehrlichia canis*-like agent isolated from a man in Venezuela: antigenic and genetic characterization. *Journal of Clinical Microbiology* 34, 2133–2139.
Perez, M., Bodor, M., Zhang, C., Xiong, Q. and Rikihisa, Y. (2006) Human infection with *Ehrlichia canis* accompanied by clinical signs in Venezuela. *Annals of the New York Academy of Sciences* 1078, 110–117.
Pritt, B.S., Allerdice, M.E.J., Sloan, L.M., Paddock, C.D., Munderloh, U.G., *et al.* (2017) Proposal to reclassify *Ehrlichia muris* as *Ehrlichia muris* subsp. *muris* subsp. nov. and description of *Ehrlichia muris* subsp. *eauclairensis* subsp. nov., a newly recognized tick-borne pathogen of humans. *International Journal of Systematic and Evolutionary Microbiology* 67, 2121–2126.
Qurollo, B.A., Balakrishnan, N., Cannon, C.Z., Maggi, R.G. and Breitschwerdt, E.B. (2014) Co-infection with *Anaplasma platys*, *Bartonella henselae*, *Bartonella koehlerae* and '*Candidatus* Mycoplasma haemominutum' in a cat diagnosed with splenic plasmacytosis and multiple myeloma. *Journal of Feline Medicine and Surgery* 16, 713–720.
Qurollo, B.A., Buch, J., Chandrashekar, R., Beall, M.J., Breitschwerdt, E.B., *et al.* (2019) Clinicopathological findings in 41 dogs (2008–2018) naturally infected with *Ehrlichia ewingii*. *Journal of Veterinary Internal Medicine* 33, 618–629.
Reeves, W.K., Loftis, A.D., Nicholson, W.L. and Czarkowski, A.G. (2008) The first report of human illness associated with the Panola Mountain *Ehrlichia* species: a case report. *Journal of Medical Case Reports* 2, 139.
Renneker, S., Abdo, J., Salih, D.E., Karagenc, T., Bilgic, H., *et al.* (2013) Can *Anaplasma ovis* in small ruminants be neglected any longer? *Transboundary and Emerging Diseases* 60(Suppl. 2), 105–112.
Ristic, M. and Holland, C.J. (1993) Canine ehrlichiosis. In: Woldehiwet, Z. and Ristic, M. (eds) *Rickettsial and Chlamydial Diseases of Domestic Animals*. Pergamon Press, Oxford, pp. 169–186.
Saito, T.B. and Walker, D.H. (2016) Ehrlichioses: an important One Health opportunity. *Veterinary Sciences* 3, 20.

Shi, K., Li, J., Yan, Y., Wang, K., Zhou, Y., *et al.* (2019) Dogs as new hosts for the emerging zoonotic pathogen *Anaplasma capra* in China. *Frontiers in Cellular and Infection Microbiology* 9, 394.

Shope, R. (1991) Global climate change and infectious diseases. *Environmental Health Perspectives* 96, 171–174.

Stuen, S., Granquist, E. and Silaghi, C. (2013) *Anaplasma phagocytophilum* – a widespread multi-host pathogen with highly adaptive strategies. *Frontiers in Cellular and Infection Microbiology* 3, 31.

Uilenberg, G. and Camus, E. (1993) Heartwater (cowdriosis). In: Woldehiwet, Z. and Ristic, M. (eds) *Rickettsial and Chlamydial Diseases of Domestic Animals*. Pergamon Press, Oxford, pp. 293–332.

Walker, D.H., Ismail, N., Olano, J.P., McBride, J.W., Yu, X.J., *et al.* (2004) *Ehrlichia chaffeensis*: a prevalent, life-threatening, emerging pathogen. *Transactions of the American Clinical and Climatological Association* 115, 375–382.

Wen, B., Rikihisa, Y., Mott, J., Fuerst, P.A., Kawahara, M., *et al.* (1995) *Ehrlichia muris* sp. nov., identified on the basis of 16S rRNA base sequences and serological, morphological, and biological characteristics. *International Journal of Systematic Bacteriology* 45, 250–254.

Zobba, R., Anfossi, A.G., Visco, S., Sotgiu, F., Dedola, C., *et al.* (2015) Cell tropism and molecular epidemiology of *Anaplasma platys*-like strains in cats. *Ticks and Tick-borne Diseases* 6, 272–280.

e044 Tick Vectors, Tick-Borne Diseases and Climate Change

Abdallah Samy[1]*, Abdelghafar Alkishe[2,3], Tatjana Pustahija[4] and Townsend Peterson[2]

[1]Ain Shams University, Cairo, Egypt; [2]University of Kansas, Lawrence, Kansas, USA; [3]University of Tripoli, Tripoli, Libya; [4]Institute of Public Health of Vojvodina, Novi Sad, Republic of Serbia

44.1 Expert Opinion

44.1.1 Introduction

The World Health Organization (WHO) has indicated that climate change threatens human health through long-term, large-scale changes in temperature and rainfall (IPCC, 2014). Ticks are sensitive to changes in climate: all arthropods are ectothermic, regulating their internal physiology by external conditions. Each life-cycle stage (e.g. larvae, nymph, adult) needs specific environmental conditions; increasing temperatures tend to increase biting rates by ticks and developmental rates of pathogens (Caminade *et al.*, 2019).

Ixodes ticks generally start searching for hosts when the temperature is at least 7°C and humidity is at least 85% (Süss *et al.*, 2008). Macro- and microclimatic requirements are not the same between tick species: some ticks are highly host- and habitat-specific, whereas others can survive in diverse habitats and use a wide variety of hosts (Petney *et al.*, 2011). Changes in climate conditions increase the risk of tick-borne diseases (TBDs) by increasing the activity, populations and ranges of numerous tick species (Bouchard *et al.*, 2019). Indeed, incidence of TBDs has multiplied globally in recent decades and continues increasing to this day. For example, the annual report of the European Centre for Disease Prevention and Control from 2014 states that the numbers of human cases of tick-borne encephalitis (TBE) in all endemic regions of Europe have increased by fourfold in just 30 years (ECDC, 2018); and the Centers for Disease Control and Prevention (CDC) reports that TBD cases have increased from 48,610 in 2016 to 59,349 in 2017 in the USA (CDC, 2019).

Factors including air temperature, humidity, soil water and vegetation type influence the physiology, which translates into the distributional potential and population abundance, of ticks and their hosts; effects on hosts may in turn affect the ticks. Temperature and humidity are the factors most important in determining the rate of tick survival and development (Estrada-Peña, 2008), but micro-scale factors (e.g. vegetation and soil type) affect tick populations and distributions as well (Schwarz *et al.*, 2009). Several studies have focused on the biology of ticks and associated pathogens (Ostfeld *et al.*, 2006; Stich *et al.*, 2008; Dantas-Torres, 2010; Estrada-Peña *et al.*, 2015). Several recent studies have focused on geographic distributions of medically important ticks (Estrada-Peña and Venzal,

*Email: samy@sci.asu.edu.eg

DOI: 10.1079/9781789249637.0044

2006; Estrada-Peña, 2008; Swart *et al.*, 2014; Alkishe *et al.*, 2017; Peterson and Raghavan, 2017; Raghavan *et al.*, 2019). Overall, the concern is that global warming will contribute to an increase in populations of ticks and hosts, dispersal of vectors and consequent range shifts and expansions.

A further, tick-mediated consequence of climate change is the emergence of new endemic foci of TBDs and the expansion of existing ones, resulting in TBD occurrence in areas where it has not been previously reported. An example is the detection of the first autochthonous human cases of Crimean–Congo haemorrhagic fever (CCHF) in Greece in 2008 (Papa *et al.*, 2010) and in Spain in 2016 (Negredo *et al.*, 2017), as well as the appearance of new TBE foci across Europe. TBD foci are also expanding to higher elevations where they have not been present before. Holzmann *et al.* (2009) described a TBE outbreak in alpine pastures at an elevation of 1564 m in 2008. In the Czech Republic, the maximum elevation at which *Ixodes ricinus* populations persist has shifted from 700 to 1100 m in just 10 years of study (Danielová *et al.*, 2008a).

Increases in average annual air temperatures in the temperate zone also prolong the period of vegetative productivity (Hatfield and Prueger, 2015), which may also extend the seasons of tick activity (Gray *et al.*, 2009). In tandem, professional or recreational activities of people in natural habitats are extended, so exposure to the natural habitats of the ticks is higher (Garcia-Marti *et al.*, 2018).

Altogether, these drivers paint a shifting picture of TBD incidence and risk. Clearly, a global response to develop and implement adequate measures to minimize the occurrence and spread of TBDs and to prevent and control outbreaks and epizootics becomes urgent. More precisely, considerable human suffering and mortality can be avoided if TBD risks and changes in that risk can be anticipated under influences of global warming.

44.1.2 Evidence for climate change impacts

Impacts of climate change vary from region to region around the world, such that some regions will likely be affected more than others. In the USA, for example, evidence of climate change manifests differently in different regions (USGCRP, 2017). According to the US Global Change Research Program, five US regions are currently facing serious impacts: heatwaves in the Northeast; sea-level rise in the East and Northwest; and heat and drought in the Midwest and Southwest (USGCRP, 2017). Those changes in climate conditions (global warming) have influenced the activity and especially biting rates of tick species (Gray *et al.*, 2009). Since the warmest known winter in 2017 (NOAA, 2020), numbers of TBD cases have increased dramatically compared with other years (CDC, 2019). Also, new reports have mentioned that several tick species have established populations in areas that did not previously hold populations (Molaei, 2020).

These climate-driven changes in TBD aetiology are increasingly gaining attention from scientists around the world (Ostfeld and Brunner, 2015). The hypothesis that climatic conditions affect the spatial distribution of ticks has been explored in numerous studies. Alkishe *et al.* (2017) assessed effects of current and future climatic conditions on the geographic distribution of *I. ricinus* in Europe, North Africa and the Middle East, using ecological niche modelling. Their models revealed potential broad expansion of suitable areas for *I. ricinus* in Northern Europe, caused by increases in average air temperature and consequently warmer winters in those areas. Such conditions allow ticks to survive longer, thus making them more likely to transmit various pathogens to humans or animals. Between the present and 2050, the potential area suitable for *I. ricinus* was predicted to increase by 10.8–11.7%, depending on the emissions scenario. Another similar study showed that an increase in temperature could affect the ecology of *I. ricinus* in Eurasian regions drastically and diseases transmitted by this tick species could emerge in previous non-endemic areas (Porretta *et al.*, 2013).

The range expansion of *I. ricinus* towards northern parts anticipated in the modelling studies has been documented in a study in Sweden over a period of 30 years. In just 18 years of the study, the area doubled in the northern part of the country where this tick species occurs (Jaenson *et al.*, 2012). Other closely related species, like *Ixodes persulcatus*, appear to be responding similarly to climate change influences; its range is also expanding and the incidence of TBE is rising in northern Russia (Tokarevich *et al.*, 2011).

Another recent modelling study (Alkishe *et al.*, 2020) for the first time mapped current and future potential distributions of the tick *Rhipicephalus sanguineus* sensu lato. Model predictions anticipated new suitable areas in the northern hemisphere in the future that are not suitable under present climatic conditions, in general suggesting possible range shifts of *R. sanguineus* towards temperate areas. A significant expansion of the geographical distribution, especially towards northern latitudes of North America, was also predicted for *Amblyomma americanum* in another recent modelling study (Raghavan *et al.*, 2019).

Other medically important tick species may have the potential to establish new natural foci at more northerly latitudes in Europe under the influence of global warming (Gray *et al.*, 2009). A possible mechanistic explanation for effects of increasing annual mean temperature on range expansions of ticks and associated diseases is via prolongation of periods of optimal temperature for survival of ticks, earlier cessation of winter diapause and earlier onset of host-seeking activity or development, all of which lead to increased abundance (Schwarz *et al.*, 2009). These reasons increase chances of transmitting pathogens transstadially, transovarially and during feeding on the same host, each of which increases numbers of infected ticks. Host activities may also increase, and highly mobile hosts may thereby aid in dispersing ticks into new areas.

Lukán *et al.* (2010) analysed tick distributions during 1961–2004 in Slovakia and found that TBE foci shifted from lowland to submontane areas in that country, in correlation with increases in annual mean temperature. At the beginning of the study period, mean elevation for TBE was 180–340 m above sea level, but the maximum elevation had risen to 832 m by the end of the study period.

Temperature and humidity impact egg hatch, moulting, and longevity of larvae and nymphs of *Dermacentor reticulatus* (Zahler and Gothe, 1995); eggs and larvae that do not complete development would not have survived the cold season of a few decades ago (Zahler and Gothe, 1995; Gray *et al.*, 2009). Zahler and Gothe (1995) noted that these climatic factors affect longevity of unfed adults and oviposition of *D. reticulatus*. Oviposition occurs at temperatures of 10–27°C. The duration of the gonotrophic cycle in ticks is also temperature-dependent; warmer conditions drive shorter gonotrophic cycles (Medlock and Leach, 2015). An important indicator of vectorial capacity of ticks is the extrinsic incubation period (i.e. time between ingestion of the pathogen by the vector and the vector becoming infectious and able to transmit the pathogen); temperature impacts this period (Ogden *et al.*, 2005; Kurtenbach *et al.*, 2006; Gubler, 2009). As this period shortens, the vectorial capacity of the tick increases.

The influence of climatic factors on the development and longevity of ticks has been explored in several other species, including *Amblyomma triguttatum triguttatum* Koch (Guglielmone, 1992), *Rhipicephalus appendiculatus* Neumann (Tukahirwa, 1976) and *Hyalomma truncatum* (Wilson *et al.*, 1993). Alahmed and Kheir (2003) showed that temperature is important in affecting the duration of pre-oviposition, incubation and moulting periods in the development of *Hyalomma dromedarii*, thus significantly affecting the survival and life cycle of this tick species.

Ixodes ricinus (sheep tick)

I. ricinus is the most widespread tick species in Europe and is a very important vector of animal and human diseases (Zahler and Gothe, 1995), including Lyme disease, TBE, anaplasmosis, babesiosis, ehrlichiosis and several rickettsial diseases. This tick lives in cool and humid habitats, such as European woodlands; in Africa, it is restricted to the cooler and more humid areas of the Mediterranean rim (e.g. Tunisia, Algeria and Morocco; Estrada-Peña *et al.*, 2004). Numerous studies have explored impacts of climate change on this species, essentially all anticipating expansion into more northern latitudes and higher elevations.

The most probable reason for this range expansion is global warming, with warmer winters and extension of spring and autumn seasons, which allows this tick to survive longer and have a longer period of activity. In Sweden and Norway, the northern distribution limit of *I. ricinus* has shifted northwards in recent decades (Tälleklint and Jaenson, 1998; Lindgren *et al.*, 2000; Skarpaas *et al.*, 2006; Jaenson and

Lindgren, 2011). This expansion of the geographic distribution of *I. ricinus* carries an increased risk of disease transmission. Indeed, in Sweden, TBE incidence has increased significantly over 1960–1998 (Lindgren and Gustafson, 2001). In the Czech Republic, the upper elevational boundary of permanent populations shifted from 700 to 1100 m above sea level (Daniel *et al.*, 2003; Materna *et al.*, 2005). TBE is also shifting to higher elevations and indeed is increasing overall in that country (Daniel *et al.*, 2004; Zeman and Beneš, 2004; Danielová *et al.*, 2008b).

Dermacentor reticulatus (ornate dog tick)

D. reticulatus is an effective vector of Omsk haemorrhagic fever, TBE, canine babesiosis, tularaemia, Q fever, anaplasmosis and several rickettsioses (Dautel *et al.*, 2006; Zivkovic *et al.*, 2007; Földvári *et al.*, 2016). It ranges across Southern and Central Europe; more precisely, the geographical distribution of this species extends from France and south-western England to Central Asia. It ranges north to northern Germany, northern Poland and Lithuania, and south to the Mediterranean coast. Its distribution is highly focal (Gray *et al.*, 2009).

Impacts of climate change on the life cycle and longevity of larvae and nymphs of this tick species have been recorded (Zahler and Gothe, 1995; Gray *et al.*, 2009). In Germany, Poland, Hungary, Slovakia, the Netherlands and Belgium, new *D. reticulatus* foci have been established, implying an expansion of its geographic distribution in recent decades (Sréter *et al.*, 2005; Földvári *et al.*, 2016; Rubel *et al.*, 2016). Changing distribution of this species was followed by an onset of canine babesiosis in new areas of Germany, Hungary and the Netherlands (Sréter *et al.*, 2005; Dautel *et al.*, 2006; Nijhof *et al.*, 2007; Gray *et al.*, 2009).

Hyalomma marginatum (Mediterranean tick)

Hyalomma marginatum is an important vector of the dangerous CCHF virus, as well as of *Rickettsia conorii, Rickettsia aeschlimannii, Anaplasma marginale, Babesia occultans, Theileria* spp., *Coxiella burnetii* and many other pathogens (Walker, 2003; Ergönül, 2006; Rumer *et al.*, 2011; Kolodziejek *et al.*, 2014; Chitimia-Dobler *et al.*, 2019). The species' geographic distribution extends from western Spain and North Africa to Ukraine and southern Russia and the Middle East (Apanaskevich and Horak, 2008; ECDC, 2018). Because it uses migratory birds as hosts, this tick species is appearing more frequently in Western and Central Europe, as migratory birds transport the ticks into the region during spring migration (Jameson *et al.*, 2012; Capek *et al.*, 2014; Cuber, 2016; Oehme *et al.*, 2017). Future climate change could affect establishment of permanent populations in these areas, as warmer autumns would reduce tick mortality (Gray *et al.*, 2009). Models of the distribution of *H. marginatum* under climate change scenarios predict that large areas of Europe would be suitable for this species, likely thanks to increasing temperature and decreasing rainfall (Estrada-Peña and Venzal, 2007).

CCHF is a dangerous zoonotic disease that is transmitted mainly by *H. marginatum*. The present foci of this disease are in parts of Africa, Asia, the Middle East and South-Eastern Europe (Hoogstraal, 1979); recent years have recorded numerous range extensions and new populations (Negredo *et al.*, 2017). A substantial increase in CCHF cases has been documented (Ergönül, 2006). The first case in Turkey was reported in 2002 and an outbreak there is ongoing (Ergönül *et al.*, 2004). First autochthonous cases were registered recently also in Greece (2008) and in Spain (2010). Seroprevalence studies indicate that this virus circulates also in Hungary, Portugal, Egypt and France (Ergönül, 2006).

Amblyomma maculatum (Gulf Coast tick)

Amblyomma maculatum is the main vector of *Rickettsia parkeri* and *Francisella tularensis* in North America (Sonenshine, 2018). Early reports on the geographic distribution of this species indicated that it was restricted to the southern coastal areas of the USA; then, the expanded ranges were observed toward western states in Kansas, Oklahoma and Tennessee (Paddock and Goddard, 2015). Between 1963 and 2000, *A. maculatum* was known as a rare species in the Carolinas, Virginia, eastern Maryland and Delaware, but it became better known in those states in recent decades (Sonenshine, 2018). Recently, the species has expanded into new areas toward northern states, particularly Connecticut

(Molaei, 2020). Factors like mild winters during recent years in Northeast states and the bird migration across those areas helped this species to arrive and establish populations. Rickettsial disease cases caused by *R. parkeri* occurred mostly in Oklahoma, Arkansas, North Carolina, Missouri, Virginia and Tennessee. The CDC reported that rickettsial disease cases have increased dramatically to 6248 cases as the warmest winter appeared in 2017 compared with an average of 4269 in previous years (CDC, 2019).

Ixodes scapularis (black-legged tick)

Ixodes scapularis is the vector of several important pathogens that cause disease in humans, including bacteria (*Borrelia burgdorferi* sensu stricto (s.s.), *Anaplasma phagocytophilum*), viruses (Powassan encephalitis virus) and parasites (*Babesia microti*) (Eisen *et al.*, 2016). In the past, this species was found only in the eastern USA; then it spread more widely across the eastern USA and southern Canada; and recently its geographic distribution has expanded to include mid-western states (North Dakota, South Dakota and Minnesota) (Russart *et al.*, 2014; Maestas *et al.*, 2018) and even higher latitudes in Canada (Clow *et al.*, 2017; Bouchard *et al.*, 2019). Based on the reports from the CDC, human cases of Lyme disease, anaplasmosis, Powassan virus infection, babesiosis and other diseases have increased significantly in response to climate changes and mild winters; Lyme disease (42,743 cases), anaplasmosis (7718 cases), babesiosis (2368 cases) and Powassan virus infections (33 cases) reached the highest case numbers in 2017 (CDC, 2019).

Haemaphysalis longicornis (Asian long-horned tick)

Haemaphysalis longicornis can transmit an impressive 30 pathogens to humans, including seven species of spotted fever group rickettsiae, seven species in the family Anaplasmataceae, four genospecies in *B. burgdorferi* s.s., two *Babesia* species, six viruses, and *Francisella*, *Bartonella*, *Coxiella* and *Toxoplasma* (Zhao *et al.*, 2020). *H. longicornis* was native to East Asia, and then invaded Australia and Oceania (Rochlin, 2019). Recently, it was discovered in the USA, first in New Jersey in 2017 (Rainey *et al.*, 2018). Since then, it has spread to 14 new states (Virginia, West Virginia, Pennsylvania, North Carolina, Tennessee, New York, Maryland, Connecticut, Kentucky, Delaware, Arkansas, Ohio, Rhode Island and South Carolina) (USDA, 2020). Only a single case of a human bite by this species was recorded in 2018, but no disease transmission has yet been reported in the USA (Wormser *et al.*, 2020).

44.1.3 Evidence against climate change impact

Only a few studies rule out the influence of climatic factors on ticks and TBDs. For example, Randolph (2010) claimed that the recent temporal patterns in TBD epidemiology result more from political transitions and human behavioural changes, which determine exposure to infected ticks, than from changes in abundance of infected ticks; however, the author did not dispute the potential impact of climate change on the geographical distributions of TBDs.

Sumilo *et al.* (2007) have argued that the worsening epidemiological situation of TBDs arose thanks to the socio-economic transitions resulting from the collapse of the Soviet Union, 30 years ago. The idea is that the collapse of industry and agriculture led to reduced air pollution and changes in land cover from agricultural and pasture to wooded land, which likely increased abundances of rodents and other hosts. One previous study (Gilbert *et al.*, 2014) suggested that *I. ricinus* has potential to be resilient to future climate change. Barbour and Fish (1993) proposed that expansion of ticks and TBDs has nothing to do with climate: afforestation of large areas cleared during the 18th and 19th centuries and the consequent return of deer populations and tick populations, in this case *I. scapularis*, are argued to have triggered the current outbreak of Lyme disease.

44.2 Future Predictions

Knowledge of climate change impacts depends on the detail available on climate dynamics, population growth, emissions-control technology and economic development (Kareiva and

Marvier, 2011). The Intergovernmental Panel on Climate Change (IPCC) has created several scenarios of future greenhouse gas (GHG) emission, in the form of four 'Representative Concentration Pathways' (RCPs): RCP2.6 (stringent mitigation scenario), RCP4.5 and RCP6.0 (intermediate scenarios), and RCP8.5 (high GHG emissions) (IPCC, 2014). These scenarios are intended to bracket the suite of possible climate futures globally.

The scenarios most frequently used to assess the potential geographic distribution of species are RCP4.5 and RCP8.5, as RCP2.6 appears to be increasingly unlikely. RCP4.5 and RCP8.5 provide estimates of less (RCP4.5) and more (RCP8.5) extreme changes in environmental conditions that would affect suitable areas of a species. Several studies have used those future climate scenarios to assess likely changes in tick species' potential geographic distributions. However, almost all previous studies have highlighted the northern hemisphere and high latitudes as newly suitable areas for tick species in North America and Europe (Williams *et al.*, 2015; Alkishe *et al.*, 2017, 2020; Raghavan *et al.*, 2019; Boorgula *et al.*, 2020); a notable exception is Marques *et al.* (2020), which focuses on a South American species.

Substantial scientific evidence indicates that global warming leads to increases in incidence of TBDs by the prolongation of tick activity even during the winter, accelerating their life cycle and increasing abundance (Süss *et al.*, 2008), then creating favourable conditions for the spread of foci of these vectors and their occurrence at higher elevations and higher latitudes. A notable gap in the literature is a focus on species that may be limited by drought rather than temperature, such as species distributed across humid-to-arid gradients in tropical regions.

Given that the annual mean temperature is expected to rise globally by 1.1°C in the 21st century (Patz *et al.*, 2000), a most likely scenario for TBDs would include the following characteristics:

1. The geographical distribution of ticks will expand, at least in some directions (i.e. upward in elevation and poleward in latitude). The potentially retracting, equatorial side of species distributions remains little studied and woefully under characterized.

2. Tick abundances will increase in some regions, as conditions become more ideal for the particular species (Martínez-Meyer *et al.*, 2013; Osorio-Olvera *et al.*, 2020).

3. Prevalence of pathogens causing TBDs will increase, at least in some regions, as conditions become more ideal for the tick species and possibly for pathogen transmission as well.

4. TBDs will likely increase in prevalence and incidence in many parts of the world.

The above insights are not new. Given that the reported TBD incidence rates have been increasing each decade (Dahlgren *et al.*, 2016), the epidemiological situation is likely to worsen in the next 10 years. The WHO has developed a Global Strategy for Response to Vector Control 2017–2030, which aims to strengthen vector control worldwide, as a main method for preventing and controlling many infectious diseases (WHO, 2017). This strategy should at least reduce the large increase in the incidence of TBDs, which appears to be otherwise inevitable in the next 10 years.

Since modelling algorithms are used widely to predict suitable areas for disease vectors, pathogens and hosts, it is important to assess uncertainty that may be introduced by transferring models to future conditions (Alkishe *et al.*, 2020). These model outputs include diverse sources of variation and uncertainty (Peterson *et al.*, 2018), so explaining those projections can be difficult. Thus, assessing uncertainty by detecting strict extrapolation areas offers a guide to where model predictions may be less robust (Owens *et al.*, 2013). Exploring sources of variation coming from different factors, such as model replicates, parameter values, general circulation models (GCMs; mathematical models representing physical processes of the atmosphere and ocean to simulate responses of global climate to increasing GHG emissions) and RCPs (four GHG concentration trajectories adopted by the IPCC's Fifth Assessment Report) in model transfers (i.e. models developed on a calibration area applied to conditions manifested in other times and places) (Peterson *et al.*, 2018; Cobos *et al.*, 2019), is important in avoiding misinterpretations of potential geographic distributions (Alkishe *et al.*, 2020). For example, Alkishe *et al.* (2017) showed areas of uncertainty in Northern Europe where interpretation of model predictions regarding the future distributional potential of the tick *I. ricinus* should be done with caution. Moreover, recent approaches use hierarchical partitioning of variance to represent the magnitude of variation coming from different sources (Alkishe *et al.*, 2020) – if most of the variation comes from 'minor' methodological choices such as parameter values and replicate re-samplings of occurrence data, the models may be overall less robust.

References

Alahmed, A.M. and Kheir, S.M. (2003) Life cycle and survival of *Hyalomma dromedarii* (Acari: Ixodidae) under laboratory conditions. *Journal of Agricultural and Marine Sciences* 8, 11–14.

Alkishe, A.A., Peterson, A.T. and Samy, A.M. (2017) Climate change influences on the potential geographic distribution of the disease vector tick *Ixodes ricinus*. *PLoS ONE* 12, e0189092.

Alkishe, A., Cobos, M.E., Peterson, A.T. and Samy, A.M. (2020) Recognizing sources of uncertainty in disease vector ecological niche models: an example with the tick *Rhipicephalus sanguineus* sensu lato. *Perspectives in Ecology and Conservation* 18, 91–102.

Apanaskevich, D.A. and Horak, I.G. (2008) The genus *Hyalomma* Koch, 1844: V. Re-evaluation of the taxonomic rank of taxa comprising the *H.* (*Euhyalomma*) *marginatum* Koch complex of species (Acari: Ixodidae) with redescription of all parasitic stages and notes on biology. *International Journal of Acarology* 34, 13–42.

Barbour, A.G. and Fish, D. (1993) The biological and social phenomenon of Lyme disease. *Science* 260, 1610–1616.

Boorgula, G.D., Peterson, A.T., Foley, D.H., Ganta, R.R. and Raghavan, R.K. (2020) Assessing the current and future potential geographic distribution of the American dog tick, *Dermacentor variabilis* (Say) (Acari: Ixodidae) in North America. *PLoS ONE* 15, e0237191.

Bouchard, C., Dibernardo, A., Koffi, J., Wood, H., Leighton, P. and Lindsay, L. (2019) Climate change and infectious diseases: the challenges. Increased risk of tick-borne diseases with climate and environmental changes. *Canada Communicable Disease Report* 45, 83–89.

Caminade, C., McIntyre, K.M. and Jones, A.E. (2019) Impact of recent and future climate change on vector-borne diseases. *Annals of the New York Academy of Sciences* 1436, 157–173.

Capek, M., Literak, I., Kocianova, E., Sychra, O., Najer, T., *et al.* (2014) Ticks of the *Hyalomma marginatum* complex transported by migratory birds into Central Europe. *Ticks and Tick-borne Diseases* 5, 489–493.

CDC (2019) Tickborne disease surveillance data summary. Centers for Disease Control and Prevention, Atlanta, Georgia. Available at: https://www.cdc.gov/ticks/data-summary/index.html (accessed 11 December 2020).

Chitimia-Dobler, L., Schaper, S., Ries, R., Bitterwolf, K., Frangoulidis, D., *et al.* (2019) Imported *Hyalomma* ticks in Germany in 2018. *Parasites & Vectors* 12, 134.

Clow, K.M., Leighton, P.A., Ogden, N.H., Lindsay, L.R., Michel, P., *et al.* (2017) Northward range expansion of *Ixodes scapularis* evident over a short timescale in Ontario, Canada. *PLoS ONE* 12, e0189393.

Cobos, M.E., Peterson, A.T., Barve, N. and Osorio-Olvera, L. (2019) kuenm: an R package for detailed development of ecological niche models using Maxent. *PeerJ* 7, e6281.

Cuber, P. (2016) Ticks (Ixodida) from the collection of the Natural History Department, Museum of Upper Silesia in Bytom, Poland – a contribution to knowledge on tick fauna and the first record of *Hyalomma marginatum* presence in Poland. *Annals of Agricultural and Environmental Medicine* 23, 379–381.

Dahlgren, F.S., Paddock, C.D., Springer, Y.P., Eisen, R.J. and Behravesh, C.B. (2016) Expanding range of *Amblyomma americanum* and simultaneous changes in the epidemiology of spotted fever group rickettsiosis in the United States. *American Journal of Tropical Medicine and Hygiene* 94, 35–42.

Daniel, M., Danielová, V., Kříž, B., Jirsa, A. and Nožička, J. (2003) Shift of the tick *Ixodes ricinus* and tick-borne encephalitis to higher altitudes in central Europe. *European Journal of Clinical Microbiology and Infectious Diseases* 22, 327–328.

Daniel, M., Danielová, V., Kříž, B. and Kott, I. (2004) An attempt to elucidate the increased incidence of tick-borne encephalitis and its spread to higher altitudes in the Czech Republic. *International Journal of Medical Microbiology Supplements* 293, 55–62.

Danielová, V., Kliegrová, S., Daniel, M. and Beneš, C. (2008a) Influence of climate warming on tick-borne encephalitis expansion to higher altitudes over the last decade (1997–2006) in the Highland Region (Czech Republic). *Central European Journal of Public Health* 16, 4–11.

Danielová, V., Schwarzová, L., Materna, J., Daniel, M., Metelka, L., *et al.* (2008b) Tick-borne encephalitis virus expansion to higher altitudes correlated with climate warming. *International Journal of Medical Microbiology* 298, 68–72.
Dantas-Torres, F. (2010) Biology and ecology of the brown dog tick, *Rhipicephalus sanguineus*. *Parasites & Vectors* 3, 26.
Dautel, H., Dippel, C., Oehme, R., Hartelt, K. and Schettler, E. (2006) Evidence for an increased geographical distribution of *Dermacentor reticulatus* in Germany and detection of *Rickettsia* sp. RpA4. *International Journal of Medical Microbiology* 296, 149–156.
ECDC (2018) *Hyalomma marginatum* – current known distribution: May 2020. European Centre for Disease Prevention and Control, Stockholm. Available at: https://www.ecdc.europa.eu/en/publications-data/hyalomma-marginatum-current-known-distribution-may-2020 (accessed 11 December 2020).
Eisen, R.J., Eisen, L., Ogden, N.H. and Beard, C.B. (2016) Linkages of weather and climate with *Ixodes scapularis* and *Ixodes pacificus* (Acari: Ixodidae), enzootic transmission of *Borrelia burgdorferi*, and Lyme disease in North America. *Journal of Medical Entomology* 53, 250–261.
Ergönül, Ö. (2006) Crimean–Congo haemorrhagic fever. *Lancet Infectious Diseases* 6, 203–214.
Ergönül, Ö., Çelikbs, A., Dokuzoğuz, B., Eren, Ş., Baykam, N. and Esener, H. (2004) Characteristics of patients with Crimean–Congo hemorrhagic fever in a recent outbreak in Turkey and impact of oral ribavirin therapy. *Clinical Infectious Diseases* 39, 284–287.
Estrada-Peña, A. (2008) Climate, niche, ticks, and models: what they are and how we should interpret them. *Parasitology Research* 103, 87–95.
Estrada-Peña, A. and Venzal, J. (2006) High-resolution predictive mapping for *Boophilus annulatus* and *B. microplus* (Acari: Ixodidae) in Mexico and Southern Texas. *Veterinary Parasitology* 142, 350–358.
Estrada-Peña, A. and Venzal, J.M. (2007) Climate niches of tick species in the Mediterranean region: modeling of occurrence data, distributional constraints, and impact of climate change. *Journal of Medical Entomology* 44, 1130–1138.
Estrada-Peña, A., Bouattour, A., Camicas, J.L. and Walker, A.R. (2004) *Ticks of Domestic Animals in the Mediterranean Region. A Guide to Identification of Species*. University of Zaragoza, Zaragoza, Spain.
Estrada-Peña, A., de la Fuente, J., Ostfeld, R.S. and Cabezas-Cruz, A. (2015) Interactions between tick and transmitted pathogens evolved to minimise competition through nested and coherent networks. *Scientific Reports* 5, 10361.
Földvári, G., Široký, P., Szekeres, S., Majoros, G. and Sprong, H. (2016) *Dermacentor reticulatus*: a vector on the rise. *Parasites & Vectors* 9, 314.
Garcia-Marti, I., Zurita-Milla, R., Harms, M.G. and Swart, A. (2018) Using volunteered observations to map human exposure to ticks. *Scientific Reports* 8, 15435.
Gilbert, L., Aungier, J. and Tomkins, J.L. (2014) Climate of origin affects tick (*Ixodes ricinus*) host-seeking behavior in response to temperature: implications for resilience to climate change? *Ecology and Evolution* 4, 1186–1198.
Gray, J., Dautel, H., Estrada-Peña, A., Kahl, O. and Lindgren, E. (2009) Effects of climate change on ticks and tick-borne diseases in Europe. *Interdisciplinary Perspectives on Infectious Diseases* 2009, 593232.
Gubler, D.J. (2009) Vector-borne diseases. *Revue Scientifique et Technique* 28, 583–588.
Guglielmone, A.A. (1992) The effect of temperature and humidity on development and longevity of *Amblyomma triguttatum triguttatum* (Acarina: Ixodidae). *Bulletin of Entomological Research* 82, 203–208.
Hatfield, J.L. and Prueger, J.H. (2015) Temperature extremes: effect on plant growth and development. *Weather and Climate Extremes* 10, 4–10.
Holzmann, H., Aberle, S.W., Stiasny, K., Werner, P., Mischak, A., *et al.* (2009) Tick-borne encephalitis from eating goat cheese in a mountain region of Austria. *Emerging Infectious Diseases* 15, 1671.
Hoogstraal, H. (1979) The epidemiology of tick-borne Crimean–Congo hemorrhagic fever in Asia, Europe, and Africa. *Journal of Medical Entomology* 15, 307–417.
IPCC (2014) *Climate Change 2014: Synthesis Report. Contribution of Working Groups I, II and III to the Fifth Assessment Report of the Intergovernmental Panel on Climate Change* [Core Writing Team, Pachauri, R.K. and Meyer, L.A. (eds)]. Intergovernmental Panel on Climate Change, Geneva, Switzerland.
Jaenson, T.G. and Lindgren, E. (2011) The range of *Ixodes ricinus* and the risk of contracting Lyme borreliosis will increase northwards when the vegetation period becomes longer. *Ticks and Tick-borne Diseases* 2, 44–49.

Jaenson, T.G., Jaenson, D.G., Eisen, L., Petersson, E. and Lindgren, E. (2012) Changes in the geographical distribution and abundance of the tick *Ixodes ricinus* during the past 30 years in Sweden. *Parasites & Vectors* 5, 8.

Jameson, L.J., Morgan, P.J., Medlock, J.M., Watola, G. and Vaux, A.G. (2012) Importation of *Hyalomma marginatum*, vector of Crimean–Congo haemorrhagic fever virus, into the United Kingdom by migratory birds. *Ticks and Tick-borne Diseases* 3, 95–99.

Kareiva, P.M. and Marvier, M. (2011) *Conservation Science: Balancing the Needs of People and Nature*. Roberts and Company, Greenwood Village, Colorado.

Kolodziejek, J., Marinov, M., Kiss, B.J., Alexe, V. and Nowotny, N. (2014) The complete sequence of a West Nile virus lineage 2 strain detected in a *Hyalomma marginatum marginatum* tick collected from a song thrush (*Turdus philomelos*) in eastern Romania in 2013 revealed closest genetic relationship to strain Volgograd 2007. *PLoS ONE* 9, e109905.

Kurtenbach, K., Hanincová, K., Tsao, J.I., Margos, G., Fish, D. and Ogden, N.H. (2006) Fundamental processes in the evolutionary ecology of Lyme borreliosis. *Nature Reviews Microbiology* 4, 660–669.

Lindgren, E. and Gustafson, R. (2001) Tick-borne encephalitis in Sweden and climate change. *The Lancet* 358, 16–18.

Lindgren, E., Tälleklint, L. and Polfeldt, T. (2000) Impact of climatic change on the northern latitude limit and population density of the disease-transmitting European tick *Ixodes ricinus*. *Environmental Health Perspectives* 108, 119–123.

Lukán, M., Bullova, E. and Petko, B. (2010) Climate warming and tick-borne encephalitis, Slovakia. *Emerging Infectious Diseases* 16, 524.

Maestas, L.P., Mays, S.E., Britten, H.B., Auckland, L.D. and Hamer, S.A. (2018) Surveillance for *Ixodes scapularis* (Acari Ixodidae) and *Borrelia burgdorferi* in eastern South Dakota state parks and nature areas. *Journal of Medical Entomology* 55, 1549–1554.

Marques, R., Krüger, R.F., Peterson, A.T., De Melo, L.F., Vicenzi, N. and Jiménez-García, D. (2020) Climate change implications for the distribution of the babesiosis and anaplasmosis tick vector, *Rhipicephalus* (*Boophilus*) *microplus*. *Veterinary Research* 51, 81.

Martínez-Meyer, E., Díaz-Porras, D., Peterson, A.T. and Yáñez-Arenas, C. (2013) Ecological niche structure and rangewide abundance patterns of species. *Biology Letters* 9, 20120637.

Materna, J., Daniel, M. and Danielová, V. (2005) Altitudinal distribution limit of the tick *Ixodes ricinus* shifted considerably towards higher altitudes in central Europe: results of three years monitoring in the Krkonose Mts. (Czech Republic). *Central European Journal of Public Health* 13, 24–28.

Medlock, J.M. and Leach, S.A. (2015) Effect of climate change on vector-borne disease risk in the UK. *Lancet Infectious Diseases* 15, 721–730.

Molaei, G. (2020) *Populations of Gulf Coast Tick, a New Species to the Northeast and a Vector of Spotted Fever to Humans, Discovered in Connecticut*. The Connecticut Agricultural Experiment Station, New Haven, Connecticut.

Negredo, A., de la Calle-Prieto, F., Palencia-Herrejón, E., Mora-Rillo, M., Astray-Mochales, J., *et al.* (2017) Autochthonous Crimean–Congo hemorrhagic fever in Spain. *New England Journal of Medicine* 377, 154–161.

Nijhof, A.M., Bodaan, C., Postigo, M., Nieuwenhuijs, H., Opsteegh, M., *et al.* (2007) Ticks and associated pathogens collected from domestic animals in the Netherlands. *Vector-Borne and Zoonotic Diseases* 7, 585–596.

NOAA (2020) Climate at a glance: global rankings. National Oceanic and Atmospheric Administration, National Centers for Environmental Information, Asheville, North Carolina. Available at: https://www.ncdc.noaa.gov/cag/ (accessed 11 December 2020).

Oehme, R., Bestehorn, M., Wölfel, S. and Chitimia-Dobler, L. (2017) *Hyalomma marginatum* in Tübingen, Germany. *Systematic and Applied Acarology* 22, 1–6.

Ogden, N.H., Bigras-Poulin, M., O'Callaghan, C.J., Barker, I.K., Lindsay, L.R., *et al.* (2005) A dynamic population model to investigate effects of climate on geographic range and seasonality of the tick *Ixodes scapularis*. *International Journal for Parasitology* 35, 375–389.

Osorio-Olvera, L., Yañez-Arenas, C., Martínez-Meyer, E. and Peterson, A.T. (2020) Relationships between population densities and niche-centroid distances in North American birds. *Ecology Letters* 23, 555–564.

Ostfeld, R.S. and Brunner, J.L. (2015) Climate change and *Ixodes* tick-borne diseases of humans. *Philosophical Transactions of the Royal Society B* 370, 20140051.

Ostfeld, R.S., Canham, C.D., Oggenfuss, K., Winchcombe, R.J. and Keesing, F. (2006) Climate, deer, rodents, and acorns as determinants of variation in Lyme-disease risk. *PLoS Biology* 4, e145.

Owens, H.L., Campbell, L.P., Dornak, L.L., Saupe, E.E., Barve, N., *et al.* (2013) Constraints on interpretation of ecological niche models by limited environmental ranges on calibration areas. *Ecological Modelling* 263, 10–18.
Paddock, C.D. and Goddard, J. (2015) The evolving medical and veterinary importance of the Gulf Coast tick (Acari: Ixodidae). *Journal of Medical Entomology* 52, 230–252.
Papa, A., Dalla, V., Papadimitriou, E., Kartalis, G. and Antoniadis, A. (2010) Emergence of Crimean–Congo haemorrhagic fever in Greece. *Clinical Microbiology and Infection* 16, 843–847.
Patz, J.A., Graczyk, T.K., Geller, N. and Vittor, A.Y. (2000) Effects of environmental change on emerging parasitic diseases. *International Journal for Parasitology* 30, 1395–1405.
Peterson, A.T. and Raghavan, R.K. (2017) The geographic distribution of *Ixodes scapularis* (Acari: Ixodidae) revisited: the importance of assumptions about error balance. *Journal of Medical Entomology* 54, 1080–1084.
Peterson, A.T., Cobos, M.E. and Jiménez-García, D. (2018) Major challenges for correlational ecological niche model projections to future climate conditions. *Annals of the New York Academy of Sciences* 1429, 66–77.
Petney, T.N., Robbins, R.G., Guglielmone, A.A., Apanaskevich, D.A., Estrada-Peña, A., *et al.* (2011) A look at the world of ticks. In: Mehlhorn, H. (ed.) *Progress in Parasitology*. Springer, Berlin/Heidelberg, Germany, pp. 283–296.
Porretta, D., Mastrantonio, V., Amendolia, S., Gaiarsa, S., Epis, S., *et al.* (2013) Effects of global changes on the climatic niche of the tick *Ixodes ricinus* inferred by species distribution modelling. *Parasites & Vectors* 6, 271.
Raghavan, R.K., Peterson, A.T., Cobos, M.E., Ganta, R. and Foley, D. (2019) Current and future distribution of the lone star tick, *Amblyomma americanum* (L.) (Acari: Ixodidae) in North America. *PLoS ONE* 14, e0209082.
Rainey, T., Occi, J.L., Robbins, R.G. and Egizi, A. (2018) Discovery of *Haemaphysalis longicornis* (Ixodida: Ixodidae) parasitizing a sheep in New Jersey, United States. *Journal of Medical Entomology* 55, 757–759.
Randolph, S.E. (2010) To what extent has climate change contributed to the recent epidemiology of tick-borne diseases? *Veterinary Parasitology* 167, 92–94.
Rochlin, I. (2019) Modeling the Asian longhorned tick (Acari: Ixodidae) suitable habitat in North America. *Journal of Medical Entomology* 56, 384–391.
Rubel, F., Brugger, K., Pfeffer, M., Chitimia-Dobler, L., Didyk, Y.M., *et al.* (2016) Geographical distribution of *Dermacentor marginatus* and *Dermacentor reticulatus* in Europe. *Ticks and Tick-borne Diseases* 7, 224–233.
Rumer, L., Graser, E., Hillebrand, T., Talaska, T., Dautel, H., *et al.* (2011) *Rickettsia aeschlimannii* in *Hyalomma marginatum* ticks, Germany. *Emerging Infectious Diseases* 17, 325.
Russart, N.M., Dougherty, M.W. and Vaughan, J.A. (2014) Survey of ticks (Acari: Ixodidae) and tick-borne pathogens in North Dakota. *Journal of Medical Entomology* 51, 1087–1090.
Schwarz, A., Maier, W.A., Kistemann, T. and Kampen, H. (2009) Analysis of the distribution of the tick *Ixodes ricinus* L. (Acari: Ixodidae) in a nature reserve of western Germany using geographic information systems. *International Journal of Hygiene and Environmental Health* 212, 87–96.
Skarpaas, T., Golovljova, I., Vene, S., Ljøstad, U., Sjursen, H., *et al.* (2006) Tickborne encephalitis virus, Norway and Denmark. *Emerging Infectious Diseases* 12, 1136.
Sonenshine, D.E. (2018) Range expansion of tick disease vectors in North America: implications for spread of tick-borne disease. *International Journal of Environmental Research and Public Health* 15, 478.
Sréter, T., Széll, Z. and Varga, I. (2005) Spatial distribution of *Dermacentor reticulatus* and *Ixodes ricinus* in Hungary: evidence for change? *Veterinary Parasitology* 128, 347–351.
Stich, R., Schaefer, J.J., Bremer, W.G., Needham, G.R. and Jittapalapong, S. (2008) Host surveys, ixodid tick biology and transmission scenarios as related to the tick-borne pathogen, *Ehrlichia canis*. *Veterinary Parasitology* 158, 256–273.
Sumilo, D., Asokliene, L., Bormane, A., Vasilenko, V., Golovljova, I. and Randolph, S.E. (2007) Climate change cannot explain the upsurge of tick-borne encephalitis in the Baltics. *PLoS ONE* 2, e500.
Süss, J., Klaus, C., Gerstengarbe, F.W. and Werner, P.C. (2008) What makes ticks tick? Climate change, ticks, and tick-borne diseases. *Journal of Travel Medicine* 15, 39–45.
Swart, A., Ibañez-Justicia, A., Buijs, J., Van Wieren, S.E., Hofmeester, T.R., *et al.* (2014) Predicting tick presence by environmental risk mapping. *Frontiers in Public Health* 2, 238.

Tälleklint, L. and Jaenson, T.G. (1998) Increasing geographical distribution and density of *Ixodes ricinus* (Acari: Ixodidae) in central and northern Sweden. *Journal of Medical Entomology* 35, 521–526.

Tokarevich, N.K., Tronin, A.A., Blinova, O.V., Buzinov, R.V., Boltenkov, V.P., *et al.* (2011) The impact of climate change on the expansion of *Ixodes persulcatus* habitat and the incidence of tick-borne encephalitis in the north of European Russia. *Global Health Action* 4, 8448.

Tukahirwa, E.M. (1976) The effects of temperature and relative humidity on the development of *Rhipicephalus appendiculatus* Neumann (Acarina, Ixodidae). *Bulletin of Entomological Research* 66, 301–312.

USDA (2020) National *Haemaphysalis longicornis* (Asian longhorned tick) situation report. US Department of Agriculture, Washington, DC. Available at: https://www.aphis.usda.gov/animal_health/animal_diseases/tick/downloads/longhorned-tick-sitrep.pdf (accessed 11 December 2020).

USGCRP (2017) *Climate Science Special Report: Fourth National Climate Assessment*, vol. I [Wuebbles, D.J., Fahey, D.W., Hibbard, K.A., Dokken, D.J., Stewart, B.C. and Maycock, T.K. (eds)]. US Global Change Research Program, Washington, DC. Available at: https://science2017.globalchange.gov/chapter/front-matter-about/ (accessed 11 December 2020).

Walker, A.R. (2003) *Ticks of Domestic Animals in Africa: A Guide to Identification of Species*. Bioscience Reports, Edinburgh.

WHO (2017) Global vector control response 2017–2030. World Health Organization, Geneva, Switzerland. Available at: https://apps.who.int/iris/bitstream/handle/10665/259002/WHO-HTM-GVCR-2017.01-eng.pdF;jsessionid=11FDB2573335384698022C22AFF730D?sequence=1 (accessed 12 December 2020).

Williams, H.W., Cross, D.E., Crump, H.L., Drost, C.J. and Thomas, C.J. (2015) Climate suitability for European ticks: assessing species distribution models against null models and projection under AR5 climate. *Parasites & Vectors* 8, 440.

Wilson, M.L., Dykstra, E.A. and Schmidt, B.A. (1993) Temperature- and humidity-dependent longevity of unfed adult *Hyalomma truncatum* (Acari: Ixodidae). *Journal of Medical Entomology* 30, 467–471.

Wormser, G.P., Mckenna, D., Piedmonte, N., Vinci, V., Egizi, A.M., *et al.* (2020) First recognized human bite in the United States by the Asian longhorned tick, *Haemaphysalis longicornis*. *Clinical Infectious Diseases* 70, 314–316.

Zahler, M. and Gothe, R. (1995) Effect of temperature and humidity on egg hatch, moulting and longevity of larvae and nymphs of *Dermacentor reticulatus* (Ixodidae). *Applied Parasitology* 36, 53–65.

Zeman, P. and Beneš, C. (2004) A tick-borne encephalitis ceiling in Central Europe has moved upwards during the last 30 years: possible impact of global warming? *International Journal of Medical Microbiology Supplements* 293, 48–54.

Zhao, L., Li, J., Cui, X., Jia, N., Wei, J., *et al.* (2020) Distribution of *Haemaphysalis longicornis* and associated pathogens: analysis of pooled data from a China field survey and global published data. *Lancet Planetary Health* 4, e320–e329.

Zivkovic, Z., Nijhof, A.M., de la Fuente, J., Kocan, K.M. and Jongejan, F. (2007) Experimental transmission of *Anaplasma marginale* by male *Dermacentor reticulatus*. *BMC Veterinary Research* 3, 32.

eo45 Climate and Other Global Factors at the Zoonotic Interface in America: Influence on Diseases Caused by Tick-Borne Pathogens

Ulrike Munderloh* and Timothy Kurtti
University of Minnesota, St. Paul, Minnesota, USA

45.1 Expert Opinion

45.1.1 Introduction to the players

Many insightful studies examining the impact of climate change on tick-borne diseases have been published that have focused on important aspects such as spatial and temporal changes in tick and host distribution, discovery of new pathogens, dynamics of disease incidence, and climate or other drivers of increasing prevalence (reviewed in Gilbert, 2021). While the body of this research has contributed immensely to our understanding of the many causes propelling the rise in tick-borne disease cases and helped guide the development of control and treatment measures, attempts to model future trends have not reached perfect agreement (Fischhoff *et al.*, 2019). A quote attributed to the American journalist and critic Henry Louis Mencken states that 'For every complex problem, there is an answer that is clear, simple, and wrong'. This nicely sums up the problems of oversimplifying our current understanding of the direct and indirect effects of climate change on human diseases caused by tick-borne pathogens. To begin with, the ticks in question are three-host ixodid ticks, because one- and two-host ticks rarely, if ever, attack humans. Although argasid (soft) ticks are known vectors of human pathogens, including relapsing fever spirochaetes and arboviruses, the diseases are relatively rare, although modelling predicts their range may change due to climate shifts (Labuda and Nuttall, 2004; Lopez *et al.*, 2016; Sage *et al.*, 2017). Three-host hard ticks feed on a variety of vertebrates, including humans, but humans are incidental hosts. Therefore, the population dynamics of three-host ticks are shaped both by the availability and characteristics (quality as a blood meal source, mobility) of their animal hosts, as well as by the ability of ticks to find them (Fig. 45.1). Second, hosts for the two most notorious species in North America, *Ixodes scapularis* (black-legged ticks) and *Amblyomma americanum* (lone star ticks), are wild animals that each respond to climate changes and human environmental disruptions in their own way, some with expanding and others with contracting or regionally shifting populations. These processes are likely to affect the host-seeking success of immature ticks in particular as they are the least mobile stage and most vulnerable to the impacts of adverse conditions. For example, although the immune

*Email: munde001@umn.edu

DOI: 10.1079/9781789249637.0045

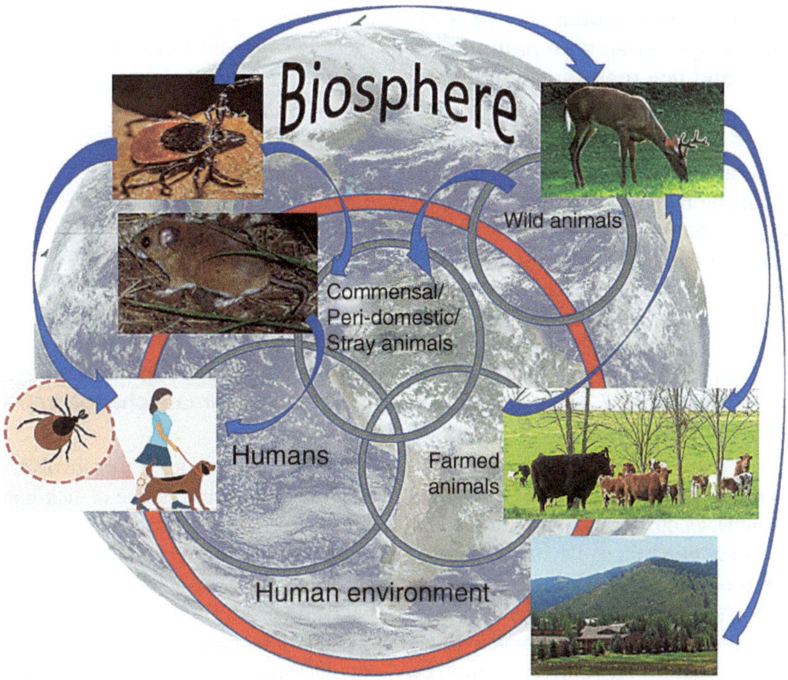

Fig. 45.1. Interactions of biotic elements within the biosphere of planet Earth. Complex interactions connect living organisms with each other within the envelope of the biosphere. Humans intentionally and unwittingly build connections by keeping domestic animals that themselves extend connections further out into the ecosphere. Humans also penetrate into the outer realms with homesteads set into, or near, wildlife habitat and through activities such as hunting and fishing. Within this system, each organism is a node receiving and sending connectivity to multiple biosphere members directly or indirectly. Each member, by virtue of its ability to affect other members directly (e.g. killing) or indirectly (e.g. diverting resources), has a certain weight that is additionally governed by numbers and activities such as resource acquisition. Thus, disturbances or removal of one node will affect other nodes in multiple ways, positively as well as negatively. (**Background diagram.** 'NASA GOES-13 Full Disk view of Earth November 24, 2010' by NASA Goddard Photo and Video is licensed under CC BY 2.0; Creator: NOAA/NASA GOES Project. **Female black-legged tick.** Creator: Lennart Tange; CC BY 2.0 license. ***Odocoileus virginianus*** **(white-tailed deer – buck in velvet).** (17 July 2018) (Newark, Ohio, USA); Creator: James St. John; CC BY 2.0 license. **Cattle in the pasture.** Creator: United Soybean Board, or the Soybean Checkoff; CC BY 2.0 license. **Home near mountains.** Between Snyderville and Park City, Utah; Creator: Ken Lund; CC BY-SA 2.0 license. **Insect–dog with a tick.** Creator: insightpest; CC BY-SA 2.0 license. **White-footed mouse.** Image info: 'P199 Either *Peromyscus leucopus* (a white footed mouse) or its very similar close cousin, *P. maniculatus* (deer mouse)' by wyldvision is licensed under CC BY 2.0; Creator: Royce Milam.)

response of Californian lizards is capable of inactivating Lyme disease borreliae during tick feeding, their support of immature tick populations may outweigh their borreliacidal effects in the transmission cycle of *Borrelia burgdorferi* (Swei *et al.*, 2011). And third, those wild hosts comprise a wide variety of animals, including reptiles, birds and mammals of every size and shape, all of which have strikingly different mobility characteristics that affect tick dispersal (Keesing *et al.*, 2009; Robinson *et al.*, 2015). There is also evidence of regionally distinct endemic transmission cycles that promote or hinder pathogen transmission to humans and these are often not considered in models of diseases caused by tick-borne pathogens (Oliver *et al.*, 2003; Johnson *et al.*, 2011).

45.1.2 Evidence for global change and climate change impacts on pathogen reservoirs and tick maintenance hosts

Recordings of terrestrial as well as of ocean water temperatures across the globe provide evidence that the Earth has been warming beginning approximately with the Industrial Revolution. Although prehistoric Earth has undergone dramatic shifts in climate, the current pace of climate change is occurring at an unprecedented rate that supports a causal rather than a coincidental relationship (Mgbemene *et al.*, 2016; Naafs *et al.*, 2016), which imposes significant stress on ecosystems. Wild animals that are blood meal sources for the different tick life stages and those that are competent pathogen reservoirs are not always the same, leading to complex interactions between these systems that can dynamically alter disease risk (Couper *et al.*, 2021). Some of the reservoirs and hosts have high reproductive rates and are highly adaptable generalists able to exploit a range of habitats and different resources, which provides them a selective advantage. This category includes small rodents such as white-footed mice that are frequently parasitized by tick larvae and nymphs and that, importantly, may harbour pathogens for extended periods and be infectious for ticks, even for their lifetime (Adler and Wilson, 1987; Shih *et al.*, 1995; Lindsay *et al.*, 1997). Their role is, however, balanced by a short survival span that may not even reach a year for most individuals in regions with harsh winters, e.g. the northern USA and Canada (Schug *et al.*, 1991). For pathogens that are not passed on transovarially from female to offspring ticks, this poses a challenge when they rely on such hosts for year-to-year survival. However, as winters become less severe in these regions, predictive models indicate that increasing numbers of small hosts will have a chance to live through winters and serve as sources of infection for newly hatched larvae early in the following summer. Indeed, trends in disease incidence show the greatest increases in regions such as the Northeast and Upper Midwest of the USA, indicating that the correlation between winter survival of small pathogen reservoirs and human disease actually holds. The northern ranges of North America, including southern Canada, have experienced notable increases in annual average temperatures. In areas where there is sufficient forest to buffer concomitant changes in humidity from either decreased or increased rainfall, the abundance of ticks has increased along with rising temperatures (Bouchard *et al.*, 2019). Habitat and behaviour shelter ticks from the effects of climate change, rendering them less susceptible than more delicate arthropods, such as mosquitoes, to events like heavy rains and resulting flushing out of breeding sites.

45.1.3 Disease incidence: evidence against climate change impacts?

There are many more actors involved in the environmental maintenance of tick-borne pathogens than ticks and mice. In areas where mice rarely survive winter, other small rodents, notably chipmunks that live multiple years in the wild, and possibly larger hosts such as wolves or coyotes, may play important roles in maintaining tick-borne pathogens through the cold season, although some have challenged this assumption (Johnson *et al.*, 2011; Levi *et al.*, 2012; Way and White, 2013). Importantly, because these animals will be infested by multiple cohorts of ticks over their lifetime, they have a chance to develop immune responses to tick-borne pathogens, and possibly ticks themselves, which could limit their reservoir potential, but also might potentially exert selective pressure on these agents to drive pathogenicity traits. Although the most dramatic changes in tick-borne illness incidence are predicted to occur in the Northeast USA (Couper *et al.*, 2021), the question begs itself why in areas where small rodents are already better able to survive winters, such as the central eastern states (e.g. the Virginias, Rhode Island, Maryland), cases of Lyme disease and human anaplasmosis have nevertheless been rising significantly (CDC, 2021). Additionally, in southern states such as Georgia, diseases caused by pathogens transmitted by *I. scapularis* are less common than in the north-east, despite the presence of related genotypes. Reasons for this presumably are that ticks are less active questing in the open in warmer climates, where most people also spend much of their time in air-conditioned indoor areas, reducing exposure (Arsnoe *et al.*, 2015, 2019). Also, well-established endemic pathogen strains, that are not infectious for humans, circulate among wildlife in these locations and may have competitively interfered with human-pathogenic strains in the past to prevent their establishment (Oliver *et al.*, 2003). This suggests that warm, humid weather by itself cannot explain heightened tick-borne disease risk. The local community structure of vectors, pathogen strains, reservoir and maintenance hosts, their habitat and human use of it, are all factors that shape disease ecology and thereby epidemiology in concert.

45.1.4 Human land use impacting tick host availability, distribution and abundance

Even within states of the USA where tick-borne pathogens causing human diseases are established, the current situation, and therefore the future, is complex across landscapes. Human land use and its impact on wildlife may more profoundly affect human disease epidemiology than climate change alone. Human land use is not necessarily driven by climate change, because of humankind's ability to alter landscapes with the help of technology. Instead, human land use may be dictated largely by factors that cause climate change: in particular, human population increases and movement that lead to human encroachment into wildlife habitat (Jones *et al.*, 2008), such as when clearing undeveloped land to plant crops or build homes. These expansions and movements have been driven by continued exponential growth of the world's human population: in fact, just 50 years ago, there were fewer than half the current figure of nearly eight billion (Population Matters, 2021). In these scenarios, the number, composition and distribution of hosts available to vectors change, often dramatically, and this happens both in rural and peri-urban areas of less developed as well as industrialized countries. Well-studied examples concern diseases transmitted by insects in the tropics, such as leishmaniasis, when natural sand fly hosts are driven away and their habitat is destroyed (Oliveira-Neto *et al.*, 1988; Jerônimo *et al.*, 1994, Lima *et al.*, 2017). Urban outbreaks of this disease are increasingly more common as epidemiological patterns change due to a combination of factors that accompany expanding city boundaries where the natural ecosystem is thrown out of balance.

The emergence of diseases caused by tick-borne pathogens can similarly be linked to ecological destruction, at regional or even much larger scales, tied to land use for agriculture. The increase in incidence of Brazilian spotted fever comes on the heels of regional deforestation leading to changes in abundance and availability of tick hosts, i.e. capybaras, that are also competent pathogen reservoirs. This combination precipitates human exposure to host-seeking ticks that spill over from capybara populations taking advantage of ready food supplies in newly established croplands (Luz *et al.*, 2019). In the USA, the situation initially was different. Here, European settlers acquired vast stretches of hunting land from Native Americans (Banner, 2005), much of which was subsequently used for growing crops. Early farmers also were avid hunters of white-tailed deer (*Odocoileus virginianus*), important maintenance hosts for ticks that transmit human-infectious agents, while at the same time being reservoirs for at least one of these. White-tailed deer populations were radically decimated during those times, but when American cities grew to dominate the economy, farmers flocked to urban areas, leaving their land to be overtaken by the secondary growth forest that deer favour. Relieved of hunting pressure and provided with ample habitat, deer populations rebounded to record numbers, along with their cargo of ticks that are vectors of important emerging human disease agents, notably Lyme disease borreliae but also other tick-borne pathogens (Paddock and Yabsley, 2007; Kilpatrick *et al.*, 2017). Forested areas also provide habitat and food for rodents that are both reservoirs of these agents as well as hosts for immature ticks to complete the intertwined pathogen–vector cycles. Indeed, predictive models have highlighted the contribution of deciduous forests to rising tick-borne disease incidence (Robinson *et al.*, 2015). Subsequent to these spatially large changes that affected vast stretches of land in the USA, suburban development and recreational land use began to encroach on existing wildlife habitat, precipitating a rise in tick-borne illnesses, as seen in South America, but affecting affluent suburbanites seeking homes close to nature as well as rural communities

(Falco and Fish, 1988; Lane *et al.*, 1992). A specialized variant of human land use is military camps where soldiers are trained for combat in natural settings. Civilian activities are largely absent, turning these camps into wildlife refuges that support tick-borne pathogen transmission cycles. Under these conditions, soldiers become active participants in sampling these environments for vector-borne pathogens and this has led to the discovery of previously unknown tick-borne disease agents (Anderson *et al.*, 1991; Dawson *et al.*, 1991).

45.1.5 Human-assisted movement of ticks

Animal trade has been known for centuries as a means for moving ticks over long distances. In subtropical and tropical climate regions, tick-borne pathogens have long been associated with diseases of domestic animals, e.g. the old scourge of canine health, *Ehrlichia canis*, transmitted by the now globally widely distributed brown dog tick, *Rhipicephalus sanguineus*, that was likely brought to the Americas by European settlers and their dogs (Dantas-Torres, 2008). Similarly, cattle fever ticks, notably *Rhipicephalus* (*Boophilus*) *microplus*, have been transported around the world with their bovine hosts accompanying humans, or with the cattle trade, and are now considered one of the most widely established livestock ectoparasites (Kanduma *et al.*, 2020). In these cases, the distribution, availability and density of hosts are facilitated by humans and their movements and animal husbandry practices. While cattle fever ticks have not been linked to human disease, brown dog ticks may feed on humans, especially when they are in close contact with infested dogs or when dogs are removed from tick-infested premises. In those cases, brown dog ticks pose a threat to human health through transmission of spotted fever rickettsiae that circulate among dogs (Demma *et al.*, 2005; Parola *et al.*, 2009; Levin *et al.*, 2012). The detection of a focus of Rocky Mountain spotted fever in Arizona is of particular interest, as it highlights the ability of an introduced tick species to pick up and transmit an endemic pathogen previously associated with tick species native to the area. This is highly significant, as it potentiates the threat posed by introduced ticks or vector arthropods in general, a scenario currently played out by reports of the invasive Asian long-horned ticks, *Haemaphysalis longicornis*, in the USA (Oakes *et al.*, 2019; Egizi *et al.*, 2020). This species feeds on a very wide range of hosts, including reptiles, birds and mammals of different sizes, including humans. It was repeatedly identified infesting animals imported into the USA before it was found in inland regions and clearly established (Beard *et al.*, 2018). The small size of the tick likely helped it escape detection for many years until it caused noticeable infestations, while its ability to reproduce asexually was a major factor contributing to its ability to gain a foothold in its new home.

45.2 Future Predictions

The spread of ticks into new habitats near and far cannot solely be attributed to climate change. A major contributor to the rising risk of tick-borne illnesses is the rapid human population growth that expands the boundaries of land used by people in various ways, be it for homes, cropland or recreation. In addition, human energy demands have resulted in ecosystem disruptions through alteration of streams to build dams, or for exploitation of fossil fuel resources, which in turn has changed the communities and distribution of tick hosts. Burning of fossil fuels has driven greenhouse gas emissions, resulting in a warming world. Ticks have taken advantage of these different human activities, e.g. by hitching rides on animals traded around the globe, or by using blood meal hosts that have multiplied, and expanded or shifted their range in adaptation to human presence and a warming climate. The northward push of white-footed mice, an important player in the tick-borne pathogen cycle, has implications for tick-borne disease risk (Roy-Dufresne *et al.*, 2013). In regions with diverse geographic landscapes and ecosystems, climate change will have diverse effects. Some species will be more severely affected than others, e.g. specialists are expected to have greater difficulties adapting to changes in availability of resources than will generalists like white-footed mice and *I. scapularis*. Additionally, mountain regions at higher elevations can provide refuge from rising temperatures as do areas near large bodies of water that act as heat sinks (Fink *et al.*, 2014; Elsen *et al.*, 2020). With these considerations in mind, it seems reasonable to predict that effects of global climate change will manifest in locally different patterns of transmission cycles and intensity. At the same time, the animal reservoirs that maintain tick populations and the pathogens they transmit will change in composition, which will impact pathogen presence and therefore human disease prevalence. Human disease risk is linked to tick abundance (Fischhoff *et al.*, 2019), thus management of factors that favour high environmental loads of ticks is important for risk mitigation. Much research remains to be done to elucidate use of altered habitat by wildlife that has been displaced or accommodated by humans, because humans themselves are incidental rather than maintenance hosts. However, it is likely that the upward trend of tick-borne human disease incidence will continue for the coming two or three decades, especially in areas at the current northern borders of tick habitat, where human land use and climate change add to each other's effects.

References

Adler, G.H. and Wilson, M.L. (1987) Demography of a habitat generalist, the white-footed mouse, in a heterogeneous environment. *Ecology* 68, 1785–1796.

Anderson, B.E., Dawson, J.E., Jones, D.C. and Wilson, K.H. (1991) *Ehrlichia chaffeensis*, a new species associated with human ehrlichiosis. *Journal of Clinical Microbiology* 29, 2838–2842.

Arsnoe, I.M., Hickling, G.J., Ginsberg, H.S., McElreath, R. and Tsao, J.I. (2015) Different populations of blacklegged tick nymphs exhibit differences in questing behavior that have implications for human Lyme disease risk. *PLoS ONE* 10, e0127450.

Arsnoe, I., Tsao, J.I. and Hickling, G.J. (2019) Nymphal *Ixodes scapularis* questing behavior explains geographic variation in Lyme borreliosis risk in the eastern United States. *Ticks and Tick-borne Diseases* 10, 553–563.

Banner, S. (2005) *How the Indians Lost Their Land – Law and Power on the Frontier*. Harvard University Press, Cambridge, Massachusetts.

Beard, C.B., Occi, J., Bonilla, D.L., Egizi, A.M., Fonseca, D.M., *et al.* (2018) Multistate infestation with the exotic disease-vector tick *Haemaphysalis longicornis* – United States, August 2017–September 2018. *Morbidity and Mortality Weekly Report* 67, 1310–1313.

Bouchard, C., Dibernardo, A., Koffi, J., Wood, H., Leighton, P.A., *et al.* (2019) Increased risk of tick-borne diseases with climate and environmental changes. *Canada Communicable Disease Report* 45, 81–89.

CDC (2021) Lyme disease data tables: historical data. Reported cases of Lyme disease by state or locality, 2010–2019. Centers for Disease Control and Prevention, Atlanta, Georgia. Available at: https://www.cdc.gov/lyme/stats/tables.html (accessed 9 June 2021).

Couper, L.I., MacDonald, A.J. and Mordecai, E.A. (2021) Impact of prior and projected climate change on US Lyme disease incidence. *Global Change Biology* 27, 738–754.

Dantas-Torres, F. (2008) The brown dog tick, *Rhipicephalus sanguineus* (Latreille, 1806) (Acari: Ixodidae): from taxonomy to control. *Veterinary Parasitology* 152, 173–185.

Dawson, J.E., Anderson, B.E., Fishbein, D.B., Sanchez, J.L., Goldsmith, C.S., *et al.* (1991) Isolation and characterization of an *Ehrlichia* sp. from a patient with human ehrlichiosis. *Journal of Clinical Microbiology* 29, 2741–2745.

Demma, L.J., Traeger, M.S., Nicholson, W.L., Paddock, C.D., Blau, D.M., *et al.* (2005) Rocky Mountain spotted fever from an unexpected tick vector in Arizona. *New England Journal of Medicine* 353, 587–594.

Egizi, A., Bulaga-Seraphin, L., Alt, E., Bajwa, W.I., Bernick, J., *et al.* (2020) First glimpse into the origin and spread of the Asian longhorned tick, *Haemaphysalis longicornis*, in the United States. *Zoonoses and Public Health* 67, 637–650.

Elsen, P.R., Monahan, W.B. and Merenlender, A.M. (2020) Topography and human pressure in mountain ranges alter expected species responses to climate change. *Nature Communications* 11, 1974.

Falco, R.C. and Fish, D. (1988) Prevalence of *Ixodes dammini* near the homes of Lyme disease patients in Westchester County, New York. *American Journal of Epidemiology* 127, 826–830.

Fink, G., Schmid, M. and Wüest, A. (2014) Large lakes as sources and sinks of anthropogenic heat: capacities and limits. *Water Resources and Research* 50, 7285–7301.
Fischhoff, I.R., Bowden, S.E., Keesing, F. and Ostfeld, R.S. (2019) Systematic review and meta-analysis of tick-borne disease risk factors in residential yards, neighborhoods, and beyond. *BMC Infectious Diseases* 19, 861.
Gilbert, L. (2021) The impacts of climate change on ticks and tick-borne disease risk. *Annual Review of Entomology* 66, 373–388.
Jerônimo, S.M.B., Oliveira, R.M., Mackay, S., Costa, R.M., Sweet, J., *et al.* (1994) An urban outbreak of visceral leishmaniasis in Natal, Brazil. *Transactions of the Royal Society of Tropical Medicine and Hygiene* 88, 386–388.
Johnson, R.C., Kodner, C., Jarnefeld, J., Eck, D.K. and Xu, Y. (2011) Agents of human anaplasmosis and Lyme disease at Camp Ripley, Minnesota. *Vector-Borne and Zoonotic Diseases* 11, 1529–1534.
Jones, K.E., Patel, N.G., Levy, M.A., Storeygard, A., Balk, D., *et al.* (2008) Global trends in emerging infectious diseases. *Nature* 451, 990–993.
Kanduma, E.G., Emery, D., Githaka, N.W., Nguu, E.K., Bishop, R.P., *et al.* (2020) Molecular evidence confirms occurrence of *Rhipicephalus microplus* Clade A in Kenya and sub-Saharan Africa. *Parasites & Vectors* 13, 432.
Keesing, F., Brunner, J., Duerr, S., Killilea, M., Logiudice, K., *et al.* (2009) Hosts as ecological traps for the vector of Lyme disease. *Proceedings of the Royal Society B* 276, 3911–3919.
Kilpatrick, A.M., Dobson, A.D.M., Levi, T., Salkeld, D.J., Swei, A., *et al.* (2017) Lyme disease ecology in a changing world: consensus, uncertainty and critical gaps for improving control. *Philosophical Transactions of the Royal Society B* 372, 20160117.
Labuda, M. and Nuttall, P.A. (2004) Tick-borne viruses. *Parasitology* 129(Suppl.), S221–S245.
Lane, R.S., Manweiler, S.A., Stubbs, H.A., Lennette, E.T., Madigan, J.E., *et al.* (1992) Risk factors for Lyme disease in a small rural community in northern California. *American Journal of Epidemiology* 136, 1358–1368.
Levi, T., Kilpatrick, A.M., Mangel, M. and Wilmers, C.C. (2012) Deer, predators, and the emergence of Lyme disease. *Proceedings of the National Academy of Sciences USA* 109, 10942–10947.
Levin, M.L., Killmaster, L.F. and Zemtsova, G.E. (2012) Domestic dogs (*Canis familiaris*) as reservoir hosts for *Rickettsia conorii*. *Vector-Borne and Zoonotic Diseases* 12, 28–33.
Lima, A.L.M., de Lima, I.D., Coutinho, J.F.V., de Sousa, U.P.S.T., Rodrigues, M.A.G., *et al.* (2017) Changing epidemiology of visceral leishmaniasis in northeastern Brazil: a 25-year follow-up of an urban outbreak. *Transactions of the Royal Society of Tropical Medicine and Hygiene* 111, 440–447.
Lindsay, L.R., Barker, I.K., Surgeoner, G.A., McEwen, S.A. and Campbell, G.D. (1997) Duration of *Borrelia burgdorferi* infectivity in white-footed mice for the tick vector *Ixodes scapularis* under laboratory and field conditions in Ontario. *Journal of Wildlife Diseases* 33, 766–775.
Lopez, J.E., Krishnavajhala, A., Garcia, M.N. and Bermudez, S. (2016) Tick-borne relapsing fever spirochetes in the Americas. *Veterinary Sciences* 3, 16.
Luz, H.R., Costa, F.B., Benatti, H.R., Ramos, V.N., de A. Serpa, M.C., *et al.* (2019) Epidemiology of capybara-associated Brazilian spotted fever. *PLoS Neglected Tropical Diseases* 13, e0007734.
Mgbemene, C.A., Nnaji, C.C. and Nwozor, C. (2016) Industrialization and its backlash: focus on climate change and its consequences. *Journal of Environmental Science and Technology* 9, 301–316.
Naafs, B.D.A, Castro, J.M., De Gea, G.A., Quijano, M.L., Schmidt, D.N., *et al.* (2016) Gradual and sustained carbon dioxide release during Aptian Oceanic Anoxic Event 1a. *Nature Geosciences* 9, 135–139.
Oakes, V.J., Yabsley, M.J., Schwartz, D., LeRoith, T., Bissett, C., *et al.* (2019) *Theileria orientalis* Ikeda genotype in cattle, Virginia, USA. *Emerging Infectious Diseases* 9, 1653–1659.
Oliveira-Neto, M.P., Pirmez, C., Rangel, E., Schubach, A. and Grimaldi, G. Jr (1988) An outbreak of American cutaneous leishmaniasis (*Leishmania braziliensis braziliensis*) in a periurban area of Rio de Janeiro city, Brazil: clinical and epidemiological studies. *Memórias do Instituto Oswaldo Cruz* 83, 427–435.
Oliver, J.H. Jr, Lin, T., Gao, L., Clark, K.L., Banks, C.W., *et al.* (2003) An enzootic transmission cycle of Lyme borreliosis spirochetes in the southeastern United States. *Proceedings of the National Academy of Sciences USA* 100, 11642–11645.
Paddock, C.D. and Yabsley, M.J. (2007) Ecological havoc, the rise of white-tailed deer, and the emergence of *Amblyomma americanum*-associated zoonoses in the United States. *Current Topics in Microbiology and Immunology* 315, 289–324.
Parola, P., Socolovschi, C. and Raoult, D. (2009) Deciphering the relationships between *Rickettsia conorii conorii* and *Rhipicephalus sanguineus* in the ecology and epidemiology of Mediterranean spotted fever. *Annals of the New York Academy of Sciences* 1166, 49–54.

Population Matters (2021) Population: the numbers. Population Matters, London. Available at: https://populationmatters.org/the-facts/the-numbers (accessed 13 June 2021).

Robinson, S.J., Neitzel, D.F., Moen, R.A., Craft, M.E., Hamilton, K.E., *et al.* (2015) Disease risk in a dynamic environment: the spread of tick-borne pathogens in Minnesota, USA. *EcoHealth* 12, 152–163.

Roy-Dufresne, E., Logan, T., Simon, J.A., Chmura, G.L. and Millien, V. (2013) Poleward expansion of the white-footed mouse (*Peromyscus leucopus*) under climate change: implications for the spread of Lyme disease. *PLoS ONE* 8, e80724.

Sage, K.M., Johnson, T.L., Teglas, M.B., Nieto, N.C. and Schwan, T.G. (2017) Ecological niche modeling and distribution of *Ornithodoros hermsi* associated with tick-borne relapsing fever in western North America. *PLoS Neglected Tropical Diseases* 11, e0006047.

Schug, M.D, Vessey, S.H. and Korytko, A.I. (1991) Longevity and survival in a population of white-footed mice (*Peromyscus leucopus*). *Journal of Mammalogy* 72, 360–366.

Shih, C.-M., Liu, L.-P. and Spielman, A. (1995) Differential spirochetal infectivities to vector ticks of mice chronically infected by the agent of Lyme disease. *Journal of Clinical Microbiology* 33, 3164–3168.

Swei, A., Ostfeld, R.S., Lane, R.S. and Briggs, C.J. (2011) Impact of the experimental removal of lizards on Lyme disease risk. *Proceedings of the Royal Society B* 278, 2970–2978.

Way, J.G. and White, B.N. (2013) Coyotes, red foxes, and the prevalence of Lyme disease. *Northeastern Naturalist* 20, 655–665.

eo46 Microclimatic Conditions and RNA Viruses in Ticks

Yan-Jang Huang, Stephen Higgs* and Dana Vanlandingham
Kansas State University, Manhattan, Kansas, USA

46.1 Expert Opinion

Ticks are vectors for a diverse range of human pathogens including both DNA and RNA viruses. Viruses with RNA genomes across five different virus families are known to be transmitted by ticks. The incidence of many tick-borne viral infections has been increasing in multiple geographic regions during the first two decades of the 21st century. Some notable examples include tick-borne encephalitis (TBE) in Europe, caused by TBE virus (TBEV), a flavivirus endemic in Eurasia, and the emergence of severe fever with thrombocytopaenia syndrome virus (SFTSV) in Asia (Kunze, 2016; Bopp *et al.*, 2020). SFTSV is a phlebovirus in the *Phenuiviridae* family which was first discovered during an outbreak of unknown febrile illness in Hubei, China, in 2009 and has been extensively detected in East Asia (Yu *et al.*, 2011). Additionally, the detection and autochthonous transmission of Crimean–Congo haemorrhagic fever virus (CCHFV) was reported for the first time in South-West Europe, marking its northward spread presumably from Africa (Estrada-Peña *et al.*, 2012).

Climate change has often been implicated in the surge of disease burden caused by the re-emergence of various arboviruses, including those transmitted by ticks. Although the dispersal of African swine fever virus, a large tick-borne DNA virus in the *Asfarviridae* family, has also led to numerous outbreaks since the latter half of the 20th century, our discussion will focus on how changes in microclimatic factors have been demonstrated to affect the infection process of medically important tick-borne RNA viruses, especially flaviviruses and bunyaviruses.

Tick development through the various life stages is sensitive to microclimatic conditions that are influenced by meteorological factors. Rising temperature and precipitation have been linked to an increased risk of transmission of tick-borne viruses using epidemiological data and mathematical models (Nah *et al.*, 2020). These studies often use increasing numbers of infections reported as markers to demonstrate more intensive transmission of TBEV in given regions. For instance, numbers of reported TBE cases have risen steadily in several European countries in the past three decades (Kunze, 2016). This trend includes countries that were historically perceived to have low or moderate risks for TBE and coincides with increased temperature (Jaenson *et al.*, 2012; Palo, 2014; Smura *et al.*, 2019). However, the increase in reported TBE cases has only been proven to be related to higher frequencies of human exposure to ticks. It is also important to recognize that the growth

*Email: shiggs@bri.ksu.edu

DOI: 10.1079/9781789249637.0046

of human populations and changes of landscape are contributing factors that result in increased populations of vectors and higher numbers of TBE cases (Palo, 2014; Hartemink *et al.*, 2019). A good example is the reports of higher exposure to *Ixodes ricinus* nymphs that often come from areas that are more densely populated. Predictive models on the distribution of *I. ricinus* nymphs which transmit TBEV show a correlation between increasing temperature and the expansion of tick geographic distribution. In the same study, it is noted that these regions are occupied by more than 70% of the Scandinavian countries' population (Kjaer *et al.*, 2019). In Central and Eastern Europe, the projected rise in temperature is expected to extend the season for questing activities and increase the overall tick population (Nah *et al.*, 2020). In the Alpine region, recent changing climatic conditions and abandonment of farmlands have led to the expansion of wooded areas at higher elevations, where the detection of *I. ricinus* has been increasingly reported in both deciduous and coniferous woods (Chemini and Rizzoli, 2003; Garcia-Vozmediano *et al.*, 2020). Similarly, it is also well documented that climatic conditions favouring the reproduction of *Hyalomma marginatum* ticks often preceded outbreaks of Crimean–Congo haemorrhagic fever (Vescio *et al.*, 2012; Duygu *et al.*, 2018).

Intriguingly, higher risk for human exposure to ticks does not fully explain the increase in the number of reported infections with tick-borne RNA viruses. The most notable example is the temporal difference that reflects the highest risk for tick exposure and the peak of TBE case numbers in the same year. Questing activities of nymphs, which transmit TBEV, increase as the temperature increases in the spring–summer period. However, the number of human TBE cases does not peak until much later in the summer–autumn period. If the elevated risk of human exposure to ticks is the only factor that drives the increase of TBE cases, one would expect the peaks of nymph questing activities and human infections to be much closer, resembling the incubation period for the disease (Daniel *et al.*, 2018). Similarly, the impact created by changes in ambient temperatures on case numbers of severe fever with thrombocytopaenia syndrome reported in three central provinces of China between 2011 and 2015 also does not match the incubation period of the disease (Sun *et al.*, 2018). This indicates that ecological and behavioural factors contribute to only a subset of mechanisms that drive the increased disease burden caused by tick-borne RNA viruses. Changes in environmental conditions clearly alter transmission of tick-borne RNA viruses. One hypothesis that explains the delayed onset of diseases caused by tick-borne RNA viruses after an increase of ambient temperature is that the higher temperature promotes viral replication, giving rise to a higher dose of virus transmitted to humans (Daniel *et al.*, 2018). Although temperature and humidity have a clear impact on the development of ticks and replication of arboviruses, there have been a very limited number of published studies addressing the kinetics of arbovirus infections in ticks.

Temperature and humidity as factors that modulate the infection process of arboviruses have been demonstrated in a wide variety of disease vectors (Turell *et al.*, 1985; Watts *et al.*, 1987; Tesh *et al.*, 1992; Reisen *et al.*, 2006; Carpenter *et al.*, 2011; Ruder *et al.*, 2015). The most widely used approach is to conduct experimental infections followed by maintenance of infected arthropods under specific environmental conditions. Higher temperatures simulated under laboratory conditions often shorten the extrinsic incubation period for transmission, leading to the conclusion that intensive transmission is more likely to occur. However, the same technique and study design, which maintains temperature and humidity at a specific level, has been proven to be too simplistic in the study of transmission dynamics of RNA viruses by ticks under different climatic conditions. Tick-borne viruses must undergo transstadial transmission in their transmission cycles, a phenomenon that is distinctive from many other arboviruses. Most arthropod vectors have the ability to transmit viruses only during the adult stage, when repeated feeding on vertebrate hosts occurs. Therefore, the impact of microclimatic conditions such as temperature and relative humidity on the transmission of most arboviruses has been mainly investigated by the quantitative analysis of viral infections and occasionally examined arthropod immune responses in adults (Adelman *et al.*, 2013). With the occurrence of transstadial transmission, the impact of microclimatic conditions on tick-borne RNA viruses is more complex. The longer extrinsic incubation period for tick-borne arboviruses than for mosquito-borne arboviruses presents challenges that are not present in other systems. With the exception of occasional host switching caused by premature host death during feeding (Wang and Nuttall, 2001), hard ticks that transmit flaviviruses and bunyaviruses do not repeatedly feed on vertebrate hosts during the same developmental stage. Most RNA viruses vectored by ticks are transmitted to the next vertebrate host after metamorphosis to the next life stage, when the infection has been established for several months. As a consequence, extrinsic incubation periods for tick-borne arboviruses can often be prolonged. Successful transstadial transmission requires not only replication of viruses but also the moulting of ticks into the next developmental stage. The process is often extended through the winter with harsh climatic conditions. Under laboratory conditions, investigating how temperature and humidity modulate arbovirus infections in ticks essentially needs to address the impact on two aspects of transstadial transmission – the development of ticks and the infection process of arboviruses.

To date, very few studies have rigorously examined these complex processes (summarized in Table 46.1). Most of the published studies focused on the effect of temperature, ignoring potential humidity effects that are clearly an important factor that influences tick development. Available experimental data have shown that the kinetics of arbovirus infections in ticks can be modulated by temperature. Published studies, albeit a relatively small number, suggest that the infection process of tick-borne flaviviruses is temperature-dependent in both hard and soft ticks. For instance, infection rates of TBEV in nymphs of *I. ricinus* were positively correlated with the increase of extrinsic temperature from 15 to 24°C (Danielová, 1990). In the same study, higher relative humidity also facilitated the establishment of TBEV in ticks. Similarly, incubation of *Ornithodoros sonrai* at 26°C showed an earlier onset of Karshi virus transmission than for those ticks incubated at 20°C (Turell *et al.*, 2004). As observed with two genetically related tick-borne phleboviruses, SFTSV and heartland virus (HRTV), in their endemic vectors, elevated temperature also facilitates the establishment of bunyavirus infections in ticks. The most well-studied example is the transstadial transmission of SFTSV in its endemic vector, *Haemaphysalis longicornis* (Luo *et al.*, 2015). Incubation of *H. longicornis* larvae infected with SFTSV at 26°C maintained the infection rate at 100% during the subsequent nymph and adult stages (Hu *et al.*, 2020), whereas the infected larvae maintained at 22°C led to a gradual reduction of infection rates in nymphs (92%) and adults (40%) (Zhuang *et al.*, 2018). Intriguingly,

Table 46.1. Published studies on the experimental challenge of ticks with arboviruses under different microclimatic conditions. (Authors' own table.)

Virus	Tick species	Microclimatic conditions tested	Outcomes	References
Tick-borne encephalitis virus	*Ixodes ricinus*	Temperature (15 versus 24°C) Relative humidity (75 versus 97%)	Higher temperature and humidity increased infection rates between 2 and 16 weeks post-infection	Danielová (1990)
Karshi virus	*Ornithodoros sonrai*	Temperature (20 versus 26°C)	Early onset of transmission at the higher temperature	Turell *et al.* (2004)
Severe fever with thrombocytopaenia syndrome virus	*Haemaphysalis longicornis*	Temperature (22 versus 26°C)	100% infection rate in larvae, nymphs and adults at the higher temperature	Luo *et al.* (2015); Hu *et al.* (2020)
Heartland virus	*Amblyomma americanum*	Temperature (22–23°C)	<20% infection rate	Godsey *et al.* (2016)

susceptibility of *Amblyomma americanum* to HRTV evaluated at 22–23°C also led to low infection rates in larvae (14–50%) and nymphs (25–50%) (Godsey *et al.*, 2016). Collectively, the relationship between extrinsic temperatures and the kinetics of arbovirus infections in ticks was established using percentages of infection at constant temperatures. Although it suggests that higher temperature favours the establishment of arbovirus infection in ticks, several key questions remain unanswered. Viral loads in the tissues and saliva of infected ticks have not been discussed in published studies.

46.2 Future Predictions

The quantities of infectious viruses secreted in the saliva of infected ticks under different microclimatic conditions is critical for assessing the transmission efficiency from infected ticks to vertebrate hosts. In the past two decades, the field has had a remarkable increase in the ability to quantify viral genomes in different types of biological samples. These same molecular virology assays have been successfully applied to determining the infection status of ticks challenged with emerging tick-borne arboviruses (Hu *et al.*, 2020). It is reasonable to assume that quantitative analysis on tissue and salivary viral loads of infected ticks maintained in different microclimatic conditions should provide much-needed information on how temperature, humidity and other climatic factors affect the infection of arboviruses in ticks.

In the past two decades, other more complex factors such as fluctuations in daily temperatures and temperature-dependent modulation of antimicrobial responses have also been proven to affect the susceptibility of disease vectors to other arboviruses (Lambrechts *et al.*, 2011; Alto *et al.*, 2018). The findings suggest that the constantly changing ecological conditions may have a more profound impact on both tick-borne arboviruses and ticks, which warrants further investigations.

References

Adelman, Z.N., Anderson, M.A., Wiley, M.R., Murreddu, M.G., Samuel, G.H., *et al.* (2013) Cooler temperatures destabilize RNA interference and increase susceptibility of disease vector mosquitoes to viral infection. *PLoS Neglected Tropical Diseases* 7, e2239.

Alto, B.W., Wiggins, K., Eastmond, B., Ortiz, S., Zirbel, K. and Lounibos, L.P. (2018) Diurnal temperature range and chikungunya virus infection in invasive mosquito vectors. *Journal of Medical Entomology* 55, 217–224.

Bopp, N.E., Kaiser, J.A., Strother, A.E., Barrett, A.D.T., Beasley, D.W.C., *et al.* (2020) Baseline mapping of severe fever with thrombocytopenia syndrome virology, epidemiology and vaccine research and development. *NPJ Vaccines* 5, 111.

Carpenter, S., Wilson, A., Barber, J., Veronesi, E., Mellor, P., *et al.* (2011) Temperature dependence of the extrinsic incubation period of orbiviruses in Culicoides biting midges. *PLoS ONE* 6, e27987.

Chemini, C. and Rizzoli, A. (2003) Land use change and biodiversity conservation in the Alps. *Journal of Mountain Ecology* 7, 1–7.

Daniel, M., Danielová, V., Fialová, A., Maly, M., Križ, B. and Nuttall, P.A. (2018) Increased relative risk of tick-borne encephalitis in warmer weather. *Frontiers in Cellular and Infection Microbiology* 8, 90.

Danielová, V. (1990) Experimental infection of ticks *Ixodes ricinus* with tick-borne encephalitis virus under different microclimatic conditions. *Folia Parasitologia* 37, 279–282.

Duygu, F., Sari, T., Kaya, T., Tavsan, O. and Naci, M. (2018) The relationship between Crimean–Congo hemorrhagic fever and climate: does climate affect the number of patients? *Acta Clinica Croatia* 57, 443–448.

Estrada-Peña, A., Palomar, A.M., Santibanez, P., Sanchez, N., Habela, M.A., *et al.* (2012) Crimean–Congo hemorrhagic fever virus in ticks, Southwestern Europe, 2010. *Emerging Infectious Diseases* 18, 179–180.

Garcia-Vozmediano, A., Krawczyk, A.I., Sprong, H., Rossi, L., Ramassa, E. and Tomassone, L. (2020) Ticks climb the mountains: ixodid tick infestation and infection by tick-borne pathogens in the Western Alps. *Ticks and Tick-borne Diseases* 11, 101489.

Godsey, M.S., Savage, H.M., Burkhalter, K.L., Bosco-Lauth, A.M. and Delorey, M.J. (2016) Transmission of heartland virus (*Bunyaviridae*: *Phlebovirus*) by experimentally infected *Amblyomma americanum* (Acari: Ixodidae). *Journal of Medical Entomology* 53, 1226–1233.

Hartemink, N., van Vliet, A., Sprong, H., Jacobs, F., Garcia-Marti, I., *et al.* (2019) Temporal-spatial variation in questing tick activity in the Netherlands: the effect of climatic and habitat factors. *Vector-Borne and Zoonotic Diseases* 19, 494–505.

Hu, Y.Y., Zhuang, L., Liu, K., Sun, Y., Dai, K., *et al.* (2020) Role of three tick species in the maintenance and transmission of severe fever with thrombocytopenia syndrome virus. *PLoS Neglected Tropical Diseases* 14, e0008368.

Jaenson, T.G., Hjertqvist, M., Bergstrom, T. and Lundkvist, A. (2012) Why is tick-borne encephalitis increasing? A review of the key factors causing the increasing incidence of human TBE in Sweden. *Parasites & Vectors* 5, 184.

Kjaer, L.J., Soleng, A., Edgar, K.S., Lindstedt, H.E.H., Paulsen, K.M., *et al.* (2019) Predicting and mapping human risk of exposure to *Ixodes ricinus* nymphs using climatic and environmental data, Denmark, Norway and Sweden, 2016. *Eurosurveillance* 24, 1800101.

Kunze, U. (2016) The International Scientific Working Group on Tick-Borne Encephalitis (ISW TBE): review of 17 years of activity and commitment. *Ticks and Tick-borne Diseases* 7, 399–404.

Lambrechts, L., Paaijmans, K.P., Fansiri, T., Carrington, L.B., Kramer, L.D., *et al.* (2011) Impact of daily temperature fluctuations on dengue virus transmission by *Aedes aegypti*. *Proceedings of the National Academy of Sciences USA* 108, 7460–7465.

Luo, L.M., Zhao, L., Wen, H.L., Zhang, Z.T., Liu, J.W., *et al.* (2015) *Haemaphysalis longicornis* ticks as reservoir and vector of severe fever with thrombocytopenia syndrome virus in China. *Emerging Infectious Diseases* 21, 1770–1776.

Nah, K., Bede-Fazekas, A., Trajer, A.J. and Wu, J. (2020) The potential impact of climate change on the transmission risk of tick-borne encephalitis in Hungary. *BMC Infectious Diseases* 20, 34.

Palo, R.T. (2014) Tick-borne encephalitis transmission risk: its dependence on host population dynamics and climate effects. *Vector-Borne and Zoonotic Diseases* 14, 346–352.

Reisen, W.K., Fang, Y. and Martinez, V.M. (2006) Effects of temperature on the transmission of West Nile virus by *Culex tarsalis* (Diptera: Culicidae). *Journal of Medical Entomology* 43, 309–317.

Ruder, M.G., Stallknecht, D.E., Howerth, E.W., Carter, D.L., Pfannenstiel, R.S., *et al.* (2015) Effect of temperature on replication of epizootic hemorrhagic disease viruses in *Culicoides sonorensis* (Diptera: Ceratopogonidae). *Journal of Medical Entomology* 52, 1050–1059.

Smura, T., Tonteri, E., Jaaskelainen, A., von Troil, G., Kuivanen, S., *et al.* (2019) Recent establishment of tick-borne encephalitis foci with distinct viral lineages in the Helsinki area, Finland. *Emerging Microbes & Infections* 8, 675–683.

Sun, J., Lu, L., Yang, J., Liu, K., Wu, H. and Liu, Q. (2018) Association between severe fever with thrombocytopenia syndrome incidence and ambient temperature. *American Journal of Tropical Medicine and Hygiene* 98, 1478–1483.

Tesh, R.B., Lubroth, J. and Guzman, H. (1992) Simulation of arbovirus overwintering: survival of Toscana virus (*Bunyaviridae*: *Phlebovirus*) in its natural sand fly vector, *Phlebotomus perniciosus*. *American Journal of Tropical Medicine and Hygiene* 47, 574–581.

Turell, M.J., Rossi, C.A. and Bailey, C.L. (1985) Effect of extrinsic incubation temperature on the ability of *Aedes taeniorhynchus* and *Culex pipiens* to transmit Rift Valley fever virus. *American Journal of Tropical Medicine and Hygiene* 34, 1211–1218.

Turell, M.J., Mores, C.N., Lee, J.S., Paragas, J.J., Shermuhemedova, D., *et al.* (2004) Experimental transmission of Karshi and Langat (tick-borne encephalitis virus complex) viruses by *Ornithodoros* ticks (Acari: Argasidae). *Journal of Medical Entomology* 41, 973–977.

Vescio, F.M., Busani, L., Mughini-Gras, L., Khoury, C., Avellis, L., *et al.* (2012) Environmental correlates of Crimean–Congo haemorrhagic fever incidence in Bulgaria. *BMC Public Health* 12, 1116.

Wang, H. and Nuttall, P.A. (2001) Intra-stadial tick-borne Thogoto virus (*Orthomyxoviridae*) transmission: accelerated arbovirus transmission triggered by host death. *Parasitology* 122, 439–446.

Watts, D.M., Burke, D.S., Harrison, B.A., Whitmire, R.E. and Nisalak, A. (1987) Effect of temperature on the vector efficiency of *Aedes aegypti* for dengue 2 virus. *American Journal of Tropical Medicine and Hygiene* 36, 143–152.

Yu, X.J., Liang, M.F., Zhang, S.Y., Liu, Y., Li, J.D., *et al.* (2011) Fever with thrombocytopenia associated with a novel bunyavirus in China. *New England Journal of Medicine* 364, 1523–1532.

Zhuang, L., Sun, Y., Cui, X.M., Tang, F., Hu, J.G., *et al.* (2018) Transmission of severe fever with thrombocytopenia syndrome virus by *Haemaphysalis longicornis* ticks, China. *Emerging Infectious Diseases* 24, 868–871.

eo47 Climate, Ticks and Tick-Borne Encephalitis in Central Europe

Vlasta Danielová and Milan Daniel*
National Institute of Public Health, Prague, Czech Republic

47.1 Expert Opinion

47.1.1 Introduction

Tick-borne encephalitis (TBE) is a serious human disease of the central nervous system (CNS) caused by the European subtype of tick-borne encephalitis virus (TBEV) – genus *Flavivirus* of the family *Flaviviridae* – in the European area of its distribution. Its major vector is the tick, *Ixodes ricinus*. TBE afflicts humans in all age groups, manifesting from asymptomatic to serious cases involving the CNS, sometimes with long-lasting sequelae or ending in death. TBE is a zoonosis characterized by a natural focality, circulating among vertebrate hosts in nature without human participation. Different species of animals are involved in TBEV circulation, forming a local biocenosis that is tied to a certain type of landscape. These places are referred to as TBE natural foci. When people visit such a place, and are attacked by an infected tick, they may become ill.

I. ricinus is a three-host tick species, which means that every active developmental stage (larva, nymph and female) feeds on a particular group of animals. In every stage a tick feeds only once, which means that the imbibed virus can be transmitted to another host only in the next developmental tick stage. Such transmission is called horizontal. Before transmission, the virus must overcome tick metamorphosis from a lower to a higher developmental stage, so-called transstadial transmission (or transstadial survival), or transovarial transmission in the case of virus transmission from female to the next filial generation (referred to as vertical transmission). Virus circulation in nature thus takes place through a combination of horizontal and vertical transmission. In addition to local biocenological conditions, the dynamics of this process are strongly influenced by abiotic factors, mainly by meteorological conditions. Under specific circumstances, a special form of horizontal transmission, so-called non-viraemic transmission, may occur between *I. ricinus* individuals feeding in close proximity on the same host in the absence of detectable viraemia, as in the case of larvae feeding en masse on small mammals.

The first successful isolation of TBEV from patients in central and western parts of Europe was performed in 1948 in former Czechoslovakia by Gallia and co-workers (Gallia *et al.*, 1949), when local outbreaks of meningoencephalitis occurred in several regions of Bohemia and Moravia (Krejčí, 1949). Approximately 80% of TBE patients recorded a tick bite before the illness, suggesting the role of ticks as vectors of the virus. This was confirmed in 1949 by Rampas and Gallia (1949) and subsequently by Krejčí (1949), when TBEV was isolated from different

*Email: midaniel@seznam.cz

DOI: 10.1079/9781789249637.0047

stages of *I. ricinus* ticks collected in the endemic areas. However, Pelnář in 1922, and later other Czech authors, described clinical cases corresponding to TBE (Hubálek, 1986); and in 1927, Schneider in Austria described meningoencephalitis of unknown aetiology (Růžek, 2015). Several synonyms were known for TBE, such as 'Central European encephalitis' and 'Czechoslovak encephalitis'.

In 1951, a large outbreak of milk-borne TBE was reported in Slovakia, when at least 660 people became infected and 271 of them were hospitalized. During this outbreak, extensive scientific investigation was performed, and the aetiology and mode of transmission – by goats' milk – were successfully identified (Blaškovič, 1954). Transmission of TBEV by milk of infected dairy animals or by milk products was experimentally confirmed by Grešíková (1957, 1958). Historically, most of the pioneering research work on TBE in Central Europe was performed in former Czechoslovakia (Růžek *et al.*, 2013).

Minor local epidemics in Bohemia and Moravia in 1948, and the knowledge that the vector of this new disease is *I. ricinus*, gave impetus to studies of TBE, first of its distribution and later of its ecology in Czechoslovakia. In 1960–1962, the first comprehensive study of *I. ricinus* distribution and seasonal abundance was undertaken in all regions of Bohemia and Moravia by means of regional and district public health centres (Černý *et al.*, 1965). Ticks were monitored by flagging 114 plots (10 × 10 m) – on 68 plots for 2 years, or on 46 plots for 3 years – at an altitude of 200–700 m above sea level (ASL). Another comprehensive field study was performed, although 50 years later (2011–2013), when the distribution, abundance and TBEV infection rate of *I. ricinus* ticks were monitored on 13 areas of three Czech and Moravian regions with high TBE incidence reported in the previous decade (Daniel *et al.*, 2016).

47.1.2 *Ixodes ricinus* tick: ecology

Many studies of the ecology of *I. ricinus* have been performed, including its role in pathogen transmission. However, they have always been narrowly specified and focused on regional issues (Hubálek, 1986; Stanko and Slovák, 2019). On a wider European scale, the main results were summarized by Randolph (2008). A comprehensive publication by Daniel and Dusbábek (1994) was devoted to the influence of landscape climate and microclimate of the tick habitat, which included results of a 3-year field experiment to monitor the entire *I. ricinus* development cycle under the conditions of thermophilic oak forest in the Czech Republic (Daniel *et al.*, 1976, 1977; Daniel, 1978). All developmental stages were observed in special microcages placed in microhabitats situated one close to another, each with a distinctive microclimate, namely: (i) inside thermophilic oak wood; (ii) at its margin; and (iii) in the adjoining open meadow. The air temperature and relative humidity of the microhabitat inside and outside the microcages were continuously recorded. All other environmental conditions were identical, including that all stages had the same optimal possibility of blood-feeding on vertebrate hosts. Abiotic factors (temperature and relative humidity) manifested as driving forces, especially for speed and success of the development cycle.

In 1981–1982, a similar methodology was used to monitor the altitudinal distribution of *I. ricinus* and verify the vertical limits of its occurrence in the Czech Republic (Daniel *et al.*, 1988; Daniel, 1993). Five experimental sites were chosen in the Krkonoše Mountains (the highest mountains in the Czech Republic), from 630 m ASL (submontaneous zone) to 1550 m ASL in the alpine zone (below the top of Sněžka Mountain, at 1603 m the highest peak of the Krkonoše Mountains). This experiment demonstrated that the vertical distribution of *I. ricinus* is dependent on the local microclimate (mainly the temperature of the air layer near the ground surface). Ticks (especially engorged females) introduced occasionally into higher altitudes by their hosts can survive there for a relatively long time, and sometimes even oviposit, but they cannot give rise to a new stable local population. This was the theoretical explanation for the upper limit of *I. ricinus* distribution in Central Europe (700 m ASL) which had been previously set according to field data.

The altitudinal study has become an important basis for assessing the impact of climate change (Materna *et al.*, 2008). In 2002–2006, host-questing *I. ricinus* were investigated by flagging along a vertical transect in the Krkonoše Mountains: (i) at 620–1270 m ASL (with six flagging sites) in the central part of Krkonoše; and (ii) at 600–1020 m ASL (with five flagging sites) in the eastern part. A comparison with the historical data revealed that the upper limit of the distribution of *I. ricinus* had shifted at least to 1100 m ASL in two decades. The likely cause of tick expansion was found when analysing long-term (1961–2005) climate data from the Krkonoše Mountains, which showed a systematic and significant rise in mean annual air temperature by 1.4°C at 1000 m ASL. All ticks collected on both vertical transects in the Krkonoše Mountains were examined individually by PCR or RT-PCR for the presence of TBEV and *Borrelia burgdorferi* sensu lato (s.l.) complex. A total of 5999 *I. ricinus* were tested (Danielová *et al.*, 2006, 2008a, 2010). TBEV RNA was detected in 26 ticks collected at up to 1140 m ASL. Demonstration of TBEV in two unfed larvae of *I. ricinus* indicates transovarial transmission. The similarity of infection rates in unfed larvae and unfed nymphs suggests that vertical transmission in TBEV circulation is important under these mountainous conditions.

Results from the Krkonoše Mountains were verified by the research in other mountainous areas (Daniel *et al.*, 2009), foremost being the northern Moravian Jeseníky Mountains, the second highest mountain range in the Czech Republic (highest summit, Praděd, 1491 m ASL). The northern Moravian region was of particular interest because, during 1970–2000, cases of TBE were recorded in localities close to mountainous areas (up to 780 m). Altogether 1253 ticks of all stages were collected at 990–1300 m covering 12 collection sites. All ticks were examined for TBEV and *B. burgdorferi* s.l. complex. *Borrelia afzelii*, *Borrelia garinii*, *Borrelia burgdorferi* sensu stricto (s.s.) and *Borrelia valaisiana* were detected in 35 ticks, but TBEV was not demonstrated. Later, in 2012, TBEV was found in an *I. ricinus* female collected at 1020 m in the area of the Jeseníky Mountains (Daniel *et al.*, 2016).

From 1979 to 2007, a significant rise in mean annual temperature of 0.5°C/10 years was recorded at 11 monitoring stations in the area of Jeseníky Mountains (658, 775 and 1490 m ASL) and other mountainous areas in the Czech Republic. The most obvious and significant trends were in spring (March–May; 0.5–0.6°C/10 years) and in summer (June–August; 0.5–0.9°C/10 years). The trends were most pronounced in the months with the highest activity of *I. ricinus* ticks. However, the trends were not significant in autumn and winter. Fluctuations in the average amount of precipitation did not show a long-term trend although there was high inter-annual variability.

Suitable climatic conditions are very important, but they are not the only factor governing the existence and development of local tick populations. The entire biocenosis, of which *I. ricinus* ticks (and also TBEV) are an integral part, is impacted by the climate; this includes vertebrate tick hosts and vegetation cover, which have a crucial influence on the survival of tick populations and TBEV circulation in nature. The existence of habitats with the potential to ensure the long-term realization of these conditions, together with the appropriate climate, is one of the basic preconditions for the emergence of active foci of TBE virus infection (Rosický, 1954). An atlas of TBE in the Czech Republic (Daniel and Kříž, 2002) containing predictive maps of *I. ricinus* high-occurrence habitats and TBE risk sites identified by satellite data (Landsat 5 TM with spatial resolution 30 m) over a territory of 52,000 km^2, and distribution maps of TBE cases (1971–2000), have been published using remote sensing and geographical system technologies (Daniel and Kolář, 1990; Daniel *et al.*, 1998, 2004, 2008a).

47.1.3 Tick-borne encephalitis: epidemiology

As for the epidemiology of TBE, as early as 1954, the first European monograph on TBE described the seasonal bimodal curve of weekly registered TBE cases, tick *I. ricinus* occurrence was recorded and the influence of meteorological factors was recognized (Raška, 1954). Also considered was the possibility of predicting TBE incidence in the following year based on the abundance of small mammals (Havlík's theory of 'mouse years'), although this was not later verified (Raška and Bárdoš, 1954).

In 1970, TBE was designated a notifiable disease by Czechoslovak ministerial decree; laboratory-confirmed cases have been reported to the central surveillance centre. Since 1993, the Czech reporting system (formerly EPIDAT, now

ISIN) has been computerized. This unique database includes the total number of TBE cases recorded since 1971 in the Czech Republic (24,023 cases for 1971–2019). It was not until 2012 that TBE became a notifiable disease in all countries of the European Union (2012/506/EU: Commission Implementing Decision of 8 August 2012), with a unified case definition (Amato-Gauci and Zeller, 2012).

According to Kříž *et al.* (2012), data registered in the Czech Republic show several important features with the pattern of TBE incidence changing over time. Until the end of the 1970s, the incidence of TBE was characterized by periods of alternately higher and lower incidence, but large year-to-year variations (from 180 to 595 cases, annually). The 1980s were a period of low occurrence and slight variability. At the beginning of the 1990s, there was a sharp rise in registered cases (since 1993) and this situation, with some year-to-year variation (maximum 1029 cases in 2006), continues (774 cases in 2019). These changes appear to be linked with changes in meteorological conditions (Daniel *et al.*, 2008b).

These data also present the following features of the changing epidemiology, which have been described previously: (i) an increased number of cases in regularly manifested natural foci of TBE infection (Danielová *et al.*, 2004); (ii) a renewed appearance of TBE in locations of former occurrence after 20 or more years (Danielová *et al.*, 2004); (iii) the appearance of new regions where TBE did not previously occur (or occurred only sporadically), mainly at higher altitudes (Danielová *et al.*, 2004, 2008a,b); and (iv) changes in the seasonal appearance of TBE (shift to early spring and late autumn months) (Danielová *et al.*, 2006).

An analysis of meteorological effects on the incidence of TBE cases in the Czech Republic over a 30-year period (1982–2011) was carried out by Brabec *et al.* (2017) in three adjacent, TBE high-incidence regions of the Czech Republic that differed considerably in baseline incidence and timing of the higher incidence period due to local environmental conditions (mainly average landscape altitude ASL and orographical character). A statistically significant relationship was found between the ambient air temperature (standard mean daily temperature) and the incidence of clinical cases of TBE, in three consecutive decades: the low-incidence decade (1982–1991), the decade recording a sharp rise in incidence change in climate (1992–2001) and the persistently high-incidence decade (2002–2011). These observations support the contention that air temperature is a driving force in the increase of TBE incidence. The effect of air temperature increases manifestly from freezing to 12–14°C, is neutral at higher temperature and slightly declines close to 20°C. This concurs with the results of long-term monitoring of the effect of air temperature on host-questing activity of *I. ricinus* (Daniel *et al.*, 2018). There is a several-year time lag in the increase in TBE cases between the highlands region and two other regions monitored (in central Bohemia and south Bohemia); however, the relationship between air temperature and the incidence of TBE is evident in all three regions. The effect of precipitation or TBE incidence–precipitation interaction in the models was not statistically significant in any region of study, but this does not prove that precipitation does not affect TBE incidence.

In 2001–2006, a detailed study of the dynamics of *I. ricinus* host-questing activity was carried out in central Bohemia within the framework of the World Health Organization/European Commission project, Climate Change and Adaptation Strategies for Human Health (Daniel *et al.*, 2006). The primary aim was to determine how well routine weather forecasts could be used to predict the risk to humans of being bitten by *I. ricinus* ticks. The study site was in an area where ticks were shown to be infected with TBEV. *B. burgdorferi* s.s., *B. afzelii* and *B. garinii*, and cases of human Lyme borreliosis (caused by *B. afzelii*), had been reported from the area. Conditions were strictly defined for the monitoring site, time course and meteorological observations (Daniel *et al.*, 2015). The concept and methodology matched (and/or exceeded) those recommended by others to reduce sampling error and bias in field studies of *I. ricinus* questing activity (Eisen, 2008; Estrada-Peña *et al.*, 2013; Medlock *et al.*, 2013). The protocol adopted enabled statistically robust comparison of tick questing activity and abiotic conditions, and hence to determine that weather conditions (and their standard forecast) can be used to predict the risk of tick bite to humans.

A computer program (acronym TICKPRO), prepared in cooperation with the Czech Hydrometeorological Institute (CHMI), forecasts the level of *I. ricinus* questing activity within a 1- to 4-day horizon (Daniel *et al.*, 2010). The program is based on routine weather forecasts produced at the CHMI as well as mathematical models describing the correlation of meteorological factors with the questing activity of *I. ricinus*. From 2008 until now (i.e. 2020), 4-day tick activity predictions are produced twice weekly for the period from March to November with a resolution of ten levels of questing activity of *I. ricinus* and hence of risk of tick bite during outdoor activities. This prediction is publicly accessible on the websites of CHMI, the National Institute of Public Health and the Ministry of Health of the Czech Republic. It is accompanied by instructions on preventive measures at various levels of the risk of tick attack. This prediction was officially included in the legal 'Decree on the system of epidemiological vigilance for selected infections' (No. 275/2010 Coll.) in Annex 28, Article 7 ('Anti-epidemic measures'). In high-risk situations it is occasionally broadcast on television news, radio and in newspapers. During the 13 years of its existence, TICKPRO has also become a significant means of increasing interest and awareness in general about tick-borne infections.

The ratio of symptomatic human TBE cases to the number of questing *I. ricinus* nymphs – the stage by which people are mostly attacked – was found to differ significantly between spring–summer and summer–autumn periods. This problem was potentially resolved in a publication by Daniel *et al.* (2018). The higher average temperature during the summer–autumn period may lead to a higher level of TBEV in ticks, reaching virus loads high enough to cause clinical disease in humans following an infected tick bite. For a forecast specially targeted at TBE, meteorological conditions provide a proxy for virus replication to levels overcoming the infection threshold for development of the clinical disease.

Circulation of TBEV in nature can be affected in various ways by several biotic factors that have one thing in common: they are influenced by abiotic factors (temperature and relative humidity), acting with different intensities and different time shifts of the changes caused by them. The need to use both abiotic meteorological factors as basic predictors of expected future changes is further supported by the fact they are precisely defined, measurable and allow quantitative assessment.

As for the risk of human TBE infection and its impact by climate change, another important factor needs to be considered, namely human behaviour, which is significantly influenced by the socio-economic situation of the human population under study. Analysis of these conditions and other anthropogenic ones in the Czech Republic (Kříž *et al.*, 2004) showed that the TBE is a recreation-related infectious disease associated with outdoor activities, which are in turn influenced by climate. Another important factor is vaccination coverage of the population, which is generally low in the Czech Republic – 29% (Orlíková *et al.*, 2020) – and consequently has little impact on TBE incidence. Hunting of wild boar (*Sus scrofa*) and roe deer (*Capreolus capreolus*), the main hosts for *I. ricinus* female ticks in the Czech Republic, potentially influences TBE incidence (Kříž *et al.*, 2014). The size of local wild boar populations may have contributed to the rising trend in TBE incidence during the 1990s, whereas the regulated population of roe deer does not seem to be implicated in variations in TBE incidence.

Spoil banks resulting from brown coal surface mining represent another type of landscape changed by anthropogenic factors. The occurrence of *I. ricinus* ticks in such regions was studied in north-west Bohemia (Černý and Daniel, 1986; Daniel and Černý, 1990). Our observations in different spoil banks about 20 years after their formation regularly found *I. ricinus* in both wide and limited banks. Particularly the reclaimed parts with artificially planted trees and bushes are invaded by them because they are regularly visited by the hosts of *I. ricinus* imagoes.

47.2 Future Predictions

Our view on future predictions is summarized in the publication titled 'Climate change and tick-borne diseases: a research field in need of long-term empirical field studies' (Eisen, 2008). We believe results of longitudinal and continuous detailed monitoring of the influence of changes in landscape macroclimate and *I. ricinus* habitat microclimate on the biology of individual tick stages (including their questing activity and vector potential), together with repeated interrogation of the long time series of registered TBE cases (accompanied by anamnestic data), prove that:

1. Climate and its changes affect the horizontal and vertical distribution of *I. ricinus*, the speed and success of tick development, population abundance, and thus the risk of human attack.
2. Because the tick does not have its own temperature regulation, its internal temperature, physiological processes and replication of the virus within it are directly affected by the temperature of the external environment. Thus, TBEV circulation within the tick vector is also affected (including vertical transmission), as is tick viral load which determines the risk of transmission of the infection to humans.
3. An analysis of the 50-year time series of TBE clinical cases shows a correlation of changes in disease incidence with changes in the overall climate of Central Europe. The driving factor for these changes is mainly air temperature together with relative air humidity and the level of their interaction, which can change in various ways both year-to-year and during individual seasons. When assessing the role of climate change, it is necessary to judge not only the effect of maximum/minimum values, but also the duration and timing of the air temperature–relative humidity interaction impacting *I. ricinus* ecology and TBEV replication in host-questing ticks during the season of tick activity.

TBE has been a serious health problem in the Czech Republic since the middle of the last century, which has aroused constant interest in the main vector, *I. ricinus*. Therefore, even before the period when signs of climate change began to appear, long-term continuous field experiments were performed to evaluate the influence of meteorological factors (both landscape macroclimate and microclimate of tick microhabitats) on the occurrence and ecology of *I. ricinus*. Some projects are briefly mentioned above. Publications for the period to 2000 have been collated by Slovák and Stanko (2020).

A summary of the results achieved, together with research experience, has provided a robust basis for assessing current climate change. Detailed analysis of the results of long-term observation of the influence of abiotic predictors on *I. ricinus* activity gave rise to a program of weather-based forecasts of the risk to humans of being bitten by *I. ricinus* ticks (TICK-PRO). This analysis allows other aspects to be addressed:

1. Supplement the general risk warning prediction 'to be attacked by an *I. ricinus* tick' and specify 'to be at risk of TBE infection'.
2. Examine the influence of abiotic predictors on the seasonal dynamics of TBEV load in ticks (mainly nymphs) with regard to the seasonality of asymptomatic and symptomatic infections.
3. Determine the influence of abiotic conditions on the level of TBEV load in wintering ticks and its effect on the occurrence of the disease in early spring or during mild winter.
4. Investigate differences in TBEV load in different altitudinal zones of the landscape, differing in local climate. Mountainous areas are an increasingly preferred zone for outdoor activities even in summer. The risk of TBE infection thus becomes a problem of travel medicine because the Czech mountains are often visited by foreign tourists (Gillingham *et al.*, 2020).

Mandatory registration of laboratory-confirmed TBE cases (since 1971), accompanied by a brief analysis of anamnestic data (history of vaccination, tick bite and location, consumption of unpasteurized milk/milk products, etc.), created a unique database to document and quantify changes in the incidence of TBE taking place simultaneously with the manifestation of climate change. The first human cases of TBE of demonstrably local origin above the previously known altitudinal limit were recorded in 1996 and 2001 in the Šumava mountain area above 900 m (Daniel *et al.*, 2003, 2005). Subsequent research on the occurrence of *I. ricinus* (2001–2002) showed an increase to 1020 m ASL in the Šumava Mountains (Fig. 47.1).

The process of *I. ricinus* penetration to higher altitudes due to climate change is conditioned by the fact that mountain ecosystems are able to provide the conditions for the entire development of *I. ricinus*. Above all, the presence of vertebrate hosts of all *I. ricinus* developmental stages is necessary. After three decades of continuous observation, it is clear that *I. ricinus* has become fully established in the local biocenoses of the Czech mountains. This means that further changes in the altitudinal distribution of *I. ricinus* (and associated pathogens) can be

Fig. 47.1. Šumava Mountains, Luzenský stream valley (1130–1150 m above sea level) with Luzný Mountain (1373 m) on the horizon, Czech Republic. Questing *Ixodes ricinus* ticks were found on the edges of forest stands of spruce (*Picea excelsa*), on the ground within clumps of spruce and under mature solitary spruce trees. (Authors' own photograph.)

expected only within changes of the whole biocenosis. The time when infected ticks first appear is conditioned by the gradual increase in *I. ricinus* abundance allowing a higher intensity of TBEV circulation up to the level of the real risk of TBE in human cases.

In the Czech Republic, there has been a general increase in the altitudinal limit of *I. ricinus* from the original 700 m ASL to 1000–1200 m ASL. After two decades, *I. ricinus* has become an integral component of the mountain biocenosis as evidenced by the findings of all developmental stages. However, it is still true that the areas of highest risk occur up to approximately 600 m ASL (Daniel *et al.*, 2016).

When studying the influence of abiotic predictors, attention should be given to the effects of extraordinary meteorological situations. For example, catastrophic floods, storms devastating forests and subsequent bark beetle calamities, and changes in the water regime of the landscape as a result of long-term precipitation deficits affect large areas and exert long-term modifications to the occurrence of *I. ricinus* and thus transmitted diseases, including TBE.

Moreover, the incidence of TBE appears to be reduced by higher ambient temperatures. A statistically significant levelling off followed by a decrease in TBE incidence was documented by Brabec *et al.* (2017). Thus, it can be assumed that changes in climate warming will to some extent lead to a reduction in the incidence of TBE not only in the southern areas of the current distribution of TBE but also in other lowland areas. It is possible that TBE will appear in them only sporadically, or natural foci of TBEV will completely disappear there.

The considered process will have a global character. Its course will be regionally modified by both the different configurations of the terrain and the diversity of vegetation cover as well as the occurrence of habitats with the potential for tick existence. Modifications also will be influenced both by the results of human activities (land use) and by human behaviour, especially by the level of TBE vaccination.

It should be emphasized that this is a highly complex problem, the solution of which will require interdisciplinary cooperation and coordinated action by specialists in fieldwork, laboratory, clinical and epidemiological practices, as well as experience in biostatistics. Another basic requirement is the long-term continuity of observations, which is driven by the long-term complicated development of *I. ricinus* and thus the long-term succession of filial generations.

References

Amato-Gauci, A. and Zeller, H. (2012) Tick-borne encephalitis joins the diseases under surveillance in the European Union. *Eurosurveillance* 17, 20299.

Blaškovič, D. (ed.) (1954) *Epidémia Encefalitidy v Rožňavskom Prírodnom Ohnisku Nákaz [An Epidemic of Encephalitis in the Rožňava Natural Focus of Infections]*. Publishing House of the Slovak Academy of Sciences, Bratislava (in Slovak and Czech).

Brabec, M., Daniel, M., Malý, M., Danielová, V., Kříž, B., *et al.* (2017) Analysis of meteorological effects on the incidence of tick-borne encephalitis in the Czech Republic over a thirty-year period. *Virology: Research & Reviews* 1. https://doi.org/10.15761/VRR.1000103

Černý, V. and Daniel, M. (1986) Occurrence of the tick *Ixodes ricinus* in a landscape strongly influenced by human activities. *Wiadomości parazytologiczne* 32, 351–353.

Černý, V., Rosický, B., Ašmera, J., Kadlčík, K., Kobík, V., *et al.* (1965) Results of investigations of phenology of the common tick *Ixodes ricinus* (L.) in the Czech lands in the years 1960–1962. *Československá parasitologie* 12, 125–131 (in Czech).

Daniel, M. (1978) Microclimate as a determining element in the distribution of ticks and their developmental cycles. *Folia Parasitologica* 25, 91–94.

Daniel, M. (1993) Influence of the microclimate on the vertical distribution of the tick *Ixodes ricinus* (L.) in Central Europe. *Acarologia* 34, 105–113.

Daniel, M. and Černý, V. (1990) Occurrence of the tick *Ixodes ricinus* (L.) under conditions of anthropopressure. *Folia Parasitologica* 37, 183–186.

Daniel, M. and Dusbábek, F. (1994) Micrometeorological and microhabitat factors affecting maintenance and dissemination of tick-borne diseases in the environment. In: Sonenshine, D.E. and Mather, T.N. (eds) *Ecological Dynamics of Tick-Borne Zoonoses*. Oxford University Press, Oxford, pp. 91–138.

Daniel, M. and Kolář, J. (1990) Using satellite data to forecast the occurrence of the common tick *Ixodes ricinus* (L.). *Journal of Hygiene, Epidemiology, Microbiology, and Immunology* 34, 243–252.

Daniel, M. and Kříž, B. (2002) Tick-borne encephalitis in the Czech Republic: I. Predictive maps of *Ixodes ricinus* tick high-occurrence habitats and tick-borne encephalitis risk assessment in Czech regions; II. Maps of tick-borne encephalitis incidence in the Czech Republic in 1971–2000. *Project WHO/EC Climate Change and Adaptation Strategies for Human Health in Europe (cCASHh), EVK-2-2000-0070*. National Institute of Public Health, Prague.

Daniel, M., Černý, V., Dusbábek, F., Honzáková, E., and Olejníček, J. (1976) Influence of microclimate on the life cycle of the common tick *Ixodes ricinus* (L.) in thermophilic oak forest. *Folia Parasitologica* 23, 327–342.

Daniel, M., Černý, V., Dusbábek, F., Honzáková, E. and Olejníček, J. (1977) Influence of microclimate on the life cycle of the common tick *Ixodes ricinus* (L.) in an open area in comparison with forest habitats. *Folia Parasitologica* 24, 149–160.

Daniel, M., Černý, V., Albrecht, V. and Honzáková, E. (1988) Mikroklima v různé nadmořské výšce Krkonoš a jeho vliv na existenci klíštěte *Ixodes ricinus* (L.) [The microclimate at different altitudes of the Krkonoše Mountains and its effect on the existence of the tick *Ixodes ricinus* (L.)]. *Opera Corcontica* 25, 76–110 (in Czech).

Daniel, M., Kolář, J., Zeman, P., Pavelka, K. and Sádlo, J. (1998) Predictive map of *Ixodes ricinus* high-incidence habitats and a tick-borne encephalitis risk assessment using satellite data. *Experimental and Applied Acarology* 22, 417–433.

Daniel, M., Danielová, V., Kříž, B., Jirsa, A. and Nožička, J. (2003) Shift of the tick *Ixodes ricinus* and tick-borne encephalitis to higher altitudes in Central Europe. *European Journal of Clinical Microbiology and Infectious Diseases* 22, 327–328. https://doi.org/10.1007/s10096-003-0918-2

Daniel, M., Kolář, J. and Zeman, P. (2004) GIS tools for tick and tick-borne diseases occurrence. *Parasitology* 129(Suppl.), S329–S352.

Daniel, M., Kříž, B., Danielová. V., Materna, J., Rudenko, N., *et al.* (2005) Occurrence of ticks infected by tick-borne virus and *Borrelia* genospecies in mountains of the Czech Republic. *Eurosurveillance* 10(13), 2672. https://doi.org/10.2807/esw.10.13.02672-en

Daniel, M., Danielová, V., Kříž, B. and Beneš, Č. (2006) Tick-borne encephalitis. In: Menne, B. and Ebi, K.L. (eds) *Climate Change and Adaptation Strategies for Human Health*. Steinkopff, Darmstadt, Germany, pp. 189–205.

Daniel, M., Kolář, J. and Zeman, P. (2008a) Analysing and predicting the occurrence of ticks and tick-borne diseases using GIS. In: Bowman, A.S. and Nuttall, P.A. (eds) *Ticks: Biology, Disease and Control*. Cambridge University Press, Cambridge, pp. 377–407.

Daniel, M., Kříž, B., Danielová, V., Valter, J. and Kott, I. (2008b) Correlation between meteorological factors and tick-borne encephalitis in the Czech Republic. *Parasitology Research* 103(Suppl. 1), S97–S107. https://doi.org/10.1007/s00436-008-1061-x

Daniel, M., Materna, J., Hönig, V., Metelka, L., Danielová, V., *et al.* (2009) Vertical distribution of the tick *Ixodes ricinus* and tick-borne pathogens in the Northern Moravian mountains correlated with climate warming (Jeseníky Mts., Czech Republic). *Central European Journal of Public Health* 17, 131–145.

Daniel, M., Vráblík, T., Valter, J., Kříž, B. and Danielová, V. (2010) The TICKPRO computer program for predicting *Ixodes ricinus* host-seeking activity and the warning system published on websites. *Central European Journal of Public Health* 18, 230–236.

Daniel, M., Malý, M., Danielová, V., Kříž, B. and Nuttall, P.A. (2015) Abiotic predictors and annual seasonal dynamics of *Ixodes ricinus*, the major disease vector of Central Europe. *Parasites & Vectors* 8, 478. https://doi.org/10.1186/s13071-015-1092-y

Daniel, M., Danielová, V., Kříž, B., Růžek, D., Fialová, A., *et al.* (2016) The occurrence of *Ixodes ricinus* ticks and important tick-borne pathogens in areas with high tick-borne encephalitis prevalence in different altitudinal levels of the Czech Republic, Part I. *Ixodes ricinus* ticks and tick-borne encephalitis virus. *Epidemiologie, Mikrobiologie, Imunologie* 65, 118–128.

Daniel, M., Danielová, V., Fialová, A., Malý, M., Kříž, B. and Nuttall, P.A. (2018) Increased relative risk of tick-borne encephalitis in warmer weather. *Frontiers in Cellular Infection and Microbiology* 8, 90. https://doi.org/10.3389/fcimb.2018.00090

Danielová, V., Kříž, B., Daniel, M., Beneš, Č., Valter, J. and Kott, I. (2004) Vliv změn klimatu na výskyt klíšťové encefalitidy v České republice v uplynulých dvaceti letech [Effects of climate changes on the incidence of tick-borne encephalitis in the Czech Republic in the past two decades]. *Epidemiologie, Mikrobiologie, Imunologie* 53, 174–181 (in Czech).

Danielová, V., Rudenko, N., Daniel, M., Holubová, J., Golovchenko, M. and Schwarzová, L. (2006) Extension of *Ixodes ricinus* ticks and agents of tick-borne diseases to mountain areas in the Czech Republic. *International Journal of Medical Microbiology* 296(Suppl. 1), 48–53. https://doi.org/10.1016/j.ijmm.2006.02.007 z

Danielová, V., Schwarzová, L., Materna, J., Daniel, M., Metelka, L., *et al.* (2008a) Tick-borne encephalitis virus expansion to higher altitudes correlated with climate warming. *International Journal of Medical Microbiology* 298(Suppl. 1), 68–72. https://doi.org/10.1016/j.ijmm.2008.02.005

Danielová, V., Kliegrová, S., Daniel, M. and Beneš, Č. (2008b) Influence of climate warming on tick-borne encephalitis expansion to higher altitudes over the last decade (1997–2006) in the Highland Region (Czech Republic). *Central European Journal of Public Health* 16, 4–11.

Danielová, V., Daniel, M., Schwarzová, L., Materna, J., Rudenko, N., *et al.* (2010) Integration of the tick-borne encephalitis virus and *Borrelia burgdorferi* sensu lato into mountain ecosystems, following a shift in the altitudinal limit of distribution of their vector, *Ixodes ricinus* (Krkonoše Mts., Czech Republic). *Vector-Borne and Zoonotic Diseases* 10, 223–230. https://doi.org/10.1089/vbz.2009.0020

Eisen, L. (2008) Climate change and tick-borne diseases: a research field in need of long-term empirical field studies. *International Journal of Medical Microbiology* 298(Suppl. 1), 12–18. https://doi.org/10.1016/j.ijmm.2007.10.004

Estrada-Peña, A., Gray, J.S., Kahl, O., Lane, R.S. and Nijhof, A.M. (2013) Research on the ecology of ticks and tick-borne pathogens – methodological principles and caveats. *Frontiers in Cellular Infection and Microbiology* 3, 29. https://doi.org/10.3389/fcimb.2013.00029

Gallia, F., Rampas, J. and Hollender, L. (1949) Laboratory infection with encephalitis virus. *Časopis Lékařů Českých* 88, 224–225.

Gillingham, E.L., Cull, B., Pietzsch, M.E., Phipps, L.P., Medlock, J.M. and Hansford, K. (2020) The unexpected holiday souvenir: the public health risk to UK travellers from ticks acquired overseas.

International Journal of Environmental Research and Public Health 17, 7957. https://doi.org/10.3390/ijerph17217957
Grešíková, M. (1957) Secretion of tick-borne encephalitis virus in goat milk. *Veterinární Časopis* 5, 177–182.
Grešíková, M. (1958) Excretion of the tick-borne encephalitis virus in the milk of subcutaneously infected cows. *Acta Virologica* 2, 188–192.
Hubálek, Z. (1986) *Czechoslovak Bibliography on Tick-Borne Encephalitis 1949–1985. Czechoslovak Bibliography on Epidemiology and Microbiology*, vol. 15, suppl. Institute of Hygiene and Epidemiology, Prague.
Krejčí, J. (1949) Isolement d'un virus nouveau en course d'une epidemie de meningoencephalitides dans la region de Vyškov (Moravie). *La Presse Médicale (Paris)* 74, 1084.
Kříž, B., Beneš, Č., Danielová, V. and Daniel, M. (2004) Socio-economic conditions and other anthropogenic factors influencing tick-borne encephalitis incidence in the Czech Republic (1997–2008). *International Journal of Medical Microbiology* 293(Suppl. 37), 63–68.
Kříž, B., Malý, M., Beneš, Č. and Daniel, M. (2012) Epidemiology of tick-borne encephalitis in the Czech Republic 1970–2008. *Vector-Borne and Zoonotic Diseases* 12, 994–999. https://doi.org/10.1089/vbz.2011.0900
Kříž, B., Daniel, M., Beneš, Č. and Malý, M. (2014) The role of game (wild boar and roe deer) in the spread of tick-borne encephalitis in the Czech Republic. *Vector-Borne and Zoonotic Diseases* 14, 801–807. https://doi.org/10.1089/vbz.2013.1569
Materna, J., Daniel, M., Metelka, L. and Harčarik, J. (2008) The vertical distribution, density and development of the tick *Ixodes ricinus* in mountain areas influenced by climatic change (the Krkonoše Mts., Czech Republic). *International Journal of Medical Microbiology* 296(Suppl. 1), 25–37.
Medlock, J.M., Hansford, K.M., Bormane, A., Derdakova, M., Estrada-Peña, A., *et al.* (2013) Driving forces for changes in geographical distribution of *Ixodes ricinus* ticks in Europe. *Parasites & Vectors* 6, 1. https://doi.org/10.1186/1756-3305-6-1
Orlíková, H., Lenz, P. and Kynčl, J. (2020) Klíšťová encefalitida v České republice v roce 2019 – zpráva o epidemiologické situaci v kontextu předcházejících let [Tick-borne encephalitis in the Czech Republic in 2019 – a report on the epidemiological situation in the context of previous years]. *Reports of the Centre of Epidemiology and Microbiology (NIPH Prague)* 29, 211–219 (in Czech).
Rampas, J. and Gallia, F. (1949) Isolation of encephalitis virus from *Ixodes ricinus* ticks. *Časopis Lékařů Českých* 88, 1179–1180.
Randolph, S.E. (2008) The impact of tick ecology on pathogen transmission dynamics. In: Bowman, A.S. and Nuttall, P.A. (eds) *Ticks: Biology, Disease and Control*. Cambridge University Press, Cambridge, pp. 40–72.
Raška, K. (ed.) (1954) *Československá klíšťová encefalitis [Czechoslovak Tick-Borne Encephalitis]*. Státní zdravotnické nakladatelství Publishing House, Prague (in Czech).
Raška, K. and Bárdoš, V. (1954) Epidemiologie Československé klíšťové encefalitidy [Epidemiology of the Czechoslovak tick-borne encephalitis]. In: Raška, K. (ed.) *Československá klíšťová encefalitis [Czechoslovak Tick-Borne Encephalitis]*. Státní zdravotnické nakladatelství Publishing House, Prague, pp. 65–77 (in Czech).
Rosický, B. (1954) Poznámky k ekologii klíštěte *Ixodes ricinus* L. ve střední Evropě se zřetelem na přírodní ohniska nákaz [Notes on the ecology of *Ixodes ricinus* L. tick in Central Europe with regard to the natural foci of diseases]. *Acta Societatis Zoologicae Bohemoslovacae* 18, 41–70 (in Czech).
Růžek, D. (2015) Virus klíšťové encefalitidy – obecná charakteristika [Tick-borne encephalitis virus – general characteristics]. In: Růžek, D. (ed.) *Klíšťová encefalitida [Tick-Borne Encephalitis]*. Grada Publishing House, Prague, pp. 19–39 (in Czech).
Růžek, D., Bilski, B. and Günther, G. (2013) Tick-borne encephalitis. In: Singh, S.K. and Růžek, D. (eds) *Neuroviral Infections*. CRC Press, Boca Raton, Florida, pp. 211–237.
Slovák, M. and Stanko, M. (2020) Bibliography of studies on ticks and tick-borne pathogens published by Slovak (until 2015) and Czech (until 1993) authors. *Entomofauna Carpatica Supplementum* 1, 1–733.
Stanko, M. and Slovák, M. (2019) *History of the Research on Tick Ecology in the Czech Republic and Slovakia Until the Year 2000*. Veda, Publishing House of the Slovak Academy of Sciences, Bratislava.

eo48 Tick-Borne Viral Haemorrhagic Fever Infections

Dennis Bente
University of Texas Medical Branch, Galveston National Laboratory, Texas, USA

48.1 Expert Opinion

48.1.1 Introduction

This expert opinion will discuss a group of tick-transmitted viruses that can cause a bona fide haemorrhagic fever in the human incidental host. Haemorrhagic fever is a life-threatening syndrome for which one of the pathognomonic hallmarks is the increased permeability of blood vessels resulting in bleeding into the skin, internally, and from puncture sites and orifices (Peters and Zaki, 2002). These viruses only cause haemorrhagic fever in the human host; there are no pathological changes in the natural vertebrate host or tick vector. Exceptions are monkeys in Kyasanur forest disease (KFD) and musk rats in Omsk haemorrhagic fever (OHF), although they might not be considered natural hosts of the virus. Besides Crimean–Congo haemorrhagic fever virus (CCHFV), no human-to-human transmission has been described for tick-borne haemorrhagic fever viruses (TBHFVs) (Bente *et al.*, 2013). The classification of viruses in the TBHFV group is purely based on the human patient's clinical presentation rather than viral taxonomy, geographic distribution or transmission dynamics. Thus, these viruses grouping into the family *Flaviviridae* and the order *Bunyavirales* are found in diverse parts of the world and are transmitted by a range of tick species (see Table 48.1).

The pathogenesis of how the viruses cause the haemorrhagic syndrome is only partly understood owing to the need to work with the viruses in a high or maximum biocontainment laboratory setting as well as the fact they cause sporadic cases in mostly remote areas. Much has been extrapolated from other haemorrhagic fever viruses, such as filoviruses. As a result, countermeasures such as antivirals or vaccines are not widely studied. Crimean–Congo haemorrhagic fever (CCHF) is the most prominent member of the TBHFV group owing to its wide geographic distribution. Studies in the last two decades have shown that more human cases are subclinical than previously expected (Fig. 48.1a) (Bodur *et al.*, 2012). If the disease progresses to the haemorrhagic stage, coagulopathy with broad ecchymosis and bleeding from orifices and injection sites is very common.

Viruses in the tick-borne encephalitis (TBE) complex are unique in that they show a spectrum of clinical presentations. OHF presents strictly as a haemorrhagic fever (LaSala and Holbrook, 2010). In comparison, Alkhurma haemorrhagic fever (AHF) and KFD can present as a haemorrhagic fever or with neurological signs

*Email: dabente@utmb.edu

DOI: 10.1079/9781789249637.0048

Table 48.1. Tick-borne viruses of medical importance causing haemorrhagic fever with severe disease. (Adapted from de la Fuente *et al.*, 2017 (Labuda and Nuttall, 2004; Johnson *et al.*, 2012).)

Virus (abbreviation)	Family/genus	Principal vector	Endemic presence	Clinical presentation
Omsk haemorrhagic fever virus (OHFV)	*Flaviviridae/ Flavivirus*	*Dermacentor reticulatus*	Asia	Bona fide haemorrhagic fever
Alkhurma haemorrhagic fever virus (AHFV)	*Flaviviridae/ Flavivirus*	*Ornithodoros savigny*, *Hyalomma dromedarii*	Saudi Arabia, East Africa	The spectrum of clinical presentation from haemorrhagic fever to neurological presentation
Kyasanur forest disease virus (KFDV)	*Flaviviridae/ Flavivirus*	*Haemaphysalis spingera*	India	The spectrum of clinical presentation from haemorrhagic fever to neurological presentation
Tick-borne encephalitis virus (TBEV)	*Flaviviridae/ Flavivirus*	*Ixodes ricinus/Ixodes persulcatus*	Europe/Asia	Strictly neurological presentation
Powassan virus (POWV)	*Flaviviridae/ Flavivirus*	*Ixodes cookei*	North America/ Russia	Strictly neurological presentation
Crimean–Congo haemorrhagic fever virus (CCHFV)	*Bunyavirales/ Nairoviridae*[b]	*Hyalomma* spp.	Africa/Asia/Middle East/Southern Europe	Bona fide haemorrhagic fever
Dabie bandavirus[a]	*Bunyavirales/ Phenuiviridae*[b]	*Haemaphysalis longicornis*	Eastern Asia	Thrombocytopaenia, elevated liver enzymes but haemorrhagic presentation rare

[a]Severe fever with thrombocytopaenia syndrome virus.
[b]Order/family.

and symptoms, or as an intermediate phenotype (Holbrook, 2012; Sawatsky *et al.*, 2014). Other viruses in the TBE complex, such as Powassan virus or TBE viruses (subtypes), have a solely neurological presentation. In addition, some Kyasanur forest disease virus (KFDV) infections may be subclinical (Munivenkatappa *et al.*, 2018). Viruses in the TBE complex, therefore, offer a unique research opportunity to correlate genotype with clinical phenotype.

There are other tick-borne viruses that cause severe disease in humans but, *per definitionem*, do not fit the clinical definition of haemorrhagic fever, e.g. Dabie bandavirus (formerly called severe fever with thrombocytopaenia virus) and Dugbe virus (Liang *et al.*, 2019). This expert opinion will focus only on CCHFV, Alkhurma haemorrhagic fever virus (AHFV), KFDV and Omsk haemorrhagic fever virus (OHFV) (although related viruses might have similar epidemiology) and how climate change might influence the tick–virus–host dynamics, although very little is known.

48.1.2 Virus dynamics

A common feature of TBHFVs is their occurrence in defined geographical foci; even within endemic areas, viruses are found in a circumscribed focus. Regularly, TBHFVs circulate in nature in silent cycles, meaning they are transmitted from ticks to their vertebrate hosts with no overt signs of disease and no contact with humans and, therefore, unbeknownst to humans (Fig. 48.1a, iceberg model). If human cases occur, they might go unnoticed because the disease is mild or, more likely, because the condition is not suspected and therefore not adequately diagnosed. As a result, we still map the distribution of TBHFVs according to human cases, ignoring the sites where silent circles

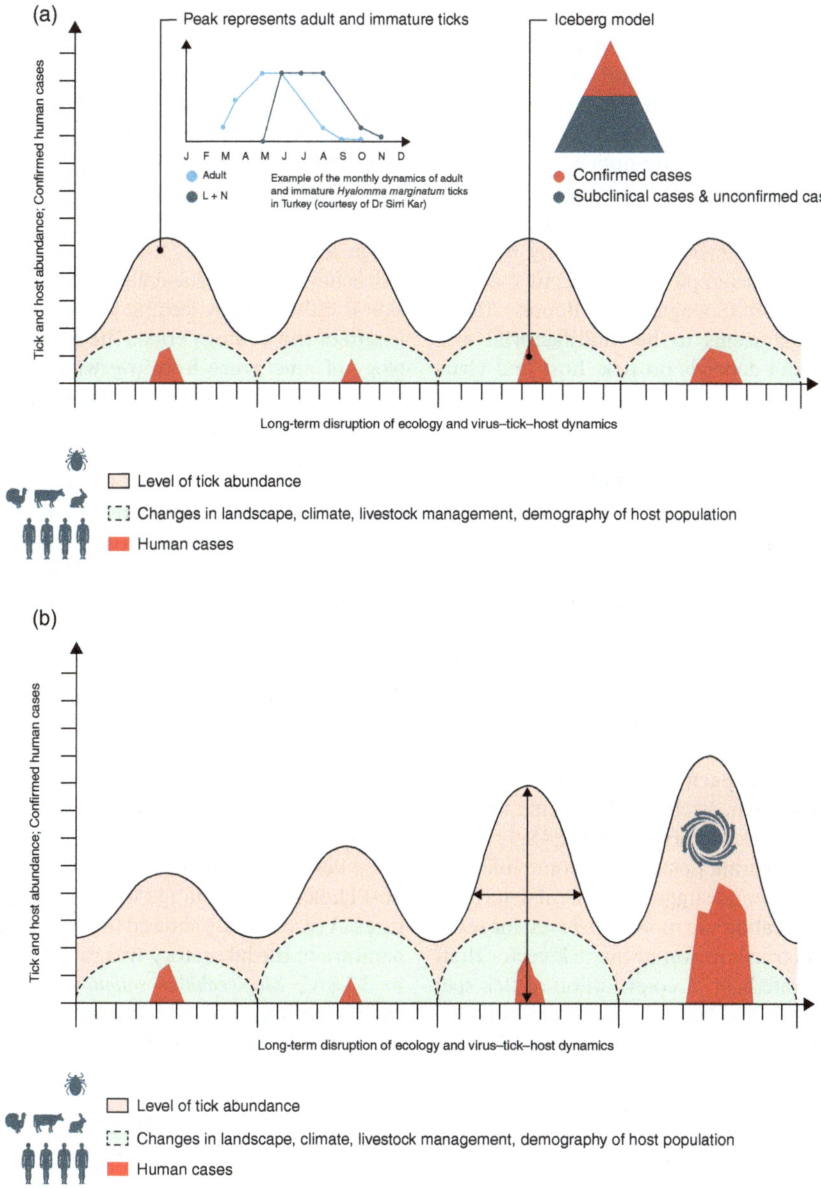

Fig. 48.1. (a) Seasonal tick, host and human case dynamics of tick-borne haemorrhagic fever viruses (TBHFVs) during undisturbed abiotic conditions (resting cycle). Peaks represent the tick abundance (orange), host abundance (green) and human case numbers (red) over four years with seasonal changes per year. The orange peaks reflect the activity of immature (larvae and nymphs, L + N) and adult life stages (see insert: 'Example of *Hyalomma marginatum* tick seasonality in Turkey'). Confirmed human cases (red) are only the tip of the iceberg (see insert: 'Iceberg model') and do not reflect asymptomatic or unconfirmed cases. (b) Potential shifts in seasonal tick, host and human case dynamics of TBHFVs due to climate change (escalated cycle). Climate change can increase host abundance as well as tick phenology (dependent or independent of host abundance changes). Arrows indicate that climate change can extend the feeding period of immature and mature tick life stages, as well as increase the tick abundance per season. This in turn will lead to a greater TBHFV transmission risk to humans and ultimately higher case numbers (vortex symbol reflects amplification of human cases). (Author's own figure.)

could exist (Kar *et al.*, 2020). When anthropogenic factors occur, such as climate change or land fragmentation, that influence host and tick behaviour, abundance and diversity, ultimately more human exposure and clinical cases result in the disease being noticed (Fig 48.1b). This is sometimes misrepresented as a TBHFV 'emerging' in a geographic area although the virus has been previously circulating there silently. A good example is the recent cases of CCHF in Spain (Spengler and Bente, 2017). This process seems to be multifactorial and complex, and depends on tick, host and virus properties.

48.1.3 Tick vector

The expansion of tick species owing to climate change into new geographic areas has been discussed in other parts of the book and shall not be discussed here.

The tick vector's role in maintaining TBHFVs in nature, the transmission from and to the host, and the role in the pathogenesis of TBHFVs have been significantly understudied; thus, the vectorial capacity of TBHFV vectors is largely unknown. Each TBHFV seems to have very few key tick vectors (see Table 48.1), but a range of vertebrate hosts can become infected. This finding would suggest that viral determinants of replication are more conserved for replication and transmission in the tick vector than the vertebrate host. A co-evolution of tick species and virus has been shown in phylogenetic studies (Honig *et al.*, 2004), suggesting that the tick shapes the virus's evolution (Xia *et al.*, 2016). It is unknown whether geographically localized 'genotypes' of the same tick species differ in vector competence. Another key question is what role other tick species play. For instance, CCHFV has been isolated from >30 tick species (Gargili *et al.*, 2017). Could they serve as bridge hosts or are there cryptic transmission cycles (Kar *et al.*, 2020)? The epidemiological theory has largely ignored the potential effects of multiple vectors on disease dynamics. As temperature changes, it is conceivable that main vector switching occurs and other tick species play a more prominent role in the transmission of the virus and dynamics of the disease.

Vector diversity is defined by species richness and species identity and can influence viral host range. Diversity in tick vector species potentially results in different pathogen transmission patterns, host preference, aggregation behaviour, movement behaviour, length of feeding period and feeding behaviour. Tick vectors vary in their seasonal phenology of emergence, host range and preference, transmission efficiency and behaviour. The probability of virus transmission is influenced by feeding behaviour and the length of the feeding period. The seasonal phenology of emergence from overwintering stages can strongly shape virus dynamics and phenology is probably influenced the most by climate change (Fig. 48.1b) (Randolph *et al.*, 2002). In general, any effects on vector fitness that lead to higher rates of population growth may significantly impact the rates of virus spread.

Hoogstraal (1979) speculated that, in addition to viral vector competence, tick behaviour can play a crucial role in transmission as certain tick species prefer to feed on humans compared with other species. A good example of host choice based on behaviour is for CCHFV in the Eastern Mediterranean: *Hyalomma marginatum* ticks are very aggressive in feeding on humans and are considered the principal tick vector, whereas CCHFV has been found in *Rhipicephalus bursa* ticks which prefer livestock as hosts. The host choice might change with rising temperatures. A recent study showed that when the temperature in the laboratory was raised from 23.3 to 37.8°C, *Rhipicephalus sanguineus* ticks that carry a pathogen were 2.5 times more likely to prefer humans over dogs (Backus *et al.*, 2020).

48.1.4 Host

All TBHFVs involve two- or three-host tick species that feed on multiple hosts during their life cycle. The risk of human exposure is most significant at the ecotone because different host species overlap at this transition area. Disease dynamics are controlled by host abundance and diversity, as well as the role of the host in the tick life cycle and virus transmission cycle. All TBHFVs seem to be stably circulating in tick–host cycles (Fig. 48.1a). Further studies need to investigate if the concept of 'endemic stability' described for tick-borne diseases of veterinary concern also applies for TBHFVs (Jonsson *et al.*, 2012). All three factors mentioned above are heavily influenced by anthropogenic factors, including climate change, which ultimately disrupt the stable cycle (Fig. 48.1b).

48.1.5 Role of the host

The vertebrate host can play multiple roles in the maintenance, amplification and dispersal of TBHFVs in nature. First, hosts serve as a blood meal source for ticks, and ticks prefer specific hosts over others (see above). Thus, individual species can serve as an amplifying host for tick vector species, resulting in increased tick abundance and density. Some of these tick–host preferences have been investigated, e.g. Lagomorpha were shown to be a critical host for immature stages of *Hyalomma* ticks (Spengler and Estrada-Peña, 2018).

Second, animal hosts serve as maintenance hosts (enzootic hosts) or amplifying hosts (epizootic hosts) for the virus; or are dead-end hosts with no transmission of the virus to uninfected ticks. Which animal species serve as what host is largely undefined for TBHFVs and deserves much greater attention to define the influence of climate change. The increase in the virus in an amplifying host versus a maintenance host can be measured in the number of TBHFV-infected ticks that complete feeding on that host. Threshold viraemia defines the quantity of virus present when mosquito feeding is required for mosquito-borne viruses to be acquired. Since ticks feed for significantly longer than do mosquitoes, threshold viraemia must be redefined for tick-borne viruses, especially since studies have shown that non-viraemic transmission of tick-borne viruses is possible. Lastly, animals that migrate over long distances, such as birds, or livestock transported by humans over considerable distances, can also serve as transport hosts for ticks. In turn, these ticks can be infected. As a result, TBHFVs can be introduced into new geographical areas, augmented by rising temperature that supports the survival of non-native tick species in these areas. *H. marginatum* and *Hyalomma rufipes* can now be found north as far as Sweden (Grandi *et al.*, 2020). More effort is needed to define the roles of different vertebrate hosts for TBHFVs to find the lynchpins in the transmission that can be targeted in sustainable One Health solutions.

48.1.6 Host abundance

Livestock (particularly local domestic cattle breeds) have played a crucial role in the spread of KFDV in south-west and south India. The forest tick, *Haemaphysalis spinigera*, is considered the primary vector and reservoir for the virus. Immature stages of *H. spinigera* feed on small mammals such as rodents and bats (Pattnaik, 2006). Larvae and nymphs can also feed on forest-dwelling monkeys, causing epizootics with high fatality rates, and the primates serve as an amplifying host of the virus. Transmission to humans can occur after contact with a sick or recently dead primate, although mainly after a tick bite from adults or immatures. Adult *H. spinigera* ticks rarely feed on monkeys but attach to large mammals such as cattle and other livestock, and also humans. It is thought that deforestation owing to activities such as agriculture, road development and mining has caused the emergence and spikes in KFD cases over the years (Purse *et al.*, 2020). The expanding human population together with deforestation has resulted in more cattle, which are grazed at the edge of forests during the wet season. The cattle can become heavily infested with ticks and bring ticks in close association with villagers. Thus, cattle undoubtedly play an essential role in tick reproduction and in maintaining high tick population densities. Since 2012, KFDV distribution in India has expanded to multiple new states. It has been speculated that the long-distance spread of KFDV occurred through KFDV-infected ticks feeding on cattle that serve as transport hosts. AHFV is genetically closely related to KFDV and was first described in Saudi Arabia; it is not found in India. The import of AHFV-infected ticks on goats or sheep has been attributed to the introduction of the virus to Saudi Arabia. A similar phenomenon has been described for CCHFV. The 1994–1995 CCHF epidemic in the United Arab Emirates was possibly associated with importation of CCHFV-infected livestock and ticks from eastern Africa (Rodriguez *et al.*, 1997). Both instances demonstrate the role of animal transport in the introduction of TBHFVs.

48.1.7 Host diversity

The diversity of hosts can have possible contrasting effects on virus dynamics, either favouring virus persistence in an environment or high virus prevalence (amplification) or reducing virus prevalence primarily through a dilution effect. At this point, there is little evidence for an amplification effect owing to host richness. The dilution effect occurs when the transmission cycle, maintained by one competent vertebrate host species, is reduced by the presence of a second host species that is less competent (Sorvillo *et al.*, 2020). Adding a second host species to the community of ticks effectively increases the number of vertebrate hosts available to feeding ticks, so the tick population expands. If this second vertebrate host is less competent for the virus, the virus is lost from the system: the second host's presence dilutes the virus. The dilution effect has been demonstrated for Lyme disease but is still somewhat controversial for other tick-borne diseases, although there is evidence for louping ill (Laurenson *et al.*, 2003). No field or laboratory data on a potential dilution effect are available for the TBHFVs. Lastly, the animal husbandry style and production systems of livestock, which play a key role in the amplification of adult ticks, are often overlooked. Livestock that is managed extensively will bring animals closer to the ecotone transition zone where hosts of the immature tick stages will deposit ticks. As livestock production systems change in response to changing climate conditions, so will the tick burden.

48.1.8 Virus

Work with mosquito-borne viruses has shown that an increased environmental temperature leads to faster virus replication in tissues, higher virus titres and quicker dissemination in the mosquito vector and mammalian hosts (Liu *et al.*, 2017). Compared with mosquitoes, the extrinsic incubation period and, therefore, the speed of virus dissemination within the vector does not play a role for virus replication within ticks and temperature has less of an impact in this respect (Nuttall *et al.*, 1994). When the extrinsic incubation period is not limiting (which is mostly the case for tick-borne viruses), increased environmental temperature may enhance virus replication and consequently increase the amount of virus transmitted during tick feeding (Daniel *et al.*, 2018). However, the effect of temperature on replication of TBHFVs in their tick vectors has not been investigated. Another difference from mosquito-borne viruses is threshold viraemia, i.e. level of virus in the bloodstream of the host required to be acquired by the mosquito. Threshold viraemia (see Section 48.1.5) is crucial for transmission from the host to the mosquito; threshold viraemia does not play a significant role for TBHFVs since ticks feed over a much more extended period and imbibe a greater volume (Nguyen *et al.*, 2013). Therefore, it is unclear if a greater viraemia level in the animal host will lead to a more significant number of ticks acquiring the infection while feeding on the host. On the contrary, the extension of the viraemia in an animal host could lead to many ticks becoming infected. The role of non-viraemic/non-systemic transmission of an infected tick to non-infected ticks through co-feeding has not been very well studied for TBHFVs and could also play a significant role in their transmission. Both higher viraemia levels and more prolonged viraemia could lead to an increased risk of infection for humans in the animal slaughter process. Nevertheless, studies have also shown that increased temperature also leads to modulation of the mammalian host immune system, which could be detrimental to the virus.

The situation might be different in the invertebrate host. No studies are available for TBHFVs on the influence of temperature on replicating kinetics in ticks, only for tick-borne encephalitis virus (TBEV). Kalenchuk and Hodko (1980) showed that female *Dermacentor reticularis* ticks parenterally infected with TBEV and kept at 23°C showed a 100-fold higher virus titre than ticks held at 37°C. It is unclear how a constant temperature of 37°C mimics climate change and the optimal temperature could have been between 23 and 37°C. Studies have suggested a co-evolution of tick-borne viruses and their tick vector species. This might also mean that the viruses evolve along with the climate niches their tick vectors colonize. A good example is the subtypes of TBEV.

48.2 Future Predictions

The broad lack of understanding of TBHFV–vector–host dynamics makes the prediction on how climate change will affect the distribution and intensity of disease very difficult and complex. Traditionally, research efforts have focused on developing countermeasures for human disease. In order to predict expansion and find sustainable long-term solutions, much greater research effort has to go into understanding the vector–virus–host interactions.

References

Backus, L., López-Pérez, A.M. and Foley, J.E. (2020) Effect of increased temperature on host selection by the brown dog tick. In: *Proceedings of the 69th Annual Meeting of the American Society of Tropical Medicine and Hygiene*, Arlington, Virginia. Available at: https://astmhpressroom.wordpress.com/annual-meeting-2020/brown-dog-ticks/ (accessed 13 June 2021).

Bente, D.A., Forrester, N.L., Watts, D.M., McAuley, A.J., Whitehouse, C.A. and Bray, M. (2013) Crimean–Congo hemorrhagic fever: history, epidemiology, pathogenesis, clinical syndrome and genetic diversity. *Antiviral Research* 100, 159–189. https://doi.org/10.1016/j.antiviral.2013.07.006

Bodur, H., Akinci, E., Ascioglu, S., Öngürü, P. and Uyar, Y. (2012) Subclinical infections with Crimean–Congo hemorrhagic fever virus, Turkey. *Emerging Infectious Diseases* 18, 640–642. https://doi.org/10.3201/eid1804.111374

Daniel, M., Danielová, V., Fialová, A., Maly, M., Kríž, B. and Nuttall, P.A. (2018) Increased relative risk of tick-borne encephalitis in warmer weather. *Frontiers in Cellular and Infection Microbiology* 8, 90. http://doi.org/10.3389/fcimb.2018.00090

de la Fuente, J., Antunes, S., Bonnet, S., Cabezas-Cruz, A., Domingos, A.G., *et al.* (2017) Tick–pathogen interactions and vector competence: identification of molecular drivers for tick-borne diseases. *Frontiers in Cellular and Infection Microbiology* 7, 114. https://doi.org/10.3389/fcimb.2017.00114

Gargili, A., Estrada-Peña, A., Spengler, J.R., Lukashev, A., Nuttall, P.A. and Bente, D.A. (2017) The role of ticks in the maintenance and transmission of Crimean–Congo hemorrhagic fever virus: a review of published field and laboratory studies. *Antiviral Research* 144, 93–119. https://doi.org/10.1016/j.antiviral.2017.05.010

Grandi, G., Chitimia-Dobler, L., Choklikitumnuey, P., Strube, C., Springer, A., *et al.* (2020) First records of adult *Hyalomma marginatum* and *H. rufipes* ticks (Acari: Ixodidae) in Sweden. *Ticks and Tick-borne Diseases* 11, 101403. https://doi.org/10.1016/j.ttbdis.2020.101403

Holbrook, M.R. (2012) Kyasanur forest disease. *Antiviral Research* 96, 353–362. https://doi.org/10.1016/j.antiviral.2012.10.005

Honig, J.E., Osborne, J.C. and Nichol, S.T. (2004) The high genetic variation of viruses of the genus *Nairovirus* reflects the diversity of their predominant tick hosts. *Virology* 318, 10–16. https://doi.org/10.1016/j.virol.2003.09.021

Hoogstraal, H. (1979) The epidemiology of tick-borne Crimean–Congo hemorrhagic fever in Asia, Europe, and Africa. *Journal of Medical Entomology* 5, 307–417. https://doi.org/10.1093/jmedent/15.4.307

Johnson, N., Voller, K., Phipps, L.P., Mansfield, K.L. and Fooks, A.R. (2012) Rapid molecular detection methods for arboviruses of livestock of importance in Northern Europe. *Journal of Biomedicine and Biotechnology* 2012, 719402. https://doi.org/10.1155/2012/719402

Jonsson, N.N., Bock, R.E., Jorgensen, W.K., Morton, J.M. and Stear, M.J. (2012) Is endemic stability of tick-borne disease in cattle a useful concept? *Trends in Parasitology* 28, 85–89. https://doi.org/10.1016/j.pt.2011.12.002

Kalenchuk, V.U. and Hodko, L.P. (1980) Effect of various temperatures on the reproduction of tick-borne encephalitis virus in ixodid ticks. *Meditsinskaya Parazitologiya i Parazitarnye Bolezni* 49, 23–25.

Kar, S., Rodriguez, S.E., Akyildiz, G., Cajimat, M.N.B., Bircan, R., *et al.* (2020) Crimean–Congo hemorrhagic fever virus in tortoises and *Hyalomma aegyptium* ticks in East Thrace, Turkey: potential of a cryptic transmission cycle. *Parasites & Vectors* 13, 201. https://doi.org/10.1186/s13071-020-04074-6

Labuda, M. and Nuttall, P.A. (2004) Tick-borne viruses. *Parasitology* 129(Suppl. 1), S221–S245. https://doi.org/10.1017/S0031182004005220

LaSala, P.R. and Holbrook, M. (2010) Tick-borne flaviviruses. *Clinics in Laboratory Medicine* 30, 221–235. https://doi.org/10.1016/j.cll.2010.01.002

Laurenson, M.K., Norman, R.A., Gilbert, L., Reid, H.W. and Hudson, P.J. (2003) Identifying disease reservoirs in complex systems: mountain hares as reservoirs of ticks and louping-ill virus, pathogens of red grouse. *Journal of Animal Ecology* 72, 177–185. https://doi.org/10.1046/j.1365-2656.2003.00688.x

Liang, M., Yu, X. and Li, D. (2019) Rename the genus *Banyangvirus* and the species *Huaiyangshan banyangvirus (Bunyavirales: Phenuiviridae)*. Available at: https://ictv.global/ICTV/proposals/2019.015M.zip (accessed 13 June 2021).

Liu, Z., Zhang, Z., Lai, Z., Zhou, T., Jia, Z., *et al.* (2017) Temperature increase enhances *Aedes albopictus* competence to transmit dengue virus. *Frontiers in Microbiology* 8, 2337. https://doi.org/10.3389/fmicb.2017.02337

Munivenkatappa, A., Sahay, R.R., Yadav, P.D., Viswanathan, R. and Mourya, D.T. (2018) Clinical and epidemiological significance of Kyasanur forest disease. *Indian Journal of Medical Research* 148, 145–150. https://doi.org/10.4103/ijmr.IJMR_688_17

Nguyen, N.M., Kien, D.T.H., Tuan, T.V., Quyen, N.T.H., Tran, C.N.B., *et al.* (2013) Host and viral features of human dengue cases shape the population of infected and infectious *Aedes aegypti* mosquitoes. *Proceedings of the National Academy of Sciences USA* 110, 9072–9077. https://doi.org/10.1073/pnas.1303395110

Nuttall, P.A., Jones, L.D., Labuda, M. and Kaufman, W.R. (1994) Adaptations of arboviruses to ticks. *Journal of Medical Entomology* 31, 1–9. https://doi.org/10.1093/jmedent/31.1.1

Pattnaik, P. (2006) Kyasanur forest disease: an epidemiological view in India. *Reviews in Medical Virology* 16, 151–165. https://doi.org/10.1002/rmv.495

Peters, C.J. and Zaki, S.R. (2002) Role of the endothelium in viral hemorrhagic fevers. *Critical Care Medicine* 30, S268–S273. https://doi.org/10.1097/00003246-200205001-00016

Purse, B.V., Darshan, N., Kasabi, G.S., Gerard, F., Samrat, A., *et al.* (2020) Predicting disease risk areas through co-production of spatial models: the example of Kyasanur forest disease in India's forest landscapes. *PLoS Neglected Tropical Diseases* 14, e0008179. https://doi.org/10.1371/journal.pntd.0008179

Randolph, S.E., Green, R.M., Hoodless, A.N. and Peacey, M.F. (2002) An empirical quantitative framework for the seasonal population dynamics of the tick *Ixodes ricinus*. *International Journal for Parasitology* 32, 979–989. https://doi.org/10.1016/S0020-7519(02)00030-9

Rodriguez, L.L., Maupin, G.O., Ksiazek, T.G., Rollin, P.E., Khan, A.S., *et al.* (1997) Molecular investigation of a multisource outbreak of Crimean–Congo hemorrhagic fever in the United Arab Emirates. *American Journal of Tropical Medicine and Hygiene* 57, 512–518. https://doi.org/10.4269/ajtmh.1997.57.512

Sawatsky, B., McAuley, A.J., Holbrook, M.R. and Bente, D.A. (2014) Comparative pathogenesis of Alkhumra hemorrhagic fever and Kyasanur forest disease viruses in a mouse model. *PLoS Neglected Tropical Diseases* 8, e2934. https://doi.org/10.1371/journal.pntd.0002934

Sorvillo, T.E., Rodriguez, S.E., Hudson, P., Carey, M., Rodriguez, L.L., *et al.* (2020) Towards a sustainable One Health approach to Crimean–Congo hemorrhagic fever prevention: focus areas and gaps in knowledge. *Tropical Medicine and Infectious Disease* 5, 113. https://doi.org/10.3390/tropicalmed5030113

Spengler, J. and Bente, D.A. (2017) Crimean–Congo hemorrhagic fever in Spain – new arrival or silent resident? *New England Journal of Medicine* 377, 106–108. https://doi.org/10.1056/NEJMp1707436

Spengler, J.R. and Estrada-Peña, A. (2018) Host preferences support the prominent role of *Hyalomma* ticks in the ecology of Crimean–Congo hemorrhagic fever. *PLoS Neglected Tropical Diseases* 12, e0006248. https://doi.org/10.1371/journal.pntd.0006248 z

Xia, H., Beck, A.S., Gargili, A., Forrester, N., Barrett, A.D.T. and Bente, D.A. (2016) Transstadial transmission and long-term association of Crimean–Congo hemorrhagic fever virus in ticks shapes genome plasticity. *Scientific Reports* 6, 35819. https://doi.org/10.1038/srep35819

eo49 Climate Impact on Lyme Borreliosis and Its Causative Agents

Natalie Rudenko[1]*, Libor Grubhoffer[1,2] and Maryna Golovchenko[1]
[1]*Czech Academy of Sciences, České Budějovice, Czech Republic;*
[2]*University of South Bohemia, České Budějovice, Czech Republic*

49.1 Expert Opinion

49.1.1 Introduction

Lyme borreliosis (LB) is a multisystem disorder with a diverse spectrum of clinical manifestations. It is one of the most frequently recorded tick-borne diseases in the northern hemisphere, caused by selected species of spirochaetes from the *Borrelia burgdorferi* sensu lato (s.l.) complex. The complex includes 22 named species recognized internationally and countless numbers of unnamed species and strains that are not fully characterized. An accurate estimation of the importance of LB to human and animal health has not been made because of multiple reasons, including significant complications in the disease diagnosis and inadequate surveillance activities (Lindgren and Jaenson, 2006). For several years, available epidemiological data and rather conservative estimates kept the official numbers of diagnosed cases at about 85,000 in Europe and approximately 20,000 in the USA. The problem of under-reporting of the real incidence of LB was acknowledged by many experts and was justified by differences in diagnostics of LB cases and diagnostic pitfalls, the lack of epidemiological data from states where LB is not a notifiable disease and the lack of awareness and failure of notification systems. Taking into consideration all these limitations, the total annual number of LB cases was estimated at as many as 255,000 worldwide (Hubálek, 2009).

Year 2013 brought a shocking announcement and became a turning point in the history of this controversial disease. Recognition of the impact of LB on public health worldwide came when, at the International Conference on Lyme Borreliosis in Boston, Massachusetts, the Centers for Disease Control and Prevention (CDC, USA), a major world-leading epidemiological centre, released the statement that about 329,000 new LB cases were diagnosed annually in the USA between 2005 and 2010, ten times more than the CDC had previously reported (Kuehn, 2013; Nelson *et al.*, 2015). According to the most recently released estimation, approximately 476,000 LB cases were diagnosed and treated in the USA annually during 2010–2018 (Kugeler *et al.*, 2021). The increase of LB cases in Europe was recognized in a resolution of the European Parliament on Lyme disease (borreliosis) (2018/2774(RSP)) as well, estimating approximately 650,000–850,000 LB cases every year (European Parliament, 2018).

LB is increasing in incidence and spreading geospatially. Recent dramatic increases in diagnosed LB cases might be due to several factors: changes in the distributional range of tick

*Email: natasharudenko@hotmail.com

DOI: 10.1079/9781789249637.0049

vectors; dispersal of infected ticks due to host animal migration; recent urbanization and an increasing overlap of human and *Borrelia* habitats; and, possibly, involvement of other blood-sucking arthropod vectors or other routes of transmission of the causative agent, for example human-to-human transmission. Additionally, increased numbers of LB cases may be the result of gene transfer between strains that lead to the appearance of recombinant genotypes new to a region. The spreading or exchange of highly pathogenic spirochaete clones between the continents might be supported by transoceanic migration, intensive colonization of new areas and establishment of new enzootic LB foci. It is now apparent that, in addition to human-related factors that affect in part the epidemiological statistics of LB, there are many natural factors that influence the transmission, incidence and worldwide distribution of LB. Climate changes that are anthropogenic in their origin are among these key drivers that might play an important role in increasing the potential risk of LB.

LB is a curious natural event, the existence of which has a few prerequisites: pathogen, competent vectors and an array of reservoir hosts. The causative agent of LB is the host-propagated bacteria that shuttle between a vertebrate host and a tick vector; hence, it cannot be affected directly by climate changes. However, the impact of climate changes on LB could be unexpectedly significant through the influence that climate has on tick vector abundance and distribution, infection prevalence in vectors and transmission of pathogens, vertebrate host population distribution, land cover and human activities that can alter the functions of ecosystems (Li *et al.*, 2019).

Model-based studies on risk prediction of LB under different climate change scenarios remain low in number. Agent-based modelling of LB dynamics, the LYMERISK model, includes four parameters that were defined based on data availability, literature-driven evidence and previous analyses of model sensitivity: they are temperature, forest land cover, distribution of deer population and pathogen transmission hosts (Li *et al.*, 2016). These four parameters were selected as key drivers of future LB risk because: (i) temperature has a big impact on activity, diapause, interstadial development, fertility and mortality of ticks, as well as on habitat suitability for their hosts; (ii) forest land cover serves as the major habitat type for both ticks and host animals; (iii) deer are considered the most important tick reproduction host, making the distribution of the deer population an important consideration; and (iv) pathogen transmission hosts can harbour the pathogenic agents of LB and transmit them to ticks. Earlier conclusions that climate warming will increase LB risk were based on climate-driven models in which the drivers like land cover or host population distribution were considered minor. Disputes on whether climate change and temperature increases may or may not lead to amplification of LB risk in the future have received much attention recently. To answer this question, LB risk must be linked to the bigger coherence of global environmental changes.

49.1.2 Evidence for climate change impacts

Transmission of spirochaetes of the *B. burgdorferi* s.l. complex between vertebrate hosts is the result of a complex and multifactorial system. Hard ticks are the main recognized arthropod vectors of the causative agent of LB. However, LB spirochaetes have been detected in *Aedes* and *Culex* mosquitoes, with evidence of transstadial and/or transovarial transmission, although as yet no records of spirochaete transmission to vertebrates (Melaun *et al.*, 2016). The life cycles of four tick vector species (*Ixodes ricinus*, *Ixodes persulcatus*, *Ixodes scapularis* and *Ixodes pacificus*), all belonging to the *I. ricinus* complex and of medical importance to humans, are strongly dependent on weather, climate and abundance of a variety of vertebrate hosts. Tick activities depend on temperatures being above a certain minimal threshold. To be able to survive in nature, the tick vectors need a temperature above 7°C and humidity over 85%. Weather defines a condition of the atmosphere during a short period of time, from a couple of weeks to a couple of months. Increasing seasonal temperature and/or lack of precipitation that are not beneficial for ticks will drive them to stay closer to the ground and to change their host-seeking behaviour. Ticks will adjust to hot and dry weather by switching to alternative, ground-foraging hosts if the changes are short. In contrast to weather, climate change refers to long-term atmospheric 'performance' over a relatively long time period. Climate changes likely cause long-term effects on vector-borne disease transmission and infection patterns, affecting both seasonal risk and broad geographic changes in disease occurrence over decades. These changes can affect tick vectors by expanding or contracting their habitats, affecting their survival rates and shifting their strategies for finding hosts.

The role of climate in maintenance of vectors and pathogens in nature is well documented and redistribution of many vector-borne diseases due to the present global warming was predicted (Brownstein *et al.*, 2005). Survival of ixodid ticks is closely linked to the abiotic environment. Because 98% of the tick life cycle takes place off the host, climate plays a vital role in maintenance of tick populations in nature and affects vector geographical redistribution by two major trends – expansion and contraction of suitable habitats. Increasing minimal seasonal temperature leads to tick expansion into higher altitudes (Danielová *et al.*, 2010) and north where the temperatures were traditionally lower than now. Such spatial expansion is shown in Europe, with *I. ricinus* now found on the west coast of Norway. Similarly, in North America, the northward progression of *I. scapularis* has led to the establishment of dense tick populations in large areas of southern Canada (Fernández-Ruiz and Estrada-Peña, 2020). Such expansion could be the result of regular host migration or, most probably, due to the migration of bird hosts, which have a range of migration significantly exceeding that of terrestrial vertebrates. By contrast, increasing maximum temperatures result in the establishment of unfavourable conditions for off-host tick populations that will lead to retraction of vectors from regions with high temperature and low humidity and a decrease in their survival rate (Kilpatrick and Randolph, 2012).

The effect of climate change on LB, from the perspective of tick vectors, will lead to a shorter developmental cycle, an increase in egg production, enlargement of tick population density and expansion of distribution. Global warming will negate the need for winter hibernation of ticks, triggering their earlier activity in spring or even in the warmer winter days. On the other hand, high temperature makes the cuticle of arthropods more permeable to weather, reducing their resistance to desiccation, which may result in elimination of vector populations in some localities with increased temperature or could potentially disrupt pathogen transmission. A key question is, to what extent could climate warming affect the intensity of *Borrelia* transmission in regions where LB spirochaete populations are already well established? Climate change might impact the intensity of enzootic spirochaete transmission via effects on the life stage-specific phenologies of infecting nymphs and infection-receiving tick larvae. The efficiency of transmission cycles increases when infected nymphs feed before larvae, enabling infection to establish in the hosts before the time of larval feeding. The earlier in the year nymphs have their blood meal, due to the mild winter temperature, the larger the gap between the peaks of nymphal and larval activities. This could reduce the efficiency of transmission due to possible mortality of older populations of infected hosts before the larval peak occurs. Another reducing factor might be greater involvement of non-infected juvenile hosts that contribute to larval blood-feeding after the peak of nymphal activity has passed (Eisen *et al.*, 2016).

The impact of climate change on LB depends as well on the impact of climate changes on the range of reservoir hosts of *Borrelia* that defines the local prevalence of *Borrelia* spp. in ticks. As spirochaetes from *B. burgdorferi* s.l. complex represent multi-host pathogens, their vertebrate host specialization becomes an issue of public health importance. Further expansion of host-specialized *Borrelia* spp. (e.g. *Borrelia afzelii* and *Borrelia garinii*) in different geographic regions results in geographical structuring of this bacterium dependent on host distribution. In contrast, spirochaete species that lack host specialization and are able to infect phylogenetically distant hosts, in some cases covering as many as three phylogenetic orders (e.g. *B. burgdorferi* sensu stricto and *Borrelia bissettii*), generate a spatially uniform population structure (Hanincova *et al.*, 2006). Established foci of pathogens should possess an appropriate environment for vector survival, an adequate density of reservoirs and a temporal and spatial overlap between them. When the stability of this connection is endangered at any point, the persistence of the foci in nature is at risk. LB risk is closely associated with fluctuations in the reservoir host population. Landscape modification, vegetation management and deforestation in urban and suburban areas negatively affect tick density due to establishment of hostile environments for their hosts. Changes in land use affect the abundance and interactions of wildlife and domestic hosts, affecting in this way tick-transmitted pathogens. The presence of host species that play key roles in feeding adult ticks, but that are not competent reservoirs for *Borrelia*, is profitable for the support of dense tick populations but will negatively affect the local prevalence of the pathogen. At the same time, deforestation and hunting might alter the host community for ticks and spirochaetes by changing the predator communities, increasing the relative abundance of small vertebrates, mostly rodents, that are principal transmission hosts for *Borrelia* species. Tick density is dependent on vertebrate hosts; however, recent evidence shows that vertebrate population fluctuations are more affected by land use, wildlife management and societal factors than by climate-associated drivers (Kilpatrick and Randolph, 2012).

If we can overlook the recently popular and widely accepted idea that climate change impacts everything, including LB and its causative agents, withstanding the pressure of model studies that often lack important parameters, we may realize there is not much that is definite and predictable. Even though climate change affects LB in some way, the results of its influence are variable, and directions are opposing. For example, as mentioned above, despite increasing the suitable areas of tick distribution, the rise in temperature might, at the same time, cause decreasing tick survival. If climate changes cause the shift in geographic distribution of tick vectors (by migratory birds), this impact most probably will not affect the distribution of key species which have a smaller range of movement (rodents or deer). If ticks disappear from localities where climate changes cause unfavourable conditions, is it possible that other blood-sucking arthropods less sensitive to the climate changes will take over the role of vectors of *Borrelia*, feeding on already infected hosts that are physically unable to leave the area? Is it possible that areas previously known as endemic for LB will remain endemic without the ticks' presence, while alternative vectors and infected hosts resilient to climate changes will be present there? If in addition to the recognized vector transmission route of *Borrelia*, the congenital and sexual human-to-human transmission route of LB spirochaetes exists (as in the case of its close relative, *Treponema pallidum*), the impact of climate changes on LB in this case will be negligible. Too many 'ifs' to answer. Human-triggered land changes, deforestation and wildlife management as well as other social factors have a crucial impact on climate. Most probably in the future LB and its causative agents will be more affected by human-associated factors (social, economic, political) than by climate changes.

49.2 Future Predictions

Even though 'global warming' and 'climate change' are often used as synonyms, 'climate change' defines the complex changes in the planet's weather and climate system, including the rise of average temperature, glaciers melting and sea-level rising, shifting wildlife populations and habitats, as well as other impacts that appear as the result of human activities. The connection between people, animals, plants and the environment we all share, all that represents the 'One Health' concept, is actually endangered by the development of humankind that leads to continuous adding of heat-trapping greenhouse gases to the atmosphere. Other natural factors like solar radiation or volcanic eruption affect the climate too, leading to the rise in temperature on the Earth and contributing about 2% to the global warming effect. The rest is coming from human-triggered factors like changes in land use or greenhouse gases, particularly carbon dioxide and methane. Global warming might cause irreversible changes; if greenhouse gas emissions are not reduced to keep the warming below 2°C, effects on the planet will be catastrophic (Ogden *et al.*, 2020).

Transmission of tick-borne diseases in nature is associated with a broad range of factors that include not only changes of temperature, but density of ticks, hosts' populations and habitats as well. While the warming climate might increase tick activity and pathogen transmission, the major regulatory role will be played by socio-economic factors that restrict the habitats by changes in land use and wildlife management.

The authors of the LYMERISK model study came to the conclusion that 'our projections suggest a highly uncertain spatiotemporal pattern for LB risk due to large projected variations in land use under the different socioeconomic scenarios' (Li *et al.*, 2019). Taking into consideration potential impacts of climate change and socio-economic factors, most probably the highest increase of LB risk might not necessarily always emerge due to climate change. However, aside from the models used for prediction of LB risk (agent-based, climate-based or environmental models), the impact of climate change on public health is indisputable.

References

Brownstein, J.S., Holford, T.R. and Fish, D. (2005) Effect of climate change on Lyme disease risk in North America. *EcoHealth* 2, 38–46.

Danielová, V., Daniel, M., Schwarzova, L., Materna, J., Rudenko, N., *et al.* (2010) Integration of a tick-borne encephalitis virus and *Borrelia burgdorferi* sensu lato into mountain ecosystems, following a shift in the altitudinal limit of distribution of their vector, *Ixodes ricinus* (Krkonoše Mountains, Czech Republic). *Vector-Borne and Zoonotic Diseases* 10, 223–230.

Eisen, R., Eisen, L., Ogden, N.H. and Beard, C.B. (2016) Linkages of weather and climate with *Ixodes scapularis* and *Ixodes pacificus* (Acari: Ixodidae), enzootic transmission of *Borrelia burgdorferi*, and Lyme disease in North America. *Journal of Medical Entomology* 53, 250–261.

European Parliament (2018) Procedure: 2018/2774(RSP). Available at: https://www.europarl.europa.eu/doceo/document/B-8-2018-0514_EN.html (accessed 14 June 2021).

Fernández-Ruiz, N. and Estrada-Peña, A. (2020) Could climate trends disrupt the contact rates between *Ixodes ricinus* (Acari, Ixodidae) and the reservoirs of *Borrelia burgdorferi* s.l.? *PLoS ONE* 15, e0233771.

Hanincova, K., Kurtenbach, K., Diuk-Wasser, M., Brei, B. and Fish, D. (2006) Epidemic spread of Lyme borreliosis, northeastern United States. *Emerging Infectious Diseases* 12, 604–611.

Hubálek, Z. (2009) Epidemiology of Lyme borreliosis. In: Lipsker, D. and Jaulhac, B. (eds) *Lyme Borreliosis. Current Problems in Dermatology*. Karger, Basel, Switzerland, pp. 31–50.

Kilpatrick, M. and Randolph, S.E. (2012) Drivers, dynamics, and control of emerging vector-borne zoonotic diseases. *The Lancet* 380, 1946–1955.

Kuehn, M.B. (2013) CDC estimates 300,000 US cases of Lyme disease annually. *Journal of the American Medical Association* 310, 1110.

Kugeler, K.J., Schwartz, A.M., Delorey, M.J., Mead, P.S. and Hinckley, A.F. (2021) Estimating the frequency of Lyme disease diagnoses, United States, 2010–2018. *Emerging Infectious Diseases* 27, 616–619.

Li, S., Gilbert, L., Harrison, P.A. and Rounsevell, M. (2016) Modelling the seasonality of Lyme disease risk and the potential impacts of a warming climate within the heterogeneous landscapes of Scotland. *Journal of the Royal Society, Interface* 13, 20160140.

Li, S., Gilbert, L., Vanwambeke, S.O., Bethan, J.Y., Purse, V. and Harrison, P.A. (2019) Lyme disease risks in Europe under multiple uncertain drivers of change. *Environmental Health Perspectives* 127, 067010.

Lindgren, E. and Jaenson, T.G.T. (2006) Lyme borreliosis in Europe: influences of climate and climate change, epidemiology and adaptation measures. In: Menne, B. and Ebi, K.L. (eds) *Climate Change and Adaptation Strategies for Human Health*. Steinkopff, Darmstadt, Germany, pp. 157–188.

Melaun, C., Zotzmann, S., Santaella, V.G., Werblow, A., Zumkowski-Xylander, H., *et al.* (2016) Occurrence of *Borrelia burgdorferi* s.l. in different genera of mosquitoes (Culicidae) in Central Europe. *Ticks and Tick-borne Diseases* 7, 256–263.

Nelson, C.A., Saha, S., Kugeler, K.J., Delorey, M.J., Shankar, M.B., *et al.* (2015) Incidence of clinician-diagnosed Lyme disease, United States, 2005–2010. *Emerging Infectious Diseases* 21, 1625–1631.

Ogden, N.H., Beard, C.B., Ginsberg, H.S. and Tsao, J.I. (2020) Possible effects of climate change on ixodid ticks and the pathogens they transmit: predictions and observations. *Journal of Medical Entomology* 2020, tjaa220. https://doi.org/10.1093/jme/tjaa220

eo50 Climate Change and Tick-Borne Encephalitis in the Greater Alpine Region

Franz Rubel*
University of Veterinary Medicine, Vienna, Austria

50.1 Expert Opinion

50.1.1 Introduction

Tick-borne encephalitis (TBE) is a viral tick-borne disease. The distribution of human TBE cases ranges from the French departments bordering Germany through Central and Eastern Europe, the Caucasus and Kazakhstan to the Far East of Russia and China. Here, TBE is described in the greater area of the European Alps, denoted as the Greater Alpine Region (GAR).

TBE vectors in the GAR are predominantly ticks of the genus *Ixodes*, especially *Ixodes ricinus*, the castor bean tick (Kahl *et al.*, 2020). However, the experimentally proven TBE vector *Dermacentor reticulatus* (Rubel *et al.*, 2020a) is also frequently found in the GAR. The latest TBE virus findings in Germany, with similar minimum infection rates in *D. reticulatus* and *I. ricinus*, indicate that *D. reticulatus* plays the same role in virus circulation as *I. ricinus* when both tick species are sympatric (Chitimia-Dobler *et al.*, 2019). A Central European map showing the distribution of TBE virus in ticks and endothermic mammals was compiled by Walter *et al.* (2020).

TBE virus is a flavivirus persisting in a natural transmission cycle between small mammals and ticks. Humans can be infected, but they are ecologically dead-end hosts (Kahl *et al.*, 2020). Since TBE can be a serious disease in humans (Ruzek *et al.*, 2019), it is notifiable in almost all endemic TBE areas. This applies in particular to the countries discussed here, Austria, Germany and Switzerland, for which reconstructed time series of the population's vaccination coverage are also available (Rubel and Brugger, 2020).

The effects of vaccination coverage and climate change on the TBE case numbers in the GAR are of the same order of magnitude, but they are opposite. Vaccination coverage reduces the national TBE case numbers, while global warming leads to an increasing number of cases, particularly in the GAR. Only if one can separate these two effects is it possible to objectively estimate the impact of climate change.

50.1.2 Reported tick-borne encephalitis cases

The accumulated TBE cases officially reported in Austria, Germany and Switzerland are shown in Fig. 50.1. A distinction is made between reported TBE cases of the partially vaccinated population (red bars) and hypothetical TBE cases, as would be expected in a completely

*Email: franz.rubel@vetmeduni.ac.at

DOI: 10.1079/9781789249637.0050

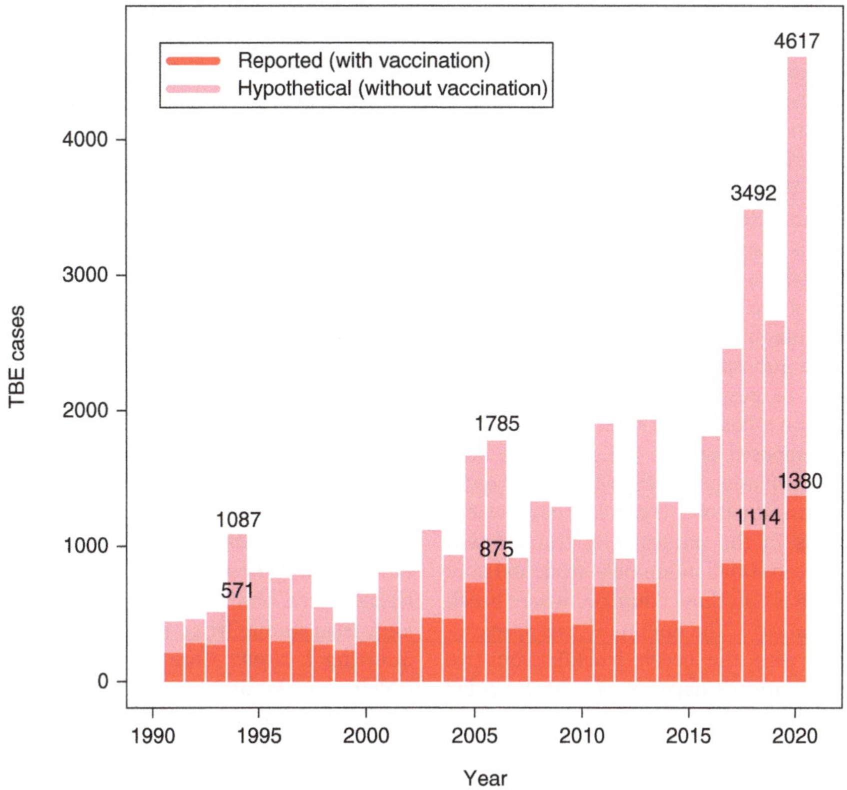

Fig. 50.1. Tick-borne encephalitis (TBE) in the Greater Alpine Region (GAR), 1991–2020. Accumulated TBE cases of the partially vaccinated residents in Austria, Germany and Switzerland as reported (red bars) and hypothetically assumed without vaccination (pink bars). TBE case numbers for the COVID-19 year of 2020 indicate a new high. (Author's own figure.)

susceptible population without any vaccination against TBE virus (pink bars). Details on the calculation of the hypothetical TBE case numbers can be found in Rubel *et al.* (2020b) and the data used (supplemented here by the years 2019 and 2020) were documented by Rubel and Brugger (2020). Exceptionally high TBE case numbers are presented for the COVID-19 year of 2020. It can be seen that despite the availability of efficient vaccines (Heinz *et al.*, 2013), TBE cases in the GAR have risen sharply in recent decades. This applies to the real TBE cases, but especially to the hypothetical TBE cases without vaccination against TBE virus.

The TBE time series begins in 1991 with a minimum of 209 TBE cases, including 128 in Austria, 44 in Germany and 37 in Switzerland. In 2020, historical maximum values of 707 cases in Germany and 454 cases in Switzerland were registered. In Austria, 219 cases were the highest reported since 1994, although 82% of the population is vaccinated against TBE virus (Kunze and Böhm, 2015). This means that a total of 1380 cases were reported in the three countries in 2020. Without vaccination, however, probably a total of 4617 TBE cases would have occurred in 2020. These would consist of 1217 Austrian, 1779 German and 1621 Swiss cases (Fig. 50.1). Similar increases in TBE cases were also recorded in the other countries of the GAR, i.e. France, Italy and Slovenia. However, the overall number of TBE cases in these countries is much lower as only a few provinces in eastern France and a few in northern Italy are affected. No data on the vaccination coverage of the population are available for these countries, which is why they were not included in Fig. 50.1.

In addition to the trend, three TBE maxima can be seen in 1994, 2006 and 2020. They indicate that the trend is overlaid by low-frequency oscillations with a period of 12–14 years. Spectral analysis of TBE time series from Austria, Germany (the federal states of Bavaria and Baden-Wuerttemberg), Slovenia, the Czech Republic and Switzerland revealed that this oscillation is synchronized over Central Europe with a periodicity of 10–16 years, depending on the length of the time series considered (Zeman, 2017a). This is due to fluctuations in the large-scale atmospheric circulation patterns, which can be best described by the so-called Scandinavian index (Rubel *et al.*, 2020b).

Superimposed on the trend and the low-frequency oscillations are high-frequency oscillations with a period of 2–3 years. They represent fluctuations in the natural TBE transmission cycle between small mammals (rodents) and ticks, which is – in those countries where beech (*Fagus sylvatica*) is the dominant tree species – associated with the beech fructification two years previously (Brugger *et al.*, 2018; Rubel *et al.*, 2020b).

50.1.3 Evidence for climate change impacts

The observed increase in the number of TBE cases in the GAR is due to three effects of climate change:

1. Increasing vector densities in TBE-endemic areas. With the exception of the high alpine regions, the Central European countries Austria, Switzerland and Slovenia are predominantly covered with deciduous forest, of which beech is the main tree species. This also applies to TBE-endemic areas in the east of France, in the south and west of Germany and in the north of Italy, where the distribution limit of the beech tree (Skjøth *et al.*, 2008) corresponds almost exactly to the TBE-endemic regions (Walter *et al.*, 2020). In addition, there are many high-performance industrial spruce forests in these countries, which are currently being replaced by deciduous or mixed forests in the course of climate adaptation strategies. A landscape-scale study in the German federal state of Baden-Wuerttemberg indicates that the density of host-seeking *I. ricinus* nymphs is significantly higher in deciduous forests than in coniferous forests (Boehnke *et al.*, 2015). This means that the fraction of deciduous trees, and with them the regions with high *I. ricinus* density, will successively increase in the coming decades. This process is already under way.

2. Extension of TBE-endemic areas. It is currently observed, especially in the Western Alps, i.e. in Switzerland (Desgrandchamps and Posfay-Barbe, 2020) and France (Hansmann and Velay, 2020), that TBE-endemic areas are expanding. Global warming is responsible for this, which has led to a significant increase in air temperature in the mountains. This shifts the forest boundaries (especially the deciduous and mixed forest boundaries important in this context) upwards, so that both TBE hosts and TBE vectors find better living conditions. The advance of *I. ricinus* into higher altitudes has already been observed. Although there are no long-term studies from the GAR in this regard, questing *I. ricinus* were recently found in the Italian Alps by dragging the vegetation up to 1884 m above sea level (Garcia-Vozmediano *et al.*, 2020). In Austria, alimentary transmission of TBE virus through the consumption of cheese made from unpasteurized milk from an infected goat has been described in an alpine pasture at an altitude of 1564 m (Holzmann *et al.*, 2009).

3. More frequent years with favourable conditions for the TBE virus transmission cycle. As a result of global warming, hot, dry summers are more common. Forest trees react to this with increased seed formation. This leads to an increased frequency of years during which the forest trees develop many seeds (e.g. beechnuts). A good supply of beechnuts means that small mammals (rodents) reproduce strongly the following year and, in turn, offer unfed tick larvae greater chances to find a host. Two years after a year of mast seeding, the number of *I. ricinus* nymphs (Brugger *et al.*, 2018) and with them the number of TBE cases (Rubel *et al.*, 2020b) increases. This explains the more frequent years with a high number of TBE cases after 2005 (Fig. 50.1).

Various socio-economic factors also play a role, only partially related to climate change. For example, up to four times as many people were counted in the recreational areas around Vienna,

Austria, during the first COVID-19 lockdown in spring 2020 than in normal years. This phenomenon is also known from other large cities in TBE-endemic areas and has certainly contributed to the strong increase in TBE case numbers in 2020 (Fig. 50.1).

50.2 Future Predictions

50.2.1 Short-range tick-borne encephalitis forecast (the next 2 years)

Forecasts of TBE case numbers for Austria, Germany and Switzerland were recently presented by Rubel and Brugger (2020). Therefore, the TBE time series were subjected to spectral analysis (Rubel *et al.*, 2020b). The resulting power spectra indicate which predictors are responsible for the trend, the low-frequency and the high-frequency oscillations. Increasing trend can be mainly associated with the demography of the increasing human population and their vaccination coverage. The low-frequency oscillations (10 years) are associated with the decadal changes of the large-scale climate in Central Europe reinforced by the socio-economic predictor net migration. The high-frequency oscillations (2–3 years) are associated with fluctuations of the natural TBE transmission cycle between small mammals and ticks, which are driven by beech fructification. As described in detail by Rubel *et al.* (2020b), peaks in the TBE incidence series are mainly caused by high beech fructification indices (mast seeding) two years previously. This means that a 2-year forecast can be made by applying generalized linear models (GLMs), which explained 80–90% of the variances in the national TBE time series. A limiting factor of the forecasts of the next 2 years' TBE incidences is the real-time availability of TBE case numbers, demographic data and the beech fructification index. These predictors are usually updated in the spring of the following year, which is why the forecasts can only be made afterwards. Although the results of the first TBE forecasts for 2019 and 2020 are encouraging (Rubel and Brugger, 2021), it is still too early for a final verification of these short-range TBE forecasts.

50.2.2 Medium-range tick-borne encephalitis forecast (the next 10 years)

Zeman (2017a) also analysed the observed TBE oscillations. However, he did not use GLMs for TBE forecasts, but rather fitted harmonic regression models to the TBE time series of Austria, the Czech Republic, the German federal state of Bavaria and Switzerland. He demonstrated that four superimposed oscillations, with periods between 2 and 10 years, can account for 55% of the variation in TBE case numbers. These TBE oscillations are relatively stable and synchronous over large areas, which makes them well suited for a forecast. Using harmonic regression models, TBE forecasts up to 10 years ahead were compiled by Zeman (2017b). Predictors such as climate parameters, beech fructification indices or host densities are not needed for this type of model. A disadvantage of this modelling approach is that trends have to be specified externally and changes in the frequency of TBE oscillations due to climate change are not taken into consideration. However, the two disadvantages mentioned can be accounted for in extended harmonic regression models.

50.2.3 Long-range tick-borne encephalitis forecast (the next 80 years)

So far only one study concerning long-range TBE forecasts has been published (Randolph and Rogers, 2000). It is based on statistical methods that are today summarized in species distribution models. That study used statistical associations between TBE cases and climate parameters from the period 1960–1990 to represent a TBE map. Future TBE maps for 2020, 2050 and 2080 were compiled with the help of the Hadley Centre for Climate Prediction and Research's 'medium-high' scenarios (HadCM2). The HadCM2 climate projections from the 1990s show warmer, drier summers for 2020 as they are actually observed today. In contrast to the correct climate prediction, the future TBE scenarios based on it did not occur. Randolph and Rogers (2000) predicted that, in the 2020s, France, Switzerland, Croatia, Slovenia, Hungary and much of Austria appear to be cleared of TBE virus.

Today, we know that not only did this not happen, but also that TBE cases in these countries have actually risen sharply. Exceptions are Slovenia (Simonović and Vuković-Janković, 2020) and Hungary (Lakos *et al.*, 2020), where extensive vaccination campaigns are responsible for the decline in TBE case numbers. This example shows that the wrong interpretation of statistical relationships can lead to significant misjudgements.

If one takes account of the above-mentioned relationships between climate change, climate adaptation strategies and TBE, it can be assumed that TBE case numbers in the GAR will remain high over the next two decades. However, a look into the more distant future seems difficult. One way is to look at the future expansion of the climatic zone responsible for the spread of beech trees as the main deciduous species. As mentioned above, there is a good correlation between the areas with beech as the most common tree species and TBE-endemic areas in the GAR. The so-called 'beech climate (Cfb)' was defined by the German-Russian climatologist Wladimir Köppen at the beginning of the 20th century as a warm temperate climate with precipitation in all seasons and warm summers. It is currently replacing the boreal coniferous forest climates in the Alpine region (Rubel *et al.*, 2017). In the scenario with the lowest global warming, the Cfb climate will occupy large areas in the GAR by 2100 and ensure sustained high TBE case numbers. In the worst-case scenario with the maximum possible global warming, the Cfb climate will be replaced by the Cfa climate with its hot summers (Rubel *et al.*, 2017), which could lead to a decrease in TBE case numbers in the GAR from the middle of the century.

References

Boehnke, D., Brugger, K., Pfäffle, M., Sebastian, P., Norra, S., *et al.* (2015) Estimating *Ixodes ricinus* densities on the landscape scale. *International Journal of Health Geography* 14, 23.

Brugger, K., Walter, M., Chitimia-Dobler, L., Dobler, G. and Rubel, F. (2018) Forecasting next season's *Ixodes ricinus* nymphal density: the example of southern Germany 2018. *Experimental and Applied Acarology* 75, 281–288.

Chitimia-Dobler, L., Lemhöfer, G., Król, N., Bestehorn, M., Dobler, G. and Pfeffer, M. (2019) Repeated isolation of tick-borne encephalitis virus from adult *Dermacentor reticulatus* ticks in an endemic area in Germany. *Parasites & Vectors* 12, 90.

Desgrandchamps, D. and Posfay-Barbe, M.K. (2020) TBE in Switzerland and Liechtenstein. In: Dobler, G., Erber, W., Bröker, M. and Schmitt, H.J. (eds) *The TBE Book*, 3rd edn. Global Health Press, Singapore, pp. 338–343.

Garcia-Vozmediano, A., Krawczyk, A.I., Sprong, H., Rossi, L., Ramassa, E. and Tomassone, L. (2020) Ticks climb the mountains: ixodid tick infestation and infection by tick-borne pathogens in the Western Alps. *Ticks and Tick-borne Diseases* 11, 101489.

Hansmann, Y. and Velay, A. (2020) TBE in France. In: Dobler, G., Erber, W., Bröker, M. and Schmitt, H.J. (eds) *The TBE Book*, 3rd edn. Global Health Press, Singapore, pp. 215–219.

Heinz, F.X., Stiasny, K., Holzmann, H., Grgic-Vitek, M., Kriz, B., *et al.* (2013) Vaccination and tick-borne encephalitis, Central Europe. *Emerging Infectious Diseases* 19, 69–76.

Holzmann, H., Aberle, S.W., Stiasny, K., Werner, P., Mischak, A., *et al.* (2009) Tick-borne encephalitis from eating goat cheese in a mountain region of Austria. *Emerging Infectious Diseases* 15, 1671–1672.

Kahl, O., Pogodina, V.V., Poponnikova, T., Süss, J. and Zlobin, V.I. (2020) A short history of TBE. In: Dobler, G., Erber, W., Bröker, M. and Schmitt, H.J. (eds) *The TBE Book*, 3rd edn. Global Health Press, Singapore, pp. 10–15.

Kunze, U. and Böhm, G. (2015) Tick-borne encephalitis (TBE) and TBE vaccination in Austria: update 2014. *Wiener Medizinische Wochenschrift* 165, 290–295 (in German).

Lakos, A., Schneider, F., Nagy, A. and Mezei, E. (2020) TBE in Hungary. In: Dobler, G., Erber, W., Bröker, M. and Schmitt, H.J. (eds) *The TBE Book*, 3rd edn. Global Health Press, Singapore, pp. 225–230.

Randolph, S.E. and Rogers, D.J. (2000) Fragile transmission cycles of tick-borne encephalitis virus may be disrupted by predicted climate change. *Philosophical Transactions of the Royal Society of London B* 267, 1741–1744.

Rubel, F. and Brugger, K. (2020) Tick-borne encephalitis incidence forecasts for Austria, Germany, and Switzerland. *Ticks and Tick-borne Diseases* 11, 101437.
Rubel, F. and Brugger, K. (2021) Operational TBE incidence forecasts for Austria, Germany, and Switzerland 2019–2021. *Ticks and Tick-borne Diseases* 12, 101579.
Rubel, F., Brugger, K., Haslinger, K. and Auer, I. (2017) The climate of the European Alps: shift of very high resolution Köppen–Geiger climate zones 1800–2100. *Meteorologische Zeitschrift* 26, 115–125.
Rubel, F., Brugger, K., Belova, O.A., Kholodilov, I.S., Didyk, Y.M., *et al.* (2020a) Vectors of disease at the northern distribution limit of the genus *Dermacentor* in Eurasia: *D. reticulatus* and *D. silvarum*. *Experimental and Applied Acarology* 82, 95–123.
Rubel, F., Walter, M., Vogelgesang, J.R. and Brugger, K. (2020b) Tick-borne encephalitis (TBE) cases are not random: explaining trend, low- and high-frequency oscillations based on the Austrian TBE time series. *BMC Infectious Diseases* 20, 448.
Ruzek, D., Županc, T.A., Borde, J.P., Chrdle, A., Eyer, L., *et al.* (2019) Tick-borne encephalitis in Europe and Russia: review of pathogenesis, clinical features, therapy, and vaccines. *Antiviral Research* 164, 23–51.
Simonović, Z. and Vuković-Janković, T. (2020) TBE in Slovenia. In: Dobler, G., Erber, W., Bröker, M. and Schmitt, H.J. (eds) *The TBE Book*, 3rd edn. Global Health Press, Singapore, pp. 321–327.
Skjøth, C.A., Geels, C., Hvidberg, M., Hertel, O., Brandt, J., *et al.* (2008) An inventory of tree species in Europe – an essential data input for air pollution modelling. *Ecological Modelling* 217, 292–304.
Walter, M., Vogelgesang, J.R., Rubel, F. and Brugger, K. (2020) Tick-borne encephalitis virus and its European distribution in ticks and endothermic mammals. *Microorganisms* 8, 1065.
Zeman, P. (2017a) Cyclic patterns in the central European tick-borne encephalitis incidence series. *Epidemiology and Infection* 145, 358–367.
Zeman, P. (2017b) Predictability of tick-borne encephalitis fluctuations. *Epidemiology and Infection* 145, 2781–2786.

eo51 The Expansion of Japanese Spotted Fever and the Complex Group of Spotted Fever Group Rickettsiae in Japan

Megumi Sato[1]*, Reiko Arai[2] and Marcello Otake Sato[3]
[1]Niigata University, Niigata City, Japan; [2]Niigata Prefectural Institute of Public Health and Environmental Sciences, Niigata City, Japan; [3]Dokkyo Medical University, Mibu, Japan

51.1 Expert Opinion

51.1.1 Japan and spotted fever group rickettsiae infections

Infection with spotted fever group rickettsiae (SFGR) was not documented in Japan until 1984 when the first cases were described from human patients (Mahara *et al.*, 1985; Uchida *et al.*, 1985). They presented symptoms after a history of tick bites in Tokushima and Kochi Prefectures, both on Shikoku Island located south of Honshu (the main island of Japan). After the first cases were described, the disease was reported in the southern part of Honshu and *Rickettsia japonica* was identified as the aetiological agent of the disease, named Japanese spotted fever (JSF) (Uchida *et al.*, 1992). However, despite a history of contact with ticks and clinical symptoms of JSF, confirmation of the aetiological agent was not possible in some patients. This led to the search in ticks for different agents that might be pathogenic for humans. Nowadays, 11 types of rickettsial agent comprise the SFGR group in Japan (Table 51.1), mainly detected from ticks and putatively causing disease in humans (Ando and Fujita, 2013).

51.1.2 Diagnosis: strategy and limitations

For diagnosis of rickettsial diseases, a procedures manual was released by the National Institute of Infectious Disease of Japan for use by local governmental institutes for health. The usual diagnosis is by serology (indirect immunofluorescence or indirect immunoperoxidase assay) and by direct techniques such as PCR and DNA sequencing (Suto, 1985; Fournier *et al.*, 2002; Gaowa *et al.*, 2013). Reference laboratories are located in each prefectural area and it is possible to consult if there are any enquiries about the tests. If one patient is suspected of having a rickettsial disease by serology, the diagnosis is confirmed by additional testing for detection and identification of the aetiological agent. For example, in 2008, a patient with Far Eastern spotted fever caused by *Rickettsia heilongjiangiensis* was first identified in Japan and the natural cycle confirmed with identification of its vector, *Haemaphysalis concinna* (Ando *et al.*, 2010). Spotted fever cases caused by *Rickettsia helvetica* and *Rickettsia tamurae* also have been reported. Additionally, imported cases of other spotted fevers, such as African tick bite fever and Mediterranean

*Email: satomeg@clg.niigata-u.ac.jp

DOI: 10.1079/9781789249637.0051

Table 51.1. Ticks and spotted fever rickettsial agents isolated and/or detected in Japan. (Modified from Ando and Fujita, 2013. Authors' own table.)

Rickettsial agents	Tick species
Rickettsia asiatica	*Ixodes ovatus*
Rickettsia heilongjiangiensis (Far Eastern spotted fever)	*Haemaphysalis concinna*
Rickettsia helvetica	*Ixodes columnae*, *Ixodes persulcatus*, *Ixodes monospinosus*, *Ixodes turdus*, *Ixodes pavlovskyi*
Rickettsia honei-like	*Ixodes granulatus*
Rickettsia japonica (Japanese spotted fever)	*Dermacentor taiwanensis*, *Haemaphysalis hystricis*, *Haemaphysalis cornigera*, *Haemaphysalis longicornis*, *Haemaphysalis flava*, *Haemaphysalis formosensis*, *Haemaphysalis megaspinosa*, *I. ovatus*
Rickettsia monacensis	*Ixodes nipponensis*
Rickettsia tamurae	*Amblyomma testudinarium*, *H. megaspinosa*
Rickettsia sp. HKT-1	*Haemaphysalis kitaokai*
Rickettsia sp. IA/IN (= In56)	*Ixodes asanumai*, *I. nipponensis*
Rickettsia sp. IT	*I. turdus*
Rickettsia sp. LON	*H. longicornis*

spotted fever, have been reported in Japan (Ando and Fujita, 2013). According to several dermatologists in Niigata Prefecture, there are a few to a dozen cases of tick-bite patients annually. Some of the patients removed the ticks themselves and for others the ticks were removed by medical doctors. However, in almost all cases, tick identification was not done. Therefore, the relationship between tick-borne diseases (including rickettsia) and tick species is unclear in most clinical cases.

51.1.3 Japanese spotted fever

In Japan, JSF is a reportable disease (Infectious Diseases Category 4). From 1999 to 2007, <100 cases were reported annually; however, since 2017, JSF cases reached >300 annually, with 44 deaths caused by *R. japonica* between 1999 and 2019. Case numbers are increasing year by year with 318 cases reported in 2019 (National Institute of Infectious Diseases, 2020). Human cases of JSF have been reported mainly in southern and western mainland Japan; however, the case report area is expanding to east and northern regions of the country, with new endemic areas reported in Toyama and Niigata (located above latitude 37°N), areas previously considered free of JSF. The increase in human cases of JSF is probably caused by exposure of more people to ticks. The expanding distribution area and population of natural tick hosts, such as deer and wild pigs, to northern Japan (Sato *et al.*, 2021) expose people living in areas previously free of ticks to tick-borne diseases such as JSF. Several factors are related with the changing population distribution and abundance of wild animals in Japan; some are related to infrastructure improvements like roads facilitating the migration of wild animals (Wu *et al.*, 2011). Another plausible factor is climate change, where previously colder areas have become suitable for these animals to live and reproduce (Gitay *et al.*, 2002). Another factor probably related with the increased occurrence of tick-borne diseases is the number of animals in urban areas, due to an unbalanced augmentation of wild animal populations (Sato *et al.*, 2021).

Surveillance studies isolated *R. japonica* from a variety of tick species, with detection not restricted to warm-area ticks: *Haemaphysalis hystricis* (species found from Nansei Island to Shikoku), *Haemaphysalis formosensis* (common in Kyushu and Shikoku), *Haemaphysalis cornigera* (found in Boso Peninsula and Shimame Peninsula) and *Dermacentor taiwanensis* (from Nansei Island to Kanagawa Prefecture, on the Pacific side of Japan); and also from ticks having a countrywide distribution, *Haemaphysalis longicornis* and *Haemaphysalis flava*, with the two species harbouring *R. japonica* in Shikoku Prefecture and Shimane Peninsula (Fournier *et al.*, 2002; Gaowa *et al.*, 2013).

51.1.4 Rickettsiosis and tick species vectors in Japan

When SFGR positivity in adult ticks is calculated without discrimination for tick species, positivity rates vary by area, such as 5.6% in Niigata Prefecture (Arai *et al.*, 2021), 3.3% in Toyama Prefecture (Yamauchi *et al.*, 2009), 22.0% in Fukui Prefecture (Ishiguro *et al.*, 2008) and 21.6% in six western prefectures including Shizuoka (Gaowa *et al.*, 2013). However, despite these rates providing indications of pathogen exposure and infection risk, if tick species is not considered, the specific threat is unknown. For example, if SFGR positivity rate is calculated for a specific species such as *Ixodes monospinosus*, the rate becomes 50.0% in Toyama (Yamauchi *et al.*, 2009), 43.8% in Fukui (Ishiguro *et al.*, 2008) and 43.8% in Niigata (Arai *et al.*, 2021). These differences show the necessity for new, more detailed studies on tick diversity and rickettsial infection for the different ecotopes of Japan.

51.1.5 Recent findings in Niigata Prefecture, the border line of Japanese spotted fever endemicity in Japan coastal region

Several *Rickettsia* spp. have strong tick specificity, such as *R. heilongjiangiensis* (Far Eastern spotted fever) detected in only *H. concinna*, and *Rickettsia asiatica* only in *Ixodes ovatus* (Table 51.1). In a recent study in Niigata Prefecture, SFGR detection confirmed this apparent specificity: *R. asiatica* from *I. ovatus*, *R. helvetica* from *I. monospinosus* and *Rickettsia monacensis* from *Ixodes nipponensis* (Arai *et al.*, 2021).

By contrast, *R. japonica* has been isolated and/or detected in numerous ixodid species (Table 51.1) (Ando and Fujita, 2013).

Tick diversity and distribution data in Niigata Prefecture were updated after 60 years, showing a dramatic change from the 1950s. Only four species were collected from 1956 to 1957 in Niigata Prefecture – *H. flava*, *Ixodes persulcatus*, *I. ovatus* and *Haemaphysalis dieentzi* – in a total of 11,404 ticks from 38 field surveys in 15 locations (Saito, 1959). However, from 2016 to 2018, 12 tick species were identified: *D. taiwanensis*, *H. flava*, *Haemaphysalis histricis*, *Haemaphysalis japonica*, *H. longicornis*, *Haemaphysalis megaspinosa*, *I. ovatus*, *I. nipponensis*, *I. persulcatus*, *I. monospinosus*, *Ixodes columnae* and *Ixodes turdus* (Sato *et al.*, 2021). If one thinks about the likelihood of being in contact with ticks and the chance of rickettsial infection, the risk has increased remarkably, a warning signal of JSF and other SFGR in the region.

The observed tick species shift in Niigata Prefecture was probably caused by multifactorial environmental changes, both natural and due to human activity (e.g. construction of highways, rural exodus). Another major reason for tick species abundance variation might be climate change. Climate changes include alterations in climate variables and may impact the survival, reproduction and distribution of disease pathogens and hosts, as well as human exposure to environments in which transmission occurs (Wu *et al.*, 2016). Besides Niigata Prefecture, this phenomenon could be happening in several areas where ticks exist, with tick species' composition shifting and new endemic areas of different tick species and their pathogens arising.

51.2 Future Predictions

For Japan, mean annual temperature has increased by about 1.0°C over the last century (Cruz *et al.*, 2007) with more frequent hot days (days with maximum temperatures >35°C) and fewer extreme cold days (JMA, 2005). Temperature projections for Japan indicate annual mean temperatures expected to rise by 2 to 3°C within the next 100 years (MOE, 2006) while, for all of East Asia, the projected increase is 5°C by 2080 (Lal *et al.*, 2001; Alam *et al.*, 2007). Increased temperatures may facilitate the spread of vector-borne and water-borne diseases by a shift in the geographic distribution and range of marine and terrestrial species (WHO, 2002; MOE, 2006). For example, the expanding habitat of mosquitoes in Japan could bring dengue fever to Hokkaido in the near future (MOE, 2017).

It is difficult to predict what exactly may happen in the next years, but changes in the environment and tick-borne diseases are foreseeable and represent an emerging issue, especially in regions of Japan where JSF/SFGR previously was not endemic. The health system and surveillance should be aware and able to respond in a collaborative, multisectoral and transdisciplinary approach to tackle the identification of tick species occurring in an area, detection of pathogens and fast diagnosis for effective treatment of patients. This is particularly relevant for the revitalization of Satoyama landscapes (SGS, 2010) where people are returning to a more natural lifestyle with increased exposure to ticks and tick-borne pathogens.

References

Alam, M., Rahman, A., Rashid, M., Rabbani, G., Bhandary, P., *et al.* (2007) Impacts, vulnerability and adaptation to climate change in Asia. Paper commissioned by Secretariat of the United Nations Framework Convention on Climate Change, Beijing, China, 11–13 April 2007. Available at: http://unfccc.int/files/adaptation/methodologies_for/vulnerability_and_adaptation/application/pdf/unfccc_asian_workshop_background_paper.pdf (accessed 23 December 2020).

Ando, S. and Fujita, H. (2013) Diversity between spotted fever group rickettsia and ticks as vector. *Medical Entomology and Zoology* 64, 5–7. https://doi.org/10.7601/mez.64.5

Ando, S., Kurosawa, M., Sakata, A., Fujita, H. and Sakai, K. (2010) Human *Rickettsia heilongjiangensis* infection, Japan. *Emerging Infectious Diseases* 16, 1306–1308. https://doi.org/10.3201/eid1608.100049

Arai, R., Sato, M., Kato, M., Aoki, J., Nishida, A., *et al.* (2021) Spotted fever group rickettsiae (SFGR) detection in ticks following reported human case of Japanese spotted fever in Niigata Prefecture, Japan. *Scientific Reports* 11, 2595. https://doi.org/10.1038/s41598-021-81587-9

Cruz, R.V., Harasawa, H., Lal, M., Wu, S., Anokhin, Y., *et al.* (2007) Asia. In: Parry, M.L., Canziani, O.F., Palutikof, J.P., van der Linden, P.J. and Hanson, C.E. (eds) *Climate Change 2007: Impacts, Adaptation and Vulnerability. Contribution of Working Group II to the Fourth Assessment Report of the Intergovernmental Panel on Climate Change*. Cambridge University Press, Cambridge, pp. 469–506.

Fournier, P.E., Fujita, H., Takada, N. and Raoult, D. (2002) Genetic identification of rickettsiae isolated from ticks in Japan. *Journal of Clinical Microbiology* 40, 2176–2181. https://doi.org/10.1128/jcm.40.6.2176-2181.2002

Gaowa, Ohashi, N., Aochi, M., Wuritu, D., Wu, Y., *et al.* (2013) Rickettsiae in ticks, Japan, 2007–2011. *Emerging Infectious Diseases* 19, 338–340. https://doi.org/10.3201/eid1902.120856

Gitay, H., Suárez, A., Dokken, D.J. and Watson, R.T. (eds) (2002) *Climate Change and Biodiversity. IPCC Technical Paper V*. Intergovernmental Panel on Climate Change, Geneva, Switzerland. Available at: https://archive.ipcc.ch/pdf/technical-papers/climate-changes-biodiversity-en.pdf (accessed 23 December 2020).

Ishiguro, F., Takada, N., Fujita, H., Noji, Y., Yano, Y., *et al.* (2008) Survey of the vectorial competence of ticks in an endemic area of spotted fever group rickettsioses in Fukui Prefecture, Japan. *Microbiology and Immunology* 52, 305–309. https://doi.org/10.1111/j.1348-0421.2008.00042.x

JMA (2005) *Ijyou Kishou Report 2005. Report on Recent Climate Change 2005 – Reviews and Outlook for the Future*. Japan Meteorological Agency, Tokyo (in Japanese).

Lal, M., Harasawa, H., Murdiyarso, D., Adger, W.N., Adhikary, S., *et al.* (2001) Asia. In: McCarthy, J.J., Canziani, O.F., Leary, N.A., Dokken, D.J. and White, K.S. (eds) *Climate Change 2001: Impacts, Adaptation, and Vulnerability. Contribution of Working Group II to the Third Assessment Report of the Intergovernmental Panel on Climate Change*. Cambridge University Press, Cambridge, pp. 533–590.

Mahara, F. (1997) Japanese spotted fever: report of 31 cases and review of the literature. *Emerging Infectious Diseases* 3, 105–111. https://doi.org/10.3201/eid0302.970203

Mahara, F., Koga, K., Sawada, S., Taniguchi, T., Shigemi, F., *et al.* (1985) The first report of the rickettsial infections of spotted fever group in Japan: three clinical cases. *Kansenshogaku Zasshi* 59, 1165–1171 (in Japanese). https://doi.org/10.11150/kansenshogakuzasshi1970.59.1165

MOE (2006) Vulnerability assessment, climate change impacts and adaptation measures. In: *Japan's Fourth National Communication Under the United Nations Framework Convention on Climate Change*. Japanese Ministry of the Environment, Government of Japan, Tokyo, pp. 191–202.

MOE (2017) Global warming – past and future. In: *Stop Global Warming*. Japanese Ministry of the Environment, Government of Japan, Tokyo, pp. 3–18. Available at: https://www.env.go.jp/earth/ondanka/knowledge/Stop2017e.pdf (accessed 4 January 2021).

National Institute of Infectious Diseases (2020) *Rickettsia japonica* 1999–2019. *Infectious Agents Surveillance Reports* 41, 133–135.

Saito, Y. (1959) Studies on ixodid ticks Part I. On ecology, with reference to distribution and seasonal occurrence of ixodid ticks in Niigata Prefecture, Japan. *Acta Medica et Biologica* 7, 193–209.
Sato, M., Ikeda, S., Arai, R., Kato, M., Aoki, J., *et al.* (2021) Diversity and distribution of ticks in Niigata prefecture, Japan (2016–2018): changes since 1950. *Ticks and Tick-borne Diseases* 12, 101683. https://doi.org/10.1016/j.ttbdis.2021.101683
SGS (2010) Towards sustainable use of *satoyama/satoumi* landscapes and regional revitalization. Satoyama Governors' Summit. Available at: https://i.unu.edu/media/unu.edu/news/854/cop10_governors_summit_message_20101026.pdf (accessed 4 January 2021).
Suto, T. (1985) Evidence of spotted fever rickettsial infection in Japan as demonstrated by the indirect immunoperoxidase test. *Microbiology and Immunology* 29, 1243–1246. https://doi.org/10.1111/j.1348-0421.1985.tb00914.x
Uchida, T., Mahara, F., Tsuboi, Y. and Oya, A. (1985) Spotted fever group rickettsiosis in Japan. *Japanese Journal of Medical Sciences and Biology* 38, 151–153. https://doi.org/10.7883/yoken1952.38.151
Uchida, T., Uchiyama, T., Kumano, K. and Walker, D.H. (1992) *Rickettsia japonica* sp. nov., the etiological agent of spotted fever group rickettsiosis in Japan. *International Journal of Systematic Bacteriology* 42, 303–305. https://doi.org/10.1099/00207713-42-2-303
WHO (2002) *Reducing Risks, Promoting Healthy Life. The World Health Report 2002*. World Health Organization, Geneva, Switzerland.
Wu, N., Abril, C., Hinić, V., Brodard, I., Thür, B., *et al.* (2011) Free-ranging wild boar: a disease threat to domestic pigs in Switzerland? *Journal of Wildlife Diseases* 47, 868–879. https://doi.org/10.7589/0090-3558-47.4.868
Wu, X., Lu, Y., Zhou, S., Chen, L. and Xu, B. (2016) Impact of climate change on human infectious diseases: empirical evidence and human adaptation. *Environment International* 86, 14–23. https://doi.org/10.1016/j.envint.2015.09.007
Yamauchi, T., Obara, M., Watanabe, M., Ando, S., Ishikura, M., *et al.* (2009) Survey of tick fauna possessing the ability to act as vectors of rickettsiosis in Toyama Prefecture, Japan. *Medical Entomology and Zoology* 60, 23–31. https://doi.org/10.7601/mez.60.23

eo52 Spatiotemporal and Demographic Patterns of Transmission of Kyasanur Forest Disease Virus in India

Neha Pandey and Sunit K. Singh*
Banaras Hindu University, Varanasi, India

52.1 Expert Opinion

52.1.1 Introduction

In the year 1957, cases of severe haemorrhagic fever were reported from the forested Shimoga district of south-western India's Karnataka state (Work and Trapido, 1957). High mortality was observed in human and non-human primates, including red-faced bonnet macaques (*Macaca radiata*) and black-faced langurs (*Semnopithecus entellus*) in the vicinity of Kyasanur forest (Holbrook, 2012). The disease was named as Kyasanur forest disease (KFD) and the causative pathogen as Kyasanur forest disease virus (KFDV). It was locally known as monkey fever or monkey disease (Holbrook, 2012). KFDV is a tick-borne flavivirus that circulates between tick vectors, mainly *Haemaphysalis spinigera*, and a vertebrate host (Holbrook, 2012; Rastogi and Singh, 2017). KFDV is characterized as a neglected and category C priority pathogen by the US National Institute of Allergy and Infectious Diseases (NIAID) and is handled exclusively in biosafety level 4 (BSL4) laboratories (Rastogi and Singh, 2017). The seasonal outbreaks of KFDV coincide with the period of maximum activity of tick vectors. Humans get infected with KFDV through bites of infected ticks or during interaction with infected mammalian and avian species (Pattnaik, 2006; Rastogi and Singh, 2017). KFDV infection causes haemorrhagic fever in humans (Holbrook, 2012). It has an incubation period of 2–8 days (Pattnaik, 2006). The symptoms include high fever (~40°C), headache, body ache, diarrhoea, anorexia, decreased blood pressure and heart rate, abnormal blood biochemistry, altered haematology, conjunctivitis that promotes photophobia, and many more symptoms (Rastogi and Singh, 2017). Histopathological findings show increased erythrophagocytosis in the spleen, nephrosis and hepatomegaly with parenchymatic degeneration (Pattnaik, 2006). Parenchymal degeneration of the kidney and expression of mononuclear phagocyte tissues in the liver, spleen, lungs, heart and skeletal muscles have been reported in fatal cases of KFDV infection (Pattnaik, 2006). About 20% of patients suffer from a relapse. These patients develop neurological manifestations including mental disturbance, confusion, convulsion, tremors, drowsiness, transient disorientation and loss of consciousness (Nayak and Srinivasa Rao, 1983; Shah *et al.*, 2018).

*Email: sunitsingh2000@gmail.com

DOI: 10.1079/9781789249637.0052

52.1.2 Kyasanur forest disease virus epidemiology and transmission

Since its first emergence in 1957, KFD outbreaks were reported to occur annually in Shimoga district with 200–500 cases. A total of 9594 cases have been reported from around 16 districts during 1957–2017 (Chakraborty *et al.*, 2019). The disease extended towards the south and south-western areas near Shimoga district. During this period approximately 320 human cases with around 4% mortality were reported from these areas (NCDC, 2019). Cases were also reported during 1971–2000 from Chikmagalur, Udupi, Uttara Kannada and Dakshina Kannada (Pattnaik, 2006). The largest outbreak of KFD was reported from Dakshina Kannada in 1983 with 1984 cases (NCDC, 2019). Over 1900 cases were recorded between 2001 and 2003, which also included Gulbarga, Chamarajanagar and Belgaun areas of Karnataka (Pattnaik, 2006; Sudarshan Pai and Harsha Kumar, 2017). During 2003–2012, 823 cases were reported from Karnataka, which included 23 deaths (Sudarshan Pai and Harsha Kumar, 2017).

KFDV has expanded its endemic area, as cases have been reported from neighbouring states of Karnataka. KFDV-infected monkeys were identified in Nilgiri district while human cases were recorded from Wayanad, Malappuram and Alappuzha districts of Kerala (Mourya and Yadav, 2016). Recently, KFDV outbreaks have been reported from Sattari taluk, Parnem and Dharbhandora taluk areas of Goa in 2015, and from Sindhudurg district, Maharashtra state, in 2016 (Awate *et al.*, 2016; Patil *et al.*, 2017). Approximately 3075 human cases have been reported between 2000 and 2019 with 53 deaths (NCDC, 2019). Increased animal–human interaction in the forest areas of Western Ghats might be responsible for this spillover of KFDV (https://www.monkeyfeverrisk.ceh.ac.uk/kfd, accessed 14 June 2021). The local population and the forest officials and staff frequently visit these areas and are at high risk of acquiring KFDV infection. The areas affected with KFD and recent KFDV infections are shown in Fig. 52.1.

KFDV is mainly transmitted by *Haemaphysalis* ticks, particularly *H. spinigera* (Holbrook, 2012; Rastogi and Singh, 2017). KFDV has also been isolated from 16 other ticks including *Ixodes*, *Argas*, *Ornithodoros*, *Hyalomma*, *Dermacentor* and *Rhipicephalus* species (Pattnaik, 2006). KFDV is transmitted transovarially from the mature tick to the progeny and transstadially to the subsequent stages in the tick life cycle. The female adult tick lays eggs on the forest litter in May/June, during the monsoon. KFDV transmission occurs through the infected tick larvae and nymph feeding on monkeys (*S. entellus* and *M. radiata*) and other small animals. Transmission to humans occurs via an infected tick bite, or through contact with an infected animal. Humans are the terminal host of KFDV; human-to-human transmission has not been reported for KFDV (Pattnaik, 2006; Holbrook, 2012; Shah *et al.*, 2018). *Haemaphysalis* ticks are the most abundant host-seeking ticks in the Western Ghat region of India. Their number is high in KFD-affected areas and their transportation to human settlements increases the risk of KFD being spread to other areas.

52.1.3 Impact of climate change on Kyasanur forest disease virus emergence and transmission

The prevailing climatic conditions of an area exert a deep influence on the spread of vector-borne diseases. Climate includes precipitation (rainfall/drought), temperature, humidity and wind. Changes in these factors influence the life cycle of the vector, vector–pathogen interaction, transmission, distribution of the pathogen by vectors, and many other factors. The increasing human population and encroachment have led to deforestation, changes in the landscape and global warming. These factors affect the precipitation and local weather of an area and influence the host–vector distributions. Deforestation promotes the migration of host species to new forest habitats. This influences the population dynamics of vectors by increasing their habitat range. The local patterns of precipitation and microclimate of a region also suffer from deforestation disturbing the vector distribution and life cycle (Ajesh *et al.*, 2017). Changes in climatic factors affect the development, seasonal activity and mortality of ticks (Ogden and Lindsay, 2016). It is also a critical determinant of

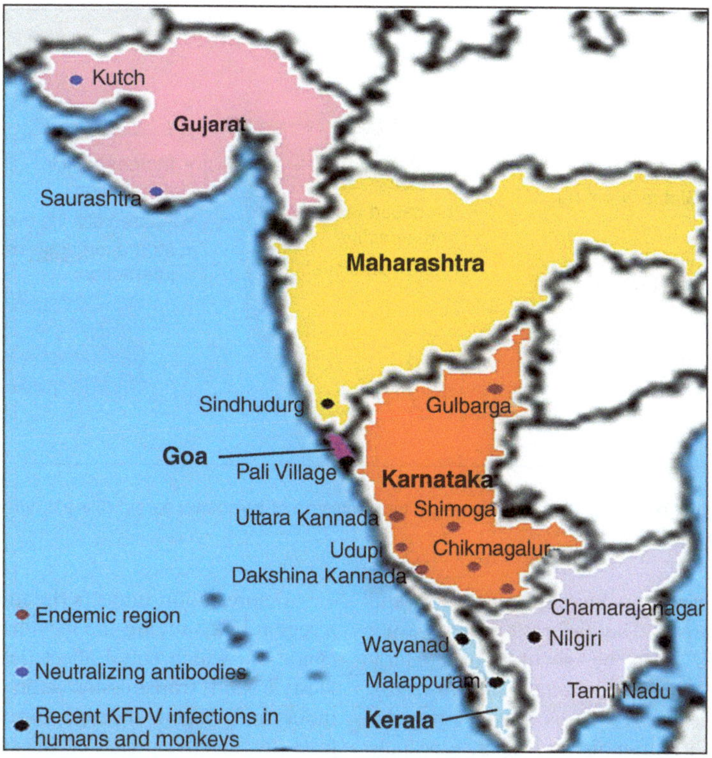

Fig. 52.1. Map showing areas affected with Kyasanur forest disease and recent Kyasanur forest disease virus (KFDV) infections. (Adapted from Rastogi and Singh, 2017, with permission.)

their distribution, life cycle and transmission. An increase in temperature promotes faster development in ticks and shortens their life cycle. The activity of ticks increases with increased humidity while increasing rainfall inhibits their activity (Süss *et al.*, 2008; Ogden and Lindsay, 2016). The impact of anthropogenic activities and climate change on the tick life cycle and KFDV transmission is shown in Fig. 52.2.

In general, ticks require high humidity of ~85% and air temperatures >7°C to complete their life cycle (Süss *et al.*, 2008). These conditions are required for transmission of the pathogen during the active period, i.e. March to May. The developmental cycle of ticks relies greatly on temperature, seasonal fluctuations and availability of hosts (Süss *et al.*, 2008; Ogden and Lindsay, 2016). Increased temperature leads to extended and accelerated development of ticks, increased production of eggs and increased population (Süss *et al.*, 2008; Ogden and Lindsay, 2016). On finding the conditions suitable for growth, these ticks shift to new areas in search of blood meals. Studies have reported that the *Ixodes* ticks require relative humidity >80% to avoid fatal desiccation (Gray *et al.*, 2009).

KFD was originally endemic to Karnataka state due to the favourable ecological and geographical conditions. Vegetation in the Shimoga district is comprised of diverse biotypes of forests. The area receives a mean rainfall of 85 inches (216 cm) per year during the rainy season that lasts from June to September (Pattnaik, 2006). The humid and fertile soil of the area is suitable for the cultivation of paddy fields and the neighbouring forests can sustain a large number of wild monkeys. The area thus provides ideal topographical and climatic conditions for the prevalence of ticks and the emergence and resurgence of tick-borne diseases like KFD. Over 26% of forest cover has been cleared from the Western Ghats in

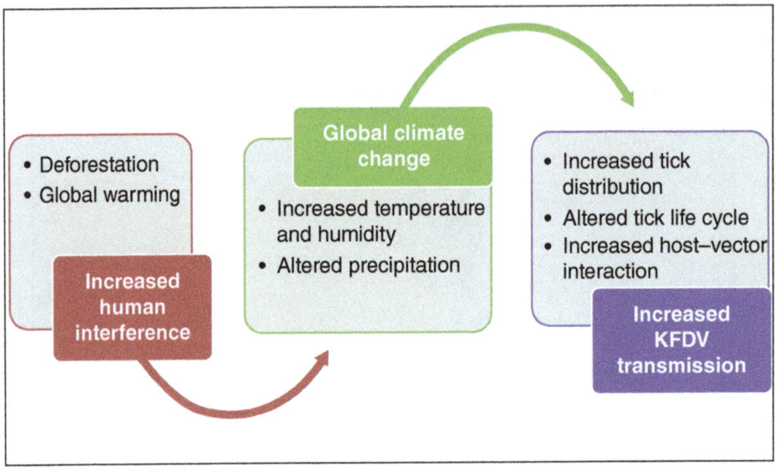

Fig. 52.2. Impact of global climate change on transmission of Kyasanur forest disease virus (KFDV). (Authors' own figure.)

the past 22 years, to increase plantation and agriculture areas. The deforested areas were rapidly taken over by shrubs, particularly *Chromolaena odorata* and *Lantana camara* (Ajesh *et al.*, 2017). These shrubs support populations of small mammals and birds, especially rodents, which in turn become hosts of larvae and nymphs. These conditions increase the transmission of KFDV. The humidity that arises from the waterlogged paddy fields might impact the tick vector life cycle (Süss *et al.*, 2008; Ajesh *et al.*, 2017).

An infected tick remains infected throughout its life cycle. Infected ticks act as reservoirs that sustain KFDV from year to year and pass the virus from adults to larvae through eggs. The activity level of ticks is stage-specific; nymphs and larvae are more energetic than adults. The nymphs and larvae are mostly active during the dry season (January to May) and their activity coincides with outbreaks of KFDV infections in human and non-human primates (Holbrook, 2012). With the onset of monsoons, human cases of KFDV infection decline. During the monsoons, the population of larvae matures under the cover of forest litter and later becomes active after the monsoon ends and forest litter dries (Pattnaik, 2006; Shah *et al.*, 2018). The association of deforestation and species richness correlates to increases in KFD (Walsh *et al.*, 2019).

Deforestation impacts the microclimate of a region by changing the local patterns of precipitation, which could affect tick distribution, lifestyle and transmission efficiency. Delayed monsoon or below-average annual rainfall affects the infective period of larvae (Ajesh *et al.*, 2017). Therefore, changes in existing climate conditions affect the emergence and re-emergence of KFD through effects on vector and host–parasite relationships.

52.2 Future Predictions

KFDV, previously endemic to Karnataka, has now spread southward and northward along the Western Ghats in India. Detection of antibodies against KFDV in humans from Gujarat, West Bengal and the Andaman and Nicobar Islands suggests its prevalence in these areas. However, there is still lack of data that could relate the seasonal fluctuations in temperature and precipitation to the tick vector life cycle and KFDV emergence. Long-term population- and molecular-based studies are required on ticks of the Western Ghats to establish the correlation of KFDV infection and transmission with the change in local weather conditions. This would help in understanding and preventing the cycles of KFDV transmission by predicting potential impacts of climate change.

References

Ajesh, K., Nagaraja, B.K. and Sreejith, K. (2017) Kyasanur forest disease virus breaking the endemic barrier: an investigation into ecological effects on disease emergence and future outlook. *Zoonoses and Public Health* 64, e73–e80.

Awate, P., Yadav, P., Patil, D., Shete, A., Kumar, V., *et al.* (2016) Outbreak of Kyasanur forest disease (monkey fever) in Sindhudurg, Maharashtra State, India, 2016. *Journal of Infection* 72, 759–761.

Chakraborty, S., Andrade, F.C.D., Ghosh, S., Uelmen, J. and Ruiz, M.O. (2019) Historical expansion of Kyasanur forest disease in India from 1957 to 2017: a retrospective analysis. *GeoHealth* 3, 44–55.

Gray, J.S., Dautel, H., Estrada-Pena, A., Kahl, O. and Lindgren, E. (2009) Effects of climate change on ticks and tick-borne diseases in Europe. *Interdisciplinary Perspectives in Infectious Diseases* 2009, 593232.

Holbrook, M.R. (2012) Kyasanur forest disease. *Antiviral Research* 96, 353–362.

Mourya, D.T. and Yadav, P.D. (2016) Recent scenario of emergence of Kyasanur forest disease in India and public health importance. *Current Tropical Medicine Reports* 3, 7–13.

Nayak, A.J. and Srinivasa Rao, P.N. (1983) Ocular changes in Kyasanur forest disease. *Journal of the Indian Medical Association* 81, 158–160.

NCDC (2019) *Kyasanur Forest Disease: A Compendium of Scientific Literature* [Department of Health & Family Welfare Services (ed.)]. National Centre for Disease Control, New Delhi. Available at: https://www.monkeyfeverrisk.ceh.ac.uk/sites/default/files/NCDC%20Consultation%20book_3rd_update.pdf (accessed 24 June 2021).

Ogden, N.H. and Lindsay, L.R. (2016) Effects of climate and climate change on vectors and vector-borne diseases: ticks are different. *Trends in Parasitology* 32, 646–656.

Patil, D.Y., Yadav, P.D., Shete, A.M., Nuchina, J., Meti, R., *et al.* (2017) Occupational exposure of cashew nut workers to Kyasanur forest disease in Goa, India. *International Journal of Infectious Diseases* 61, 67–69.

Pattnaik, P. (2006) Kyasanur forest disease: an epidemiological view in India. *Reviews in Medical Virology* 16, 151–165.

Rastogi, M. and Singh, S.K. (2017) Kyasanur forest disease. In: Singh, S.K. (ed.) *Neglected Tropical Diseases – South Asia*. Springer, Cham, Switzerland, pp. 373–386.

Shah, S.Z., Jabbar, B., Ahmed, N., Rehman, A., Nasir, H., *et al.* (2018) Epidemiology, pathogenesis, and control of a tick-borne disease – Kyasanur forest disease: current status and future directions. *Frontiers in Cellular and Infection Microbiology* 8, 149.

Sudarshan Pai, U. and Harsha Kumar, H.N. (2017) Epidemiological trends in Kyasanur forest disease in Karnataka State, South India. *Journal of Pharmacy Practice and Community Medicine* 3, 103–107.

Süss, J., Klaus, C., Gerstengarbe, F.W. and Werner, P.C. (2008) What makes ticks tick? Climate change, ticks, and tick-borne diseases. *Journal of Travel Medicine* 15, 39–45.

Walsh, M.G., Mor, S.M., Maity, H. and Hossain, S. (2019) Forest loss shapes the landscape suitability of Kyasanur forest disease in the biodiversity hotspots of the Western Ghats, India. *International Journal of Epidemiology* 48, 1804–1814.

Work, T.H. and Trapido, H. (1957) Summary of preliminary report of investigations of the Virus Research Centre on an epidemic disease affecting forest villagers and wild monkeys of Shimoga District, Mysore. *Indian Journal of Medical Science* 11, 341–342.

eo53 Argasid Ticks, Relapsing Fever and a Changing Climate

Mike Teglas*
University of Nevada, Reno, Nevada, USA

53.1 Expert Opinion

53.1.1 Introduction

Epidemics of relapsing fever, described by Hippocrates and the missionary-explorer David Livingstone, have been recorded since 1744 (Reynolds, 1937). The term 'relapsing fever' was first used to describe the symptoms associated with outbreaks in the 1800s. Since then, relapsing fever has been diagnosed in human populations on every continent except Australia and Antarctica (Dunn and Clark, 1933; Legge, 1933; Birkhaug, 1942). Outbreaks involving millions of individuals have been reported in North Africa, Europe and Russia, especially during famine or global conflicts such as World Wars. Clinical manifestations of relapsing fever include periods of high fever (40–41.7°C), headache, general malaise, chills and sweats, and myalgia/arthralgia lasting 3 to 7 days followed by a respite from clinical signs for about a week when the spirochaetes that cause the disease are reduced to undetectable numbers in the blood. Relapse cycles of spirochaetaemia and resultant symptoms can occur multiple times following infection until the host becomes immune, receives appropriate medical treatment or dies.

The disease is caused by multiple species of spirochaetes in the genus *Borrelia* that are often categorized taxonomically based on the vector species involved in their transmission. Epidemic or louse-borne relapsing fever, caused by *Borrelia recurrentis*, is transmitted by the human body louse (*Pediculus humanus humanus*) and has historically been associated with the majority of relapsing fever outbreaks worldwide. *B. recurrentis* infections cause the most severe form of relapsing fever with the highest fatality rates, but the disease has all but been eliminated due to drastic reductions in the number of people globally infested with body lice (Warrell, 2019).

Endemic or tick-borne relapsing fever (TBRF) is caused by a wider variety of *Borrelia* species and is further broken down based on tick vector species. *Borrelia miyamotoi*, transmitted by ixodid (hard) ticks in the genus *Ixodes*, is prevalent across a wide area of Europe, Northern Asia and North America. Other TBRF group borreliae have been identified in *Amblyomma americanum*, *Dermacentor variabilis* and *Rhipicephalus* spp. ticks, but only *B. miyamotoi* is associated with human disease (Talagrand-Reboul *et al.*, 2018). The effects of climate change on the vectors that transmit *B. miyamotoi* (*Ixodes persulcatus* and *Ixodes ricinus* in Eurasia, *Ixodes scapularis* and *Ixodes pacificus* in North America) are covered in

*Email: mteglas@unr.edu

DOI: 10.1079/9781789249637.0053

detail by other authors in this publication. The remainder of this expert opinion will focus on TBRF *Borrelia* species vectored by argasid (soft) ticks.

Argasid ticks that vector TBRF group spirochaetes pathogenic to humans belong in the genus *Ornithodoros*. These tick species have a worldwide distribution and can be found in both temperate and tropical climates. The spirochaete pathogens they transmit show a strong specificity for their unique species of *Ornithodoros* and are typically maintained in enzootic cycles with domestic and wild animals serving as reservoirs and humans serving mainly as opportunistic hosts (Talagrand-Reboul *et al.*, 2018). The exception, *Borrelia duttonii* in Africa, associated with the most severe and fatal form of TBRF, is strictly confined to human hosts and does not have an identified animal reservoir (Lopez *et al.*, 2020). Ticks in the genus *Ornithodoros* are characterized by their rapid feeding (usually <30 min), multiple nymphal stages (typically five to seven), indiscriminate host preferences and general longevity (up to a decade) (Cutler, 2010). The ticks often feed on their hosts while they sleep, and most species in the genera are nidicolous, spending the majority of their lifespan living in or near the nests or burrows of their vertebrate hosts, including structures inhabited by humans and their livestock. These lifestyle characteristics contribute to making collection and identification of the tick vectors difficult. Researchers are often forced to rely on other information – such as historical data, disease prevalence in the human population or the presence of the pathogen circulating in animal reservoir populations – to make predictions on the distribution and habitat requirements of the tick species involved in TBRF transmission.

53.1.2 Current knowledge

Argasid ticks can survive in habitats with much lower-humidity microenvironments than most ixodid tick species can. Components such as cementum and oilskin in the argasid cuticle provide them greater resistance to water evaporation and allow *Ornithodoros* species to survive in regions with high temperatures and xeric conditions (Sonenshine and Roe, 2014). For example, two African species, *Ornithodoros moubata* and *Ornithodoros savignyi*, can remain active even when air temperatures reach 63 and 75°C, respectively (Vial, 2009). Yet cooler, moister conditions are required to enable these ticks to complete components of their developmental cycle. These microclimatic factors are often found within the tick host's shelters (i.e. nests, burrows, buildings used for human or domestic animal shelters, caves) and the body heat and moisture excreted from their hosts can provide the conditions required for tick development (Sonenshine, 1993). The success and duration of *Ornithodoros* development are dependent on climatic conditions (both inside and outside their harbourages) and their ability to obtain sufficient blood meals to provide them with the energy and nutrients to reproduce or to complete development into a new instar or life stage (Olmedo *et al.*, 1995). For example, *Ornithodoros tholozani* in Central Asia stops feeding and undergoes diapause in the winter, whereas the same species remains active and feeds all year round in parts of its range where the Mediterranean Sea moderates winter temperatures. In Europe, *Ornithodoros erraticus* will not feed on its domestic pig hosts until the temperature inside the pigsties reaches 13 to 15°C (Vial, 2009).

The majority of *Ornithodoros* species involved in the transmission of TBRF display indiscriminate feeding habits. Their variation in host preference may simply be a result of host availability within the tick's appropriate microclimate. The ticks feed on small mammals, reptiles or domestic animals when available; when absent, they will readily feed on humans (Nieto and Teglas, 2014; Sonenshine and Roe, 2014). With the exception of *O. moubata*, humans do not constitute an important blood meal source for these tick vectors and most cases of human transmission are due to opportunistic feeding by infected ticks.

Our ability to make accurate predictions on the distribution and habitat requirements of argasid ticks involved in TBRF transmission is hampered by their nidicolous lifestyle. Collection and identification of tick species typically requires removal of burrow contents or building materials within a structure. Identification using morphological characteristics alone is difficult for many species and can require molecular techniques to make them accurately. Studies investigating ixodid vector distribution can incorporate climate and environmental data as important variables when developing species distribution models (Li *et al.*, 2019). However, argasid ticks are buffered from their external environments by the stable microclimates offered inside their harbourages. This buffering effect must then be accounted for in the development of soft tick distribution models and for modelling the associated diseases they vector. The population dynamics of important host species for ixodid ticks, such as deer and rodents, are used to predict the incidence of Lyme disease successfully but the host preferences of Ornithodorine ticks are less distinct (Brisson and Dykhuizen, 2006; Hacker *et al.*, 2018). This makes comparisons between host species abundance and disease distribution a less reliable predictive variable when estimating TBRF prevalence across a landscape. As a result, only a few studies have assessed the role that climate and host diversity play in the distribution of *Ornithodoros* spp. tick vectors and the epidemiology of TBRF over a large geographic scale (Trape *et al.*, 2013; Donaldson *et al.*, 2016; Sage *et al.*, 2017; Vial *et al.*, 2018).

To make more accurate predictions regarding the epidemiology of endemic TBRF, investigators have incorporated tick field collection data, the presence of TBRF-infected human cases and the locations of TBRF *Borrelia*-positive animal hosts into habitat suitability function analyses and species distribution models to define suitable habitat for *Ornithodoros* ticks in North America, Europe, Northern Africa and Western Asia. Although climate and precipitation variables used for the analyses differed by study, a few notable trends become apparent. In North America, the two most common species of *Borrelia* involved in human cases of endemic relapsing fever are *Borrelia hermsii* in the western USA and Canada, and *Borrelia turicatae* in the southern USA and northern Central America. *Ornithodoros hermsi* serves as the primary vector in the western USA, while *Ornithodoros turicata* and *Ornithodoros talaje* are the primary vectors in the southern USA and Central America, respectively. The distribution of *O. hermsi* is restricted to mountain ranges at elevations above 1000 m, while *O. turicata* is found in arid regions of the south-western USA through to semi-tropical areas of the Gulf Coast. For both species, one of the most predictive climate variables associated with their presence is the maximum temperature during the warmest month, typically July or August (Donaldson *et al.*, 2016; Sage *et al.*, 2017). Mean spring (April–June) temperatures had the most predictive value for determining the occurrence of nine *Ornithodoros* species involved in the transmission of endemic relapsing fever in the Western Palaearctic region (Vial *et al.*, 2018). For most tick species, feeding and questing activities stop during the winter when temperatures drop below a certain threshold and resume again when average temperatures rise above it. For *Ornithodoros* species in Europe and the Middle East, this corresponds with average temperatures in April through June. Afterwards, physiological activities such as feeding, moulting and reproduction require a time interval of consistently higher temperatures to complete the developmental cycle. For *Ornithodoros* species found in cold climates, such as *O. hermsi*, the maximum temperatures during the summer months also become an important factor. Human cases of TBRF often peak on a seasonal basis, during the summer in North America and spring in Northern Africa, reflecting the period of highest tick activity and also the time of the year when people are most likely to be participating in outdoor activities and are at higher risk of contact with ticks (Vial *et al.*, 2006; Dworkin *et al.*, 2008).

Precipitation, or the lack thereof, is likely a much more important driver of *Ornithodoros* distribution than temperature. With one exception, the distribution of all TBRF vector species studied occurs in regions of the northern hemisphere where average precipitation is 500 mm per year or less. In Europe and Western Asia, mean precipitation during the driest quarter and annual precipitation of >500 mm were limiting factors for the distribution of *Ornithodoros*. The same trend also holds for *Ornithodoros* species in Africa, except for *Ornithodoros sonrai* in West Africa. This species can tolerate regions with an average precipitation of up to 750 mm per year (Trape *et al.*, 1996). A period of prolonged drought in western sub-Saharan Africa is thought to have allowed for the successful colonization of *O. sonrai* into the Sudan savannah, extending the southern limit of TBRF on the continent (Lecompte and Trape, 2003). For *O. hermsi* in western North America, the minimum temperature during the coldest month, elevation and average precipitation of 25–75 mm/month during the wettest months also had the highest predictive values in determining the distribution of these tick vectors. As most precipitation within the range of *O. hermsi* occurs during the winter months, the amount and duration of the snowpack might play the largest role in limiting this species' distribution. The distribution of *O. turicata* in the southern USA is also heavily influenced by precipitation patterns and their range is restricted to regions of the country that receive precipitation in the warmest quarter of the year, such as the summer monsoons that occur in the desert south-west and the hurricane-prone south-eastern states. Regions with <200 mm average precipitation during the driest quarter had the highest predictive value for determining the presence of this tick species. Conversely, a gap in the distribution of *O. turicata* exists between the states of Texas and Florida that appears correlated with climate. The same animal hosts that *O. turicata* primarily feeds on are present in this region of the country, but it is characterized by high temperatures and average precipitation >200 mm during the driest quarter (Donaldson *et al.*, 2016; Lopez *et al.*, 2016).

53.2 Future Predictions

Few studies have investigated how potential climatic changes will affect argasid tick distribution and the prevalence of TBRF. In western North America, the distribution of *O. hermsi* will likely change due to warmer, drier winters leading to contraction in much of its range to higher elevations (Sage *et al.*, 2017). This may be one of the few examples where 'global warming' will lead to a decrease in the risk of acquiring a vector-borne disease, but it may also open up new habitats in British Columbia and the Rocky Mountains for this tick species to expand into. Globally, precipitation is projected to increase with climate change, much of it falling back to Earth in the form of more severe weather events such as typhoons and hurricanes. The south-eastern USA is an area where such weather events are predicted to occur with increasing frequency. As a result, suitable habitats for *O. turicata* in the eastern half of its ranges (areas receiving less than an average 200 mm of precipitation in the driest quarter) will likely diminish or cease to exist, driving the distribution of this tick species westwards where the climate is expected to become hotter and drier in the south-western region of the USA. The same scenario may also play out for important TBRF vector species in Africa. Regions of North-eastern Africa where *O. moubata* currently exist may become unsuitable in the future due to predicted increases in annual precipitation for this part of the continent. On the other hand, *O. sonrai* and the other *Ornithodoros* species involved in endemic relapsing fever transmission in western Africa may be able to expand their range in sub-Saharan Africa as droughts become more and more frequent. The distribution of *Ornithodoros* species involved in TBRF transmission in Southern Europe and the Middle East is unlikely to suffer from additional precipitation in the coming decades as these parts of the world are expected to become drier under current climate models (European Environment Agency, 2019). In this expert's opinion, endemic relapsing fever is an aptly named disease. Although the geographic regions that we consider endemic for this disease will change with the climate, it is highly unlikely that TBRF will cease being a global public health concern in the next century.

References

Birkhaug, K. (1942) Otto H.F. Obermeier. In: Moulton, F.R. (ed.) *A Symposium of Relapsing Fever in the Americas*. American Association for the Advancement of Science, Washington, DC, pp. 7–9.

Brisson, D. and Dykhuizen, D.E. (2006) A modest model explains the distribution and abundance of *Borrelia burgdorferi* strains. *American Journal of Tropical Medicine and Hygiene* 74, 615–622.

Cutler, S.J. (2010) Relapsing fever – a forgotten disease revealed. *Journal of Applied Microbiology* 108, 1115–1122.

Donaldson, T.G., Pèrez de León, A.A., Li, A.Y., Castro-Arellano, I., Wozniak, E., *et al.* (2016) Assessment of the geographic distribution of *Ornithodoros turicata* (Argasidae): climate variation and host diversity. *PLoS Neglected Tropical Diseases* 10, e0004383.

Dunn, L.H. and Clark, H.C. (1933) Notes on relapsing fever in Panama with special reference to animal hosts. *American Journal of Tropical Medicine* 13, 201–209.
Dworkin, M.S., Schwan, T.G., Anderson, D.E. Jr and Borchardt, S.M. (2008) Tick-borne relapsing fever. *Infectious Disease Clinics of North America* 22, 449–468.
European Environment Agency (2019) Heavy precipitation in Europe. Available at: https://www.eea.europa.eu/data-and-maps/indicators/precipitation-extremes-in-europe-3/assessment-1 (accessed 16 December 2020).
Hacker, G.M., Brown, R.N., Fedorova, N., Girard, Y.A., Higley, M., *et al.* (2018) Spatial clustering of *Borrelia burgdorferi* sensu lato within populations of Allen's chipmunks and dusky-footed woodrats in northwestern California. *PLoS ONE* 13, e0195586.
Lecompte, Y. and Trape, J.F. (2003) West African tick-borne relapsing fever. *Annales de Biologie Clinique* 61, 541–548.
Legge, R.T. (1933) Relapsing fever – a new etiological observation. *California and Western Medicine* 38, 380.
Li, S., Gilbert, L., Vanwambeke, S.O., Yu, J., Purse, B.V. and Harrison, P.A. (2019) Lyme disease risks in Europe under multiple uncertain drivers of change. *Environmental Health Perspectives* 127, 67010.
Lopez, J.E., Krishnavajhala, A., Garcia, M.N. and Bermudez, S. (2016) Tick-borne relapsing fever spirochetes in the Americas. *Veterinary Sciences* 3, 16.
Lopez, J., Hovius, J.W. and Bergström, S. (2020) Pathogenesis of relapsing fever. *Current Issues in Molecular Biology* 42, 519–550.
Nieto, N.C. and Teglas, M.B. (2014) Relapsing fever group *Borrelia* in southern California rodents. *Journal of Medical Entomology* 51, 1029–1034.
Olmedo, F.F., Estrada-Peña, A. and Espuny, J.C. (1995) Variations in the size and weight of adults of *Ornithodoros marocanus* Velu (Acari: Argasidae) according to nymphal blood-intake. *Acarologia* 36, 195–201.
Reynolds, F.C. (1937) Relapsing fever: comments on its incidence in Nevada. *California and Western Medicine* 47, 170–174.
Sage, K.M., Johnson, T.L., Teglas, M.B., Nieto, N.C. and Schwann, T.G. (2017) Ecological niche modeling of the distribution of tick-borne relapsing fever in western North America. *PLoS Neglected Tropical Diseases* 11, e0006047.
Sonenshine, D.E. (1993) *The Biology of Ticks*, vol. 2. Oxford University Press, New York.
Sonenshine, D.E. and Roe, R.M. (2014) *The Biology of Ticks*, vol. 2, 2nd edn. Oxford University Press, New York.
Talagrand-Reboul, E., Boyer, P.H., Bergström, S., Vial, L. and Boulanger, N. (2018) Relapsing fevers: neglected tick-borne diseases. *Frontiers in Cellular and Infection Microbiology* 8, 98.
Trape, J.F., Godeluck, B., Diatta, G., Rogier, C., Legros, F., *et al.* (1996) The spread of tick-borne borreliosis in West Africa and its relationship to sub-Saharan drought. *American Journal of Tropical Medicine and Hygiene* 54, 289–293.
Trape, J.F., Diatta, G., Arnathau, C., Bitam, I., Sarih, M., *et al.* (2013) The epidemiology and geographic distribution of relapsing fever borreliosis in West and North Africa, with a review of the *Ornithodoros erraticus* complex (Acari: Ixodida). *PLoS ONE* 8, 11.
Vial, L. (2009) Biological and ecological characteristics of soft ticks (Ixodida: Argasidae) and their impact for predicting tick and associated disease distribution. *Parasite* 16, 191–202.
Vial, L., Diatta, G., Tall, A., Ba, E.H., Bouganali, H., *et al.* (2006) Incidence of tick-borne relapsing fever in west Africa: longitudinal study. *The Lancet* 368, 37–43.
Vial, L., Ducheyne, E., Filatov, S., Gerilovych, A., McVey, D.S., *et al.* (2018) Spatial multi-criteria decision analysis for modelling suitable habitats of *Ornithodoros* soft ticks in the Western Palearctic region. *Veterinary Parasitology* 249, 2–16.
Warrell, D.A. (2019) Louse-borne relapsing fever (*Borrelia recurrentis* infection). *Epidemiology and Infection* 147, e106.

eo54 The Potential Effects of Climate Change on Lyme Borreliosis in East-Central Europe

Attila Trájer*
University of Pannonia, Veszprém, Hungary

54.1 Expert Opinion

The aim of this expert opinion is to consider the potential future effects of the changing climate on the risk patterns of Lyme borreliosis (LB) in East-Central Europe. This geographical domain comprises the Baltic states, the countries of the Visegrád Group, the Balkan states excluding Greece, and East European countries excluding the Russian Federation (East Central European Center, 2020). LB is the most important human vector-borne disease in East-Central Europe and has the highest annual incidence values among tick-borne diseases. While in the Baltic states the incidence of LB is about twice that of tick-borne encephalitis (Lindgren *et al.*, 2006), in such Central European countries as Hungary the ratio is an order of magnitude greater in favour of LB (Zöldi *et al.*, 2013). There is a connection between the observed changing climatic conditions and the altering incidence and distribution patterns of the disease (Daniel *et al.*, 2009; Trájer *et al.*, 2013a). For example, in the Czech Republic, an inverse relationship was found between the incidence of LB and winter severity (Daniel *et al.*, 2008). In Hungary, a significant increase in the LB incidence rates was observed between 1998 and 2010 corresponding to a shift in the start of the vegetation season (Trájer *et al.*, 2013a). Regional differences at the start of the vegetation season can also influence LB incidence and the annual pattern of the incidence curve (Trájer *et al.*, 2013b). Another analysis showed that the annual incidence pattern is a function of human presence in nature and tick activity, factors that are also strongly related to meteorological conditions (Trájer *et al.*, 2014). The increasing residential proximity of humans to high-LB-risk habitats is also an important factor (Zeman *et al.*, 2015). It is very likely that climate change will substantially alter LB incidence in East-Central Europe. A model designed to forecast the potential future annual pattern of LB incidence in Hungary showed a longer disease season for 2081–2100 compared with 1998–2008. However, the summer dormancy period of adult *Ixodes* ticks was predicted to be longer in the future than currently (Bede-Fazekas and Trájer, 2019). A study considering the different possibilities of tick infection found that the likelihood of tick-borne disease transmission from ticks to humans would increase in the 21st century in Hungary (Nah *et al.*, 2020). Of course, these studies cannot consider the effects of future healthcare and socio-economic changes, which could greatly change the observed incidence in the future.

What makes it difficult to predict the future incidence patterns of LB in East-Central Europe

*Email: trajer.attila@mk.uni-pannon.hu

DOI: 10.1079/9781789249637.0054

is that the border of the forest and forest–steppe biomes stretches through the south-east part of the area. Zöldi *et al.* (2013) showed that most LB cases are tied to forested regions. Already in the 20th century, warming of the climate has caused the decline of forests in the lowlands of the Carpathian Basin (Kotroczó *et al.*, 2012) and it is plausible that this trend will continue in the 21st century (Móricz *et al.*, 2013). It is also known that tick-borne diseases are frequent among foresters and hunters (Li *et al.*, 2018), which raises the problem of changes in population-level risks in an ecosystem that loses its forests. Urbanized areas today may still pose unpredictable tick-borne disease risks to both humans and pets (Rizzoli *et al.*, 2014). Despite these theoretical uncertainties, recent observations suggest that climate change accounts for the upsurge of LB in East-Central Europe, as was observed in Ukraine (e.g. Rogovskyy *et al.*, 2020), and the seasonal shift of tick-borne diseases as found in Estonia (Lindgren *et al.*, 2006). Climate change could have a somewhat heterogeneous effect on the vectors due to the different seasonality of different tick species. In Latvia, *Ixodes ricinus* is active from spring to autumn whereas *Ixodes persulcatus* is found only in spring and the first half of summer (Lindgren *et al.*, 2006). Because climate change results in varying degrees of temperature rise across the seasons, the specific seasonality of tick vectors should be considered in the development of forecasting models. What is less certain are the factors limiting the southern distribution of LB. Under the hot and dry summer climate of Sicily, the first LB case was not observed until the early 1990s (Rinaldi *et al.*, 1991). However, the important vector *I. ricinus* is prevalent in such Mediterranean regions like Cádiz, Málaga and Granada in Spain as well as Palermo and Catania in Sicily, which indicates that the hot and dry summer climates of the Mediterranean region are tolerated by the main European vector of LB (Beugnet and Chalvet-Monfray, 2013). It is plausible that warming will not generate disadvantageous conditions for the vectors of LB in the southern regions of East-Central Europe.

54.2 Future Predictions

Based on the 5°C lower thermal threshold for the LB season (Trájer *et al.*, 2014), the potential maximum length of the LB season is estimated to vary between 7 and 9 months in a major part of East-Central Europe during 1960–1990; in north Latvia and Estonia, this value may be only 6 months. The longest LB seasons probably occur on the Adriatic coast of Albania, Croatia and Montenegro, where, theoretically, the seasonal duration could reach 11–12 months. Concerning the continental regions, the longest seasonal duration probably occurs in the Carpathian Basin and the north Balkans, including the lowland areas of Bosnia and Herzegovina, Croatia, North Macedonia, Romania and Serbia. In these regions and countries, the potential duration of the LB season is 8–9 months, whereas in the eastern part of East-Central Europe, the potential LB season may be less than 8 months. In Belarus, the Czech Republic, Lithuania, Moldova, Poland and a major part of Slovakia and Ukraine, the most frequent maximum LB season is estimated as 7 months for 1960–1990. In the Carpathian Mountains and the Balkan Mountains ranges, the potential seasonal duration may vary between 6 and 4 months, excluding the highest elevations.

These patterns are predicted to change for 2041–2060 when estimated using Community Climate System Model 4 (CCSM4) climate scenarios (Gent *et al.*, 2011) and Representative Concentration Pathways (RCPs) for greenhouse gases and aerosols (Van Vuuren *et al.*, 2011) (Fig. 54.1). The models show that in the western areas of Poland and Slovakia, Transylvania, south Moldova, and in major areas of the north Balkan states, the annual seasonal duration of LB will reach 9 months. This is also true for central and south-east Czech Republic. Throughout the lowland regions of Hungary and Croatia, the future LB season is predicted to be 9 months.

The models predict the smallest changes in seasonal duration between the reference period-based (1960–1990) and the RCP4.5-based models (Fig. 54.2a). In Belarus, Lithuania, the major part of Ukraine, the south and central part of Hungary, western areas of the Czech Republic, the lowlands of Croatia, Romania and Serbia, as well as in the highest ranges of the Carpathian Mountains, the models do not show notable changes in the LB season (>1 month).

The greatest changes in geographical distribution of the LB season are predicted when comparing the reference period and the RCP8.5-based seasonality models (Fig. 54.2c). In the middle

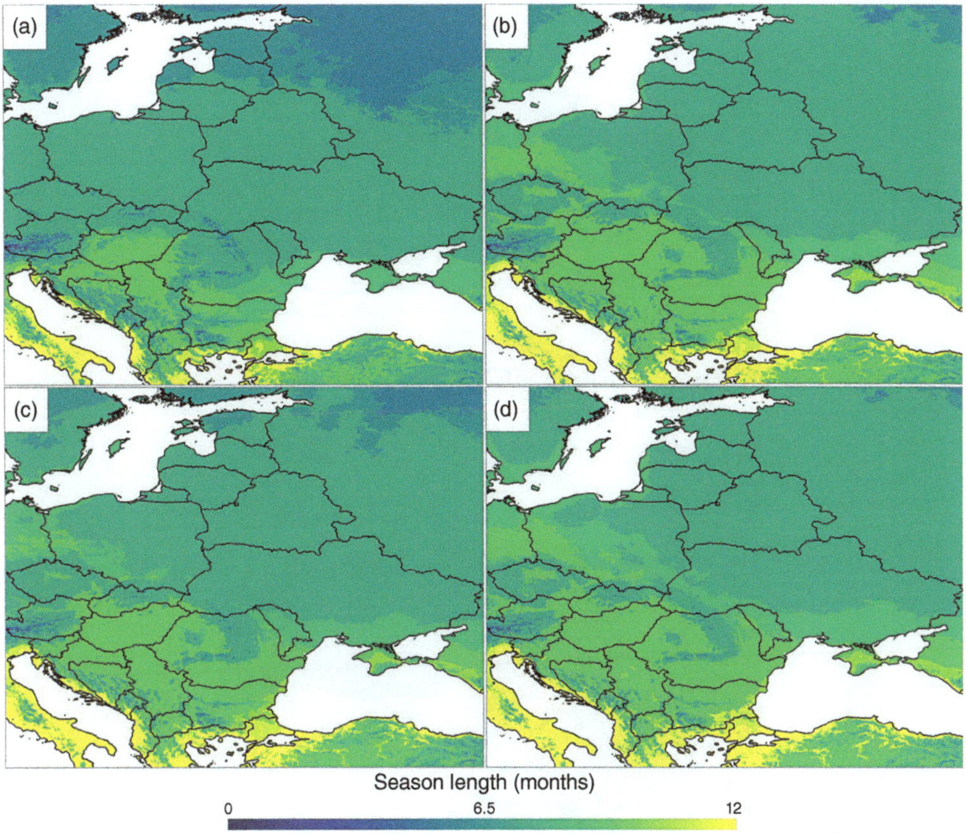

Fig. 54.1. The modelled maximum seasonal duration of Lyme borreliosis (LB) in East-Central Europe for 1960–1990 (a) and 2041–2060 (b, c, d), according to the 5°C thermal threshold of the LB season (Trájer *et al.*, 2014) and Representative Concentration Pathway (RCP)-based Community Climate System Model 4 (CCSM4) climate projections: RCP4.5 (b), RCP6.0 (c) and RCP8.5 (d). (Author's own figure.)

mountain regions of the Carpathian Basin, the foothills of the Carpathian Mountains in Romania, the mid-elevation regions of the north Balkan Mountains and in some western regions of Latvia and Moldova, the change in the seasonal duration may reach 2 months. In north Latvia, Estonia, large parts of Albania, Bulgaria and Moldova, and in the westernmost areas of Poland, the change in LB season may reach 1–2 months based on comparisons of the modelled values for 1960–1990 and 2041–2060 (Fig. 54.2).

Based on the algorithms of Trájer *et al.* (2014), the correlation between the mean monthly temperature and the relative incidence of LB – the relative risk or, in other words, the temperature-based component of the risk of LB acquisition – was calculated. Annual relative incidence values were estimated by addition of the modelled relative incidence values. The Hungarian incidence values of 1998–2010 formed the base of the comparison. Highest annual relative incidence values occur on the Adriatic coasts of Albania, Croatia and Montenegro in the period 1960–1990, reaching about 150–200% of the Hungarian incidence values in the period of 1998–2010. In the lowlands of Macedonia, Bulgaria, Romania, Serbia and Ukraine, the temperature-based maximum relative incidence theoretically reaches 90–110%; in Slovakia, the highest values are 80–90%. In Poland and the Czech Republic, the maximum values could be less than 72% but, in general, the modelled relative LB values range between 40 and 70%. In the Baltic states, the modelled relative incidence values are about 40% of the base incidence values. The climate models predict a notable increase in the risk of LB in the future. The temperature-based relative risk of LB could reach 180–240%

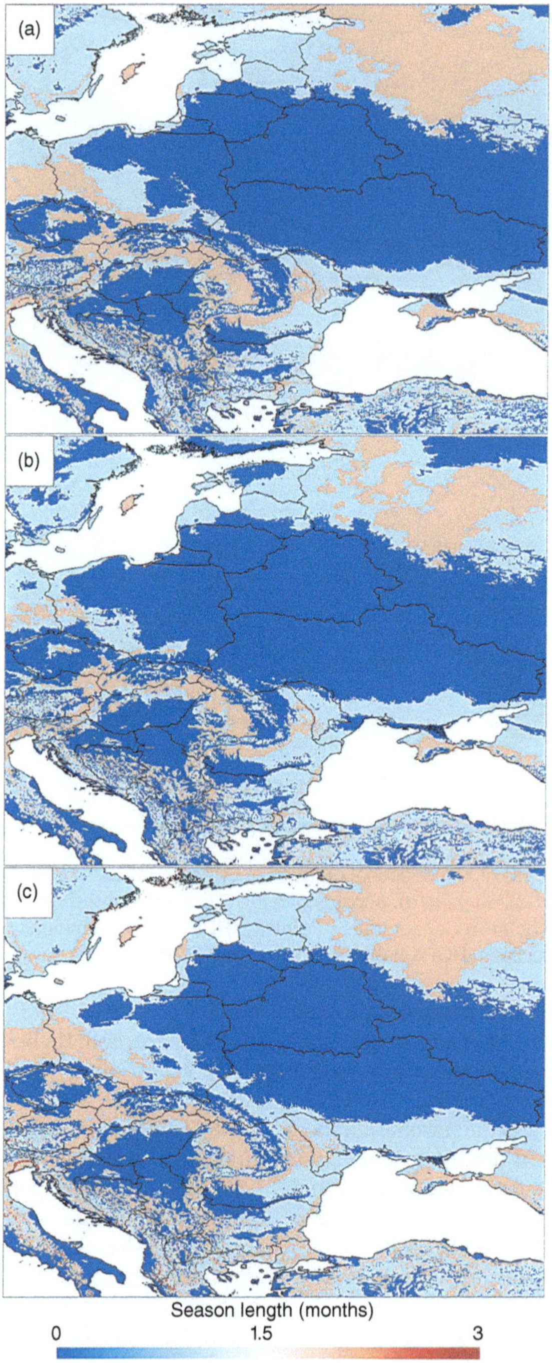

Fig. 54.2. Projected changes in the modelled maximum seasonal duration of Lyme borreliosis (LB) in East-Central Europe between 1960–1990 and 2041–2060, according to the 5°C thermal threshold of the LB season (Trájer *et al.*, 2014) and Representative Concentration Pathway (RCP)-based Community Climate System Model 4 (CCSM4) climate projections: RCP4.5 (a), RCP6.0 (b) and RCP8.5 (c). (Author's own figure.)

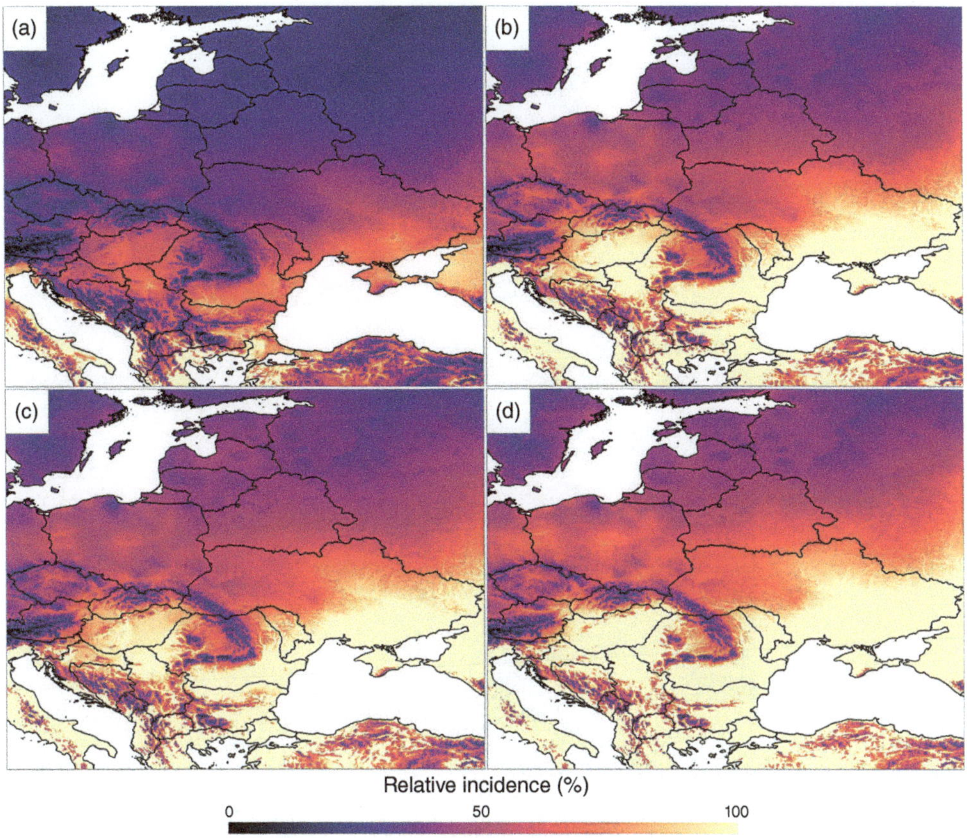

Fig. 54.3. Modelled relative incidence of Lyme borreliosis (LB) (%) in East-Central Europe for 1960–1990 (a) and 2041–2060 (b, c, d), according to the monthly mean temperature-based relative LB incidence function of Trájer *et al.* (2014) and Representative Concentration Pathway (RCP)-based Community Climate System Model 4 (CCSM4) climate projections: RCP4.5 (b), RCP6.0 (c) and RCP8.5 (d). (Author's own figure.)

relative risk values in the north Balkan lowlands, the Carpathian Basin and south Ukraine and Moldova. In the Czech Republic and north-west Poland, highest values reach 70–100%. In the Baltic states, the highest predicted values vary around 70–80%. In Belarus, the maximum values could reach about 100%. In the Adriatic territories of Albania, Croatia and Montenegro, the theoretical maximum could overwhelm the 230% values for 2041–2060 (Fig. 54.3).

Concerning the modelled alterations in the relative risk patterns of LB, it can be said that the highest increases are expected in the lowland regions of the north Balkans, south Ukraine and Moldova, the South Pannonian Biogeographical Region and the Adriatic coasts. In these regions, the range of the modelled alterations is between 170 and 200% or more, depending on the different RCP-scenario-based CCSM4 climate models. In the non-mountainous regions of the Czech Republic and Poland, this value ranges between 40 and 170%. In the Baltic states, the increase is 20–30%. In Belarus, the projected range of the changing relative incidences is between 30 and 40% (Fig. 54.4).

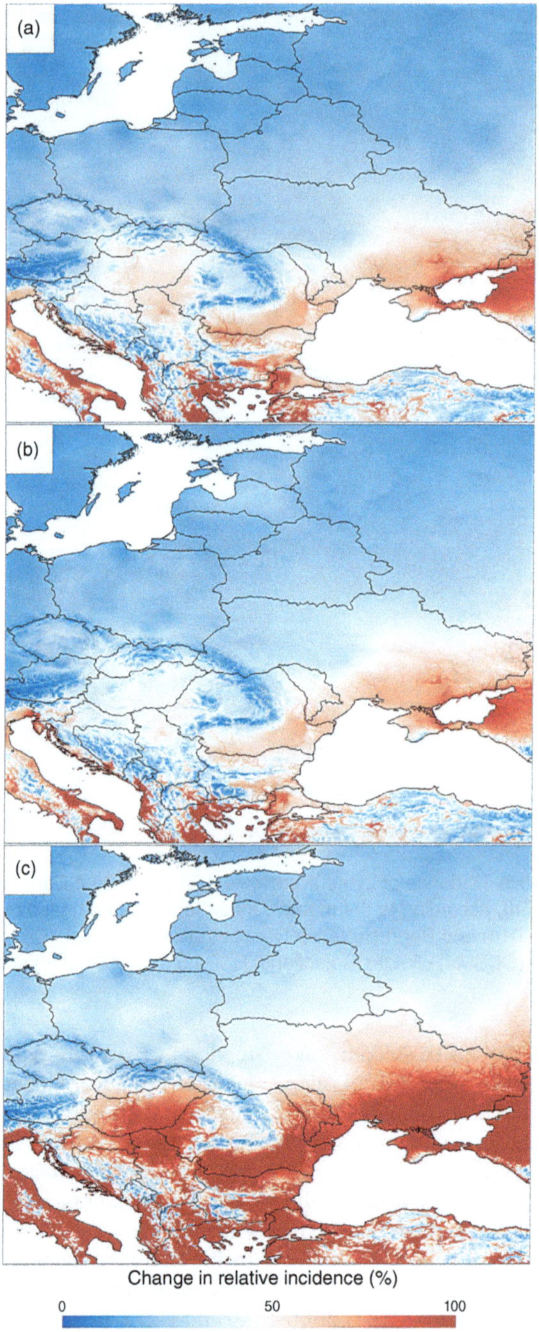

Fig. 54.4. Projected changes in the modelled relative incidence of Lyme borreliosis (LB) (%) in East-Central Europe between 1960–1990 and 2041–2060 according to the monthly mean temperature-based relative LB incidence function of Trájer *et al.* (2014) and Representative Concentration Pathway (RCP)-based Community Climate System Model 4 (CCSM4) climate projections: RCP4.5 (a), RCP6.0 (b) and RCP8.5 (c). (Author's own figure.)

References

Bede-Fazekas, Á. and Trájer, A.J. (2019) A framework for predicting the effects of climate change on the annual distribution of Lyme borreliosis incidences. *International Journal of Global Warming* 18, 81–102.
Beugnet, F. and Chalvet-Monfray, K. (2013) Impact of climate change in the epidemiology of vector-borne diseases in domestic carnivores. *Comparative Immunology, Microbiology and Infectious Diseases* 36, 559–566.
Daniel, M., Kříz, B., Valter, J., Kott, I. and Danielová, V. (2008) The influence of meteorological conditions of the preceding winter on the incidences of tick-borne encephalitis and Lyme borreliosis in the Czech Republic. *International Journal of Medical Microbiology* 298(Suppl. 1), 60–67.
Daniel, M., Materna, J., Hönig, V., Metelka, L., Danielová, V., *et al.* (2009) Vertical distribution of the tick *Ixodes ricinus* and tick-borne pathogens in the northern Moravian mountains correlated with climate warming (Jeseníky Mts., Czech Republic). *Central European Journal of Public Health* 17, 139–145.
East Central European Center (2020) Institute on East Central Europe at Columbia University. Available at: https://ece.columbia.edu/ (accessed 15 June 2021).
Gent, P.R., Danabasoglu, G., Donner, L.J., Holland, M.M., Hunke, E.C., *et al.* (2011) The Community Climate System Model version 4. *Journal of Climate* 24, 4973–4991.
Kotroczó, Z., Veres, Z., Fekete, I., Papp, M. and Tóth, J.A. (2012) Effects of climate change on litter production in a *Quercetum petraeae-cerris* forest in Hungary. *Acta Silvatica et Lignaria Hungarica* 8, 31–38.
Li, S., Juhász-Horváth, L., Trájer, A., Pintér, L., Rounsevell, M.D.A. and Harrison, P.A. (2018) Lifestyle, habitat and farmers' risk of exposure to tick bites in an endemic area of tick-borne diseases in Hungary. *Zoonoses and Public Health* 65, e248–e253.
Lindgren, E., Jaenson, T.G., Menne, B. and World Health Organization (2006) *Lyme Borreliosis in Europe: Influences of Climate and Climate Change, Epidemiology, Ecology and Adaptation Measures (EUR/04/5046250).* WHO Regional Office for Europe, Copenhagen.
Móricz, N., Rasztovits, E., Gálos, B., Berki, I., Eredics, A. and Loibl, W. (2013) Modelling the potential distribution of three climate zonal tree species for present and future climate in Hungary. *Acta Silvatica et Lignaria Hungarica* 9, 85–96.
Nah, K., Bede-Fazekas, Á., Trájer, A.J. and Wu, J. (2020) The potential impact of climate change on the transmission risk of tick-borne encephalitis in Hungary. *BMC Infectious Diseases* 20, 34.
Rinaldi, R., Gabellini, A.S., Procaccio, L., Benassi, G. and D'Alessandro, R. (1991) Lyme disease. *First reported case in Sicily. Italian Journal of Neurological Sciences* 12, 105–107.
Rizzoli, A., Silaghi, C., Obiegala, A., Rudolf, I., Hubálek, Z., *et al.* (2014) *Ixodes ricinus* and its transmitted pathogens in urban and peri-urban areas in Europe: new hazards and relevance for public health. *Frontiers in Public Health* 2, 251.
Rogovskyy, A.S., Biatov, A.P., Davis, M.A., Liu, S. and Nebogatkin, I.V. (2020) Upsurge of Lyme borreliosis in Ukraine: a 20-year survey. *Journal of Travel Medicine* 27, 6.
Trájer, A., Bobvos, J., Páldy, A. and Krisztalovics, K. (2013a) Association between incidence of Lyme disease and spring–early summer season temperature changes in Hungary – 1998–2010. *Annals of Agricultural and Environmental Medicine* 20, 245–251.
Trájer, A., Bobvos, J., Krisztalovics, K. and Páldy, A. (2013b) Regional differences between ambient temperature and incidence of Lyme disease in Hungary. *Quarterly Journal of the Hungarian Meteorological Service* 117, 175–186.
Trájer, A., Bede-Fazekas, Á., Hufnagel, L., Bobvos, J. and Páldy, A. (2014) The paradox of the binomial *Ixodes ricinus* activity and the observed unimodal Lyme borreliosis season in Hungary. *International Journal of Environmental Health Research* 24, 226–245.
Van Vuuren, D.P., Edmonds, J., Kainuma, M., Riahi, K., Thomson, A., *et al.* (2011) The representative concentration pathways: an overview. *Climatic Change* 109, 5.
Zeman, P., Benes, C. and Markvart, K. (2015) Increasing residential proximity of Lyme borreliosis cases to high-risk habitats: a retrospective study in central Bohemia, the Czech Republic, 1987–2010. *EcoHealth* 12, 519–522.
Zöldi, V., Juhász, A., Nagy, C., Papp, Z. and Egyed, L. (2013) Tick-borne encephalitis and Lyme disease in Hungary: the epidemiological situation between 1998 and 2008. *Vector-Borne and Zoonotic Diseases* 13, 256–265.

eo55 Epidemiology of Severe Fever with Thrombocytopaenia Syndrome in China

Wei Liu*, Li-Qun Fang and Hao Li
Beijing Institute of Microbiology and Epidemiology, Beijing, People's Republic of China

55.1 Expert Opinion

55.1.1 Introduction

Severe fever with thrombocytopaenia syndrome (SFTS) is a novel emerging virus infection that was first reported in the People's Republic of China in 2010. It is most likely that SFTS had been prevalent in China for some time, but it had not been identified. The aetiological pathogen is a novel phlebovirus, severe fever with thrombocytopaenia syndrome virus (SFTSV), in the family *Phenuiviridae* of the order *Bunyavirales* (Yu *et al.*, 2011). The disease was identified first in China in 2010 and subsequently in South Korea and Japan in 2013 (Yu *et al.*, 2011; Kim *et al.*, 2013; Takahashi *et al.*, 2014). So far, the number of SFTS cases has increased and the geographical distribution has expanded consistently, with the cumulative case numbers attaining 7721 (laboratory-confirmed) in China by 2018, 866 in South Korea by 2018 and 467 in Japan by August 2019, according to the most recent updates (Miao *et al.*, 2020).

The national guideline for prevention and control of SFTS was released by the Ministry of Health in 2010. A probable SFTS case is defined as a patient with acute fever, epidemiological risk factors (residing, farming or travelling in an endemic region during the endemic season or exposure to ticks 2 weeks before disease onset) and laboratory-identified abnormalities consisting of thrombocytopaenia and leukocytopaenia (Ding *et al.*, 2014). A confirmed case is defined by laboratory findings meeting one or more of the following criteria (Sun *et al.*, 2017): a positive SFTSV culture, or a positive result for SFTSV RNA by molecular detection, or seroconversion or a fourfold increase in specific immunoglobulin G antibody titres to SFTSV between the acute and the convalescent serum samples collected at least 2 weeks apart. Approximately 30% of probable cases of SFTS cannot be confirmed by laboratory tests (Yu *et al.*, 2011; Wen *et al.*, 2014) and the differential diagnosis should be made for other diseases, including human granulocytic anaplasmosis, rickettsiosis, haemorrhagic fever with renal syndrome, dengue fever, sepsis, typhoid fever and thrombocytopaenic purpura.

55.1.2 Epidemiological characteristics of severe fever with thrombocytopaenia syndrome in China

Geographical distribution

By the end of 2018, laboratory-confirmed cases had been found in 25 provinces (Miao *et al.*,

*Email: liuwei@bmi.ac.cn

DOI: 10.1079/9781789249637.0055

2020) of which more than 95% of cases have been reported in the seven provinces of Henan, Shandong, Hubei, Anhui, Liaoning, Zhejiang and Jiangsu (Li *et al.*, 2015). The geographical distribution of SFTSV infection in China has increased continuously over recent years, witnessed by increasing numbers of counties reporting SFTSV infection during 2011–2018 (98, 99, 113, 140, 153, 167, 184 and 187, respectively) (Sun *et al.*, 2021). Spatial clustering of SFTS cases revealed four hotspots in China (Fig. 55.1a). Cluster I is located at the Changbai Mountain area in Liaoning Province of north-eastern China. Two clusters reside in Shandong Province on the northern part of the east coast: Cluster II on the Jiaodong Peninsula and Cluster III surrounding the Taishan Mountain area. Cluster IV, the largest among the four clusters, centres around the Huaiyangshan Mountain area in central China and stretches over five provinces: Henan, Anhui, Hubei, Jiangsu and Zhejiang (Miao *et al.*, 2020). Within each cluster, the disease mainly spread from the centre to its surrounding areas (Fig. 55.1b). These four endemic foci are situated in four different ecogeographical zones, but all are located in wooded or mountainous areas and share similar mountainous or hilly landscapes covered by forests, shrub or grassland, and temperate humid or subtropical climates.

Demographic characteristics

Based on the confirmed cases reported from 2010 to 2018, the median age of patients in China is 63 years old (interquartile range: 54–70 years, ranging from 1 to 93 years), and the majority are 50 years of age or above (Table 55.1). In total, 55 paediatric cases under the age of 18 were reported (X.A. Zhang, 2021, unpublished results). The median age increased from 56 years in 2010 to 64 years in 2018, probably reflecting the natural ageing of the at-risk population and the lack of replenishment of the exposed population. The incidence rate of SFTS is significantly higher in elderly people and the percentage of residents aged ≥60 years is a driving factor for both presence and high incidence of SFTS. Slightly more cases were female (52.6%), but no significant difference was found overall by sex. The specific demographic distribution of SFTS in China could mainly be attributed to social factors. In rural areas of endemic regions, young adults pursue work opportunities in urban cities most of the year and senior adults become the main exposed population through agricultural activities. Similarly, women assume more tea-farming activities than men, which may partly account for the higher SFTS incidence among females.

Seasonal distribution

SFTS epidemics at the national level are highly seasonal. The epidemic season spans from May to August with a peak in May (Table 55.1). However, the epidemic curve reveals slightly different characteristics when comparing hotspots. The epidemic season appears to be slightly later at higher latitude, e.g. July–September in Cluster I (40°N) versus April–July in Cluster IV (30°N), and the peak months occur in July, June and May in clusters I, II and IV, respectively. Cluster III has two peaks, one in May and the other in August. Seasonality of SFTS occurrence may be attributed to the activity level of both the natural hosts and the vectors. From spring to summer, both outdoor activity of farmers as well as tick density and activity in the endemic regions increase, leading to enhanced exposure to vectors and competent natural hosts.

55.1.3 Epidemiological risk factors of severe fever with thrombocytopaenia syndrome disease in China

The primary source of SFTSV infection is zoonotic, e.g. direct/indirect contact with ticks, wild mammals or livestock; all might serve as the reservoir (Li *et al.*, 2014; Liang *et al.*, 2014; Yun *et al.*, 2015; Xing *et al.*, 2017). Therefore, all factors that might contribute to increased outdoor activities of humans, increasing the chances of contact with ticks or animals, could lead to increased odds of SFTSV infection. Epidemiological studies on SFTS in China indicate that occupation (agricultural tea pickers, farmers, forest workers) and personal behaviour (farming, farm work time, grass mowing) are related to increased risk of SFTSV infection. Ecological and meteorological factors that provide a comfortable environment for ticks to grow and reproduce (acidic soils,

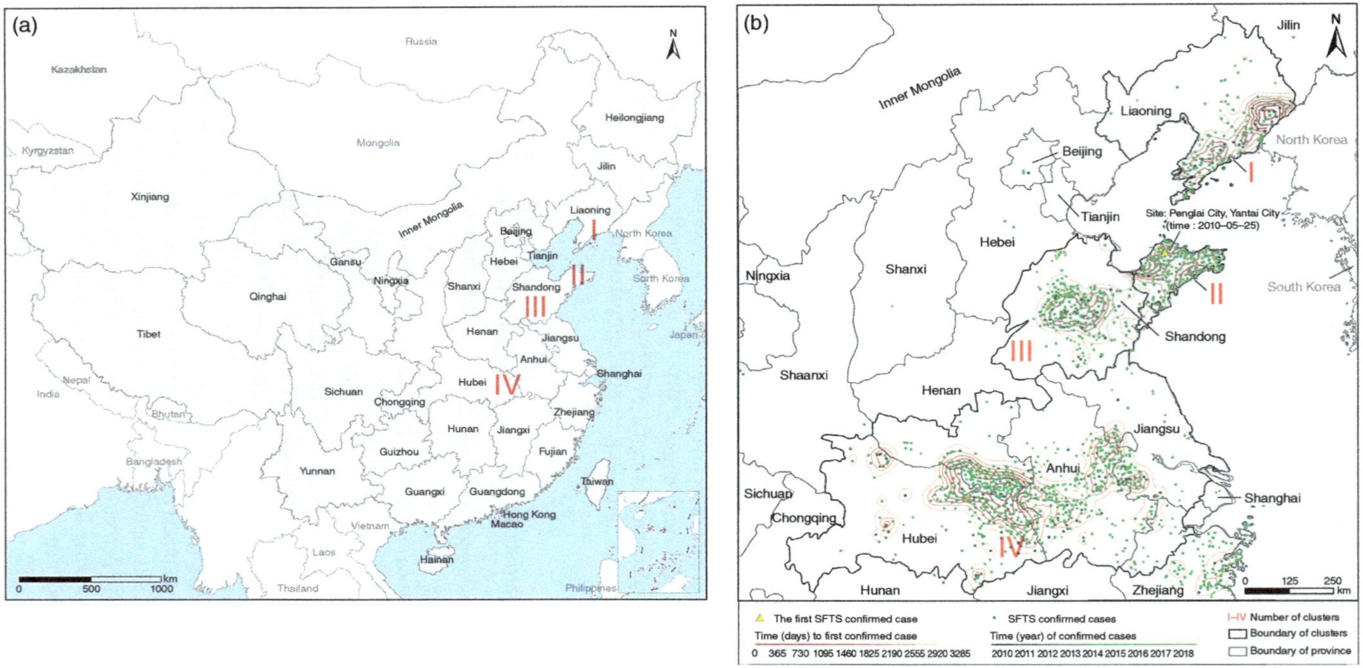

Fig. 55.1. The spatial dispersion of four geographical clusters of cases with severe fever with thrombocytopaenia syndrome (SFTS) in China, 2010–2018. (a) Geographical location of the four clusters (I–IV). (b) Confirmed SFTS cases are marked as dots with the colour indicating the year of reporting. The four clusters (I–IV) are delineated according to the spatial trend surface analysis and the geographical aggregation of cases. In each cluster, contour lines of the time intervals (in days) from the first case reported in China to the first case reported in each SFTS-affected township reflect the diffusion direction of the disease. (Authors' own figure.)

Table 55.1. Baseline demographic characteristics of patients with laboratory-confirmed severe fever with thrombocytopaenia syndrome (SFTS) in China from 2010 to 2018, for the whole nation and stratified by cluster. (Authors' own table.)

	Total	Cluster I	Cluster II	Cluster III	Cluster IV	Other
No. of confirmed cases	7721	386	1510	1128	4286	411
Annual incidence (/10^5)	0.064	0.099	0.92	0.16	0.20	0.0048
Average annual % change (95% CI)	18.5 (3.5, 35.6)	14 (2.3, 27)	18.9 (10, 28.5)	31.6 (27, 36.4)	11.2 (−4.5, 29.4)	25.1 (10.2, 42.2)
No. of deaths	810	20	185	107	450	48
Case-fatality ratio % (95% CI)	10.5 (9.8, 11.2)	5.2 (3.3, 8)	12.3 (10.7, 14)	9.5 (7.9, 11.4)	10.5 (9.6, 11.5)	11.7 (8.8, 15.3)
Age in years, median (IQR)*	63 (54–70)	62 (55–70)	64 (55–71)	63 (55–71)	62 (52–70)	65 (55–73)
No. (%) of female cases*	4063 (52.6)	165 (42.7)	730 (48.3)	567 (50.3)	2383 (55.6)	218 (53)
Days from disease onset to admission, median (IQR)	6 (4–9)	6 (3–9)	6 (4–9)	6 (4–8)	6 (4–9)	5 (2–8)
Seasonality	May–August	July–September	May–August	May–August	April–July	May–August

IQR, interquartile range.
*Significant difference between the four clusters ($P < 0.05$).

percentage coverage of waterbodies, altitude, yearly average temperature, yearly accumulated precipitation, monthly mean pressure, mean two-minute wind speed, duration of sunshine, yearly average relative humidity) and animal density (domestic and wild animals) can affect the risk of human infection with SFTSV (Liu *et al.*, 2015; Wang, G.S. *et al.*, 2017; Wang *et al.*, 2020; Wu *et al.*, 2020). Person-to-person transmission of SFTSV, mainly through excretions containing high levels of SFTSV, have been recorded (Bao *et al.*, 2011). Previous studies on a family cluster of SFTS cases suggested direct contact with patient blood and/or bloody secretions is a risk factor; those without effective personal protection were associated with increased odds ratio of secondary infection (Ye and Qi, 2021). So far, there is no evidence suggesting air-borne or faecal–oral transmission between humans.

55.1.4 Epidemiological distribution in animal vectors

The virus life cycle and natural mechanisms of sustained transmission remain poorly understood, but transmission to humans via tick bites is deemed the most probable route (Zhang *et al.*, 2012; Xing *et al.*, 2016; Wang, T. *et al.*, 2017). The public health concern of SFTSV is primarily posed by the wide distribution of its vectors and the broad spectrum of its vertebrate hosts (Yu *et al.*, 2011). *Haemaphysalis longicornis*, a tick species with a wide distribution and broad host range in China, was confirmed to be a competent vector of SFTSV, evidenced by successful SFTSV transmission through bites on mice in an experimental study (Hu *et al.*, 2020). On the other hand, there are other tick species that co-circulate with *H. longicornis* in the SFTS-endemic region; some of them were found to carry SFTSV by molecular evidence, e.g. *Rhipicephalus microplus*, *Amblyomma testudinarium*, *Haemaphysalis flava* and *Ixodes nipponensis*, however their vector competence has not been clarified (Yu *et al.*, 2011; Zhang *et al.*, 2012; Yun *et al.*, 2014; Meng *et al.*, 2015; Wang *et al.*, 2015). There is no strong evidence of SFTSV vectors other than ticks in the endemic region, such as mosquitoes or mites (Yu *et al.*, 2011; Xiong *et al.*, 2012).

The natural reservoir hosts of SFTSV have not yet been fully clarified, although SFTSV-specific antibody has been detected in a wide variety of animals, including mice, dogs, goats, cattle, chickens, wild birds, weasels, cats, pigs, sheep, geese, hedgehogs, spotted doves, shrews, deer, minks, rodents and yaks, with the possibility of some animals acting as reservoir hosts (Jiao *et al.*, 2012; Zhao *et al.*, 2012; Cui *et al.*, 2013; Niu *et al.*, 2013; Diang *et al.*, 2014; Li *et al.*, 2014; Liu *et al.*, 2014; Jin *et al.*, 2015; Sun *et al.*, 2017; Tian *et al.*, 2017; Wang, G.S. *et al.*, 2017; Chen *et al.*, 2019; Huang *et al.*, 2019; Li *et al.*, 2019; Matsuu *et al.*, 2019; Yu *et al.*, 2019). Many epidemiological studies supported the risk of human infection through contact with animals, such as cattle, goats and cats (Ding *et al.*, 2014; Li *et al.*, 2014; Liang *et al.*, 2014; Xing *et al.*, 2017). Ecological studies have revealed associations between the migration route and transmission paths of SFTSV, suggesting migratory birds may play a role in spreading SFTSV (Yun *et al.*, 2015; Miao *et al.*, 2020). However, it is difficult to specify which species might be the true amplifiers or reservoirs. No clinical infection has ever been reported from the aforementioned animals, except for cats. In Japan, SFTS-like clinical signs were reported for infected domestic cats (Matsuu *et al.*, 2019). Experimental evidence also verified their high susceptibility to SFTSV (Hwang *et al.*, 2017; Kang *et al.*, 2020). In addition, Japan has reported two cases of cat-to-human transmission, causing clinical infection in three veterinary personnel, raising the potential of SFTSV transmission without a tick as vector (Kida *et al.*, 2019; Yamanaka *et al.*, 2020). Whether cats might be an overlooked animal reservoir in China remains to be investigated.

The spatial-temporal distribution and spread of SFTSV in the recent epidemic foci of China likely depend on complex interactions between multiple ecological, environmental and societal variables. As environmental variables may affect ticks, their hosts and habitat, environmental data can potentially be used directly and as proxies in predictive models. Meteorological factors and risk of SFTS have been intensively explored. Ecological niche models (ENMs) have been applied to analyse the association between meteorological factors, environmental factors and epidemics of SFTS in Shandong and Jiangsu provinces, in eastern China. Based on the data from January 2010 to April 2013 in Shandong Province, Du *et al.* (2014) revealed that the air temperature, precipitation, land cover area, normalized difference vegetation index and duration of sunshine were significantly related to the spatial distribution of SFTS. The suitable ranges for SFTS occurrence were annual mean temperature of 11.6 to 12.8°C and January mean precipitation of 5.5 to 7.8 mm (Du *et al.*, 2014). Another study in Jiangsu Province indicated that slope and maximum temperature of the warmest month were key environmental factors for SFTS occurrence (Zhang *et al.*, 2019). However, a study in which an ENM was similarly applied on the national reported data from 2011 to 2018 revealed that altitude, yearly average temperature, yearly accumulated precipitation and yearly average relative humidity accounted for 94.1% of the total effect in the model. SFTS occurrence probability was high where the yearly average temperature, yearly accumulated precipitation and yearly relative humidity were 12.5–17.5°C, 700–2250 mm and 63–82%, respectively (Sun *et al.*, 2021).

Poisson regression analysis has been applied in other studies. Wang, T. *et al.* (2017) reported that the spatial distribution of SFTS in Hubei Province was significantly associated with cattle density, farmland coverage ratio, temperature and relative humidity. When a boosted regression tree model was applied on the national surveillance data for SFTS from 2010 to 2013, the occurrence of SFTSV human infection was found to be significantly associated with temperature, rainfall, relative humidity, sunshine hours, elevation, distribution of *H. longicornis* ticks, cattle density and coverage of forest, with most of the factors showing non-linear relationships with the risk of SFTS. Temperature was found to have the maximum correlation with the monthly incidence of SFTS, followed by rainfall, sunshine hours and relative humidity (Liu *et al.*, 2015).

In most studies, the percentage contribution of the yearly average temperature to the spatial distribution of SFTS is independently significant. In some studies, this effect is higher than most other factors, and the relationship between the yearly average temperature and the risk of SFTS occurrence is non-linear. Heterogeneity of spatial-temporal transmission, as well as significant risk factors, were often observed across studies.

Most of the suitable environmental factors are conducive to tick growth and reproduction, resulting in an increasing density of ticks, as well as being suitable for human outdoor activities and hence increasing the risk to humans of tick exposure. Yearly accumulated precipitation also had a non-linear relationship with the spatial distribution of SFTS. With these findings, it is also inferred that a certain range of precipitation is beneficial to growth and reproduction of SFTSV's competent vectors.

55.2 Future Predictions

The complex enzootic cycles of SFTSV are made up of a combination of interactions between multiple factors, which include: (i) competent tick species implicated as vectors; (ii) reservoir hosts for immature and adult stages of the ticks; (iii) landscape and vegetation types; (iv) meteorological factors; and (v) socio-economic factors, among others. All the assessments of published studies have only touched on the basic aspects of the ecoepidemiology of the disease, mostly on habitat, landscape and climate, while lacking in-depth investigation on the combination of interactions.

H. longicornis ticks have been incriminated in the transmission of SFTSV under field conditions and by experimental transmission assays, but the role of other ticks found infected in nature remains to be tested under experimental conditions. The involvement of these species in the epidemiology of the virus under natural conditions is unclear. The survival of SFTSV in ticks under natural conditions is important from an epidemiological point of view. However, there is still a dearth of knowledge regarding the infection rates and related factors that might affect the transmission rate and harbouring rate of the virus under different environmental factors. Ticks are highly sensitive to small changes in climate; long-term, sustained and small differences in key climate variables may drive a serious change. Furthermore, transmission potential by ticks is very sensitive to climatically induced variations in tick development rates (Randolph, 1998).

According to Miao *et al.* (2020), it seems reasonable to conclude that SFTSV has already reached its full geographic extent where suitable conditions currently exist for its survival. However, we cannot be sure that the virus is not existing in hidden enzootic cycles, where there is seroprevalence but not human case reports. Most of the previous studies are also subject to such limitation, in that case reports are obtained from passive surveillance of patients presenting clinical signs who sought medical care in hospitals. The inclusion of serological data from both human beings and animal hosts might enhance the predictive power of models in the future. The role of birds in the establishment of ticks in new areas, the rates of antibody production in animals and the impact of climate or host abundance on the epidemiological chain must be included before their impact can be fully elucidated.

References

Bao, C.-J., Guo, X.-L., Qi, X., Hu, J.L., Zhou, M.H., *et al.* (2011) A family cluster of infections by a newly recognized bunyavirus in eastern China, 2007: further evidence of person-to-person transmission. *Clinical Infectious Diseases* 53, 1208–1214.

Chen, C., Li, P., Li, K.F., Wang, H.L., Dai, Y.X., *et al.* (2019) Animals as amplification hosts in the spread of severe fever with thrombocytopenia syndrome virus: a systematic review and meta-analysis. *International Journal of Infectious Diseases* 79, 77–84.

Cui, F., Cao, H.X., Wang, L., Zhang, S.F., Ding, S.J., *et al.* (2013) Clinical and epidemiological study on severe fever with thrombocytopenia syndrome in Yiyuan County, Shandong Province, China. *American Journal of Tropical Medicine and Hygiene* 88, 510–512.

Diang, S., Yin, H., Xu, X., Liu, G., Jiang, S., *et al.* (2014) A cross-sectional survey of severe fever with thrombocytopenia syndrome virus infection of domestic animals in Laizhou City, Shandong Province, China. *Japanese Journal of Infectious Diseases* 67, 1–4.

Ding, F., Guan, X.H., Kang, K., Ding, S.J., Huang, L.Y., *et al.* (2014) Risk factors for bunyavirus-associated severe fever with thrombocytopenia syndrome, China. *PLoS Neglected Tropical Diseases* 8, e3267.

Du, Z., Wang, Z., Liu, Y., Wang, H., Xue, F., *et al.* (2014) Ecological niche modeling for predicting the potential risk areas of severe fever with thrombocytopenia syndrome. *International Journal of Infectious Diseases* 26, 1–8.

Hu, Y.Y., Zhuang, L., Liu, K., Sun, Y., Dai, K., *et al.* (2020) Role of three tick species in the maintenance and transmission of severe fever with thrombocytopenia syndrome virus. *PLoS Neglected Tropical Diseases* 14, e0008368.

Huang, X.Y., Du, Y.H., Wang, H.F., You, A.G., Li, Y., *et al.* (2019) Prevalence of severe fever with thrombocytopenia syndrome virus in animals in Henan Province, China. *Infectious Diseases of Poverty* 8, 56.

Hwang, J., Kang, J.G., Oh, S.S., Chae, J.B., Cho, Y.K., *et al.* (2017) Molecular detection of severe fever with thrombocytopenia syndrome virus (SFTSV) in feral cats from Seoul, Korea. *Ticks and Tick-borne Diseases* 8, 9–12.

Jiao, Y., Zeng, X., Guo, X., Qi, X., Zhang, X., *et al.* (2012) Preparation and evaluation of recombinant severe fever with thrombocytopenia syndrome virus nucleocapsid protein for detection of total antibodies in human and animal sera by double-antigen sandwich enzyme-linked immunosorbent assay. *Journal of Clinical Microbiology* 50, 372–377.

Jin, C., Jiang, H., Liang, M., Han, Y., Gu, W., *et al.* (2015) SFTS virus infection in nonhuman primates. *Journal of Infectious Diseases* 211, 915–925.

Kang, J.G., Cho, Y.K., Han, S.W., Jeon, K., Choi, H., *et al.* (2020) Molecular and serological investigation of severe fever with thrombocytopenia syndrome virus in cats. *Vector-Borne and Zoonotic Diseases* 20, 916–920.

Kida, K., Matsuoka, Y., Shimoda, T., Matsuoka, H., Yamada, H., *et al.* (2019) A case of cat-to-human transmission of severe fever with thrombocytopenia syndrome virus. *Japanese Journal of Infectious Diseases* 72, 356–358.

Kim, K.H., Yi, J., Kim, G., Choi, S.J., Jun, K.I., *et al.* (2013) Severe fever with thrombocytopenia syndrome, South Korea, 2012. *Emerging Infectious Diseases* 19, 1892–1894.

Li, D.X. (2015) Severe fever with thrombocytopenia syndrome: a newly discovered emerging infectious disease. *Clinical Microbiology and Infection* 21, 614–620.

Li, Y., Zhou, H., Mu, D., Yin, W. and Yu, H. (2015) Epidemiological analysis on severe fever with thrombocytopenia syndrome under the national surveillance data from 2011 to 2014, China. *Zhonghua Liu Xing Bing Xue Za Zhi* 36, 598–602.

Li, Z., Hu, J., Bao, C., Li, P., Qi, X., *et al.* (2014) Seroprevalence of antibodies against SFTS virus infection in farmers and animals, Jiangsu, China. *Journal of Clinical Virology* 60, 185–189.

Li, Z.F., Bao, C.J., Hu, J.L., Gao, C.F., Zhang, N., *et al.* (2019) Susceptibility of spotted doves (*Streptopelia chinensis*) to experimental infection with the severe fever with thrombocytopenia syndrome phlebovirus. *PLoS Neglected Tropical Diseases* 13, 14.

Liang, S., Bao, C., Zhou, M., Hu, J., Tang, F., *et al.* (2014) Seroprevalence and risk factors for severe fever with thrombocytopenia syndrome virus infection in Jiangsu Province, China, 2011. *American Journal of Tropical Medicine and Hygiene* 90, 256–259.

Liu, J.W., Wen, H.L., Fang, L.Z., Zhang, Z.T., He, S.T., *et al.* (2014) Prevalence of SFTSV among Asian house shrews and rodents, China, January–August 2013. *Emerging Infectious Diseases* 20, 2126–2128.

Liu, K., Zhou, H., Sun, R.X., Yao, H.W., Li, Y., *et al.* (2015) A national assessment of the epidemiology of severe fever with thrombocytopenia syndrome, China. *Scientific Reports* 5, 9679.

Matsuu, A., Momoi, Y., Nishiguchi, A., Noguchi, K., Yabuki, M., *et al.* (2019) Natural severe fever with thrombocytopenia syndrome virus infection in domestic cats in Japan. *Veterinary Microbiology* 236, 108346.

Meng, K., Sun, W., Cheng, Z., Guo, H., Liu, J., *et al.* (2015) First detection of severe fever with thrombocytopenia syndrome virus in the tick species *Haemaphysalis concinna* in Shandong Province, China. *Parasitology Research* 114, 4703–4707.

Miao, D., Liu, M.J., Wang, Y.X., Ren, X., Lu, Q.B., *et al.* (2020) Epidemiology and ecology of severe fever with thrombocytopenia syndrome in China, 2010–2018. *Clinical Infectious Diseases* 2020, ciaa1561.

Niu, G., Li, J., Liang, M., Jiang, X., Jiang, M., *et al.* (2013) Severe fever with thrombocytopenia syndrome virus among domesticated animals, China. *Emerging Infectious Diseases* 19, 756–763.

Randolph, S.E. (1998) Ticks are not insects: consequences of contrasting vector biology for transmission potential. *Parasitology Today* 14, 186–192.

Sun, J., Lu, L., Wu, H., Yang, J., Ren, J., *et al.* (2017) The changing epidemiological characteristics of severe fever with thrombocytopenia syndrome in China, 2011–2016. *Scientific Reports* 7, 9236.

Sun, J.M., Wu, H., Lu, L., Liu, Y., Mao, Z.Y., *et al.* (2021) Factors associated with spatial distribution of severe fever with thrombocytopenia syndrome. *Science of the Total Environment* 750, 141522.

Takahashi, T., Maeda, K., Suzuki, T., Ishido, A., Shigeoka, T., *et al.* (2014) The first identification and retrospective study of severe fever with thrombocytopenia syndrome in Japan. *Journal of Infectious Diseases* 209, 816–827.

Tian, H., Yu, P., Chowell, G., Li, S., Wei, J., *et al.* (2017) Severe fever with thrombocytopenia syndrome virus in humans, domesticated animals, ticks, and mosquitoes, Shaanxi Province, China. *American Journal of Tropical Medicine and Hygiene* 96, 1346–1349.

Wang, G.S., Wang, J.B., Tian, F.L., Zhang, H.J., Yin, F.F., *et al.* (2017) Severe fever with thrombocytopenia syndrome virus infection in minks in China. *Vector-Borne and Zoonotic Diseases* 17, 596–598.

Wang, S., Li, J., Niu, G., Wang, X., Ding, S., *et al.* (2015) SFTS virus in ticks in an endemic area of China. *American Journal of Tropical Medicine and Hygiene* 92, 684–689.

Wang, T., Li, X.L., Liu, M., Song, X.J., Zhang, H., *et al.* (2017) Epidemiological characteristics and environmental risk factors of severe fever with thrombocytopenia syndrome in Hubei Province, China, from 2011 to 2016. *Frontiers in Microbiology* 8, 387.

Wang, X., Qi, C., Zhang, D.D., Li, C.Y., Zheng, Z.L., *et al.* (2020) Epidemic character and environmental factors in epidemic areas of severe fever with thrombocytopenia syndrome in Shandong Province. *Ticks and Tick-borne Diseases* 12, 101593.

Wen, H.L., Zhao, L., Zhai, S., Chi, Y., Cui, F., *et al.* (2014) Severe fever with thrombocytopenia syndrome, Shandong Province, China, 2011. *Emerging Infectious Diseases* 20, 1–5.

Wu, H., Wu, C., Lu, Q., Ding, Z., Xue, M., *et al.* (2020) Spatial-temporal characteristics of severe fever with thrombocytopenia syndrome and the relationship with meteorological factors from 2011 to 2018 in Zhejiang Province, China. *PLoS Neglected Tropical Diseases* 14, e0008186.

Xing, X., Guan, X., Liu, L., Zhan, J., Jiang, H., *et al.* (2016) Natural transmission model for severe fever with thrombocytopenia syndrome bunyavirus in villages of Hubei Province, China. *Medicine* 95, e2533.

Xing, X., Guan, X., Liu, L., Xu, J., Li, G., *et al.* (2017) A case–control study of risk sources for severe fever with thrombocytopenia syndrome in Hubei Province, China. *International Journal of Infectious Diseases* 55, 86–91.

Xiong, W.Y., Feng, Z.J., Matsui, T. and Foxwell, A.R. (2012) Risk assessment of human infection with a novel bunyavirus in China. *Western Pacific Surveillance Response Journal* 3, 61–66.

Yamanaka, A., Kirino, Y., Fujimoto, S., Ueda, N., Himeji, D., *et al.* (2020) Direct transmission of severe fever with thrombocytopenia syndrome virus from domestic cat to veterinary personnel. *Emerging Infectious Diseases* 26, 2994–2998.

Ye, C. and Qi, R. (2021) Risk factors for person-to-person transmission of severe fever with thrombocytopenia syndrome. *Infection Control and Hospital Epidemiology* 42, 582–585.

Yu, K.M., Jeong, H.W., Park, S.J., Kim, Y.I., Yu, M.A., *et al.* (2019) Shedding and transmission modes of severe fever with thrombocytopenia syndrome phlebovirus in a ferret model. *Open Forum Infectious Diseases* 6, ofz309.

Yu, X.J., Liang, M.F., Zhang, S.Y., Liu, Y., Li, J.D., *et al.* (2011) Fever with thrombocytopenia associated with a novel bunyavirus in China. *New England Journal of Medicine* 364, 1523–1532.

Yun, S.M., Lee, W.G., Ryou, J., Yang, S.C., Park, S.W., *et al.* (2014) Severe fever with thrombocytopenia syndrome virus in ticks collected from humans, South Korea, 2013. *Emerging Infectious Diseases* 20, 1358–1361.

Yun, Y., Heo, S.T., Kim, G., Hewson, R., Kim, H., *et al.* (2015) Phylogenetic analysis of severe fever with thrombocytopenia syndrome virus in South Korea and migratory bird routes between China, South Korea, and Japan. *American Journal of Tropical Medicine and Hygiene* 93, 468–474.

Zhang, D., Sun, C., Yu, H., Li, J., Liu, W., *et al.* (2019) Environmental risk factors and geographic distribution of severe fever with thrombocytopenia syndrome in Jiangsu Province, China. *Vector-Borne and Zoonotic Diseases* 19, 758–766.

Zhang, Y.Z., Zhou, D.J., Qin, X.C., Tian, J.H., Xiong, Y., *et al.* (2012) The ecology, genetic diversity, and phylogeny of Huaiyangshan virus in China. *Journal of Virology* 86, 2864–2868.

Zhao, L., Zhai, S., Wen, H., Cui, F., Chi, Y., *et al.* (2012) Severe fever with thrombocytopenia syndrome virus, Shandong Province, China. *Emerging Infectious Diseases* 18, 963–965.

eo56 Climate Change and Debilitating Symptom Complexes Attributed to Ticks in Australia

Mona Dehhaghi[1,4], Hamed Kazemi Shariat Panahi[1], Richard Schloeffel[1,2], Bernard Hudson[3,4], Benjamin Ruiwen[1] and Gilles Guillemin[1,4]*

[1]Macquarie University, Macquarie Park, NSW, Australia; [2]Grove Health Centre, Gordon, NSW, Australia; [3]Royal North Shore Hospital, Sydney, NSW, Australia; [4]PANDIS.org

56.1 Expert Opinion

56.1.1 Climate change and temperatures in Australia

As a consequence of global climate warming, Australia's climate has warmed on average by 1.44 ± 0.24°C since national records began in 1910, leading to an increase in the frequency of extreme heat events (Fig. 56.1). Most regions of the continent have experienced warming (Fig. 56.2).

56.1.2 Tick-borne diseases in Australia

Vector-borne diseases (VBDs) account for 17% of all infectious diseases, with more than 700,000 deaths annually worldwide (WHO, 2020). Over 40% of emerging zoonotic illnesses are vector-borne and over half the world's population is exposed to VBDs. It is thought that one-sixth of human illnesses and disability suffered worldwide is from VBDs, leading to significant morbidity and mortality globally (RACGP, 2018). The socio-economic and health impacts to individuals and society are a concern and, with global warming and an increasing risk of contracting a VBD, the world's population (including that of Australia) is vulnerable.

According to various reports, six of the eight subfamilies of ticks – Amblyomminae, Argasinae, Bothriocrotinae, Haemaphysalinae, Ixodinae and Ornithodorinae – have been recognized in Australia (Fig. 56.3). However, only 72 tick species (almost 8% of all valid species) are endemic in Australia; of these, 17 species are known to feed on humans and other animals (Dehhaghi *et al.*, 2019). Infectious agents found in tick-borne disease patients in Australia include *Borrelia*, *Rickettsia*, *Bartonella*, *Babesia* and *Coxiella burnetii*. A large number of viruses also have been found in ticks (Harvey *et al.*, 2019). The dilemma arises that not all ticks in Australia carry these pathogens, whereas in many parts of the northern hemisphere, particularly in North America, Europe, China, Africa and throughout Asia, ticks are known to carry a number of pathogens.

Q fever caused by *C. burnetii*, which can be transmitted by ticks, has been studied extensively. This illness may cause acute fevers, fatigue, cough, nausea, night sweats, headaches and myalgic

*Email: gilles.guillemin@mq.edu.au

DOI: 10.1079/9781789249637.0056

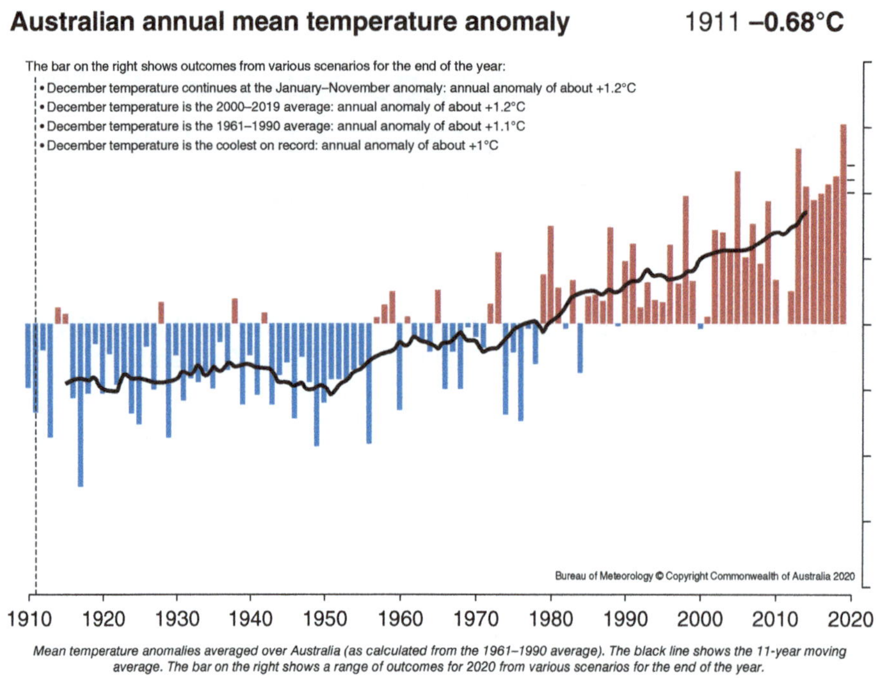

Fig. 56.1. Australian mean annual temperature anomaly. The black line depicts the 11-year moving average. (© Australian Bureau of Meteorology, reproduced with permission.)

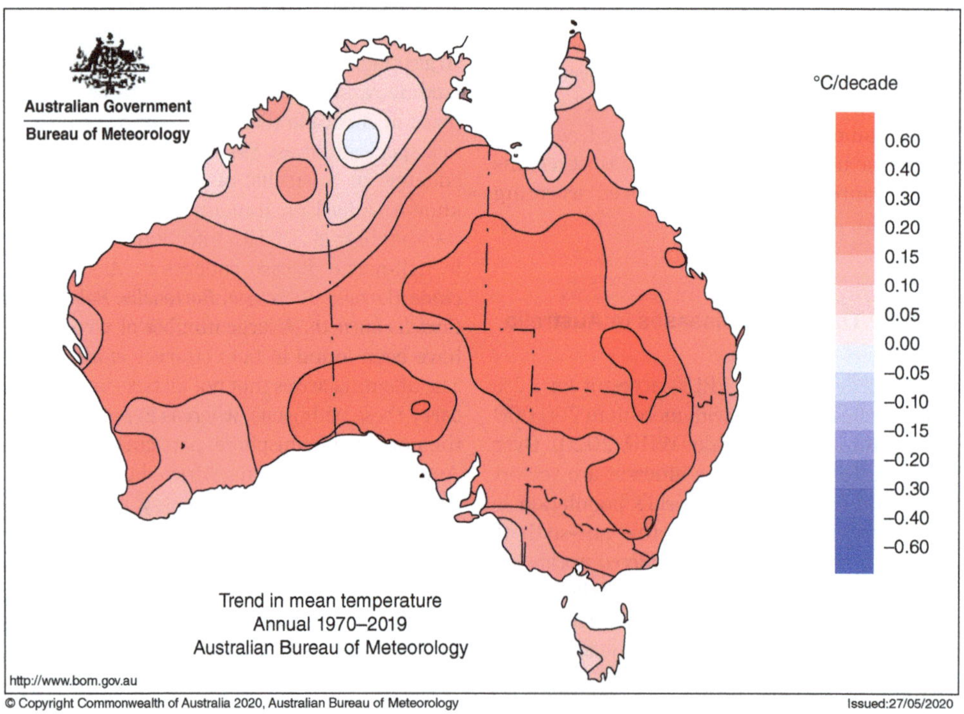

Fig. 56.2. Map of Australian mean temperatures between 1970 and 2019 showing that temperatures have increased in most of the continent over the last 50 years. (© Australian Bureau of Meteorology, reproduced with permission.)

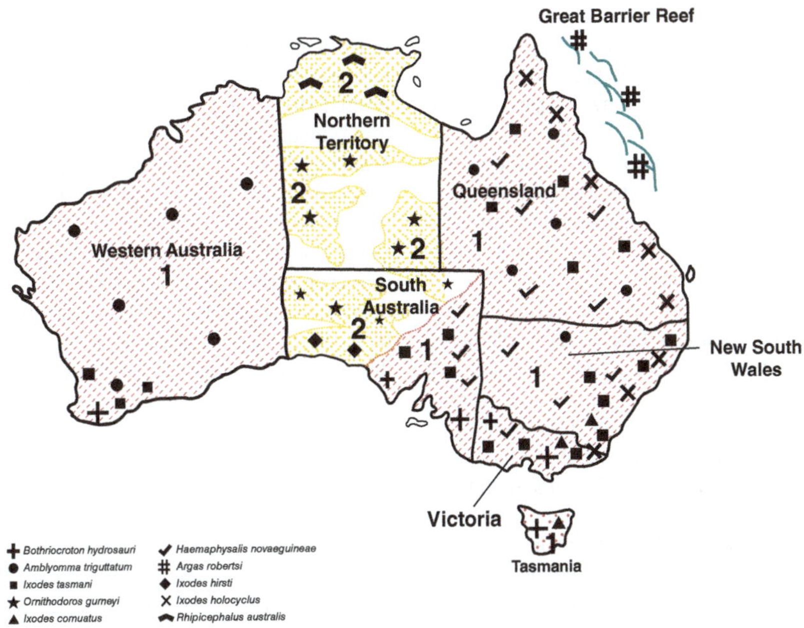

Fig. 56.3. Distribution of the most important potentially human-biting ticks of Australia, showing tick-borne diseases (red and 1) and potential tick-borne diseases (yellow and 2). (From Dehhaghi *et al.*, 2019, reproduced with permission.)

pain, often resulting in a chronic fatigue syndrome. Australian tick typhus (*Rickettsia australis* and Flinders' Island rickettsia) can cause acute and some forms of chronic illness. *Bartonella* infections (*Bartonella henselae*) may be acquired from cat fleas and ticks, causing an acute illness with cat scratch fever, striae, fevers and lymphadenopathy (Stenos and Graves, 2018). This infection is now recognized as a cause of tick-borne illness and associated with chronic neurological disorders.

Borreliosis infections like Lyme disease and Lyme-like illnesses are present in many other continents of the world. Research into Australian ticks has identified minimal evidence of *Borrelia* species. Many Australian patients are reporting acute and chronic symptoms suggestive of a locally acquired borreliosis (similar to patients overseas) but the lack of diagnostic pathology and absence of *Borrelia* identification are hampering diagnosis and appropriate treatment. Consequently, in Australia, Lyme disease is not recognized as such by the medical associations and the government. Lyme disease, originally described in North America, is specifically caused by the spirochaete *Borrelia burgdorferi* sensu stricto. Scientific literature dating back to 1959 has demonstrated that other species of *Borrelia* are present in Australia (Mackerras and Mackerras, 1960; Dehhaghi *et al.*, 2019). *Borrelia theileri* and *Borrelia anserina*, which are the causative agents of bovine borreliosis and avian spirochaetosis, respectively, have been reported in Queensland, New South Wales and Northern Territory (Mulhearn, 1946; Dehhaghi *et al.*, 2019). *Borrelia queenslandica* was isolated from *Rattus villosissimus* in north-western Queensland (Carley and Pope, 1962). Unfortunately, since the original reports, only limited and sometimes controversial research has been performed (Chalada *et al.*, 2016; Collignon *et al.*, 2016; Newby, 2019). The Australian Government has acknowledged that there is a group of Australian patients suffering from the symptoms of a 'chronic debilitating illness', which many associate with a tick bite (DoH, 2020).

The Australian Government recently released a guideline for Debilitating Symptom Complexes Attributed to Ticks (DSCATT) which is not supportive of a Lyme-like illness being present in Australia (DoH, 2020). The guidelines fall short in assisting Australians with tick-borne illnesses similar to other parts of the world from being diagnosed. In the USA, the Centers for Disease Control and Prevention has reported 30,000 to 40,000 patients diagnosed with Lyme disease and co-infections each year (Mayer, 2021), with evidence of climate change impacts increasing disease incidence (CDC, 2019; DoH, 2020). Doctors working in Western Australia have reported cases of Lyme-like illnesses which have yet to be scientifically confirmed.

The two most common ticks that bite humans are the paralysis tick (*Ixodes holocyclus*) and, in Western Australia, the kangaroo tick (*Amblyomma triguttatum*), which are known particularly to cause infectious diseases such as rickettsioses (Barker and Barker, 2018). Attachment of an adult female *I. holocyclus*, usually in spring and summer, can induce paralysis (Barker and Walker, 2014; Dehhaghi *et al.*, 2019). Although the larvae cannot feed on humans and are only associated with localized dermatitis, nymphs of *I. holocyclus* can attach and feed on humans and cause acute anaphylactic shock and paralysis due to injection of neurotoxins (i.e. holocyclotoxin) (Dehhaghi *et al.*, 2019). It is important to note that humans are usually accidental hosts for many of Australia's ticks; however, tick-borne diseases can negatively affect human health.

Several non-infectious illnesses from tick bites include allergic reactions with localized inflammation to anaphylaxis, generalized urticarias, and paralysis reactions if bitten around the neck, especially in young children. More recently, Professor Sheryl van Nunen has identified the mammalian meat allergy antigens which are transmitted by ticks and cause patients to have significant reactions to beef, pork and lamb. Often a severe and chronic condition, patient diagnoses are supported by the recently developed anti-alpha-gal test (van Nunen, 2018).

56.1.3 Climate change, Australian ticks and associated human diseases

Ticks are relatively sensitive to temperature and humidity. They do not tolerate well conditions that are too cold or too hot and dry. Australian mild winters help them survive; their eggs hatch sooner, lengthening the feeding and moulting season. Australasian migratory birds help them move across Australia into areas where cooler temperatures would have once significantly limited their survival and proliferation (Barker and Walker, 2014; Barker *et al.*, 2014; Kwak and Madden, 2017). Ticks become active when temperatures reach 7–8°C and they thrive in areas with at least 85% humidity. Warming temperatures associated with climate change are likely to increase the range of suitable tick habitat in Australia and are therefore one of multiple factors driving the observed spread of tick-borne diseases. Interestingly, it has been reported that rising temperature could switch the host preference of dog ticks (*Rhipicephalus sanguineus*) to bite humans rather than dogs. However, it is unclear why dog ticks change their host preference at higher temperatures (Pratap, 2020).

Climate change can affect human health in various ways, including by facilitating the spread of infectious diseases. In September 2019, the President of the Australian Medical Association, Dr Tony Bartone, released a document declaring 'there is clear scientific evidence indicating severe impacts for our patients and communities now and into the future due to climate change affecting our health. The warming of the planet will affect human health and well-being by increasing the environment and situation in which infectious disease can be transmitted.' He stated that 'climate change will cause increases in the transmission of vector-borne diseases' (Murphy, 2019). The concept of VBD permeates through human history with plagues, malaria, dengue fever, chikungunya virus and more recently Zika virus, and in Australia, Ross River fever and Barmah Forest virus disease (DoH, 2019; Murphy, 2019).

VBDs may be linked to several factors including increasing global temperature, deforestation, alteration in demographics and industrialization (Semenza and Suk, 2018; Caminade *et al.*, 2019). According to a government report, more than 80% of Australia's people live within 50 km of coastal regions and two-fifths of the total population reside in Sydney and Melbourne (Cechet *et al.*, 2011) (Fig. 56.4a). The Australian coast supports diverse species, ecosystems and habitats. Human migration to coastal regions has revealed a continuing demographic trend of depopulation of interior areas (Smith *et al.*, 2016). High humidity and rainfall in coastal

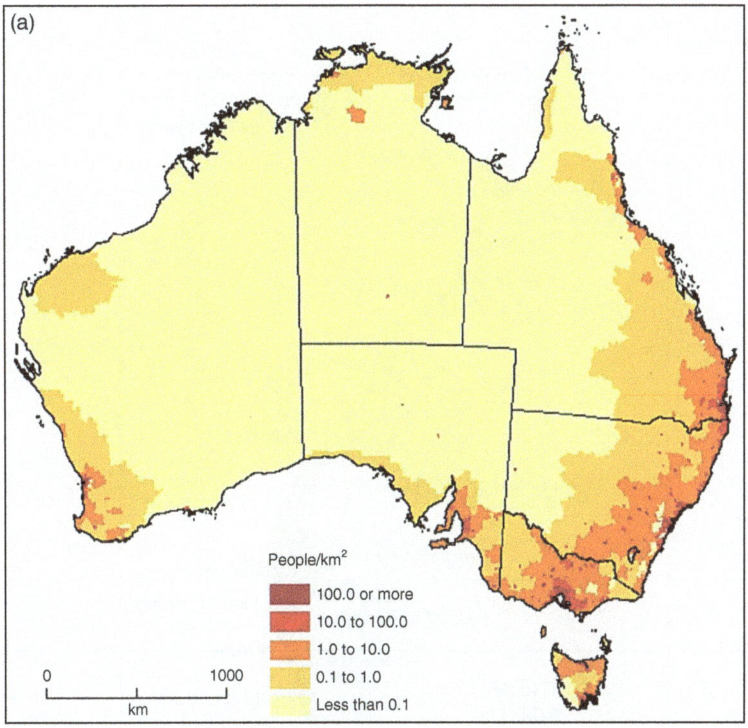

Source: Regional Population Growth, Australia (3218.0)

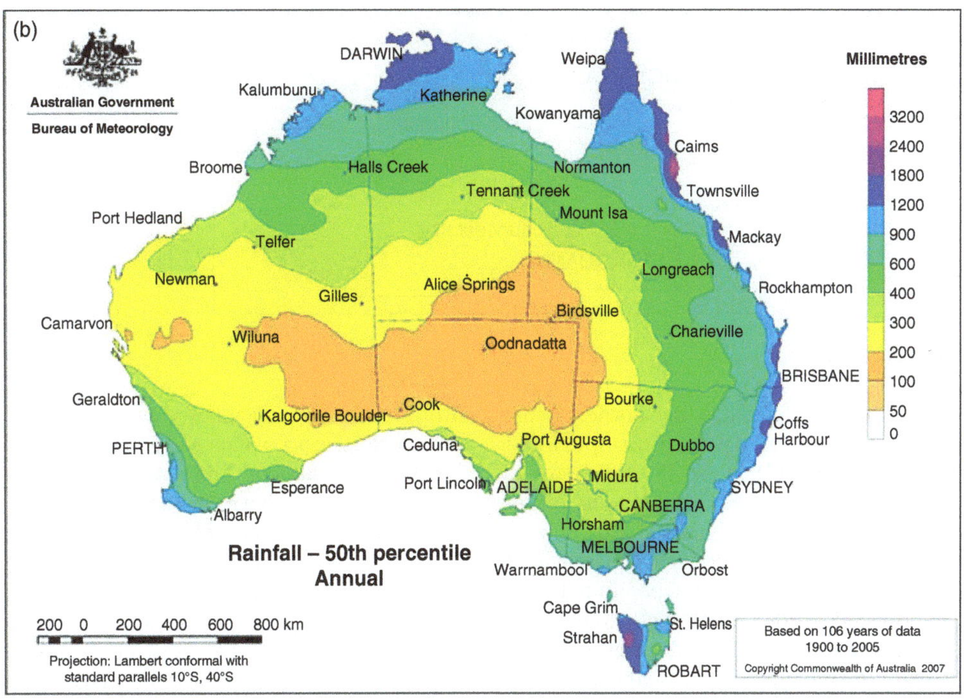

Fig. 56.4. (a) Population density in Australia in 2010 (Australian Bureau of Statistics, 2010); (b) average annual rainfall (Bureau of Meteorology, 2000); and (c) distribution of *Ixodes holocyclus* in coastal regions (from Teong *et al*., 2015). (Reproduced with permission from the cited sources.)

Fig. 56.4. Continued.

regions possibly influence tick-borne disease transmission and tick development (Bureau of Meteorology, 2000) (Fig. 56.4b). As shown in Fig. 56.4c, the paralysis tick, *I. holocyclus*, one of the medically important ticks in Australia, is usually found in the eastern coastline region of Australia, which is the most populated area (Jackson *et al.*, 2007; Teong *et al.*, 2015).

Moreover, warm winters, as well as lingering warmth in summer, increase the activity of hosts for several tick species, facilitating tick dispersal. Accordingly, scientists forecast that future increases in weather temperature may increase the tick population and enhance the risk of tick-borne infections. Professor Peter Irwin, a researcher into tick-borne diseases in Australia, suggests that the impacts of climate change on the habitat suitability to ticks will change greatly over the next few decades (eo61: Oskam *et al.*). Professor Irwin has also looked at tick populations, from all around Australia, seeking pathogens that may well infect humans. It has been known for at least the past 25 years that many Australians are reporting development of acute and chronic illnesses associated with tick bites (Irwin *et al.*, 2017).

56.2 Future Predictions

Much is known in regard to most VBDs, with general acceptance by the medical fraternity throughout Australia that these illnesses are present. Mosquito-borne infections are notifiable but tick-borne infections are not reportable Australia-wide. VBDs are being identified as they move into different populated areas, moving south with changes in the environment and with changes in climate, precipitation and global warming. The difficulty lies primarily in the tick-borne illnesses which have lacked scientific research in both ticks and patients. Climate change and VBDs will require in-depth scientific investigations into ticks and patients, especially those who have a history of an acute or chronic illness. Research using novel technologies, including highly specific serology to identify pathogens, and the adoption of metagenomics to elucidate viruses, bacteria and other parasites, is needed for diagnosis of patient illness. Advanced biomarker research into inflammatory and immune changes in the body will improve the understanding of patients with tick-borne illnesses. It is known in Europe that ticks represent 'nature's dirty needle'. There is no reason to think that Australia is any different from other parts of the world inhabited by ticks in respect to pathogens carried by these vectors. It is only with hindsight and good research that we will delineate the types of pathogens that are present in tick populations. This is becoming a more urgent need in Australia as the climate is changing and more people already are, and will be even more, affected by vector bites, including those of ticks. One issue the authors have in regard to tick-borne illnesses in Australia is that they are not notifiable. As illness related to tick bite is of significant public concern, and there exists considerable controversy about infections that may be transmitted by ticks, it is very important that notification of illness caused by known tick-borne pathogens should be mandatory for all Federal and State Health Departments of Australia. Agreed case definitions are required, just as they are for other notifiable diseases. Apparent illness related to tick bite that does not fit the agreed case definitions could also be monitored nationally, to facilitate detection of infections caused by novel agents. Climate change will increase the need for education and prevention programmes to assist human populations dealing with increasing VBDs. The challenges of global warming are present here and now. The way forward is not only to reduce the human impact on the environment but also to meet the challenges of emerging illnesses impacting all nations, including Australia, with the science of climate change and VBDs.

Acknowledgements

We thank Peter Owen from the Tick-borne Illness Community Network Australia Inc. (TICNA) for his help with the manuscript. Ms Mona Dehhaghi and Mr Hamed Kazemi Shariat Panahi are both supported by an International PhD Scholarship from Macquarie University. Professor Guillemin is supported by the National Health and Medical Research Council (NHMRC), the Handbury Foundation, the Snow Foundation, PANDIS.org and Macquarie University. Tick-borne disease research by Dr Hudson is supported by the National Health and Medical Research Council (NHMRC), the Snow Foundation and PANDIS.org.

References

Australian Bureau of Statistics (2010) Regional population 2010. Available at: https://www.abs.gov.au/ausstats/abs@.nsf/Lookup/by%20Subject/1301.0~2012~Main%20Features~Geographic%20distribution%20of%20the%20population~49 (accessed 23 June 2021).

Barker, S.C. and Barker, D. (2018) Ticks in Australia: endemics; exotics; which ticks bite humans? *Microbiology Australia* 39, 194–199.

Barker, S.C. and Walker, A.R. (2014) Ticks of Australia. The species that infest domestic animals and humans. *Zootaxa* 3816, 1–144.

Barker, S.C., Walker, A.R. and Campelo, D. (2014) A list of the 70 species of Australian ticks; diagnostic guides to and species accounts of *Ixodes holocyclus* (paralysis tick), *Ixodes cornuatus* (southern paralysis tick) and *Rhipicephalus australis* (Australian cattle tick); and consideration of the place of Australia in the evolution of ticks with comments on four controversial ideas. *International Journal for Parasitology* 44, 941–953.

Bureau of Meteorology (2000) Climatic Atlas of Australia – Rainfall. Bureau of Meteorology, Commonwealth of Australia. Available at: http://www.bom.gov.au/climate/how/newproducts/images/rainatlas_sample2.shtml (accessed 23 June 2021).

Caminade, C., McIntyre, K.M. and Jones, A.E. (2019) Impact of recent and future climate change on vector-borne diseases. *Annals of the New York Academy of Sciences* 1436, 157–173.

Carley, J. and Pope, J. (1962) A new species of *Borrelia* (*B. queenslandica*) from *Rattus villosissimus* in Queensland. *Australian Journal of Experimental Biology and Medical Science* 40, 255–261.

CDC (2019) Division of Vector-Borne Diseases (DVBD). Centers for Disease Control and Prevention, Atlanta, Georgia. Available at: https://www.cdc.gov/ncezid/dvbd/index.html (accessed 15 June 2021).

Cechet, B., Taylor, P., Griffin, C. and Hazelwood, M. (2011) Australia's coastline: adapting to climate change. *AUSGEO News*, March 2011, issue 101. Geoscience Australia, Commonwealth of Australia. Available at: http://www.ga.gov.au/ausgeonews/ausgeonews201103/climate.jsp (accessed 15 June 2021).

Chalada, M.J., Stenos, J. and Bradbury, R.S. (2016) Is there a Lyme-like disease in Australia? Summary of the findings to date. *One Health* 2, 42–54.

Collignon, P.J., Lum, G.D. and Robson, J.M. (2016) Does Lyme disease exist in Australia? *Medical Journal of Australia* 205, 413–417.

Dehhaghi, M., Panahi, H.K.S., Holmes, E.C., Hudson, B.J., Schloeffel, R. and Guillemin, G.J. (2019) Human tick-borne diseases in Australia. *Frontiers in Cellular and Infection Microbiology* 9, 3.

DoH (2019) Ross River virus and Barmah Forest virus. Department of Health, Tasmanian Government. Available at: https://www.dhhs.tas.gov.au/publichealth/communicable_diseases_prevention_unit/infectious_diseases/ross_river (accessed 15 June 2021).

DoH (2020) Debilitating Symptom Complexes Attributed to Ticks (DSCATT). Australian Government, Department of Health. Available at: https://www1.health.gov.au/internet/main/publishing.nsf/Content/ohp-lyme-disease.htm (accessed 15 June 2021).

Harvey, E., Rose, K., Eden, J.-S., Lo, N., Abeyasuriya, T., *et al.* (2019) Extensive diversity of RNA viruses in Australian ticks. *Journal of Virology* 93, e01358-18.

Irwin, P.J., Robertson, I.D., Westman, M.E., Perkins, M. and Straubinger, R.K. (2017) Searching for Lyme borreliosis in Australia: results of a canine sentinel study. *Parasites & Vectors* 10, 114.

Jackson, J., Beveridge, I., Chilton, N. and Andrews, R. (2007) Distributions of the paralysis ticks *Ixodes cornuatus* and *Ixodes holocyclus* in south-eastern Australia. *Australian Veterinary Journal* 85, 420–424.

Kwak, M.L. and Madden, C. (2017) The first record of infestation by a native tick (Acari: Ixodidae) on the Australian emu (*Dromaius novaehollandiae*) and a review of tick paralysis in Australian birds. *Experimental and Applied Acarology* 73, 103–107.

Mackerras, M.J. and Mackerras, I. (1960) The haematozoa of Australian birds. *Australian Journal of Zoology* 8, 226–260.

Mayer, M. (2021) Mapping Lyme: CDC reveals distribution of Lyme disease-causing bacteria by county. *Entomology Today*, 22 January 2021. Available at: https://entomologytoday.org/2021/01/22/mapping-lyme-cdc-reveals-distribution-of-lyme-disease-causing-bacteria-county-ticks/ (accessed 23 June 2021).

Mulhearn, C. (1946) A note on two blood parasites of cattle (*Spirochaeta theileri* and *Bartonella bovis*) recorded for the first time in Australia. *Australian Veterinary Journal* 22, 118.

Murphy, K. (2019) Australian Medical Association declares climate change a health emergency. *The Guardian*, 2 September 2019. Available at: https://www.theguardian.com/australia-news/2019/sep/03/australian-medical-association-declares-climate-change-a-health-emergency (accessed 23 June 2021).

Newby, K. (2019) *Bitten: The Secret History of Lyme Disease and Biological Warfare*. HarperCollins, New York.

Pratap, A. (2020) Dog ticks carrying deadly Rocky Mountain spotted fever twice as likely to choose humans over canines when temperatures rise. *ScienceNews*, 30 November 2020. Available at: https://www.sciencenews.org/article/dog-ticks-bite-human-blood-disease-climate-change (accessed 23 June 2021).

RACGP (2018) The impact of climate change on human health. Royal Australian College of General Practitioners. Available at: https://www.racgp.org.au/advocacy/position-statements/view-all-position-statements/clinical-and-practice-management/the-impact-of-climate-change-on-human-health (accessed 15 June 2021).

Semenza, J.C. and Suk, J.E. (2018) Vector-borne diseases and climate change: a European perspective. *FEMS Microbiology Letters* 365, fnx244.

Smith, T., Leitch, A. and Thomsen, D. (2016) Climate change impacts on coastal communities. In: *Community Engagement. Information Manual 9*. National Climate Change Adaptation Research Facility, Gold Coast, Australia. Available at: https://nccarf.edu.au/wp-content/uploads/2019/03/IM09_community_engagement.pdf (accessed 15 June 2021).

Stenos, J. and Graves, S.R. (2018) Laboratory diagnosis of human infections transmitted by ticks, fleas, mites and lice in Australia. *Microbiology Australia* 39, 182–184.

Teong, J.M., Adler, P.A., Doggett, S.L., Daneshvar, D. and Shields, M. (2015) Conjunctival attachment of a live paralysis tick, *Ixodes holocyclus*, in a child: a case report. *Case Reports in Ophthalmology* 6, 120–126.

van Nunen, S.A. (2018) Tick-induced allergies: mammalian meat allergy and tick anaphylaxis. *Medical Journal of Australia* 208, 316–321.

WHO (2020) Vector-borne diseases. World Health Organization, Geneva, Switzerland. Available at: https://www.who.int/news-room/fact-sheets/detail/vector-borne-diseases#:~:text=Vector%2Dborne%20diseases%20account%20for,infection%20transmitted%20by%20Anopheline%20mosquitoes (accessed 15 June 2021).

eo57 Effect of Climate Change on Mosquito-Borne Pathogens

Michael Turell*
VectorID LLC, Frederick, Maryland, USA

57.1 Expert Opinion

57.1.1 Effect of environmental temperature on pathogen transmission

Environmental temperature is known to affect the ability of mosquitoes to transmit a variety of pathogens (Davis, 1932; Bates and Roca-Garcia, 1946; Chamberlain and Sudia, 1955; Hurlbut, 1973; Turell *et al.*, 1985, 2020; Watts *et al.*, 1987; Dohm and Turell, 2001; Dohm *et al.*, 2002; Reisen *et al.*, 2006). For many of these mosquito-transmitted pathogens, increased temperature enhances, or is even critical, for the mosquito to be able to transmit the pathogen. Various studies have shown an association between increased temperature and increased amount of disease caused by mosquito-borne pathogens (Hess *et al.*, 1963; Paz, 2006; Ruiz *et al.*, 2010; Hahn *et al.*, 2015; Cheng *et al.*, 2020). In addition, many mosquitoes are unable to survive in cold climates. Thus, there is concern that with increased global warming many tropical diseases will be able to extend their range into more temperate climates (Baylis, 2017; Rossati, 2017; Ogden and Gachon, 2019; Ryan *et al.*, 2019; El-Sayed and Kamel, 2020; Fischer *et al.*, 2020). While there have been recent widespread outbreaks of viruses in regions where these viruses have not previously been reported, e.g. West Nile virus (WNV), chikungunya virus (CHIKV) and Zika virus (ZIKV) in the western hemisphere, the question that remains is, are these recent outbreaks due to climate change or to some other factor?

57.1.2 A brief history lesson

Despite being considerably cooler in the 18th and 19th centuries in Europe and North America, mosquito-borne diseases such as malaria and yellow fever were significant causes of human disease in both these places. For example, during this time period, there were >100,000 deaths in the USA from yellow fever alone, with major outbreaks involving thousands of deaths in Philadelphia, New York and Boston (Patterson, 1992). Yellow fever also occurred in Europe and an outbreak in Barcelona in 1821 was responsible for about 20,000 deaths (Chastel, 1999). Similarly, malaria (or relapsing fever or ague, as it was often called) was a common cause of illness and death in Europe up until the 20th century, including hundreds of thousands of cases of malaria in Finland alone during the 19th century (Hulden and Hulden, 2009). Likewise, malaria caused massive numbers of cases and deaths in the USA and southern Canada up until the 20th

*Email: mturell@erols.com

 Climate, Ticks and Disease (ed. P. Nuttall)
DOI: 10.1079/9781789249637.0057

century (Patterson, 1992; Kakkilaya, 2016). Malaria was responsible for 1.3 million cases and 10,000 deaths in soldiers in the Union Army during the 4 years of the American Civil War (Kakkilaya, 2016). Therefore, despite the much cooler temperatures in Europe and the USA than those in the tropics or even those occurring in Europe and the USA today, malaria and yellow fever were significant causes of disease and death. Outbreaks of both of these diseases essentially disappeared from Europe and the USA during the 20th century. It was not cooler temperatures that prevented mosquitoes from transmitting these pathogens in cooler climates, but rather socio-economic changes (i.e. screening, air conditioning, piped water, television) that explain why these pathogens essentially disappeared from Europe and the USA but yet remain a problem in tropical areas (Moreno-Madriñán and Turell, 2017, 2018).

57.1.3 Zoonotic versus anthroponotic diseases

Many of the more common mosquito-borne human pathogens (e.g. malaria *Plasmodium* spp., yellow fever virus (YFV), dengue virus (DENV), CHIKV and ZIKV) are anthroponotic pathogens, i.e. during outbreaks, the primary vertebrate involved is a human. In contrast, for pathogens such as WNV, Japanese encephalitis, St. Louis encephalitis, eastern equine encephalitis and Rift Valley fever viruses, vertebrates other than humans serve as the primary vertebrate amplifying host (i.e. the animal from which the mosquito obtains an infectious blood meal). This has major implications for the persistence and expansion of range of these pathogens and the diseases that they cause. For example, as mentioned above, yellow fever and malaria were once both very common causes of disease and death in both Europe and North America, even when temperatures were considerably cooler than they are now. Because in outbreaks, humans are the source of virus or parasite, the mosquito that transmits these pathogens must take two separate blood meals on a human, one to pick up the pathogen and one to transmit it. In most of Europe and North America, socio-economic conditions (e.g. screening, air conditioning, piped water, etc.) have greatly reduced contact between mosquitoes and humans. We still get a lot of mosquito bites, but it has become extremely unlikely for a single individual mosquito to obtain two separate blood meals on a human. This has led to the essential disappearance of major outbreaks of these anthroponotic pathogens in North America and Europe. However, birds, cattle, horses and other wild vertebrates do not live in air-conditioned houses and spend hours each day/night serving as a blood source for mosquitoes. Therefore, increases in socio-economic status do not protect these vertebrates from becoming infected and serving as a source of pathogens for mosquitoes. For example, the natural cycle for WNV is between birds and a variety of mosquito species (dependent on the geographic region). An infected mosquito bites a bird, it becomes viraemic and is bitten by another mosquito, and the cycle goes on. Every once in a while, that infected mosquito bites a human (or a horse) and the mammal can develop West Nile fever or even die. However, neither the horse nor the human develops a viraemia sufficient to infect a mosquito, so they are referred to as a 'dead-end host' and are not involved in the transmission cycle. For these zoonotic pathogens, increased temperature is often associated with increased transmission rates (Hahn *et al.*, 2015). While increased temperature increases the likelihood of transmission of many mosquito-borne pathogens, other factors, primarily socio-economic status, may be even more important, at least for anthroponotic pathogens. However, for zoonotic pathogens, higher environmental temperature is likely to lead to increased pathogen transmission and increased disease.

57.1.4 Risk of the spread of exotic viruses

The continued increase in rapid transportation of people and merchandise has led to the introduction of novel viruses into regions where they were previously unknown. One only has to look at the Americas to see what happens when a novel virus such as WNV, CHIKV or ZIKV is introduced. The explosive outbreaks of chikungunya and Zika were not due to climate change, but rather to a virus being introduced into an area with essentially no immunity. Think of the introduction of SARS-CoV-2 virus. As known by the persistence of DENV in much of tropical America, the mosquito vector, *Aedes aegypti*, is well established and socio-economic conditions allow for close contact between people and this vector. Therefore, when a novel virus is introduced that is primarily transmitted by *A. aegypti*, the results can be explosive, with millions of cases. On the other hand, in areas with higher socio-economic status like the USA, despite over 10,000 documented imported cases of chikungunya and Zika (CDC, 2020a,b), there were only a few hundred locally acquired cases, and all of them in southern Florida or along the Texas–Mexico border, areas known to be infested with *A. aegypti*. Therefore, despite the warmer climate today than in the 1700s and 1800s when yellow fever caused major outbreaks as far north as Boston, current socio-economic conditions prevent large outbreaks. In contrast, when the zoonotic WNV was introduced into New York in 1999 (CDC, 1999), it rapidly became established and spread throughout the Americas. As mentioned earlier, this virus does better in warmer conditions, so a general increase in average temperatures would allow this virus to expand its distribution northward and allow for more severe outbreaks in areas where it is already established.

57.1.5 Summary

Because anthroponotic arthropod-borne pathogens require the same individual mosquito to feed twice on a human, the current socio-economic status of the USA and most of Europe (i.e. screening, air conditioning, piped water, etc.) make it unlikely for a new anthroponotic pathogen to become established. However, warmer temperatures increase the risk of the introduction and severity of zoonotic pathogens.

57.2 Future Predictions

Anthroponotic pathogens (e.g. DENV, YFV, ZIKV or malaria parasites) are unlikely to cause large outbreaks in Europe or the USA, due to the current use of screening, air conditioning, piped water, etc. However, localized outbreaks of a few cases might occur, similar to what is happening with malaria in the USA (i.e. about 1500 imported cases and 0–3 locally transmitted cases each year). In contrast, as illustrated by the introduction of WNV, higher socio-economic status does not protect against the introduction and establishment of zoonotic pathogens. Two pathogens, Japanese encephalitis virus (JEV) and Rift Valley fever virus (RVFV), are of particular concern. WNV is in the JEV complex and both viruses have a similar transmission cycle. Both vertebrate amplifying hosts (e.g. birds and swine) and North American mosquitoes have been shown to be competent vectors for JEV (Nemeth *et al.*, 2012; Huang *et al.*, 2015; Park *et al.*, 2018). Therefore, if JEV is introduced into the Americas (or Europe), it is likely that, like WNV, JEV would become established. Unfortunately, the disease in humans caused by JEV is much more severe than that caused by WNV, and despite there being an effective vaccine for JEV, about 10,000 people die every year from infection with JEV (Mansfield *et al.*, 2017). Of even greater concern is Rift Valley fever. RVFV has been expanding its range. Prior to 1977, when it caused a major outbreak in Egypt (Meegan, 1979), this virus was only known to exist in sub-Saharan Africa. Then in 2000, it was detected on the Arabian Peninsula for the first time (Jupp *et al.*, 2002). Although primarily a disease of domestic animals (cattle, goats and sheep) with nearly a 100% mortality rate in very young animals and almost a 100% abortion rate in pregnant animals, it can also cause disease in humans (Easterday, 1965; Meegan and Bailey, 1988; Bird *et al.*, 2009). To further complicate matters, unlike most other zoonotic viruses that do not produce a high viraemia in humans, RVFV can produce a viraemia $>10^8$ infectious units (Meegan, 1979), more than enough to infect a mosquito. It is possible for a family to visit Kenya and go on a safari. On their last day, while seeing the elephants, one of the family members is bitten by an infected mosquito (note: mosquito bites are common in the area). The family leaves to fly home the next day and arrives about 40 h later. The person bitten by the infected mosquito feels miserable, but attributes it to a combination of jet lag and spending nearly 40 h travelling, so just rests in a hammock and feeds the local mosquitoes for a day or two. Because they feel so miserable, they do not even notice the mosquitoes biting them, and the virus is now introduced. Unfortunately, many North American and European mosquito species are competent laboratory vectors for RVFV (Gargan *et al.*, 1988; Turell *et al.*, 2008, 2010; Iranpour *et al.*, 2011; Birnberg *et al.*, 2019) and the vertebrate amplifying hosts, cattle, goats and sheep, are also common. There are also other zoonotic viruses, and we must be prepared to deal with any novel viruses that might be introduced.

References

Bates, M. and Roca-Garcia, M. (1946) The development of the virus of yellow fever in *Haemagogus* mosquitoes. *American Journal of Tropical Medicine and Hygiene* 26, 585–605.

Baylis, M. (2017) Potential impact of climate change on emerging vector-borne and other infections in the UK. *Environmental Health* 16, 112.

Bird, B.H., Ksiazek, T.G., Nichol, S.T. and Maclachlan, N.J. (2009) Rift Valley fever virus. *Journal of the American Veterinary Medical Association* 234, 883–893.

Birnberg, L., Talavera, S., Aranda, C., Núñez, A.I., Napp, S. *et al.* (2019) Field-captured *Aedes vexans* (Meigen, 1830) is a competent vector for Rift Valley fever phlebovirus in Europe. *Parasites & Vectors* 12, 484.

CDC (Centers for Disease Control and Prevention) (1999) Outbreak of West Nile-like viral encephalitis – New York, 1999. *Morbidity and Mortality Weekly Report* 48, 845–849.

CDC (2020a) Chikungunya virus in the United States. Centers for Disease Control and Prevention, Atlanta, Georgia. Available at: https://www.cdc.gov/chikungunya/geo/united-states.html (accessed 14 November 2020).

CDC (2020b) Zika virus, statistics and maps. Centers for Disease Control and Prevention, Atlanta, Georgia. Available at: https://www.cdc.gov/zika/reporting/case-counts.html (accessed 14 November 2020).

Chamberlain, R.W. and Sudia, W.D. (1955) The effects of temperature upon the extrinsic incubation of eastern equine encephalitis in mosquitoes. *American Journal of Hygiene* 62, 295–305.

Chastel, C. (1999) The 'plague' of Barcelona. Yellow fever epidemic of 1821. *Bulletin de la Societe de Pathologie Exotique* 92, 405–407.

Cheng, J., Bambrick, H., Yakob, L., Devine, G., Frentiu, F.D., *et al.* (2020) Heatwaves and dengue outbreaks in Hanoi, Vietnam: new evidence on early warning. *PLoS Neglected Tropical Diseases* 14, e0007997.

Davis, N.C. (1932) The effect of various temperatures in modifying the extrinsic incubation period of yellow fever virus in *Aedes aegypti*. *American Journal of Hygiene* 16, 163–176.

Dohm, D.J. and Turell, M.J. (2001) Effect of incubation at overwintering temperatures on the replication of West Nile virus in New York *Culex pipiens* (Diptera: Culicidae). *Journal of Medical Entomology* 38, 462–464.

Dohm, D.J., O'Guinn, M.L. and Turell, M.J. (2002) Effect of environmental temperature on the ability of *Culex pipiens* (Diptera: Culicidae) to transmit West Nile virus. *Journal of Medical Entomology* 39, 221–225.

Easterday, B.C. (1965) Rift Valley fever. *Advances in Veterinary Science and Comparative Medicine* 10, 65–127.

El-Sayed, A. and Kamel, M. (2020) Climatic changes and their role in emergence and re-emergence of diseases. *Environmental Science and Pollution Research* 27, 22336–22352.

Fischer, L., Gültekin, N., Kaelin, M.B., Fehr, J. and Schlagenhauf, P. (2020) Rising temperature and its impact on receptivity to malaria transmission in Europe: a systematic review. *Travel Medicine and Infectious Disease* 36, 101815.

Gargan, T.P. 2nd, Clark, G.G., Dohm, D.J., Turell, M.J. and Bailey, C.L. (1988) Vector potential of selected North American mosquito species for Rift Valley fever virus. *American Journal of Tropical Medicine and Hygiene* 38, 440–446.

Hahn, M.B., Monaghan, A.J., Hayden, M.H., Eisen, R.J., Delorey, M.J., *et al.* (2015) Meteorological conditions associated with increased incidence of West Nile virus disease in the United States, 2004–2012. *American Journal of Tropical Medicine and Hygiene* 92, 1013–1022.

Hess, A.D., Churubin, C.E. and Lamotte, L.C. (1963) Relationship of temperature to activity of western and St. Louis encephalitis viruses. *American Journal of Tropical Medicine and Hygiene* 12, 657–667.

Huang, Y.J., Harbin, J.N., Hettenbach, S.M., Maki, E., Cohnstaedt, L.W., *et al.* (2015) Susceptibility of a North American *Culex quinquefasciatus* to Japanese encephalitis virus. *Vector-Borne and Zoonotic Diseases* 15, 709–711.

Hulden, L. and Hulden, L. (2009) The decline of malaria in Finland – the impact of the vector and social variables. *Malaria Journal* 8, 94.

Hurlbut, H.S. (1973) The effect of environmental temperature upon the transmission of St. Louis encephalitis virus by *Culex pipiens quinquefasciatus*. *Journal of Medical Entomology* 10, 1–12.

Iranpour, M., Turell, M.J. and Lindsay, L.R. (2011) Potential for Canadian mosquitoes to transmit Rift Valley fever virus. *Journal of the American Mosquito Control Association* 27, 363–369.

Jupp, P.G., Kemp, A., Grobbelaar, A., Lema, P., Burt, F.J., *et al.* (2002) The 2000 epidemic of Rift Valley fever in Saudi Arabia: mosquito vector studies. *Medical and Veterinary Entomology* 16, 245–252.

Kakkilaya, B.S. (2016) Malaria in wars and victims. Available at: https://www.malariasite.com/tag/civil-war/ (accessed 29 October 2020).

Mansfield, K.L., Hernández-Triana, L.M., Banyard, A.C., Fooks, A.R. and Johnson, N. (2017) Japanese encephalitis virus infection, diagnosis and control in domestic animals. *Veterinary Microbiology* 201, 85–92.

Meegan, J.M. (1979) The Rift Valley fever epizootic in Egypt 1977–78. 1. Description of the epizootic and virological studies. *Transactions of the Royal Society of Tropical Medicine and Hygiene* 73, 618–623.

Meegan, J.M. and Bailey, C.L. (1988) Rift Valley fever. In: Monath, T.P. (ed.) *The Arboviruses: Epidemiology and Ecology*, vol. IV. CRC Press, Boca Raton, Florida, pp. 61–76.

Moreno-Madriñán, M.J. and Turell, M. (2017) Factors of concern regarding Zika and other *Aedes aegypti*-transmitted viruses in the United States. *Journal of Medical Entomology* 54, 251–257.

Moreno-Madriñán, M.J. and Turell, M. (2018) History of mosquito-borne diseases in the United States and implications for new pathogens. *Emerging Infectious Diseases* 24, 821–826.

Nemeth, N., Bosco-Lauth, A., Oesterle, P., Kohler, D. and Bowen, R. (2012) North American birds as potential amplifying hosts of Japanese encephalitis virus. *American Journal of Tropical Medicine and Hygiene* 87, 760–767.

Ogden, N.H. and Gachon, P. (2019) Climate change and infectious diseases: what can we expect? *Canada Communicable Disease Report* 45, 76–80.

Park, S.L., Huang, Y.S., Lyons, A.C., Ayers, V.B., Hettenbach, S.M., *et al.* (2018) North American domestic pigs are susceptible to experimental infection with Japanese encephalitis virus. *Scientific Reports* 8, 7951.

Patterson, K.D. (1992) Yellow fever epidemics and mortality in the United States, 1693–1905. *Social Science & Medicine* 34, 855–865.

Paz, S. (2006) The West Nile virus outbreak in Israel (2000) from a new perspective: the regional impact of climate change. *International Journal of Environmental Health Research* 16, 1–13.

Reisen, W.K., Fang, Y. and Martinez, V.M. (2006) Effects of temperature on the transmission of West Nile virus by *Culex tarsalis* (Diptera: Culicidae). *Journal of Medical Entomology* 43, 309–317.

Rossati, A. (2017) Global warming and its health impact. *International Journal of Occupational and Environmental Medicine* 8, 7–20.

Ruiz, M.O., Chaves, L.F., Hamer, G.L., Sun, T., Brown, W.M., *et al.* (2010) Local impact of temperature and precipitation on West Nile virus infection in *Culex* species mosquitoes in northeast Illinois, USA. *Parasites & Vectors* 3, 19.

Ryan, S.J., Carlson, C.J., Mordecai, E.A. and Johnson, L.R. (2019) Global expansion and redistribution of *Aedes*-borne virus transmission risk with climate change. *PLoS Neglected Tropical Diseases* 13, e0007213.

Turell, M.J., Rossi, C.A. and Bailey, C.L. (1985) Effect of extrinsic incubation temperature on the ability of *Aedes taeniorhynchus* and *Culex pipiens* to transmit Rift Valley fever virus. *American Journal of Tropical Medicine and Hygiene* 34, 1211–1218.

Turell, M.J., Dohm, D.J., Mores, C.N., Terracina, L., Wallette, D.L. Jr, *et al.* (2008) Potential for North American mosquitoes to transmit Rift Valley fever virus. *Journal of the American Mosquito Control Association* 24, 502–507.

Turell, M.J., Wilson, W.C. and Bennett, K.E. (2010) Potential for North American mosquitoes (Diptera: Culicidae) to transmit Rift Valley fever virus. *Journal of Medical Entomology* 47, 884–889.

Turell, M.J., Cohnstaedt, L.W. and Wilson, W.C. (2020) Effect of environmental temperature on the ability of *Culex tarsalis* and *Aedes taeniorhynchus* (Diptera: Culicidae) to transmit Rift Valley fever virus. *Vector-Borne and Zoonotic Diseases* 20, 454–460.

Watts, D.M., Burke, D.S., Harrison, B.A., Witmire, R.E. and Nisalak, A. (1987) Effect of temperature on the vector efficiency of *Aedes aegypti* for dengue 2 virus. *American Journal of Tropical Medicine and Hygiene* 36, 143–152.

eo58 *Ornithodoros* Tick Vectors and African Swine Fever Virus

Armanda Bastos[1]* and Fernando Boinas[2]
[1]University of Pretoria, Pretoria, South Africa; [2]University of Lisbon, Lisbon, Portugal

58.1 Expert Opinion

58.1.1 Introduction

African swine fever (ASF) virus, currently the only recognized DNA arbovirus, is the causative agent of a highly virulent and frequently lethal haemorrhagic fever of pigs and wild boar. The virus is readily transmitted between infected *Sus scrofa* (wild and domestic), as well as through ingestion of infected pork, fomites and some haematophagous vectors. Although several arthropod species have been evaluated for virus competence, argasids (soft ticks) of the genus *Ornithodoros* are the only known invertebrates that are reservoirs of the virus. *Ornithodoros* ticks are also biological vectors of several species of *Borrelia*, which cause tick-borne relapsing fever in humans (Boinas *et al*., 2014). In contrast to hard ticks (Ixodidae), soft ticks (Argasidae) do not have a dorsal scutum and they inhabit the host's resting/nesting sites. It is this endophilic (nidicolous) lifestyle, and relatively narrow host range, that are at the centre of predictions regarding the impact of climate change and other anthropogenic factors on ASF virus–tick persistence.

ASF virus was initially confined to an ancient sylvatic cycle in Africa involving warthogs, *Phacochoerus africanus*, and long-lived, soft ticks of the *Ornithodoros moubata* species complex, which inhabit warthog burrows (Fig. 58.1). The disease only became recognized following investigation of acute mortalities in domestic pigs (*S. scrofa*) occurring in close proximity to warthogs (Montgomery, 1921). While the importance of the *Ornithodoros* soft tick as both reservoir and vector of the virus is now readily appreciated, its central role in the epidemiology of ASF was not clarified until the disease was introduced to the Iberian Peninsula in the mid-20th century and researchers in Spain confirmed virus presence and persistence in the European *Ornithodoros* soft tick counterpart, *Ornithodoros erraticus*, that is associated with pigsties (Sanchez Botija, 1963). This resulted in renewed efforts to evaluate ASF virus presence in sylvatic cycles involving soft ticks, which, up until that point, had failed to detect the virus. These activities culminated in the first report of ASF virus in soft ticks sampled from warthog burrows in northern Tanzania (Plowright *et al*., 1969). ASF virus-infected tick populations have since been identified throughout eastern and southern Africa, albeit at very low levels: 0.06–5.1% (Table 58.1), the only exception being a recent report from Tanzania (18%; Peter *et al*., 2021). With the concern that accompanied the expanding global footprint of the genotype I ASF virus introduced to Iberia in the 1960s, a number of successive experimental infection studies were conducted to determine

*Email: adbastos@zoology.up.ac.za

DOI: 10.1079/9781789249637.0058

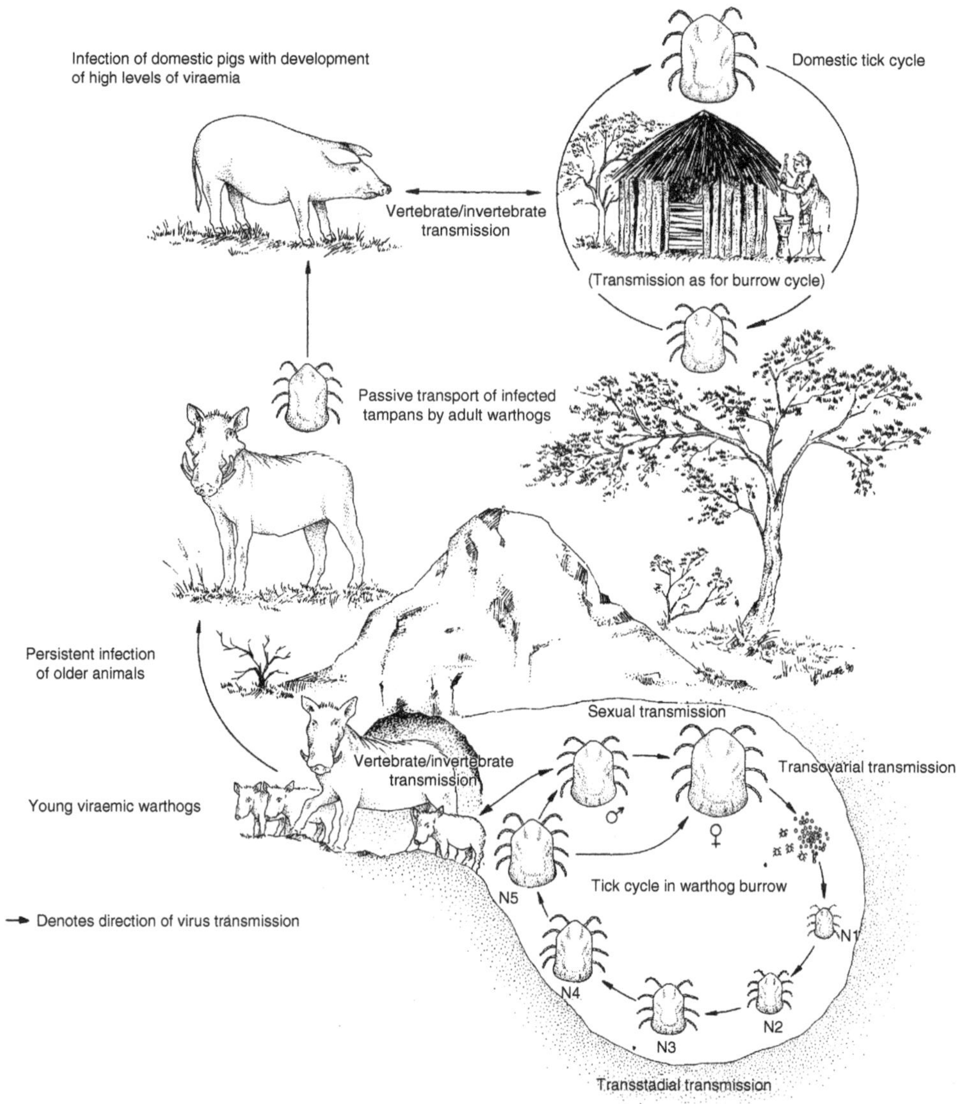

Fig. 58.1. Depiction of the African swine fever (ASF) virus cycles in sub-Saharan Africa, i.e. the sylvatic (warthog–tick) cycle and domestic (pig–tick) cycle. Cyclical amplification of the virus in the sylvatic cycle occurs when neonatal warthogs are infected for the first time and in turn infect naïve ticks. Transmission of ASF virus from the sylvatic/domestic cycle to domestic pigs is via infected *Ornithodoros moubata* complex ticks. Once introduced to domestic pigs, there is efficient horizontal transmission of the ASF virus between pigs. (From Penrith *et al.*, 2019b, with permission.)

vector competence and virus persistence in New World *Ornithodoros* soft tick species (Table 58.2). To date, just eight of the >190 potential soft tick vectors of ASF virus have been experimentally assessed, of which six have been shown capable of ASF virus transmission to pigs. In each case, different degrees of vector competence for ASF virus transmission were identified, including transstadial, sexual (male to female) and transovarial, within tick populations. Ticks of the *O. moubata* and *O. erraticus* species complexes were the most efficient, transmitting the virus by

Table 58.1. Summary of studies investigating the African swine fever (ASF) virus infection status of sylvatic cycle *Ornithodoros* tick populations in Africa. (Authors' own table.)

Country	Sampling locality	Sampling date	No. of infested burrows/ total (%)	No. of ASF-positive burrows	No. of ticks evaluated	Tick pooling	No. of positives (detection method)	Prevalence[a] (%): total (adult)	Reference/s
Uganda	Queen Elizabeth NP	NS	NK	0	>1100	Yes	0 (PBM culture)	0	Plowright *et al.* (1969)
Tanzania	Kirawira, Serengeti NP	1968	15/15 (100)	4	3,554	Yes	15 (PBM culture)	0.42 (3.8)	Plowright *et al.* (1969)
South Africa	Rooibokkraal	NS	NK	NK	14,023	Yes	45 (PBL culture)	0.3 (1.6)	Thomson *et al.* (1983)
South Africa	Mkuze NP	1978	13/40 (32.5)	NK	5,018	Yes	3 (PBL culture)	0.06 (NK)	Thomson *et al.* (1983); Thomson (1985)
South Africa	Kruger NP	1979 & 1981	10/18 (55.6)	NK	1,026	Yes	14 (PBL culture)	1.4 (4.2)	Thomson *et al.* (1983); Thomson (1985)
Namibia	Windhoek District	1980	NK	NK	2,349	Yes	11 (PBL culture)	0.5 (1.8)	Thomson *et al.* (1983)
Malawi	Liwonde NP	1985	1[b]	0	1,400	Yes	0 (PBM culture)	0	Haresnape *et al.* (1988)
Zambia	South Luangwa NP	1981–1982	13/36 (36.1)	5	1,492	Yes	6 (PBM culture)	0.4 (5.2)	Wilkinson *et al.* (1988)
Zambia	Kakumbe GMA	1982	5/11 (45.5)	3	696	Yes	11 (PBM culture)	1.6 (5.3)	Wilkinson *et al.* (1988)
Zambia	Kafue NP & Nalusanga GMA	1981–1983	41/70 (58.6)	7	1,476	Yes	21 (PBM culture)	1.4 (5.0)	Wilkinson *et al.* (1988)
Zambia	Livingstone GP	1982–1983	9/60 (15.0)	12	1,133	Yes	58 (PBM culture)	5.1 (13.2)	Wilkinson *et al.* (1988)
Zambia	Sumbu NP	1983	22/35 (62.9)	3	481	Yes	10 (PBM culture)	2.1 (15.1)	Wilkinson *et al.* (1988)
South Africa	Mkuze NP	2002	59/98 (60.2)	0	348	Yes	PCR	0	Arnot *et al.* (2009)
Mozambique	Gorongosa NP	2006–2007	29/32 (90.6)	21	1,662	Yes	27 (PBMC culture), 57 (nested PCR)	1.6 (NS)	Quembo *et al.* (2018)
Swaziland	Various GMAs ($n = 7$)	2012–2013	23/58 (39.7)	0	562	Yes	0 (PCR)	0	Boshoff *et al.* (2014)
Tanzania	Saadani NP	NS	5/12 (41.7)	5	111	No	20 (PCR)	18.0 (NS)	Peter *et al.* (2021)

GMA, Game Management Area; GP, Game Park; NK, not known; NP, National Park; NS, not specified; PBL, porcine blood leucocyte; PBM, pig bone marrow; PBMC, peripheral blood mononuclear cells.

[a]Prevalence for pooled samples assumes one positive tick per tick pool (following Plowright *et al.*, 1969).

[b]Culvert used by warthogs.

Table 58.2. *Ornithodoros* tick species experimentally evaluated for African swine fever virus vector competence. (From Boinas, 1994; with updates from Pereira de Oliveira *et al*., 2019.)

				Virus transmission			
					Within tick populations		
Ornithodoros species/complex	Distribution	Main hosts	Laboratory estimate of viral persistence (field-based estimate)	To pigs	Transovarial	Transstadial	Sexual
New World							
Ornithodoros coriaceus	Americas	Cattle, deer, pigs	502 days	Yes	No	Yes	–
Ornithodoros turicata	Americas	Prairie dogs, ground squirrels, snakes, cattle, pigs, tortoises	23 days	Yes	–	–	–
Ornithodoros puertoricensis	Americas	Humans	506 days	Yes	Yes	Yes	–
Ornithodoros parkeri	Americas	Rodents, owls, goats	70 days	No	–	–	–
Old World							
Ornithodoros erraticus complex	Europe/North Africa	Pigs, rodents, birds	655 days (1915 days)	Yes	Yes	Yes	Yes
Ornithodoros verrucosus	Europe/Caucasus	Mammals, birds	240 days	No	–	–	–
Ornithodoros moubata complex	Africa	Warthogs, pigs, humans	239 days (1095 days)	Yes	Yes	Yes	Yes
Ornithodoros savignyi	Africa and Asia	Livestock, wild animals, humans	106 days	Yes	No	Yes	–

all three routes. However, in some cases, transmission to pigs did not occur after ticks infected with ASF virus fed on pigs (Pereira de Oliveira *et al.*, 2019). The large disparities in the results under different experimental settings is likely due to a range of factors including tick developmental stage, virus titre and tick–virus diversity (Pereira de Oliveira *et al.*, 2020).

While ASF virus vector competence has been confirmed for a number of *Ornithodoros* species under laboratory settings (Table 58.2), reports of naturally infected ticks are limited to three species, namely *Ornithodoros sonrai* sampled from rodent burrows in Senegal (Vial *et al.*, 2007) and the two suid-associated tick species, *O. moubata* and *O. erraticus* (Boinas *et al.*, 2011; Jori and Bastos, 2009). It is unlikely that other Old World *Ornithodoros* species play a substantial role in virus maintenance and spread due to the distinct host and habitat preferences of these nidicolous ticks. Of the New World species, *Ornithodoros turicata* in the USA is of concern for potential establishment of the virus on introduction, as it has a relatively broad host range, inclusive of suids (Table 58.2), and occurs in states in which feral swine densities are particularly high (Brown and Bevins, 2018). While acknowledging that additional competent vectors may yet be found, current knowledge, based on experimental infection studies and on multiple records of naturally infected tick populations, indicate that there are just two Old World species complexes, *O. erraticus* in the Iberian Peninsula and *O. moubata* in sub-Saharan Africa, that are biologically relevant vectors of ASF virus.

58.1.2 Ticks of the *Ornithodoros erraticus* species complex

O. erraticus is found in patchy areas in Central and Southern Iberia in the generally hotter and drier environments of their temperate climate. This soft tick has only been associated with pigs produced in the traditional type of free-range pig herds of the Iberian/Alentejano extensive production system. Pigs are considered to be the main hosts of *O. erraticus* and other hosts, including humans, rodents, bovids, sheep and birds, are considered accidental. The habitat of *O. erraticus* in the Iberian Peninsula is in traditional pigsties and, occasionally, in animal burrows and bird nests in the vicinity. These ticks are usually found in the cracks and crevices of the traditional old buildings made of adobe and/or loose stone and occur only rarely in modern buildings or outside the sties (Boinas *et al.*, 2014). In contrast to ixodid ticks, which remain attached to the host for long periods of time, *O. erraticus* are rapid feeders (10–30 min to engorgement) after which they detach from the host and drop to the ground. This feeding strategy severely limits tick dispersal, although movement could feasibly occur over short distances through transfer of utensils contaminated with the parasite or by the passive transfer of ticks feeding on animals when they are being moved.

The biological cycle of *O. erraticus* includes the stages of eggs, larvae, nymphs (three to five stages) and adults. While there is a blood meal requirement to move through successive stages and reproduce, *Ornithodoros* ticks have a remarkable resistance to fasting: several reports confirm that the larger developmental stages can withstand fasting periods of >5 years. Moreover, these long-lived ticks have lifespans of 15–20 years (Boinas *et al.*, 2011). These features, together with a preference for hiding in fissures of buildings, make acaricide control difficult (Boinas *et al.*, 2014). In Iberia, the field activity of the tick varies with the season. The ticks are inactive in winter and restart their activity when the temperature inside the pigsties reaches 13–15°C. Although it takes 2–3 years to complete the cycle under field conditions (Encinas Grandes *et al.*, 1993), the biological cycle can be completed within 154 days in the laboratory under optimal temperature and humidity conditions (El Shoura, 1987).

Towards the end of the ASF-endemic period in Portugal and Spain (1980s and 1990s), sporadic outbreaks and seropositive pig herds were detected in areas in which the soft tick occurs. Surveys suggest that *O. erraticus* was responsible for ~5% of the outbreaks in Spain and Portugal, and an important source of repeat outbreaks following culling and restocking of premises with pigs (Boinas *et al.*, 2011). Legislation prohibiting pig repopulation of ASF-positive infested premises, for a period of 5 years, was introduced in 1999 after a sporadic ASF outbreak occurred in pigs that were housed in *O. erraticus*-infested premises that had been depopulated for several years (Boinas *et al.*, 2011). This legislation led to higher rates of abandonment of premises and traditional pig farming practices in many areas.

58.1.3 Ticks of the *Ornithodoros moubata* complex – the African sylvatic cycle

On the African continent, three ASF epizootiological cycles occur: (i) the sylvatic cycle in which the virus cycles between warthogs and ticks; (ii) a domestic pig–tick cycle that is reliant on traditional pig housing conducive to tick establishment; and (iii) a domestic pig cycle involving horizontal transmission between pigs without the agency of ticks or warthogs (Penrith *et al.*, 2019a). Historically, outbreaks were associated with the sylvatic cycle (Fig. 58.1) with control being reliant on the physical separation of domestic pigs from sylvatic cycle hosts. However, a review of the causes of outbreaks on the African continent over the last few decades revealed a discernible shift in the drivers of outbreaks, with the sylvatic cycle playing a minor role compared with anthropogenic/socio-economic factors (Penrith *et al.*, 2019a).

While *O. moubata* is known to be highly resilient to increased temperature and decreased relative humidity under laboratory conditions (Brett, 1939), and is predicted to have the capacity to expand its distributional range under a number of climate change scenarios (Cumming and van Vuuren, 2006), these abiotic/macroclimatic factors overlook the complex interdependence of biotic factors that influence tick survival and distribution. First, these eyeless ticks, like their *O. erraticus* counterparts, are confined to the resting/sleeping place of their preferred host (in this case, the common warthog) and, being rapid feeders, are rarely found on warthogs outside the burrow. Second, burrows offer a buffered microclimate to both the warthog and the ticks that inhabit them, making the availability of these burrows in the landscape one of the most critical considerations for predicting climate change impacts on the persistence of warthogs, ticks and virus in the sylvatic cycle. Third, the availability of these burrows is dependent on the aardvarks (*Orycteropus afer*) that dig them; aardvarks, being myrmecophagous, are in turn dependent on the termites and ants they feed upon. The persistence of the warthog–tick–ASF virus sylvatic cycle is thus reliant on the persistence of another vertebrate–invertebrate relationship.

Aardvarks are considered to be a keystone species because the burrows they excavate provide shelter to a broad range of animals, including warthogs. Although some burrows are relatively long-lived and can persist for months to years (Cumming, 1975), many are not permanent fixtures in the landscape; thus any negative impacts on aardvark survival will have substantial knock-on effects for all animals that are reliant on the persistence of burrows. Recent research from the Kalahari in South Africa suggests that climate change impacts will be particularly severe for aardvarks (Rey *et al.*, 2018). The significance of aardvarks and burrow loss will impact neonate warthog survival as they are unable to adequately thermoregulate and easily succumb to weather fluctuations if they do not have access to a sheltered burrow environment (Cumming, 1975).

58.2 Future Predictions

We predict that both *Ornithodoros* species complexes (*O. erraticus* and *O. moubata*) are likely to undergo localized extinctions in the future; the European soft ticks due to changing pig husbandry practices and their African counterparts due to climate change. The former prediction is based on the results of four surveys conducted in Portugal over a 25-year period which suggest that farm management practices, such as the availability of traditional pigsties, is more important than climate for *O. erraticus* persistence as both tick prevalence and suitable habitat decreased over the study period (Wilson *et al.*, 2013). This declining trend in suitable tick habitat is set to continue as initiatives to re-establish free-range pig production reliant on traditional premises were shown to be unprofitable, in part due to the impact of the ticks on pig health as well as the costly biosecurity restrictions aimed at limiting contact between free-ranging domestic herds and wild boar. This has led to a high rate of abandonment of the old traditional type of premises, or to their replacement with modern facilities in which soft ticks cannot persist because there is no place to hide.

The prediction regarding the fate of *O. moubata* and the ASF virus it maintains within a sylvatic cycle, in which neonate warthogs play an important role in cyclical amplification of the virus, is educated conjecture that requires investigation. While both ticks and warthogs will likely be shielded from climate change impacts in the short term, due to the buffered microclimate offered by the burrow environment, in the longer term this refuge is likely to become a finite resource in affected areas, at which point both warthogs and ticks will be negatively impacted. The study on aardvarks in the Kalahari, in which drought-induced starvation was reported from a study area that provides a window to future climate change impacts, highlights the vulnerability of this species to global warming. A reduction or absence of aardvarks from areas in which they previously occurred will have a cascading effect on all burrow-dependent species, including warthogs, *O. moubata* ticks and the ASF virus that these ticks maintain and vector.

References

Arnot, L.F., Du Toit, J.T. and Bastos, A.D.S. (2009) Molecular monitoring of African swine fever virus using surveys targeted at adult *Ornithodoros* ticks: a re-evaluation of Mkuze Game Reserve, South Africa. *Onderstepoort Journal of Veterinary Research* 76, 385–392.

Boinas, F.J.S. (1994) The role of *Ornithodoros erraticus* in the epidemiology of African swine fever in Portugal. PhD Thesis, University of Reading, Reading, UK.

Boinas, F.S., Wilson, A.J., Hutchings, G.H., Martins, C. and Dixon, L.J. (2011) The persistence of African swine fever virus in field-infected *Ornithodoros erraticus* during the ASF endemic period in Portugal. *PLoS ONE* 6, e20383.

Boinas, F., Ribeiro, R., Madeira, S., Palma, M., de Carvalho, I.L., *et al.* (2014) The medical and veterinary role of *Ornithodoros erraticus* complex ticks (Acari: Ixodida) on the Iberian Peninsula. *Journal of Vector Ecology* 39, 238–248.

Boshoff, C.I., Bastos, A.D.S., Dube, M.M. and Heath, L. (2014) First molecular assessment of the African swine fever virus status of *Ornithodoros* ticks from Swaziland. *Onderstepoort Journal of Veterinary Research* 81, 846.

Brett, G.A. (1939) On the period of survival of the egg, larva and first nymph stages of the argasid tick *Ornithodoros moubata*, Murray, at different relative humidities. *Bulletin of Entomological Research* 30, 247–253.

Brown, V.R. and Bevins, S.N. (2018) A review of African swine fever and the potential for introduction into the United States and the possibility of subsequent establishment in feral swine and native ticks. *Frontiers in Veterinary Science* 5, 11.

Cumming, D. (1975) *A Field Study of the Ecology and Behaviour of Warthog*. Trustees of the National Museums and Monuments of Rhodesia, Salisbury, Rhodesia.

Cumming, G.S. and van Vuuren, D.P. (2006) Will climate change affect ectoparasite species ranges? *Global Ecology and Biogeography* 16, 486–497.

El Shoura, S.M. (1987) The life cycle of *Ornithodoros* (*Pavlovskyella*) *erraticus* (Acari: Ixodoidea: Argasidae) in the laboratory. *Journal of Medical Entomology* 24, 229–234.

Encinas Grandes, A., Oleaga Perez, A., Prez Sanchez, R. and Astirraga, A. (1993) Datos sobre el reservorio y vector de la peste porcina Africana, *Ornithodoros erraticus*. *Anaporc* 121, 38–47.

Haresnape, J.M., Wilkinson, P.J. and Mellor, P.S. (1988) Isolation of African swine fever virus from ticks of the *Ornithodoros moubata* complex (Ixodoidea: Argasidae) collected within the African swine fever enzootic area of Malawi. *Epidemiology and Infection* 101, 173–185.

Jori, F. and Bastos, A.D.S. (2009) Role of wild suids in the epidemiology of African swine fever. *EcoHealth* 6, 296–310.

Montgomery, R.E. (1921) On a form of swine fever occurring in British East Africa (Kenya Colony). *Journal of Comparative Pathology* 34, 159–191.

Penrith, M.-L., Bastos, A.D.S., Etter, E. and Beltrán-Alcrudo, D. (2019a) Epidemiology of African swine fever in Africa today: sylvatic cycle versus socio-economic imperatives. *Transboundary and Emerging Diseases* 66, 672–686.

Penrith, M.-L., Thomson, G.R., Bastos, A.D.S. and Etter, E.M.C. (2019b) African swine fever, 2019. In: Coetzer, J.A.W., Thomson, G.R., Maclachlan, N.J. and Penrith, M.-L. (eds) *Infectious Diseases of*

Livestock, pp. 1–110. Anipedia, J.A.W. Coetzer and P. Oberem (Directors). Available at: https://anipedia.org/resources/infectious-diseases-of-livestock/196 (accessed 16 June 2021).

Pereira de Oliveira, R.P., Hutet, E., Paboeuf, F., Duhayon, M., Boinas, F., *et al.* (2019) Comparative vector competence of the Afrotropical soft tick *Ornithodoros moubata* and Palearctic species, *O. erraticus* and *O. verrucosus*, for African swine fever virus strains circulating in Eurasia. *PLoS ONE* 14, e0225657.

Pereira de Oliveira, R.P., Hutet, E., Lancelot, R., Paboeuf, F., Duhayon, M., *et al.* (2020) Differential vector competence of *Ornithodoros* soft ticks for African swine fever virus: what if it involves more than just crossing organic barriers in ticks? *Parasites & Vectors* 13, 618.

Peter, E., Machuka, E., Githae, D., Okoth, E., Cleaveland, S., *et al.* (2021) Detection of African swine fever virus genotype XV in a sylvatic cycle in Saadani National Park, Tanzania. *Transboundary and Emerging Diseases* 68, 813–823.

Plowright, W., Parker, J. and Peirce, M.A. (1969) African swine fever virus in ticks (*Ornithodoros moubata*, Murray) collected from animal burrows in Tanzania. *Nature* 221, 1071–1073.

Quembo, C.J., Jori, F., Vosloo, W. and Heath, L. (2018) Genetic characterization of African swine fever virus isolates from soft ticks at the wildlife/domestic interface in Mozambique and identification of a novel genotype. *Transboundary and Emerging Diseases* 65, 420–431.

Rey, B., Fuller, A., Mitchell, D., Meyer, L.C.R. and Hetem, R.S. (2018) Drought-induced starvation of aardvarks in the Kalahari: an indirect effect of climate change. *Biology Letters* 13, 20170301.

Sanchez Botija, C. (1963) Reservorios del virus de la peste porcina Africana. *Bulletin Office International des Epizooties* 6, 895–899.

Thomson, G.R. (1985) The epidemiology of African swine fever: the role of free-living hosts in Africa. *Onderstepoort Journal of Veterinary Research* 52, 201–209.

Thomson, G.R., Gainaru, M., Lewis, A., Biggs, H., Nevill, E., *et al.* (1983) The relationship between African swine fever virus, the warthog and *Ornithodoros* species in southern Africa. In: Wilkinson, P.J. (ed.) *African Swine Fever*. Commission of the European Communities, Luxembourg, pp. 85–100.

Vial, L., Wieland, B., Jori, F., Etter, E., Dixon, L. and Roger, F. (2007) African swine fever virus DNA in soft ticks, Senegal. *Emerging Infectious Diseases* 13, 1928–1931.

Wilkinson, P.J., Pegram, R.G., Perry, B.D., Lemche, J. and Schels, H.F. (1988) The distribution of African swine fever virus isolated from *Ornithodoros moubata* in Zambia. *Epidemiology and Infection* 101, 547–564.

Wilson, A.J., Ribeiro, R. and Boinas, F. (2013) Use of a Bayesian network model to identify factors associated with the presence of the tick *Ornithodoros erraticus* on pig farms in southern Portugal. *Preventive Veterinary Medicine* 110, 45–53.

eo59 Tick-Borne Diseases of Livestock in the UK

Nicholas Johnson[1,2*] and L. Paul Phipps[1]
[1]*Animal and Plant Health Agency, Weybridge, UK;* [2]*University of Surrey, Guildford, UK*

59.1 Expert Opinion

Tick-borne diseases of livestock endemic within the UK are caused by a small number of pathogens (see Table 59.1), with almost all transmitted by the common sheep tick, *Ixodes ricinus* (Folly *et al.*, 2020). This tick species is ubiquitous in the UK and a generalist that feeds on a wide range of vertebrates. It is also the vector for a number of human diseases, most importantly Lyme borreliosis (Medlock *et al.*, 2018a). *I. ricinus* is abundant in areas of the UK that are dominated by cattle and sheep farming, including Scotland, Wales, and northern and south-west England. These are also areas most affected by tick-borne diseases of livestock (Lihou *et al.*, 2020).

A similar picture is observed in Northern Ireland, where cattle farming is particularly affected by babesiosis (Taylor, 1983). Two other tick species, *Haemaphysalis punctata* and *Dermacentor reticulatus*, are present in the UK and feed on livestock but their known distribution is limited to a number of coastal locations in England and Wales.

The key questions to address when considering the effects of climate change on tick-borne disease are: (i) what are the likely climatic changes expected; (ii) what impact will these have on the tick population; and (iii) how will these changes influence disease transmission by ticks? The most recent report from the Intergovernmental Panel on Climate Change (IPCC) categorizes the British Isles (consisting of the Republic of Ireland, the UK and the British Crown dependencies) as having an Atlantic climate based on the classification developed by Metzger *et al.* (2005). The Fifth Assessment Report (IPCC, 2014a) suggests that the decadal average temperature increased by 1.3°C between 2002 and 2011 compared with the previous century and if no action is taken to curb carbon emissions, this trend will continue. There is also evidence for increases in high temperature extremes and annual precipitation in Northern Europe. The Report predicts that temperatures and precipitation will continue to increase, particularly in the winter months. In addition, extreme weather events such as heatwaves, droughts and heavy precipitation leading to flooding will increase in frequency. Based on the Fifth Assessment Report, IPCC predicts that vector-borne diseases will increase: 'Risks from vector-borne diseases are projected to generally increase with warming, due to the extension of the infection area and season, despite reductions in some areas that become too hot for disease vectors (medium confidence)' (IPCC, 2014b).

Tick-borne disease transmission is seasonal and intimately linked to the periods of the year

*Email: nick.johnson@apha.gov.uk

DOI: 10.1079/9781789249637.0059

Table 59.1. Endemic tick-borne diseases of veterinary importance in the UK. (Authors' own table.)

Disease (common name)	Livestock affected	Pathogen	Vector
Encephalomyelitis (louping ill)	Sheep, cattle	Louping ill virus (LIV)	Common sheep tick (*Ixodes ricinus*)
Anaplasmosis (tick-borne fever)	Cattle, sheep	*Anaplasma phagocytophilum*	Common sheep tick (*I. ricinus*)
Babesiosis (redwater fever)	Cattle	*Babesia divergens*	Common sheep tick (*I. ricinus*)
Tick pyaemia	Sheep	*Staphylococcus aureus*	Common sheep tick (*I. ricinus*), red sheep tick (*Haemaphysalis punctata*)
Ovine theileriosis	Sheep	*Theileria luwenshuni*	Red sheep tick (*H. punctata*)

when *I. ricinus* is active. An analysis of the effect of weather and seasonality on bovine babesiosis in southern England between 1955 and 1967 suggested that the incidence of disease was bimodal, with a large 'spring' peak in May followed by a lower 'late summer' peak in September (Donnelly and MacKellar, 1970). Disease was rarely reported during the winter months between December and March. This seasonality corresponded closely with the bimodal activity of *I. ricinus* (Milne, 1945), which was observed to begin questing during the spring months. It was concluded that rising temperatures early in the year correlated with the appearance of disease (Donnelley and MacKellar, 1970).

An increase in the decadal average temperature of 1–2°C alone will not affect tick activity in itself, although increased temperature may extend the period of time when ticks become active which will, in turn, extend the periodicity of disease transmission to earlier and later in the year. Increases in precipitation are expected to have a positive effect on tick abundance and disease transmission as ticks spend most of their life within the moist vegetative layer at the base of herbage in order to remain hydrated, only moving to higher levels to seek a vertebrate host on which to feed. However, extreme weather events such as drought are likely to be detrimental to ticks. If the vegetative layer dries out, dehydration leads to tick death and overall the tick population will decline. In addition, changes in land management by farmers that maintains relatively shorter grass may also reduce the suitability of grazing land for ticks. However, in fields with hedge boundaries, adjacent areas will support populations of *I. ricinus*, confounding efforts to fully eliminate populations of the tick. Flooding may have short-term negative effects but usually flooding is likely to affect a relatively small area adjacent to rivers and for a short period of time. Overall, the impact of flooding on the wider tick population will be correspondingly small.

Surveillance for ticks in continental Europe and North America has suggested that *Ixodes* spp. are now active at higher elevations and at higher latitudes than in previous years (Medlock *et al.*, 2013; Jore *et al.*, 2014; Clow *et al.*, 2017), thus increasing the area over which they are found. For the UK, ticks are often abundant on upland areas where sheep are extensively grazed and correspond to the areas affected by louping ill virus (LIV). This suggests that the current distribution of *I. ricinus* will not change dramatically. Other factors such as deer presence and landscape management are likely to have a more direct impact on tick abundance. The range and density of the deer population that provides a major host for nymph and adult tick stages have been shown to directly increase tick abundance (Gilbert *et al.*, 2012). Also, human activity, for example farming practices, will also influence exposure of livestock to ticks. Tick pyaemia in lambs, a disease caused by systemic bacterial abscesses due to *Staphylococcus aureus* infection and exacerbated by co-infection with *Anaplasma phagocytophilum* following a tick bite, can result when lambs are first turned out with their dams on to tick-infested pasture (Macrelli *et al.*, 2020). A further factor challenging control is the chronic nature of some tick-borne diseases leading to recrudescence independent of tick activity (Suarez *et al.*, 2019).

59.2 Future Predictions

The distribution of LIV seroprevalence in Scottish sheep appears to be associated with geographical location (Gilbert *et al.*, 2020). Higher seroprevalence levels are observed in western areas, associated with a milder climate. Ticks and seropositive sheep are present in eastern areas but at lower levels. If the climate becomes milder in these eastern areas, then more transmission of LIV is possible and levels of seroprevalence will rise. It also seems likely that if winter temperatures rise over time, ticks will become active earlier in the year and opportunities for disease transmission will happen during the winter months. Intensive measurement of climatic variables and tick activity have shown that maximum temperatures over 7°C for 5 days will trigger tick questing (Perret *et al.*, 2000). However, this is difficult to assess in a single year where short-term extremes of temperature occur and can in themselves lead to short periods when ticks begin to quest, leading to cases of tick-borne disease transmission, for example babesiosis (Johnson *et al.*, 2020).

A significant problem facing future predictions is the limited surveillance for ticks and tick-borne diseases of livestock in the UK. The Animal and Plant Health Agency (APHA), as the main governmental delivery body of research into diseases of livestock in the UK, publishes data on all positive cases identified from diagnostic tests conducted by the Agency. However, this is clearly not all the cases of disease that occur throughout the country, and while it may give an indication of the distribution of disease, for example the distribution of louping ill cases in Great Britain (Carson *et al.*, 2020), it is not an indication of incidence. Surveys of livestock farmers have repeatedly corroborated the regions of the country that are most affected by tick-borne diseases (Barton *et al.*, 2006; Lihou *et al.*, 2020). However, this at best identifies the number of properties affected, not the number of animals infected. Surveillance for tick bites in humans takes place through the Public Health England Tick Surveillance Scheme (PHE-TSS). This periodically reports on submissions, and observed that between January 2013 and December 2018, tick bites were dominated by *I. ricinus* and occurred on humans during every month of the year (Cull *et al.*, 2019). However, the vast majority occurred between April and September with a distinct peak for bites by nymphs and adults in the month of June. Bites from larvae peaked in July.

The south-east of England has resident populations of the red sheep tick, *H. punctata*. This species has been associated with nuisance biting of humans and is a vector of two piroplasmid parasites, *Babesia motasi* and *Theileria luwenshuni*; both are relatively mild pathogens of sheep but capable of causing disease (Phipps *et al.*, 2016). However, they have been associated as concurrent infections with cases of tick pyaemia in lambs following severe tick infestation (Macrelli *et al.*, 2020). The current distribution of *H. punctata* is limited but appears to be increasing in a westerly direction within the South Downs National Park (Medlock *et al.*, 2018b). This trend is likely to continue and could lead to more instances of disease in livestock. Tick control that restricts the infestation will be needed to prevent disease establishment and transmission.

Changes in the climate may make the UK more hospitable to tick species normally present in countries of Southern Europe. *Rhipicephalus* and *Hyalomma* spp. are regularly detected in Northern Europe and brought into the UK on domestic animals (Hansford *et al.*, 2016, 2017) or potentially by migrating birds (England *et al.*, 2016; Hansford *et al.*, 2019). Until now there is no evidence for these species establishing breeding populations in the UK but changes in climate may enable them to establish and provide an opportunity for exotic pathogens to establish.

Currently there is no equivalent scheme investigating the infestation of livestock analogous to the PHE-TSS despite the range of tick-borne diseases that are endemic or the possibility of ingress of exotic vectors and infectious hazards. This points to a clear need for more detailed surveillance of tick infestations and tick-borne diseases of livestock, wildlife and at the human–animal interface in the UK, as well as increased options for tick control.

Recommendations are:

1. Increased surveillance for tick infestation and endemic tick-borne diseases within the UK over time, to monitor changes in incidence and distribution in response to changing climatic conditions.

2. Surveillance and reporting of exotic ticks and exotic tick-borne diseases associated with livestock.

3. Improved diagnostic ability for detecting tick-borne diseases and strategies for preventing (vaccination, tick control), ameliorating (landscape and livestock management) and mitigation of tick-borne diseases, including zoonoses.
4. Better understanding of the effect of co-infection with multiple pathogens on the severity of tick-borne diseases.

Unless surveillance for tick-borne disease in livestock improves, there will be uncertainty around the status of individual diseases in the UK beyond anecdotal observations. Any trends will not be detected or documented and the role of factors such as climate will remain within the realm of modellers and expert opinion rather than robust evidential data. Who delivers this surveillance presents a critical challenge. Government-funded bodies will no doubt need to play a coordinating role but other actors/stakeholders such as the veterinary profession, academic organizations and the livestock industry also can play a role in this. A cross-sector coordinated approach (wildlife–livestock–companion animal–human) would be the most robust and effective way to address the challenges of tick-borne disease.

Acknowledgements

Nicholas Johnson and Paul Phipps are funded by the Department for Environment, Food and Rural Affairs (Defra) through grants SV3045 and ED1600.

References

Barton, C.H., Dale, E.F., Dixon, C. and Coles, C.G. (2006) Survey of parasite control on beef farms in south-west England. *Veterinary Record* 159, 682–684.

Carson, A., Mitchell, S., Phipps, P., Macrelli, M. and Dunnett, E. (2020) Tickborne diseases of sheep. *Veterinary Record* 186, 408–409.

Clow, K.M., Leighton, P.A., Ogden, N.H., Lindsay, L.R., Michel, P., *et al.* (2017) Northward range expansion of *Ixodes scapularis* evident over a short timescale in Ontario, Canada. *PLoS ONE* 12, e0189393.

Cull, B., Pietzsch, M.E., Gillingham, E.L., McGinley, L., Medlock, J.M. and Hansford, K.M. (2019) Seasonality and anatomical location of human tick bites in the United Kingdom. *Zoonoses and Public Health* 67, 112–121.

Donnelly, J. and MacKellar, J.C. (1970) The effect of weather and season on the incidence of redwater fever in cattle in Britain. *Agricultural Meteorology* 7, 5–17.

England, M.E., Phipps, P., Medlock, J.M., Atkinson, P.M., Atkinson, B., *et al.* (2016) *Hyalomma* ticks on northward migrating birds in southern Spain: implications for the risk of entry of Crimean–Congo haemorrhagic fever virus to Great Britain. *Journal of Vector Ecology* 41, 128–134.

Folly, A.J., Dorey-Robinson, D., Hernández-Triana, L.M., Phipps, L.P. and Johnson, N. (2020) Emerging threats to animals in the United Kingdom by arthropod-borne diseases. *Frontiers in Veterinary Science* 7, 20.

Gilbert, L., Maffey, G.L., Ramsay, S.L. and Hester, A.J. (2012) The effect of deer management on the abundance of *Ixodes ricinus* in Scotland. *Ecological Applications* 22, 658–667.

Gilbert, L., Brülisauer, F., Willoughby, K. and Cousens, C. (2020) Identifying environmental risk factors for louping ill virus seroprevalence in sheep and the potential to inform wildlife management policy. *Frontiers in Veterinary Science* 7, 377.

Hansford, K.M., Medlock, J.M., Atkinson, B. and Santos-Silva, M.M. (2016) Importation of a *Hyalomma lusitanicum* tick into the UK on a dog. *Veterinary Record* 179, 415.

Hansford, K.M., Phipps, L.P., Cull, B., Pietzsch, M.E. and Medlock, K.M. (2017) *Rhipicephalus sanguineus* importation into the UK: surveillance, risk, public health awareness and One Health response. *Veterinary Record* 180, 119.

Hansford, K.M., Carter, D., Gillingham, E.L., Hernández-Triana, L.M., Chamberlain, J., *et al.* (2019) *Hyalomma rufipes* on an untraveled horse: is this the first evidence of *Hyalomma* nymphs successfully moulting in the United Kingdom? *Ticks and Tick-borne Diseases* 10, 704–708.

IPCC (2014a) Kovats, R.S., Valentini, R., Bouwer, L.M., Georgopoulou, E., Jacob, D., *et al.* (2014) Europe. In: Barros, V.R., Field, C.B., Dokken, D.J., Mastrandrea, M.D., Mach, K.J., *et al.* (eds) *Climate Change 2014: Impacts, Adaptation, and Vulnerability. Part B: Regional Aspects. Contribution of Working Group*

II to the Fifth Assessment Report of the Intergovernmental Panel on Climate Change. Cambridge University Press, Cambridge and New York, pp. 1267–1326. Available at: https://www.ipcc.ch/site/assets/uploads/2018/02/WGIIAR5-Chap23_FINAL.pdf (accessed 30th July 2020).

IPCC (2014b) *Climate Change 2014: Synthesis Report. Contribution of Working Groups I, II and III to the Fifth Assessment Report of the Intergovernmental Panel on Climate Change* [Core Writing Team, Pachauri, R.K. and Meyer, L.A. (eds)]. Intergovernmental Panel on Climate Change, Geneva, Switzerland. Available at: https://www.ipcc.ch/site/assets/uploads/2018/05/SYR_AR5_FINAL_full_wcover.pdf (accessed 7 August 2020).

Johnson, N., Phipps, L.P., McFazdean, H. and Barlow, A.M. (2020) An outbreak of bovine babesiosis in February, 2019, triggered by above average winter temperatures in southern England and co-infection with *Babesia divergens* and *Anaplasma phagocytophilum*. *Parasites & Vectors* 13, 305.

Jore, S., Vanwambeke, O., Viljugrein, H., Isaksen, K., Kristoffersen, A.B., *et al.* (2014) Climate and environmental change drives *Ixodes ricinus* geographical expansion at the northern range margin. *Parasites & Vectors* 7, 11.

Lihou, K., Rose Vineer, H. and Wall, R. (2020) Distribution and prevalence of ticks and tick-borne disease on sheep and cattle farms in Great Britain. *Parasites & Vectors* 13, 406.

Macrelli, M., Phipps, P., McGinley, L., Medlock, J. and Johnson, N. (2020) First report of fatal tick pyaemia caused by heavy infestation with the red sheep tick, *Haemaphysalis punctata* and co-infection with *Babesia* and *Theileria* spp. *Veterinary Record Case Report* 8, e001267.

Medlock, J.M., Hansford, K.M., Bormane, A., Derdakova, M., Estrada-Peña, A., *et al.* (2013) Driving forces for changes in geographical distribution of *Ixodes ricinus* ticks in Europe. *Parasites & Vectors* 6, 1.

Medlock, J.M., Hansford, K.M., Vaux, A.G.C., Cull, B., Gillingham, E. and Leach, S. (2018a) Assessment of the public health threats posed by vector-borne disease in the United Kingdom. *International Journal of Environmental Health* 15, 2145.

Medlock, J.M., Hansford, K.M., Vaux, A.G.C., Cull, B., Pietzsch, M.E., *et al.* (2018b) Has the red sheep tick, *Haemaphysalis punctata*, recently expanded its range in England? *Medical and Veterinary Entomology* 32, 473–480.

Metzger, M.J., Bunce R.G.H., Jongman, R.H.G., Mücher, C.A. and Watkins, J.W. (2005) A climatic stratification of the environment of Europe. *Global Ecology and Biogeography* 14, 549–563.

Milne, A. (1945) The ecology of the sheep tick, *Ixodes ricinus* L. The seasonal activity in Britain with particular reference to northern England. *Parasitology* 36, 142–150.

Perret, J.-L., Guigoz, E., Rais, O. and Gern, L. (2000) Influence of saturation deficit and temperature on *Ixodes ricinus* tick questing activity in a Lyme borreliosis-endemic area (Switzerland). *Parasitology Research* 86, 554–557.

Phipps, L.P., Hernández-Triana, L.M., Goharriz, H., Welchman, D. and Johnson, N. (2016) Detection of *Theileria luwenshuni* in sheep from Great Britain. *Parasites & Vectors* 9, 203.

Suarez, C.E., Alzan, H.F., Silva, M.G., Rathinasamy, V., Poole, W.A. and Cooke, B.M. (2019) Unravelling the cellular and molecular pathogenesis of bovine babesiosis: is the sky the limit? *International Journal for Parasitology* 49, 183–197.

Taylor, S.M. (1983) Assessment of prevalence of clinical babesiosis in cattle in Northern Ireland. *Veterinary Record* 112, 247–250.

eo60 Impact of Climate Change on Tick-Borne Diseases of Livestock in Pakistan – Looking Ahead

Abdul Ghafar, Robin Gasser and Abdul Jabbar*
The University of Melbourne, Werribee, Australia

60.1 Expert Opinion

60.1.1 Introduction

Climate change has profound and long-term implications for animal and human health globally, with a direct impact on the occurrence and spread of infectious diseases, particularly those transmitted by ectoparasites (e.g. Caminade *et al.*, 2019; Rocklöv and Dubrow, 2020). Ticks and tick-borne diseases (TTBDs) are no exception; climate change is expected to result in alterations of tick biology, host distribution and vegetation, as well as the behaviour and features of pathogens transmitted by ticks (Reisen *et al.*, 2006; Ogden and Lindsay, 2016; Caminade *et al.*, 2019; Gilbert, 2021). For instance, the establishment and growth of tick species are mainly determined by a set of temperature and precipitation conditions, known as climatic niche (Estrada-Peña, 2008). Various tick species require different levels of specialization in their niche, which can lead to varying distributions, ranging from generalist and ubiquitous occurrence (e.g. *Ixodes ricinus* and *Rhipicephalus microplus*) to narrow geographic distributions (such as *Hyalomma hussaini*) (Kaiser and Hoogstraal, 1964; Soberón, 2007; Porretta *et al.*, 2013; Gilbert, 2021). Significant changes in such conditions could either expand or contract the size and distribution of tick populations, thereby also impacting the prevalence and intensity of tick-borne diseases (TBDs). However, determining such impact is not as simple as it seems, as it requires information about complex tick–pathogen–host–environment interactions as well as the individual and cumulative effects to likely events of climate change (Gilbert, 2021). This expert opinion aims to summarize the current status of TTBDs of ruminants in Pakistan and to assess evidence of climate change and its likely impact on TTBDs.

60.1.2 Climate and livestock in Pakistan

Pakistan lies in the subtropical to temperate zone, having a diverse climate that varies substantially across topographically distinct locations. Four seasons are recognized: (i) cool, dry winter (December–February); (ii) hot, dry spring (March–May); (iii) summer (also known as the south-west monsoon period; June–September); and (iv) the retreating monsoon (October–November). Annual rainfall is very limited across most parts of the country, and climatic variations are quite common, particularly during the El Niño cycle (World Bank Group, 2021). Like many other agriculture-based economies, the livestock industry plays

*Email: jabbara@unimelb.edu.au

DOI: 10.1079/9781789249637.0060

critical roles in the livelihoods of approximately eight million farming families, the animal traction for transport and vital power for cultivation and, importantly, the national gross domestic product (Government of Pakistan, 2020). Milk, the most important agricultural commodity in Pakistan, is produced by a large population of cattle and buffaloes (~200 million) (Government of Pakistan, 2020). Although the national cattle herd mainly comprises domestic breeds, there has been a sizeable increase in the population of exotic and cross-bred cattle in recent decades. Estimates have shown that more than 90% of dairy herds are small, with fewer than ten animals per herd (Umm e Zia, 2011).

60.1.3 Ticks and tick-borne diseases of ruminants in Pakistan

In Pakistan, tropical to temperate climate, availability of millions of ruminants, poor husbandry practices and a lack of access to proper control measures collectively provide favourable conditions for TTBDs (Jabbar *et al.*, 2015; Ghafar *et al.*, 2020a). Several epidemiological investigations have demonstrated that ruminant populations are exposed to heavy tick infestation, particularly during the summer season (Ghafar *et al.*, 2020b,c,d, 2021). Predominant tick species include *Haemaphysalis* (mainly *Haemaphysalis bispinosa*, *Haemaphysalis punctata* and *Haemaphysalis sulcata*), *Hyalomma* (*Hyalomma anatolicum*, *Hyalomma detritum*, *Hyalomma dromedarii* and *Hyalomma excavatum*) and *Rhipicephalus* (*Rhipicephalus microplus*, *Rhipicephalus sanguineus* and *Rhipicephalus turanicus*) (Karim *et al.*, 2017; Ghafar *et al.*, 2020b,c,d, 2021). Besides causing significant, direct economic losses associated with reduced production and control-associated costs, these ticks also transmit pathogens of veterinary and public health significance.

Initial molecular studies of the prevalence of TBDs in ruminants in Pakistan (Jabbar *et al.*, 2015; Ghafar *et al.*, 2020d,e, 2021) suggest that more than half of the total population is affected by at least one of the three major TBDs – theileriosis, babesiosis and anaplasmosis. Theileriosis is considered the most important TBD of ruminants in this country and is caused by *Theileria annulata* (in large ruminants), *Theileria ovis* and *Theileria lestoquardi* (small ruminants). Studies have shown that up to 72.5 and 85.8% of apparently healthy populations of small and large ruminants, respectively, could be the carriers of *Theileria* spp. (Ghafar *et al.*, 2020d, 2021). The second most important TBD is anaplasmosis and is caused by *Anaplasma centrale*, *Anaplasma marginale* (large ruminants) and *Anaplasma ovis* (small ruminants), with an overall prevalence of up to 55.3 and 53.7%, respectively (Ghafar *et al.*, 2020d). Babesiosis, caused by *Babesia bovis*, *Babesia bigemina* (large ruminants) and *Babesia ovis* (small ruminants), is also a major TBD with high mortality rates, particularly in large ruminants, and has an overall reported prevalence of 55.0 and 56.8%, respectively (Jabbar *et al.*, 2015; Ghafar *et al.*, 2021).

60.2 Future Predictions

Climate data suggest that the annual average temperature of Pakistan has increased by 0.6°C during 1947–2016 (Fig. 60.1a). However, this change is not uniform across different seasons and/or different geographical locations where domestic ruminants are kept. The highest increase in temperature has been recorded in the winter season and in the southern parts of Pakistan, including the provinces of Punjab, Sindh and Balochistan. Similarly, annual rainfall has decreased by 10–15% in the arid plains and coastal areas since 1960, while it has increased in most other parts of Pakistan (Fig. 60.1b) (World Bank Group, 2021). These changes in climate could have a significant impact on the occurrence and distribution of TTBDs in Pakistan; however, data are lacking. Such information is also very scarce in other parts of the world, because climate change is a long-term process, and large, longitudinal data sets are needed to establish possible or likely impacts on TTBDs. Additionally, even if such data are available, it is challenging to separate the impact(s) of climate alone from that of other factors, such as variation in host and habitat, because they can also change, irrespective of climate (Gilbert, 2021).

In Pakistan, like many other developing countries, there are several constraints on projecting the impact of climate change on TTBDs.

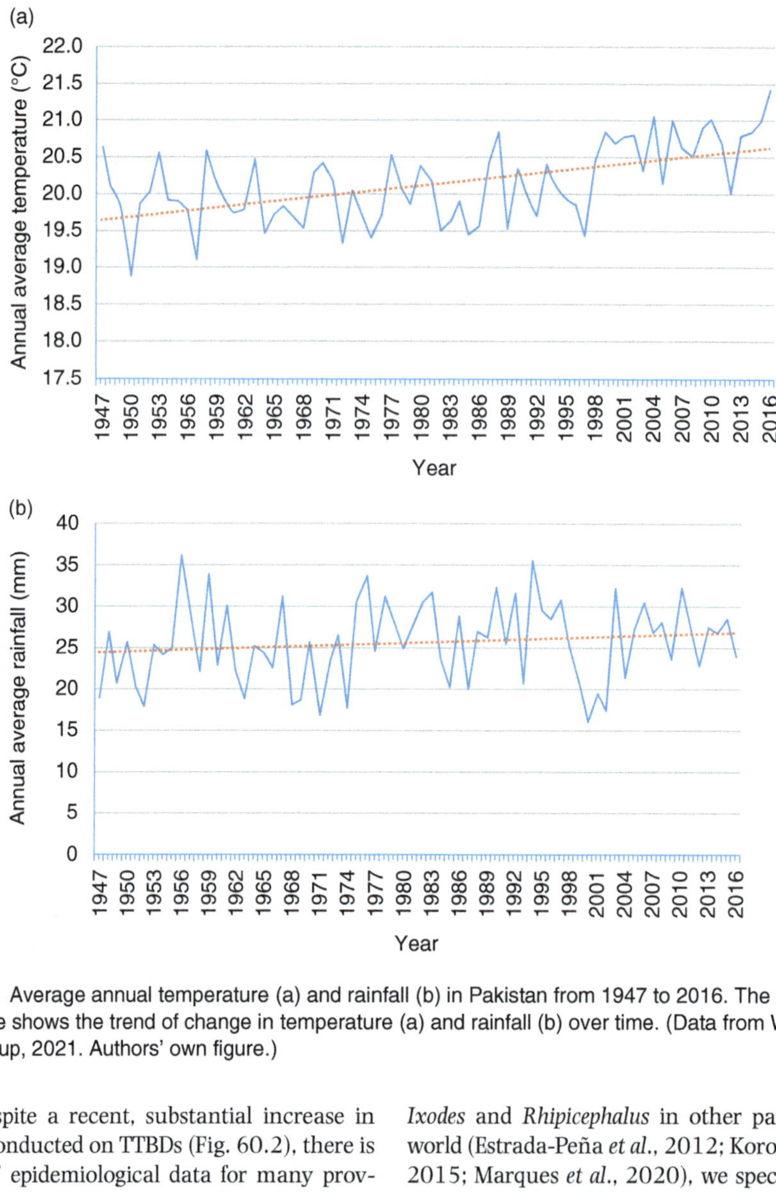

Fig. 60.1. Average annual temperature (a) and rainfall (b) in Pakistan from 1947 to 2016. The orange dotted line shows the trend of change in temperature (a) and rainfall (b) over time. (Data from World Bank Group, 2021. Authors' own figure.)

First, despite a recent, substantial increase in studies conducted on TTBDs (Fig. 60.2), there is a lack of epidemiological data for many provinces/regions of this country. Second, even if such data are available, they are not accurate, at least in some cases, due to issues relating to study design and/or a lack of use of reliable molecular diagnostic tools in epidemiological studies. Third, animal movement, from one region to another, for migration or marketing purposes, is uncontrolled and there is no national livestock animal identification system in place. Nevertheless, by taking into consideration the predictions of distributional range expansion of *Hyalomma*, *Ixodes* and *Rhipicephalus* in other parts of the world (Estrada-Peña *et al.*, 2012; Korotkov *et al.*, 2015; Marques *et al.*, 2020), we speculate that climate change is likely to alter the prevalence and distribution of two key tick genera, *Hyalomma* and *Rhipicephalus*, which mainly infest ruminants in Pakistan (Ghafar *et al.*, 2020b,d).

Hyalomma ticks are found on all ruminant species in Pakistan and are known vectors of the aetiological agents of theileriosis and Crimean–Congo haemorrhagic fever (CCHF) in ruminants and humans, respectively (Ghafar *et al.*, 2020b,d). Ticks of the genus *Rhipicephalus* include some of the economically most important ticks – they

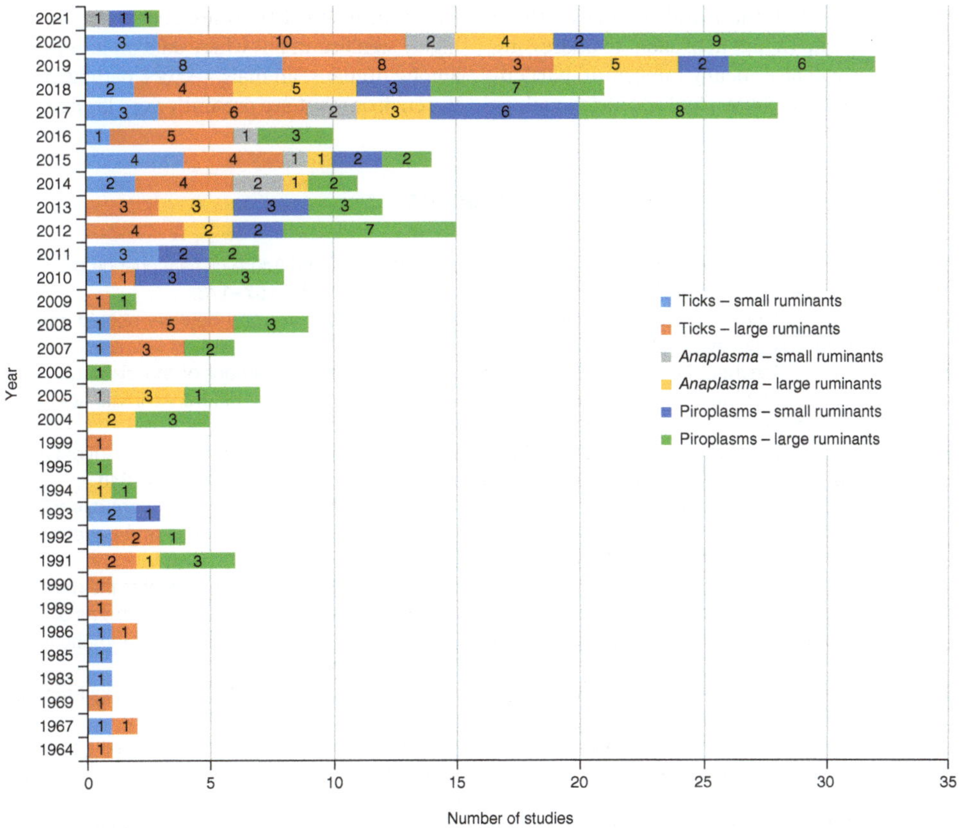

Fig. 60.2. The numbers of studies of ticks and tick-borne diseases of ruminants in Pakistan from 1964 to January 2021. Different coloured bars indicate particular groups of ticks and tick-borne pathogens of small and large ruminants. (Authors' own figure.)

transmit the causative agents of anaplasmosis and babesiosis in Pakistan and elsewhere (Ghafar *et al.*, 2020b). Additionally, several potentially zoonotic species of rickettsiae (including *Rickettsia massiliae* and *Rickettsia aeschlimannii*) have also been detected in both tick genera (Ghafar *et al.*, 2020e). With the rise in environmental temperature and in rainfall in some regions, it could be postulated that the distribution and abundance of TTBDs will increase because the distribution of ruminant populations remains relatively stable in Pakistan, except if there were major climatic calamities, such as floods, which would likely displace human and animal populations. Tick-borne zoonotic pathogens will also, potentially, continue to increase, as indicated by increasing reports of CCHF in recent years in Pakistan (Umair *et al.*, 2020). A similar trend may develop for other tick-borne zoonotic pathogens (such as rickettsiae), but information and epidemiological data are presently sparse due to a lack of surveillance and clinical testing in this country (Ghafar *et al.*, 2020e). Clearly, these hypotheses need testing through detailed spatial and temporal studies, focused on accurately estimating the prevalence of TTBDs in different climatic zones of Pakistan utilizing sensitive and specific surveillance tools. Subsequently, such data could be explored using mathematical modelling methods to assess changes in the occurrence, distribution and/or abundance of TTBDs due to climate change.

There is a clear need for coordinated surveillance programmes for TTBDs in Pakistan to underpin critical analyses of field and laboratory findings using *in silico* models, to inform us about emerging or re-emerging TTBDs and to predict the effects of a changing climate on the prevalence, distribution and impact of these diseases. A 'One Health approach' to addressing such issues will be critically important, particularly in tropical or subtropical countries such as Pakistan, where TTBDs are socio-economically highly significant, but where data on interactions among humans, animals, vectors, pathogens and the environment are presently very limited.

References

Caminade, C., McIntyre, K.M. and Jones, A.E. (2019) Impact of recent and future climate change on vector-borne diseases. *Annals of the New York Academy of Sciences* 1436, 157–173.

Estrada-Peña, A. (2008) Climate, niche, ticks, and models: what they are and how we should interpret them. *Parasitology Research* 103, 87–95.

Estrada-Peña, A., Sánchez, N. and Estrada-Sánchez, A. (2012) An assessment of the distribution and spread of the tick *Hyalomma marginatum* in the western Palearctic under different climate scenarios. *Vector-Borne and Zoonotic Diseases* 12, 758–768.

Ghafar, A., McGill, D., Stevenson, M.A., Badar, M., Kumbher, A., *et al.* (2020a) A participatory investigation of bovine health and production issues in Pakistan. *Frontiers in Veterinary Science* 7, 248.

Ghafar, A., Gasser, R.B., Rashid, I., Ghafoor, A. and Jabbar, A. (2020b) Exploring the prevalence and diversity of bovine ticks in five agro-ecological zones of Pakistan using phenetic and genetic tools. *Ticks and Tick-borne Diseases* 11, 101472.

Ghafar, A., Khan, A., Cabezas-Cruz, A., Gauci, C.G., Niaz, S., *et al.* (2020c) An assessment of the molecular diversity of ticks and tick-borne microorganisms of small ruminants in Pakistan. *Microorganisms* 8, 1428.

Ghafar, A., Abbas, T., Rehman, A., Sandhu, Z.-U.-D., Cabezas-Cruz, A., *et al.* (2020d) Systematic review of ticks and tick-borne pathogens of small ruminants in Pakistan. *Pathogens* 9, 937.

Ghafar, A., Cabezas-Cruz, A., Galon, C., Obregon, D., Gasser, R.B., *et al.* (2020e) Bovine ticks harbour a diverse array of microorganisms in Pakistan. *Parasites & Vectors* 13, 1.

Ghafar, A., Koehler, A.V., Hall, R.S., Gauci, C.G., Gasser, R.B., *et al.* (2021) Targeted next-generation sequencing and informatics as an effective tool to establish the composition of bovine piroplasm populations in endemic regions. *Microorganisms* 9, 21.

Gilbert, L. (2021) The impacts of climate change on ticks and tick-borne disease risk. *Annual Review of Entomology* 66, 373–388.

Government of Pakistan (2020) Pakistan Economic Survey 2019–2020. Ministry of Finance, Islamabad. Available at: http://www.finance.gov.pk/survey_1920.html (accessed 20 January 2021).

Jabbar, A., Abbas, T., Sandhu, Z.D., Saddiqi, H.A., Qamar, M.F., *et al.* (2015) Tick-borne diseases of bovines in Pakistan: major scope for future research and improved control. *Parasites & Vectors* 8, 283.

Kaiser, M. and Hoogstraal, H. (1964) The *Hyalomma* ticks (Ixodoidea, Ixodidae) of Pakistan, India, and Ceylon, with keys to subgenera and species. *Acarologia* 6, 257–286.

Karim, S., Budachetri, K., Mukherjee, N., Williams, J., Kausar, A., *et al.* (2017) A study of ticks and tick-borne livestock pathogens in Pakistan. *PLoS Neglected Tropical Diseases* 11, e0005681.

Korotkov, Y., Kozlova, T. and Kozlovskaya, L. (2015) Observations on changes in abundance of questing *Ixodes ricinus*, castor bean tick, over a 35-year period in the eastern part of its range (Russia, Tula region). *Medical and Veterinary Entomology* 29, 129–136.

Marques, R., Krüger, R.F., Peterson, A.T., De Melo, L.F., Vicenzi, N., *et al.* (2020) Climate change implications for the distribution of the babesiosis and anaplasmosis tick vector, *Rhipicephalus* (*Boophilus*) *microplus*. *Veterinary Research* 51, 81.

Ogden, N.H. and Lindsay, L.R. (2016) Effects of climate and climate change on vectors and vector-borne diseases: ticks are different. *Trends in Parasitology* 32, 646–656.

Porretta, D., Mastrantonio, V., Amendolia, S., Gaiarsa, S., Epis, S., *et al.* (2013) Effects of global changes on the climatic niche of the tick *Ixodes ricinus* inferred by species distribution modelling. *Parasites & Vectors* 6, 271.

Reisen, W., Fang, Y. and Martinez, V. (2006) Effects of temperature on the transmission of West Nile virus by *Culex tarsalis* (Diptera: Culicidae). *Journal of Medical Entomology* 43, 309–317.

Rocklöv, J. and Dubrow, R. (2020) Climate change: an enduring challenge for vector-borne disease prevention and control. *Nature Immunology* 21, 479–483.

Soberón, J. (2007) Grinnellian and Eltonian niches and geographic distributions of species. *Ecology Letters* 10, 1115–1123.

Umair, M., Khurshid, A., Alam, M.M., Akhtar, R., Salman, M., *et al.* (2020) Genetic diversity and phylogenetic analysis of Crimean–Congo hemorrhagic fever viruses circulating in Pakistan during 2019. *PLoS Neglected Tropical Diseases* 14, e0008238.

Umm e Zia, Mahmood, T. and Ali, M.R. (2011) *Dairy Development in Pakistan*. Food and Agriculture Organization of the United Nations, Rome. Available at: http://www.fao.org/3/a-al750e.pdf (accessed 20 January 2021).

World Bank Group (2021) Climate Change Knowledge Portal: Pakistan. World Bank Group, Washington, DC. Available at: https://climateknowledgeportal.worldbank.org/country/pakistan (accessed 20 January 2021).

eo61 The Emergence of Tick-Borne Diseases in Domestic Animals in Australia

Charlotte Oskam[1]*, Isobel Ronai[2] and Peter Irwin[1]
[1]*Murdoch University, Perth, Australia;* [2]*The University of Sydney, Sydney, Australia*

61.1 Expert Opinion

61.1.1 Background

Australia currently has 74 recognized tick species: 60 hard tick (family, Ixodidae) and 14 soft tick (family, Argasidae) species (Barker, 2019; Evans *et al.*, 2019). The majority of these species are native to Australia and have co-evolved with their diverse native vertebrate hosts, including marsupials, reptiles and birds. Notably however, five tick species have been introduced into the Australian continent following the arrival of Europeans in 1788 due to the movement of domestic animals: poultry, *Argas persicus*; horses, *Otobius megnini*; cattle, *Haemaphysalis longicornis* and *Rhipicephalus australis* (syn. *Boophilus microplus*); and dogs, *Rhipicephalus sanguineus*, although *R. sanguineus* may have been introduced approximately 3500 years ago with dingoes (Jackson *et al.*, 2017). The establishment and spread of these introduced tick species, together with the presence of susceptible hosts, place Australia in a precarious situation – a situation by which the introduction of new tick-borne infectious agents could establish and spread rapidly throughout Australia, putting livestock, companion animals and possibly wildlife at risk.

The climate of Australia is one of extremes and variability, with climate change exacerbating these characteristics. In particular, the number of extreme heat events is increasing, the fire season has extended, droughts are longer, rainfalls are heavier and rainfall patterns have changed across the continent (CSIRO and BoM, 2020). The effects of climate change also vary across the large land mass of Australia, impacting its many climatic zones. Therefore, the impact of climate change on habitat suitability for ticks in Australia will be region-dependent.

In this expert opinion, two case studies of tick-borne diseases of domestic animals in Australia have been selected to illustrate how climate change can potentially influence the spread of introduced tick species and the pathogens they transmit. The first reports an incursion of canine monocytic ehrlichiosis (CME) into northern regions of Australia in April/June 2020 and the second describes the geographical expansion of bovine anaemia due to *Theileria orientalis* group (BATOG) across Australia since 2006. The introduction of novel pathogens into ecosystems exposed to climate change provides a unique opportunity to study disease dynamics in real time.

*Email: c.oskam@murdoch.edu.au

DOI: 10.1079/9781789249637.0061

61.1.2 Case study: *Ehrlichia canis* in Australia

One of the most serious vector-borne diseases of dogs worldwide is CME, caused by *Ehrlichia canis*, a Gram-negative obligate intracellular bacterium that resides within membrane-bound vacuoles ('morulae') in the cytoplasm of canine monocytes and endothelial cells and is transmitted by the brown dog tick (*R. sanguineus*). This disease has a global distribution, reflecting the extensive geographical range of the vector; however, CME was considered to be absent from Australia until recently. One targeted surveillance study conducted approximately 20 years ago in dogs from major urban centres of northern Australia (where *R. sanguineus* is enzootic) concluded there was no evidence of *E. canis* in any of the dogs ($n = 316$) examined (Mason *et al.*, 2001).

E. canis organisms are transmitted in the tick's saliva relatively quickly – within a few hours – after attachment and infected dogs develop clinical signs within 4 weeks (Fourie *et al.*, 2013). The disease is recognized to develop through three phases: acute, subclinical and chronic/end stage. Clinical illness in dogs with *E. canis* infections includes fever, lethargy, anorexia, bleeding diatheses (including nasal haemorrhage), enlarged lymph nodes and spleen, weight loss and ocular signs. Some dogs die of sepsis due to an inability to fight infection, while others succumb to uncontrollable haemorrhage (de Castro *et al.*, 2004). The diagnosis of CME is confirmed by PCR conducted on blood samples and serological testing (Harrus and Waner, 2011).

In Australia, CME is a notifiable disease, meaning that veterinarians are obliged to report suspected cases to state and national veterinary authorities. In April 2020, veterinarians in northern Western Australia reported examining dogs with a severe illness. Cases were confirmed as CME by state and national laboratories and a nationwide disease response was coordinated by the Australian Government's Consultative Committee on Emergency Animal Disease (CCEAD; https://www.agriculture.gov.au/animal/health/committees/ccead, accessed 16 June 2021). Within 12 months of the discovery, over 500 dogs had been confirmed in Western Australia and in the Northern Territory, 2500 km to the east of the index cases, and sporadic reports of CME have been made in travelled dogs in every Australian State and Territory. Restriction orders for the movement of dogs were imposed; however, the area where cases of CME have been reported is enormous, spanning sparsely populated country with remote towns and indigenous communities.

R. sanguineus is an endophilic tick (adapted to indoor living) that can complete up to four generations per year in warmer climatic conditions (Dantas-Torres, 2010). Worryingly, *R. sanguineus* ticks are capable of survival in the environment and even infest dogs during winter in temperate climates (e.g. south-eastern Oklahoma and north-western Arkansas, USA; Northern Europe) (Dantas-Torres, 2010). Within Australia, *R. sanguineus* is well established in areas with tropical and subtropical climates, north to the Tropic of Capricorn. There has also been a southern extension of *R. sanguineus* into more populous regions due to human migration and urbanization, and interstate and transregional travel, with consequent movement of companion dogs (Greay *et al.*, 2018; Chandra *et al.*, 2020). The establishment of this tick in more southern latitudes of Australia will likely be supported by the poleward expansion of tropical and subtropical zones due to climate change.

Questions have been raised about how the disease arrived despite Australia's strict biosecurity control measures, and about how it is likely to play out into the future with the current and anticipated changes to Australia's climate. How did the infection enter Australia? Was it carried, undetected, by a dog travelling from an endemic country, or did infected ticks arrive from overseas? When exactly did the infection arrive? Lastly, what are the implications of this disease in other animals – and humans – with regard to CME in Australia? It would seem that our native marsupials are in no danger from this disease, however the potential impact on eutherian dingoes (*Canis lupus dingo*) is unknown and is of concern. Furthermore, given public interest in tick-associated illness in humans in Australia, it is pertinent that *E. canis* has been reported in a small number of people in South America since 1996 (Perez *et al.*, 1996; Bouza-Mora *et al.*, 2017). Research will be required to answer these questions.

61.1.3 Case study: bovine anaemia due to *Theileria orientalis* group

An emerging tick-borne disease for cattle in Australia is BATOG, also known as oriental theileriosis.

The causative agent of BATOG is the intracellular haemoprotozoan parasite, *T. orientalis* (previously *Theileria sergenti*). *T. orientalis* is transmitted principally by ticks and the competent vectors in Australia are three species in the genus *Haemaphysalis* (Stewart *et al.*, 1996; Watts *et al.*, 2016). Both the tick vector and bovine host are needed for the protozoal parasite to progress through its life cycle. While other *Theileria* spp. cause more severe bovine diseases, such as East Coast fever and tropical theileriosis, they are absent in Australia.

T. orientalis invades the host's red blood cells (erythrocytes) and clinical disease in infected cattle can include severe anaemia characterized by pale mucous membranes, weakness, lethargy and loss of appetite (Izzo *et al.*, 2010). While cattle develop functional immunity to *T. orientalis*, they can remain infected and, under stressful conditions such as calving or peak milk production, the disease re-emerges (Gebrekidan *et al.*, 2015; Watts *et al.*, 2016). Furthermore, young calves in endemic areas and naïve cattle that are moved to endemic areas are susceptible to the disease as they have not developed immunity to the parasite. As there is no vaccine, BATOG has been estimated to cost the Australian cattle industry AUS$20 million per annum due to the increase in mortality, reduction in fertility and loss of milk production (Perera *et al.*, 2014; GHD Pty Ltd, 2015). Despite this, BATOG remains a non-notifiable disease in Australia.

Cattle with signs of BATOG are traditionally diagnosed via morphological identification of the parasite within erythrocytes (Forshaw *et al.*, 2020). Serological and molecular methods are also commonly used to detect the presence of *T. orientalis*. Importantly, the sequence of the major piroplasm surface protein (*mpsp*) gene (Kamau *et al.*, 2011) distinguishes 11 genotypes in the *T. orientalis* complex: type 1 (chitose), type 2 (ikeda), type 3 (buffeli), types 4–8 and N1–N3.

The distribution and diversity of *T. orientalis* have changed dramatically in Australia. Since the early 1900s, cattle infected with *T. orientalis* were generally asymptomatic (Stewart *et al.*, 1996) and, retrospectively, presumed to be infected with the apathogenic *T. orientalis* type 3 (Eamens *et al.*, 2013b; Perera *et al.*, 2014). In the 1960s, there were three fatal cases linked to *T. orientalis* (Rogers and Callow, 1966) and then in the 1990s, cattle exported from Australia to Japan became unwell and *T. orientalis* type 1, a type that causes clinical disease, was detected for the first time (Kubota *et al.*, 1996). Since 2006, there have been over 500 major outbreaks of BATOG in the eastern states of New South Wales and Queensland (Izzo *et al.*, 2010; Perera *et al.*, 2014), which were linked with *T. orientalis* types previously undetected in Australia: pathogenic type 1 (Kamau *et al.*, 2011; Eamens *et al.*, 2013a,b); pathogenic type 2 (Kamau *et al.*, 2011; Eamens *et al.*, 2013a,b); and type 4 (Kamau *et al.*, 2011). Since the early 2010s, there have also been outbreaks of BATOG in Victoria and South Australia with pathogenic type 1, pathogenic type 2 and type 5 (Islam *et al.*, 2011; Perera *et al.*, 2013, 2014; Gebrekidan *et al.*, 2015) and also in Western Australia with pathogenic type 2 (Forshaw *et al.*, 2020). BATOG disease dynamics are dependent on the host and vector, both of which will be affected by climate change.

A susceptible host is important for BATOG outbreaks. The importation of cattle from endemic regions has preceded the BATOG outbreak in Victoria (Islam *et al.*, 2011), South Australia (Gebrekidan *et al.*, 2015) and possibly Western Australia (Forshaw *et al.*, 2020). However, there have been outbreaks of BATOG in closed herds, so the role of sympatric wildlife in the movement of *Haemaphysalis* spp. (Perera *et al.*, 2013; Loh *et al.*, 2018) requires further investigation. Furthermore, climate change is increasing the frequency of extreme climate events in Australia, so cattle are more likely to be affected by heat stress and dehydration. These stressful environmental conditions will make cattle more susceptible to the re-emergence of BATOG.

An established population of a competent tick vector is important for BATOG outbreaks. The native *Haemaphysalis bremneri* (formerly *Haemaphysalis humerosa*) (Stewart *et al.*, 1987a,b; Fujisaki *et al.*, 1994) and *Haemaphysalis bancrofti* (Riek, 1982; Stewart *et al.*, 1989) transmit *T. orientalis*, most likely the apathogenic type 3; *H. longicornis*, also known as the bush tick in Australia, the New Zealand cattle tick or the Asian long-horned tick, can transmit the pathogenic *T. orientalis* type 2 (Greay *et al.*, 2018; Marendy *et al.*, 2020) but is likely an inefficient vector for apathogenic type 3 (Riek, 1982; Uilenberg *et al.*, 1985; Stewart *et al.*, 1987a,b). *H. longicornis* was first identified in eastern Australia in 1897, likely introduced from Japan (Roberts, 1963; Hoogstraal *et al.*, 1968), and was first detected in Western Australia in 1983 (Besier and Wroth, 1985). Outbreaks of BATOG are restricted to the range of *H. longicornis* in Australia (Riek, 1982; Hammer *et al.*, 2015; Jenkins, 2018). There is also a close association of this disease and the recent introduction of *H. longicornis* in the USA (Oakes *et al.*, 2019). Therefore, habitat suitability for *H. longicornis* is likely a key driver of BATOG outbreaks. For example, the outbreak of BATOG in Victoria was preceded by favourable environmental conditions for *H. longicornis* (Perera *et al.*, 2013). Wherever *H. longicornis* becomes established in Australia, there will be the threat of BATOG outbreaks. Climate change will likely alter the habitat suitability for *H. longicornis* in Australia.

61.1.4 Tick-borne diseases in Australia and climate change

Climate change is likely having an impact on the introduced tick vectors associated with CME and BATOG. *R. sanguineus* has been introduced into more populous, southern locations in Australia (Greay *et al.*, 2016; Chandra *et al.*, 2019) and *H. longicornis* is also expanding its range, now being found in multiple locations in Western Australia (Forshaw *et al.*, 2020); favourable environmental conditions will support the establishment of both tick species. For example, in Western Australia, climate change is increasing the average annual temperature, increasing summer rainfall and decreasing the frequency of frosts (Sudmeyer *et al.*, 2016). However, the effect of climate change on tick habitat suitability, and therefore the potential distribution for these two introduced tick species in Australia, is yet to be properly explored. An approach similar to a recent study of the effect of climate change on *H. longicornis* in New Zealand would be a good starting point (Heath, 2016; Lawrence *et al.*, 2017). The distribution of these two introduced tick species also needs to be correlated with the potential distribution of the domesticated animals, cattle and wild dogs, and climate change. Studying the effect of climate change on ticks and tick-borne diseases in Australia is an urgent priority.

61.2 Future Predictions

There is little doubt that the recent detection of *E. canis* and *T. orientalis* in Australia will have significant ramifications for domestic animal health in this country. There is concern that CME and BATOG will become increasingly widespread as, to date, the tick vectors are unlikely to have reached their full geographical range in Australia, especially if climate change leads to favourable environmental conditions for ticks, such as higher temperatures and increasing summer rainfall. The relative abundance of the two tick vectors is also likely to change in regions where they are established.

In the short term, the enzootic range of *H. longicornis* will likely continue to expand as ticks are dispersed inadvertently with cattle transport and spread is further facilitated through wildlife movements (Loh *et al.*, 2018). However, in the long term it is hypothesized that *H. longicornis* in Australia will encounter unfavourable environmental conditions due to climate change and, being parthenogenetic (Bremner, 1959), tick populations will be unable to adapt as a result of restricted genetic diversity. Similarly, the movement of infected dogs from northern to southern regions of Australia, in combination with favourable local environments, will see the continued establishment of *R. sanguineus* in new locations (Greay *et al.*, 2018; Chandra *et al.*, 2020). However, unlike in remote regions of northern Australia, where *R. sanguineus* burdens are high as a consequence of the favourable interaction between climate, large dog populations and limited acaricidal usage, it is unlikely that the brown dog tick will become similarly abundant in the south if pet owners apply effective tick prophylaxis.

In conclusion, the interplay of climate change, ticks and tick-borne diseases in Australia is a dynamic situation. Australia urgently needs better surveillance of tick-borne diseases. In particular, we propose a system be developed to provide real-time updates on the presence and geographical distribution of pathogenic organisms and associated vectors in Australia. These real-time reports will be invaluable for informing key stakeholders, such as veterinarians, the farming community and government departments. Furthermore, long-term tick surveillance data will enable the impact of climate change on ticks in Australia to be quantified. Surveillance is essential to developing strategies to control the spread and prevalence of ticks and tick-borne diseases in a changing world.

References

Barker, D. (2019) *Ixodes barkeri* n. sp. (Acari: Ixodidae) from the short-beaked echidna, *Tachyglossus aculeatus*, with a revised key to the male *Ixodes* of Australia, and list of the subgenera and species of *Ixodes* known to occur in Australia. *Zootaxa* 4658, 331–342.

Besier, R. and Wroth, R. (1985) Discovery of the tick *Haemaphysalis longicornis* in Western Australia. *Australian Veterinary Journal* 62, 205–206.

Bouza-Mora, L., Dolz, G., Solórzano-Morales, A., Romero-Zuñiga, J.J., Salazar-Sánchez, L., *et al.* (2017) Novel genotype of *Ehrlichia canis* detected in samples of human blood bank donors in Costa Rica. *Ticks and Tick-borne Diseases* 8, 36–40.

Bremner, K. (1959) Observations on the biology of *Haemaphysalis bispinosa* Neumann (Acarina: Ixodidae) with particular reference to its mode of reproduction by parthenogenesis. *Australian Journal of Zoology* 7, 7–12.

Chandra, S., Ma, G.C., Burleigh, A., Brown, G., Norris, J.M., *et al.* (2020) The brown dog tick *Rhipicephalus sanguineus* sensu Roberts, 1965 across Australia: morphological and molecular identification of *R. sanguineus* s.l. tropical lineage. *Ticks and Tick-borne Diseases* 11, 101305.

CSIRO and BoM (2020) *State of the Climate 2020*. Commonwealth Scientific and Industrial Research Organisation and The Bureau of Meteorology, Melbourne, Australia. Available at: https://www.csiro.au/en/Showcase/state-of-the-climate (accessed 6 December 2020).

Dantas-Torres, F. (2010) Biology and ecology of the brown dog tick, *Rhipicephalus sanguineus*. *Parasites & Vectors* 3, 26.

de Castro, M.B., Machado, R.Z., de Aquino, L.P.C.D., Alessi, A.C. and Costa, M.T. (2004) Experimental acute canine monocytic ehrlichiosis: clinicopathological and immunological findings. *Veterinary Parasitology* 119, 73–86.

Eamens, G.J., Bailey, G., Jenkins, C. and Gonsalves, J.R. (2013a) Significance of *Theileria orientalis* types in individual affected beef herds in New South Wales based on clinical, smear and PCR findings. *Veterinary Parasitology* 196, 96–105.

Eamens, G.J., Gonsalves, J.R., Jenkins, C., Collins, D. and Bailey, G. (2013b) *Theileria orientalis* MPSP types in Australian cattle herds associated with outbreaks of clinical disease and their association with clinical pathology findings. *Veterinary Parasitology* 191, 209–217.

Evans, M., Egan, S., Irwin, P. and Oskam, C. (2019) Automatic barcode gap discovery reveals large COI intraspecific divergence in Australia Ixodidae. *Zootaxa* 4656, 393–396.

Forshaw, D., Alex, S.M., Palmer, D.G., Cotter, J., Roberts, W.D., *et al.* (2020) *Theileria orientalis* Ikeda genotype infection associated with anaemia, abortion and death in beef cattle in Western Australia. *Australian Veterinary Journal* 98, 290–297.

Fourie, J.J., Luus, H.G., Stanneck, D. and Jongejan, F. (2013) The efficacy of Advantix to prevent transmission of *Ehrlichia canis* to dogs by *Rhipicephalus sanguineus* ticks. *Parasite* 20, 36.

Fujisaki, K., Kawazu, S. and Kamio, T. (1994) The taxonomy of the bovine *Theileria* spp. *Parasitology Today* 10, 31–33.

Gebrekidan, H., Gasser, R.B., Perera, P.K., McGrath, S., McGrath, S., *et al.* (2015) Investigating the first outbreak of oriental theileriosis in cattle in South Australia using multiplexed tandem PCR (MT-PCR). *Ticks and Tick-borne Diseases* 6, 574–578.

GHD Pty Ltd (2015) *Priority List of Endemic Diseases for the Red Meat Industries*. Meat & Livestock Australia Limited, North Sydney, Australia.

Greay, T.L., Oskam, C.L., Gofton, A.W., Rees, R.L., Ryan, U.M. and Irwin, P.J. (2016) A survey of ticks (Acari: Ixodidae) of companion animals in Australia. *Parasites & Vectors* 9, 207.

Greay, T.L., Zahedi, A., Krige, A.-S., Owens, J.M., Rees, R.L., *et al.* (2018) Endemic, exotic and novel apicomplexan parasites detected during a national study of ticks from companion animals in Australia. *Parasites & Vectors* 11, 197.

Hammer, J.F., Emery, D., Bogema, D.R. and Jenkins, C. (2015) Detection of *Theileria orientalis* genotypes in *Haemaphysalis longicornis* ticks from southern Australia. *Parasites & Vectors* 8, 229.

Harrus, S. and Waner, T. (2011) Diagnosis of canine monocytic ehrlichiosis (*Ehrlichia canis*): an overview. *Veterinary Journal* 187, 292–296.

Heath, A.C.G. (2016) Biology, ecology and distribution of the tick, *Haemaphysalis longicornis* Neumann (Acari: Ixodidae) in New Zealand. *New Zealand Veterinary Journal* 64, 10–20.

Hoogstraal, H., Roberts, F.H., Kohls, G.M. and Tipton, V.J. (1968) Review of *Haemaphysalis* (*Kaiseriana*) *longicornis* Neumann (resurrected) of Australia, New Zealand, New Caledonia, Fiji, Japan, Korea, and northeastern China and USSR, and its parthenogenetic and bisexual populations (Ixodoidea, Ixodidae). *Journal of Parasitology* 54, 1197–1213.

Islam, M.K., Jabbar, A., Campbell, B.E., Cantacessi, C. and Gasser, R.B. (2011) Bovine theileriosis – an emerging problem in south-eastern Australia? *Infection, Genetics and Evolution* 11, 2095–2097.

Izzo, M., Poe, I., Horadagoda, N., De Vos, A. and House, J. (2010) Haemolytic anaemia in cattle in NSW associated with *Theileria* infections. *Australian Veterinary Journal* 88, 45–51.

Jackson, S.M., Groves, C.P., Fleming, P.J.S., Aplin, K.P., Eldridge, M.D.B., *et al.* (2017) The wayward dog: is the Australian native dog or dingo a distinct species? *Zootaxa* 4317, 201–224.

Jenkins, C. (2018) Bovine theileriosis in Australia: a decade of disease. *Microbiology Australia* 39, 215.

Kamau, J., de Vos, A.J., Playford, M., Salim, B., Kinyanjui, P. and Sugimoto, C. (2011) Emergence of new types of *Theileria orientalis* in Australian cattle and possible cause of theileriosis outbreaks. *Parasites & Vectors* 4, 22.

Kubota, S., Sugimoto, C., Kakuda, T. and Onuma, M. (1996) Analysis of immunodominant piroplasm surface antigen alleles in mixed populations of *Theileria sergenti* and *T. buffeli*. *International Journal for Parasitology* 26, 741–747.

Lawrence, K.E., Summers, S.R., Heath, A.C.G., McFadden, A.M.J., Pulford, D.J., *et al.* (2017) Using a rule-based envelope model to predict the expansion of habitat suitability within New Zealand for the tick *Haemaphysalis longicornis*, with future projections based on two climate change scenarios. *Veterinary Parasitology* 243, 226–234.

Loh, S.-M., Egan, S., Gillett, A., Banks, P.B., Ryan, U.M., *et al.* (2018) Molecular surveillance of piroplasms in ticks from small and medium-sized urban and peri-urban mammals in Australia. *International Journal for Parasitology: Parasites and Wildlife* 7, 197–203.

Marendy, D., Baker, K., Emery, D., Rolls, P. and Stutchbury, R. (2020) *Haemaphysalis longicornis*: the life-cycle on dogs and cattle, with confirmation of its vector status for *Theileria orientalis* in Australia. *Veterinary Parasitology* 3, 100022.

Mason, R.J., Lee, J.M., Curran, J.M., Moss, A., van der Heide, B. and Daniels, P.W. (2001) Serological survey for *Ehrlichia canis* in urban dogs from the major population centres of northern Australia. *Australian Veterinary Journal* 79, 559–562.

Oakes, V.J., Yabsley, M.J., Schwartz, D., LeRoith, T., Bissett, C., *et al.* (2019) *Theileria orientalis* Ikeda genotype in cattle, Virginia, USA. *Emerging Infectious Diseases* 25, 1653–1659.

Perera, P.K., Gasser, R.B., Anderson, G.A., Jeffers, M., Bell, C.M. and Jabbar, A. (2013) Epidemiological survey following oriental theileriosis outbreaks in Victoria, Australia, on selected cattle farms. *Veterinary Parasitology* 197, 509–521.

Perera, P.K., Gasser, R.B., Firestone, S.M., Anderson, G.A., Malmo, J., *et al.* (2014) Oriental theileriosis in dairy cows causes a significant milk production loss. *Parasites & Vectors* 7, 73.

Perez, M., Rikihisa, Y. and Wen, B. (1996) *Ehrlichia canis*-like agent isolated from a man in Venezuela: antigenic and genetic characterization. *Journal of Clinical Microbiology* 34, 2133–2139.

Riek, R. (1982) Epidemiology and transmission of *Theileria* sp. of cattle in Australia. *Australian Veterinary Journal* 59, 89–92.

Roberts, F. (1963) A systematic study of the Australian species of the genus *Haemaphysalis* Koch (Acarina: Ixodidae). *Australian Journal of Zoology* 11, 35–80.

Rogers, R. and Callow, L. (1966) Three fatal cases of *Theileria mutans* infection. *Australian Veterinary Journal* 42, 42–46.

Stewart, N., De Vos, A., McGregor, W. and Shiels, I. (1987a) *Haemaphysalis humerosa*, not *H. longicornis*, is the likely vector of *Theileria buffeli* in Australia. *Australian Veterinary Journal* 64, 280–282.

Stewart, N., De Vos, A., Shiels, I. and McGregor, W. (1987b) The experimental transmission of *Theileria buffeli* of cattle in Australia by *Haemaphysalis humerosa*. *Australian Veterinary Journal* 64, 81–83.

Stewart, N., Uilenberg, G. and De Vos, A. (1996) Review of Australian species of *Theileria*, with special reference to *Theileria buffeli* of cattle. *Tropical Animal Health and Production* 28, 81–90.

Stewart, N.P., Devos, A.J., Shiels, I.A. and Jorgensen, W.K. (1989) Transmission of *Theileria buffeli* to cattle by *Haemaphysalis bancrofti* fed on artificially infected mice. *Veterinary Parasitology* 34, 123–127.

Sudmeyer, R., Edward, A., Fazakerley, V., Simpkin, L. and Foster, I. (2016) *Climate Change: Impacts and Adaptation for Agriculture in Western Australia. Bulletin No. 4870*. Department of Agriculture and Food, Western Australia, Perth, Australia.

Uilenberg, G., Perie, N., Spanjer, A. and Franssen, F. (1985) *Theileria orientalis*, a cosmopolitan blood parasite of cattle: demonstration of the schizont stage. *Research in Veterinary Science* 38, 352–360.

Watts, J., Playford, M. and Hickey, K. (2016) *Theileria orientalis*: a review. *New Zealand Veterinary Journal* 64, 3–9.

eo62 Tick-Borne Infections in Central Europe

Maria Kazimirova*
Institute of Zoology, Slovak Academy of Sciences, Bratislava, Slovakia

62.1 Expert Opinion

62.1.1 Introduction

In Central Europe (CE), Lyme borreliosis (LB) and tick-borne encephalitis (TBE) are the most important tick-borne infections in humans. Their causative agents, spirochaetes of the *Borrelia burgdorferi* sensu lato (s.l.) complex and tick-borne encephalitis virus (TBEV), respectively, are transmitted by the castor bean tick, *Ixodes ricinus* (Medlock *et al.*, 2013). *I. ricinus* is also a competent vector for other pathogenic microorganisms of medical and veterinary importance, such as *Borrelia miyamotoi*, *Anaplasma phagocytophilum*, *Neoehrlichia mikurensis*, *Ehrlichia* and *Rickettsia* spp., and the protozoan parasites, *Babesia divergens*, *Babesia microti* and *Babesia venatorum*. *I. ricinus* is associated with a multitude of vertebrate host species and all three developmental stages (larvae, nymphs, adults) can bite humans. The wide range of vertebrate hosts (including those amplifying pathogens) that live in different habitats, the peculiar population ecology, phenology and physiology of *I. ricinus*, together with abiotic factors (mainly temperature and relative humidity), determine: (i) its widespread occurrence across Europe; (ii) its high abundance in many locations; and (iii) its ability to harbour and transmit multiple pathogens.

The ornate dog tick, *Dermacentor reticulatus*, and the ornate sheep tick, *Dermacentor marginatus*, are other important tick species in CE that transmit pathogens of diseases in humans and domestic animals, including the causative agents of rickettsioses (*Rickettsia slovaca*, *Rickettsia raoultii*), tularaemia (*Francisella tularensis*), Q fever (*Coxiella burnetii*) and canine babesiosis (*Babesia canis*) (Földvári *et al.*, 2016; Walter *et al.*, 2016). There is growing evidence that *D. reticulatus* may also be implicated in transmission of TBEV in regions where it occurs sympatrically with *I. ricinus* (Chitimia-Dobler *et al.*, 2019). Adults of both *Dermacentor* spp. are able to infest a relatively broad range of larger vertebrate hosts, including domestic animals and pets, whereas their immature stages are endophilic and parasitize mainly rodents and other small mammals (Földvári *et al.*, 2016; Rubel *et al.*, 2016).

Haemaphysalis spp. ticks occurring in CE (*Haemaphysalis concinna*, *Haemaphysalis punctata*, *Haemaphysalis inermis*) are also involved in the transmission of pathogens (e.g. TBEV, *F. tularensis*, *C. burnetii*, *Babesia* spp., *Theileria* spp.), but due to their focal distribution and low abundance, their overall medical and veterinary impact is lower than of the tick species mentioned above

*Email: maria.kazimirova@savba.sk

DOI: 10.1079/9781789249637.0062

(Hornok *et al.*, 2015; Rubel *et al.*, 2018). They are exophilic, three-host species associated with a number of vertebrate hosts and occasionally bite humans.

The occurrence in CE of a new species, *Ixodes inopinatus*, and its infection with several pathogens transmitted by *I. ricinus*, have been confirmed recently (Chitimia-Dobler *et al.*, 2018; Hauck *et al.*, 2019). Local occurrence of the brown dog tick, *Rhipicephalus sanguineus* – vector of the zoonotic pathogens *Rickettsia conorii* and *Rickettsia massiliae* and pathogens of canines such as *Ehrlichia canis*, *Anaplasma platys*, *Babesia vogeli*, *Babesia gibsoni*, *Babesia vulpes*, and *Hepatozoon canis* in Southern Europe – has repeatedly been reported in CE (Dantas-Torres and Otranto, 2015; Buczek and Buczek, 2020). Moreover, several cases of the introduction of *Hyalomma* spp. (e.g. *Hyalomma marginatum*, *Hyalomma rufipes*), vectors of the important zoonotic pathogens Crimean–Congo haemorrhagic fever virus (CCHFV) and *Rickettsia aeschlimannii*, by migratory birds to CE, have been reported as well (Chitimia-Dobler *et al.*, 2016; Duscher *et al.*, 2018; Hubálek *et al.*, 2020).

During the past decades, there has been growing evidence of: (i) geographic expansion of indigenous tick species northward and to higher altitudes in CE; (ii) increasing tick survival and abundance in certain areas including urban parks and gardens; (iii) prolonged seasonal tick questing activities, increasing the number of infected hosts and the probability of pathogen transmission; (iv) emergence of new tick-borne disease foci; and (v) changing epidemiological patterns and increasing incidence of tick-borne diseases in endemic foci. As in the rest of the northern hemisphere, global warming and climate change in CE have been attributed to these changes, acting through multiple pathways, i.e. through direct effects on ticks, the pathogens they carry and their hosts, but also through alteration of whole ecosystems. However, non-climate drivers such as globalization, socio-economic and demographic changes, changing agricultural and hunting practices and public health interventions interact with climatic factors, and it is difficult or even impossible to isolate solely the role of climate change on the changing epidemiological patterns of tick-borne diseases (Rocklöv and Dubrow, 2020).

62.1.2 Expansion of indigenous tick species and diseases to new geographical areas and habitats, and emergence of new pathogens

I. ricinus is widely distributed in various habitat types all over CE. During the last few decades, the species has been emerging in new sites located in mountainous areas at altitudes about 300–400 m higher than previously recorded, as well as in urban and suburban sites (Materna *et al.*, 2008; Gray *et al.*, 2009; Medlock *et al.*, 2013; Rizzoli *et al.*, 2014). *I. ricinus* occurs mainly in broad-leaf and mixed forests at altitudes up to 700–800 m above sea level (ASL) (Daniel *et al.*, 2003; Medlock *et al.*, 2013). It prefers habitats with a relative humidity of at least 80% and sufficient vegetation cover and leaf litter which provide shelter for the ticks as well as for their hosts. Nymphs and adults are usually active in spring and autumn. Data published from the Czech Republic, Slovakia and Switzerland, but also from the Alpine region in north-western Italy, provide evidence for expansion of *I. ricinus* to higher altitudes (Materna *et al.*, 2005, 2008; Daniel *et al.*, 2009; Peťko *et al.*, 2011; Rusňáková Tarageľová *et al.*, 2016; Garcia-Vozmediano *et al.*, 2020). For example, monitoring of climatic conditions and the altitudinal distribution of *I. ricinus* in different mountain areas in the Czech Republic revealed a similar trend of increasing mean temperatures and the occurrence of tick populations up to the timberline (at about 1250 m ASL) (Daniel *et al.*, 2003, 2009; Materna *et al.*, 2005, 2008). The shift of the upper altitudinal limit of *I. ricinus* has been explained by the continuous rise of the mean annual air temperature at higher altitudes, especially in May to August, which is connected with prolonged questing activity, increased survival and the ability of the tick to establish new local populations (Materna *et al.*, 2005, 2008; Daniel *et al.*, 2009). Prolonged questing activity of *I. ricinus* extending to winter months in non-mountainous regions is another consequence of the changing climate, as demonstrated e.g. in Germany (Dautel *et al.*, 2008; Gray *et al.*, 2009). Increasing abundance of *I. ricinus* in different habitats, including urban parks and gardens, is another phenomenon observed during recent years (Rizzoli *et al.*, 2014). Climate change may have, in part, a direct effect on changing tick abundance in endemic areas by influencing tick phenology (Randolph, 2004). However, climate also affects the biodiversity, phenology and production of plants, and thus the food availability for herbivorous vertebrate hosts, and consequently their population densities and the densities of predators (Bregnard *et al.*, 2020). Anthropogenic factors such as urbanization, changing patterns of land use, deforestation, reforestation, hunting practices, etc. affect the biodiversity and population density of vertebrate hosts maintaining tick populations (Rizzoli *et al.*, 2014; Rocklöv and Dubrow, 2020).

Along with expansion of *I. ricinus* to new areas, increasing abundance and prolonged seasonal questing activity, the presence of pathogens transmitted by this tick in CE has been reported in new locations, together with increases in case numbers and the emergence of new foci of the notifiable diseases, LB and TBE (Lindgren *et al.*, 2006; Beauté *et al.*, 2018). TBE is endemic in all CE countries, with the highest incidence rate in the Czech Republic (Beauté *et al.*, 2018). According to the European Centre for Disease Prevention and Control, the overall mean prevalence of *B. burgdorferi* s.l. in ticks has been estimated at about 12% in Europe, with the highest rates (>10% in nymphs and >20% in adults) in CE, mainly in Austria, Czech Republic, southern Germany, Switzerland, Slovakia and Slovenia (https://www.ecdc.europa.eu/en/borreliosis/facts/factsheet, accessed 17 June 2021). According to Rizzoli *et al.* (2011), LB incidence is also the highest in CE. However, recent reports, based on more exact surveillance data, indicate higher incidence rates in more northern latitudes (Vandekerckhove *et al.*, 2019). Notwithstanding, there is evidence that, as a result of climate change, TBE foci and the range of LB are expanding into higher altitudes (Gray *et al.*, 2009). Moreover, there is increasing evidence of the presence of infected ticks in urban and suburban areas (Rizzoli *et al.*, 2014). In the Czech Republic, together with the expansion of *I. ricinus* in mountainous areas, the shift of TBEV (and human TBE cases) and *B. burgdorferi* s.l. into higher altitudes was reported (Daniel *et al.*, 2009; Danielová *et al.*, 2006, 2010). In addition to changes in geographical distribution, changing patterns of seasonal TBE incidence have been registered since the late 1990s, with increasing proportions of cases in spring and autumn as a consequence of prolonged seasonal tick activity (Kriz *et al.*, 2012). In Austria, an outbreak of alimentary TBE infection in an alpine pasture at 1500 m ASL was recorded in 2008 (Holzmann *et al.*, 2009). Furthermore, new foci of TBE in the western part of the country emerged at higher altitudes than previously recorded (Heinz *et al.*, 2015). In Slovakia, a shift of active TBE foci from lowlands to highlands and mountainous areas has been observed since the 1980s (Lukan *et al.*, 2010). *Borrelia*-infected *I. ricinus* were found in a mountain region in northern Slovakia up to 1070 m ASL. Interestingly, the genospecies *Borrelia lusitaniae*, known to be associated with warmer xerothermic habitats, dominated up to 660 m ASL and was present in ticks up to 750 m ASL (Rusňáková Tarageľová *et al.*, 2016). Diseases in humans, caused by *A. phagocytophilum*, *Babesia divergens*, *B. microti*, *B. venatorum*, *B. miyamotoi*, *N. mikurensis*, *Rickettsia helvetica* and *R. monacensis*, are rare in CE and data on the pathogenicity of these microorganisms are limited (Azagi *et al.*, 2020). Thus, it is difficult to relate their emergence and incidence to changing climate.

Geographic range expansion has been observed for *D. reticulatus* and to a lesser extent for *D. marginatus* (Földvári *et al.*, 2016; Rubel *et al.*, 2016; Drehmann *et al.*, 2020). Their distribution areas overlap between 41°N and 51°N (Rubel *et al.*, 2016). Generally, *D. reticulatus* prefers the moist and colder climate of more northern latitudes and is able to overcome unfavourable conditions (e.g. survival underwater for several weeks during floods), while *D. marginatus* prefers warmer, steppe-like habitats in more southern latitudes (Földvári *et al.*, 2016; Rubel *et al.*, 2016); however, *D. marginatus* was recently recorded in Alpine valleys of north-western Italy (Garcia-Vozmediano *et al.*, 2020). Northward expansion of *D. reticulatus* and its emergence in new locations including urban sites have been documented in Germany (Drehmann *et al.*, 2020), Poland (Karbowiak, 2014; Mierzejewska *et al.*, 2016; Kloch *et al.*, 2017; Kubiak *et al.*, 2018), Slovakia (Bullová *et al.*, 2009) and the Czech Republic (Široký *et al.*, 2011). *D. marginatus* was found in new locations north-west of its previous range in North Rhine-Westphalia in Germany (Drehmann *et al.*, 2020). Factors influencing the spread of *D. reticulatus* probably include global warming, although habitat fragmentation, due to changing patterns of land use, plays a role by creating patches of green areas along water courses and reservoirs, which lead to increases in number and diversity of vertebrate hosts (Karbowiak, 2014). In addition, anthropogenic factors, mainly travelling with pets and transport of livestock, are suggested to contribute to the spread of *D. reticulatus* (Kloch *et al.*, 2017). In contrast to *I. ricinus*, the medical importance of *Dermacentor* species in CE is lower. As larvae and nymphs are endophilic, only adults attack humans, but human bites are rare. They may transmit TBEV in endemic areas (Chitimia-Dobler *et al.*, 2019) and are vectors of the causative agents of tick-borne lymphadenopathy (*R. slovaca*, *R. raoultii*), tularaemia and Q fever (Földvári *et al.*, 2016; Rubel *et al.*, 2016; Drehmann *et al.*, 2020). In contrast, the veterinary importance of *D. reticulatus* due to transmission of *B. canis*, the causative agent of canine babesiosis, is increasing coincident with the expansion of *D. reticulatus* into new areas (Solano-Gallego *et al.*, 2016).

Except for the relict tick, *H. concinna*, data on the spread of *Haemaphysalis* spp. to new geographical areas in CE are limited, probably due to their fragmented and focal distribution and low abundance. *H. concinna* inhabits mainly deciduous and mixed forests of the temperate zone of Europe and can also occur in urban parks and forests. In some regions, it forms abundant populations, so-called clusters. One such cluster is located in an area along the Danube and Morava rivers on the borders of the Czech Republic, Slovakia, Austria and Hungary, where *H. concinna* is locally the second or third most abundant tick species after *I. ricinus* and *D. reticulatus* (Rubel *et al.*, 2018). Recently, a local population of *H. concinna* was recorded in a new site in western Poland (Dwużnik *et al.*, 2019). In CE, *Haemaphysalis* spp. harbour zoonotic pathogens (TBEV, *A. phagocytophilum*, *C. burnetii*) and pathogens that infect domestic animals and wildlife (*Babesia* spp., *Theileria* spp.) (Hornok *et al.*, 2015; Rubel *et al.*, 2018). However, the relevance of these ticks for public health is relatively low. Recently, *Babesia crassa* (known to infect small ruminants in the Middle East) was detected in *H. inermis*. *Babesia motasi* known from *H. punctata* from Northern, Western and Southern Europe, and genotypes identical with *Babesia* known from Far Eastern Russia and East Siberia, were detected in *H. concinna* from Hungary. In addition, *Theileria* sp. OT3, known from the Mediterranean region, was detected in *H. punctata* (Hornok *et al.*, 2015). Two *Babesia* genotypes related to species known from Far Eastern Russia were detected in *H. concinna* in south-western Slovakia (Hamšíková *et al.*, 2016). The presence of diverse *Babesia* spp. in *Haemaphysalis* spp. in CE is probably connected with the wide geographical distribution of *H. concinna* in Eurasia and introduction to new habitats by migratory birds, rather than as a consequence of climate change.

62.1.3 Introduction of new tick species and risk of establishment of diseases in non-endemic areas

The newly described species, *I. inopinatus* (Estrada-Peña *et al.*, 2014), is closely related to *I. ricinus* and its distribution area has been restricted to Southern Europe and Northern Africa, but a few specimens have been found in Rhineland-Palatinate (Germany). Sympatric occurrence of *I. inopinatus* and *I. ricinus* has been reported recently from south-eastern and northern Germany, western Austria and Romania (Chitimia-Dobler *et al.*, 2018; Hauck *et al.*, 2019). It remains to be seen if *I. inopinatus* is a hitherto unrecognized but established member of the tick fauna of CE occurring sympatrically with *I. ricinus*, or if it was introduced from Southern Europe. DNA of *Borrelia* spp. (*Borrelia afzelii*, *B. garinii*, *B. valaisiana*, *B. burgdorferi* sensu stricto, *B. bissettiae*, *B. miyamotoi*), *R. helvetica* and *A. phagocytophilum* was identified in questing *I. inopinatus* from northern Germany, indicating its potential vector role for these pathogens (Hauck *et al.*, 2019).

Species of the *R. sanguineus* group have been frequently imported by companion animals to several countries in CE, mainly from southern parts of the continent, but also from other endemic areas (Buczek and Buczek, 2020). These ticks are endophilic and associated with kennels or human dwellings. Their main hosts are dogs, but human bites are quite frequent. Under favourable conditions inside kennels and houses, or during extremely warm summers, the ticks can reproduce and establish abundant temporal populations as reported e.g. from Poland, Germany and Hungary. Pathogens such as the causative agent of Mediterranean spotted fever, *R. conorii*, and *R. massiliae* have been detected in the exotic ticks, which pose a risk of introduction and spread of new diseases (Buczek and Buczek, 2020). An increase in mean temperature by about 2–3°C from April to September could lead to permanent *R. sanguineus* populations in CE (Gray *et al.*, 2009). Hungary is the first country outside the original endemic area where increasing autochthonous cases of dog infestations and new foci of *R. sanguineus* s.l. have been reported recently, suggesting establishment of the species in the southern part of the country (Hornok *et al.*, 2020). Detections of *E. canis* and *R. massiliae* in *D. marginatus* nymphs collected from dogs, and of *E. canis* in *Ixodes canisuga* larvae feeding on red foxes in southern Hungary, are likely to confirm this assumption (Hornok *et al.*, 2013b).

In Europe, species of the *H. marginatum* complex are endemic in the south and southeast, including countries in the Mediterranean region. Immature stages (larvae, nymphs) have often been recorded on migratory birds arriving in CE in spring. Such records are known, e.g., from the Czech Republic, Slovakia, Slovenia, Germany, Switzerland and Hungary (Hubálek *et al.*, 2020). Adult specimens of the *H. marginatum* complex have been reported on various mammalian hosts (including humans) in the Czech Republic, Slovakia, Poland, Germany, Hungary and Austria (Hubálek *et al.*, 2020). These ticks were probably imported by birds and moulted to the adult stage under the climatic conditions of CE. In addition, data from southern Hungary suggest that overwintering of *Hyalomma* spp. could also take place (Hornok *et al.*, 2020). *R. aeschlimannii* was detected in subadult *H. marginatum* introduced by birds to Germany (Rumer *et al.*, 2011) and Hungary (Hornok *et al.*, 2013a). The same rickettsial species was detected in an adult *H. marginatum* in Austria (Duscher *et al.*, 2018). However, recent data on the climatic requirements of *Hyalomma* spp. suggest that currently they are not able to establish permanent populations in CE and the probability of introduction of these ticks and the risk of spread of pathogens they transmit is low (Gray *et al.*, 2009; Estrada-Peña *et al.*, 2021).

62.2 Future Predictions

Future scenarios of the geographical spread and habitat expansion of ticks and associated diseases in CE must consider not only changes in climate, but the whole complexity of drivers of such changes including socio-economic factors, introduction of effective tick-control measures and of vaccines against LB, increased vaccine coverage against TBE and increased public awareness of ticks. It is also important to bear in mind that finding a new, potentially pathogenic microorganism is not necessarily the consequence of climate change or globalization but may be the result of improved diagnostic methods and more intensive research. Spreading of ticks and diseases to higher altitudes and northwards is likely to continue in CE under the current trend of global warming; however, this trend can be altered by taking effective measures to reduce the impact of anthropogenic factors on the environment.

References

Azagi, T., Hoornstra, D., Kremer, K., Hovius, J.W.R. and Sprong, H. (2020) Evaluation of disease causality of rare *Ixodes ricinus*-borne infections in Europe. *Pathogens* 9, 150.

Beauté, J., Spiteri, G., Warns-Petit, E. and Zeller, H. (2018) Tick-borne encephalitis in Europe, 2012 to 2016. *Eurosurveillance* 23, 1800201.

Bregnard, C., Rais, O. and Voordouw, M.J. (2020) Climate and tree seed production predict the abundance of the European Lyme disease vector over a 15-year period. *Parasites & Vectors* 13, 408.

Buczek, A. and Buczek, W. (2020) Importation of ticks on companion animals and the risk of spread of tick-borne diseases to non-endemic regions in Europe. *Animals* 11, E6.

Bullová, E., Lukáň, M., Stanko, M. and Peťko, B. (2009) Spatial distribution of *Dermacentor reticulatus* tick in Slovakia in the beginning of the 21st century. *Veterinary Parasitology* 165, 357–360.

Chitimia-Dobler, L., Rieß, R., Kahl, O., Wölfel, S., Dobler, G., *et al.* (2018) *Ixodes inopinatus* – occurring also outside the Mediterranean region. *Ticks and Tick-borne Diseases* 9, 196–200.

Chitimia-Dobler, L., Lemhöfer, G., Król, N., Bestehorn, M., Dobler, G. and Pfeffer, M. (2019) Repeated isolation of tick-borne encephalitis virus from adult *Dermacentor reticulatus* ticks in an endemic area in Germany. *Parasites & Vectors* 12, 90.

Daniel, M., Danielová, V., Kříž, B., Jirsa, A. and Nožička, J. (2003) Shift of the tick *Ixodes ricinus* and tick-borne encephalitis to higher altitudes in Central Europe. *European Journal of Clinical Microbiology and Infectious Diseases* 22, 327–328.

Daniel, M., Materna, J., Hönig, V., Metelka, L., Danielová, V., *et al.* (2009) Vertical distribution of the tick *Ixodes ricinus* and tick-borne pathogens in the northern Moravian mountains correlated with climate warming (Jeseníky Mts., Czech Republic). *Central European Journal of Public Health* 17, 139–145.

Danielová, V., Rudenko, N., Daniel, M., Holubova, J., Materna, J., *et al.* (2006) Extension of *Ixodes ricinus* ticks and agents of tick-borne diseases to mountain areas in the Czech Republic. *International Journal of Medical Microbiology* 296, 48–53.

Danielová, V., Daniel, M., Schwarzova, L., Materna, J., Rudenko, N., *et al.* (2010) Integration of a tick-borne encephalitis virus and *Borrelia burgdorferi* sensu lato into mountain ecosystems, following a shift in the altitudinal limit of distribution of their vector, *Ixodes ricinus* (Krkonose mountains, Czech Republic). *Vector-Borne and Zoonotic Diseases* 10, 223–230.

Dantas-Torres, F. and Otranto, D. (2015) Further thoughts on the taxonomy and vector role of *Rhipicephalus sanguineus* group ticks. *Veterinary Parasitology* 208, 9–13.

Dautel, H., Dippel, C., Kämmer, D., Werkhausen, A. and Kahl, O. (2008) Winter activity of *Ixodes ricinus* in a Berlin forest. *International Journal of Medical Microbiology* 298, 50–54.

Drehmann, M., Springer, A., Lindau, A., Fachet, K., Mai, S., *et al.* (2020) The spatial distribution of *Dermacentor* ticks (Ixodidae) in Germany – evidence of a continuing spread of *Dermacentor reticulatus*. *Frontiers in Veterinary Science* 7, 578220.

Duscher, G.G., Hodžić, A., Hufnagl, P., Wille-Piazzai, W., Schotta, A.-M., *et al.* (2018) Adult *Hyalomma marginatum* tick positive for *Rickettsia aeschlimannii* in Austria. *Eurosurveillance* 23, 1800595.

Dwużnik, D., Mierzejewska, E.J., Alsarraf, M. and Bajer, A. (2019) A new focus of the tick *Haemaphysalis concinna* in Western Poland. *Experimental and Applied Acarology* 78, 93–112.

Estrada-Peña, A., Nava, S. and Petney, T. (2014) Description of all the stages of *Ixodes inopinatus* n. sp. (Acari: Ixodidae). *Ticks and Tick-borne Diseases* 5, 734–743.

Estrada-Peña, A., D'Amico, G. and Fernández-Ruiz, N. (2021) Modelling the potential spread of *Hyalomma marginatum* ticks in Europe by migratory birds. *International Journal for Parasitology* 21, 1–11.

Földvári, G., Široký, P., Szekeres, S., Majoros, G. and Sprong, H. (2016) *Dermacentor reticulatus*: a vector on the rise. *Parasites & Vectors* 9, 314.

Garcia-Vozmediano, A., Krawczyk, A.I., Sprong, H., Rossi, L., Ramassa, E. and Tomassone, L. (2020) Ticks climb the mountains: ixodid tick infestation and infection by tick-borne pathogens in the Western Alps. *Ticks and Tick-borne Diseases* 11, 101489.

Gray, J.S., Dautel, H., Estrada-Peña, A., Kahl, O. and Lindgren, E. (2009) Effects of climate change on ticks and tick-borne diseases in Europe. *Interdisciplinary Perspectives on Infectious Diseases* 2009, 593232.

Hamšíková, Z., Kazimírová, M., Haruštiaková, D., Mahríková, L., Slovák, M., *et al.* (2016) *Babesia* spp. in ticks and wildlife in different habitat types of Slovakia. *Parasites & Vectors* 9, 292.

Hauck, D., Springer, A., Pachnicke, S., Schunack, B., Fingerle, V. and Strube, C. (2019) *Ixodes inopinatus* in northern Germany: occurrence and potential vector role for *Borrelia* spp., *Rickettsia* spp., and *Anaplasma phagocytophilum* in comparison with *Ixodes ricinus*. *Parasitology Research* 118, 3205–3216.

Heinz, F.X., Stiasny, K., Holzmann, H., Kundi, M., Sixl, W., *et al.* (2015) Emergence of tick-borne encephalitis in new endemic areas in Austria: 42 years of surveillance. *Eurosurveillance* 20, 9–16.

Holzmann, H., Aberle, S.W., Stiasny, K., Werner, P., Mischak, A., *et al.* (2009) Tick-borne encephalitis from eating goat cheese in a mountain region of Austria. *Emerging Infectious Diseases* 15, 1671–1673.

Hornok, S., Csörgő, T., de la Fuente, J., Gyuranecz, M., Privigyei, C., *et al.* (2013a) Synanthropic birds associated with high prevalence of tick-borne rickettsiae and with the first detection of *Rickettsia aeschlimannii* in Hungary. *Vector-Borne and Zoonotic Diseases* 13, 77–83.

Hornok, S., de la Fuente, J., Horváth, G., Fernández de Mera, I.G., Wijnveld, M., *et al.* (2013b) Molecular evidence of *Ehrlichia canis* and *Rickettsia massiliae* in ixodid ticks of carnivores from South Hungary. *Acta Veterinaria Hungarica* 61, 42–50.

Hornok, S., Takács, N., Kontschán, J., György, Z., Micsutka, A., *et al.* (2015) Diversity of *Haemaphysalis*-associated piroplasms of ruminants in Central-Eastern Europe, Hungary. *Parasites & Vectors* 9, 627.

Hornok, S., Kováts, D., Horváth, G., Kontschán, J. and Farkas, R. (2020) Checklist of the hard tick (Acari: Ixodidae) fauna of Hungary with emphasis on host-associations and the emergence of *Rhipicephalus sanguineus*. *Experimental and Applied Acarology* 80, 311–328.

Hubálek, Z., Sedláček, P., Estrada-Peña, A., Vojtíšek, J. and Rudolf, I. (2020) First record of *Hyalomma rufipes* in the Czech Republic, with a review of relevant cases in other parts of Europe. *Ticks and Tick-borne Diseases* 11, 101421.

Karbowiak, G. (2014) The occurrence of the *Dermacentor reticulatus* tick: its expansion to new areas and possible causes. *Annals of Parasitology* 60, 37–47.

Kloch, A., Mierzejewska, E.J., Karbowiak, G., Slivinska, K., Alsarraf, M., *et al.* (2017) Origins of recently emerged foci of the tick *Dermacentor reticulatus* in Central Europe inferred from molecular markers. *Veterinary Parasitology* 237, 63–69.

Kriz, B., Maly, M., Benes, C. and Daniel, M. (2012) Epidemiology of tick-borne encephalitis in the Czech Republic 1970–2008. *Vector-Borne and Zoonotic Diseases* 12, 994–999.

Kubiak, K., Sielawa, H., Dziekońska-Rynko, J., Kubiak, D., Rydzewska, M. and Dzika, E. (2018) *Dermacentor reticulatus* ticks (Acari: Ixodidae) distribution in north-eastern Poland: an endemic area of tick-borne diseases. *Experimental and Applied Acarology* 75, 289–298.

Lindgren, E., Jaenson, T.G.T. and Menne, B. (2006) *Lyme Borreliosis in Europe: Influences of Climate and Climate Change, Epidemiology, Ecology and Adaptation Measures*. WHO Regional Office for Europe, Copenhagen. Available at: https://apps.who.int/iris/handle/10665/107800 (accessed 17 June 2021).

Lukan, M., Bullova, E. and Petko, B. (2010) Climate warming and tick-borne encephalitis, Slovakia. *Emerging Infectious Diseases* 16, 524–526.

Materna, J., Daniel, M. and Danielová, V. (2005) Altitudinal distribution limit of the tick *Ixodes ricinus* shifted considerably towards higher altitudes in Central Europe: results of three years monitoring in the Krkonoše Mts. (Czech Republic). *Central European Journal of Public Health* 13, 24–28.

Materna, J., Daniel, M., Metelka, L. and Harčarik, J. (2008) The vertical distribution, density and the development of the tick *Ixodes ricinus* in mountain areas influenced by climate changes (the Krkonoše Mts., Czech Republic). *International Journal of Medical Microbiology* 298, 25–37.

Medlock, J.M., Hansford, K.M., Bormane, A., Derdakova, M., Estrada-Peña, A., *et al.* (2013) Driving forces for changes in geographical distribution of *Ixodes ricinus* ticks in Europe. *Parasites & Vectors* 6, 1.

Mierzejewska, E.J., Estrada-Peña, A., Alsarraf, M., Kowalec, M. and Bajer, A. (2016) Mapping of *Dermacentor reticulatus* expansion in Poland in 2012–2014. *Ticks and Tick-borne Diseases* 7, 94–106.

Pet'ko, B., Bullová, E. and Lukáň, M. (2011) Changes in the distribution of *Ixodes ricinus* tick in Slovakia in the past three decades and the assessment of its causes. *Folia Veterinaria* 55(Suppl. 1), 12–13.

Randolph, S.E. (2004) Tick ecology: processes and patterns behind the epidemiological risk posed by ixodid ticks as vectors. *Parasitology* 129(Suppl.), S37–S65.

Rizzoli, A., Hauffe, H.C., Carpi, G., Vourc'h, G.I., Neteler, M. and Rosà, R. (2011) Lyme borreliosis in Europe. *Eurosurveillance* 16, 19906.

Rizzoli, A., Silaghi, C., Obiegala, A., Rudolf, I., Hubálek, Z., *et al.* (2014) *Ixodes ricinus* and its transmitted pathogens in urban and peri-urban areas in Europe: new hazards and relevance for public health. *Frontiers in Public Health* 2, 251.

Rocklöv, J. and Dubrow, R. (2020) Climate change: an enduring challenge for vector-borne disease prevention and control. *Nature Immunology* 21, 479–483.

Rubel, F., Brugger, K., Pfeffer, M., Chitimia-Dobler, L., Didyk, Y.M., *et al.* (2016) Geographical distribution of *Dermacentor marginatus* and *Dermacentor reticulatus* in Europe. *Ticks and Tick-borne Diseases* 7, 224–233.

Rubel, F., Brugger, K., Walter, M., Vogelgesang, J.R., Didyk, Y.M., *et al.* (2018) Geographical distribution, climate adaptation and vector competence of the Eurasian hard tick *Haemaphysalis concinna*. *Ticks and Tick-borne Diseases* 9, 1080–1089.

Rumer, L., Graser, E., Hillebrand, T., Talaska, T., Dautel, H., *et al.* (2011) *Rickettsia aeschlimannii* in *Hyalomma marginatum* ticks, Germany. *Emerging Infectious Diseases* 17, 325–326.

Rusňáková Tarageľová, V., Mahríková, L., Selyemová, D., Václav, R. and Derdáková, M. (2016) Natural foci of *Borrelia lusitaniae* in a mountain region of Central Europe. *Ticks and Tick-borne Diseases* 7, 350–356.

Široký, P., Kubelová, M., Bednář, M., Modrý, D., Hubálek, Z. and Tkadlec, E. (2011) The distribution and spreading pattern of *Dermacentor reticulatus* over its threshold area in the Czech Republic: how much is range of this vector expanding? *Veterinary Parasitology* 183, 130–135.

Solano-Gallego, L., Sainz, Á., Roura, X., Estrada-Peña, A. and Miró, G. (2016) A review of canine babesiosis: the European perspective. *Parasites & Vectors* 9, 336.
Vandekerckhove, O., De Buck, E. and Van Wijngaerden, E. (2019) Lyme disease in Western Europe: an emerging problem? A systematic review. *Acta Clinica Belgica* 18, 1–9.
Walter, M., Brugger, K. and Rubel, F. (2016) The ecological niche of *Dermacentor marginatus* in Germany. *Parasitology Research* 115, 2165–2174.

eo63 Impact of Climate Change on Ticks and Tick-Borne Infections in Russia

Eduard Korenberg*
Gamaleya National Research Centre for Epidemiology and Microbiology, Moscow, Russia

63.1 Expert Opinion

63.1.1 Introduction

The alternation of two climate types, warm and cold, in the Late Pleistocene and especially in the Holocene had a strong effect on the ecology and formation of recent ranges of ixodid ticks and the disease agents they transmit, with zonal boundaries gradually shifting northward or, at the next phase, southward. Against this background, minor climate fluctuations occurred with a periodicity of several years to several centuries. According to climatological data, the current climate warming accompanied by increasing precipitation began in the late 1970s. In Russia, the year 2019 was the warmest on record. However, the effect of these changes varies over the vast territory of Russia, depending on differences in zonal climatic conditions and the level of climate continentality. It is predicted that the rate of climate warming in the next few decades will be higher in the Arctic and north-western regions, the Urals and Far East than in other regions of the country (http://www.meteorf.ru/, accessed 17 June 2021).

The fauna of Russia includes approximately 70 ixodid tick species of the subfamilies Ixodinae and Amblyomminae (Filippova, 1997) as well as several species of argasid ticks (the family Argasidae) (Filippova, 1966). These ticks differ in geographic distribution and ecosystem connections, which depend on species-specific requirements for environmental conditions. Due to space limitations, this expert opinion considers only the most significant aspects of the problem denoted in its title. Attention is focused on the species that are most widespread in the extratropical part of the northern hemisphere: the taiga tick, *Ixodes persulcatus*, and the European forest tick, *Ixodes ricinus*, which are hosts and vectors of tick-borne encephalitis (TBE) virus and *Borrelia* spirochaetes, agents of ixodid tick-borne borrelioses (ITBBs), i.e. of Lyme and Lyme-like diseases.

63.1.2 Preconditions for the habitation of certain tick species in a given area

Species-specific adaptations of ixodid ticks are aimed at ensuring their life cycle (i.e. reproduction) under given conditions, which must satisfy the requirements of ticks at all phases of development. The total duration of the life cycle of a tick generation may vary from 3 to 5–6 years. All blood-sucking tick phases become engorged within no more than 25 days. Their mode of life

*Email: edkornberg@yandex.ru

DOI: 10.1079/9781789249637.0063

before contact with a feeding host and after detaching from it (about 98% of the life cycle) is non-parasitic. The existence of ticks is usually limited not by the scarcity of feeding hosts but rather by the adverse conditions of the environment to which their non-parasitic stages are exposed (Balashov, 2009). The hydrothermal regime of the air has an influence on the activity of ticks but usually does not determine the very possibility of survival for unfed individuals. Therefore, it appears ecologically pointless to attempt to reveal a relationship between probable shifts in the range boundaries of ticks 'pasture-waylaying' or their population density within the range, on the one hand, and changes in the meteorological parameters of the air over certain periods, on the other hand.

When moisture in the upper soil layers is sufficient, the main limiting factor for tick development is the possibility to receive from the soil, within a definite period of time, sufficient warmth necessary for the development of eggs and the metamorphosis of fed larvae and nymphs, i.e. for accomplishing each consecutive stage in the life cycle of a tick generation (Korenberg *et al.*, 2013). These thermal constants of development are an essential characteristic of the fundamental ecological niche that has been formed in the course of tick speciation. Other niche characteristics (timing of activation and contact of unfed individuals with host, synchronization of transition to the next life cycle stage, diapauses in the life cycle, their regulation depending on photoperiod, etc.) ensure the maintenance of these constants under different conditions. The mechanism responsible for synchronization of the complex life cycle of ticks is based on their response to daylight length. This should be sufficient for supplying ticks with the amount of warmth necessary for completing each stage of development within a certain calendar period, which varies depending on the geographic location of a given area. The range of species' adaptive features manifested in a particular ecosystem (the realized ecological niche) may be markedly narrower than the potential range in the fundamental ecological niche.

The taiga tick, *I. persulcatus*, has a broad ecological niche that supposedly was formed during the relatively cold Pliocene epoch as a result of adaptation to diverse landscape–climatic conditions in forest biocenoses similar to recent relict forests of the Ussuri type and dark taiga forests in the mountains of southern Primorye, southern Siberia and the Korean Peninsula (Filippova, 1977). Especially characteristic of this species are morphophysiological diapauses (delays) during the transition to the next developmental phase, which are induced by a short-daylight photoperiod, and the obligate diapause in adult ticks. Adult diapause occurs at the end of the warm season, is independent of photoperiod and prevents the appearance of fed females or egg laying doomed to failure because of insufficient warmth (Balashov, 2009). According to our approximate calculations based on published data on the duration of embryonic development and metamorphosis of fed ticks at specified temperature, the thermal constant of development (the sum of temperatures in Celsius) is 570–620°C for *I. persulcatus* eggs and larvae, and 800–920°C for nymphs. Such parameters have contributed to the expansion of this species, which currently has a vast range in the boreal belt of Eurasia.

The European forest tick, *I. ricinus*, is an autochthon of the Mediterranean–European forest fauna (Filippova, 1977). Its fundamental ecological niche had been formed as late as the Holocene, under relatively dry and mild conditions of the Boreal period, in the zone with long and warm winters. As a result, this species is less susceptible to moisture deficit compared with *I. persulcatus*. Different phases of *I. ricinus* developing in the range optimum may receive the necessary amount of warmth during most of the year. The life cycle of this species is less synchronized than that of *I. persulcatus*, and the periods of activation and seasonal activity of all blood-sucking phases are longer. *I. ricinus* starts laying eggs or metamorphoses without strict dependence on photoperiodic regulation, and autumn egg-laying and late-feeding females successfully overwinter under optimal climatic conditions. *I. ricinus* is widespread in moderately hygrophilic and mesophilic (less frequently, xerophilic) lowland and highland broadleaf forests. The thermal constants of development for different phases of this species are approximately 1.5 times higher than in *I. persulcatus*.

63.1.3 Effects of climate change on ixodid ticks

Climate fluctuations can have an effect on the boundaries of tick species' range and on populations within it. This range resembles a patchwork formed by ecosystems inhabited or not inhabited by ticks. The distribution and population abundance of ticks in mountain areas depend on their geographic coordinates, elevation above sea level and slope aspect. The patchiness of the range usually increases as its boundaries are approached, which is due to general reduction in the size of the area where conditions are suitable for accomplishing the life cycle (see above). Single ticks periodically occur far beyond the limits of their reproductive distribution. Such findings themselves are not proof of a shift in the range boundaries under the impact of climate change; they only indicate that ticks have been brought to a new area by their migrating hosts. Information on changes in the boundaries of permanently inhabited range may be considered reliable only if it is based on the results of a representative census of all tick phases confirming that they were previously absent (or present) in the locality with specified coordinates.

The range boundaries of TBE virus vectors depend on the presence of hydrothermal conditions necessary and sufficient for their development. However, almost no data are available on the impact of climate change on the boundaries of the main (continental) range of the most widespread Eurasian tick species during the past few decades.

As shown in our long-term studies in European southern taiga forests (1960–1972), Far Eastern middle taiga forests (1976–1985) and East European mountain taiga forests of Russia (1993–2011), changes in the abundance of unfed adult *I. persulcatus* ticks show no definite trend attributable to global climate change over at least the past 40 years. These changes are characterized by distinct cyclicity related to the population cycles of small forest mammals, the main hosts for tick larvae and nymphs. This is confirmed by observations of other authors in Far Eastern conifer–broadleaf forests and in different landscapes of Central Siberia. Synchronous fluctuations in the abundance of *I. persulcatus* populations observed in certain parts of its range appear to be explained primarily by the effect of regional climate variations.

In estimating the possible influence of global climate warming on the abundance of tick populations, it is necessary to take into account the consequences of their long-term exposure to the impact of anthropogenic factors. For example, the northern distribution limits of *I. persulcatus* in several Russian regions were determined in the 1960s–1970s and have not changed since then. However, clear-cut forest harvesting over several hundred thousand square kilometres triggered the formation of ecosystems where the abundance of ticks has increased sharply. A serious anthropogenic impact on ecosystems against the background of long-term climate fluctuations in the south of central European Russia and in Western Siberia in the second half of the 20th century has led to changes in the abundance and landscape distribution of ticks from the genera *Ixodes* and *Dermacentor*.

As a consequence of global warming, many parts of the world, including large regions of Eurasia, are increasingly suffering from extreme climate events and related disasters (serious shifts in seasonal hydrothermal regimes, overabundant precipitation leading to inundation of large land areas, droughts, forest and steppe fires, etc.). Such events can certainly cause increased mortality of ticks in the period of their seasonal activity. This is especially true of larvae, since this phase of the tick life cycle is the most vulnerable to abiotic factors. Due to warming, the onset and completion of adult *I. persulcatus* activation after wintering have already shifted to earlier dates in some regions, but the total length of the activity season remains approximately the same or even decreases because of limited nutrient resources in individual ticks. However, the effects of climate anomalies on the abundance and phenology of ticks in different parts of their species ranges have not yet been studied sufficiently.

63.1.4 Effects of abiotic factors on disease agents ecologically connected with ticks

Agents of obligate transmissible diseases cannot survive in natural ecosystems without ticks, their hosts and vectors. TBE virus and *Borrelia* spirochaetes reproduce in infected ticks. The rate of their reproduction under laboratory conditions varies depending on temperature and humidity; in nature, this rate depends on the regional and local range of hydrothermal conditions in the upper soil layer where ticks continue their development after sucking blood and detaching from the host. The reproduction rate of microorganisms (i.e. the infection level in individual ticks) may also vary depending on the conditions within the bodies of ticks at different phases of development.

According to long-term observations, the relative abundance of infected adult unfed ticks in the natural foci of diseases is largely determined by the prevalence of infection at the preceding phase (i.e. among nymphs), which depends on cyclic fluctuations in the abundance of small mammals (see above), the feeding hosts for nymphs and the main reservoir hosts for *Borrelia*. There are as yet no facts confirming the direct effect of the current change in global climate on the agents of obligate transmissible diseases adapted to ixodid ticks. Changes in the range of these agents in any direction are conditioned by the ecological plasticity and adaptive properties of their main vectors.

63.1.5 Climate change and morbidity of obligatory transmissible infections

The incidence of diseases with natural focality is a function of two interacting variables: (i) the rate of pathogen circulation in the foci, which depends mainly on ecological factors; and (ii) the rate of human contact with the foci, which depends on social factors. Discrimination between their effects on the dynamics of morbidity is possible only if one of the variables has remained relatively stable during the period under analysis.

The hypothesis of global warming as the main cause of increased TBE morbidity in a considerable part of the southern taiga subzone was advanced more than 55 years ago. Its discussion showed, however, that there was no definite cause-and-effect relationship between climate warming by the mid-20th century and an abrupt increase in TBE morbidity in the 1950s and early 1960s. This increase was explained by the effect of several anthropogenic and sociodemographic factors as well as by the improvement in quality of laboratory diagnostics and medical records (Korenberg *et al.*, 2013).

The effects of climate change on the components of parasitic systems formed by the agents of tick-borne infections (i.e. the biological aspects of the problem at issue) have been considered above (Sections 63.1.3 and 63.1.4). Today, the epidemic manifestation of natural disease foci is largely conditioned by urbanization and economic activities in non-urbanized areas, and the consequent increase in frequency of human contacts with ticks and pathogens. These interrelated parallel processes are intensifying. For example, the growth of morbidity in several Eurasian countries in the late 1980s to the late 1990s was explained by radical changes in the socio-economic pattern of life. In Russia, this occurred after an approximately 20-year period of relatively low TBE morbidity (on average, fewer than 2 cases per 100,000 population). Then in the 1990s, changes in legislation provided the opportunity to use undeveloped land for individual housing and garden plots. According to official data, the total number of such plots around Russian cities with populations above 300,000–400,000 increased three- to fivefold between 1980 and 1995. This resulted in unprecedented mass contacts of city people with ticks and explosive growth of morbidity, which reached a peak of 9955 cases, or 6.8 cases per 100,000 population, in 1999 (the highest value over the history of TBE case records in Russia). It has become popular in publications and comments of Russian authors to explain this situation by global warming, but this explanation is speculative and baseless. In our opinion, there were also no grounds for the World Health Organization's conclusion that the observed increase in morbidity may be regarded as a consequence of climate change (McMichael *et al.*, 2003). The point is that TBE morbidity began to decrease in 2000 and has been falling since then, despite climate warming. In particular, the morbidity rate per 100,000 population in 2006 was 2.44 cases, which is 2.8–2.9 times lower than in 1999 and approximately similar to that in 1989; about 2800 cases (1.95 per 100,000 population) were recorded in 2008. On the whole, the incidence of TBE in Russia has shown a downward trend over the past two decades. Although the annual number of recorded cases increased slightly every 3 years, each increase was smaller than the previous one. In 2019, the TBE morbidity rate was slightly above one case per 100,000 population.

The countrywide programme for organizing clinical laboratory diagnosis of ITBBs in Russia was on the whole accomplished by 1993. Beginning from that year, a parallelism has been observed between the dynamics of annually recorded

ITBB cases and of TBE cases. According to the official statistics over the past decade (2010–2019), the incidence of ITBB was the highest in 2011 and the lowest in 2013 (6.96 versus 3.99 cases per 100,000 population), varying within this range (increasing or decreasing) in other years. For example, this parameter in 2019 was 5.48 cases (a total of about 8050 cases were recorded). Similar but not completely synchronous changes in the statistical parameters of annual ITBB morbidity were also observed in some large Russian regions with the highest epidemic risk.

Reliable time series data on the parameters of morbidity over many years are in most cases too short for mathematical analysis aimed at robust attribution of changes in these parameters to the impact of global climatic factors. The data presented above show that conclusive evidence for the directional effect of climate change on the morbidity of obligate transmissible tick-borne infections in Russia has not been obtained to date (Korenberg *et al.*, 2013).

63.2 Future Predictions

Changes in ecosystems in the near and more distant future will be determined by the magnitude and interaction of differently directed natural factors and the increasing impact of human activities on the environment. These changes in different regions of Eurasia will differ in intensity and even in direction, under the influence of typical zonal, intrazonal and extrazonal phenomena and their complex combinations (Korenberg *et al.*, 2013). They will largely depend on the type, frequency, duration and magnitude of climate anomalies (see Section 63.1.3) resulting from the general disbalance in the global climatic system, and even short-term prediction of these changes is problematic.

The data considered here suggest that the expansion of *I. persulcatus* to continental areas at latitudes above 62–63°N, especially in the north-western part of its range, is unlikely during the next decade, and the same applies to *I. ricinus* in the north-eastern part of its range. The main limiting factor for *I. persulcatus* is that the period with soil temperatures above 10°C, the threshold temperature for tick metamorphosis (Filippova, 1977; Balashov, 2009), is too short at these latitudes and the obligate behavioural diapause starts earlier; in *I. ricinus*, the main factor is that late egg-laying and females engorged in autumn are at high risk of freezing. However, there are isolated populations of *I. persulcatus* on the northern Gulf of Bothnia coast, in Sweden, at a latitude above 65°N, and of *I. ricinus* on the west coast of Norway at 66°N. These facts indicate the minimum extent to which the boundaries of the main ranges of these ticks (and agents of obligate transmissible diseases) may shift northward in the more distant future under continuing climate warming, which will be accompanied by changes in the zonal landscape boundaries and hydrothermal regimes of soils. The northward expansion of *I. persulcatus* may lead to broadening of the zone of sympatry between this species and *I. ricinus*.

The natural foci of diseases can withstand significant changes in their external environment. The problem of revealing the influence of climate on the parasitic systems formed by disease agents eventually comes down to analysis of their stability. A sharp increase in the frequency of human contact with the foci, which is conditioned by socio-economic factors and simultaneously affects a considerable proportion of the Russian population, may lead to a new significant rise of morbidity level.

References

Balashov, Yu.S. (2009) *The Acari and Insect Parasitism on Terrestrial Vertebrates*. Nauka, St. Petersburg, Russia (in Russian).

Filippova, N.A. (1966) Argasid ticks (Argasidae). In: *The Fauna of the Soviet Union: Arachnida*, vol. IV, issue 3. Nauka, Moscow/Leningrad, Russia (in Russian).

Filippova, N.A. (1977) Ixodid ticks of the subfamily Ixodinae. In: *The Fauna of the Soviet Union: Arachnida*, vol. IV, issue 4. Nauka, Leningrad, Russia (in Russian).

Filippova, N.A. (1997) Ixodid ticks of the subfamily Ambliomminae. In: *The Fauna of Russia and Neighboring Countries*, vol. IV, issue 5. Nauka, St. Petersburg, Russia (in Russian).
Korenberg, E.I., Pomelova, V.G. and Osin, N.S. (2013) *Infections with Natural Focality Transmitted by Ixodid Ticks*. Kommentariy, Moscow (in Russian).
McMichael, A.J., Campbell-Lendrum, D.H., Corvalán, C.F., Ebi, K.L., Githeko, A.K. and Scheraga, J.D. (eds) (2003) *Climate Change and Human Health: Risks and Responses*. World Health Organization, Geneva, Switzerland.

eo64 Is Climate Change Affecting Ticks and Tick-Borne Diseases in Taiwan?

Chi-Chien Kuo*
National Taiwan Normal University, Taiwan

64.1 Expert Opinion

64.1.1 Introduction

Taiwan is an island of similar land area to Switzerland, located off the south-eastern coast of China. The climate in lowland regions is primarily subtropical, with monthly temperatures averaging from 15.2 to 29.6 °C (Taiwan Weather Bureau; https://www.cwb.gov.tw/V8/C/, accessed 17 June 2021). However, mountains comprise a large part of the island, with the highest peak close to 4000 m in elevation. The temperature can drop to below 0°C at high altitudes during the winter. For the past 100 years, Taiwan has experienced a 1.0 to 1.4°C increase in temperature, which is much higher than the global average increasing rate of 0.07°C per decade.

In Taiwan, a total of 32 tick species were recorded, based on checklists published before 1984 (Robbins, 2005). Recently, five species have been newly recorded from hosts seldom inspected previously, including birds (three tick species), sea snakes (one tick species), and bears and sambar deer (*Rusa unicolor*) (one tick species). Particularly significant is the discovery of the Asian long-horned tick (*Haemaphysalis longicornis*) collected from bears and sambar deer (Tu *et al.*, 2019). This species typically occurs in temperate regions and is the primary vector of severe fever with thrombocytopaenia syndrome (SFTS) virus.

SFTS and Lyme disease are the only notifiable tick-borne diseases in Taiwan, despite other tick-borne pathogens having been identified in ticks, vertebrate hosts and humans. The first and only SFTS human case to date was identified in northern Taiwan in late 2019 (Peng *et al.*, 2020). In early 2020, SFTS was designated as a notifiable disease by Taiwan Centres for Disease Control. Lyme disease became a notifiable disease in October 2007. Since then, a total of 14 imported but no autochthonous cases have been documented. However, *Borrelia burgdorferi*, the causative agent of Lyme disease, has been identified in ticks and rodents as well as patients with no recent travel history abroad (Chao *et al.*, 2011). Other tick-borne pathogens identified in Taiwan include *Borrelia valaisiana*, *Borrelia turdi*, tick-borne spotted fever group rickettsiae (SFGR) (*Rickettsia* spp.), *Anaplasma phagocytophilum*, *Ehrlichia chaffeensis* and several *Babesia* species.

The impacts of climate change on ticks and tick-borne diseases can only be determined with appropriate long-term spatial and temporal data, after controlling for other confounding factors such as land-use change and disease-control measures. Long-term data sets are necessary to distinguish long-term trend from short-term

*Email: cckuo@ntnu.edu.tw

DOI: 10.1079/9781789249637.0064

noise. Conclusions based on a comparison of only two data sets separated by an extended period of time can be misleading if, for example, the first sampling occurred in a particularly cold year while the second was in a notably warm year while the temperature remained similar over time despite the temperature fluctuations. Ideally, the sampling areas would include edges of distribution ranges of ticks or tick-borne pathogens so that their emergence can be more readily detected. Likewise, the impact of climate change is more likely to manifest by focusing on seasons when ticks start to emerge and quest for hosts. Lastly, an observed expansion or contraction in distribution can be due to land-use change, for example the emergence of Lyme disease following reforestation in New England, USA, rather than a consequence of climate change.

Therefore, not surprisingly, solid evidence for the impact of climate change on ticks is generally lacking in most countries, including Taiwan. This is especially apparent in developing or least developed countries where research spending and rigorous studies are typically limited. Most of these countries are in tropical and subtropical regions where the tick response to climate change may be less obvious and therefore receive less attention from local researchers. Indeed, a representative (although not comprehensive) survey of literature in the database PubMed published between 2000 and 2020 using the keywords 'climate change' and 'ticks or tick-borne' revealed 93 relevant empirical studies (editorial or review papers excluded), of which nearly 90% were focused on Europe (46 studies; 49.5%) or North America (37; 40.0%). Other continents constituted only a small proportion (Fig. 64.1). It should be stressed, however, that not all of the 93 studies involved an explicit investigation of potential climate change impacts. For example, the only two relevant studies in Asia (both in South Korea) reported seasonal variation in tick abundance over a period of 4 years (Kim-Jeon *et al.*, 2019; Park *et al.*, 2020) without examining its association with climate change. Similarly, about 60% (40/67) of studies focusing on ticks concerned the temperate *Ixodes ricinus* distributed in Europe and *Ixodes scapularis* in the USA and Canada, while 80% (25/31) of studies focusing on tick-borne diseases centred on Lyme disease (13) or tick-borne encephalitis (12), likewise in Europe and North America. Again, this demonstrates that research has concentrated on a limited number of temperate tick species and tick-borne diseases distributed in more developed countries.

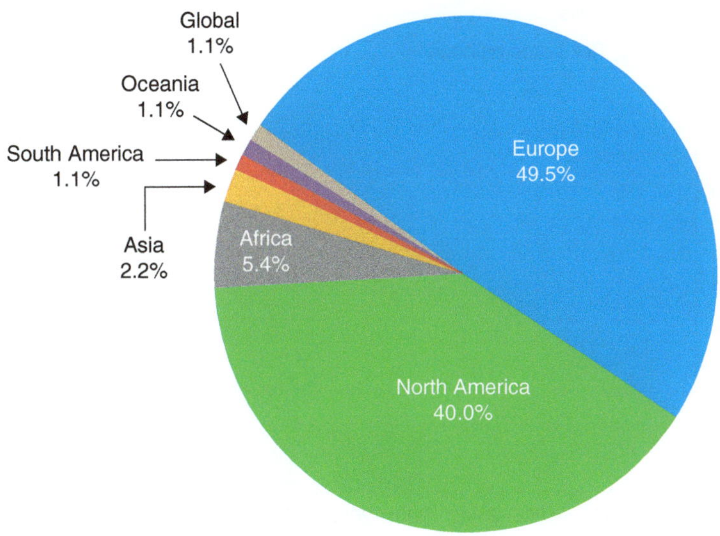

Fig. 64.1. Percentage of research literature from different continents concerning climate change and ticks or tick-borne pathogens, 2000–2020. (Author's own figure.)

64.1.2 Evidence for and against climate change impacts

Despite the increase in publications during the past decade (Fig. 64.2), most studies in Taiwan have been sporadic rather than systematic and long-term. Surveys were predominantly in confined areas instead of across the whole island and usually at different time points. Hence latitudinal or altitudinal boundaries cannot be delineated at a given time period. Moreover, ticks collected in Taiwan were commonly retrieved from a limited range of hosts (rodents, pets or livestock), which can result in misleading and restricted distributions when not all potential host animals are thoroughly sampled. By contrast, a dragging or flagging sampling approach for surveying questing ticks is less prone to such bias. Likewise, a given site was typically sampled at a certain period of time without covering the whole year or at least including the start of the questing period. This makes comparison of the phenology of ticks across years difficult to assess and questions such as whether ticks appear earlier in more recent years cannot be addressed. Furthermore, in studies that did extend for several seasons, ticks were found all year round, which is not unexpected given the mild climate of subtropical regions. Relative to temperate regions, where ticks become dormant when the temperature drops below a threshold, validating any warming effect on tick phenology requires particularly intensive sampling given the continuous appearance of ticks in subtropical regions. Beyond that, the long-term research essential for climate change studies is lacking for ticks. In fact, the effect of climate change (including temperature and precipitation) on ticks has never been explicitly investigated in Taiwan.

In addition to tick studies, there is also an increase in tick-borne pathogen studies in recent years (Fig. 64.2). However, this is not the result of a surge in tick-borne disease incidence in Taiwan. Instead, it is the outcome of greater research interest in emerging infectious diseases. For example, the identification of the first human SFTS case was an accidental discovery and unlikely to be a true record of the first appearance of this pathogen in Taiwan (Peng *et al.*, 2020). SFTS virus and other tick-borne pathogens are not tested for routinely even though aetiological agents of many febrile diseases cannot be identified. Some research has investigated infection rates in several corners of Taiwan (e.g. Kuo *et al.*,

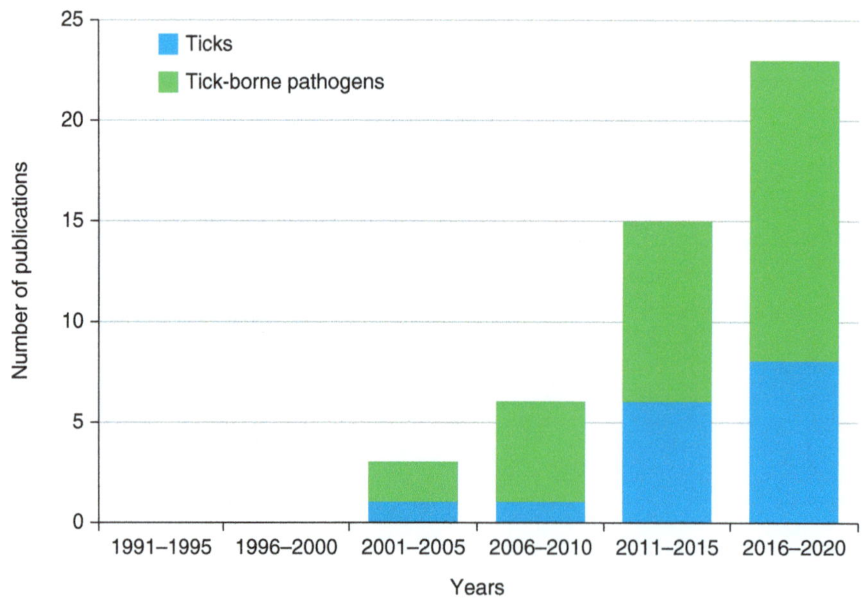

Fig. 64.2. Number of research publications concerning ticks or tick-borne pathogens in Taiwan between 1991 and 2020. (Author's own figure.)

2015) but none has been repeated long term to enable evaluations of the potential impact of climate change. Similarly, time series data on the incidence of tick-borne diseases in humans are not available in Taiwan.

64.2 Future Predictions

With the lack of long-term studies on ticks and tick-borne diseases in Taiwan, any current impact or lack of impact of climate change is impossible to validate, thus rendering future predictions challenging. Dengue fever transmitted by mosquitoes and scrub typhus transmitted by chigger mites are by far the most important vector-borne diseases and receive most attention in Taiwan. By contrast, there are very few human cases of tick-borne diseases, and tick-borne pathogens have not yet been a serious health concern. It is therefore expected that tick-borne diseases will not be a research priority in the near future.

Given the mild subtropical climate and few human cases, is Taiwan exempt from the impact of climate change on tick-borne diseases, which is more evident globally in temperate regions with high incidence? *Aedes aegypti* is the most important mosquito vector in transmitting dengue fever in Taiwan. This species is currently limited to southern Taiwan (south of the Tropic of Cancer) with higher temperature compared with northern Taiwan, where winter temperature is much lower (Chen, 2018). It would be unsurprising if global warming facilitates the northward expansion of *A. aegypti* and dengue fever in Taiwan. The example of *A. aegypti* demonstrates that tropical or subtropical regions can still encompass distribution limits of some vector species and therefore are not free from the impact of climate change. It also supports the need for better knowledge on the geographical ranges of tropical/subtropical tick species, especially those with high potential for pathogen transmission, which generally are not well known worldwide, including in Taiwan. What is also lacking in Taiwan is an inclusive assessment of all potential tick-borne diseases besides Lyme disease and SFTS and therefore the real human exposure to tick-borne pathogens. For example, SFGR related to *Rickettsia rickettsii* may occur in southern Taiwan (Lai *et al.*, 2014) but the extent of SFGR infection in Taiwan remains unclear and confirmation of the causative agent and tick vector is still lacking. Not until the majority of tick species and tick-borne diseases are well documented can the impact of climate change be reliably and comprehensively evaluated. As a first step, further studies can focus on tick species with significant disease threat (e.g. *H. longicornis* and SFTS) and track their distributional and seasonal changes in the long term to help unravel the potential effects of climate change on tick-borne diseases in Taiwan.

References

Chao, L.L., Chen, Y.J. and Shih, C.M. (2011) First isolation and molecular identification of *Borrelia burgdorferi* sensu stricto and *Borrelia afzelii* from skin biopsies of patients in Taiwan. *International Journal of Infectious Diseases* 15, e182–e187.

Chen, W.J. (2018) Dengue outbreaks and the geographic distribution of dengue vectors in Taiwan: a 20-year epidemiological analysis. *Biomedical Journal* 41, 283–289.

Kim-Jeon, M.D., Jegal, S., Jun, H., Jung, H., Park, S.H., *et al.* (2019) Four year surveillance of the vector hard ticks for SFTS, Ganghwa-do, Republic of Korea. *Korean Journal of Parasitology* 57, 691–698.

Kuo, C.C., Shu, P.Y., Mu, J.J., Lee, P.L., Wu, Y.W., *et al.* (2015) Widespread *Rickettsia* spp. infections in ticks (Acari: Ixodoidea) in Taiwan. *Journal of Medical Entomology* 52, 1096–1102.

Lai, C.H., Chang, L.L., Lin, J.N., Tsai, K.H., Hung, Y.C., *et al.* (2014) Human spotted fever group rickettsioses are underappreciated in southern Taiwan, particularly for the species closely-related to *Rickettsia felis*. *PLoS ONE* 9, e95810.

Park, J.W., Lee, S.H., Lee, G.S., Seo, J.J. and Chung, J.K. (2020) Epidemiological characteristics of field tick-borne pathogens in Gwang-ju Metropolitan Area, South Korea, from 2014 to 2018. *Osong Public Health and Research Perspectives* 11, 177–184.

Peng, S.H., Yang, S.L., Tang, S.E., Wang, T.C., Hsu, T.C., *et al.* (2020) Human case of severe fever with thrombocytopenia syndrome virus infection, Taiwan, 2019. *Emerging Infectious Diseases* 26, 1612–1614.

Robbins, R.G. (2005) The ticks (Acari: Ixodida: Argasidae, Ixodidae) of Taiwan: a synonymic checklist. *Proceedings of the Entomological Society of Washington* 107, 245–253.

Tu, W.C., Lin, Y.H., Wu, H.H. and Chi, W.L. (2019) *Distribution and Risk Assessment of Emerging Tick-Borne Diseases and Tick Vectors in Taiwan. Research Report*. Taiwan Centers for Disease Control, Taipei (in Chinese with an English abstract).

eo65 Ticks and Tick-Borne Pathogens in the Caribbean Region in the Context of Climate Change

Dasiel Obregón Alvarez[1]*, Roxanne Albertha Charles[2] and Agustín Estrada-Peña[3,4]
[1]University of Guelph, Guelph, Canada; [2]University of the West Indies, St. Augustine, Trinidad and Tobago; [3]University of Zaragoza, Zaragoza, Spain; [4]Instituto Agroalimentario de Aragón, Zaragoza, Spain

65.1 Expert Opinion

65.1.1 Introduction

The Caribbean region, located south-east of the Gulf of Mexico, consists of more than 700 islands, islets, reefs and cays distributed over nearly 4 million km^2 in the Caribbean Sea, as well as the surrounding coastlines along Central and South America. The Caribbean is a biodiversity hotspot due to the geography and climate of tropical and semi-tropical islands (BirdLife International, 2019). It is expected that climate change will have highly negative consequences for the Caribbean basin as a whole, with some differentiated effects on some territories. In particular, small Caribbean islands are highly vulnerable to climate change (Rhiney, 2015). Consistent changes in temperature and precipitation indices have been observed during the past 50 years in the countries and territories of the region (Stephenson *et al.*, 2014).

The tick fauna in the Caribbean region is diverse, encompassing several species of ticks from different genera (i.e. *Ornithodoros*, *Antricola*, *Amblyomma*, *Argas*, *Ixodes*, *Rhipicephalus*, *Haemaphysalis* and *Dermacentor*), some of them being serious pests of domestic animals or humans (Barros-Battesti *et al.*, 2009; Basu and Charles, 2017). Several tick species are endemic to this region, while exotic species have been naturally introduced by migratory birds from continental America. Other exotic tick species were introduced through animal movements from Europe, Africa and Asia. Some of these tick species are implicated in the transmission of tick-borne pathogens (TBPs), including members of the genera *Ehrlichia*, *Anaplasma*, *Babesia*, *Hepatozoon*, *Rickettsia* and *Borrelia*, which pose a constraint for public and animal health (Gondard *et al.*, 2017; Obregón *et al.*, 2020). Surveillance and control of ticks and TBPs in the Caribbean countries are highly costly and the measures in place are not always effective. Due to their importance, countries and territories in the region are working together through the Caribbean animal health network (CaribVET) to anticipate risks of (re)-emergence of tick-borne disease pathogens, among others (CaribVET, 2012).

Some tick species of the region pose serious threats to human and animal health. For example, the tropical bont tick, *Amblyomma variegatum*, imported from Africa, is a vector of *Ehrlichia ruminantium* and *Rickettsia africae*, the causative

*Email: dasieloa@uoguelph.ca

DOI: 10.1079/9781789249637.0065

agents of heartwater in ruminants (Camus and Barré, 1990) and African tick-bite fever (ATBF) in humans, respectively (Kelly *et al.*, 2010). The presence of the tick complex *Amblyomma cajennense* sensu lato (s.l.) represents a special problem in the region due to its role as a vector of spotted fever group rickettsiae (SFGR). The recent reclassification of *A. cajennense* into six species (Nava *et al.*, 2014), separated by allopatric speciation, raises new questions about the ecoepidemiological significance of these species. As far as we know, the only study in the area (Cuba) identified *Amblyomma mixtum* as the species of the complex colonizing the Caribbean region (Encinosa *et al.*, 2021). This expert opinion refers to the most important ticks and TBPs in the region and how climate change may influence their distribution in the next decades.

65.1.2 Ticks and tick-borne pathogens of importance in the Caribbean region

Despite the broad tick species diversity in the region, the most important species transmitting diseases to animals and humans are *Rhipicephalus microplus*, *A. mixtum*, *A. variegatum*, *Dermacentor nitens* and *Rhipicephalus sanguineus* s.l. Of them, *R. microplus* is the most widespread and is the vector of *Anaplasma marginale* as well as the protozoan parasites, *Babesia bovis* and *Babesia bigemina*, in cattle and buffaloes. These pathogens are responsible for the diseases known as the cattle tick fever complex (CTF), which causes significant annual losses in livestock in the Caribbean. The climatic conditions (i.e. high temperature and humidity) support the survival of free-living stages throughout the year, with up to four generations completed in a single year. In that context, tick eradication is impracticable, and the epidemiological goal is to maintain endemic stability (i.e. epidemiological status in which clinical disease is unusual despite high infection prevalence) of CTF in the herds. However, this situation limits the genetic improvement of herds through the inclusion of more productive *Bos taurus* genotypes that are more susceptible to infection by these parasites.

A. variegatum, commonly referred to as the tropical bont tick, is an emergent problem in the Caribbean region. It was introduced to Guadeloupe from Senegal and was restricted to three islands (Guadeloupe, Antigua and Martinique) until the 1960s. However, by the end of the 1980s, the species was widely disseminated among 18 Caribbean islands, most likely spread by migrating cattle egrets (*Bubulcus ibis*) in the region (Barré *et al.*, 1995). Interestingly, *E. ruminantium* has not followed the expansion of *A. variegatum* and is only present in the first three islands. Furthermore, *A. variegatum* is also a vector of *R. africae*, the pathogen responsible for ATBF which is a common disease of locals and tourists in sub-Saharan Africa. Recent studies revealed that *R. africae* is present in some Caribbean islands and it poses a risk for the health of indigenous people and tourists in the region, with potential expansion to America (Kelly *et al.*, 2010). In addition, *A. variegatum* supports development of the acute form of dermatophilosis, a skin disease caused by the bacterium, *Dermatophilum congolensis*. For these reasons, the eradication of *A. variegatum* is proposed as a goal across the Caribbean region. Surprisingly, the tick has never been found on the continental land mass. Even if cattle egrets transport immatures of the tick (and it may be transferred in some unregulated movement of animals between the islands and the mainland), efforts to locate the tick on the mainland have been unsuccessful. The reasons for this are unknown.

Another latent concern in the region is the presence of *A. cajennense* s.l. and its potential as a vector of pathogens affecting animals and humans. *A. cajennense* s.l. was redescribed by Nava *et al.* (2014) as a complex of six tick species with different ecological preferences and distributions across America. According to this classification, *A. mixtum* is the species present in Mesoamerica, and is also thought to be the species present in the Caribbean region since the two specimens analysed by Nava *et al.* (2014) from Cuba and Jamaica belonged to *A. mixtum*. However, after reclassification of the *A. cajennense* complex, no further studies have addressed which species is present in the Caribbean islands and highlands, except in Cuba where recently Encinosa *et al.* (2021) confirmed that *A. mixtum* was the species present in all the localities they studied. On the other hand, little is known about the vectorial capacity of this species in the region. Studies in Cuba indicated that *A. mixtum* is the vector of *Coxiella burnetii*, the agent of the zoonotic disease Q fever, as well as the vector of *Rickettsia amblyommatis*, a member of the SFGR that may cause mild symptoms in humans (Noda *et al.*, 2016). The presence of *R. amblyommatis* in *A. mixtum* biting humans is a public health concern in Cuba (Chitimia-Dobler *et al.*, 2020), but also in Panama where *R. amblyommatis* was detected in *A. mixtum* at high (91%) prevalence (Bermúdez *et al.*, 2021).

The brown dog tick, *R. sanguineus* s.l., is a heteroxenous tick widely distributed within the Caribbean region. It is the most common tick species affecting dogs and cats in several Caribbean islands (Basu and Charles, 2017; Obregón *et al.*, 2020). This species is the vector of *Ehrlichia canis*, *Anaplasma platys*, *Babesia canis*, *Babesia gibsoni* and *Hepatozoon canis* (Gondard *et al.*, 2017) that are common in dogs. For humans, a public health concern has been proposed recently in the region related to *R. sanguineus* s.l. (Díaz-Sánchez *et al.*, 2020), since pathogens very similar to *E. canis* were detected infecting humans in Venezuela and Costa Rica, with clinical signs (Perez *et al.*, 2006; Bouza-Mora *et al.*, 2017). However, the scientific evidence is insufficient; further studies should examine the transmission of pathogens from dogs to humans through *R. sanguineus* considering our coexistence with dogs.

Also important in most of the territories of the region is the monoxene tick, *D. nitens*, a parasite of horses that constitutes the main vector of *Babesia caballi* and *Theileria equi*, the pathogens of equine piroplasmosis. *D. nitens* may also be an emergent threat to public health in the region, since studies conducted in Panama revealed its potential as a vector of the highly pathogenic *Rickettsia rickettsii*, the causative agent of Rocky Mountain spotted fever (Bermúdez *et al.*, 2009).

65.1.3 Climate change in the Caribbean region of significance for ticks and tick-borne infections

The climate of the Caribbean region is described as dry-winter tropical with a bimodal seasonal rainfall, with an early rainfall season in April–July and a late one in August–November. Differences in climatology are also found from south to north islands, and between high- and lowland islands. Vast scientific evidence has demonstrated that the Caribbean region is already being affected by climate change. These effects have been recorded across the Caribbean region during the period 1961–2010, with significant warming of 0.19°C per decade and 0.28°C per decade in daily maximum and minimum temperatures, respectively. Likewise, changes in rainfall indices were observed in the region in the last years. However, rainfall trends were not very consistent in the period 1961–2010, specifically during 1986–2010 when an increase was observed in annual total precipitation, daily intensity, number of consecutive dry days and heavy rainfall events (Stephenson *et al.*, 2014). For the purposes of this expert opinion, we included an analysis of the anomalies from the annual mean temperature and precipitation in the Caribbean region. A comparison of the annual mean temperature and precipitation between the quinquennia 1980–1985 and 2015–2019 confirms an increase of approximately 1°C in mean temperature for the last 40 years, whereas the precipitation tended to increase in some areas and decrease in others (Fig. 65.1).

65.2 Future Predictions

Several climatological studies have provided estimates of end-of-century climatic changes in the Caribbean region. From these analyses, we will consider the recent projections presented by McLean *et al.* (2015), based on the PRECIS (Providing Regional Climates for Impacts Studies) climate model, since they validated the models on time-series parameters from several meteorological stations across the region. In that study, future projections on changes of mean climate and extremes are presented for the Caribbean in two scenarios (A2 and B2) in the future period 2071–2099. Temperatures are expected to rise by 3.4 and 2.6°C for those scenarios, respectively, relative to the model baseline validated on 1979–1989. The projections predict greater warming in the wet season (May–November) and towards the Eastern Caribbean islands (i.e. Virgin Islands, St. Maarten and Puerto Rico). A slight increase in the annual rainfall of 0.9% (scenario A2) and 2.9% (scenario B2) is also expected, with particularly intense rainfall events predicted for the northern and eastern zones. On the other hand, drier conditions are projected for the southern zone (e.g. Tobago, Guyana) with

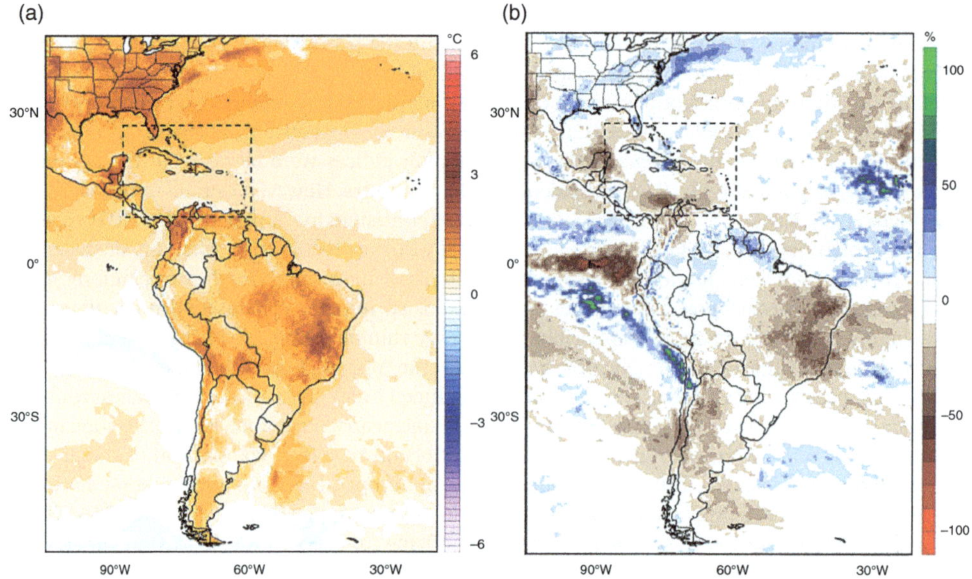

Fig. 65.1. Anomalies of temperature and precipitation in the Caribbean region for 1980–2019. (a) Anomalies in the annual mean temperature in the Caribbean region (area within dashed lines) and parts of America, calculated as the difference in annual mean temperatures for 2015–2019 compared with 1980–1985. (b) Anomalies in the annual mean precipitation, calculated in the same way and for the same time. (Data, images and calculations were performed using the Climate Reanalyzer webtool; https://ClimateReanalyzer.org, accessed 21 November 2020. Authors' own figure.)

increases in consecutive dry days and less intense rainfall events. According to these authors, these changes can lead to changes in the spatiotemporal occurrences of vector-borne diseases.

Concerning the *A. variegatum*/*E. ruminantium* system, the experience achieved in the Caribbean region suggests that *A. variegatum* eradication cannot be achieved in the Caribbean because of socio-economic and ecological reasons. Two eradication programmes were launched in ten islands of the English Lesser Antilles and the two French islands in 1994 and 1995, respectively (Pegram *et al.*, 2007). Despite *A. variegatum* being eliminated from eight islands at that time, some tick-free islands were repopulated with these ticks from unknown origins and the ticks were also discovered in some new islands, likely through the movement of immature life stages (i.e. through migratory birds) (Bournez *et al.*, 2011). These situations persist as a threat for the next decades, even though the tick has never been found on the South American mainland. It seems that after invasion of the islands, and the spread among islands due to cattle egrets, the situation may be stable. In the Caribbean, *A. variegatum* has coexisted with cattle egrets for over 40 years yet spread to the mainland has never been reported. If this situation occurs, however, there is a possibility of *E. ruminantium* adapting to local South American tick fauna.

No adequate data exist regarding the distribution of *A. mixtum* in the territory and therefore it is impossible to confirm its presence (of note, it was a member of the former *A. cajennense* s.l. and the only way to confirm its presence is rechecking already reported specimens). Therefore, it is risky to overview how climate could affect its distribution, especially as *A. mixtum* and *A. cajennense* stricto sensu (s.s.) are expected to have similar large areas of environmentally suitable niches in regions of the Caribbean and Mesoamerica (Estrada-Peña *et al.*, 2014). Official organizations involved in animal health should promote coordinated cooperation (e.g. coordinated through CaribVET) in the region to define the presumed large range of this tick.

Probably the best studied tick in the region is *R. microplus*. Studies suggest that *R. microplus* could either expand its known range in the target area and/or speed up its development cycle

(Estrada-Peña *et al.*, 2006). Since it is a one-host species, completing its feeding period in 21 days, most of its life cycle is spent in the ground, subjected to the effects of environmental stressors. The only 'free' stage is the ovipositing, engorged female, laying eggs on the ground. While the increase of temperature is expected to have a direct impact on a faster developmental period for the eggs, it is not known how the changes in precipitation or soil surface moisture can shape the mortality while oviposition takes place. This would increase the number of generations per year and therefore lead to an increase in the frequency of chemical treatments, with negative effects on tick resistance to acaricides. Complete knowledge of the biology of the tick under new conditions of climate is necessary to fit new strategies of control. This looks like a promising field of research on basic tick ecology, specifically for modelling new patterns of tick behaviour and the ecoepidemiology of TBPs.

References

Barré, N., Garris, G. and Camus, E. (1995) Propagation of the tick *Amblyomma variegatum* in the Caribbean. *Revue Scientifique et Technique (International Office of Epizootics)* 14, 841–855. https://doi.org/10.20506/rst.14.3.883

Barros-Battesti, D., Reyes, M.H., Onofrio, V.C., Beati, L., Famada, K. and Guglielmone, A.A. (2009) The ixodid ticks (Acari: Ixodidae) of Cuba. *Systematic and Applied Acarology* 14, 101–128. https://doi.org/10.11158/saa.14.2.3

Basu, A.K. and Charles, R. (2017) *Ticks of Trinidad and Tobago – An Overview*. Academic Press, London.

Bermúdez, C.S., Zaldívar, Y., Domínguez, L., Hernández, M., Barnett de Antinori, M.E. and Krawczak, F.S. (2021) *Rickettsia amblyommatis* isolated from *Amblyomma mixtum* (Acari: Ixodida) from two sites in Panama. *Ticks and Tick-borne Diseases* 12, 101597. https://doi.org/10.1016/j.ttbdis.2020.101597

Bermúdez, S.E., Zaldívar, Y., Domínguez, A.L., Hernández, M., de Antinori, M.E.B. and Krawczak, F.S. (2009) Detection and identification of rickettsial agents in ticks from domestic mammals in eastern Panama. *Journal of Medical Entomology* 46, 856–861. https://doi.org/10.1603/033.046.0417

BirdLife International (2019) The Caribbean islands biodiversity hotspot. Available at: https://www.cepf.net/sites/default/files/final_caribbean_ep.pdf (accessed 18 June 2021).

Bournez, L., Pradel, J., Hackshaw, K.H. and Lefrançois, T. (2011) Tropical bont tick surveillance and control in the Caribbean: how research outputs can help decisions. In: Evans, E.A., Davis, C.G. and Ballen, F. (eds) *Assuring Caribbean Food and Nutrition Security in the Context of Climate Change*. Caribbean Food Crops Society, Bridgetown, Barbados, pp. 109–120.

Bouza-Mora, L., Dolz, G., Solórzano-Morales, A., Romero-Zuñiga, J.J., Salazar-Sánchez, L., *et al.* (2017) Novel genotype of *Ehrlichia canis* detected in samples of human blood bank donors in Costa Rica. *Ticks and Tick-borne Diseases* 8, 36–40. https://doi.org/10.1016/j.ttbdis.2016.09.012

Camus, E. and Barré, N. (1990) *Amblyomma variegatum* and associated diseases in the Caribbean: strategies for control and eradication in Guadeloupe. *Parassitologia* 32, 185–193.

CaribVET (2012) *Caribbean Animal Health Network Charter*. CaribVET Steering Committee, Guyana. Available at: https://www.caribvet.net/about-the-network/charter (accessed 18 June 2021).

Chitimia-Dobler, L., Schaper, S., Mansfeld, P., Gonschorrek, J., Bröker, M. and Nava, S. (2020) Detection of *Amblyomma mixtum* (Acari: Ixodidae) in Germany on a human traveler returning from Cuba. *Journal of Medical Entomology* 57, 962–964. https://doi.org/10.1093/jme/tjz225

Díaz-Sánchez, A.A., Corona-González, B., Meli, M.L., Roblejo-Arias, L., Fonseca-Rodríguez, O., *et al.* (2020) Molecular diagnosis, prevalence and importance of zoonotic vector-borne pathogens in Cuban shelter dogs – a preliminary study. *Pathogens* 9, 901. https://doi.org/10.3390/pathogens9110901

Encinosa, P.E., García, Y., Lleonart, R., Aliaga, D., Fernández, Y., *et al.* (2021) Morphological and molecular characterization supporting *Amblyomma mixtum* presence in Cuba. *Ticks and Tick-borne Diseases* 12, 101602. https://doi.org/10.1016/j.ttbdis.2020.101602

Estrada-Peña, A., García, Z. and Sánchez, H.F. (2006) The distribution and ecological preferences of *Boophilus microplus* (Acari: Ixodidae) in Mexico. *Experimental and Applied Acarology* 38, 307–316. https://doi.org/10.1007/s10493-006-7251-2

Estrada-Peña, A., Tarragona, E.L., Vesco, U., Meneghi, D., de Mastropaolo, M., *et al.* (2014) Divergent environmental preferences and areas of sympatry of tick species in the *Amblyomma cajennense* complex (Ixodidae). *International Journal for Parasitology* 44, 1081–1089. https://doi.org/10.1016/j.ijpara.2014.08.007

Gondard, M., Cabezas-Cruz, A., Charles, R.A., Vayssier-Taussat, M., Albina, E. and Moutailler, S. (2017) Ticks and tick-borne pathogens of the Caribbean: current understanding and future directions for more comprehensive surveillance. *Frontiers in Cellular and Infection Microbiology* 7, 490. https://doi.org/10.3389/fcimb.2017.00490

Kelly, P., Lucas, H., Beati, L., Yowell, C., Mahan, S. and Dame, J. (2010) *Rickettsia africae* in *Amblyomma variegatum* and domestic ruminants on eight Caribbean Islands. *Journal of Parasitology* 96, 1086–1088. https://doi.org/10.1645/GE-2552.1

McLean, N.M., Stephenson, T.S., Taylor, M.A. and Campbell, J.D. (2015) Characterization of future Caribbean rainfall and temperature extremes across rainfall zones. *Advances in Meteorology* 2015, 425987. https://doi.org/10.1155/2015/425987

Nava, S., Beati, L., Labruna, M.B., Cáceres, A.G., Mangold, A.J. and Guglielmone, A.A. (2014) Reassessment of the taxonomic status of *Amblyomma cajennense* (Fabricius, 1787) with the description of three new species, *Amblyomma tonelliae* n. sp., *Amblyomma interandinum* n. sp. and *Amblyomma patinoi* n. sp., and reinstatement of *Amblyomma mixtum* Koch, 1844, and *Amblyomma sculptum* Berlese, 1888 (Ixodida: Ixodidae). *Ticks and Tick-borne Diseases* 5, 252–276. https://doi.org/10.1016/j.ttbdis.2013.11.004

Noda, A.A., Rodríguez, I., Miranda, J., Mattar, S., and Cabezas-Cruz, A. (2016) First report of spotted fever group *Rickettsia* in Cuba. *Ticks and Tick-borne Diseases* 7, 1057–1058. https://doi.org/10.1016/j.ttbdis.2016.06.004

Obregón, D., Corona-González, B., Rodríguez-Mallón, A., Rodríguez-Gonzalez, I., Alfonso, P., *et al.* (2020) Ticks and tick-borne diseases in Cuba, half a century of scientific research. *Pathogens* 9, 616. https://doi.org/10.3390/pathogens9080616

Pegram, R.G., Wilsmore, A.J., Lockhart, C., Pacer, R.E. and Eddi, C.S. (2007) The Carribean *Amblyomma variegatum* eradication programme: success or failure? In: Vreysen, M.J.B., Robinson, A.S. and Hendrichs, J. (eds) *Area-Wide Control of Insect Pests: From Research to Field Implementation*. Springer, Dordrecht, the Netherlands, pp. 709–720. https://doi.org/10.1007/978-1-4020-6059-5_65

Perez, M., Bodor, M., Zhang, C., Xiong, Q. and Rikihisa, Y. (2006) Human infection with *Ehrlichia canis* accompanied by clinical signs in Venezuela. *Annals of the New York Academy of Sciences* 1078, 110–117. https://doi.org/10.1196/annals.1374.016

Rhiney, K. (2015) Geographies of Caribbean vulnerability in a changing climate: issues and trends. *Geography Compass* 9, 97–114. https://doi.org/10.1111/gec3.12199

Stephenson, T.S., Vincent, L.A., Allen, T., Van Meerbeeck, C.J., Mclean, N., *et al.* (2014) Changes in extreme temperature and precipitation in the Caribbean region, 1961–2010. *International Journal of Climatology* 34, 2957–2971. https://doi.org/10.1002/joc.3889

eo66 The Strange Case of Tick-Borne Viruses in Turkey

Sırrı Kar[1] and Koray Ergünay[2]*
[1]*Tekirdağ Namık Kemal University, Tekirdağ, Turkey;* [2]*Hacettepe University, Ankara, Turkey*

66.1 Expert Opinion

66.1.1 Tick-borne viruses

With its diverse ecological niches, vegetation and wildlife supporting a rich fauna of arthropods and vertebrates, Turkey can be considered a haven for ticks and tick-borne diseases. It is located amidst Old World continents, has diverse regions under the influence of different climatic zone characteristics (Black Sea, Mediterranean, Caucasus and Middle East) and is on the Black Sea Flyway of wild birds. This flyway, which runs along South Africa, Madagascar, the eastern part of sub-Saharan Africa, the Rift Valley, the Nile Delta, the Middle East, Anatolia, Thrace, Europe and the northern Black Sea, is used by millions of birds (Kirwan *et al*., 2014; Najdenski *et al*., 2018). Migratory birds on the Black Sea Flyway facilitate the transfer of several tick species (such as immature stages of *Hyalomma marginatum* and *Hyalomma aegyptium*) in both directions between continents (Hoogstraal *et al*., 1963). In Turkey, around 30 tick species are distributed, most of which differ significantly in biological and ecological features. Therefore, Turkey exhibits significant potential in monitoring possible effects of climate change on tick-borne diseases as well as evaluating forward-looking scenarios for pathogen spread and disease emergence.

Crimean–Congo haemorrhagic fever virus (CCHFV) is currently the notable tick-borne viral infection occurring in Turkey (Ergünay *et al*., 2020a). Moreover, several other tick-associated viruses, either tick-borne or tick-specific (i.e. not transmitted to vertebrates), are in circulation or have recently been introduced. Although some cause human and other animal diseases in various parts of the world, such an association has not been reported yet in Turkey. For instance, Tamdy virus (TAMV) has been detected on numerous occasions in *H. aegyptium* ticks from Central Anatolia (Brinkmann *et al*., 2018; Ergünay *et al*., 2020b). Analyses of the prototype genome of TAMV revealed potential recombination events, suggesting interactions with identical or closely related viruses. A tentative nairovirus, named Meram virus, was also reported in *H. aegyptium* ticks from Central Anatolia, with closest relatives being Hazara virus and Tofla virus (Ergünay *et al*., 2020b). Jingmen tick virus (JMTV), a recently described novel human pathogen, was also isolated from ticks (Dinçer *et al*., 2019). The prototype strain of JMTV in Turkey was isolated during a CCHFV screening and a subsequent survey detected the virus in 3.9% of tick pools

*Email: ekoray@hacettepe.edu.tr

DOI: 10.1079/9781789249637.0066

collected from eastern Thrace and from Mediterranean and eastern Anatolia. Various tick species of the *Rhipicephalus* and *Haemaphysalis* genera were found to be infected, as well as *H. marginatum*. The initial findings suggest a ubiquitous distribution of JMTV in ticks. Screening field-collected ticks also revealed novel virus sequences related to tick-associated members of the family *Phenuiviridae* (Dinçer *et al*., 2017; Brinkmann *et al*., 2018; Emanet *et al*., 2019). These tentative viruses exhibit considerable genetic diversity, with several probable clades or genotypes, suggested to represent a novel tick-associated phlebovirus species. They also appear as widespread in Turkey, with positive tick pools originating from Thrace as well as Mediterranean, south-eastern and eastern Anatolia. Metagenome-based assessments have revealed other less-explored viruses in ticks of Turkey. These include flavivirus-like viruses, arthropod-associated rhabdoviruses and chuviruses (Brinkmann *et al*., 2018; Ergünay *et al*., 2020b). Near complete or partial genome sequences closely related to Tacheng tick virus 8, Wuhan tick virus 1, Bole tick virus 2 and Bole tick virus 3 have been reported in several tick species in Turkey, from various regions (Brinkmann *et al*., 2018; Ergünay *et al*., 2020b).

66.1.2 The case of tick-borne encephalitis

Despite the plethora of viruses and the rich tick fauna comprising many species with high vector capacity, the number of reported cases of tick-borne viral diseases from Turkey is relatively low. In addition, well-known agents such as tick-borne encephalitis virus (TBEV), transmitted by *Ixodes ricinus* and prevalent in Europe and Northern Asia, are hardly ever documented. The prevalence and density of *I. ricinus* are high in coastal regions of the Black Sea, in the Marmara Sea basin as well as in some other foci in Turkey, and thousands of individuals are bitten by different stages of this species annually in Istanbul (Kar *et al*., 2017). However, confirmed human or vertebrate exposure has been documented in only a few instances and tick surveys throughout Anatolia and Thrace have so far been unable to detect TBEV (Ergünay *et al*., 2020a). Another related virus – Turkish tick-borne encephalitis or Turkish sheep encephalitis virus – was isolated during an outbreak of acute encephalomyelitis in sheep in the Marmara region during the 1960s, which was later classified among mammalian tick-borne flaviviruses (Hartley *et al*., 1969; Gao *et al*., 1993; Grard *et al*., 2007). No further data on vertebrate or tick exposure are available.

The emergence of tick-borne encephalitis (TBE) in an area is influenced by various drivers such as meteorological parameters, wildlife dynamics, vector seasonality and human-related factors such as socio-economic structure and outdoor activity (Randolph *et al*., 2008; Borde *et al*., 2019). In North America and Europe, a shifting of areas affected by tick-borne diseases (including TBE) to higher altitudes and northern latitudes has been documented, which is primarily associated with changing climatic conditions in the last two decades (Esteve-Gassent *et al*., 2016; Beauté *et al*., 2018; Hellenbrand *et al*., 2019). These findings indicate that temperature values below or above a certain level negatively affect the natural dynamics of TBEV, which likely seems to be related to temperature-mediated changes in the natural dynamics of the vector ticks. TBEV infection in wildlife hosts exhibits a short-term character and co-feeding of infected nymphs and uninfected larvae is essential for maintenance (Randolph *et al*., 1999). Immature or adult forms of *I. ricinus* are capable of using rodents as hosts, but the lack of large non-susceptible ungulates such as deer (dilution hosts) in the environment with vector ticks and susceptible rodents has been documented to increase infection risk (Allan *et al*., 2003; Perkins *et al*., 2006). Milder climatic conditions also support virus maintenance by enhancing host-seeking activity of the nymphs and larvae, which presumably contribute to the observed shifts in the TBEV-affected areas to higher altitude and northern latitudes. Therefore, virus maintenance is negatively affected below or above a particular temperature threshold, in parallel with temperature-mediated changes in vector dynamics.

In Istanbul and many other localities on the Black Sea coast, dense *I. ricinus* populations are present in rural and peri-urban forests, including in the peri-urban fragmented forest patches with very low or no wild ungulate animal populations. Although there are no detailed studies on the seasonal dynamics of *I. ricinus*, our observations and some studies (e.g. Kar *et al*., 2017) indicate that all three stages of this tick species can be active for most of the year. Therefore, the milder climatic conditions in Turkey, compared with endemic regions in Europe or the Baltics, might be among the drivers for the observed paucity of TBEV infections. High prevalence, high density, high rate of human biting, long-term co-activation of the stages during the year and associated high probability of co-feeding of *I. ricinus* have so far been insufficient to produce TBEV-endemic foci under the conditions present in Turkey. It is also possible that consequent tick infestations trigger protective mechanisms in the host, as well as inter- or intraspecific competition between closely related viruses in vectors and hosts. Hence, it can be hypothesized that the predicted climate changes are unlikely to increase the risk related to TBEV and other *Ixodes*-transmitted pathogens with similar ecological and geographical patterns, as exemplified by the situation in Turkey.

66.1.3 The case of Crimean–Congo haemorrhagic fever

Since the emergence of CCHFV infections in the early 2000s, Turkey has become the epicentre of the disease. Although infections are reported from almost all parts of the country, more than 95% are documented in the north-east plateau of Central Anatolia, especially in the Kelkit Valley. The virus has been detected in various tick species, but *H. marginatum* is considered the principal vector responsible for transmission (Ergünay *et al.*, 2020a). However, contradicting patterns are observed in disease prevalence, *H. marginatum* phenology, climatic parameters and geographical characteristics between the Kelkit Valley and other parts of the country, such as eastern Thrace and the European part of Turkey. The main features of ecology, wildlife, livestock management and human population are comparable in these two regions, with meteorological data indicating milder conditions in Thrace. However, prevalence and density of *H. marginatum* feeding on cattle and possibly other livestock are higher in the mildly endemic area of Thrace, with greater tick species diversity, longer *H. marginatum* annual activity and higher virus prevalence and diversity in ticks (Gargili *et al.*, 2011; Gunes *et al.*, 2011). Thus, within the climatic range of Turkey, higher temperature and milder conditions are insufficient to explain ongoing regional variations in disease prevalence. Other routes for virus introduction, such as via wild bird migration, are also possible. However, any contribution of this route in regional CCHFV activity should have been evident in the Thrace region, located in close vicinity to the Black Sea Flyway (Kirwan *et al.*, 2014; Najdenski *et al.*, 2018). Overall, wild bird migration can be considered to have a negligible impact, at least in the particular circumstances covering the last 20 years in Turkey.

When agent–vector–host interaction and human behaviour are taken into consideration, the following observations are crucial for understanding transmission dynamics in CCHFV-hyperendemic areas in Turkey: (i) increase in the population of cattle in recent decades, which are the main host for adult stages of *H. marginatum*; (ii) shift of the cattle population from endemic breeds to non-endemic European breeds, with probable variations in susceptibility to parasites and infectious diseases (Shyma *et al.*, 2015; Robbertse *et al.*, 2017). Unauthorized importation and other transborder activities may further contribute to host and virus diversity; (iii) reduced number of wild ungulates that can host *H. marginatum*; (iv) decrease in the predator population in the wild and increase in the number of small and medium-sized animals such as hares and partridges (major hosts for immature stages of *H. marginatum*) due to hunting, deforestation, agricultural applications and other human-induced factors; and (v) increase in the lifespan of infected and tick-infested animals due to reduced predator pressure. It is interesting to note the low prevalence of CCHFV activity in the Thrace region where many parameters appear to be in favour of the virus. Possibly, some kind of endemic stability has been developed in this region, involving the diversity of tick species present, the multitude of closely and distantly related tick-associated viruses and the vertebrate hosts.

66.2 Future Predictions

The interplay of many factors associated directly or indirectly with climate – such as wildlife demography, host community, pathogen diversity, human demography and human behaviour – affects ticks and tick-borne diseases in non-linear and complex ways. This complexity makes definitive assessment and prediction difficult (Randolph *et al.*, 2008; Stefanoff *et al.*, 2012; Dantas-Torres, 2015; Essl *et al.*, 2020; Seebens *et al.*, 2020). However, given the parameters of possible climate and ecological changes, and the ecoclimatic characteristics of Turkey, infections due to new invasions or already existing tick-borne pathogens will likely occur as regional, sporadic and/ or periodic outbreaks. It also seems possible that the ongoing environmental changes may not allow for the formation of effective endemic stability, that there may be no long-term and widespread predominance of a particular virus and that the agents responsible for possible outbreaks may vary.

To become endemic, tick-borne agents have to adapt to the local drivers such as host population, climatic features and other environmental parameters, which may alter major features including biology, virulence, vector competence and ability to colonize (McCoy *et al.*, 2013; Araya-Anchetta *et al.*, 2015; Esteve-Gassent *et al.*, 2016). Such processes are usually long-term and require relatively stable environmental conditions. However, the progressive character of the predicted changes seems unlikely to allow for stabilization. It is also doubtful that expected alterations in climate and ecosystems will enable all existing viruses to adapt successfully in local environments. Therefore, we hypothesize that climate change will not overtly increase the impact of tick-borne pathogens already established in Turkey, such as CCHFV. However, particular environmental conditions might provide opportunities for expansion into previously unaffected areas or emergence of other, less prevalent viruses.

Due to its geographical location, Turkey is expected to be significantly affected by the predicted changes in climate. Being in a critical terrestrial transit route between Europe, Asia and Africa, and on the Black Sea Flyway, Turkey is prone to the introduction of several vector-borne agents, which has undoubtedly occurred throughout history. Climate change may potentially allow for agents that would have been previously unsuccessful to adapt due to the regional conditions. However, it is difficult to predict how new intruders will interact with already existing viruses and will behave in environments with unfamiliar vectors and hosts. A holistic, One Health-based approach is required to understand the impact and consequences of climate change on vector-borne agents in complex environments like Turkey.

References

Allan, B.F., Keesing, F. and Ostfeld, R.S. (2003) Effect of forest fragmentation on Lyme disease risk. *Conservation Biology* 17, 267–272.

Araya-Anchetta, A., Busch, J.D., Scoles, G.A. and Wagner, D.M. (2015) Thirty years of tick population genetics: a comprehensive review. *Infection, Genetics and Evolution* 29, 164–179.

Beauté, J., Spiteri, G., Warns-Petit, E. and Zeller, H. (2018) Tick-borne encephalitis in Europe, 2012 to 2016. *Eurosurveillance* 23, 1800201.

Borde, J.P., Kaier, K., Hehn, P., Bohmer, M.M., Kreusch, T.M. and Dobler, G. (2019) Tick-borne encephalitis virus infections in Germany. Seasonality and in-year patterns. A retrospective analysis from 2001–2018. *PLoS ONE* 14, e0224044.

Brinkmann, A., Dinçer, E., Polat, C., Hekimoğlu, O., Hacıoğlu, S., *et al.* (2018) A metagenomic survey identifies Tamdy orthonairovirus as well as divergent phlebo-, rhabdo-, chu- and flavi-like viruses in Anatolia, Turkey. *Ticks and Tick-borne Diseases* 9, 1173–1183.

Dantas-Torres, F. (2015) Climate change, biodiversity, ticks and tick-borne diseases: the butterfly effect. *International Journal for Parasitology: Parasites and Wildlife* 4, 452–461.

Dinçer, E., Brinkmann, A., Hekimoğlu, O., Hacloğlu, S., Földes, K., *et al.* (2017) Generic amplification and next generation sequencing reveal Crimean–Congo hemorrhagic fever virus AP92-like strain and distinct tick phleboviruses in Anatolia, Turkey. *Parasites & Vectors* 10, 335.

Dinçer, E., Hacioğlu, S., Kar, S., Emanet, N., Brinkmann, A., *et al.* (2019) Survey and characterization of Jingmen tick virus variants. *Viruses* 11, 1071.

Emanet, N., Kar, S., Dinçer, E., Brinkmann, A., Hacıoglu, S., Farzani, T.A., *et al.* (2019) Novel tick phlebovirus genotypes lacking evidence for vertebrate infections in Anatolia and Thrace, Turkey. *Viruses* 11, 703.

Ergünay, K., Polat, C. and Özkul, A. (2020a) Vector-borne viruses in Turkey: a systematic review and bibliography. *Antiviral Research* 183, 104934.
Ergünay, K., Dinçer, E., Kar, S., Emanet, N., Yalçınkaya, D., *et al.* (2020b) Multiple orthonairoviruses including Crimean–Congo hemorrhagic fever virus, Tamdy virus and the novel Meram virus in Anatolia. *Ticks and Tick-borne Diseases* 11, 101448.
Essl, F., Lenzner, B., Bacher, S., Bailey, S., Capinha, C., *et al.* (2020) Drivers of future alien species impacts: an expert-based assessment. *Global Change Biology* 26, 4880–4893.
Esteve-Gassent, M.D., Castro-Arellano, I., Feria-Arroyo, T.P., Patino, R., Li, A.Y., *et al.* (2016) Translating ecology, physiology, biochemistry and population genetics research to meet the challenge of tick and tick-borne diseases in North America. *Archives of Insect Biochemistry and Physiology* 92, 38–64.
Gao, G.F., Hussain, M.H., Reid, H.W. and Gould, E.A. (1993) Classification of a new member of the TBE flavivirus subgroup by its immunological, pathogenetic and molecular characteristics: identification of subgroup-specific pentapeptides. *Virus Research* 30, 129–144.
Gargili, A., Midilli, K., Ergonul, O., Ergin, S., Alp, H.G., *et al.* (2011) Crimean–Congo hemorrhagic fever in European part of Turkey: genetic analysis of the virus strains from ticks and a seroepidemiological study in humans. *Vector-Borne and Zoonotic Diseases* 11, 747–752.
Grard, G., Moureau, G., Charrel, R.N., Lemasson, J.J., Gonzalez, J.P., *et al.* (2007) Genetic characterization of tick-borne flaviviruses: new insights into evolution, pathogenetic determinants and taxonomy. *Virology* 361, 80–92.
Gunes, T., Poyraz, O. and Vatansever, Z. (2011) Crimean–Congo hemorrhagic fever virus in ticks collected from humans, livestock, and picnic sites in the hyperendemic region of Turkey. *Vector-Borne and Zoonotic Diseases* 11, 1411–1416.
Hartley, W., Martin, W., Hakiolu, F. and Chifney, S. (1969) A viral encephalitis of sheep in Turkey. *Pendik Veterinary Control Institute Journal* 2, 89–100.
Hellenbrand, W., Kreusch, T., Bohmer, M.M., Wagner-Wiening, C., Dobler, G., *et al.* (2019) Epidemiology of tick-borne encephalitis (TBE) in Germany, 2001–2018. *Pathogens* 8, 42.
Hoogstraal, H., Kaiser, M.N., Traylor, M.A., Guindy, E. and Gaber, S. (1963) Ticks (Ixodidae) on birds migrating from Europe and Asia to Africa. *Bulletin of the World Health Organization* 28, 235–262.
Kar, S., Yilmazer, N., Akyildiz, G. and Gargili, A. (2017) The human infesting ticks in the city of Istanbul and its vicinity with reference to a new species for Turkey. *Systematic and Applied Acarology* 22, 2245–2255.
Kirwan, G.M., Ozen, M., Ertuhan, M. and Atahan, A. (2014) Turkey bird report 2007–2011. *Sandgrouse* 36, 146–175.
McCoy, K.D., Leger, E. and Dietrich, M. (2013) Host specialization in ticks and transmission of tick-borne diseases: a review. *Frontiers in Cellular and Infection Microbiology* 3, 57.
Najdenski, H., Dimova, T., Zaharieva, M.M., Nikolov, B., Petrova-Dinkova, G., *et al.* (2018) Migratory birds along the Mediterranean–Black Sea Flyway as carriers of zoonotic pathogens. *Canadian Journal of Microbiology* 64, 915–924.
Perkins, S.E., Cattadori, I.M., Tagliapietra, V., Rizzoli, A.P. and Hudson, P.J. (2006) Localized deer absence leads to tick amplification. *Ecology* 87, 1981–1986.
Randolph, S.E., Miklisová, D., Lysy, J., Rogers, D.J. and Labuda, M. (1999) Incidence from coincidence: patterns of tick infestations on rodents facilitate transmission of tick-borne encephalitis virus. *Parasitology* 118, 177–186.
Randolph, S.E., Asokliene, L., Avsic-Zupanc, T., Bormane, A., Burri, C., *et al.* (2008) Variable spikes in tickborne encephalitis incidence in 2006 independent of variable tick abundance but related to weather. *Parasites & Vectors* 1, 44.
Robbertse, L., Richards, S.A. and Maritz-Olivier, C. (2017) Bovine immune factors underlying tick resistance: integration and future directions. *Frontiers in Cellular and Infection Microbiology* 7, 522.
Seebens, H., Bacher, S., Blackburn, T.M., Capinha, C., Dawson, W., *et al.* (2020) Projecting the continental accumulation of alien species through to 2050. *Global Change Biology* 27, 970–982.
Shyma, K.P., Gupta, J.P. and Singh, V. (2015) Breeding strategies for tick resistance in tropical cattle: a sustainable approach for tick control. *Journal of Parasitic Diseases* 39, 1–6.
Stefanoff, P., Rosinska, M., Samuels, S., White, D.J., Morse, D.L. and Randolph, S.E. (2012) A national case-control study identifies human socio-economic status and activities as risk factors for tick-borne encephalitis in Poland. *PLoS ONE* 7, e45511.

eo67 Melting, Melting Pot – Climate Change and Its Impact on Ticks and Tick-Borne Pathogens in the Arctic

Jiří Černý[1]*, Jana Elsterová[2,3] and Lauren Culler[4]
[1]Czech University of Life Sciences, Prague, Czech Republic; [2]Czech Academy of Sciences, České Budějovice, Czech Republic; [3]Veterinary Research Institute, Brno, Czech Republic; [4]Dartmouth College, Hanover, New Hampshire, USA

67.1 Expert Opinion

67.1.1 Introduction

Ticks are an important arthropod in marine and terrestrial Arctic ecosystems and are known to transmit significant pathogens to humans and wildlife. Of the three families of ticks, only hard ticks are known from the Arctic. But since recent decades, the Arctic has been undergoing rapid rates of warming and ecological change. This is restructuring marine and terrestrial ecosystems in which ticks are present, with consequences for tick and pathogen abundance and distribution. Warming is also creating novel environmental conditions that permit the establishment of species currently restricted to more southern latitudes. Natural processes such as bird migration as well as human-caused species introductions render the melting Arctic a melting pot for ticks and their pathogens. Significantly more baseline data, monitoring and surveillance are needed to anticipate future change and limit the health and economic impacts of tick-borne diseases in Arctic wildlife and humans.

The low diversity of ticks is directly related to the extreme conditions, specifically to the low temperatures and short growing seasons that characterize the Arctic (Hahn *et al.*, 2020). To survive such harsh conditions numerous ecophysiological adaptations, which we do not fully understand yet, are required. For example, the seabird tick, *Ixodes uriae*, has an extremely low supercooling point (almost −30°C) and is very resistant to desiccation by aggregating in moist rocky microhabitats (Lee and Baust, 1987; Benoit *et al.*, 2007). Despite this, *I. uriae* does not show any metabolic cold adaptation (Lee and Baust, 1982), nor were researchers able to BLAST homologues of any known tick anti-freeze proteins (Neelakanta *et al.*, 2010; Perner *et al.*, 2018) in *I. uriae* transcriptome libraries.

A few tick species well adapted to the Arctic conditions are well-known parasites of seabirds as well as terrestrial birds and mammals. Given that the Arctic is undergoing some of the most rapid rates of change in the world, and that it serves as a melting pot based on migrating birds and increases in human activity (e.g. tourism), we expect that ticks will increase in prominence in the future Arctic, bringing along with them pathogens and diseases that could have significant health and economic impacts on Arctic residents and wildlife. Further studies are needed to increase our understanding of the complex interactions between ticks and Arctic ecosystems (Revich

*Email: jiricerny@ftz.czu.cz

 Climate, Ticks and Disease (ed. P. Nuttall)
DOI: 10.1079/9781789249637.0067

et al., 2012). Further, it was shown that emerging disease outbreaks in the last 30 years have been either zoonotic or vector-borne (Ruscio *et al.*, 2015; Waits *et al.*, 2018). Indeed, a recent analysis found that ticks were the arthropod vector most often associated with climate-sensitive infections in the Arctic (Omazic *et al.*, 2019).

67.1.2 Defining the Arctic and its ecological conditions for ticks

Despite our intuitive understanding of what the Arctic is, its precise definition is quite complicated and ambiguous, but important to discuss in the context of this expert opinion. The Arctic can be defined most simply as the region north of the Arctic Circle (66°33′N, Fig. 67.1). Geographically, the Arctic consists of the Arctic Ocean, the northernmost parts of the Atlantic and Pacific Oceans, the northernmost parts of continental Europe, Asia and North America, and surrounding seas and islands. Biogeographically, the Arctic consists of the northernmost parts of the Palaearctic and Nearctic regions. There is no sharp division between the Arctic and remaining parts of these biogeographic regions, which allows migration of many species from lower altitudes into the Arctic (Vincent *et al.*, 2011; Clairbaux *et al.*, 2019; Joly *et al.*, 2019).

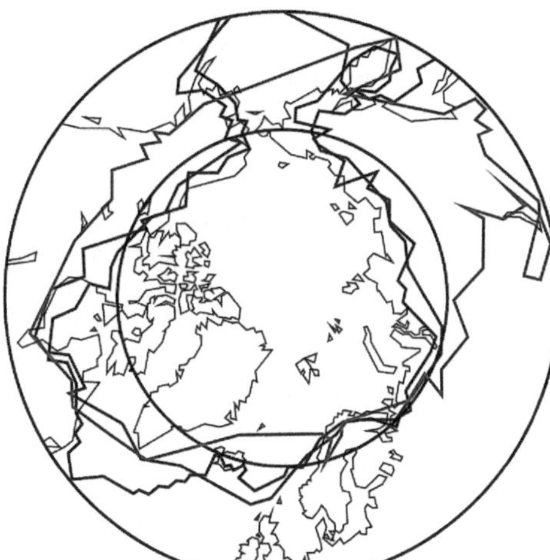

Fig. 67.1. Current geographic range of the Arctic. The Arctic can be defined by latitude 66.5°N (red line), 10°C isotherm (blue line), treeline (green line) and the area covered by permafrost (grey line). (Authors' own figure.)

From the ecological point of view, which is most relevant to ticks and their pathogens, the Arctic is best defined as the 10°C isotherm, i.e. the region northward from the isotherm on which the average temperature for July (the warmest month) is 10°C (Fig. 67.1). This is approximately identical to the region north from the northernmost treeline and the region covered by permafrost, i.e. soil that has been frozen continuously for two or more years (Fig. 67.1). It is important to stress, however, that the 10°C isotherm is moving northwards due to climate change. In this expert opinion, when we talk about the Arctic in the future, we mean the region which is understood to be the Arctic today, but which can be situated south from the 10°C isotherm in the future (Vincent *et al.*, 2011).

During the short Arctic summer, ticks are active in both the sea-ice-dominated marine ecosystem and the terrestrial ecosystem, the latter of which consists mostly of Arctic tundra or, in some parts, boreal forest. A uniting feature of both the marine and terrestrial Arctic is cold temperatures and the presence of frozen water throughout the year, in the form of sea ice in the oceans and in the form of glaciers in both marine and terrestrial ecosystems. Nevertheless, the surface layer (at least in some parts of the marine and terrestrial ecosystems) melts each summer, creating liquid water. The limiting conditions for life in the Arctic are thus: (i) temperature; and (ii) availability of liquid water. Ecosystems in the Arctic are usually quite simple with short food chains and low species richness. Nevertheless, numbers of individuals of some species can be very high due to high summer productivity of Arctic ecosystems. This is true especially for aquatic ecosystems, which are rich in oxygen due to low water temperatures and saturated by sunlight during the long polar summer days. These conditions allow high biomass production by primary producers, which can subsequently serve as food source for primary consumers (e.g. fish). Finally, high abundance of prey allows summer migration of numerous secondary consumer species (e.g. seabirds) from lower latitudes to the Arctic (Vincent *et al.*, 2011; Bölter and Müller, 2016; Ravolainen *et al.*, 2016).

These animals can then serve as food source for numerous blood-feeding arthropods, enabling transmission and circulation of various vector-borne pathogens.

67.1.3 Climate change in the Arctic

The Arctic underwent more dramatic climate changes in past years than the rest of the world, warming at more than twice the rate of other regions. This is a result of polar amplification, a phenomenon caused by ice–albedo positive feedbacks. Briefly, increased temperatures lead to melting of snow and ice, which uncovers darker soil or surface water. Darker surfaces then absorb more heat and cause more melting. Ice and snow albedo can be also increased via accumulation of dark-coloured dust particles on melting ice or snow (Kim *et al.*, 2016). The implications of polar amplification are global. Melting of Arctic glaciers leads to the rise of global sea levels, changes in ocean salinity and alterations to global ocean currents (Milner *et al.*, 2017). Melting of permafrost affects global climate due to release of fixed organic carbon into the inorganic forms of carbon dioxide and methane (Anthony *et al.*, 2018; Biskaborn *et al.*, 2019).

In the Arctic, temperatures have increased on average by about 0.5–1°C during the past 50 years. The increases are much higher in some geographical parts of the Arctic and during some seasons. The most significant increase in temperature can be observed in winter and the most affected part of the Arctic is Alaska and western Canada, where the average winter temperatures rose by 3–4°C in the past 50 years (ACIA, 2006; Sulikowska *et al.*, 2019). Noticeable also are the temperature extremes from recent years. On 20 June 2020, a record high temperature of 38°C was measured north from the Arctic Circle in the Russian town of Verkhoyansk, where the normal maxima are around 20°C (WMO, 2020). Similarly, in 2012, most of the Greenland ice sheet experienced surface melting due to an extreme heatwave (Nghiem *et al.*, 2012).

The above-mentioned changes strongly affect the whole Arctic ecosystem. Increased temperatures in lower parts of the Arctic allow introduction of numerous species typical for the temperate climate zone (Henttonen *et al.*, 2001). It also leads to increased ecological pressure on Arctic species that lose the optimal ecosystems on which they were adapted and that have to compete for resources with newly arriving species (Kausrud *et al.*, 2008; Tyler *et al.*, 2008; Hunter *et al.*, 2010; Vincent *et al.*, 2011). Finally, it also leads to introduction of new parasites and pathogens into the Arctic. This can be very dangerous for many animal species adapted for the Arctic, as the level of pathogen pressure was traditionally rather low there and these species are not adapted to deal with the newly arriving parasites and pathogens (Piersma, 1997; Omazic *et al.*, 2019).

67.1.4 Ticks and tick-borne pathogens in the Arctic

Several tick and tick-borne pathogen (TBP) species have been described in localities up to the 80th parallel. Recently published research articles have described increased tick activity and TBP prevalence in some Arctic regions as well as the introduction of new tick species. Nevertheless, compared with temperate and tropical zones, we have very limited data about actual abundances of ticks in the Arctic and we almost lack any data that describe the situation in the more distant past. The lack of baseline data makes any interpretations of actual observations very hard as it is often impossible to definitely state if current data about new tick species and increased abundance of previously established tick species, in particular Arctic areas, are real phenomena or just the consequence of increased research and public awareness. Our opinion is that both effects play some role. Therefore, we further describe only examples for which there are at least precise data for several recent decades.

In the marine Arctic, the dominant tick species are those which infest seabirds. These birds usually live in dense colonies, creating an ideal environment for blood-feeding arthropods including ticks. The most important tick species feeding on seabirds in the Arctic is *I. uriae*. This species has a circumpolar distribution in both hemispheres (Table 67.1), but the results of population genetics studies show that these two populations do not mix (Dietrich *et al.*, 2014) despite evidence that *I. uriae*-borne viruses do

Table 67.1. Tick species in the Arctic. (Authors' own table.)

Tick species	Current distribution	Possible future distribution
Ixodes uriae	Circumpolar	Population density increase
Ixodes ricinus	Arctic regions of Sweden, Norway, Faroes, Iceland	Northern Norway and Sweden, south-eastern Greenland?
Ixodes persulcatus	Boreal forests in Russia and Fennoscandia	Arctic Russia and Fennoscandia
Rhipicephalus sanguineus	Sporadic introductions to Alaska	Established population in Alaska

mix (Pettersson *et al.*, 2020). *I. uriae* is a nidicolous tick species that feeds on its hosts in their nests and overwinters in biological debris. *I. uriae* ticks are present up to almost 80°N in Svalbard. They were not common in Svalbard at the beginning of this century (McCoy, 2001), but their numbers there are increasing (Coulson *et al.*, 2009) in relation to increases in winter temperatures, which explains almost 90% of the average prevalence of ticks infesting an Arctic seabird species. Moreover, an increase of 1°C in average winter temperature at the nesting colony site was associated with a 5% increase in the number of birds infested by these ectoparasites in the subsequent breeding season (Descamps, 2013). Despite the fact these ticks transmit numerous TBPs in lower latitudes, such as *Borrelia* spirochaetes (Olsen *et al.*, 1995; Larsson *et al.*, 2007) as well as various arboviruses (Pettersson *et al.*, 2020), no TBPs were detected in *I. uriae* ticks collected at Jan Mayen, Bjornoya and Svalbard (Elsterova *et al.*, 2015).

Colonization of the Arctic by ticks, obligatory haematophagous parasites, depends mostly on the availability of hosts. In isolated areas, such as Svalbard or Antarctica, in which the distribution of ticks depends on migrating birds, the seabird tick (*I. uriae*) is the most prevalent species (Bergström *et al.*, 1999). On the other hand, in the continental Arctic, a wider diversity of ticks can be expected, as these ticks can be introduced during natural migration of their avian (seabirds) and mammalian (rodents and ungulates) hosts. In the North American Arctic, hare tick (*Haemaphysalis leporispalustris*) and vole tick (*Ixodes angustus*) are prevalent together with seabird ticks (*Ixodes auritulus*, *Ixodes howelli*, *Ixodes signatus* and *I. uriae*) (Hahn *et al.*, 2020). Migratory birds nesting in the Arctic or subarctic region but overwintering in lower latitudes are ideal vectors for spreading new tick species. For example, *Hyalomma rufipes*, a tick species typical for Africa, was reported from southern Norway where it was collected on migrating passerine birds (Hasle *et al.*, 2009). Ticks introduced by this way are usually unable to survive winter and therefore they do not pose a significant risk.

Ixodes ricinus, the most medically important European tick species, has increased its range northwards, having been reported from within the Polar Circle in Norway and Sweden where it is an important vector of tick-borne encephalitis virus (TBEV), Lyme borreliosis spirochaetes and other TBPs (Jaenson *et al.*, 2012; Hvidsten *et al.*, 2014, 2015; Larsson *et al.*, 2018; Soleng *et al.*, 2018; Jenkins *et al.*, 2019; Vikse *et al.*, 2020). *I. ricinus* was also reported from islands in the northern Atlantic such as the Faroe Islands (Medlock *et al.*, 2017) and Iceland (Alfredsson *et al.*, 2017), where they can transmit TBPs to local animals such as foxes (Hornok *et al.*, 2020). Moreover, geographically isolated populations of *I. ricinus* show clinal variation in questing activity in response to temperature, suggesting that a physiological threshold is not fixed in this species (Gilbert *et al.*, 2014).

Similarly, *Ixodes persulcatus*, which is a tick species typical of forests in Eastern Europe and southern Siberia, has invaded the boreal forests of Fennoscandia where they serve as important vectors for *Borrelia* spirochaetes (Jaenson *et al.*, 2016; Laaksonen *et al.*, 2017; Pakanen *et al.*, 2020). In Komi Republic and Arkhangelsk Oblast, situated in the north of the European subarctic zone of Russia, increases of mean annual air temperatures and temperatures during the tick active season resulted in the northward expansion of *I. persulcatus* (Tokarevich *et al.*, 2011, 2017). Further, a study investigating epidemiological changes in tick-borne encephalitis incidence in the Russian Federation over 20 years (1997–2006 and 2007–2016) found that changes are occurring most intensively in subarctic parts of Siberia and the Far East (Waits *et al.*, 2018).

Novel tick species can be introduced into the Arctic also by human activities. Several introductions of the brown dog tick, *Rhipicephalus sanguineus*, to Alaska were recently described (Hahn *et al.*, 2020). As these ticks are strongly synanthropic and can successfully survive and reproduce indoors, there is a high chance that this originally tropical tick can establish a stable population in Alaska (Hahn *et al.*, 2020).

67.2 Future Predictions

67.2.1 Future prediction for climate in Arctic

Almost all climatological models agree that the current trend of rapid warming connected with deglaciation and decrease in snow and ice cover will continue in the following century. According to the even most conservative models, all summer sea ice in the Arctic will melt before 2100, but most probably around 2050 (Holland *et al.*, 2006; Kim *et al.*, 2016). Several models also state that the current melting of the Greenland ice sheet is an irreversible process that will lead to complete deglaciation of Greenland (Huybrechts *et al.*, 1991; Bevis *et al.*, 2019; King *et al.*, 2020).

All these changes will have a massive impact on Arctic animals. They will suffer through loss of habitats and by increased competition from the related species migrating to the Arctic from the south. Introduction of new species may also lead to interspecies hybrids which could lead to the disruption of integrity of the original species' gene pool (Vincent *et al.*, 2011). Finally, it can be expected that numerous new pathogens and parasites will be introduced into the area of the Arctic, which would create an additional pressure to the already stressed species and be potentially dangerous for human populations living there (Omazic *et al.*, 2019).

67.2.2 Future predictions for ticks and tick-borne pathogens in the Arctic

Tick populations in the Arctic and the prevalence of TBPs will likely increase in the future due to the increased availability of suitable animal hosts resulting from increased temperatures (Parkinson *et al.*, 2014). Populations of tick species that are already established will grow in numbers and new foci will appear in areas that are currently tick-free. The current distribution of four tick species present in the Arctic and their possible future expansion are shown in Fig. 67.2. *I. uriae* is already well established in the Arctic, having a circumpolar distribution aided by its ability to survive extremely cold temperatures. In future decades we may expect increased population densities of this species in existing foci. *I. ricinus* and *I. persulcatus* are typical of temperate zones, but they were recently found at the southern borders of the Arctic. In future decades, we may expect introduction of these species to higher altitudes of Eurasia, as well as expansion of *I. ricinus* in northern Norway and Sweden and of *I. persulcatus* in Arctic Russia and Fennoscandia. *R. sanguineus* is typical for tropical and subtropical zones. As it is a synanthropic tick species, able to reproduce indoors, potentially it may establish stable populations in some parts of the Arctic.

Hand in hand with climate change and introduction of new host species, new tick species will be introduced. Most of these introductions will not be successful and the populations of newly introduced ticks will collapse soon after introduction, but at least some of them might establish (Hahn *et al.*, 2020). All these changes will also lead to increased prevalence of already established TBPs and introduction of new ones. These changes can have very negative impacts on local animal species, domestic animals, as well as on humans (Omazic *et al.*, 2019).

This process can be slowed down by careful public health and veterinary controls that would limit introductions caused by human activities. But as new tick and TBP species can be introduced to the Arctic by natural processes (such as bird migrations) we cannot stop this process completely.

We also have to realize that the Arctic is a kind of melting pot for many important pathogens as it is the place where birds following different flyways meet during the nesting season and where they can share parasites and pathogens that would otherwise never meet, then further introduce these 'bugs' to localities where they

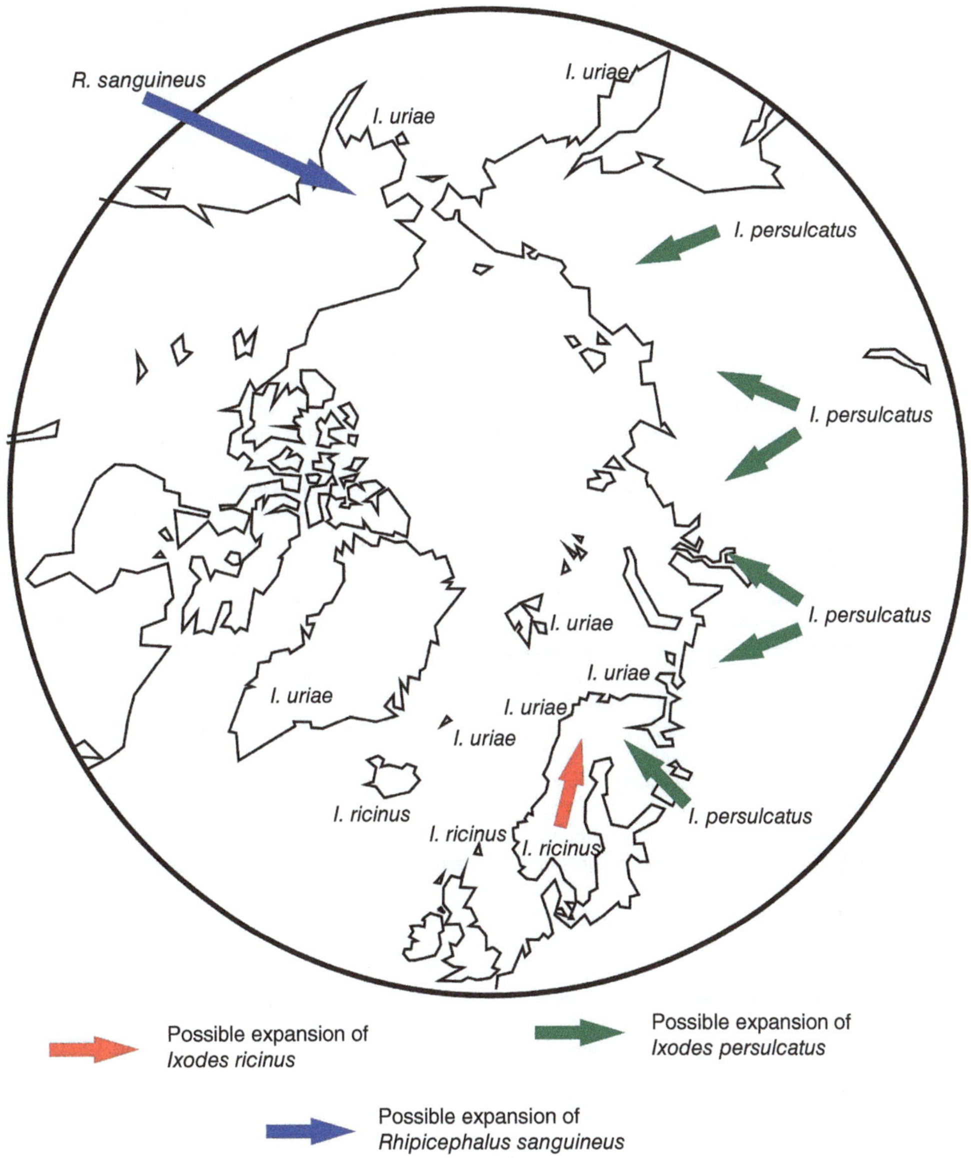

Fig. 67.2. Current distribution of selected tick species in the Arctic and their expansion potential. (Authors' own figure.)

overwinter (Wahlgren *et al.*, 2008; Lee *et al.*, 2015; Ramey *et al.*, 2016; Morin *et al.*, 2018). Climate change followed by massive melting in the Arctic can theoretically allow new tick and TBP species to reach this melting pot and then disseminate to other parts of the world. This route was most probably used when Powassan virus invaded from North America to North Asia via Beringia (Leonova *et al.*, 2009; Deardorff *et al.*, 2013). Theoretically, in future, similar routes could lead to introduction of other TBPs such as TBEV from Asia to North America.

From the scientific point of view, and as with all Arctic arthropods (Gillespie *et al.*, 2020), we need to precisely describe the current distribution of ticks and TBPs in the Arctic to create a baseline that we can use as a reference to understand future changes (Hahn *et al.*, 2020). It will require an extensive amount of work in the field and careful analysis of already published data, which is unfortunately beyond the scope of this expert opinion.

Acknowledgement

The work was supported by the Ministry of Education, Youth and Sports of the Czech Republic (ECOPOLARIS, Project No. CZ.02.1.01/0.0/0.0/16_013/0001708).

References

ACIA (2006) *Arctic Climate Impact Assessment – Scientific Report*. Cambridge University Press, Cambridge.

Alfredsson, M., Olafsson, E., Eydal, M., Unnsteinsdottir, E.R., Hansford, K., *et al.* (2017) Surveillance of *Ixodes ricinus* ticks (Acari: Ixodidae) in Iceland. *Parasites & Vectors* 10, 466.

Anthony, K.W., von Deimling, T.S., Nitze, I., Frolking, S., Emond, A., *et al.* (2018) 21st-century modeled permafrost carbon emissions accelerated by abrupt thaw beneath lakes. *Nature Communications* 9, 3262.

Benoit, J.B., Yoder, J.A., Lopez-Martinez, G., Elnitsky, M.A., Lee, R.E. Jr and Denlinger, D.L. (2007) Habitat requirements of the seabird tick, *Ixodes uriae* (Acari: Ixodidae), from the Antarctic Peninsula in relation to water balance characteristics of eggs, nonfed and engorged stages. *Journal of Comparative Physiology B* 177, 205–215.

Bergström, S., Haemig, P.D. and Olsen, B. (1999) Increased mortality of black-browed albatross chicks at a colony heavily-infested with the tick *Ixodes uriae*. *International Journal for Parasitology* 29, 1359–1361.

Bevis, M., Harig, C., Khan, S.A., Brown, A., Simons, F.J., *et al.* (2019) Accelerating changes in ice mass within Greenland, and the ice sheet's sensitivity to atmospheric forcing. *Proceedings of the National Academy of Sciences USA* 116, 1934–1939.

Biskaborn, B.K., Smith, S.L., Noetzli, J., Matthes, H., Vieira, G., *et al.* (2019) Permafrost is warming at a global scale. *Nature Communications* 10, 264.

Bölter, M. and Müller, F. (2016) Resilience in polar ecosystems: from drivers to impacts and changes. *Polar Science* 10, 52–59.

Clairbaux, M., Fort, J., Mathewson, P., Porter, W., Strøm, H. and Grémillet, D. (2019) Climate change could overturn bird migration: transarctic flights and high-latitude residency in a sea ice free Arctic. *Scientific Reports* 9, 17767.

Coulson, S.J., Lorentzen, E., Strom, H. and Gabrielsen, G.W. (2009) The parasitic tick *Ixodes uriae* (Acari: Ixodidae) on seabirds from Spitsbergen, Svalbard. *Polar Research* 28, 399–402.

Deardorff, E.R., Nofchissey, R.A., Cook, J.A., Hope, A.G., Tsvetkova, A., *et al.* (2013) Powassan virus in mammals, Alaska and New Mexico, USA, and Russia, 2004–2007. *Emerging Infectious Diseases* 19, 2012–2016.

Descamps, S. (2013) Winter temperature affects the prevalence of ticks in an Arctic seabird. *PLoS ONE* 8, e65374.

Dietrich, M., Kempf, F., Boulinier, T. and McCoy, K.D. (2014) Tracing the colonization and diversification of the worldwide seabird ectoparasite *Ixodes uriae*. *Molecular Ecology* 23, 3292–3305.

Elsterova, J., Cerny, J., Mullerova, J., Sima, R., Coulson, S., *et al.* (2015) Search for tick-borne pathogens in the Svalbard Archipelago and Jan Mayen. *Polar Research* 34, 27466.

Gilbert, L., Aungier, J. and Tomkins, J.L. (2014) Climate of origin affects tick (*Ixodes ricinus*) host-seeking behavior in response to temperature: implications for resilience to climate change? *Ecology and Evolution* 4, 1186–1198.

Gillespie, M.A.K., Alfredsson, M., Barrio, I.C., Bowden, J., Convey, P., *et al.* (2020) Circumpolar terrestrial arthropod monitoring: a review of ongoing activities, opportunities and challenges, with a focus on spiders. *Ambio* 49, 704–717.

Hahn, M.B., Disler, G., Durden, L.A., Coburn, S., Witmer, F., *et al.* (2020) Establishing a baseline for tick surveillance in Alaska: tick collection records from 1909–2019. *Ticks and Tick-borne Diseases* 11, 101495.

Hasle, G., Bjune, G., Edvardsen, E., Jakobsen, C., Linnehol, B., *et al.* (2009) Transport of ticks by migratory passerine birds to Norway. *Journal of Parasitology* 95, 1342–1351.

Henttonen, H., Fuglei, E., Gower, C.N., Haukisalmi, V., Ims, R.A., *et al.* (2001) *Echinococcus multilocularis* on Svalbard: introduction of an intermediate host has enabled the local life-cycle. *Parasitology* 123, 547–552.

Holland, M.M., Bitz, C.M. and Tremblay, L.B. (2006) Future abrupt reductions in the summer Arctic sea ice. *Geophysical Research Letters* 33, L23503.

Hornok, S., Mühldorfer, K., Takács, N., Hofmann-Lehmann, R., Meli, M.L., *et al.* (2020) Broad range screening of vector-borne pathogens in Arctic Foxes. *Animals* 10, 2031.

Hunter, C.M., Caswell, H., Runge, M.C., Regehr, E.V., Amstrup, S.C. and Stirling, I. (2010) Climate change threatens polar bear populations: a stochastic demographic analysis. *Ecology* 91, 2883–2897.

Huybrechts, P., Letreguilly, A. and Reeh, N. (1991) The Greenland ice sheet and greenhouse warming. *Palaeogeography, Palaeoclimatology, Palaeoecology* 89, 399–412.

Hvidsten, D., Stuen, S., Jenkins, A., Dienus, O., Olsen, R.S., *et al.* (2014) *Ixodes ricinus* and *Borrelia* prevalence at the Arctic Circle in Norway. *Ticks and Tick-borne Diseases* 5, 107–112.

Hvidsten, D., Stordal, F., Lager, M., Rognerud, B., Kristiansen, B.E., *et al.* (2015) *Borrelia burgdorferi* sensu lato-infected *Ixodes ricinus* collected from vegetation near the Arctic Circle. *Ticks and Tick-borne Diseases* 6, 768–773.

Jaenson, T.G., Jaenson, D.G., Eisen, L., Petersson, E. and Lindgren, E. (2012) Changes in the geographical distribution and abundance of the tick *Ixodes ricinus* during the past 30 years in Sweden. *Parasites & Vectors* 5, 8.

Jaenson, T.G., Värv, K., Fröjdman, I., Jääskeläinen, A., Rundgren, K., *et al.* (2016) First evidence of established populations of the taiga tick *Ixodes persulcatus* (Acari: Ixodidae) in Sweden. *Parasites & Vectors* 9, 377.

Jenkins, A., Raasok, C., Pedersen, B.N., Jensen, K., Andreassen, Å., *et al.* (2019) Detection of *Candidatus* Neoehrlichia mikurensis in Norway up to the northern limit of *Ixodes ricinus* distribution using a novel real time PCR test targeting the *groEL* gene. *BMC Microbiology* 19, 199.

Joly, K., Gurarie, E., Sorum, M.S., Kaczensky, P., Cameron, M.D., *et al.* (2019) Longest terrestrial migrations and movements around the world. *Scientific Reports* 9, 15333.

Kausrud, K.L., Mysterud, A., Steen, H., Vik, J.O., Østbye, E., *et al.* (2008) Linking climate change to lemming cycles. *Nature* 456, 93–97.

Kim, K.-Y., Hamlington, B.D. and Kim, J. (2016) Mechanism of seasonal Arctic sea ice evolution and Arctic amplification. *The Cryosphere* 10, 2191–2202.

King, M.D., Howat, I.M., Candela, S.G., Noh, M.J., Jeong, S., *et al.* (2020) Dynamic ice loss from the Greenland Ice Sheet driven by sustained glacier retreat. *Nature Communications Earth & Environment* 1, 1.

Laaksonen, M., Sajanti, E., Sormunen, J.J., Penttinen, R., Hänninen, J., *et al.* (2017) Crowdsourcing-based nationwide tick collection reveals the distribution of *Ixodes ricinus* and *I. persulcatus* and associated pathogens in Finland. *Emerging Microbes and Infections* 6, e31.

Larsson, C., Comstedt, P., Olsen, B. and Bergstrom, S. (2007) First record of Lyme disease *Borrelia* in the Arctic. *Vector-Borne and Zoonotic Diseases* 7, 453–456.

Larsson, C., Hvidsten, D., Stuen, S., Henningsson, A.J. and Wilhelmsson, P. (2018) '*Candidatus* Neoehrlichia mikurensis' in *Ixodes ricinus* ticks collected near the Arctic Circle in Norway. *Parasites & Vectors* 11, 620.

Lee, D.H., Torchetti, M.K., Winker, K., Ip, H.S., Song, C.S. and Swayne, D.E. (2015) Intercontinental spread of Asian-origin H5N8 to North America through Beringia by migratory birds. *Journal of Virology* 89, 6521–6524.

Lee, R.E. Jr and Baust, J.G. (1982) Respiratory metabolism of the Antarctic tick, *Ixodes uriae*. *Comparative Biochemistry and Physiology Part A: Physiology* 72, 167–171.

Lee, R.E. Jr and Baust, J.G. (1987) Cold-hardiness in the Antarctic tick, *Ixodes uriae*. *Physiological Zoology* 60, 499–506.

Leonova, G.N., Kondratov, I.G., Ternovoi, V.A., Romanova, E.V., Protopopova, E.V., *et al.* (2009) Characterization of Powassan viruses from Far Eastern Russia. *Archives of Virology* 154, 811–820.

McCoy, K.D. (2001) Consequences of dispersal in host–parasite systems: population dynamics, genetic structure, and local adaptation of the seabird tick *Ixodes uriae*. PhD thesis, Pierre and Marie Curie University, Paris.

Medlock, J.M., Hansford, K., Vaux, A.G.C., Simonsen, W., Jensen, J.K., *et al.* (2017) Surveillance for *Ixodes ricinus* ticks (Acari, Ixodidae) on the Faroe Islands. *Ticks and Tick-borne Diseases* 8, 190–195.

Milner, A.M., Khamis, K., Battin, T.J., Brittain, J.E., Barrand, N.E., *et al.* (2017) Glacier shrinkage driving global changes in downstream systems. *Proceedings of the National Academy of Sciences USA* 114, 9770–9778.

Morin, C.W., Stoner-Duncan, B., Winker, K., Scotch, M., Hess, J.J., *et al.* (2018) Avian influenza virus ecology and evolution through a climatic lens. *Environment International* 119, 241–249.

Neelakanta, G., Sultana, H., Fish, D., Anderson, J.F. and Fikrig, E. (2010) *Anaplasma phagocytophilum* induces *Ixodes scapularis* ticks to express an antifreeze glycoprotein gene that enhances their survival in the cold. *Journal of Clinical Investigation* 120, 3179–3190.

Nghiem, S.V., Hall, D.K., Mote, T.L., Tedesco, M., Albert, M.R., *et al.* (2012) The extreme melt across the Greenland ice sheet in 2012. *Geophysical Research Letters* 39, L20502.

Olsen, B., Duffy, D., Jaenson, T., Gylfe, A., Bonnedahl, J. and Bergstrom, S. (1995) Transhemispheric exchange of Lyme-disease spirochetes by seabirds. *Journal of Clinical Microbiology* 33, 3270–3274.

Omazic, A., Bylund, H., Boqvist, S., Högberg, A., Björkman, C., *et al.* (2019) Identifying climate-sensitive infectious diseases in animals and humans in Northern regions. *Acta Veterinaria Scandinavica* 61, 53.

Pakanen, V.M., Sormunen, J.J., Sippola, E., Blomqvist, D. and Kallio, E.R. (2020) Questing abundance of adult taiga ticks *Ixodes persulcatus* and their *Borrelia* prevalence at the north-western part of their distribution. *Parasites & Vectors* 13, 384.

Parkinson, A.J., Evengard, B., Semenza, J.C., Ogden, N., Børresen, M.L., *et al.* (2014) Climate change and infectious diseases in the Arctic: establishment of a circumpolar working group. *International Journal of Circumpolar Health* 73, 25163.

Perner, J., Kropáčková, S., Kopáček, P. and Ribeiro, J.M.C. (2018) Sialome diversity of ticks revealed by RNAseq of single tick salivary glands. *PLoS Neglected Tropical Diseases* 12, e0006410.

Pettersson, J.H., Ellström, P., Ling, J., Nilsson, I., Bergström, S., *et al.* (2020) Circumpolar diversification of the *Ixodes uriae* tick virome. *PLoS Pathogens* 16, e1008759.

Piersma, T. (1997) Do global patterns of habitat use and migration strategies co-evolve with relative investments in immunocompetence due to spatial variation in parasite pressure? *Oikos* 80, 623–631.

Ramey, A.M., Reeves, A.B., TeSlaa, J.L., Nashold, S., Donnelly, T., *et al.* (2016) Evidence for common ancestry among viruses isolated from wild birds in Beringia and highly pathogenic intercontinental reassortant H5N1 and H5N2 influenza A viruses. *Infection, Genetics and Evolution* 40, 176–185.

Ravolainen, V., Soininen, E.M., Jónsdóttir, I.S., Eischeid, I., Forchhammer, M., *et al.* (2016) High Arctic ecosystem states: conceptual models of vegetation change to guide long-term monitoring and research. *Ambio* 10, 666–677.

Revich, B., Tokarevich, N. and Parkinson, A.J. (2012) Climate change and zoonotic infections in the Russian Arctic. *International Journal of Circumpolar Health* 71, 18792.

Ruscio, B.A., Brubaker, M., Glasser, J., Hueston, W. and Hennessy, T.W. (2015) One Health – a strategy for resilience in a changing Arctic. *International Journal of Circumpolar Health* 74, 27913.

Soleng, A., Edgar, K.S., Paulsen, K.M., Pedersen, B.N., Okbaldet, Y.B., *et al.* (2018) Distribution of *Ixodes ricinus* ticks and prevalence of tick-borne encephalitis virus among questing ticks in the Arctic Circle region of northern Norway. *Ticks and Tick-borne Diseases* 9, 97–103.

Sulikowska, A., Walawender, J.P. and Walawender, E. (2019) Temperature extremes in Alaska: temporal variability and circulation background. *Theoretical and Applied Climatology* 136, 955–970.

Tokarevich, N.K., Tronin, A.A., Blinova, O.V., Buzinov, R.V., Boltenkov, V.P., *et al.* (2011) The impact of climate change on the expansion of *Ixodes persulcatus* habitat and the incidence of tick-borne encephalitis in the north of European Russia. *Global Health Action* 4, 8448.

Tokarevich, N., Tronin, A., Gnativ, B., Revich, B., Blinova, O. and Evengard, B. (2017) Impact of air temperature variation on the ixodid ticks habitat and tick-borne encephalitis incidence in the Russian Arctic: the case of the Komi Republic. *International Journal of Circumpolar Health* 76, 1298882.

Tyler, N.J., Forchhammer, M.C. and Øritsland, N.A. (2008) Nonlinear effects of climate and density in the dynamics of a fluctuating population of reindeer. *Ecology* 89, 1675–1686.

Vikse, R., Paulsen, K.M., Edgar, K.S., Pettersson, J.H.-O., Ottesen, P.S., *et al.* (2020) Geographical distribution and prevalence of tick-borne encephalitis virus in questing *Ixodes ricinus* ticks and phylogeographic structure of the *Ixodes ricinus* vector in Norway. *Zoonoses and Public Health* 67, 370–381.

Vincent, W.F., Callaghan, T.V., Dahl-Jensen, D., Johansson, M., Kovacs, K.M., *et al.* (2011) Ecological implications of changes in the Arctic cryosphere. *Ambio* 40, 87–99.

Wahlgren, J., Waldenström, J., Sahlin, S., Haemig, P.D., Fouchier, R.A., *et al.* (2008) Gene segment reassortment between American and Asian lineages of avian influenza virus from waterfowl in the Beringia area. *Vector-Borne and Zoonotic Diseases* 8, 783–790.

Waits, A., Emelyanova, A., Oksanen, A., Abass, K. and Rautio, A. (2018) Human infectious diseases and the changing climate in the Arctic. *Environment International* 121, 703–713.

WMO (2020) Reported new record temperature of 38°C north of Arctic Circle. World Meteorological Organization, Geneva, Switzerland. Available at: https://public.wmo.int/en/media/news/reported-new-record-temperature-of-38%C2%B0c-north-of-arctic-circle (accessed 8 December 2020).

eo68 Ticks and Tick-Borne Diseases in the Middle East

Ali Reza Chavshin* **and Seyyed Javad Seyyed-Zadeh**
Urmia University of Medical Sciences, Urmia, Iran

68.1 Expert Opinion

68.1.1 Introduction

The Middle East, as a transcontinental region, includes some parts of Western Asia, North Africa and Turkey (partly in South-East Europe). Due to the large size of the Middle East, a lot of climatic diversity can be seen in this region (Fig. 68.1), but it is largely arid to semi-arid (Evans *et al.*, 2004). The existence of climatic and biological diversity in this vast area has provided suitable hosts and habitats for ticks. In many parts of the Middle East, the conducive overlap of ticks, hosts and climatic conditions has led to the establishment of transmission cycles of various tick-borne diseases in this region.

Many studies on ticks and their diseases have been conducted in Middle Eastern countries. Each of them has reported valuable findings based on their specific goals and their summaries, like puzzle pieces, can draw an informative picture of the ticks in the region. This expert opinion is based on published and available information. Obviously, it is impossible to claim the completeness and accuracy of all these studies and the availability of comprehensive information about ticks and their diseases. The passage of time will make the picture more complete and accurate.

Another important point to consider is the scientific challenge in identification of species and the taxonomy of ticks. Here, only a summary of published and available studies is presented, and the challenges of species identification and classification of ticks are not considered.

From 12 countries in the Middle East where reports and articles were published, a total of 136 species of ticks have been reported. Of these, 111 species belong to the family Ixodidae and 25 species belong to the Argasidae. Reported species of the Ixodidae belong to eight genera (*Hyalomma*, 25; *Rhipicephalus*, 23; *Haemaphysalis*, 22; *Ixodes*, 20; *Amblyomma*, 10; *Dermacentor*, 6; *Boophilus*, 4; and *Aponomma*, 1 species), while the Argasidae belong to five genera (*Ornithodoros*, 13; *Argas*, 8; *Alveonasus*, 2; *Carios*, 1; and *Otobius*, 1 species). Based on the results of available reports, it seems that species belonging to the genera *Hyalomma*, *Rhipicephalus*, *Haemaphysalis*, *Ixodes* and *Ornithodoros* have the highest species diversity in the Middle East. The distribution of some important species is shown in Fig. 68.2.

68.1.2 Ticks in the Middle East

Egypt

Fifty-one species of ticks (40 of Ixodidae and 11 of Argasidae) have been reported, from the genera *Hyalomma* (*aegyptium*, *anatolicum*, *detritum*, *dromedarii*, *excavatum*, *franchinii*, *impeltatum*,

*Email: chavshin@gmail.com

DOI: 10.1079/9781789249637.0068

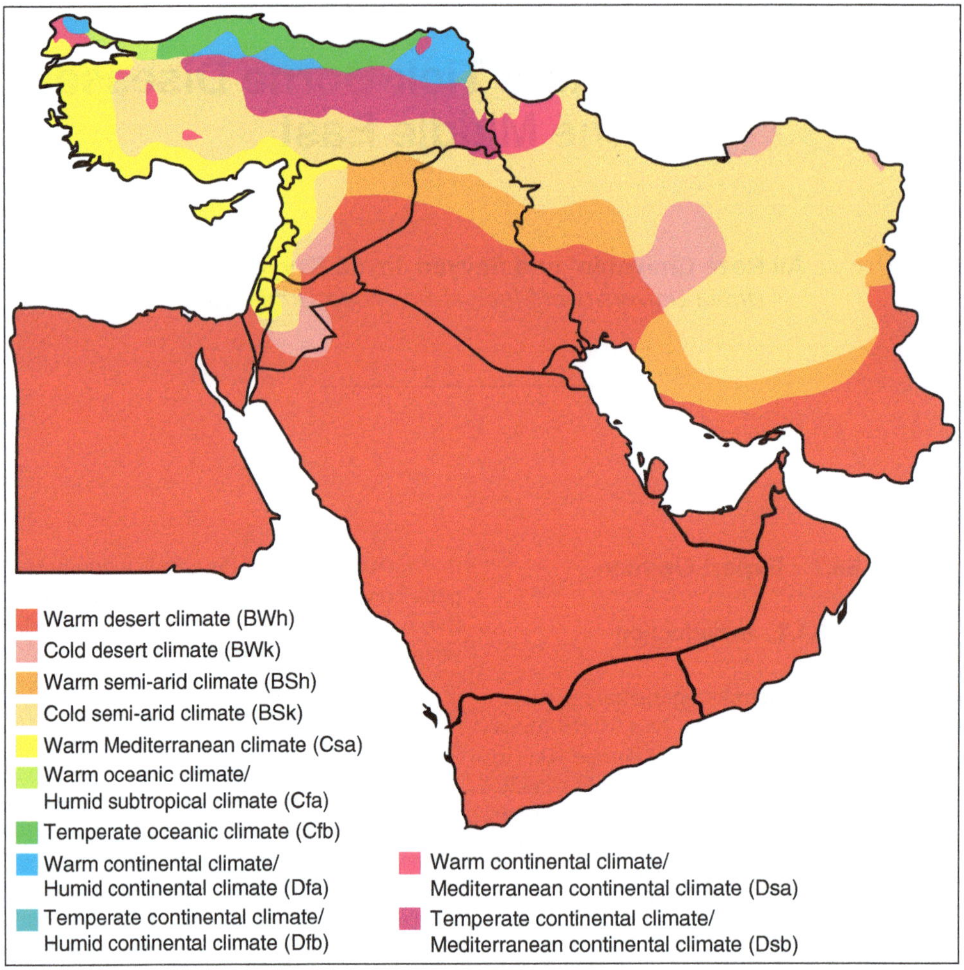

Fig. 68.1. Middle East map of Köppen–Geiger climate classification. (Map from Kottek *et al.*, 2006; enhanced, modified and vectorized by Ali Zifan. Authors' own figure.)

marginatum, rufipes, runcatim, schulzei, scupense, sulcata, truncatum), *Rhipicephalus* (*annulatus, bursa, camicasi, guilhoni, praetextatus, pulchellus, sanguineus, turanicus*), *Amblyomma* (*gemma, hebraeum, lipidum, maftnoreum, marmoreum, repidum, variegatum*), *Haemaphysalis* (*inermis, leaehi, otophila, pavlovskyi, punctata, sulcata*), *Argas* (*arboreus, boutei, hermanni, persicus, reflexus, streptopelia*), *Ornithodoros* (*coniceps, erraticus, normandi, salahi, savignyi*), *Ixodes* (*arborieola, frontalis, inopinatus, ricinus*) and *Boophilus annulatus* (Hoogstraal and Kaiser, 1958, 1961; El Kammah *et al.*, 2001; El-seify *et al.*, 2017).

Iran

Forty-six species of Iranian ticks (36 of Ixodidae and 10 of Argasidae) have been reported: *Hyalomma* (*aegyptium, anatolicum, asiaticum, dromedarii, excavatum, impeltatum, kumari, marginatum, rufipes, schulzei, scupense*), *Haemaphysalis* (*concinna, doenitzi, erinacei, caucasica, inermis, kashmirensis, kopetdaghica, parva, punctata, sulcata*), *Rhipicephalus* (*annulatus, bursa, kohlsi, rossicus, sanguineus, schulzei, turanicus*), *Ixodes* (*crenulatus, occulatus, redikorzevi, ricinus, vespertilionis*), *Ornithodoros* (*erraticus, tartakovskyi, tholozani, verrucosus*), *Dermacentor* (*marginatus, niveus, raskemensis*),

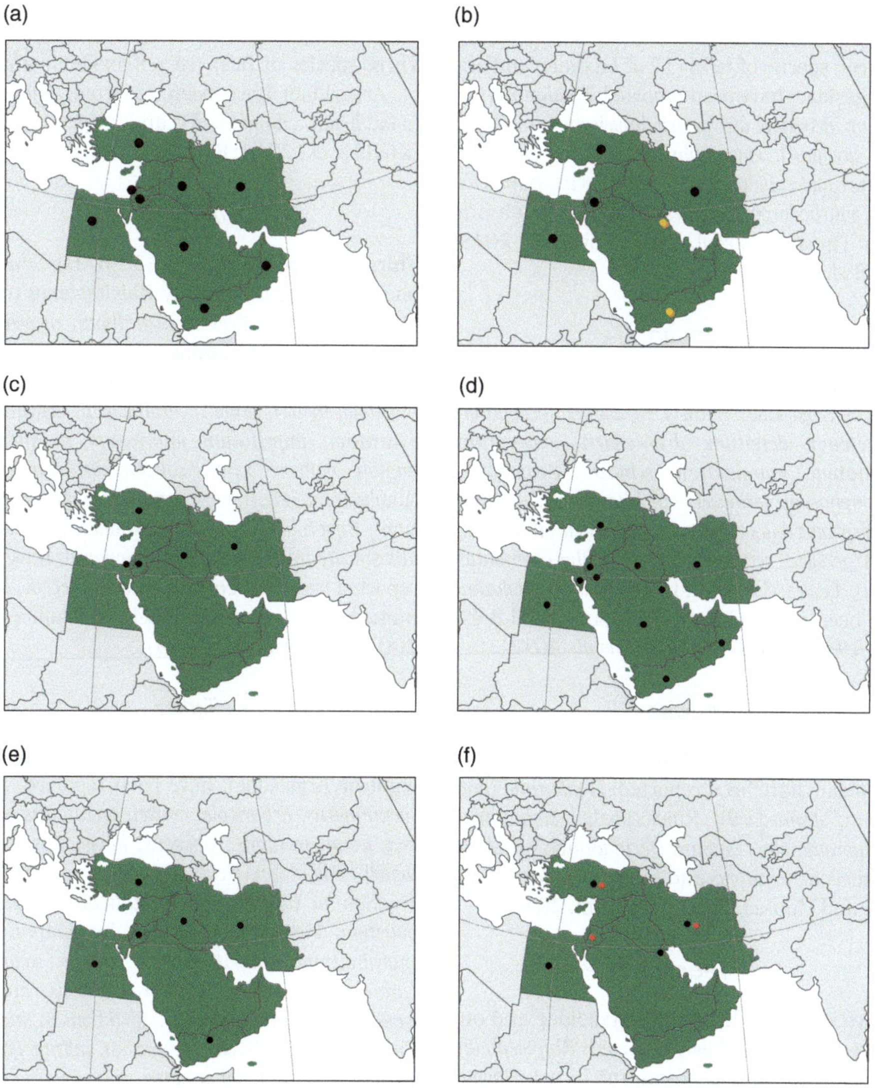

Fig. 68.2. The distribution of some of the medically important ticks in the Middle East region: (a) *Hyalomma marginatum*; (b) *Ixodes ricinus* (•) and *Ixodes scapularis* (•); (c) *Hyalomma scupense* (syn. *Hyalomma detritum*); (d) *Rhipicephalus sanguineus*; (e) *Argas persicus*; (f) *Ornithodoros erraticus* (•) and *Ornithodoros tholozani* (•). (Authors' own figure.)

Alvenasus (*canestrinii*, *lahorensis*), *Argas* (*persicus*, *reflexus*), *Carios vespertilionis* and *Otobius megnini* (Rahbari *et al.*, 2007; Ramezani *et al.*, 2014; Hosseini-Chegeni *et al.*, 2019).

Iraq

Thirty-two species of ticks (23 of Ixodidae and eight of Argasidae) have been reported: *Hyalomma* (*analoticum*, *asiaticum*, *detritum*, *dromedarii*, *excavatum*, *impeltatum*, *marginatum*, *schulzei*, *scupense*, *turanicum*), *Rhipicephalus* (*annulatus*, *appendiculatus*, *bursa*, *leporis*, *sanguineus*, *turanicus*), *Haemaphysalis* (*adleri*, *alderi*, *erinacei*, *parva*, *sulcata*), *Argas* (*confuses*, *persicus*, *reflexus*, *vespertilionis*), *Ornithodoros* (*erraticus*, *lahorensis*, *savignyi*), *Boophilus* (*annulatus*, *kohlsi*), *Amblyomma lepidum* and *Otobius megnini* (Shamsuddin and Mohammad, 1988; Shubber *et al.*, 2014; Mohammad, 2015).

Palestine and Israel

Thirteen species of ticks (12 of Ixodidae and one of Argasidae) have been reported: *Hyalomma* (*aegyptium, detritum, dromedarii, impeltatum, marginatum, savignyi*), *Haemaphysalis* (*adleri, concinna, parva*), *Rhipicephalus* (*bursa, sanguineus, turanicus*) and *Ornithodoros pipillipes* (Feldman-Muehsam, 1948; Tirosh-Levy *et al.*, 2018; Ravi *et al.*, 2019; Zaid *et al.*, 2019).

Jordan

Twenty-three species of ticks (21 of Ixodidae and three of Argasidae), namely *Hyalomma* (*aegyptium, anatolicum, detritum, dromedarii, excavatum, impeltatum, marginatum, schulzei, turanicum*), *Haemaphysalis* (*erinacei, otophila, parva, sulcata*), *Rhipicephalus* (*bursa, comicasi, sanguineus, turanicus*), *Argas* (*persicus, vespertilionis*), *Boophilus* (*annulatus, kohlsi*), *Ixodes ricinus* and *Ornithodoros tholozani*, have been reported from this country (El-Rabie *et al.*, 1990; Amr, 2000; Qablan *et al.*, 2012).

Kuwait

Seven species of ticks (six of Ixodidae and one of Argasidae) have been reported: *Hyalomma* (*anatolicum, dromedarii*), *Rhipicephalus sanguineus, Amblyomma americanum, Dermacentor variabilis, Ornithodoros erraticus* and *Ixodes scapularis* (Converse and Moussa, 1982; Ibrahium, 2018).

Lebanon

Twelve species of ticks (11 of Ixodidae and one of Argasidae) have been reported: *Rhipicephalus* (*annulatus, bursa, sanguineus, turanicus*), *Haemaphysalis* (*parva, punctata*), *Hyalomma* (*aegyptium, anatolicum*), *Ixodes* (*simplex, vespertilionis*), *Argas vespertilionis* and *Dermacentor marginatus* (Dabaja *et al.*, 2017; de Mera *et al.*, 2018).

Oman

Seventeen species of ticks (14 of Ixodidae and three of Argasidae) have been reported: *Hyalomma* (*anatolicum, arabica, dromedarii, excavatum, impeltatum, marginatum, rufipes, turanicum*), *Rhipicephalus* (*appendiculatus, evertsi, pulchellus, sanguineus*), *Ornithodoros* (*buettikeri, foleyi, sonrai*), *Amblyomma variegatum* and *Haemaphysalis indica* (Papadopoulos *et al.*, 1991; Schwarz *et al.*, 1995).

Qatar

Three species of ticks (two of Ixodidae and one of Argasidae) have been reported: *Hyalomma* (*anatolicum, dromedarii*) and *Argas hermanni* (Khalil *et al.*, 2014; Barradas *et al.*, 2020).

Saudi Arabia

Thirty-five species of ticks (34 of Ixodidae and one of Argasidae), namely *Rhipicephalus* (*appendiculatus, comicasi, evertsi, guilhoni, neavei, pravus, pulchellus, punctatus, pusillus, sanguineus, senegalensis, simus, sulcatus, turanicus*), *Hyalomma* (*anatolicum, arabica, dromedarii, erythraeum, excavatum, impeltatum, impressum, marginatum, onatoli, rufipes, schulzei, turanicum, turanicus*), *Amblyomma* (*eburneum, gemma, lepidum, variegatum*), *Boophilus* (*decoloratus, kohlsi*), *Haemaphysalis sulcate* and *Ornithodoros savignyi*, have been reported from this country (Banaja *et al.*, 1980; Banaja and Ghandour, 1994; Al-Khalifa *et al.*, 2007; Chandra *et al.*, 2019).

Turkey

Sixty-three species of ticks (55 of Ixodidae and eight of Argasidae), have been reported. *Ixodes* (*acuminatus, arboricola, cookei, crenulatus, eldaricus, festai, frontalis, gibbosus, hexagonus, kaiseri, laguri, redikorzevi, ricinus, simplex, vespertilionis*), *Hyalomma* (*aegyptium, anatolicum, asiaticum, detritum, dromedarii, excavatum, impeltatum, isacii, marginatum, mauritanicum, rufipes, scupense, turanicum*), *Haemaphysalis* (*aksarensis, aponommoides, concinna, erinacei, ibrikliensis, inermis, numidiana, otophila, parva, punctata, sulcata, yalvaci*), *Rhipicephalus* (*annulatus, bursa, calcaratus, haemaphysaloides, rossicus, sanguineus, turanicus*), *Dermacentor* (*daghestanicus, marginatus, niveus, reticulatus*), *Ornithodoros* (*coniceps, erraticus, lahorensis, tholozani*), *Argas* (*persicus, reflexus, vespertilionis*), *Boophilus* (*annulatus, calcaratus, kohlsi*), *Amblyomma variegatum* and *Otobius megnini* are from this country (Aydin and Bakirci, 2007; Karaer *et al.*, 2011; Bursali *et al.*, 2012, 2014).

Yemen

Twenty-seven species of ticks (25 of Ixodidae and two of Argasidae) have been reported. *Hyalomma* (*anatolicum, arabica, dromedarii, erythraeum, excavatum, impeltatum, rufipes, truncatum*), *Rhipicephalus* (*camicasi*, *evertsi*, *sanguineus*, *simus*, *turanicus*), *Amblyomma* (*gemma*, *gervaisi*, *variegatum*), *Haemaphysalis* (*leachi*, *muhsami*, *sulcata*), *Ixodes* (*hoogstraali*, *pacificus*, *scapularis*), *Boophilus* (*annulatus*, *kohlsi*), *Aponomma latum*, *Argas persicus* and *Ornithodoros savignyi* are from this country (Sanborn and Hoogstraal, 1953; Ba-hassan, 2018).

68.1.3 Tick-borne diseases in the Middle East

Due to the high diversity of tick species in the region and the abundance of susceptible vertebrate hosts, the presence of various tick-borne diseases is expected in this region. Among the important diseases, Crimean–Congo haemorrhagic fever (CCHF) has been reported from Iran (Chinikar *et al.*, 2010; Telmadarraiy *et al.*, 2015), Turkey (Yilmaz *et al.*, 2008; Düzlü *et al.*, 2020), Iraq (Al-Tikriti *et al.*, 1981), Kuwait (Al-Nakib *et al.*, 1984), United Arab Emirates (Suleiman *et al.*, 1980), Saudi Arabia (El-Azazy and Scrimgeour, 1997) and Oman (Schwarz *et al.*, 1995). CCHF can be considered widespread in large parts of the region and may possibly spread further and establish transmission cycles in large parts of the Middle East (Okely *et al.*, 2020). Among the widespread tick-borne diseases, ehrlichiosis (Sacchini *et al.*, 2007; M'ghirbi *et al.*, 2009; Tajedin *et al.*, 2016), anaplasmosis (Ghafar and Amer, 2012; Hosseini-Vasoukolaei *et al.*, 2014; Aktas and Özübek, 2015; Cetinkaya *et al.*, 2016), Lyme borreliosis (Elhelw *et al.*, 2014; Naddaf *et al.*, 2020), babesiosis (Hashemi-Fesharki, 1997; Adham *et al.*, 2009; Razmi *et al.*, 2013; Ghafar and Amer, 2019), Alkhurma haemorrhagic disease (Carletti *et al.*, 2010; Memish *et al.*, 2011) and Kyasanur forest disease (Madani, 2005; Mahdi *et al.*, 2011) have been reported from different parts of the region. The most important take-home message from the expression of tick-borne diseases in the Middle East is the existence of many different diseases in this region, including other diseases not mentioned here (Lange *et al.*, 1992; Hashemi-Fesharki, 1997; Keysary *et al.*, 2007; Orkun *et al.*, 2014; Barghash *et al.*, 2016; Düzlü *et al.*, 2020).

68.2 Future Predictions

Considering the diversity of tick species and the diseases transmitted by them in the Middle East, the effect of the following factors should be accounted for when considering the future:

1. Due to the notable climatic diversity of the Middle East, climate change will have different effects in different regions. Climate change in each region due to possible changes in environmental conditions such as temperature and humidity (Sweatman, 1968; Gilbert, 2010), as well as changes in the distribution pattern of ticks' hosts and vegetation (Cumming, 2002; Tack *et al.*, 2011), can change the pattern of presence and distribution of ticks and tick-borne diseases. To identify the impact of climate change on the presence and distribution of ticks, the regional effects should be determined by local studies and focusing on the bioecology of tick species in that region.

2. As the distribution of some tick species depends on migratory birds (Hoogstraal *et al.*, 1961; Björsdorff *et al.*, 2001; Palomar *et al.*, 2013), which migrate from different regions and sometimes even outside the Middle East, changes related to some important ticks and important diseases transmitted by them are not limited to climatic changes.

3. Other important and influential factors that intensify the effect of other effective factors are political unrest and crises that lead to changes in human demographics, environmental changes, human migration and even changes in the pattern of ticks' hosts. These should be considered as factors affecting ticks and the diseases transmitted by them in the Middle East.

References

Adham, F.K., Abd-El-Samie, E.M., Gabre, R.M. and Hussein, H.E. (2009) Detection of tick blood parasites in Egypt using PCR assay I – *Babesia bovis* and *Babesia bigemina*. *Parasitology Research* 105, 721–730.

Aktas, M. and Özübek, S. (2015) Bovine anaplasmosis in Turkey: first laboratory confirmed clinical cases caused by *Anaplasma phagocytophilum*. *Veterinary Microbiology* 178, 246–251.

Al-Khalifa, M.S., Khalil, G.M. and Diab, F.M. (2007) A two years study of ticks infesting goats and sheep in Abha, Saudi Arabia. *Saudi Journal of Biological Sciences* 14, 83–91.
Al-Nakib, W., Lloyd, G., El-Mekki, A., Platt, G., Beeson, A., *et al.* (1984) Preliminary report on arbovirus-antibody prevalence among patients in Kuwait: evidence of Congo/Crimean virus infection. *Transactions of the Royal Society of Tropical Medicine and Hygiene* 78, 474–476.
Al-Tikriti, S.K., Al-Ani, F., Jurji, F.J., Tantawi, H., Al-Moslih, M., *et al.* (1981) Congo/Crimean haemorrhagic fever in Iraq. *Bulletin of the World Health Organization* 59, 85–90.
Amr, Z.S. (2000) *Jordan Country Study on Biological Diversity: Arthropods of Medical Importance in Jordan*. United Nations Environment Programme, Nairobi.
Aydin, L. and Bakirci, S. (2007) Geographical distribution of ticks in Turkey. *Parasitology Research* 101, 163–166.
Ba-hassan, M.S. (2018) Study the presence of the ticks Ixodid Ticks *Camelus dromedarius* in the provinces of Aden and Lahij in Yemen. Available at: https://www.academia.edu/38723036/Study_the_presence_of_the_ticks_Ixodid_Ticks_Camelus_dromedarius_in_the_provinces_of_Aden_and_Lahij_in_Yemen (accessed 19 June 2021).
Banaja, A.A. and Ghandour, A.M. (1994) A review of parasites of camels (*Camelus dromedarius*) in Saudi Arabia. *Journal of King Abdulaziz University* 6, 75–86.
Banaja, A.A., Madbouly, M.H. and Roshdy, M.A. (1980) Ticks of Saudi Arabia. 1. Ticks (Ixodoides) infesting imported and local breeds of domestic animals at Jeddah. In: *Proceedings of the Fourth Conference on the Biological Aspects of Saudi Arabia*. University of Riyadh, Saudi Arabia, pp. 339–346.
Barghash, M.S., Hafez, A.A., Darwish, A.M. and El-Naga, T.R.A. (2016) Molecular detection of pathogens in ticks infesting camels in Matrouh Governorate, Egypt. *Journal of Bacteriology and Parasitology* 7(2), 1000269.
Barradas, P.F., Mesquita, J.R., Lima, C., Cardoso, L., Alho, A.M., *et al.* (2020) Pathogenic *Rickettsia* in ticks of spur-thighed tortoise (*Testudo graeca*) sold in a Qatar live animal market. *Transboundary and Emerging Diseases* 67, 461–465.
Bjöersdorff, A., Bergström, S., Massung, R.F., Haemig, P.D. and Olsen, B. (2001) *Ehrlichia*-infected ticks on migrating birds. *Emerging Infectious Diseases* 7, 877–879.
Bursali, A., Keskin, A. and Tekin, S. (2012) A review of the ticks (Acari: Ixodida) of Turkey: species diversity, hosts and geographical distribution. *Experimental and Applied Acarology* 57, 91–104.
Bursali, A., Tekin, S., Keskin, A., Ekici, M. and Dundar, E. (2014) Species diversity of ixodid ticks feeding on humans in Amasya, Turkey: seasonal abundance and presence of Crimean–Congo hemorrhagic fever virus. *Journal of Medical Entomology* 48, 85–93.
Carletti, F., Castilletti, C., Di Caro, A., Capobianchi, M.R., Nisii, C., *et al.* (2010) Alkhurma hemorrhagic fever in travelers returning from Egypt, 2010. *Emerging Infectious Diseases* 16, 1979–1982.
Cetinkaya, H., Matur, E., Akyazi, I., Ekiz, E.E., Aydin, L., *et al.* (2016) Serological and molecular investigation of *Ehrlichia* spp. and *Anaplasma* spp. in ticks and blood of dogs, in the Thrace region of Turkey. *Ticks and Tick-borne Diseases* 7, 706–714.
Chandra, S., Smith, K., Alanazi, A.D., Alyousif, M.S., Emery, D., *et al.* (2019) *Rhipicephalus sanguineus* sensu lato from dogs and dromedary camels in Riyadh, Saudi Arabia: low prevalence of vector-borne pathogens in dogs detected using multiplexed tandem PCR panel. *Folia Parasitologica* 66, 1–13.
Chinikar, S., Ghiasi, S.M., Hewson, R., Moradi, M. and Haeri, A. (2010) Crimean–Congo hemorrhagic fever in Iran and neighboring countries. *Journal of Clinical Virology* 47, 110–114.
Converse, J.D. and Moussa, M.I. (1982) Quaranfil virus from *Hyalomma dromedarii* (Acari: Ixodoidea) collected in Kuwait, Iraq and Yemen. *Journal of Medical Entomology* 19, 209–210.
Cumming, G.S. (2002) Comparing climate and vegetation as limiting factors for species ranges of African ticks. *Ecology* 83, 255–268.
Dabaja, M.F., Tempesta, M., Bayan, A., Vesco, G., Greco, G., *et al.* (2017) Diversity and distribution of ticks from domestic ruminants in Lebanon. *Veterinaria Italiana* 53, 147–155.
de Mera, I.G.F., Blanda, V., Torina, A., Dabaja, M.F., El Romeh, A., *et al.* (2018) Identification and molecular characterization of spotted fever group rickettsiae in ticks collected from farm ruminants in Lebanon. *Ticks and Tick-borne Diseases* 9, 104–108.
Düzlü, Ö., Inci, A., Yıldırım, A., Doğanay, M., Özbel, Y., *et al.* (2020) Vector-borne zoonotic diseases in Turkey: rising threats on public health. *Türkiye Parazitolojii Dergisi* 44, 168–175.
El-Azazy, O.M.E. and Scrimgeour, E.M. (1997) Crimean–Congo haemorrhagic fever virus infection in the western province of Saudi Arabia. *Transactions of the Royal Society of Tropical Medicine and Hygiene* 91, 275–278.

Elhelw, R.A., El-Enbaawy, M.I. and Samir, A. (2014) Lyme borreliosis: a neglected zoonosis in Egypt. *Acta Tropica* 140, 188–192.
El Kammah, K.M., Oyoun, L.M., El Kady, G.A. and Shafy, S.A. (2001) Investigation of blood parasites in livestock infested with argasid and ixodid ticks in Egypt. *Journal of the Egyptian Society of Parasitology* 31, 365–371.
El-Rabie, Y., Amr, Z.S. and Hyland, K.E. (1990) Observations on the ixodid ticks associated with domestic ungulate animals in Jordan. *Zoology in the Middle East* 4, 85–92.
El-seify, M.A., Ebrahaium, A.A., Hashem, M.A. and Kamel, Z.M. (2017) Studies on role of ticks in transmission of some blood parasites in camel. *Kafrelsheikh Veterinary Medical Journal* 15, 105–125.
Evans, J.P., Smith, R.B. and Oglesby, R.J. (2004) Middle East climate simulation and dominant precipitation processes. *International Journal of Climatology* 24, 1671–1694.
Feldman-Muehsam, B. (1948) On larvae and nymphs of some species of Palestinian *Hyalomma*. *Parasitology* 39, 138–147.
Ghafar, M.W. and Amer, S.A. (2012) Prevalence and first molecular characterization of *Anaplasma phagocytophilum*, the agent of human granulocytic anaplasmosis, in *Rhipicephalus sanguineus* ticks attached to dogs from Egypt. *Journal of Advanced Research* 3, 189–194.
Ghafar, M.W. and Amer, S.A. (2019) A preliminary molecular survey of *Babesia divergens* and first evidence of *Theileria annulata* in cattle from Saudi Arabia. *Veterinary World* 12, 266–270.
Gilbert, L. (2010) Altitudinal patterns of tick and host abundance: a potential role for climate change in regulating tick-borne diseases. *Oecologia* 162, 217–225.
Hashemi-Fesharki, R. (1997) Tick-borne diseases of sheep and goats and their related vectors in Iran. *Parassitologia* 39, 115–117.
Hoogstraal, H. and Kaiser, M.N. (1958) The ticks (Ixodoidea) of Egypt. A brief review and keys. *Journal of the Egyptian Public Health Association* 33, 51–85.
Hoogstraal, H. and Kaiser, M.N. (1961) Ticks from European-Asiatic birds migrating through Egypt into Africa. *Science* 133, 277–278.
Hoogstraal, H., Kaiser, M.N., Traylor, M.A., Gaber, S. and Guindy, E. (1961) Ticks (Ixodoidea) on birds migrating from Africa to Europe and Asia. *Bulletin of the World Health Organization* 24, 197–212.
Hosseini-Chegeni, A., Tavakoli, M. and Telmadarraiy, Z. (2019) The updated list of ticks (Acari: Ixodidae & Argasidae) occurring in Iran with a key to the identification of species. *Systematic and Applied Acarology* 24, 2133–2166.
Hosseini-Vasoukolaei, N., Oshaghi, M.A., Shayan, P., Vatandoost, H., Babamahmoudi, F., *et al.* (2014) Anaplasma infection in ticks, livestock and human in Ghaemshahr, Mazandaran Province, Iran. *Journal of Arthropod-Borne Diseases* 8, 204–211.
Ibrahium, S.M.A.A. (2018) Study of internal and external parasites in cats (*Felis catus*) in animal house veterinary hospital in Salmiya region, Kuwait. PhD thesis, Sudan University of Science & Technology, Khartoum.
Karaer, Z., Guven, E., Nalbantoglu, S., Kar, S., Orkun, O., *et al.* (2011) Ticks on humans in Ankara, Turkey. *Experimental and Applied Acarology* 54, 85–91.
Keysary, A., Massung, R.F., Inbar, M., Wallach, A.D., Shanas, U., *et al.* (2007) Molecular evidence for *Anaplasma phagocytophilum* in Israel. *Emerging Infectious Diseases* 13, 1411–1412.
Khalil, G.M., Marzouk, A.S., Mohamed, F.S. and Darwish, Z.E. (2014) Effect of 25-azacholestane on *Argas (Argas) hermanni* (Ixodoidea: Argasidae). *Journal of Medical Entomology* 33, 783–789.
Kottek, M., Grieser, J., Beck, C., Rudolf, B. and Rubel, F. (2006) World map of the Köppen–Geiger climate classification updated. *Meteorologische Zeitschrift* 15, 259–263.
Lange, J.V., El Dessouky, A.G., Manor, E., Merdan, A.I. and Azad, A.F. (1992) Spotted fever rickettsiae in ticks from the northern Sinai Governate, Egypt. *American Journal of Tropical Medicine and Hygiene* 46, 546–551.
Madani, T.A. (2005) Alkhumra virus infection, a new viral hemorrhagic fever in Saudi Arabia. *Journal of Infection* 51, 91–97.
Mahdi, M., Erickson, B.R., Comer, J.A., Nichol, S.T., Rollin, P.E., *et al.* (2011) Kyasanur forest disease virus Alkhurma subtype in ticks, Najran province, Saudi Arabia. *Emerging Infectious Diseases* 17, 945–947.
Memish, Z.A., Albarrak, A., Almazroa, M.A., Al-Omar, I., Alhakeem, R., *et al.* (2011) Seroprevalence of Alkhurma and other hemorrhagic fever viruses, Saudi Arabia. *Emerging Infectious Diseases* 17, 2316–2318.
M'ghirbi, Y., Ghorbel, A., Amouri, M., Nebaoui, A., Haddad, S., *et al.* (2009) Clinical, serological, and molecular evidence of ehrlichiosis and anaplasmosis in dogs in Tunisia. *Parasitology Research* 104, 767–774.

Mohammad, M.K. (2015) Distribution of ixodid ticks among domestic and wild animals in central Iraq. *Bulletin of the Iraq Natural History Museum* 13, 23–30.
Naddaf, S.R., Mahmoudi, A., Ghasemi, A., Rohani, M., Mohammadi, A., *et al.* (2020) Infection of hard ticks in the Caspian Sea littoral of Iran with Lyme borreliosis and relapsing fever borreliae. *Ticks and Tick-borne Diseases* 11, 101500.
Okely, M., Anan, R., Gad-Allah, S. and Samy, A.M. (2020) Mapping the environmental suitability of etiological agent and tick vectors of Crimean–Congo hemorrhagic fever. *Acta Tropica* 203, 105319.
Orkun, Ö., Karaer, Z., Çakmak, A. and Nalbantoğlu, S. (2014) Spotted fever group rickettsiae in ticks in Turkey. *Ticks and Tick-borne Diseases* 5, 213–218.
Palomar, A.M., Portillo, A., Santibáñez, P., Mazuelas, D., Arizaga, J., *et al.* (2013) Crimean–Congo hemorrhagic fever virus in ticks from migratory birds, Morocco. *Emerging Infectious Diseases* 19, 260–263.
Papadopoulos, B., Büttiker, W., Morel, P.C. and Aeschlimann, A. (1991) Ticks (Acarina, Fam. Argasidae & Ixodidae) of Oman. *Fauna of Saudi Arabia* 12, 200–208.
Qablan, M.A., Kubelová, M., Široký, P., Modrý, D. and Amr, Z.S. (2012) Stray dogs of northern Jordan as reservoirs of ticks and tick-borne hemopathogens. *Parasitology Research* 111, 301–307.
Rahbari, S., Nabian, S. and Shayan, P. (2007) Primary report on distribution of tick fauna in Iran. *Parasitology Research* 101, 175–177.
Ramezani, Z., Chavshin, A.R., Telmadarraiy, Z., Edalat, H., Dabiri, F., *et al.* (2014) Ticks (Acari: Ixodidae) of livestock and their seasonal activities, northwest of Iran. *Asian Pacific Journal of Tropical Disease* 4(Suppl. 2), S754–S757.
Ravi, A., Ereqat, S., Al-Jawabreh, A., Abdeen, Z., Abu Shamma, O., *et al.* (2019) Metagenomic profiling of ticks: identification of novel rickettsial genomes and detection of tick-borne canine parvovirus. *PLoS Neglected Tropical Diseases* 13, e0006805.
Razmi, G., Pourhosseini, M., Yaghfouri, S., Rashidi, A. and Seidabadi, M. (2013) Molecular detection of *Theileria* spp. and *Babesia* spp. in sheep and ixodid ticks from the northeast of Iran. *Journal of Parasitology* 99, 77–81.
Sacchini, F., Cessford, R.J. and Robinson, B.M. (2007) Outbreak of canine monocytic ehrlichiosis in Saudi Arabia. *Veterinary Clinical Pathology* 36, 331–335.
Sanborn, C.C. and Hoogstraal, H. (1953) Some mammals of Yemen and their ectoparasites. *Fieldiana: Zoology* 34, 229–252.
Schwarz, T., Nitschko, H., Jäger, G., Nsanze, H., Longson, M., *et al.* (1995) Crimean–Congo haemorrhagic fever in Oman. *The Lancet* 346, 1230.
Shamsuddin, M. and Mohammad, M.K. (1988) Incidence, distribution, and host relationships of some ticks (Ixodoidea) in Iraq. *Journal of the University of Kuwait (Science)* 15, 321–330.
Shubber, H.W.K., Al-Hassani, N.A.W. and Kadhim, M. (2014) Ixodid tick diversity in the Middle and South of Iraq. *International Journal of Recent Scientific Research* 5, 1518–1523.
Suleiman, M.N.E.H., Muscat-Baron, J.M., Harries, J.R., Satti, A.G.O., Platt, G.S., *et al.* (1980) Congo/Crimean haemorrhagic fever in Dubai: an outbreak at the Rashid Hospital. *The Lancet* 316, 939–941.
Sweatman, G.K. (1968) Temperature and humidity effects on the oviposition of *Hyalomma aegyptium* ticks of different engorgement weights. *Journal of Medical Entomology* 5, 429–439.
Tack, W., Madder, M., De Frenne, P., Vanhellemont, M., Gruwez, R., *et al.* (2011) The effects of sampling method and vegetation type on the estimated abundance of *Ixodes ricinus* ticks in forests. *Experimental and Applied Acarology* 54, 285–292.
Tajedin, L., Bakhshi, H., Faghihi, F. and Telmadarraiy, Z. (2016) High infection of *Anaplasma* and *Ehrlichia* spp. among tick species collected from different geographical locations of Iran. *Asian Pacific Journal of Tropical Disease* 6, 787–792.
Telmadarraiy, Z., Chinikar, S., Vatandoost, H., Faghihi, F. and Hosseini-Chegeni, A. (2015) Vectors of Crimean Congo hemorrhagic fever virus in Iran. *Journal of Arthropod-Borne Diseases* 9, 137–147.
Tirosh-Levy, S., Gottlieb, Y., Apanaskevich, D.A., Mumcuoglu, K.Y. and Steinman, A. (2018) Species distribution and seasonal dynamics of equine tick infestation in two Mediterranean climate niches in Israel. *Parasites & Vectors* 11, 546.
Yilmaz, G.R., Buzgan, T., Torunoglu, M.A., Safran, A., Irmak, H., *et al.* (2008) A preliminary report on Crimean–Congo haemorrhagic fever in Turkey, March–June 2008. *Eurosurveillance* 13, 18953.
Zaid, T., Ereqat, S., Nasereddin, A., Al-Jawabreh, A., Abdelkader, A., *et al.* (2019) Molecular characterization of *Anaplasma* and *Ehrlichia* in ixodid ticks and reservoir hosts from Palestine: a pilot survey. *Veterinary Medicine and Science* 5, 230–242.

eo69 The Emergence of Ticks and Tick-Borne Diseases in the USA

Charles Hart and Saravanan Thangamani*
SUNY Upstate Medical University, Syracuse, New York, USA

69.1 Expert Opinion

69.1.1 Introduction

Ticks are the primary source of vector-borne pathogens in the USA. The native tick species of the USA are adapted for a number of distinct ecosystems and mammalian hosts. As such, several overlapping sylvatic cycles exist within the tick populations. These occur in nature without human involvement. The incursion of humans into these environments can result in sylvatic spillover of these pathogens to humans, causing clinical disease. As tick populations shift, the risk of exposure to emerging, newly introduced and re-emerging pathogens increases, posing a direct threat to human and animal health. Several genera are the source of most tick-borne pathogens in the USA. These include *Dermacentor*, *Amblyomma* and *Ixodes*, with a lesser role for *Ornithodoros* soft ticks. This expert opinion focuses on the influence of the changing landscape of ticks in the USA and its impact on the emergence of tick-borne pathogens that occurs as a result.

69.1.2 *Dermacentor* spp. and transmitted pathogens

The Rocky Mountain wood tick (*Dermacentor andersoni*) is native to the western half of the USA, with its range localized around the Rocky Mountains (CDC, 2020f). It is the primary vector of *Rickettsia rickettsii*, the causative agent of Rocky Mountain spotted fever (RMSF). This pathogen causes severe febrile illness associated with a characteristic petechial rash; if untreated, the resulting loss of vascular integrity can lead to permanent limb damage or rapid death (Álvarez-López *et al.*, 2020). RMSF was reported in 5544 individuals in 2018, a threefold increase from 2009 and an 11-fold increase from 2000 (CDC, 2020h). The majority of these cases occur in the more north-western states and even more commonly in the American South, a region outside the range of *D. andersoni*, suggesting a secondary vector or pathogen. These ticks have also recently been associated with the Colorado tick fever reovirus (Williamson *et al.*, 2019). This virus produces biphasic febrile illness with early symptoms similar to RMSF. Although the illness may be severe, Colorado tick fever rarely results in death.

*Email: thangams@upstate.edu

DOI: 10.1079/9781789249637.0069

R. rickettsii may also be transmitted by *Dermacentor variabilis* (Kakumanu *et al.*, 2018), which is nearly ubiquitous in the region east of the Rocky Mountains and only slightly sympatric with *D. andersoni* (Dergousoff *et al.*, 2013). Since the majority of RMSF cases are diagnosed in the central-southern portions of the USA (CDC, 2020h), this tick most likely also plays a substantial role in transmission but only through a limited portion of its range. This tick's success relates to its environmental adaptability in lower-altitude environments, allowing it to exist in areas with higher human habitation than those endemic for *D. andersoni*.

An additional species of *Rickettsia* has been identified in *Dermacentor occidentalis* ticks on the Pacific Coast. These ticks are capable of transmitting a spotted fever *Rickettsia*, including *Rickettsia* 364D (Rodino *et al.*, 2020) and *Rickettsia philipii* (Rochlin and Toledo, 2020). The tick *Dermacentor albipictus* in the same region has also been associated with the apicomplexan parasite *Babesia duncani*. Together, these constitute new and emerging threats from less-often considered *Dermacentor* species.

Both *D. andersoni* and *D. variabilis* have been associated with the transmission of *Francisella tularensis*, the causative agent of tularaemia (CDC, 2020i). A total of 229 cases of tularaemia were diagnosed in 2018, with a yearly average of 195 from 2008 to 2018 (CDC, 2020j). These were detected throughout the USA but primarily occurred in a region centred in Kansas, Oklahoma, Missouri and Arkansas. In tick-derived cases, the disease manifests with a cutaneous condition that can progress to a life-threatening systematic infection if untreated. Assessing the actual rate of *F. tularensis* infection in ticks is difficult as *Dermacentor* ticks almost universally contain endosymbiotic, non-pathogenic *Francisella* (Lefcort *et al.*, 2020). Therefore, it remains unclear what proportion of *F. tularensis* cases in the USA result from tick bites versus non-tick sources.

69.1.3 *Amblyomma* spp. and transmitted pathogens

The primary *Amblyomma* species present in the USA is *Amblyomma americanum*, the lone star tick. This tick is common in the eastern USA and is generally associated with southern climates that receive infrequent frost. This species is aggressive, with both nymphs and adults attacking humans. It is a vector of several species of *Ehrlichia* and spotted fever group *Rickettsia* (Rodino *et al.*, 2020; Rochlin and Toledo, 2020; CDC, 2020i). The *Ehrlichia* spp. include *Ehrlichia ewingii*, which tends to infect granulocytes, and *Ehrlichia chaffeensis*, which infects monocytes and produces more severe disease (Rodino *et al.*, 2020). These result in febrile illnesses that may become lethal. *E. chaffeensis* was reported to cause 1799 cases in 2018, with a twofold increase since 2009 and a ninefold increase since 2000 (CDC, 2020b). *Rickettsia* may additionally cause leukopaenia and thrombocytopaenia; the resulting condition is similar to RMSF and may be partially responsible for the high incidence of spotted fever cases in the southern USA outside the range of *D. andersoni*. This has been attributed specifically to *Rickettsia parkeri* (Rodino *et al.*, 2020; Rochlin and Toledo, 2020).

A. americanum transmits Heartland virus, a phlebovirus that can cause multisystem organ failure (Fill *et al.*, 2017); it is also a suspected vector of Bourbon virus, a recently discovered thogotovirus with a similar course of illness (Kosoy *et al.*, 2015). *A. americanum* is also associated with southern tick-associated rash illness (STARI), a condition similar to Lyme disease with cutaneous manifestations and Lyme-like febrile illness; this has been tentatively attributed to the spirochaete *Borrelia lonestari* (Varela *et al.*, 2004). Like *Dermacentor*, *A. americanum* may transmit *F. tularensis* (CDC, 2020i). In areas with bobcats (*Lynx rufus*), it also serves as the vector of *Cytauxzoon felis*, a *Theileria*-like protozoan parasite that does not infect humans, but which causes fatal illness in domestic cats (Thomas *et al.*, 2018). Galactose-α-1,3-galactose (alpha-gal) allergy syndrome has also been associated with the bite of *A. americanum* (Diaz, 2020). This is a non-infectious condition that results in a semi-permanent and potentially life-threatening food allergy to red meat.

69.1.4 *Ixodes* spp. and transmitted pathogens

The most significant arthropod vector of the USA is the deer tick, *Ixodes scapularis*. This species' range is centred in the Northeast and Upper Midwest US regions but extends southward to the Gulf Coast (CDC, 2020f). These ticks are generalist feeders. Larvae feed primarily on rodents, while adults prefer large animals. Nymphs may feed on small, mid-sized or larger animals. Both nymphs and adults commonly bite humans. This means that nymphs may transmit pathogens acquired as larvae or feed on rodent pathogen reservoirs again for a second chance to become infected as adults.

I. scapularis transmits a variety of pathogens. The most common is the spirochaete *Borrelia burgdorferi*, which causes Lyme disease. Discovered in the 1970s, Lyme disease is a progressive illness that, if untreated in its early stages, can result in arthritis, carditis and neuroinvasive illness (Biesiada *et al.*, 2010). In 2018, 23,558 cases were reported (CDC, 2020c), with a higher number of undiagnosed cases likely existing. In addition to *B. burgdorferi*, the deer tick also transmits *Borrelia miyamotoi*, which results in disease more similar to relapsing fever (Cutler *et al.*, 2018), *Anaplasma phagocytophilum* and *Ehrlichia muris eauclairensis*. Infection with *A. phagocytophilum* and *E. muris eauclairensis* results in an acute febrile illness with thrombocytopaenia and leukopaenia (Johnson *et al.*, 2015; Baker *et al.*, 2020). While *E. muris eauclairensis* infection is rarely diagnosed, 4008 cases of anaplasmosis were reported in 2018 (CDC, 2020a), a threefold increase from 2009 and an 11-fold increase from 2000. *Babesia microti*, an apicomplexan parasite, is also transmitted by *I. scapularis*; it inhabits erythrocytes and results in generally self-resolving febrile illness accompanied by anaemia (Homer *et al.* 2000). This condition was reported for 2161 cases in 2018, a twofold increase since 2009 (CDC, 2020d). In immunocompromised and especially asplenic individuals, babesiosis can result in life-threatening parasitaemia. In addition to transmitting bacteria and protozoan parasites, *I. scapularis* also transmits Lineage II Powassan virus (deer tick virus, DTV) (Telford *et al.*, 1997), a version of the virus which utilizes *Peromyscus leucopus* as a reservoir instead of *Marmota monax* as with Lineage I Powassan virus (POWV). Infection with DTV leads to Powassan encephalitis, a rare but fatal condition. In 2019, 39 human cases were reported to CDC. However, this is a fourfold increase from eight human cases in 2010 (CDC, 2020e), indicating a steadily increasing number of yearly diagnoses.

I. scapularis is uniquely resistant to both warm and cold temperatures. The species has populations as far south as Florida, as west as Texas and as far north as Canada. Most of Lyme disease transmission and other *I. scapularis*-borne pathogens is focused on the northern portion of the population, specifically in the Northeast and the Upper Midwest. These areas have harsh, sub-freezing winters. During winter, the ticks enter dormancy in an unattached state, although they can temporarily become active whenever temperatures rise above freezing and snow cover is limited. As such, they are often active during winter if the temperature is not consistently low.

Other *scapularis*-like species exist throughout the range of *I. scapularis*. These morphologically similar ticks include several species, i.e. *Ixodes muris*, *Ixodes dentatus*, *Ixodes woodi* and others. Although these are closely related to *I. scapularis*, they are rarely studied directly. Their vectorial capacity for canonically *I. scapularis*-infecting pathogens and their native ranges are not well understood.

69.1.5 *Ornithodoros* spp.

Ornithodoros genus soft ticks are present throughout the USA and concentrated in more arid areas of the Southwest region (Donaldson *et al.*, 2015). They are capable of transmitting several forms of relapsing-fever *Borrelia* (*hermsii*, *turicate*, *parkeri*) depending on the tick species (CDC, 2020g). They may also possess the capacity to transmit African swine fever virus, a lethal pathogen of swine that has recently been introduced to much of Asia and Europe. Knowledge concerning the full vectorial capacity of American *Ornithodoros* with multiple pathogens is limited. Field studies are difficult due to the genus' close association with nests and burrows.

69.2 Future Predictions

The tick-borne pathogens of the USA have existed for extended periods of time in specific sylvatic cycles. Their emergence, therefore, is either the result of spillover from a natural sylvatic cycle or the introduction of a new pathogen in the existing systems. In either case, the risk of human exposure is increased by disruption of the ecosystem status quo. At present, several forces are driving changes in tick distribution and affecting the emergence of tick-borne diseases in the USA. Throughout North America, the distributions of medically consequential ticks are in flux. This is in response to land-use changes leading to reforestation and from weather effects of a changing climate. Although the temperatures of tick-endemic regions have changed only slightly, the alterations to storms and depths of winter have caused increased survival and expansion of several species formerly restricted to warmer environments. These changing distributions and increased tick numbers enhance the risk of sylvatic escape of tick-borne pathogens, thereby increasing the risk of human exposure to both well-known and emerging tick-borne diseases. What will eventually become of this changing system is unknown. It is unclear what the tick population of North America will look like when the situation finally reaches a new equilibrium. Whatever the result, the landscape of ticks and tick-borne diseases will be radically different in the near future. Since this will have far-reaching implications to human health, it is necessary to fully understand the new state of tick-borne pathogens to better inform future medical treatment and policy decisions in the face of expanding and emerging tick-borne pathogens.

69.2.1 Emergence potential of *Dermacentor* ticks

Dermacentor ticks already enjoy a consistent range throughout the USA. *D. andersoni* remains geographically restricted to the Rocky Mountains region, while the range of *D. variabilis* already covers the eastern portion of the country (CDC, 2020h). However, the density of *D. variabilis* may increase over time due to greater survival during milder winters. An increase in density can alter pathogens' sylvatic cycling by pushing the vector numbers above a critical threshold limit, thereby allowing for the maintenance and transmission of a pathogen. This may be part of the reason for increased RMSF infections in the range of *D. variabilis*, with increased populations being able to better maintain a sylvatic cycle of spotted fever group *Rickettsia*.

69.2.2 Northward expansion of *Amblyomma* ticks

Recently, *A. americanum* has undergone a dramatic migration northward (Sonenshine, 2018). This tick is associated with the American South, with consistent colonization in the northern states being a recent phenomenon. As of 2020, *A. americanum* is the dominant tick species of Long Island, New York, and has successfully colonized mainland New York and Connecticut. The amenability of *A. americanum* to northern climates is not unexpected. North America's continental centre generally has lower temperatures than the coasts, resulting in a curved band of minimum winter temperatures. Based on these minimums, the conditions of southern New York match those of Tennessee (USDA, 2020), a state well into *A. americanum*'s native range.

This is most likely not related directly to minimum temperatures, as the temperatures have not decreased as to exceed a clear survivability threshold. Instead, the driving force appears to be a trend toward overall milder winters. This has resulted in winters that include fewer sudden, precipitous drops in temperature and a lack of consistent freezing. More days with above-freezing temperatures increase the likelihood of tick survival. While intermittent *A. americanum* populations may have been established previously, their numbers were restricted by periodic harsh winters. Now, they are able to overwinter and establish extensive breeding populations. Further, the ticks are exposed to strong selective pressure to develop adaptations for greater frost tolerance. This expansion corresponds to the emergence of several diseases. The range of *E. chaffeensis* has been increased due to the new tick populations, spreading to northern regions where it was not normally endemic. Additionally, alpha-gal allergy syndrome occurrence has been observed in areas of recent introduction.

Despite this northern expansion, the southern range of *A. americanum* is also limited, although not by temperature. Instead, the populations are partially restricted by the invasive fire ant, *Solenopsis invicta*. This species of ant is a predator of ticks, with a preference for *A. americanum* over *Amblyomma maculatum* in sympatric areas (Kjeldgaard *et al.*, 2019). However, this effect has been disputed due to differences in habitat preference between the ants and *A. americanum* (Gleim *et al.*, 2013). In either event, the introduction and spread of the ant and the creation of the disturbed habitat it favours can affect the distribution of *A. americanum* in its southern range and the potential dominance of *A. maculatum* in near-tropical Gulf Coast environments.

While *A. americanum* enjoys an extensive distribution throughout the USA, its relatives, *A. maculatum* and *Amblyomma cajennense*, are restricted to the southernmost areas of the country where frost is rare or does not occur. As vectors, these serve in much the same capacity as *A. americanum*. *A. maculatum* is known to transmit *R. parkeri* (Fornadel *et al.*, 2011; CDC, 2020f), while the vectorial capacity of *A. cajennense* and other less-studied *Amblyomma* remains unclear.

In general, the range of *Amblyomma* ticks is anticipated to continue to increase with consistent northward spread. This will lead to *A. americanum* becoming a dominant tick species in the northern USA, but also indicates the potential for frost-intolerant *Amblyomma* species to undergo expansion or to emerge in North America. This will, in turn, correspond to an expansion of the range of *Amblyomma*-transmitted pathogens into areas where they were previously rare or absent.

69.2.3 Changes in range of *Ixodes* ticks

I. scapularis is already well adapted for northern environments. It is also tolerant of warm summers, making it perfectly suited for areas with both hot summers and cold winters, such as in the Northeast region. This also explains the broad range across multiple climatic conditions throughout the US East Coast. At present, *I. scapularis* is expanding into territory that it was not formerly known to inhabit, including many areas of the Northeast where it was never before observed. This corresponds with an increasing number of cases of *I. scapularis*-borne pathogens, such as Lyme disease and POWV, as well as the emergence of previously unknown ones like *B. miyamotoi*. However, this response is not driven primarily by temperature but rather by changes in land use throughout the endemic region (Sonenshine, 2018; Talbot *et al.*, 2019). Historically, much of New York and New England were deforested for agricultural development and lumber. In the last century, much of this area has undergone reforestation as agriculture has been reduced or abandoned completely. This has led to the re-establishment of the native habitat of *I. scapularis*, complete with an environment fully capable of supporting animals that serve as hosts for both the ticks and the pathogens they transmit.

Milder winters relate to *I. scapularis* as secondary factors. In the USA, the lack of severe winters results in *I. scapularis* that quest in the winter, during intermittent thaws. Increased primary productivity can also result in greater fruit and nuts production, thereby enhancing the rodent population when moist weather corresponds to longer growing seasons. Should both warmth and moisture be present, this may enhance the frequency of mast years in oaks and other nut-bearing trees, in turn resulting in extended 'boom periods' for small mammals and, thus, ticks; it additionally supplies a greater number of naïve animals for the sylvatic cycle of tick-borne pathogens.

Additional considerations involve polymicrobial transmission, where multiple pathogens are introduced by a single deer tick and may result in unique disease states, as well as other pathogens that exist within ticks but are undiscovered as of yet.

Several non-scapularis *Ixodes* species are also important but uncommon vectors. *Ixodes pacificus* serves as the primary vector of *B. burgdorferi* on the US West Coast. It may transmit other *I. scapularis* pathogens or related versions of them, although this is unclear. In the north-eastern part of the USA, *Ixodes cookei* exists sympatrically with *I. scapularis*. This is the southernmost range of *I. cookei*, which is generally associated with colder conditions north of the range of *I. scapularis* (Patterson *et al.*, 2017). This species, unlike *I. scapularis*, prefers mid-sized host animals and quests less vigorously. As such, it is less commonly encountered by humans. *I. cookei* is the vector of Lineage I POWV and a potential vector of *A. phagocytophilum*. It is a poor vector of *B. burgdorferi* and is not associated with Lyme disease transmission.

Like *I. scapularis*, *I. cookei* is advancing northward. Considering that *I. cookei* is sympatric with *I. scapularis* over its southern range, this may relate to the same reforestation effect as with *I. scapularis*. However, much of *I. cookei*'s range has historically not been as extensively cleared as more arable southern regions. Instead, the expansion appears caused by the lessening severity of winters in far northern, sub-arctic regions in Canada. However, this is difficult to ascertain as the northern range of *I. cookei* is not well documented. The southern edge of its range in Pennsylvania, New York and New England appears to be undergoing retraction. This is partially due to intolerance of warmer summer temperatures. Another proposed effect is the presence of *I. scapularis*, which is capable of forming sterile hybrids with *I. cookei* (Patterson *et al.*, 2017). This is detrimental to *I. cookei* populations in the presence of larger numbers of *I. scapularis*. Despite this, *I. cookei* is still present in New York and New England, serving as a vector of Lineage I POWV and possibly as a secondary vector of *A. phagocytophilum*.

Other species of non-scapularis *Ixodes* are thought to be expanding in parallel with *I. scapularis*, although this has not been assessed in detail.

69.2.4 Emergence potential of *Ornithodoros* ticks

The current range of *Ornithodoros* ticks is already poorly understood, specifically with regard to which biotic or abiotic factors influence their survival. Their range is heavily impacted by temperature, precipitation and available host population (Donaldson *et al.*, 2015). The greatest threat posed in the USA by this species of tick is the possible introduction of African swine fever. Several North American species have been identified as potentially competent vectors of the virus (Golnar *et al.*, 2019), meaning that local transmission may be possible. Considering the presence of feral hogs in the USA throughout much of the range of *Ornithodoros* ticks, there is a small but present risk of the formation of a persistent sylvatic cycle that could make the pathogen difficult to eradicate and contain.

69.2.5 New and emerging threats

An additional threat is the emergence of new tick species that did not previously exist in the USA. The introduction of new species is a well-understood threat in the USA and has been previously observed with *Haemaphysalis longicornis* and *Rhipicephalus microplus*. Other species may also pose a threat of introduction in the future. These can provide a set of established tick vectors that may later serve as a transmission source of introduced pathogens or transmit local pathogens with greater human and animal risk. *H. longicornis* is native to Asia and was introduced to the USA in 2017 or earlier, likely from imported livestock. The species quickly established wild populations, aided by its parthenogenic reproduction and tolerance for cold temperatures. As of 2020, *H. longicornis* is present in several states and is expanding its range rapidly, with a wide potential area of habitation (Raghavan *et al.*, 2018). The tick tends to eschew mice and small rodents that serve as reservoirs for many native pathogens, instead feeding on mid-sized animals (Tufts *et al.*, 2020). Additionally, as of now, *H. longicornis* has not been confirmed to transmit any native pathogens. However, *H. longicornis* is a vector of several pathogens in its native range, including the cattle apicomplexan, *Theileria orientalis* (Thompson *et al.*, 2020), which causes febrile disease with anaemia and has been identified recently in the USA. The Asian species is also a vector of severe fever with thrombocytopaenia syndrome virus (SFTSV), a virus which causes a febrile illness with potentially haemorrhagic thrombocytopaenia. Unlike *T. orientalis*, SFTSV has not been observed in *H. longicornis* present in North America. The exact nature of the potential sylvatic cycle of these pathogens in the USA remains unclear, although the tick is understood to be a keystone element to the process.

Due to its high rate of reproduction and the apparent amenability of North American habitats to its life cycle, *H. longicornis* will continue to breed and spread throughout the USA. This will lead to it becoming a common, ubiquitous species in many areas that are currently inhabited primarily by *I. scapularis* and *A. americanum*, resulting in a dramatic shift in the nature of the tick populations in those areas. It is therefore almost certain that the tick population encountered in various areas will drastically vary within the near future. Within a decade, the tick populations will be almost unrecognizable compared with what they are now, with a corresponding profound change in the ranges of the diseases they transmit.

References

Álvarez-López, D., Ochoa-Mora, E., Heitman, K.N., Binder, A.M., Alvarez, G. and Armstrong, P.A. (2020) Epidemiology and clinical features of Rocky Mountain spotted fever from enhanced surveillance, Sonora, Mexico: 2015–2018. *American Journal of Tropical Medicine and Hygiene* 104, 190–197. https://doi.org/10.4269/ajtmh.20-0854

Baker, A., Wang, H.H., Mogg, M., Derouen, Z., Borski, J. and Grant, W.E. (2020) Increasing incidence of anaplasmosis in the United States, 2012 through 2016. *Vector-Borne and Zoonotic Diseases* 20, 855–859. https://doi.org/10.1089/vbz.2019.2598

Biesiada, G., Czepiel, J., Lesniak, M.R., Garlicki, A. and Mach, T. (2010) Lyme disease: review. *Archives of Medical Science* 8, 978–982. https://doi.org/10.5114/aoms.2012.30948

CDC (2020a) Anaplasmosis: Epidemiology and Statistics. Centers for Disease Control and Prevention, Atlanta, Georgia. Available at: https://www.cdc.gov/anaplasmosis/stats/index.html (accessed 30 November 2020).

CDC (2020b) Ehrlichiosis: Epidemiology and Statistics, Centers for Disease Control and Prevention, Atlanta, Georgia. Available at: https://www.cdc.gov/ehrlichiosis/stats/index.html (accessed 30 November 2020).

CDC (2020c) Lyme Disease Data Tables: Historical Data. Centers for Disease Control and Prevention, Atlanta, Georgia. Available at: https://www.cdc.gov/lyme/stats/tables.html (accessed 30 November 2020).

CDC (2020d) Parasites – Babesiosis: Data & Statistics. Centers for Disease Control and Prevention, Atlanta, Georgia. Available at: https://www.cdc.gov/parasites/babesiosis/data-statistics/index.html (accessed 30 November 2020).

CDC (2020e) Powassan Virus: Statistics & Maps. Centers for Disease Control and Prevention, Atlanta, Georgia. Available at: https://www.cdc.gov/powassan/statistics.html (accessed 30 November 2020).

CDC (2020f) Ticks: Regions where ticks live. Centers for Disease Control and Prevention, Atlanta, Georgia. Available at: https://www.cdc.gov/ticks/geographic_distribution.html (accessed 25 November 2020).

CDC (2020g) Relapsing Fever: Transmission. Centers for Disease Control and Prevention, Atlanta, Georgia. Available at: https://www.cdc.gov/relapsing-fever/transmission/index.html (accessed 1 November 2020).

CDC (2020h) Rocky Mountain Spotted Fever (RMSF): Epidemiology and Statistics. Centers for Disease Control and Prevention, Atlanta, Georgia. Available at: https://www.cdc.gov/rmsf/stats/index.html (accessed 30 November 2020).

CDC (2020i) Tickborne Diseases of the United States: Tick ID. Centers for Disease Control and Prevention, Atlanta, Georgia. Available at: https://www.cdc.gov/ticks/tickbornediseases/tickID.html (accessed 25 November 2020).

CDC (2020j) Tularemia: Statistics. Centers for Disease Control and Prevention, Atlanta, Georgia. Available at: https://www.cdc.gov/tularemia/statistics/index.html (accessed 30 November 2020).

Cutler, S., Vayssier-Taussat, M., Estrada-Peña, A., Potkonjak, A., Mihalca, A.D. and Zeller, H. (2018) A new *Borrelia* on the block: *Borrelia miyamotoi* – a human health risk? *Eurosurveillance* 24, 1800170. https://doi.org/10.2807/1560-7917.ES.2019.24.18.1800170

Dergousoff, S.J., Galloway, T.D., Lindsay, L.R., Curry, P.S. and Chilton, N.B. (2013) Range expansion of *Dermacentor variabilis* and *Dermacentor andersoni* (Acari: Ixodidae) near their northern distributional limits. *Journal of Medical Entomology* 50, 510–520. https://doi.org/10.1603/ME12193

Diaz, J.H. (2020) Red meat allergies after lone star tick (*Amblyomma americanum*) bites. *Southern Medical Journal* 113, 267–274. https://doi.org/10.14423/SMJ.0000000000001102

Donaldson, T.G., Perez de Leon, A.A., Li, A.I., Castro-Areliano, I., Wozniak, E., *et al.* (2015) Assessment of the geographic distribution of *Ornithodoros* turicata (Argasidae): climate variation and host diversity. *PLoS Neglected Tropical Diseases* 10, e0004383. https://doi.org/10.1371/journal.pntd.0004383

Fill, M.M.A., Compton, M.L., McDonald, E.C., Moncayo, A.C., Dunn, J.R., *et al.* (2017) Novel clinical and pathologic findings in a Heartland virus-associated death. *Clinical and Infectious Diseases* 64, 510–512. https://doi.org/10.1093/cid/ciw766

Fornadel, C.M., Zhang, X., Smith, J.D., Paddock, C.D., Arias, J.R. and Norris, D.E. (2011) High rates of *Rickettsia parkeri* infection in Gulf Coast ticks (*Amblyomma maculatum*) and identification of '*Candidatus* Rickettsia Andeannae' from Fairfax County, Virginia. *Vector-Borne and Zoonotic Diseases* 11, 1535–1539. https://doi.org/10.1089/vbz.2011.0654

Gleim, E.R., Conner, L.M. and Yabsley, M.J. (2013) The effects of *Solenopsis invicta* (Hymenoptera: Formicidae) and burned habitat on the survival of *Amblyomma americanum* (Acari: Ixodidae) and *Amblyomma maculatum* (Acari: Ixodidae). *Journal of Medical Entomology* 50, 270–276. https://doi.org/10.1603/ME12168

Golnar, A.J., Martin, E., Wormington, J.D., Kading, R.C., Teel, P.D., *et al.* (2019) Reviewing the potential vectors and hosts of African swine fever virus transmission in the United States. *Vector-Borne and Zoonotic Diseases* 19, 512–524. https://doi.org/10.1089/vbz.2018.2387

Homer, M.J., Aguilar-Delfin, I., Telford, S.R. III, Krause, P.J. and Persing, D.H. (2000) Babesiosis. *Clinical Microbiology Review* 13, 451–469. https://doi.org/10.1128/CMR.13.3.451

Johnson, D.K.H., Schiffman, E.K., Davis, J.P., Neitzel, D.F., Sloan, L.M., *et al.* (2015) Human infection with *Ehrlichia muris*-like pathogen, United States, 2007–2013. *Emerging Infectious Diseases* 21, 1794–1799. https://doi.org/10.3201/eid2110.150143

Kakumanu, M.L., Ponnusamy, L., Sutton, H., Meshnick, S.R., Nicholson, W.L. and Apperson, C.S. (2018) Prevalence of *Rickettsia* species (Rickettsiales: Rickettsiaceae) in *Dermacentor variabilis* ticks (Acari: Ixodidae) in North Carolina. *Journal of Medical Entomology* 55, 1284–1291. https://doi.org/10.1093/jme/tjy074

Kjeldgaard, M.K., Takano, O.M., Bockhoven, A.A., Teel, P.D., Light, J.E., *et al.* (2019) Red imported fire ant (*Solenopsis invicta*) aggression influences the behavior of three hard tick species. *Experimental and Applied Acarology* 79, 87–97. https://doi.org/10.1007/s10493-019-00419-8

Kosoy, O.I., Lambert, A.J., Hawkinson, D.J., Pastula, D.M., Goldsmith, C.S., *et al.* (2015) Novel thogotovirus associated with febrile illness and death, United States, 2014. *Emerging Infectious Diseases* 21, 760–764. https://doi.org/10.3201/eid2105.150150

Lefcort, H., Tsybulnik, D.Y., Browning, R.J., Eagle, H.P., Eggleston, T.E., *et al.* (2020) Behavioral characteristics and endosymbionts of two potential tularemia and Rocky Mountain spotted fever tick vectors. *Journal of Vector Ecology* 45, 321–332. https://doi.org/10.1111/jvec.12403

Patterson, J.W., Duncan, A.M., McIntyre, K.C., Lloyd, V.K. (2017) Evidence for genetic hybridization between *Ixodes scapularis* and *Ixodes cookei*. *Canadian Journal of Zoology* 95, 527–537. https://doi.org/10.1139/cjz-2016-0134

Raghavan, R.K., Barker, S.C., Cobos, M.E., Barker, D., Teo, E.J.M., *et al.* (2018) Potential spatial distribution of the newly introduced long-horned tick, *Haemaphysalis longicornis* in North America. *Scientific Reports* 9, 498. https://doi.org/10.1038/s41598-018-37205-2

Rochlin, I. and Toledo, A. (2020) Emerging tick-borne pathogens of public health importance: a mini-review. *Journal of Medical Microbiology* 69, 781–791. https://doi.org/10.1099/jmm.0.001206

Rodino, K.G., Theel, E.S. and Pritt, B.S. (2020) Tick-borne diseases in the United States. *Clinical Chemistry* 66, 537–548. https://doi.org/10.1093/clinchem/hvaa040

Sonenshine, D.E. (2018) Range expansion of tick disease vectors in North America: implications for the spread of tick-borne disease. *International Journal of Environmental Research and Public Health* 15, 478. https://doi.org/10.3390/ijerph15030478

Talbot, B., Slatculescu, A., Thickstun, C.R., Koffi, J.K., Leighton, P.A., *et al.* (2019) Landscape determinants of density of blacklegged ticks, vectors of Lyme disease, at the northern edge of their distribution in Canada. *Scientific Reports* 9, 16652. https://doi.org/10.1038/s41598-019-50858-x

Telford, S.R. III, Armstrong, P.M., Katavolos, P., Foppa, I., Garcia, A.S., *et al.* (1997) A new tick-borne encephalitis-like virus infecting New England deer ticks, *Ixodes dammini*. *Emerging Infectious Diseases* 3, 165–170.

Thomas, J.E., Ohmes, C.M., Payton, M.E., Hostetler, J.A. and Reichard, M.V. (2018) Minimum transmission time of *Cytauxzoon felis* by *Amblyomma americanum* to domestic cats in relation to duration of infestation, and investigation of ingestion of infected ticks as a potential route of transmission. *Journal of Feline Medicine and Surgery* 20, 67–72. https://doi.org/10.1177/1098612X17691172

Thompson, A.T., White, S., Shaw, D., Egizi, A., Lahmers, K., *et al.* (2020) *Theileria orientalis* Ikeda in host-seeking *Haemaphysalis longicornis* in Virginia, USA. *Ticks and Tick-borne Diseases* 11, 101450. https://doi.org/10.1016/j.ttbdis.2020.101450

Tufts, D.M., Goodman, L.B., Benedict, M.C., Davis, A.D., VanAcker, M.C., and Diuk-Wasser, M. (2020) Association of the invasive *Haemaphysalis longicornis* tick with vertebrate hosts, other native tick vectors, and tick-borne pathogens in New York City, USA. *International Journal for Parasitology* 51, 149–157. https://doi.org/10.1016/j.ijpara.2020.08.008

USDA (2020) *Plant Hardiness Zones*. US Department of Agriculture, US Forest Service, Washington, DC. Available at: https://www.fs.fed.us/wildflowers/Native_Plant_Materials/Native_Gardening/hardinesszones.shtml (accessed 25 November 2020).

Varela, A.S., Luttrell, M.P., Howerth, E.W., Moore, V.A., Davison, W.R., *et al.* (2004) First culture isolation of *Borrelia lonestari*, putative agent of southern tick-associated rash illness. *Journal of Clinical Microbiology* 42, 1163–1169. https://doi.org/10.1128/JCM.42.3.1163-1169.2004

Williamson, B.N., Fischer, R.J., Lopez, J.E., Ebihara, H. and Schwan, T.G. (2019) Prevalence and strains of Colorado tick fever virus in Rocky Mountain wood ticks in the Bitterroot Valley, Montana. *Vector-Borne and Zoonotic Diseases* 19, 694–702. https://doi.org/10.1089/vbz.2018.2407

eo70 Role of Climate and Other Factors in Determining the Dynamics of Tick and Tick-Transmitted Pathogen Populations and Distribution in Western, Central and Eastern Africa

Naftaly Githaka[1]*, Esther Kanduma[2] and Richard Bishop[3]
[1]International Livestock Research Institute (ILRI), Nairobi, Kenya; [2]University of Nairobi, Nairobi, Kenya; [3]Washington State University, Pullman, Washington, USA

70.1 Expert Opinion

70.1.1 General perspectives

The African continent ranks highly among regions of the world that are vulnerable to the effects of climate change resulting in shifts in the range and distribution of vector-borne infectious diseases. In tropical Africa, the epidemiology of tick-borne diseases is complex, often characterized by co-infestation of animal hosts by multiple tick species and typically resulting in co-infection of livestock and wildlife with multiple tick-borne pathogens. This results in variability in disease outcomes at both the individual animal and population levels. Moreover, animal production systems across the continent vary greatly: from intensive small-scale zero-grazing in the cooler highlands, to extensive pastoral systems in the arid and semi-arid rangelands, to hybrid agropastoral systems in areas where ecological parameters, particularly rainfall, are intermediate. Such heterogeneity makes it difficult to make generic inferences regarding the potential impact of climate change on vector-borne diseases of livestock. A combination of anthropogenic factors such as livestock movement, and ecological factors including predation and competition, in addition to climatic changes such as the extent and distribution of rainfall, are key drivers of the geographical distribution of ticks and the pathogens that they transmit (Fig. 70.1). There is evidence that certain species of ticks, e.g. *Rhipicephalus microplus*, the major vector of *Babesia bovis* and *Anaplasma marginale*, are expanding their geographical ranges very rapidly within the continent, increasing the socio-economic and health burden of tick-borne diseases of animals. The current lack of modelling data and research capacity in this discipline within Africa represents a severe impediment to accurate prediction of the extent to which climate change will impact the future distribution of ticks and tick-borne diseases in the region.

Climate can affect the transmission dynamics, geographical spread, emergence and

*Email: n.githaka@cgiar.org

Climate, Ticks and Disease (ed. P. Nuttall)
DOI: 10.1079/9781789249637.0070

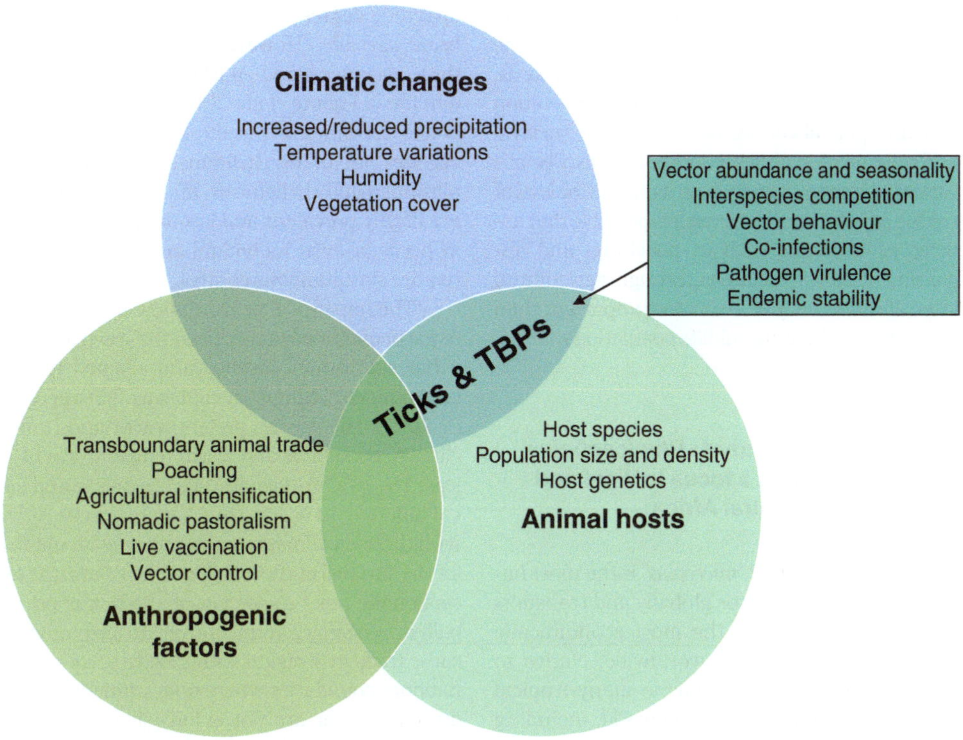

Fig. 70.1. Climatic and non-climatic drivers of tick-borne pathogens (TBPs) in tropical Africa. The effects of climate changes on the epidemiology of TBPs are augmented by human and animal factors that are largely unique to the African region, such as uncontrolled cross-border livestock trade and displacement of wildlife by human settlements. The extent of this interplay between climate and other factors is heterogeneous and, consequently, the outcomes are likely context specific. (Authors' own figure.)

re-emergence of human and animal vector-borne diseases through multiple pathways, including direct effects on vector abundance and seasonality, vector–pathogen interactions and host parameters, such as population density and diversity. Climate has direct effects on the biological development and adaptation of individual species and the pathogens they transmit, defining their prevalence in different agroecological zones and the distribution of animal and human hosts. Climate change has been shown to affect ecological niches and ecosystem parameters in which vectors and their hosts occur at different spatial ranges, influencing survival, adaptation, transmission and disease dynamics.

For tick vectors, rainfall and humidity are important climatic factors influencing abundance. Increased rainfall results in improved pasture and vegetation cover which increases the number of potential vector breeding sites. Mammalian hosts also occur at higher density, due to increased availability of forage and pasture. These higher densities in turn create the potential for enhanced pathogen transmission. Habitat degradation driven by climate change can also reduce habitat quality, resulting in reduction of vector populations and pathogen transmission. Climate change can therefore have variable impacts on vectors and tick-borne diseases depending on the specific environmental context.

A high incidence of important tick-borne diseases such as East Coast fever, babesiosis and anaplasmosis is recorded following rainy seasons when vector abundance is high. Mammalian host density is also influenced by climate through abundance of food and water sources making habitats more suitable for tick growth and survival and causing increased disease transmission. African wildlife, particularly the Cape buffalo (*Syncerus caffer*), constitute reservoirs of both tick vectors and tick-transmitted pathogens. Decreased wildlife populations and altered migration patterns, caused by factors such as decreased vegetation and drought, affect vector abundance, survival and transmission of tick-borne pathogens. Climatic variability is one parameter contributing to the level of wildlife populations. However, it is probable that anthropogenic factors such as poaching, and displacement of wildlife areas through agricultural intensification, may be even more important than climate in modulating wildlife populations.

70.1.2 Case study 1: *Rhipicephalus microplus* with a focus in West and Central Africa

The Asian blue tick, *R. microplus*, is the most important arthropod vector globally and transmits *B. bovis*, which causes the most economically important cattle disease worldwide. Native to Asia, the tick is now endemic in many tropical and subtropical regions of the world, including Australia and large parts of Latin America, decreasing the profitability of beef and dairy production. Until recently, *R. microplus* was relatively rare in Africa, with the exception of South Africa, and a limited focus at the East African coast and the Tanzanian islands of Pemba and Zanzibar (Walker *et al.*, 2003). However, the situation has changed dramatically following rapid increases of the tick in Tanzania replacing the indigenous *Rhipicephalus decoloratus* (Lynen *et al.*, 2008) and two introductions into different countries in West Africa through the importation of infested cattle from Brazil (Madder *et al.*, 2007, 2012). Although the economic and disease burden resulting from this dispersal is yet to be determined, there is already evidence (from application of the reverse line blot technique) for the presence of *B. bovis* in cattle in Nigeria (Kamani *et al.*, 2017). The impact of *R. microplus* introduction is predicted to be severe, due to the propensity of this tick species to rapidly acquire resistance against commercially available acaricides and to achieve high densities in susceptible animal hosts (Burrow *et al.*, 2019). In Australia and South America where *R. microplus* complex is the only tick species commonly found on cattle, anti-tick vaccines have been used with some apparent degree of efficacy, although this has been variable. However, the effectiveness of these vaccines based on BM86 'hidden' gut protein (de la Fuente *et al.*, 2007) in tropical Africa is less certain since multiple major tick-borne diseases, including theileriosis, babesiosis, anaplasmosis and ehrlichiosis, and their tick vectors are highly prevalent and frequently overlapping in herds or even individual animals, and these require simultaneous control.

The spread of *R. microplus* in Central Africa is of particular importance given the speed at which it has successfully invaded and adapted to new geographical regions since it was first reported in 2007. The species prefers warm and humid climatic conditions and warm temperatures of at least 15–20°C are required for egg laying and larval hatching to occur. Thus, any region in Africa with these ecoclimatic conditions may be suitable for the survival of the tick (Fig. 70.2). First, the exceptionally rapid dispersion of the tick is principally driven by largely uncontrolled transboundary cattle trade in a region with mostly porous international boundaries where quarantine measures do not exist or are not enforceable. As climate change accelerates, this trade will intensify and disseminate the tick deep into new areas that are ecologically suitable. Second, seasonal variation in precipitation is a major driver of nomadic pastoralism in countries such as Chad that border countries where *R. microplus* has become established (Silatsa *et al.*, 2019). Civil unrest in the Central African Republic, Cameroon and the Sudan have the potential to further catalyse the spread of animal diseases, including those transmitted by *R. microplus*. Moreover, we have recently shown that a specific genotype of *R. microplus* occurs in all countries where the tick has been reported in Africa (Kanduma *et al.*, 2020). The role of climatic changes and other factors in the selection of highly adaptable tick lineages, such as this one, remain to be determined.

70.1.3 Case study 2: northern expansion of *Rhipicephalus appendiculatus* into South Sudan through anthropogenic cattle movement

For most important African tick vectors that have been studied, the geographical range is

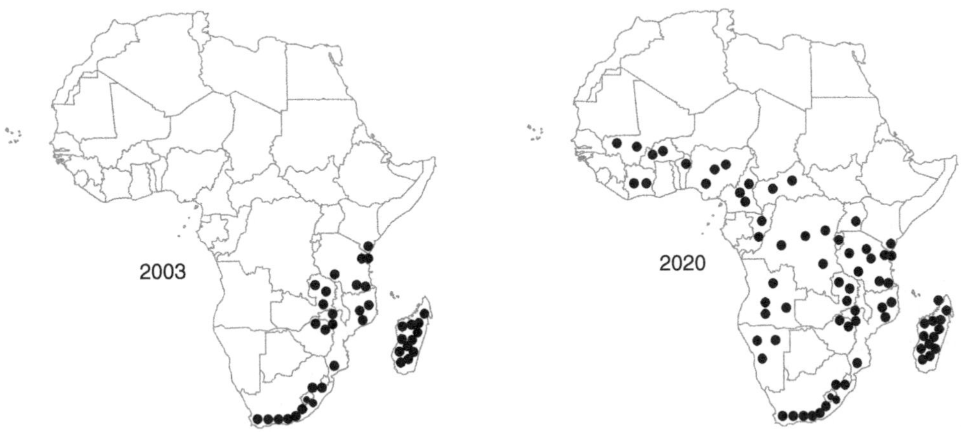

Fig. 70.2. Shift in the distribution of *Rhipicephalus microplus* across Africa over the past 17 years. The establishment of *R. microplus* into new foci follows the patterns of local climatic conditions, with prevalence varying across different agroecological zones within individual countries. There is evidence that the encroachment of *R. microplus* into new areas is influencing the distribution of native tick species through e.g. mate competition, requiring review of existing vector distribution maps. (Authors' own figure.)

influenced by ambient temperatures. For instance, *Rhipicephalus appendiculatus* is known to be limited to relatively higher, cooler and wetter regions of sub-Saharan Africa. For the last 20 years, which have seen an increase in global temperatures, *R. appendiculatus* has been predicted to expand its geographical range into new areas of Central Africa that now have suitable precipitation and temperature conditions to support tick reproduction and survival (Olwoch *et al.*, 2007). Adaptation and establishment of the tick to new suitable habitats will drive further expansion into new areas with environmental conditions suitable for tick survival. Therefore, as global temperatures continue to rise, the tick may spread to higher latitudes and altitudes. However, it is interesting to note that, by contrast to predictions of expansion, in some areas that are apparently ecologically suitable for *R. appendiculatus*, e.g. Murchison Falls and Kidepo Valley National Parks in Northern Uganda, the tick is surprisingly not present (Obara *et al.*, 2020), and this correlates with the absence of *Theileria parva* from Cape buffalo in these areas (Oura *et al.*, 2011). It was speculated that this might be due to biotic factors such as competition with other tick species or predation (Obara *et al.*, 2020).

Extensive anthropogenically mediated movement of cattle is known to occur between Northern Uganda and South Sudan and, as a consequence, both *R. appendiculatus* and *T. parva* have been detected over extensive regions of South Sudan (Salih *et al.*, 2018; Marcellino *et al.*, 2012, 2017). As a result, East Coast fever is an emerging problem in that country and the long-term impact may be considerable.

70.2 Future Predictions

One interesting issue which may be worthy of investigation is why *R. appendiculatus* and *T. parva* do not occur in West Africa when there appears to be a corridor of climatically and ecologically suitable subhumid habitat stretching from north-west Uganda through the Central African Republic and northern Democratic Republic of the Congo to Cameroon and Nigeria. In addition, the indigenous mammalian wildlife host, Cape buffalo, is present in these countries, although a different subspecies may be involved. Possibly biotic competition could be a factor. Some parts of south-west Ethiopia also appear ecologically suitable for *R. appendiculatus* and this region may be worthy of investigation for the presence of both the tick and East Coast fever.

The impact of the displacement of indigenous livestock breeds through increased interbreeding with exotic improved *Bos taurus* breeds may have a significant effect on promoting the spread of ticks and the pathogens they transmit. Such cross-bred cattle may exhibit decreased resistance to infestation by ticks and infection by tick-borne pathogens, thus resulting in an increased prevalence of tick-borne diseases. This suggestion is supported by the observation that animals are more susceptible to tick infestation and parasitic diseases during periods of nutritional restriction, which will become more frequent due to climate changes, particularly a higher frequency of droughts. Moreover, future climatic variability may amplify anthropogenic factors such as transboundary cattle movement for trade or to find grazing. Furthermore, the increased use of vaccines based on live attenuated parasites, or in the case of East Coast fever, the infection and treatment procedure (reviewed by Bishop *et al.*, 2020), may result in induction of tick-transmissible carrier states in vaccinated animals. When these vaccines are applied at the boundary of the existing distribution of the parasite, this may in turn lead to the emergence and establishment of tick-borne diseases in areas where they were previously absent.

References

Bishop, R.P., Odongo, D., Ahmed, J., Mwamuye, M., Fry, L.M., *et al.* (2020) A review of recent research on *Theileria parva*: implications for the infection and treatment vaccination method for control of East Coast fever. *Transboundary and Emerging Diseases* 67(Suppl. 1), 56–67.

Burrow, H.M., Mans, B.J., Cardoso, F.F., Birkett, M.A., Kotze, A.C., *et al.* (2019) Towards a new phenotype for tick resistance in beef and dairy cattle: a review. *Animal Production Science* 59, 1401–1427.

de la Fuente, J., Almazán, C., Canales, M., Pérez de la Lastra, J., Kocan, K., *et al.* (2007) A ten-year review of commercial vaccine performance for control of tick infestations on cattle. *Animal Health Research Reviews* 8, 23–28.

Kamani, J., Apanaskevich, D., Gutiérrez, R., Nachum-Biala, Y., Baneth, G., *et al.* (2017) Morphological and molecular identification of *Rhipicephalus* (*Boophilus*) *microplus* in Nigeria, West Africa: a threat to livestock health. *Experimental and Applied Acarology* 73, 283–296.

Kanduma, E.G., Emery, D., Githaka, N.W., Nguu, E.K., Bishop, R.P. and Šlapeta, J. (2020) Molecular evidence confirms occurrence of *Rhipicephalus microplus* Clade A in Kenya and sub-Saharan Africa. *Parasites & Vectors* 13, 432.

Lynen, G., Zeman, P., Bakuname, C., Di Giulio, G., Mtui, P., *et al.* (2008) Shifts in the distributional ranges of *Boophilus* ticks in Tanzania: evidence that a parapatric boundary between *Boophilus microplus* and *B. decoloratus* follows climate gradients. *Experimental and Applied Acarology* 44, 147–164.

Madder, M., Thys, E., Geysen, D., Baudoux, C. and Horak, I. (2007) *Boophilus microplus* ticks found in West Africa. *Experimental and Applied Acarology* 43, 233–234.

Madder, M., Adehan, S., De Deken, R., Adehan, R. and Lokossou, R. (2012) New foci of *Rhipicephalus microplus* in West Africa. *Experimental and Applied Acarology* 56, 385–390.

Marcellino, W., Salih, D., Julla, I. and El Hussein, A. (2012) Sieroprevalenza di febbre della costa orientale (theileriosi) nel Central Equatoria State, Sud Sudan. *Veterinaria Italiana* 48, 379–385. Available at: https://pubmed.ncbi.nlm.nih.gov/23277119/ (accessed 14 January 2021).

Marcellino, W.L., Salih, D., Njahira, M., Ndiwa, N., Araba, A., *et al.* (2017) The emergence of *Theileria parva* in Jonglei State, South Sudan: confirmation using molecular and serological diagnostic tools. *Transboundary and Emerging Diseases* 64, 1229–1235.

Obara, I., Githaka, N., Nijhof, A., Krücken, J., Nanteza, A., *et al.* (2020) The *Rhipicephalus appendiculatus* tick vector of *Theileria parva* is absent from Cape buffalo (*Syncerus caffer*) populations and associated ecosystems in northern Uganda. *Parasitology Research* 119, 2363–2367.

Olwoch, J., Van Jaarsveld, A., Scholtz, C. and Horak, I. (2007) Climate change and the genus *Rhipicephalus* (Acari: Ixodidae) in Africa. *Onderstepoort Journal of Veterinary Research* 74, 45–72.

Oura, C., Tait, A., Asiimwe, B., Lubega, G. and Weir, W. (2011) Haemoparasite prevalence and *Theileria parva* strain diversity in Cape buffalo (*Syncerus caffer*) in Uganda. *Veterinary Parasitology* 175, 212–219.

Salih, D., Mwacharo, J., Pelle, R., Njahira, M., Odongo, D., *et al.* (2018) Genetic diversity and population structure of *Theileria parva* in South Sudan. *Ticks and Tick-borne Diseases* 9, 806–813.
Silatsa, B.A., Simo, G., Githaka, N., Mwaura, S., Kamga, R.M., *et al.* (2019) A comprehensive survey of the prevalence and spatial distribution of ticks infesting cattle in different agro-ecological zones of Cameroon. *Parasites & Vectors* 12, 489.
Walker, A., Bouattour, A., Camicas, J., Estrada-Peña, A., Horak, I., *et al.* (2003) *Ticks of Domestic Animals in Africa: A Guide to Identification of Species*. Bioscience Reports, Edinburgh.

eo71 Ticks and Tick-Borne Pathogens in China

Hao Li, Li-Qun Fang and Wei Liu*
Beijing Institute of Microbiology and Epidemiology, Beijing, People's Republic of China

71.1 Expert Opinion

71.1.1 Introduction

Tick-borne pathogens have emerged worldwide in recent decades and become a great threat to human health. China has made great progress in the control and prevention of infectious diseases; however, emerging infectious diseases are the new challenge now facing China. Although outbreaks of severe acute respiratory syndrome and H7N9 avian influenza virus infection have attracted great attention, emerging tick-borne diseases are generally neglected by primary healthcare providers and clinicians. In China, more than 120 tick species have been documented, including over 100 species in the family Ixodidae (hard ticks) and 19 species in the family Argasidae (soft ticks); 30 tick species are reported to feed on human beings (Zhang *et al.*, 2019). Due to the rich and diverse natural habitat types, as well as the preference of different tick species for distinct biotopes or environments, the geographic distribution of ticks and the area of risk for human tick-borne infections are various in China. In the past three decades, an increasing number of tick-borne pathogens, including species of *Anaplasma*, *Babesia*, spotted fever group rickettsiae (SFGR), *Borrelia* and viruses, have been reported in China (Fang *et al.*, 2015). This expert opinion provides an overview of the type and distribution of tick species and emerging tick-borne pathogens in China and considers the potential influence of climate change.

71.1.2 Ticks and tick-borne infections

Ticks and their distributions

In total, 113 hard tick species in seven genera and 11 soft tick species in two genera have been recorded in China, together with 103 tick-associated agents detected in either ticks or humans (Zhao *et al.*, 2021). Four genera, namely *Dermacentor*, *Haemaphysalis*, *Ixodes* and *Rhipicephalus*, are reported to be the most widely distributed in more than 400 counties of China, followed by *Hyalomma*, *Argas*, *Ornithodoros*, *Amblyomma* and *Anomalohimalaya* found in 37 to 298 counties. At the species level, *Dermacentor nuttalli*, *Haemaphysalis longicornis*, *Dermacentor silvarum*, *Hyalomma scupense* and *Rhipicephalus sanguineus* are the predominant species identified in China with each located in more than 200 counties, followed by *Rhipicephalus microplus*, *Ixodes persulcatus*, *Ixodes sinensis*, *Ixodes granulatus* and *Hyalomma*

*Email: liuwei@bmi.ac.cn

DOI: 10.1079/9781789249637.0071

asiaticum, which are locally predominant species with each detected in 100–200 counties.

Geographical distributions differ among species, even for those in the same genus. *D. nuttalli* and *D. silvarum* are reported in the vast region of north-eastern and north-western China, where there are middle to high elevations, shrub grassland, strong seasonality in temperature, relatively low temperatures and low precipitation in the driest month. *H. longicornis*, *Hy. scupense* and *R. sanguineus* are mainly found in landscapes of shrub grassland and irrigated or rain-fed croplands at low to middle elevations (<1600 m) in central, eastern and north-western China. *R. microplus*, *Rhipicephalus haemaphysaloides*, *I. granulatus* and *Haemaphysalis hystricis* are mainly distributed in coniferous or broad-leaf woodland at low elevations in southern and central China, where the weather is warm and humid with low seasonality in temperature. *I. persulcatus*, *Haemaphysalis concinna* and *Haemaphysalis japonica* are mainly distributed in coniferous or broad-leaf forests as well as cropland in north-western and north-eastern China, featuring strong seasonality in temperature and low temperatures in the driest season. For *Hy. asiaticum*, *Dermacentor marginatus*, *Dermacentor daghestanicus* and *Argas persicus*, the natural habitats are meadow, desert grassland and cropland in northern and north-western China, featuring low precipitation in the wettest/warmest quarter or month of the year.

Tick-borne infections

A total of 103 tick-borne agents have been detected in China, of which 65 were newly identified in the past two decades. *H. longicornis* is the tick species harbouring the greatest variety of tick-borne agents, with as many as 44 known species including seven *Rickettsia* species, seven *Babesia*, 12 Anaplasmataceae, four *Theileria*, four *Borrelia*, nine viruses and *Francisella tularensis*. Other competent tick species that carry 20 or more agents are *I. persulcatus* (36 agents), *D. nuttalli* (32 agents), *R. microplus* (31), *D. silvarum* (30), *H. concinna* (24) and *Hy. asiaticum* (23). Agents that parasitize more than ten tick species are *Rickettsia raoultii* (in 15 tick species), *Rickettsia heilongjiangensis* (14), *Anaplasma phagocytophilum* (22), *Ehrlichia chaffeensis* (16), *Anaplasma bovis* (10), *Borrelia burgdorferi* sensu stricto (s.s.) (20), *Borrelia garinii* (18), *Borrelia afzelii* (11), *Coxiella burnetii* (14), Jingmen tick virus (12) and *Theileria annulata* (11). From 1980 to the end of 2018, a total of 2786 human cases had been confirmed for infections with *Borrelia* (five species and uncharacterized species) and 415, 215 and 129 cases were reported to be infected with Anaplasmataceae (four species), *Babesia* spp. (five species and uncharacterized species) and SFGR (six specific species and uncharacterized species), respectively. In addition, 216 human cases were infected with other bacteria (three species), including 120 with *F. tularesis*, 95 with *C. burnetii* and one with *Colpodella* spp. *annulata* (Zhao *et al.*, 2021).

Altogether, 19 tick-borne viruses have been identified in China, six of which were associated with human patients. Among them, severe fever with thrombocytopaenia syndrome bunyavirus (SFTSV) and tick-borne encephalitis virus (TBEV) are responsible for an enormous disease burden unparalleled by any other tick-borne pathogens. In addition, hundreds of human cases infected by Crimean–Congo haemorrhagic fever virus (CCHFV) have been reported in Xinjiang, with a few reported in Yunnan Province. Jingmen tick virus (JMTV) was first found in 2010 and is distributed in central, eastern and north-eastern China, with 12 human cases reported in Heilongjiang Province. Another recently discovered agent, Alongshan virus (ALSV), was detected in ticks in the northern tips of Inner Mongolia and Heilongjiang, as well as in human cases across the two provinces (Wang *et al.*, 2019).

Tick-borne diseases and climate change

The effect of worldwide climate change on the emergence of most tick-borne diseases and other vector-borne diseases is thought to be less important than changes in land use, animal host communities, human living conditions and societal factors (Kilpatrick and Randolph, 2012; Fang *et al.*, 2015; Liu *et al.*, 2017). Although the effects of climate on transmission of infectious diseases are thought to be non-linear and act in opposing directions in different climate regions, the seasonal dynamics of tick vectors are largely defined by climate conditions, which might further affect the seasonal pattern of tick-borne diseases. Climate change, such as increased temperature, promotes the development and reproduction of ticks, and also increases host abundance, which may further increase tick abundance and expand the spatial scope of their habitat (Gilbert, 2010). An increase in winter temperatures is known to cause the northward extension and increased abundance of *Ixodes ricinus*, which subsequently raises the risk of tick-borne disease dissemination (Parola, 2004; Harrus and Baneth, 2005; Altizer *et al.*, 2013). Although this topic is still under debate, climatic change could have a role in the emergence of tick-borne diseases in China. Further studies are needed to better elucidate this issue.

71.1.3 Emerging tick-borne pathogens in tick vectors and human beings

Anaplasma

Anaplasma species are zoonotic pathogens with tick vectors and mammalian reservoir hosts. Seven species in the genus *Anaplasma* have been identified in China, including *A. phagocytophilum, Anaplasma ovis, Anaplasma capra, A. bovis, Anaplasma centrale, Anaplasma platys* and *Anaplasma marginale*, of which the former three species have been identified as human pathogens. *A. phagocytophilum*, first confirmed as the causative agent of human granulocytic anaplasmosis (HGA) in the USA in 1994 (Chen *et al.*, 1994), is mainly responsible for human anaplasmosis. In China, *A. phagocytophilum* was first detected in *I. persulcatus* in Heilongjiang Province in 1997 (Cao *et al.*, 2000) and the first suspected human case was described in Anhui Province in 2006 (Zhang *et al.*, 2006). Subsequent reports cited about 100 HGA cases in Beijing, Tianjin, Shandong Province, Henan Province, Hubei Province and Inner Mongolia (Fang *et al.*, 2015). *A. phagocytophilum* has been detected in various tick species from five genera including *Dermacentor, Haemaphysalis, Hyalomma, Ixodes* and *Rhipicephalus*, and is now the most widely distributed species in the genus *Anaplasma* over broad areas where various tick species are vectors. *A. ovis* was first identified in Cyprus in 2007 (Chochlakis *et al.*, 2010), although no human *A. ovis* infection has been reported in China. *A. ovis* was detected in *Haemaphysalis qinghaiensis, Dermacentor abaensis* and *Ixodes crenulatus* ticks that are associated with the distribution of goats (Han *et al.*, 2019). Similarly, *A. bovis* and *A. platys* were detected in several tick species that are associated with the distribution of ruminants (Han *et al.*, 2019). *A. marginale* was reported in *H. asiaticum* in north-western China and *A. centrale* was detected in *H. longicornis* in north-eastern China. *A. capra* is a novel *Anaplasma* species, so called because it was originally recognized in asymptomatic goats, and the agent was subsequently identified in febrile patients in Heilongjiang Province of China (Li *et al.*, 2015). Recent studies have shown that *A. capra* is identified in several tick species such as *I. persulcatus, R. microplus, H. longicornis* and *H. qinghaiensis*, as well as in many domesticated animals such as goats, sheep, cattle, deer and dogs (Peng *et al.*, 2018; Shi *et al.*, 2019), thus indicating the agent is widely distributed throughout China. However, the tick vector competence for *A. capra* has not yet been proven. Generally, patients with *Anaplasma* infection present undifferentiated and mild clinical manifestations that are easily misdiagnosed in China.

Babesia

Tick-borne parasites usually include *Babesia, Cytauxzoon, Theileria* and *Hepatozoon*. Babesiosis, caused by *Babesia* spp., is now classified as a notifiable disease that is recognized as an emerging health risk in several parts of the world. Over 100 *Babesia* species have been reported to infect many types of wild and domestic mammals, resulting in a significant health impact on livestock; however, only a few have been confirmed to infect people. Since the 1980s, about 13 *Babesia* species have been discovered in China, including *Babesia ovis, Babesia major, Babesia ovata, Babesia orientalis, Babesia motasi, Babesia caballi, Babesia* sp. Kashi, *Babesia* sp. Xinjiang, *Babesia microti, Babesia divergens, 'Babesia venatorum', Babesia crassa*-like and *Babesia* sp. XXB/Hangzhou (Fang *et al.*, 2015; Man *et al.*, 2016; Jia *et al.*, 2018). Among the 13 species, *B. microti, B. divergens, 'B. venatorum', B. crassa*-like and *Babesia* sp. XXB/Hangzhou have been reported in human infections. Transmission of babesiosis is associated with about 13 tick species from five genera including *Dermacentor, Haemaphysalis, Hyalomma, Ixodes* and *Rhipicephalus*. *B. microti* is the most widely distributed species in China and has been detected in *I. persulcatus, H. concinna,*

H. longicornis, *D. nuttalli*, *Rhipicephalus pumilio* and *H. asiaticum* ticks. Human infections with *B. microti* diagnosed by PCR and sequencing analysis have been reported in Zhejiang, Yunnan, Heilongjiang, Guangxi and Fujian provinces (Li and Zhang, 2019). Through serological and blood morphological investigations, suspected infections with *B. microti* have been reported in Xinjiang, Heilongjiang and Taiwan of China. *B. divergens* has been detected in *I. persulcatus*, *H. concinna* and *H. japonica* ticks, as well as in two patients in Shandong Province (Qi *et al.*, 2011; Li and Zhang, 2019). '*B. venatorum*' (formerly known as *Babesia* sp. EU1), which was discovered in patients without spleens in Europe, has been detected in a paediatric patient in Xinjiang Uygur Autonomous Region and 48 immunocompetent patients in Heilongjiang Province (Sun *et al.*, 2014; Jiang *et al.*, 2015). Through active surveillance of vectors in the north-eastern region of China, '*B. venatorum*' has been identified in *I. persulcatus* and *D. silvarum* ticks. *B. crassa*-like was first proven to infect humans in Heilongjiang and Jilin provinces of China in 2018 and has been detected in *I. persulcatus* and *H. concinna* ticks (Jia *et al.*, 2018). A novel species named *Babesia* sp. XXB/Hangzhou was discovered in a patient with a history of tick bite in Zhejiang Province in 2016 (Man *et al.*, 2016). The tick vector of *Babesia* sp. XXB/Hangzhou has not been identified yet. Additionally, several cases of babesiosis caused by uncharacterized *Babesia* species have been reported in China (Li *et al.*, 1984; Wang, 2012).

Spotted fever group rickettsiae

SFGR are obligate intracellular bacteria of the genus *Rickettsia* and family Rickettsiaceae and comprise >20 species identified as human pathogens. Most SFGR are transmitted by ticks; flea-transmitted *Rickettsia felis* and mite-transmitted *Rickettsia akari* are recognized as members of the transitional group rickettsiae. In China, more than ten emerging species and uncharacterized species of SFGR are associated with over 16 tick species (Fang *et al.*, 2015). Six species including *R. heilongjiangiensis*, *Rickettsia sibirica* sp. BJ-90, *R. raoultii*, *Candidatus* Rickettsia tarasevichiae, *Rickettsia japonica* and *Candidatus* Rickettsia xinyangensis, as well as one new genotype named *Rickettsia* sp. XY99, have been reported to infect human beings. *R. heilongjiangiensis* is the most common SFGR in China, and often detected in *Dermacentor* and *Haemaphysalis* ticks. *R. heilongjiangiensis* has also been reported in *Ixodes* ticks in north-eastern China and in *Hyalomma* and *Rhipicephalus* ticks in Inner Mongolia of China. About 34 people were reported to be infected with *R. heilongjiangiensis*, including 19 diagnosed in forested areas of north-eastern China in the 1990s (Wu *et al.*, 1998, 2001) and 15 diagnosed in Hainan Province in 2008 (Jin *et al.*, 2011). *R. sibirica* sp. BJ-90 was detected in *Dermacentor sinicus* ticks from a Beijing suburb in 1990 (Yu *et al.*, 1993). Two cases of *R. sibirica* sp. BJ-90 infection were reported in Heilongjiang Province, and another one was reported in Beijing (Jia *et al.*, 2013a; Li *et al.*, 2017). *R. raoultii* was detected in *D. silvarum* and *D. marginatus* in western China, in *Haemaphysalis verticalis* collected from Inner Mongolia, in *D. silvarum*, *D. nuttalli* and *H. longicornis* from Heilongjiang Province, and in *H. longicornis* from Liaoning Province. In China, 28 patients with *R. raoultii* infection have been reported, including two in Heilongjiang Province, 11 in Inner Mongolia, two in Shandong Province and 13 in Henan Province (Fang *et al.*, 2015; Li, H. *et al.*, 2018c). *Candidatus* R. tarasevichiae, an emerging tick-borne *Rickettsia* species initially found in Russia, has been identified in human beings and *I. persulcatus* ticks in the north-eastern region of China in 2013 (Jia *et al.*, 2013b). Subsequent surveillance has shown that *Candidatus* R. tarasevichiae is endemic in the central eastern region of China, with *H. longicornis* tick as the potential vector (Liu, W. *et al.*, 2016). *R. japonica*, the causative agent of Japanese spotted fever (JSF) that was first reported in Japan in 1984, has been detected in many tick species in China such as *Haemaphysalis flava*, *H. hystricis*, *Haemaphysalis cornigera*, *H. longicornis* and *Ixodes ovatus*. Human infections of *R. japonica* have been reported in Anhui, Zhejiang and Henan provinces (Li, H. *et al.*, 2018a,b; Li, J.B. *et al.*, 2018; Lu *et al.*, 2018). *Candidatus* R. xinyangensis, a novel SFGR that is phylogenetically related to *Rickettsia vini*, *R. japonica* and *R. heilongjiangiensis*, was first identified in febrile patients in Xinyang of Henan Province in 2020 (Li *et al.*, 2020). Further field investigation revealed the detection of *Candidatus* R. xinyangensis in *H. longicornis* ticks. A novel SFGR genotype, designated

Rickettsia sp. XY99, was detected in five ill patients in the eastern central region of China (Li *et al.*, 2016); however, the potential tick vectors have not yet been identified. Additionally, four species, including *R. sibirica* sp. *mongolotimonae*, *Rickettsia monacensis*, *Rickettsia slovaca* and *Candidatus* Rickettsia hebeiii, have not been proven to infect human beings.

Borrelia

Lyme disease (borreliosis), caused by *Borrelia burgdorferi* sensu lato (s.l.), is one of the most prevalent tick-borne zoonoses in Eastern Asia. At least 22 species of *B. burgdorferi* s.l. have been described by means of PCR-restriction fragment length polymorphism of 5S–23S ribosomal rRNA gene spacer amplicons and multilocus sequence analysis. In China, six *B. burgdorferi* genospecies, including *B. garinii*, *B. afzelii*, *Borrelia valaisiana*-related strain, *B. valaisiana*, *B. burgdorferi* s.s. and *Borrelia sinica*, have been detected in about 26 tick species from over 25 provinces (Fang *et al.*, 2015). *B. garinii* is the most common genospecies and has been identified in many tick species from the genera *Amblyomma*, *Dermacentor*, *Haemaphysalis*, *Hyalomma*, *Ixodes* and *Rhipicephalus* (Hao *et al.*, 2011; Zhang *et al.*, 2015). *B. afzelii* is the second most common genospecies and has been identified in tick species from the genera *Dermacentor*, *Haemaphysalis*, *Hyalomma* and *Ixodes* (Chu *et al.*, 2011; Zhang *et al.*, 2015). *B. valaisiana*-related genospecies (a genetically related but distinct genospecies of *B. valaisiana*) was detected in *I. granulatus* and *H. longicornis* from Guizhou Province (Chu *et al.*, 2008). These three genospecies have been reported to cause human infections in China. Infections with different genospecies show slightly different clinical manifestations, including erythema migrans, arthritis or arthralgia, fever, headache and fatigue (Ni *et al.*, 2014). Generally, clinical manifestations of Lyme disease in China are mild compared with those in the USA and Europe. Human cases of *B. burgdorferi* s.l. infection have been reported frequently in almost all provinces in China, except for Tibet and Shanghai; however, the causative genospecies of most cases have not been characterized. Three other genospecies, including *B. valaisiana*, *B. burgdorferi* s.s. and *B. sinica*, have not been reported to infect human beings in China. *Borrelia miyamotoi*, first detected in *I. persulcatus* ticks and rodents from Japan in 1994, has been recognized as an emerging tick-borne pathogen that causes relapsing fever. Since the first identification of human infection in Russia in 2011, patients infected with *B. miyamotoi* have been reported in the USA, Europe and Japan. In 2018, *B. miyamotoi* was first identified in patients with febrile influenza-like illness in the north-eastern region of China (Jiang *et al.*, 2018). To date, *B. miyamotoi* has been detected in *I. persulcatus* and *H. concinna* ticks in the north-eastern region and in *H. longicornis* ticks from Jiangsu of China (Yang *et al.*, 2018); nevertheless, the public health significance as well as the potentially wider distribution of *B. miyamotoi* deserve further investigation.

Viruses

Tick-borne viruses can be divided into two orders and nine families, of which five families –*Nairoviridae*, *Phenuiviridae*, *Flaviviridae*, *Orthomyxoviridae* and *Reoviridae* – contain tick-borne viruses that can cause human diseases. CCHFV, also known as 'Xinjiang haemorrhagic fever virus' in China, is a member of the family *Nairoviridae* and is endemic in Xinjiang Uygur Autonomous Region, with *H. asiaticum* as the main vector. In 2018, CCHFV was detected in *D. nuttalli* and *Hyalomma detritum* ticks from Xinjiang (Moming *et al.*, 2018), indicating that these tick species were also vectors for CCHFV. Seroepidemiological investigation showed a potential epidemic of CCHFV in other regions such as Yunnan Province. TBEV, a member of the family *Flaviviridae*, invades the central nervous system and causes serious morbidity. TBEV is classified into the European (TBEV-Eu), Siberian (TBEV-Sib) and Far Eastern (TBEV-FE) subtypes. TBEV-FE is endemic in the north-eastern region of China and is mainly transmitted by *I. persulcatus* (Chen *et al.*, 2019). TBEV-FE was also detected in *D. silvarum* and *H. concinna* ticks. Xinjiang, Yunnan and Xizang (Tibet) may also be natural foci based on serological surveys (Gao *et al.*, 2010). TBEV-Sib was first discovered in *I. persulcatus* tick in Xinjiang in 2016 (Liu, R. *et al.*, 2016). SFTSV, a member of the family *Phenuiviridae*, was first discovered as the causative agent of severe haemorrhagic syndrome disease (SFTS) in China in 2009 (Yu *et al.*, 2011). Since then, SFTS has been listed as a nationally reportable disease in China and this disease has been found in more than 25 provinces up to now, resulting in it being the most widespread tick-borne viral illness in China (Li, H. *et al.*, 2018a). *H. longicornis* is a proven vector of SFTSV, with both transovarial and transstadial modes of transmission (Zhuang *et al.*, 2018). SFTSV was also detected in *R. microplus*, *R. sanguineus* and *H. concinna* in SFTS-endemic regions in China. An experimental transmission study has shown that *D. silvarum* and *I. persulcatus* ticks may not serve as efficient vectors of SFTSV (Hu *et al.*, 2020). JMTV, a member in the family *Flaviviridae*, was first isolated from *R. microplus* in Hubei Province in 2010, then identified in human patients in Heilongjiang Province in 2019 (Jia *et al.*, 2019). JMTV was also detected in several other tick species, such as *Amblyomma javanense*, *D. silvarum* and *I. persulcatus*. In 2019, ALSV, a novel member in the family *Flaviviridae*, was first discovered in febrile patients in Inner Mongolia and Heilongjiang, China (Wang *et al.*, 2019). *I. persulcatus* may serve as the vector of ALSV. In 2018, an outbreak of African swine fever (ASF) in pigs was reported in China. The disease is caused by infection with ASF virus (ASFV). Soft tick *Ornithodoros* spp. that can serve as vectors of ASFV have been reported in the western region of China; however, there are no studies reporting the infection of ASFV in the soft ticks from these areas. Several other tick-borne viruses that can cause human infections, such as Powassan virus, Thogoto virus, Colorado tick fever virus and Heartland bandavirus, have not been reported in China.

71.2 Future Predictions

In the past three decades in China, there is no doubt that advances in and application of molecular technologies have resulted in the discovery of new tick-borne pathogens and identification of diseases caused by agents previously detected in ticks. However, the epidemic status of tick-borne infectious diseases may still be greatly underestimated. Not only SFTS, but other tick-borne diseases should be considered as notifiable infectious diseases in China. Further application of microbiome research techniques should be made to identify tick symbionts and potential tick-borne microbes, and to construct tick-borne microbiome databases for medical and veterinary importance. Due to increasing population sizes and rapid urbanization, changes in tick geographical distributions and numbers, as well as the emergence of new tick-borne infectious agents, will continue. Integrated strategies, combining field surveillance for monitoring tick species' changes and their infesting agents with hospital surveillance for monitoring febrile patients with a tick exposure history, should be developed.

Disclaimer

The findings and conclusions of this expert opinion are by the authors and do not necessarily represent the views of the Beijing Institute of Microbiology and Epidemiology.

References

Altizer, S., Ostfeld, R.S., Johnson, P.T., Kutz, S. and Harvell, C.D. (2013) Climate change and infectious diseases: from evidence to a predictive framework. *Science* 341, 514–519.

Cao, W.C., Zhao, Q.M., Zhang, P.H., Dumler, J.S., Zhang, X.T., *et al.* (2000) Granulocytic ehrlichiae in *Ixodes persulcatus* ticks from an area in China where Lyme disease is endemic. *Journal of Clinical Microbiology* 38, 4208–4210.

Chen, S.M., Dumler, J.S., Bakken, J.S. and Walker, D.H. (1994) Identification of a granulocytotropic *Ehrlichia* species as the etiologic agent of human disease. *Journal of Clinical Microbiology* 32, 589–595.

Chen, X.J., Li, F., Yin, Q.K., Liu, W.J., *et al.* (2019) Epidemiology of tick-borne encephalitis in China, 2007–2018. *PLoS ONE* 14, e0226712.

Chochlakis, D., Ioannou, I., Tselentis, Y. and Psaroulaki, A. (2010) Human anaplasmosis and *Anaplasma ovis* variant. *Emerging Infectious Diseases* 16, 1031–1032.

Chu, C.Y., Liu, W., Jiang, B.G., Wang, D.M., Jiang, W.J., *et al.* (2008) Novel genospecies of *Borrelia burgdorferi* sensu lato from rodents and ticks in southwestern China. *Journal of Clinical Microbiology* 46, 3130–3133.

Chu, C.Y., Jiang, B.G., He, J., Gao, Y., Zhang, P.-H., *et al.* (2011) Genetic diversity of *Borrelia burgdorferi* sensu lato isolates from Northeastern China. *Vector-Borne and Zoonotic Diseases* 11, 877–882.
Fang, L.Q., Liu, K., Li, X.L., Liang, S., Yang, Y., *et al.* (2015) Emerging tick-borne infections in mainland China: an increasing public health threat. *Lancet Infectious Diseases* 15, 1467–1479.
Gao, X., Nasci, R. and Liang, G. (2010) The neglected arboviral infections in mainland China. *PLoS Neglected Tropical Diseases* 4, e624.
Gilbert, L. (2010) Altitudinal patterns of tick and host abundance: a potential role for climate change in regulating tick-borne diseases? *Oecologia* 162, 217–225.
Han, R., Yang, J.F., Mukhtar, M.U., Chen, Z., Niu, Q.L., *et al.* (2019) Molecular detection of *Anaplasma* infections in ixodid ticks from the Qinghai–Tibet Plateau. *Infectious Diseases of Poverty* 8, 12.
Hao, Q., Hou, X.X., Geng, Z. and Wan, K.L. (2011) Distribution of *Borrelia burgdorferi* sensu lato in China. *Journal of Clinical Microbiology* 49, 647–650.
Harrus, S. and Baneth, G. (2005) Drivers for the emergence and re-emergence of vector-borne protozoal and bacterial diseases. *International Journal for Parasitology* 35, 1309–1318.
Hu, Y.Y., Zhuang, L., Liu, K., Sun, Y., Dai, K., *et al.* (2020) Role of three tick species in the maintenance and transmission of severe fever with thrombocytopenia syndrome virus. *PLoS Neglected Tropical Diseases* 14, e0008368.
Jia, N., Jiang, J.F., Huo, Q.B., Jiang, B.G. and Cao, W.C. (2013a) *Rickettsia sibirica* subspecies *sibirica* BJ-90 as a cause of human disease. *New England Journal of Medicine* 369, 1176–1178.
Jia, N., Zheng, Y.C., Jiang, J.F., Ma, L. and Cao, W.C. (2013b) Human infection with *Candidatus* Rickettsia tarasevichiae. *New England Journal of Medicine* 369, 1178–1180.
Jia, N., Zheng, Y.C., Jiang, J.F., Jiang, R.R., Jiang, B.G., *et al.* (2018) Human babesiosis caused by a *Babesia crassa*-like pathogen: a case series. *Clinical Infectious Diseases* 67, 1110–1119.
Jia, N., Liu, H.B., Ni, X.B., Bell-Sakyi, L., Zheng, Y.C., *et al.* (2019) Emergence of human infection with Jingmen tick virus in China: a retrospective study. *EBioMedicine* 43, 317–324.
Jiang, B.G., Jia, N., Jiang, J.F., Zheng, Y.C., Chu, Y.L., *et al.* (2018) *Borrelia miyamotoi* infections in humans and ticks, northeastern China. *Emerging Infectious Diseases* 24, 236–241.
Jiang, J.F., Zheng, Y.C., Jiang, R.R., Li, H., Huo, Q.B., *et al.* (2015) Epidemiological, clinical, and laboratory characteristics of 48 cases of '*Babesia venatorum*' infection in China: a descriptive study. *Lancet Infectious Diseases* 15, 196–203.
Jin, Y.M., Zhang, L.J., Sun, L.Y., Lao, S.J., Han, J., *et al.* (2011) Epidemiological investigation of emerging spotted fever in Chengmai County, Hainan Province. *Disease Surveillance* 26, 18–22.
Kilpatrick, A.M. and Randolph, S.E. (2012) Drivers, dynamics, and control of emerging vector-borne zoonotic diseases. *The Lancet* 380, 1946–1955.
Li, H., Zheng, Y.C., Ma, L., Jia, N., Jiang, B.G., *et al.* (2015) Human infection with a novel tick-borne *Anaplasma* species in China: a surveillance study. *Lancet Infectious Diseases* 15, 663–670.
Li, H., Cui, X.M., Cui, N., Yang, Z.D., Hu, J.G., *et al.* (2016) Human infection with novel spotted fever group *Rickettsia* genotype, China, 2015. *Emerging Infectious Diseases* 22, 2153–2156.
Li, H., Fu, X.Y., Jiang, J.F., Liu, R.X., Li, R., *et al.* (2017) Severe illness caused by *Rickettsia sibirica* subspecies *sibirica* BJ-90 infection, China. *Emerging Microbes and Infections* 6, e107.
Li, H., Lu, Q.B., Xing, B., Zhang, S.F., Liu, K., *et al.* (2018a) Epidemiological and clinical features of laboratory-diagnosed severe fever with thrombocytopenia syndrome in China, 2011–17: a prospective observational study. *Lancet Infectious Diseases* 18, 1127–1137.
Li, H., Zhang, P.H., Du, J., Yang, Z.D., Cui, N., *et al.* (2018b) *Rickettsia japonica* infections in humans, Xinyang, China, 2014–2017. *Emerging Infectious Diseases* 25, 1719–1722.
Li, H., Zhang, P.H., Huang, Y., Du, J., Cui, N., *et al.* (2018c) Isolation and identification of *Rickettsia raoultii* in human cases: a surveillance study in 3 medical centers in China. *Clinical Infectious Diseases* 66, 1109–1115.
Li, H., Li, X.M., Du, J., Zhang, X.A., Cui, N., *et al.* (2020) *Candidatus* Rickettsia xinyangensis as cause of spotted fever group rickettsiosis, Xinyang, China, 2015. *Emerging Infectious Diseases* 26, 985–988.
Li, J.B., Hu, W., Wu, T., Li, H.B., Hu, W.F., *et al.* (2018) Japanese spotted fever in eastern China, 2013. *Emerging Infectious Diseases* 24, 2107–2109.
Li, J.F., Meng, D.B., Wang, Q.F., Dong, X.R. and Wang, J.Y. (1984) The discovery of human babesiosis. *Chinese Journal of Veterinary Medicine* 6, 19–20.
Li, L.H. and Zhang, Y. (2019) Epidemic status and control of tick-borne parasitic diseases in China. *Chinese Journal of Schistosomiasis Control* 31, 58–62.
Liu, M., Li, T., Yu, Z.J., Qiu, Z.X., Yan, P., *et al.* (2017) Seasonal dynamics of the tick *Haemaphysalis tibetensis* in the Tibetan Plateau, China. *Medical and Veterinary Entomology* 31, 449–456.

Liu, R., Zhang, G.L., Liu, X.M., Li, Y.C., Zheng, Z., *et al.* (2016) Detection of the Siberian tick-borne encephalitis virus in the Xinjiang Uygur Autonomous Region, northwestern China. *Chinese Journal of Virology* 32, 26–31.

Liu, W., Li, H., Lu, Q.B., Cui, N., Yang, Z.D., *et al.* (2016) *Candidatus* Rickettsia tarasevichiae infection in eastern central China: a case series. *Annals of Internal Medicine* 17, 641–648.

Lu, Q.Y., Yu, J.P., Yu, L.Q., Zhang, Y.J., Chen, Y.T., *et al.* (2018) *Rickettsia japonica* infections in humans, Zhejiang Province, China, 2015. *Emerging Infectious Diseases* 24, 2077–2079.

Man, S.Q., Qiao, K., Cui, J., Feng, M., Fu, Y.F. and Cheng, X.J. (2016) A case of human infection with a novel *Babesia* species in China. *Infectious Diseases of Poverty* 5, 28.

Moming, A., Yue, X.H., Shen, S., Chang, C.C., Wang, C., *et al.* (2018) Prevalence and phylogenetic analysis of Crimean–Congo hemorrhagic fever virus in ticks from different ecosystems in Xinjiang, China. *Virologica Sinica* 33, 67–73.

Ni, X.B., Jia, N., Jiang, B.G., Sun, T., Zheng, Y.C., *et al.* (2014) Lyme borreliosis caused by diverse genospecies of *Borrelia burgdorferi* sensu lato in northeastern China. *Clinical Microbiology and Infection* 20, 808–814.

Parola, P. (2004) Tick-borne rickettsial diseases: emerging risks in Europe. *Comparative Immunology, Microbiology and Infectious Diseases* 27, 297–304.

Peng, Y., Wang, K., Zhao, S., Yan, Y., Wang, H., *et al.* (2018) Detection and phylogenetic characterization of *Anaplasma capra*: an emerging pathogen in sheep and goats in China. *Frontiers in Cellular and Infection Microbiology* 8, 283.

Qi, C.H., Zhou, D., Liu, J.Z, Cheng, Z.Q., Zhang, L., *et al.* (2011) Detection of *Babesia divergens* using molecular methods in anemic patients in Shandong Province, China. *Parasitology Research* 109, 241–245.

Shi, K., Li, J.Q., Yan, Y.Q., Chen, Q., Wang, K.L., *et al.* (2019) Dogs as new hosts for the emerging zoonotic pathogen *Anaplasma capra* in China. *Frontiers in Cellular and Infection Microbiology* 9, 394.

Sun, Y., Li, S.G., Jiang, J.F., Wang, X., Zhang, Y., *et al.* (2014) *Babesia venatorum* infection in child, China. *Emerging Infectious Diseases* 20, 896–897.

Wang, H.X. (2012) A case of human babesiosis in southwest region of China. *International Journal of Medical Parasitic Diseases* 39, 190–192.

Wang, Z.D., Wang, B., Wei, F., Han, S.Z., Zhang, L., *et al.* (2019) A new segmented virus associated with human febrile illness in China. *New England Journal of Medicine* 380, 2116–2125.

Wu, Y.M., Wei, A.M., Hu, L.M., Liu, X.X., Qing, Y., *et al.* (1998) Identification of spotted fever group (SFG) H-5 strain isolated from the patient's blood samples. *Chinese Journal of Zoonoses* 14, 23–26.

Wu, Y.M., Wei, A.M., Hu, L.M., Zhang, Y.G., Zhang, J.Z., *et al.* (2001) Study on aetiology of *Rickettsia heilongianggii* tick borne spotted fever. *China Public Health* 7, 28–30.

Yang, Y., Yang, Z.P., Kelly, P., Li, J., Ren, Y.J. and Wang, C.M. (2018) *Borrelia miyamotoi* sensu lato in Père David deer and *Haemaphysalis longicornis* ticks. *Emerging Infectious Diseases* 24, 928–931.

Yu, X.J., Jin, Y., Fan, M.Y., Xu, G.M. and Raoult, D. (1993) Genotypic and antigenic identification of two new strains of spotted fever group rickettsiae isolated from China. *Journal of Clinical Microbiology* 31, 83–88.

Yu, X.J., Liang, M.F., Zhang, S.Y., Liu, Y., Li, J.D., *et al.* (2011) Fever with thrombocytopenia associated with a novel bunyavirus in China. *New England Journal of Medicine* 364, 1523–1532.

Zhang, J.J., Zhang, F., Li, L. and Liu, Z.J. (2015) *Borrelia burgdorferi* infection and their genotypes in ticks and rodents collected from Maijishan region, Gansu Province, China. *Chinese Journal of Zoonoses* 31, 357–360.

Zhang, L.J., Liu, Y., Ni, D.X., Li, Q., Yu, Y.L., *et al.* (2006) Nosocomial transmission of human granulocytic anaplasmosis in China. *Journal of the American Medical Association* 300, 2263–2270.

Zhang, Y.K., Zhang, X.Y. and Liu, J.Z. (2019) Ticks (Acari: Ixodoidea) in China: geographical distribution, host diversity, and specificity. *Archives of Insect Biochemistry and Physiology* 102, e21544.

Zhao, G.P., Wang, Y.X., Fan, Z.W., Ji, Y., Liu, M.J., *et al.* (2021) Mapping ticks and tick-borne pathogens in China. *Nature Communications* 12, 1075.

Zhuang, L., Sun, Y., Cui, X.M., Tang, F., Hu, J.G., *et al.* (2018) Transmission of severe fever with thrombocytopenia syndrome virus by *Haemaphysalis longicornis* ticks, China. *Emerging Infectious Diseases* 24, 868–871.

eo72 Tick-Borne Rickettsioses in Africa

Agatha Kolo*
Council for Scientific and Industrial Research, Pretoria, South Africa

72.1 Expert Opinion

72.1.1 Introduction

Tick-borne rickettsioses are caused by obligate intracellular Gram-negative bacteria of the spotted fever group (SFG) of rickettsiae (Cazorla *et al.*, 2008). Around the world it has been acknowledged that rickettsioses are significant models of emerging arthropod-borne diseases (Parola *et al.*, 2005). Traditionally, the identification of tick-borne rickettsial pathogens was confined to serology (La Scola and Raoult, 1997) but in contemporary times the detection of various and novel SFG rickettsiae has been made easier using cell culture approaches and the evolution of genomic tools (Jado *et al.*, 2006). It has been reported that global warming and climate change has led to a rise in human attacks and the likelihood of increased transmission of life-threatening rickettsioses by the brown dog tick, *Rhipicephalus sanguineus* (Parola *et al.*, 2008). A study conducted in south-western Tanzania found that high seropositivity to SFG rickettsiae was distinctively linked to localities with higher temperatures (Heinrich *et al.*, 2015). Nevertheless, there is currently a paucity of information on whether climate change is having an impact on the incidence, prevalence and distribution of rickettsial infections in Africa. The current knowledge of tick-borne rickettsial agents reported on the African continent is presented in this expert opinion.

72.1.2 *Rickettsia africae*

Rickettsia africae causes African tick-bite fever (ATBF) (Parola *et al.*, 2005). Cattle ticks of the *Amblyomma* genus are responsible for its transmission (Kelly *et al.*, 1996), with *Amblyomma hebraeum* in south-eastern Africa and *Amblyomma variegatum* in African countries south of the Sahara Desert considered the primary hosts and vectors of this pathogen (Jensenius *et al.*, 2003a). Around Africa, *R. africae* has been discovered in a variety of ticks including *Amblyomma lepidum*, *Amblyomma compressum*, *Rhipicephalus annulatus*, *Rhipicephalus evertsi evertsi*, *Rhipicephalus decoloratus*, *R. sanguineus*, *Rhipicephalus geigyi*, *Hyalomma dromedarii* and *A. variegatum* in Djibouti, Guinea, Senegal, Nigeria, Liberia, the Democratic Republic of Congo, Botswana, Algeria, Egypt, Cameroon, Benin, Kenya and Madagascar (Ndip *et al.*, 2004; Portillo *et al.*, 2007; Socolovschi *et al.*, 2007; Mediannikov *et al.*, 2010a,b, 2012a,b; Abdel-Shafy *et al.*, 2012; Kernif *et al.*, 2012; Ogo *et al.*, 2012; Reye *et al.*, 2012; Maina *et al.*, 2014; Adjou Moumouni *et al.*, 2016; Keller *et al.*, 2016; Vanegas

*Email: afrianimalfriend@yahoo.co.uk

DOI: 10.1079/9781789249637.0072

et al., 2018). Within South Africa, *R. africae* was found in *A. hebraeum* (Halajian *et al.*, 2016) and *Haemaphysalis elliptica* (Kolo *et al.*, 2016).

Clinical signs and symptoms of ATBF include headaches, nuchal myalgia, painful lymphadenopathy and inoculation eschars (Fig. 72.1) (Jensenius *et al.*, 2003a). *R. africae* has been identified in ticks and/or humans in 22 countries in sub-Saharan Africa (Parola *et al.*, 2013). The majority of cases of ATBF are seen in international travellers coming back from Africa (Fournier *et al.*, 1998; Jensenius *et al.*, 2003b; Althaus *et al.*, 2010). In Cameroon, *R. africae* DNA was detected in 6% of the individuals who presented with an undifferentiated fever that did not originate from malaria or typhoid fever (Ndip *et al.*, 2004). A serological survey conducted in a rural community in South Africa detected prior exposure to SFG *Rickettsia* in 92.2% of patients who presented with signs of non-malarial acute febrile illness in local clinics and 64.4% of farmers, herdsmen and veterinary personnel who were screened (Simpson *et al.*, 2018). Seropositivity to *R. africae* was found in 26.9% of individuals from rural villages in Cameroon and 20.6–45.6% in rural Senegal (Mediannikov *et al.*, 2010a; Ndip *et al.*, 2011).

72.1.3 *Rickettsia conorii*

Multilocus sequence typing proposed four subspecies of *Rickettsia conorii* as *R. conorii* subsp. *conorii*, *R. conorii* subsp. *indica*, *R. conorii* subsp. *caspia* and *R. conorii* subsp. *israelensis* (Zhu *et al.*, 2005). *R. conorii* subsp. *conorii* is the cause of Mediterranean spotted fever (MSF). The brown dog tick *R. sanguineus* is the main vector of its transmission (Parola *et al.*, 2005). *R. conorii* subsp. *conorii* has been described in the Central African Republic, Zimbabwe, Senegal, South Africa, Uganda, Swaziland, Kenya, Libya, Egypt, Algeria, Tunisia and Morocco (Socolovschi *et al.*, 2007; Mediannikov *et al.*, 2010a; Parola *et al.*, 2013). It has been identified using PCR in *R. sanguineus* in Algeria, Tunisia and Morocco (Bitam *et al.*, 2006; Boudebouch *et al.*, 2009; Sfar *et al.*, 2009). *R. conorii* subsp. *conorii* was found in *Rhipicephalus mushumae* picked off cattle within the Central African Republic and in *Rhipicephalus simus* and *Haemaphysalis leachi* ticks from dogs in Zimbabwe (Parola *et al.*, 2013). The organism was identified from *R. evertsi evertsi* removed from horses inside Senegal and in *Hyalomma punctaleachi* from a dog in Uganda (Socolovschi *et al.*, 2007; Mediannikov *et al.*, 2010a). Clinical signs of MSF include fever, headache, malaise, arthromyalgias, eschar at the location of the tick bite and a maculopapular rash (Fig. 72.2) (Rovery and Raoult, 2008). Human cases of MSF in Algeria, Tunisia and Morocco have been on the increase (Kernif *et al.*, 2012). In Algeria, a reported 49% of patients hospitalized for MSF had severe forms of the disease (Kaabia *et al.*, 2009; Mouffok *et al.*, 2009). The organism was identified from the blood sample of a feverish young patient in Senegal (Mediannikov *et al.*, 2010a) and in a South African man on a trip to Brazil who tragically died of the infection (De Almeida *et al.*, 2010). Clinical signs seen in this

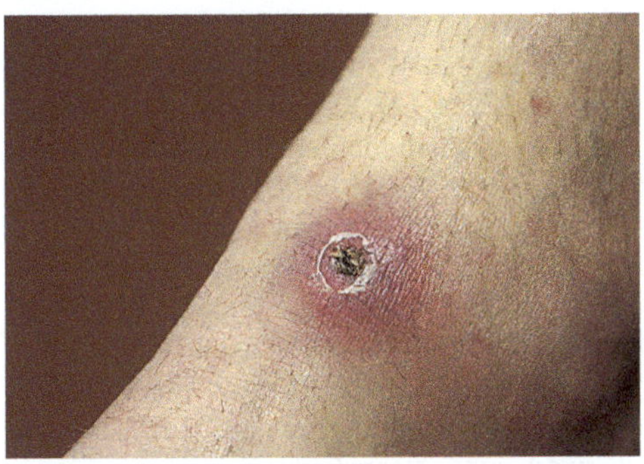

Fig. 72.1. Image showing common appearance of an eschar from a tick bite. (Photograph courtesy of Dr John Frean.)

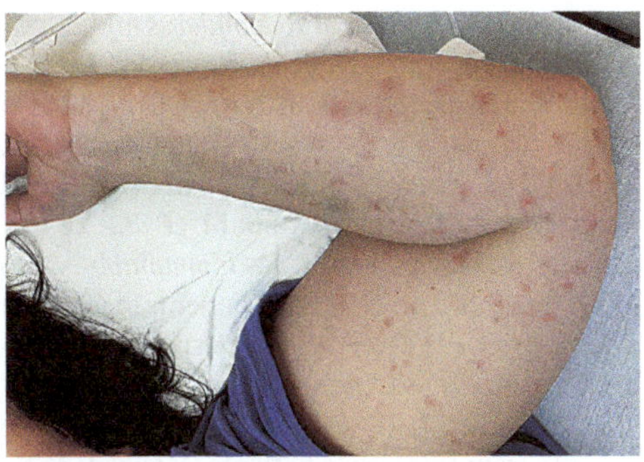

Fig. 72.2. Generalized maculopapular skin rash usually seen in *Rickettsia conorii* infection. (Photograph courtesy of Dr Jeremy Nel.)

fatal case included fever, chills, vomiting, haematuria, sore throat and a maculopapular rash (De Almeida *et al.*, 2010). Of late, *R. conorii* was reported in a survey of small mammals across seven provinces in South Africa and parts of Namibia (Essbauer *et al.*, 2018).

Other subspecies of *R. conorii* detected in Africa include *R. conorii* subsp. *israelensis* which causes Israeli spotted fever (ISF). The organism has been detected using molecular techniques in skin biopsy samples from patients in southern Tunisia (Znazen *et al.*, 2011). Astrakhan fever is the disease condition caused by *Rickettsia conorii* subsp. *caspia*. The organism was identified in a French traveller who was returning from Chad and presented with clinical signs of fever and a maculopapular rash (Fournier *et al.*, 2003).

72.1.4 *Rickettsia aeschlimannii*

Rickettsia aeschlimannii is the cause of a disease resembling MSF. The organism was first isolated from *Hyalomma marginatum marginatum* in Morocco (Beati *et al.*, 1997). Thereafter, *R. aeschlimannii* has been identified in different species of *Hyalomma* ticks collected from cattle and camels in Egypt, Sudan, Algeria and Tunisia (Kernif *et al.*, 2012; Abdelkadir *et al.*, 2019), in *Hyalomma marginatum rufipes* and *Hyalomma truncatum* in Cameroon and Senegal (Mediannikov *et al.*, 2010a; Vanegas *et al.*, 2018), in *H. truncatum* in Sudan (Morita *et al.*, 2004), in *H. marginatum rufipes* in Ethiopia, Chad and Nigeria (Mura *et al.*, 2008; Kamani *et al.*, 2015), as well as in *R. evertsi evertsi* in Nigeria and Senegal (Mediannikov *et al.*, 2010a; Reye *et al.*, 2012). The organism has also been detected in *R. annulatus* and *A. variegatum* in Nigeria (Reye *et al.*, 2012), in *H. dromedarii* picked off camels in central Tunisia and Sudan (Morita *et al.*, 2004; Demoncheaux *et al.*, 2012) and in *Hyalomma impeltatum* obtained from Nigerian camels (Kamani *et al.*, 2015). Serological exposure to *R. aeschlimannii* has also been described in a febrile patient in Tunisia (Znazen *et al.*, 2006). Other reported human infections are in a French traveller after a recent trip to Morocco (Raoult *et al.*, 2002), inside Algeria from two febrile patients (Mokrani *et al.*, 2008) and in a South African patient infected after a *Rhipicephalus appendiculatus* tick bite from an excursion (Pretorius and Birtles, 2002).

72.1.5 *Rickettsia sibirica* subsp. *mongolotimonae*

Lymphangitis-associated rickettsiosis (LAR) is the condition caused by *Rickettsia sibirica* subsp. *mongolotimonae* (Socolovschi *et al.*, 2010). The organism was detected in *H. truncatum* ticks removed from cattle in Niger and Cameroon (Parola *et al.*, 2001; Vanegas *et al.*, 2018), in *H. truncatum* picked off horses, donkeys, cattle, sheep and goats in Senegal (Mediannikov *et al.*, 2010a) and in a topi (*Damaliscus lunatus* ssp. *jimela*) in Kenya (Ndeereh *et al.*, 2017). Human infection with this organism has been reported in South Africa in a patient that had clinical signs of lymphangitis, fever and a toe eschar (Pretorius and Birtles, 2004). In other reported cases, two travellers on a recent return to France from Algeria (Fournier *et al.*, 2005; Socolovschi *et al.*, 2010) and Egypt (Fournier *et al.*, 2005; Socolovschi *et al.*, 2010) presented with fever, asthenia, headaches and arthromyalgia.

72.1.6 Other tick-borne rickettsiae detected in Africa

Other rickettsial agents have been reported around the African continent. *Rickettsia slovaca* and *Rickettsia raoultii* which cause a scalp eschar and neck lymphadenopathy (SENLAT) syndrome were identified in *Dermacentor marginatus* within Morocco and northern Algeria (Sarih *et al.*, 2008; Kernif *et al.*, 2012). *Rickettsia monacensis* and *Rickettsia helvetica* have been identified from *Ixodes ricinus* collected from livestock and dogs in Tunisia (Sfar *et al.*, 2008), Morocco (Sarih *et al.*, 2008) and Algeria (Dib *et al.*, 2009; Kernif *et al.*, 2012). *Rickettsia helvetica* was also identified in small mammals in southern Africa (Essbauer *et al.*, 2018). *Rickettsia massiliae*, implicated in causing human infection in Europe and South America (Parola *et al.*, 2013), was identified in *Rhipicephalus guilhoni* picked off donkeys and cattle inside Senegal (Mediannikov *et al.*, 2010a), in *Rhipicephalus senegalensis* ticks in Guinea (Mediannikov *et al.*, 2012b) and from *R. evertsi evertsi* obtained using tick drags in Nigeria (Reye *et al.*, 2012). *R. massiliae* was also identified in *Amblyomma sylvaticum* as well as *R. simus* ticks collected from a leopard tortoise in Limpopo

Province, South Africa (Halajian *et al.*, 2016), in *Rhipicephalus lunulatus* picked off cattle in Cameroon (Vanegas *et al.*, 2018) and in small mammals from southern Africa (Essbauer *et al.*, 2018). Three novel *Candidatus* species, *Candidatus* Rickettsia africaustralis, *Candidatus* Rickettsia rhabdomydis and *Candidatus* Rickettsia muridii, have been detected from small mammals collected across South Africa and Namibia (Essbauer *et al.*, 2018). Other rickettsial agents of undetermined disease potential reported in Africa include *Rickettsia rhipicephali* detected in the Central African Republic from *R. lunulatus* and *Rhipicephalus composites* (Parola *et al.*, 2005), *Rickettsia hoogstraalii* detected from the avian soft tick *Argas persicus* within Ethiopia (Pader *et al.*, 2012) and *Candidatus* Rickettsia liberiensis identified from *Ixodes muniensis* picked off dogs in Liberia (Mediannikov *et al.*, 2012b). The novel *Candidatus* Rickettsia barbariae was detected in *R. lunulatus* in Cameroon (Vanegas *et al.*, 2018) and in *Hyalomma excavatum* and *Rhipicephalus bursa* picked off livestock in Algeria (Abdelkadir *et al.*, 2019).

72.2 Future Predictions

As mentioned above, research has indicated that global warming and climate change will favour the proliferation of ticks, lead to an increase in human and animal tick bites, and increase the incidence of life-threatening rickettsioses. Advances in science will also result in the increased recognition of novel rickettsial agents. The potential pathogenicity of these new rickettsial agents remains to be seen.

References

Abdelkadir, K., Palomar, A.M., Portillo, A., Oteo, J.A., Ait-Oudhia, K. and Khelef, D. (2019) Presence of *Rickettsia aeschlimannii*, '*Candidatus* Rickettsia barbariae' and *Coxiella burnetii* in ticks from livestock in northwestern Algeria. *Ticks and Tick-borne Diseases* 10, 924–928.

Abdel-Shafy, S., Allam, N.A., Mediannikov, O., Parola, P. and Raoult, D. (2012) Molecular detection of spotted fever group rickettsiae associated with ixodid ticks in Egypt. *Vector-Borne and Zoonotic Diseases* 12, 346–359.

Adjou Moumouni, P.F., Terkawi, M.A., Jirapattharasate, C., Cao, S., Liu, M., *et al.* (2016) Molecular detection of spotted fever group rickettsiae in *Amblyomma variegatum* ticks from Benin. *Ticks and Tick-borne Diseases* 7, 828–833.

Althaus, F., Greub, G., Raoult, D. and Genton, B. (2010) African tick-bite fever: a new entity in the differential diagnosis of multiple eschars in travelers. Description of five cases imported from South Africa to Switzerland. *International Journal of Infectious Diseases* 14, E274–E276.

Beati, L., Meskini, M., Thiers, B. and Raoult, D. (1997) *Rickettsia aeschlimannii* sp. nov., a new spotted fever group rickettsia associated with *Hyalomma marginatum* ticks. *International Journal of Systemic Bacteriology* 47, 548–554.

Bitam, I., Parola, P., Matsumoto, K., Rolain, J.M., Baziz, B., *et al.* (2006) First molecular detection of *R. conorii*, *R. aeschlimannii*, and *R. massiliae* in ticks from Algeria. *Annals of the New York Academy of Sciences* 1078, 368–372.

Boudebouch, N., Sarih, M., Socolovschi, C., Amarouch, H., Hassar, M., *et al.* (2009) Molecular survey for spotted fever group rickettsiae in ticks from Morocco. *Clinical Microbiology and Infection* 15, 259–260.

Cazorla, C., Socolovschi, C., Jensenius, M. and Parola, P. (2008) Tick-borne diseases: tick-borne spotted fever rickettsioses in Africa. *Infectious Disease Clinics of North America* 22, 531–544.

De Almeida, D.N., Favacho, A.R., Rozental, T., Barcaui, H., Guterres, A., *et al.* (2010) Fatal spotted fever group rickettsiosis due to *Rickettsia conorii conorii* mimicking a hemorrhagic viral fever in a South African traveler in Brazil. *Ticks and Tick-borne Diseases* 1, 149–150.

Demoncheaux, J.-P., Socolovschi, C., Davoust, B., Haddad, S., Raoult, D. and Parola, P. (2012) First detection of *Rickettsia aeschlimannii* in *Hyalomma dromedarii* ticks from Tunisia. *Ticks and Tick-borne Diseases* 3, 398–402.

Dib, L., Bitam, I., Bensouilah, M., Parola, P. and Raoult, D. (2009) First description of *Rickettsia monacensis* in *Ixodes ricinus* in Algeria. *Clinical Microbiology and Infection* 15, 261–262.

Essbauer, S., Hofmann, M., Kleinemeier, C., Wolfel, S. and Matthee, S. (2018) *Rickettsia* diversity in Southern Africa: a small mammal perspective. *Ticks and Tick-borne Diseases*, 9, 288–301.

Fournier, P.E., Roux, V., Caumes, E., Donzel, M. and Raoult, D. (1998) Outbreak of *Rickettsia africae* infections in participants of an adventure race in South Africa. *Clinical Infectious Diseases* 27, 316–323.

Fournier, P.E., Xeridat, B. and Raoult, D. (2003) Isolation of a *Rickettsia* related to Astrakhan fever rickettsia from a patient in Chad. *Annals of the New York Academy of Sciences* 990, 152–157.

Fournier, P.E., Gouriet, F., Brouqui, P., Lucht, F. and Raoult, D. (2005) Lymphangitis-associated rickettsiosis, a new rickettsiosis caused by *Rickettsia sibirica mongolotimonae*: seven new cases and review of the literature. *Clinical Infectious Diseases* 40, 1435–1444.

Halajian, A., Palomar, A.M., Portillo, A., Heyne, H., Luus-Powell, W.J. and Oteo, J.A. (2016) Investigation of *Rickettsia*, *Coxiella burnetii* and *Bartonella* in ticks from animals in South Africa. *Ticks and Tick-borne Diseases* 7, 361–366.

Heinrich, N., Dill, T., Dobler, G., Clowes, P., Kroidl, I., *et al.* (2015) High seroprevalence for spotted fever group rickettsiae, is associated with higher temperatures and rural environment in Mbeya region, Southwestern Tanzania. *PLoS Neglected Tropical Diseases* 9, e0003626.

Jado, I., Escudero, R., Gil, H., Jiménez-Alonso, M.I., Sousa, R., *et al.* (2006) Molecular method for identification of *Rickettsia* species in clinical and environmental samples. *Journal of Clinical Microbiology* 44, 4572–4576.

Jensenius, M., Fournier, P.-E., Kelly, P., Myrvang, B. and Raoult, D. (2003a) African tick bite fever. *Lancet Infectious Diseases* 3, 557–564.

Jensenius, M., Fournier, P.E., Vene, S., Hoel, T., Hasle, G., *et al.* (2003b) African tick bite fever in travelers to rural sub-equatorial Africa. *Clinical Infectious Diseases* 36, 1411–1417.

Kaabia, N., Bellazreg, F., Hachfi, W., Khalifa, M., Ghanouchi, N., *et al.* (2009) Rickettsial infection in hospitalised patients in central Tunisia: report of 119 cases. *Clinical Microbiology and Infection* 15(Suppl. 2), 216–217.

Kamani, J., Baneth, G., Apanaskevich, D.A., Mumcuoglu, K.Y. and Harrus, S. (2015) Molecular detection of *Rickettsia aeschlimannii* in *Hyalomma* spp. ticks from camels (*Camelus dromedarius*) in Nigeria, West Africa. *Medical and Veterinary Entomology* 29, 205–209.

Keller, C., Krüger, A., Schwarz, N.G., Rakotozandrindrainy, R., Rakotondrainiarivelo, J.P., *et al.* (2016) High detection rate of *Rickettsia africae* in *Amblyomma variegatum* but low prevalence of anti-rickettsial antibodies in healthy pregnant women in Madagascar. *Ticks and Tick-borne Diseases* 7, 60–65.

Kelly, P.J., Beati, L., Mason, P.R., Matthewman, L.A., Roux, V. and Raoult, D. (1996) *Rickettsia africae* sp. nov., the etiological agent of African tick bite fever. *International Journal of Systematic Bacteriology*, 46, 611–614.

Kernif, T., Messaoudene, D., Ouahioune, S., Parola, P., Raoult, D. and Bitam, I. (2012) Spotted fever group rickettsiae identified in *Dermacentor marginatus* and *Ixodes ricinus* ticks in Algeria. *Ticks and Tick-borne Diseases* 3, 380–381.

Kolo, A.O., Sibeko-Matjila, K.P., Maina, A.N., Richards, A.L., Knobel, D.L. and Matjila, P.T. (2016) Molecular detection of zoonotic rickettsiae and *Anaplasma* spp. in domestic dogs and their ectoparasites in Bushbuckridge, South Africa. *Vector-Borne and Zoonotic Diseases* 16, 245–252.

La Scola, B. and Raoult, D. (1997) Laboratory diagnosis of rickettsioses: current approaches to diagnosis of old and new rickettsial diseases. *Journal of Clinical Microbiology* 35, 2715–2727.

Maina, A.N., Jiang, J., Omulo, S.A., Cutler, S.J., Ade, F., *et al.* (2014) High prevalence of *Rickettsia africae* variants in *Amblyomma variegatum* ticks from domestic mammals in rural western Kenya: implications for human health. *Vector-Borne and Zoonotic Diseases* 14, 693–702.

Mediannikov, O., Diatta, G., Fenollar, F., Sokhna, C., Trape, J.-F. and Raoult, D. (2010a) Tick-borne rickettsioses, neglected emerging diseases in rural Senegal. *PLoS Neglected Tropical Diseases* 4, e821.

Mediannikov, O., Trape, J.F., Diatta, G., Parola, P., Fournier, P.E. and Raoult, D. (2010b) *Rickettsia africae*, western Africa. *Emerging Infectious Diseases* 16, 571–573.

Mediannikov, O., Davoust, B., Socolovschi, C., Tshilolo, L., Raoult, D. and Parola, P. (2012a) Spotted fever group rickettsiae in ticks and fleas from the Democratic Republic of the Congo. *Ticks and Tick-borne Diseases* 3, 370–372.

Mediannikov, O., Diatta, G., Zolia, Y., Balde, M.C., Kohar, H., *et al.* (2012b) Tick-borne rickettsiae in Guinea and Liberia. *Ticks and Tick-borne Diseases* 3, 43–48.

Mokrani, N., Parola, P., Tebbal, S., Dalichaouche, M., Aouati, A. and Raoult, D. (2008) *Rickettsia aeschlimannii* infection, Algeria. *Emerging Infectious Diseases* 14, 1814–1815.

Morita, C., El Hussein, AR., Matsuda, E., Abdel Gabbar, K.M., Muramatsu, Y., *et al.* (2004) Spotted fever group rickettsiae from ticks captured in Sudan. *Japanese Journal of Infectious Diseases* 57, 107–109.

Mouffok, N., Parola, P., Lepidi, H. and Raoult, D. (2009) Mediterranean spotted fever in Algeria – new trends. *International Journal of Infectious Diseases* 13, 227–235.

Mura, A., Socolovschi, C., Ginesta, J., Lafrance, B., Magnan, S., *et al.* (2008) Molecular detection of spotted fever group rickettsiae in ticks from Ethiopia and Chad. *Transactions of the Royal Society of Tropical Medicine and Hygiene* 102, 945–949.

Ndeereh, D., Thaiyah, A., Muchemi, G. and Miyunga, A.A. (2017) Molecular surveillance of spotted fever group rickettsioses in wildlife and detection of *Rickettsia sibirica* in a topi (*Damaliscus lunatus* ssp. *jimela*) in Kenya. *Onderstepoort Journal of Veterinary Research* 84, e1–e7.

Ndip, L.M., Fokam, E.B., Bouyer, D.H., Ndip, R.N., Titanji, V.P.K., *et al.* (2004) Detection of *Rickettsia africae* in patients and ticks along the coastal region of Cameroon. *American Journal of Tropical Medicine and Hygiene* 71, 363–366.

Ndip, L.M., Biswas, H.H., Nfonsam, L.E., Lebreton, M., Ndip, R.N., *et al.* (2011) Risk factors for African tick-bite fever in rural central Africa. *American Journal of Tropical Medicine and Hygiene* 84, 608–613.

Ogo, I.N., Fernandez De Mera, I.G., Galindo, R.C., Okubanjo, O.O., Inuwa, H.M., *et al.* (2012) Molecular identification of tick-borne pathogens in Nigerian ticks. *Veterinary Parasitology* 187, 572–577.

Pader, V., Buniak, J.N., Abdissa, A., Adamu, H., Tolosa, T., *et al.* (2012) *Candidatus* Rickettsia hoogstraalii in Ethiopian *Argas persicus* ticks. *Ticks and Tick-borne Diseases* 3, 338–345.

Parola, P., Inokuma, H., Camicas, J., Brouqui, P. and Raoult, D. (2001) Detection and identification of spotted fever group rickettsiae and ehrlichiae in African ticks. *Emerging Infectious Diseases* 7, 1014–1017.

Parola, P., Paddock, C.D. and Raoult, D. (2005) Tick-borne rickettsioses around the world: emerging diseases challenging old concepts. *Clinical Microbiology Reviews* 18, 719–756.

Parola, P., Socolovschi, C., Jeanjean, L., Bitam, I., Fournier, P.-E., *et al.* (2008) Warmer weather linked to tick attack and emergence of severe rickettsioses. *PLoS Neglected Tropical Diseases* 2, e338.

Parola, P., Paddock, C.D., Socolovschi, C., Labruna, M.B., Mediannikov, O., *et al.* (2013) Update on tick-borne rickettsioses around the world: a geographic approach. *Clinical Microbiology Reviews* 26, 657.

Portillo, A., Perez-Martinez, L., Santibanez, S., Blanco, J.R., Ibarra, V. and Oteo, J.A. (2007) Detection of *Rickettsia africae* in *Rhipicephalus* (*Boophilus*) *decoloratus* – ticks from the Republic of Botswana, South Africa. *American Journal of Tropical Medicine and Hygiene* 77, 376–377.

Pretorius, A.M. and Birtles, R.J. (2002) *Rickettsia aeschlimannii*: a new pathogenic spotted fever group rickettsia, South Africa. *Emerging Infectious Diseases* 8, 874.

Pretorius, A.-M. and Birtles, R.J. (2004) Rickettsia *mongolotimonae* infection in South Africa. *Emerging Infectious Diseases* 10, 125–126.

Raoult, D., Fournier, P.E., Abboud, P. and Caron, F. (2002) First documented human *Rickettsia aeschlimannii* infection. *Emerging Infectious Diseases* 8, 748–749.

Reye, A.L., Arinola, O.G., Hubschen, J.M. and Muller, C.P. (2012) Pathogen prevalence in ticks collected from the vegetation and livestock in Nigeria. *Applied and Environmental Microbiology* 78, 2562–2568.

Rovery, C. and Raoult, D. (2008) Mediterranean spotted fever. *Infectious Disease Clinics of North America* 22, 515–530.

Sarih, M., Socolovschi, C., Boudebouch, N., Hassar, M., Raoult, D. and Parola, P. (2008) Spotted fever group rickettsiae in ticks, Morocco. *Emerging Infectious Diseases* 14, 1067–1073.

Sfar, N., M'ghirbi, Y., Letaïef, A., Parola, P., Bouattour, A. and Raoult, D. (2008) First report of *Rickettsia monacensis* and *Rickettsia helvetica* from Tunisia. *Annals of Tropical Medicine and Parasitology* 102, 561–564.

Sfar, N., Kaabia, N., Letaief, A., Rolain, J.M., Parola, P., *et al.* (2009) First molecular detection of *R. conorii* subsp. *conorii* 99 years after the Conor description of Mediterranean spotted fever, in Tunisia. *Clinical Microbiology and Infection* 15, 309–310.

Simpson, G.J.G., Quan, V., Frean, J., Knobel, D.L., Rossouw, J., *et al.* (2018) Prevalence of selected zoonotic diseases and risk factors at a human–wildlife–livestock interface in Mpumalanga Province, South Africa. *Vector-Borne and Zoonotic Diseases* 18, 303–310.

Socolovschi, C., Matsumoto, K., Marie, J.-L., Davoust, B., Raoult, D. and Parola, P. (2007) Identification of rickettsiae, Uganda and Djibouti. *Emerging Infectious Diseases* 13, 1508–1509.

Socolovschi, C., Barbarot, S., Lefebvre, M., Parola, P. and Raoult, D. (2010) *Rickettsia sibirica mongolitimonae* in traveler from Egypt. *Emerging Infectious Diseases* 16, 1495–1496.

Vanegas, A., Keller, C., Krüger, A., Manchang, T.K., Hagen, R.M., *et al.* (2018) Molecular detection of spotted fever group rickettsiae in ticks from Cameroon. *Ticks and Tick-borne Diseases* 9, 1049–1056.

Zhu, Y., Fournier, P.-E., Eremeeva, M. and Raoult, D. (2005) Proposal to create subspecies of *Rickettsia conorii* based on multi-locus sequence typing and an emended description of *Rickettsia conorii*. *BMC Microbiology* 5, 11.

Znazen, A., Rolain, J.-M., Hammami, N., Hammami, A., Jemaa, M.B. and Raoult, D. (2006) *Rickettsia felis* infection, Tunisia. *Emerging Infectious Diseases* 12, 138.

Znazen, A., Hammami, B., Lahiani, D., Ben Jemaa, M. and Hammami, A. (2011) Israeli spotted fever, Tunisia. *Emerging Infectious Diseases* 17, 1328–1330.

eo73 Climate and the Emergence of Tick-Borne Disease in Canada

Patrick Leighton[1]*, L. Robbin Lindsay[2] and Nicholas Ogden[1,2]
[1]*University of Montréal, Saint-Hyacinthe, Canada;* [2]*National Microbiology Laboratory, Canada*

73.1 Expert Opinion

73.1.1 The Canadian context

Geography and climate

Canada sits at the northern range limit for many North American tick species. With its cold winters and short summers, there is less time for ticks to complete their life cycle. Much of northern and western Canada is characterized by habitats that are marginal (e.g. Great Plains and Rocky Mountains) or inhospitable (e.g. Boreal Shield and Arctic Tundra) for most tick species. However, while climate conditions continue to be marginal for many tick species, climate warming is occurring in Canada at over twice the rate observed elsewhere in the world (Bush and Lemmen, 2019), allowing tick ranges to push further north and occupy new areas, thus leading to the recent emergence of tick-borne diseases as a significant public health issue across the country.

Surveillance

Being situated at the northern range limit has also provided a unique opportunity to study the process of tick-borne disease emergence in Canada. In response to the emerging threat of Lyme disease in the USA, Canada put in place high-quality, early and sustained tick surveillance that has provided a unique view of the full process of tick-borne disease emergence over a 30-year period (Ogden *et al.*, 2006b, 2010; Ripoche *et al.*, 2018; Guillot *et al.*, 2020).

73.1.2 A brief history of tick-borne disease in Canada

The recent history of tick-borne disease research in Canada has been shaped to a great extent by the emergence of Lyme disease in the early 1990s.

The pre-Lyme era (before 1990)

Prior to 1990, there were relatively few recognized tick-borne diseases that were transmitted by tick species native to Canada. During this time, the main public health threats from ticks in Canada were Powassan encephalitis (Corrin *et al.*, 2018), tularaemia (Artsob *et al.*, 1984), relapsing fever (Spiller, 1986) and Rocky Mountain spotted fever (Wood and Artsob, 2012). However, human infections with these pathogens occurred sporadically and the burden of human disease was not substantive (Lindquist *et al.*, 2016).

*Email: patrick.a.leighton@umontreal.ca

DOI: 10.1079/9781789249637.0073

Lyme disease emergence (1990 to present)

The tick-borne disease landscape began to shift in the early 1990s as the tick vector of Lyme disease in eastern and central USA, *Ixodes scapularis* (the black-legged tick), invaded eastern Canada, leading to isolated foci of ixodid tick establishment and associated tick-borne disease risk across southern Canada (Ogden *et al.*, 2009). The tick vector of Lyme disease in the western USA, *Ixodes pacificus* (the western black-legged tick), was known to have been present in southern British Columbia for some years and there is no evidence to date that the geographic range of this tick has changed (Mak *et al.*, 2010). However, *I. scapularis* began spreading rapidly in the southern parts of eastern and central Canada to occupy the limits of its climate envelope from the Atlantic coast to the eastern margin of the Great Plains in Manitoba (Bouchard *et al.*, 2015). Other native tick species of potential medical importance also expanded their ranges northward during this period as climate warmed (Dergousoff *et al.*, 2013; Gasmi *et al.*, 2018). By 2010, Lyme disease cases began to spike in areas with established *I. scapularis* populations (Gasmi *et al.*, 2017), sparking new efforts to control growing tick-borne disease risk including major information campaigns by federal and provincial health authorities, the creation of a Federal Framework for Lyme Disease (Government of Canada, 2017) and the launch of the federally funded Canadian Lyme Disease Research Network (Guillot *et al.*, 2020).

Current situation

Lyme disease is now endemic in most Canadian provinces, with over 1000 confirmed human cases per year across the country (Government of Canada, 2020). With their principal tick vectors now abundant, anaplasmosis, babesiosis and Powassan virus are also being detected more frequently in *Ixodes*-endemic regions, and novel *Borrelia* species emerging in North America (e.g. *Borrelia miyamotoi*, *Borrelia mayonii*) have been detected at low frequency in ticks from Canada (Bouchard *et al.*, 2019). As climate conditions become more favourable, other medically important tick species are poised to invade southern Canada. Passive tick surveillance now regularly detects small numbers of lone star ticks (*Amblyomma americanum*) in regions bordering the eastern USA (Gasmi *et al.*, 2019; Nelder *et al.*, 2019) and the climate has been shown to be suitable for *A. americanum* establishment in southern Canada (Ludwig *et al.*, 2016; Sagurova *et al.*, 2019). A more tick-friendly climate also makes Canada vulnerable to invasion by tick species expanding their global distribution, such as the Asian long-horned tick (*Haemaphysalis longicornis*) (Hutcheson *et al.*, 2019). This evolving situation has given rise to novel surveillance initiatives to track changing tick-borne disease risk (Clow *et al.*, 2019), including crowd-sourced tick surveillance by citizen scientists submitting photographs of ticks for identification (Koffi *et al.*, 2017) and a new national network of sentinel surveillance sites (Guillot *et al.*, 2020).

73.1.3 The role of climate

Relationships between tick-borne pathogens and climate have been most extensively studied in *I. scapularis* due to the public health importance of this tick (Bouchard *et al.*, 2019).

Tick physiology and life cycle

Canada's northern climate shapes tick population dynamics primarily through multiple effects of temperature on tick physiology. A short summer combined with cool spring and autumn conditions extend development times for each tick stage and, along with daylength-induced diapause, extend the generation times for *I. scapularis* to 3–4 years, compared with the typical 2 years seen further south (Ogden *et al.*, 2004, 2005). Counterintuitively, harsh winter conditions likely contribute less to mortality as ticks survive air temperatures of $<-20°C$ where woodland habitats provide a litter layer refuge (Ogden *et al.*, 2020). Climate change may affect phenology of the different tick stages, altering the degree of overlap in activity periods among juvenile stages (larvae and nymphs), with potential consequences for the efficiency of tick-borne pathogen transmission cycles and the frequency of co-feeding transmission of certain pathogens (Ogden *et al.*, 2007, 2008a, 2018).

Temperature and tick population growth

Climate effects on the length of the tick life cycle have strongly limited tick population growth in

Canada, with positive population growth only expected (Ogden *et al.*, 2005) and seen (Gabriele-Rivet *et al.*, 2015) in regions with more than 2860 mean annual degree-days >0°C. Limits on tick population growth have knock-on effects for pathogen transmission, since successful pathogen invasion requires an established tick population and the basic reproduction number (R_0) for most tick-borne pathogens is related to tick population density (Ogden *et al.*, 2013a).

Geographic range limits

These physiological constraints on tick development in cooler climates (see Section 'Tick physiology and life cycle' above) impose a theoretical northern range limit for all tick species in Canada. This range limit is expected to shift with climate warming, expanding suitable areas for tick population establishment to the north (Ogden *et al.*, 2006a). However, this expansion will also be limited by the distribution of suitable habitats (Lindsay *et al.*, 1995; Ogden *et al.*, 2008b), with ixodid ticks, for example, surviving poorly in the intensive agricultural landscapes of the Great Plains (Gabriele-Rivet *et al.*, 2017) and in some coniferous forests of the north (Gabriele-Rivet *et al.*, 2015, 2017). Nevertheless, courtesy of the extensive and long-timescale surveillance of *I. scapularis* in Canada, there is much evidence that recent climate warming in Canada has been a significant driver of the emergence of Lyme disease here (Ebi *et al.*, 2017).

Speed of spread

Leighton *et al.* (2012) analysed invasion speeds of *I. scapularis* ticks into Canada based on 20 years of passive tick submissions and found that the rate at which ticks were projected to expand their range within eastern Canada ranged from 55 km/year in the warmest years to 35 km/year in the coolest years observed during the study period. Pathogens invade once *I. scapularis* populations are established, but this may occur rapidly or with a significant lag; in eastern Canada, endemic cycles of *Borrelia burgdorferi* transmission appear approximately 5 years after tick population establishment (Ogden *et al.*, 2013b).

Climate and other ticks in Canada

While this section so far has focused on *I. scapularis*, many of the reported effects of climate are likely important for other medically important ticks at the northern limit of their ranges. *I. pacificus*, the main vector on Canada's west coast, is currently limited in its distribution to the coastal regions in southern British Columbia but expansion beyond this range is likely with climate warming (Mak *et al.*, 2010). *Dermacentor variabilis* has undergone a steady range expansion in western Canada since the early 1980s, pushing north and west in the Great Plains where it now shares a 200 km zone of overlap with *Dermacentor andersoni* (Dergousoff *et al.*, 2013). The northern range limit of *A. americanum* is thought to be governed by similar physiological constraints (temperature-dependent development rate and diapause) as *I. scapularis* (Ludwig *et al.*, 2016) and this species may become established in Canada in the near future. Finally, recent surveys suggest that the majority of tick species in Canada have increased in local abundance over the past decade (Nelder *et al.*, 2014; Gasmi *et al.*, 2018). While most are not frequent vectors of human pathogens, these findings illustrate how the Canadian landscape is becoming generally more suitable for ticks and their associated pathogens.

73.2 Future Predictions

Climate warming is occurring rapidly in Canada. Warmer summers and earlier snow melt are expected to both increase the potential geographic range for most tick species and accelerate the expansion of tick populations within these range limits. Increased temperature and humidity are likely to lead to greater local densities of tick populations, increasing the transmission potential for tick-borne pathogens, the abundance of infected ticks in the environment and, ultimately, the human and animal health burden of tick-borne diseases in Canada.

While most research over the past three decades in Canada has focused on ixodid ticks and the pathogens they carry, the coming decades will likely require a broadened focus as increased climate suitability facilitates the introduction and spread of invasive tick species and their associated pathogens. Modelling studies have been instrumental in translating the complex ecology and physiology of ticks into measures of tick-borne disease risk that are relevant to public health and, in combination with the ongoing collection of high-quality surveillance data, continue to provide useful projections of how climate will shape the future tick-borne disease landscape in Canada.

References

Artsob, H., Spence, L., Surgeoner, G., McCreadie, J., Thorsen, J., *et al.* (1984) Isolation of *Francisella tularensis* and Powassan virus from ticks (Acari: Ixodidae) in Ontario, Canada. *Journal of Medical Entomology* 21, 165–168.

Bouchard, C., Leonard, E., Koffi, J.K., Pelcat, Y., Peregrine, A., *et al.* (2015) The increasing risk of Lyme disease in Canada. *Canadian Veterinary Journal* 56, 693–699.

Bouchard, C., Dibernardo, A., Koffi, J., Wood, H., Leighton, P. and Lindsay, L. (2019) Increased risk of tick-borne diseases with climate and environmental changes. *Canada Communicable Disease Report* 45, 83–89.

Bush, E. and Lemmen, D.S. (2019) *Canada's Changing Climate Report*. Government of Canada, Ottawa.

Clow, K.M., Leighton, P.A., Pearl, D.L. and Jardine, C.M. (2019) A framework for adaptive surveillance of emerging tick-borne zoonoses. *One Health* 7, 100083.

Corrin, T., Greig, J., Harding, S., Young, I., Mascarenhas, M. and Waddell, L.A. (2018) Powassan virus, a scoping review of the global evidence. *Zoonoses and Public Health* 65, 595–624.

Dergousoff, S.J., Galloway, T.D., Lindsay, L.R., Curry, P.S. and Chilton, N.B. (2013) Range expansion of *Dermacentor variabilis* and *Dermacentor andersoni* (Acari: Ixodidae) near their northern distributional limits. *Journal of Medical Entomology* 50, 510–520.

Ebi, K.L., Ogden, N.H., Semenza, J.C. and Woodward, A. (2017) Detecting and attributing health burdens to climate change. *Environmental Health Perspectives* 125, 085004.

Gabriele-Rivet, V., Arsenault, J., Badcock, J., Cheng, A., Edsall, J., *et al.* (2015) Different ecological niches for ticks of public health significance in Canada. *PLoS ONE* 10, e0131282.

Gabriele-Rivet, V., Koffi, J.K., Pelcat, Y., Arsenault, J., Cheng, A., *et al.* (2017) A risk model for the Lyme disease vector *Ixodes scapularis* (Acari: Ixodidae) in the prairie provinces of Canada. *Journal of Medical Entomology* 54, 862–868.

Gasmi, S., Ogden, N., Lindsay, L., Burns, S., Fleming, S., *et al.* (2017) Emerging infections: surveillance for Lyme disease in Canada: 2009–2015. *Canada Communicable Disease Report* 43, 194–199.

Gasmi, S., Bouchard, C., Ogden, N.H., Adam-Poupart, A., Pelcat, Y., *et al.* (2018) Evidence for increasing densities and geographic ranges of tick species of public health significance other than *Ixodes scapularis* in Québec, Canada. *PLoS ONE* 13, e0201924.

Gasmi, S., Ogden, N.H., Ripoche, M., Leighton, P.A., Lindsay, R.L., *et al.* (2019) Detection of municipalities at-risk of Lyme disease using passive surveillance of *Ixodes scapularis* as an early signal: a province-specific indicator in Canada. *PLoS ONE* 14, e0212637.

Government of Canada (2017) *Lyme Disease in Canada – A Federal Framework*. Ministry of Health, Ottawa.

Government of Canada (2020) Surveillance of Lyme disease. Available at: https://www.canada.ca/en/public-health/services/diseases/lyme-disease/surveillance-lyme-disease.html (accessed 1 December 2020).

Guillot, C., Badcock, J., Clow, K., Cram, J., Dergousoff, S., *et al.* (2020) Sentinel surveillance of Lyme disease risk in Canada, 2019: results from the first year of the Canadian Lyme Sentinel Network (CaLSeN). *Canada Communicable Disease Report* 46, 354–361.

Hutcheson, H.J., Lindsay, L.R. and Dergousoff, S.J. (2019) *Haemaphysalis longicornis*: a tick of considerable importance, now established in North America. *Canadian Journal of Public Health* 110, 118–119.

Koffi, J., Savage, J., Thivierge, K., Lindsay, L., Bouchard, C., *et al.* (2017) Evaluating the submission of digital images as a method of surveillance for *Ixodes scapularis* ticks. *Parasitology* 144, 877–883.

Leighton, P.A., Koffi, J.K., Pelcat, Y., Lindsay, L.R. and Ogden, N.H. (2012) Predicting the speed of tick invasion: an empirical model of range expansion for the Lyme disease vector *Ixodes scapularis* in Canada. *Journal of Applied Ecology* 49, 457–464.

Lindquist, E.E., Galloway, T.D., Artsob, H., Lindsay, L.R., Drebot, M., *et al.* (2016) *A Handbook to the Ticks of Canada (Ixodida: Ixodidae, Argasidae)*. Biological Survey of Canada.

Lindsay, L.R., Barker, I.K., Surgeoner, G.A., McEwen, S.A., Gillespie, T.J. and Robinson, J.T. (1995) Survival and development of *Ixodes scapularis* (Acari: Ixodidae) under various climatic conditions in Ontario, Canada. *Journal of Medical Entomology* 32, 143–152.

Ludwig, A., Ginsberg, H.S., Hickling, G.J. and Ogden, N.H. (2016) A dynamic population model to investigate effects of climate and climate-independent factors on the lifecycle of *Amblyomma americanum* (Acari: Ixodidae). *Journal of Medical Entomology* 53, 99–115.

Mak, S., Morshed, M. and Henry, B. (2010) Ecological niche modeling of Lyme disease in British Columbia, Canada. *Journal of Medical Entomology* 47, 99–105.

Nelder, M.P., Russell, C., Lindsay, L.R., Dhar, B., Patel, S.N., *et al.* (2014) Population-based passive tick surveillance and detection of expanding foci of blacklegged ticks *Ixodes scapularis* and the Lyme disease agent *Borrelia burgdorferi* in Ontario, Canada. *PLoS ONE* 9, e105358.

Nelder, M.P., Russell, C.B., Clow, K.M., Johnson, S., Weese, J.S., *et al.* (2019) Occurrence and distribution of *Ambylomma americanum* as determined by passive surveillance in Ontario, Canada (1999–2016). *Ticks and Tick-borne Diseases* 10, 146–155.

Ogden, N., Lindsay, L., Beauchamp, G., Charron, D., Maarouf, A., *et al.* (2004) Investigation of relationships between temperature and developmental rates of tick *Ixodes scapularis* (Acari: Ixodidae) in the laboratory and field. *Journal of Medical Entomology* 41, 622–633.

Ogden, N., Bigras-Poulin, M., O'Callaghan, C., Barker, I., Lindsay, L., *et al.* (2005) A dynamic population model to investigate effects of climate on geographic range and seasonality of the tick *Ixodes scapularis*. *International Journal for Parasitology* 35, 375–389.

Ogden, N., Maarouf, A., Barker, I., Bigras-Poulin, M., Lindsay, L., *et al.* (2006a) Climate change and the potential for range expansion of the Lyme disease vector *Ixodes scapularis* in Canada. *International Journal for Parasitology* 36, 63–70.

Ogden, N., Trudel, L., Artsob, H., Barker, I., Beauchamp, G., *et al.* (2006b) *Ixodes scapularis* ticks collected by passive surveillance in Canada: analysis of geographic distribution and infection with Lyme borreliosis agent *Borrelia burgdorferi*. *Journal of Medical Entomology* 43, 600–609.

Ogden, N.H., Bigras-Poulin, M., O'Callaghan, C.J., Barker, I.K., Kurtenbach, K., *et al.* (2007) Vector seasonality, host infection dynamics and fitness of pathogens transmitted by the tick *Ixodes scapularis*. *Parasitology* 134, 209–227.

Ogden, N.H., Bigras-Poulin, M., Hanincova, K., Maarouf, A., O'Callaghan, C.J. and Kurtenbach, K. (2008a) Projected effects of climate change on tick phenology and fitness of pathogens transmitted by the North American tick *Ixodes scapularis*. *Journal of Theoretical Biology* 254, 621–632.

Ogden, N.H., St-Onge, L., Barker, I.K., Brazeau, S., Bigras-Poulin, M., *et al.* (2008b) Risk maps for range expansion of the Lyme disease vector, *Ixodes scapularis*, in Canada now and with climate change. *International Journal of Health Geographics* 7, 24.

Ogden, N.H., Lindsay, L.R., Morshed, M., Sockett, P.N. and Artsob, H. (2009) The emergence of Lyme disease in Canada. *Canadian Medical Association Journal* 180, 1221–1224.

Ogden, N.H., Bouchard, C., Kurtenbach, K., Margos, G., Lindsay, L.R., *et al.* (2010) Active and passive surveillance and phylogenetic analysis of *Borrelia burgdorferi* elucidate the process of Lyme disease risk emergence in Canada. *Environmental Health Perspectives* 118, 909–914.

Ogden, N., Mechai, S. and Margos, G. (2013a) Changing geographic ranges of ticks and tick-borne pathogens: drivers, mechanisms and consequences for pathogen diversity. *Frontiers in Cellular and Infection Microbiology* 3, 46.

Ogden, N.H., Lindsay, L.R. and Leighton, P.A. (2013b) Predicting the rate of invasion of the agent of Lyme disease *Borrelia burgdorferi*. *Journal of Applied Ecology* 50, 510–518.

Ogden, N.H., Pang, G., Ginsberg, H.S., Hickling, G.J., Burke, R.L., *et al.* (2018) Evidence for geographic variation in life-cycle processes affecting phenology of the Lyme disease vector *Ixodes scapularis* (Acari: Ixodidae) in the United States. *Journal of Medical Entomology* 55, 1386–1401.

Ogden, N.H., Beard, C.B., Ginsberg, H.S. and Tsao, J.I. (2020) Possible effects of climate change on ixodid ticks and the pathogens they transmit: predictions and observations. *Journal of Medical Entomology* 2020, tjaa220.

Ripoche, M., Lindsay, L.R., Ludwig, A., Ogden, N.H., Thivierge, K. and Leighton, P.A. (2018) Multi-scale clustering of Lyme disease risk at the expanding leading edge of the range of *Ixodes scapularis* in Canada. *International Journal of Environmental Research and Public Health* 15, 603.

Sagurova, I., Ludwig, A., Ogden, N.H., Pelcat, Y., Dueymes, G. and Gachon, P. (2019) Predicted northward expansion of the geographic range of the tick vector *Amblyomma americanum* in North America under future climate conditions. *Environmental Health Perspectives* 127, 107014.

Spiller, G.W. (1986) Tick-borne relapsing fever due to *Borrelia hermsii* in British Columbia. *Canadian Medical Association Journal* 134, 46–47.

Wood, H. and Artsob, H. (2012) Spotted fever group rickettsiae: a brief review and a Canadian perspective. *Zoonoses and Public Health* 59, 65–79.

eo74 Climate Change Impacts on *Ixodes ricinus* Ticks in Scotland and Implications for Lyme Disease Risk

Lucy Gilbert
University of Glasgow, Glasgow, UK

74.1 Expert Opinion

74.1.1 Changes in *Ixodes ricinus* and Lyme disease cases in Scotland

Ixodes ricinus is a generalist tick, feeding on almost any land vertebrate, and has an extremely broad distribution throughout most of Europe and into Asia. It is abundant in many areas of Scotland and is vector to a wide range of tick-borne pathogens, including *Borrelia burgdorferi* sensu lato (s.l.), the complex of bacteria that cause Lyme disease. *I. ricinus* abundance has increased in many areas of the UK over the last few decades (Scharlemann *et al.*, 2008), including Scotland (Kirby *et al.*, 2004). Lyme disease incidence has also increased over the last few decades in the UK (Cairns *et al.*, 2019), including Scotland, where reported human cases (based on positive blood samples) increased around 30-fold during a 10-year period, from fewer than 10 cases per year in the late 1990s to around 300 per year 10 years later (Health Protection Scotland, 2016). Meanwhile, a UK-wide study using clinical diagnosis of Lyme disease as well as blood testing estimated the number of diagnosed cases of Lyme disease to have increased from fewer than 1000 in 2001 to almost 8000 in 2012 (>12 per 100,000 people). Of all the UK nations, Scotland suffers the highest incidence: the estimated number of diagnosed cases was almost 2000 in 2012 (>37 per 100,000 people) (Cairns *et al.*, 2019).

It is extremely challenging to address, in retrospect, the question of what has caused these increases in *I. ricinus* tick abundance and Lyme disease cases. However, we can examine which factors have changed in these time frames that affect tick survival, fecundity or activity.

74.1.2 Scotland's changing climate

The UK in general has been getting warmer and wetter, with more frequent extreme weather events. UKCP18 (Met Office, 2019) reports that the last decade (2009–2018) was around 1°C warmer than the pre-industrial period (1850–1900), i.e. similar to the global average. The last decade in the UK was 12–13% wetter than the 1960s, 1970s and 1980s. These increases in rainfall have been strongest in Scotland, in contrast to minimal rainfall changes in the south and east of England. These trends of milder, wetter winters and warmer, drier summers are set to continue; however, the changes projected for Scotland are among the least in Europe. For example, from 1961–1999 to 2070–2099 summers are projected to be only 1.2–1.8°C warmer in Scotland compared with around 2.4°C in

*Email: lucy.gilbert@glasgow.ac.uk

 Climate, Ticks and Disease (ed. P. Nuttall)
DOI: 10.1079/9781789249637.0074

Norway and south-east England and around 5°C across much of Southern Europe (Christensen *et al.*, 2011).

74.1.3 Can we attribute Scotland's increases in *Ixodes ricinus* to climate change?

Warmer temperatures speed up interstadial development, oviposition and egg development rates of *I. ricinus* (MacLeod, 1935; Randolph *et al.*, 2002). In addition, if the air is humid enough, rising temperatures also greatly increase the proportion of *I. ricinus* that are active (Gilbert *et al.*, 2014). These mechanisms translate into population-level effects: there are higher densities of questing *I. ricinus* nymphs in warmer areas and in warmer survey conditions in Scotland (James *et al.*, 2013; Millins *et al.*, 2016). Similarly, the prevalence of *B. burgdorferi* s.l. in questing nymphs and the density of infected nymphs were higher in areas of Scotland that had warmer climates (as estimated by higher growing degree-days) (Millins *et al.*, 2016) or lower altitudes (James *et al.*, 2013). In addition, there are more Lyme disease cases reported in years with warmer winters and springs (Mavin *et al.*, 2005). These studies indicate that in a cool country such as Scotland, a warmer climate benefits *I. ricinus* and may increase Lyme disease risk.

Several studies have shown how *I. ricinus* activity requires low saturation deficit (the drying power of the air, a function of temperature and relative humidity) (Perret *et al.*, 2000; Randolph, 2004). In hot, dry areas of Southern Europe ticks are more likely to be limited by dryness rather than cold temperatures, so an increase in rainfall could dramatically aid tick activity and survival. However, Scotland is a famously wet and cool country, so ticks are probably not routinely inhibited by high saturation deficits and an increase in rainfall is much less likely to increase tick abundance or distribution. In fact, multivariate analysis (that took account of variance due to altitude, hosts, vegetation and season) of data on questing *I. ricinus* across Scotland shows that areas with higher rainfall and conditions of higher relative humidity are associated with fewer (rather than more) questing *I. ricinus* (James *et al.*, 2013). Similarly, the likelihood of *I. ricinus* nymphs being infected with *B. burgdorferi* s.l. was also negatively associated with rainfall and relative humidity (James *et al.*, 2013). This suggests that in a cool, wet country such as Scotland, it can be too wet for *I. ricinus*, with associated effects on Lyme disease hazard.

Thus, Scotland's increased rainfall is unlikely to have contributed to the observed increase in *I. ricinus* (Kirby *et al.*, 2004), but the temperature increase might have, to a limited extent. However, it is likely that other factors have had stronger impacts on *I. ricinus* in Scotland. In particular, deer, which are the main reproduction host for *I. ricinus* in Scotland (Gilbert *et al.*, 2012), have greatly increased in distribution and abundance in Scotland (Clutton-Brock *et al.*, 2004) in the last few decades.

74.1.4 Indirect climate change impacts

Climate change may have influenced ticks indirectly, such as through changes in habitat and hosts. There is currently a paucity of data to investigate this and because *I. ricinus* is a generalist, feeding on almost all terrestrial vertebrates, changes in host community composition may not have such clear impacts on this tick species as on specialist ticks that have a narrow choice of hosts. However, deer abundance is positively associated with *I. ricinus* abundance in Scotland (Gilbert, 2010; Ruiz-Fons and Gilbert, 2010; Gilbert *et al.*, 2012; James *et al.*, 2013) and this can translate to Lyme disease hazard (density of infected nymphs) both empirically (James *et al.*, 2013) and theoretically (Li *et al.*, 2016). The rise in red deer abundance in Scotland (by around 400% since the 1960s) is thought to be related to changes in winter weather conditions as well as a reduction in culling rates by land managers (Clutton-Brock *et al.*, 2004). Changes in human behaviour due to a change in climate may further influence the risk of exposure of people to tick bites and tick-borne disease. For example, Mavin *et al.* (2005) suggested that drier weather from April to July in some years was associated with more cases of Lyme disease in Scotland as a result of people spending more time outdoors.

74.2 Future Predictions

74.2.1 Predictions due to direct impacts of climate change

Not only increases in hosts, but also increased temperatures, are predicted to increase *I. ricinus* densities in the UK context (Dobson and Randolph, 2011). Mechanistic agent-based models predict peak values of Lyme disease hazard in Scotland to increase two- to sevenfold as the climate warms by 1–2°C (Li *et al.*, 2016).

Altitude

I. ricinus has a huge geographic distribution across most of Europe and into Asia, encompassing an impressively large climate envelope. Scotland lies well within its geographic distribution, but land above 550 m (the highest of Scotland's hills reach 1400 m) is currently mostly tick-free (Fig. 74.1). This is because these cold, higher-altitude areas currently lie outside this tick's climate envelope (Gilbert, 2010). As Scotland's climate is projected to warm by 1.2–1.8°C by 2070–2099 (Christensen *et al.*, 2011), we may predict that the altitudinal threshold of *I. ricinus* in Scotland will shift from the current 550 m to almost 800 m elevation (see also Li *et al.*, 2016).

Seasonality

The length of the questing season is also predicted to expand. Currently, seasonal temperatures in Scotland dictate that most *I. ricinus* are inactive in March with a steep increase in the proportion of active ticks through April and May (Gilbert, 2010). In 50–80 years' time the start of this pattern of seasonal activity in spring is predicted to advance by a month in Scotland (Gilbert, 2010; Li *et al.*, 2016).

In addition, the projected warmer summers are likely to cause a more pronounced second peak in questing tick abundance in late summer/autumn. Surveys using blanket drag transects of questing nymphs as well as studies counting

Fig. 74.1. A walker at around 800 m elevation in Scotland's West Highlands, well above the current altitudinal threshold for *Ixodes ricinus* in Scotland, which is around 550 m. However, depending on the climate warming scenario, models predict this threshold to increase to almost 800 m elevation, putting walkers like this at higher risk of exposure to ticks. (Author's own photograph.)

larvae, nymphs and adult tick burdens on hosts have mostly revealed one seasonal peak in tick abundance (in late spring or summer) in most of Scotland (L. Gilbert, 2021, unpublished results). However, in warmer conditions tick activity can start earlier in the year and interstadial development rates are faster, enabling a second cohort of ticks to emerge later in the same year, manifested as a second peak in autumn (Dobson and Randolph, 2011). Temperatures are currently too cool in most of Scotland (except the west coast) to allow time for a second cohort to develop and emerge in the same season; however, as the climate warms a second peak may become more apparent in more Scottish areas.

74.2.2 Resilience to climate change?

I. ricinus, along with *Ixodes scapularis* in North America, has perhaps the broadest geographic distribution, demonstrating that it has adapted to a wide range of environments. Experiments (Gilbert *et al.*, 2014) have shown how fewer *I. ricinus* from warm climates (south-central France) are active at cool temperatures than those from cool climates (Scotland). This clearly demonstrates adaptation to local thermal climates which might suggest that, as Scotland's temperatures increase towards those of present-day France, Scotland's *I. ricinus* may not increase their activity as much as assumed in predictive models.

74.2.3 Green recovery impacts

An important indirect impact of climate change may be anthropogenic, especially the response of people (governments and non-governmental organizations) to climate change by incentivizing and implementing 'green recovery' schemes in an attempt to reach 'net zero'. For example, current policy to create more woodlands might increase *I. ricinus* and Lyme disease risk. Woodlands in Aberdeenshire, north-east Scotland, contained double the density of questing *I. ricinus* nymphs and five times the density of nymphs infected with *B. burgdorferi* s.l. compared with adjacent open habitats (Gilbert, 2016). However, deer management (Gilbert *et al.*, 2012) and type of woodland (James *et al.*, 2013) will affect this relationship, so it is difficult to predict how future woodland creation will impact on ticks and Lyme disease risk without knowing the type, management and extent of afforestation.

References

Cairns, V., Wallenhorst, C., Rietbrock, S. and Martinez, C. (2019) Incidence of Lyme disease in the UK: a population-based cohort study. *BMJ Open* 9, e025916.

Christensen, O.B., Goodess, C.M., Harris, I. and Watkiss, P. (2011) *European and Global Climate Change: Discussion of Climate Change Model Outputs, Scenarios and Uncertainty in the EC RTD ClimateCost Project. The ClimateCost Project, Final Report, Volume 1: Europe*. Stockholm Environment Institute, Sweden. Available at: http://www.climatecost.cc/images/Policy_brief_1_Projections_05_lowres.pdf (accessed 21 June 2021).

Clutton-Brock, T.H., Coulson, T. and Milner, J.M. (2004) Red deer stocks in the highlands of Scotland. *Nature* 429, 261–262.

Dobson, A.D.M. and Randolph, S.E. (2011) Modelling the effects of recent changes in climate, host density and acaricide treatments on population dynamics of *Ixodes ricinus* in the UK. *Journal of Applied Ecology* 48, 1029–1037.

Gilbert, L. (2010) Altitudinal patterns of tick and host abundance: a potential role for climate change in regulating tick-borne diseases? *Oecologia* 162, 217–225.

Gilbert, L. (2016) How landscapes shape Lyme borreliosis risk. In: Braks, M.A.H., van Wieren, S.E., Takken, W. and Sprong, H. (eds) *Ecology and Prevention of Lyme Borreliosis. Ecology and Control of Vector-borne Diseases*, vol. 4. Wageningen Academic Publishers, Wageningen, the Netherlands, pp. 161–171. https://doi.org/10.3920/978-90-8686-838-4_11

Gilbert, L. Maffey, G.L., Ramsay, S.L. and Hester, A.J. (2012) The effect of deer management on the abundance of *Ixodes ricinus* in Scotland. *Ecological Applications* 22, 658–667.

Gilbert, L., Aungier, J. and Tomkins, J.L. (2014) Climate of origin affects tick (*Ixodes ricinus*) host-seeking behavior in response to temperature: implications for resilience to climate change? *Ecology and Evolution* 4, 1186–1198.

Health Protection Scotland (2016) Lyme Disease, Scotland, Annual Totals as at 28 July 2016. Available at: https://hpspubsrepo.blob.core.windows.net/hps-website/nss/1761/documents/1_lyme-disease.pdf (accessed 18 November 2020).

James, M.C., Bowman, A.S., Forbes, K.J., Lewis, F., McLeod, J.E. and Gilbert, L. (2013) Environmental determinants of *Ixodes ricinus* ticks and the incidence of *Borrelia burgdorferi* sensu lato, the agent of Lyme borreliosis, in Scotland. *Parasitology* 140, 237–246.

Kirby, A.D., Smith, A.A., Benton, T.G. and Hudson, P.J. (2004) Rising burden of immature sheep ticks (*Ixodes ricinus*) on red grouse (*Lagopus lagopus scoticus*) chicks in the Scottish uplands. *Medical and Veterinary Entomology* 18, 67–70.

Li, S., Gilbert, L., Harrison, P.A. and Rounsevell, M.D.A. (2016) Modelling the seasonality of Lyme disease risk and the potential impacts of a warming climate within the heterogeneous landscapes of Scotland. *Journal of the Royal Society Interface* 13, 20160140.

MacLeod, J. (1935) *Ixodes ricinus* in relation to its physical environment: II. *The factors governing survival and activity. Parasitology* 27, 123–144.

Mavin, S., Joss, A. and Ho-Yen, D. (2005) Incidence of Scottish Lyme disease appears to be related to the effects of weather on survival and human behaviour. *Journal of the Royal College of Physicians, Edinburgh* 35, 112–115.

Met Office (2019) *UK Climate Projections: Headline Findings*. Met Office, Exeter, UK. Available at: https://www.metoffice.gov.uk/binaries/content/assets/metofficegovuk/pdf/research/ukcp/ukcp-headline-findings-v2.pdf (accessed 18 November 2020).

Millins, C., Gilbert, L., Johnson, P., James, M., Kilbride, E., *et al.* (2016) Heterogeneity in the abundance and distribution of *Ixodes ricinus* and *Borrelia burgdorferi* (sensu lato) in Scotland: implications for risk prediction. *Parasites & Vectors* 9, 595.

Perret, J.L., Guigoz, E., Rais, O. and Gern, L. (2000) Influence of saturation deficit and temperature on *Ixodes ricinus* tick questing activity in a Lyme borreliosis-endemic area (Switzerland). *Parasitology Research* 86, 554–557.

Randolph, S.E. (2004) Tick ecology: processes and patterns behind the epidemiological risk posed by ixodid ticks as vectors. *Parasitology* 129(Suppl.), S37–S65.

Randolph, S.E., Green, R.M., Hoodless, A.N. and Peacey, M.F. (2002) An empirical quantitative framework for the seasonal population dynamics of the tick *Ixodes ricinus*. *International Journal for Parasitology* 32, 979–989.

Ruiz-Fons, F. and Gilbert, L. (2010) The role of deer as vehicles to move ticks, *Ixodes ricinus*, between contrasting habitats. *International Journal for Parasitology* 40, 1013–1020.

Scharlemann, J.P.W., Johnson, P.J., Smith, A.A., Macdonald, D.W. and Randolph, S.E. (2008) Trends in ixodid tick abundance and distribution in Great Britain. *Medical and Veterinary Entomology* 22, 238–247.

eo75 Possible Impact of Climate and Environmental Change on Ticks and Tick-Borne Disease in England

Jolyon Medlock* **and Kayleigh Hansford**
Public Health England, Salisbury, UK

75.1 Expert Opinion

75.1.1 Introduction

Although there are more than 20 different species of tick in the UK, here we focus on *Ixodes ricinus*, the main tick species and the most important to human health (Cull *et al.*, 2020). *Dermacentor reticulatus*, *Haemaphysalis punctata*, *Hyalomma* spp., *Rhipicephalus sanguineus* species complex (hereinafter *R. sanguineus*) are also included, as changes in distribution and/or importation of these species is already occurring and has the potential to change in the future (Medlock *et al.*, 2017, 2018; Hansford *et al.*, 2018). Other tick species are reported to bite humans in the UK, but they are not addressed here due to lower importance to human health.

The survival, density and distribution of ticks are determined by three key elements: microclimate, habitat and hosts (Randolph, 2000; Anderson and Magnarelli, 2008; Estrada-Peña, 2015), all of which can be impacted by climate change (Gray *et al.*, 2009; Medlock and Leach, 2015; Baylis, 2017). The public health risk from ticks is further influenced by human behaviour and the way in which we access and manage the environment where *I. ricinus* or key tick hosts are found (Černý *et al.*, 2020). This expert opinion considers how these factors influence tick-borne disease (TBD) transmission in a changing climate, first by discussing direct effects of climate change on ticks and TBD, and second the indirect effects and environmental changes that make direct comparisons of climate and ticks so challenging.

75.1.2 The possible direct effects of climate change on ticks and tick-borne disease

Increased temperatures, heatwaves, mild winters on survival and expansion of ticks

A warming climate can influence ectothermic invertebrate vectors by impacting their geographic range, density and seasonality. Changing temperatures or extreme weather such as heatwaves or flooding are likely to impact ticks. However, this is not a linear relationship and many other factors related to tick hosts, habitats and human behaviour might impact to compound or negate this climate influence (Mannelli *et al.*, 2012; Medlock *et al.*, 2013; Sprong *et al.*, 2018; van Oort *et al.*, 2020).

*Email: jolyon.medlock@phe.gov.uk

DOI: 10.1079/9781789249637.0075

The role of temperature on native ticks

In simplistic terms, changing temperatures are forecast to bring milder winters and hotter summers in the UK (Met Office, 2019). In England, *I. ricinus* seasonal questing activity peaks during warming months of April through to early July, before activity declines as temperature rises, with an occasional peak in autumn, depending upon the weather and locality (Randolph, 2004). Generally, variability in this phenology can be explained by milder winters initiating earlier questing activity in spring, with ticks remaining active at low densities through the winter months. Hot and dry periods in summer can lead to a decline in questing activity or density of host-seeking *I. ricinus*, due to high temperatures and declining potential evaporation. Increased rainfall can potentially have a negative impact on nymph density (Rosà *et al.*, 2018). Autumnal densities tend to be lower compared with spring (10% in southern England studies) (Public Health England, 2021, unpublished results), but in some years a peak in autumnal questing is reported, likely linked to both a suppression in spring densities and an inability to quest during hot summers, as well as unusually warm autumnal days. Changes in weather patterns, therefore, have the potential to modify the activity/exposure periods for ticks, which in turn can influence risk at the local level (Gilbert *et al.*, 2014; Li *et al.*, 2016).

Elsewhere in Europe, a warming climate has been linked to altitudinal and latitudinal changes in the distribution of *I. ricinus* (Danielová *et al.*, 2004; Jaenson *et al.*, 2012; Hvidsten *et al.*, 2020). In the UK, *I. ricinus* occurs as far north as the Northern Isles and as far south as western Cornwall (Cull *et al.*, 2018), so we do not expect latitudinal range expansion with climate change. As the UK is a predominantly lowland country, *I. ricinus* occurs throughout the lowlands below 700 m where there is suitable habitat; and although there may be some altitudinal expansion, this will be into montane habitats, where the landscape is largely tree-less and likely to be unsuitable for *I. ricinus* establishment (Gilbert, 2010).

For other native species, such as *D. reticulatus*, localized spread in distribution has been reported, both in the UK and elsewhere in Europe (Medlock *et al.*, 2017; Drehmann *et al.*, 2020; Rubel *et al.*, 2020a; Kiewra *et al.*, 2021). Although some of this has been attributed to climate change, other factors including habitat and land-use change, as well as host composition and spread by livestock, are also key factors (Gray *et al.*, 2009; Földvári *et al.*, 2016; Medlock *et al.*, 2017). *D. reticulatus* host-seeking tends to peak in cool temperatures in March and remains active during winter, but these patterns could change under future climate scenarios. *H. punctata* is another important species changing distribution in the UK, likely as a result of host movements and land management practices (Medlock *et al.*, 2018). To the best of our knowledge, there have been no studies on the potential impact of climate change on this species.

Non-native ticks and climatic change

It is a different story for the non-native tick species. *R. sanguineus* and *Hyalomma marginatum* are tick species of the Mediterranean and Africa and so prosper in warmer, often more arid conditions (Dantas-Torres, 2010; Valcárcel *et al.*, 2020). Northward expansion of these species in Europe could occur under warmer climate scenarios (Gray *et al.*, 2009) and both species are expected to benefit from warmer summers or heatwaves in the UK. *H. marginatum*, a vector of Crimean–Congo haemorrhagic fever virus (CCHFV), is regularly imported as engorged nymphs on migratory birds (Jameson *et al.*, 2012). Historical records and recent research have implicated particular migratory bird species which acquire larvae before or during their migration that moult to the nymphal stage and feed while upon the same host (Hubálek *et al.*, 2020). Upon arrival, the nymphal tick's ability to complete its moult to a host-seeking adult stage, in order to pose a disease risk, or to facilitate local establishment, has long been considered climatically limited in a UK setting (Estrada-Peña *et al.*, 2011). However, during the summer heatwave of 2018, where July mean maximum temperatures exceeded 3.5°C above the 1981–2010 climate averages, conditions were suitable for nymphal moulting of several *Hyalomma* species in Northern Europe, including both *H. marginatum* (on a human in Norfolk) and *Hyalomma rufipes* (on a horse in Dorset) in England (Hansford *et al.*, 2019; McGinley *et al.*, 2021). These two-host tick species can disperse considerable distances as they remain on their host during the larval moult. On these occasions there was no further evidence of tick establishment. Although both ticks tested negative for CCHFV, both tested positive for the spotted fever group rickettsia, *Rickettsia aeschlimannii*. These occurrences can be expected to become common under warmer climatic conditions (Grandi *et al.*, 2020) and biting of humans by infected *Hyalomma* could occur as a result. Incidences of Crimean–Congo haemorrhagic fever (CCHF) on the main bird migration routes in Iberia have been fairly rare (England *et al.*, 2016) until recently, with a small number of fatal CCHF cases in western Spain (Negredo *et al.*, 2017) associated with *Hyalomma lusitanicum* (Estrada-Peña *et al.*, 2012). This species is less likely to be imported into the UK by migratory birds and although it is imported on travelling dogs (Hansford *et al.*, 2016, 2018), it requires drier and consistently warmer conditions compared with *H. marginatum* (Valcárcel *et al.*, 2020).

R. sanguineus is regularly imported on travelling and rescued dogs in the UK (Hansford *et al.*, 2017a; Wright *et al.*, 2018). Although it does not currently occur outdoors, it likely has the potential to do so, in hotter summers (Gray *et al.*, 2009). It is able to overwinter and survive indoors and house infestations in England have been linked to travelled dogs or the adoption of rescued dogs (Hansford *et al.*, 2015). Although *H. marginatum* and *R. sanguineus* are recognized vectors of human disease (e.g. *Rickettsia conorii*), there have so far been no incidences of local human cases. Much like the endemic tick species already discussed, introduction and subsequent establishment of these non-native species will not be driven by climate change alone.

Changing temperatures on tick-borne disease

Like so many mosquito-borne viruses that are acutely affected by increases in temperature on their extrinsic incubation period, temperature changes also impact on TBDs. Pathogens such as *Borrelia burgdorferi* sensu lato (s.l.) and tick-borne encephalitis virus (TBEV) are impacted by temperature, but in complex and differing ways. *Borrelia* infection rates in *I. ricinus* are influenced by the diversity and abundance of mammal and bird reservoir hosts in the environment (Strnad *et al.*, 2017), with differing genospecies dominating and fluctuating regionally (Public Health England, 2021, unpublished results). These faunal variations are in turn affected by habitats, which are shaped by climate and land-use factors. This makes modelling impacts of climate change on TBDs such as Lyme borreliosis complicated (Li *et al.*, 2016; Fernández-Ruiz and Estrada-Peña, 2020), as discussed below (see Section 75.1.3).

In the case of TBEV, a change in temperature, whether that be through rapid cooling in autumn or rapid warming in spring, has been linked to the sporadic foci of TBEV transmission. This affects the synchronicity of tick stage-specific seasonal activity, enabling co-feeding larvae and nymphs on small mammals which facilitates subdermal virus transmission to create foci of TBEV-infected questing nymphs, creating localized public health risk (Rizzoli *et al.*, 2019). The key research question is what drives these foci. Recent evidence of TBEV in the UK in certain regional foci (currently in Thetford Forest and the New Forest) could be linked to a temperature driver (Holding *et al.*, 2019), but further research is required. Met Office climate data show considerable temperature anomalies for both areas, particularly for Thetford where 1–2°C annual/spring temperature anomalies have been recorded in recent years. In addition to temperature, aspects of soil or geology (i.e. sandy soils) may enhance the warming and cooling effects that promote co-feeding mediated by small mammals, or foci could be related to tick host densities. With increased risk from TBEV suggested for Northern and Central Europe (Semenza and Suk, 2018), improved climate/weather-based modelling, and empirical field data on questing seasonality and co-feeding on small mammals and deer, may help unravel these complexities.

75.1.3 Possible indirect effects of climate change on ticks and tick-borne disease in a changing environment

The UK Government's 25 Year Environment Plan published in 2018 (https://www.gov.uk/government/publications/25-year-environment-plan, accessed 28 June 2021) promotes a

'Green Future' with a vision to 'protect and grow the natural capital' and will likely lead to a much-changed landscape. It promotes four main areas: (i) to increase woodland to 12% cover by 2060 through the planting of 180,000 hectares of trees, including supporting the Northern Forest; (ii) to design and deliver a new Environmental Land Management system; (iii) to provide opportunities for the reintroduction of native species; and (iv) to strengthen the planning policy framework to promote 'net biodiversity gains' as part of urban development. In view of the current and proposed future environmental changes, any predictions of the impact of future climate change on vectors and vector-borne disease must consider these significant environmental changes, which are, in part, an adaptation to climate change but also to mitigate biodiversity loss. Predictions must also consider the indirect effects of a changing climate on the tick's habitat and their hosts, which are discussed below.

Changes in landscape management

Climate may be suitable for ticks; however, habitat is a key factor in mediating climate suitability, supporting both the microclimate and the assemblage of tick host mammals, birds and livestock. Although woodland is important for *I. ricinus* populations, suitability is not uniform across all woodland types. Fragmented woodland, with mosaic habitat structure of glades and rides or areas of scrub grassland, provide more favourable habitats for ticks, likely as a result of increased ecotonal habitat (Medlock *et al.*, 2012, 2020) and a wider diversity of tick hosts. Initiatives that increase woodland cover through planting, woodland regeneration and rewilding will likely increase the overall habitat availability for ticks and their hosts. Recent studies on ticks in agri-environment field margins highlight their suitability when planted adjacent to tick woodland habitats (Medlock *et al.*, 2020). Densities of ticks and more importantly the densities of infected ticks vary between habitats and within woodland depending upon biotic and abiotic factors that determine wild mammal and bird densities (Medlock *et al.*, 2013; Ehrmann *et al.*, 2018). Habitat management, therefore, can impact on the relative proportions of *Borrelia* genospecies in ticks by determining the diversity of tick hosts, which can be independent of the direct impacts of climate. The relative densities of small mammals, wild birds and game birds are also dictated by anthropogenic impacts such as wildlife and game management, which can impact on tick infection rates and genospecies dominance (Kurtenbach *et al.*, 1998; Mysterud *et al.*, 2019; Becker and Han, 2020). In open grazed habitats, sheep and cattle can act to dilute infection rates (Ostfeld and Keesing, 2000) and so stocking densities, dictated by landscape management, can impact on tick numbers and pathogen infection rates. The dilution effect is, however, debated and the potential impacts can differ locally (Randolph and Dobson, 2012; Ehrmann *et al.*, 2018; Ruyts *et al.*, 2018). Each tick host species is impacted directly by climate and habitat, which leads to an indirect effect of climate on tick numbers and pathogen infection rates (Medlock and Leach, 2015).

Woodlands are important and dynamic tick habitats (Medlock *et al.*, 2012; James *et al.*, 2013; Millins *et al.*, 2016; Hansford *et al.*, 2017b) subject to a great deal of management, e.g. for timber extraction, biodiversity management and public access, which can impact on ticks and their hosts (Medlock *et al.*, 2012). Increasing networks of rides and glades to improve biodiversity and assist with climate change adaptation for biodiversity (such as woodland butterflies) can increase ecotonal habitat for ticks and their hosts. Other non-adaptation strategies, such as woodland thinning, the impact of tree diseases (such as ash dieback) and improving public access, can change the way in which a woodland is exploited by animals, change the microclimate for invertebrates such as ticks and change the way humans access woodland.

Outside this key habitat, the provision of increased urban green space for health and well-being, or mitigation against the effects of warming cities, can provide additional habitat for ticks and provide wildlife corridors for tick hosts (Rizzoli *et al.*, 2014). The expansion of urban areas into green belt, where new housing developments are constructed among existing or new tick habitat, can increase exposure to ticks (Hansford *et al.*, 2017b). It can also increase the interaction between tick hosts and humans through gardens (Jameson and Medlock, 2011), recreational areas and school playing fields, which can have considerable implications for management and disease risk.

Climate change is also expected to disproportionately impact biodiversity in fragmented habitats (Estrada-Peña, 2003; Rizzoli *et al.*, 2019), so many strategies seek to enhance resilience of urban and rural habitats to reduce the impacts of stochastic weather events on wildlife, with knock-on effects on ticks. Rough grassland is a seral stage prior to scrub encroachment and the return to a climax woodland. It is an important habitat for grassland flora, birds and invertebrates (Hawes, 2015). Grazing livestock is the favoured conservation strategy for grassland management; however, inadvertent movement of tick-infested livestock for conservation grazing can result in the movement of tick vectors and create disease issues for wildlife, livestock and humans. Recent examples of this include *D. reticulatus* in the Essex marshes and urban grasslands (Medlock *et al.*, 2017), with subsequent cases of canine babesiosis (de Marco *et al.*, 2017), and *H. punctata* in the Sussex Downs and urban grasslands and their role in tick pyaemia in sheep (Macrelli *et al.*, 2020). These apparent emergences of tick infestation and associated disease are independent of the direct effects of climate change. However, the indirect impacts of adapting biodiverse habitats for climate change or enhancing peri-urban green space can ultimately have subsequent effects on TBDs.

Impact of climate change on tick hosts

Two important groups of mammals for *I. ricinus* are deer and small mammals (Perez *et al.*, 2016; Qviller *et al.*, 2016). Deer continue to increase in numbers in the UK in the absence of natural predators. Milder winters also increase the overwintering success rate of young deer and increased deer density has been linked to increased questing nymph densities (Gilbert, 2010). Such changes may have contributed to the increased range of tick reports to the Public Health England Tick Surveillance Scheme (Cull *et al.*, 2018). Studies in Sweden have shown that milder winters lead to fewer frost days, an extended vegetation period and an expansion in deer populations, and consequently a northerly expansion in the range of *I. ricinus* (Jaenson *et al.*, 2012). Deer management activities such as culling, deer fencing and considerations for release of large predators are likely to impact on deer densities at a local and national scale over the coming decades and may be important tick-management strategies (Sprong *et al.*, 2018; Černý *et al.*, 2020). This will be particularly important in urban areas (Telford, 2017), as deer and ticks become features of urban green space.

Small mammals are acutely responsive to the direct and indirect effects of climate. Temperature (including extremes, such as drought or severe cold) impacts on the production of tree mast, with summer drought increasing seed production and late frosts reducing flower survival. A resultant good or poor mast year drives the population density of small mammals, which, coupled with high mammal overwintering rates and elevated tick survival/continued activity in mild winters, can determine success rates of tick feeding and thus *Borrelia* (Bregnard *et al.*, 2020; Krawczyk *et al.*, 2020) and TBEV transmission (Rubel *et al.*, 2020b). High densities of fed larval *I. ricinus* can impact densities of questing infected nymphs to increase human exposure and disease transmission risk (Perez *et al.*, 2016). For other tick hosts, any changes in their current numbers, such as an increase or decrease in the number of released game birds, changing bird migration patterns, or the return of extirpated mammals such as wild boar, beaver, pine marten and lynx, will all have an impact that can be completely independent of climate change.

A change in human exposure to ticks with climate change

Tick presence, increased infection rates in ticks and an increase in the density of infected ticks do not necessarily correlate with increased human exposure or disease risk. There is a crucial and perhaps not easily quantified component in transmission risk, namely human exposure to tick bites and subsequent development of disease (Wilhelmsson *et al.*, 2013). A woodland with high densities of infected ticks with no footpaths does not necessarily present an increased public health risk compared with a woodland with extensive public access with low densities of *Borrelia*-infected ticks. Recent studies on the density of *Borrelia*-infected ticks in English cities have begun to quantify the disparity between tick presence and *Borrelia* infection rates in urban areas (Hansford *et al.*, 2017b), but further studies are required on human exposure and behaviour (Wierzbicka *et al.*, 2016) and subsequent disease development following tick exposure in peri/urban and rural green spaces.

As towns and cities expand into greener areas of the countryside, enveloping and incorporating natural habitat, or become greener, the contact between people and the countryside is enhanced. Management of tick issues in these high-footfall areas, through both direct management of habitats (fencing, mowing, habitat partitioning, directional flow of visitors) and public health awareness campaigns, will be crucial in managing risk (Rizzoli *et al.*, 2014). Improving knowledge, education and awareness of tick and TBD risk will be particularly important in a changing climate, especially if more people dress for warmer temperatures and access cool green spaces for exercise and recreation. Increased visitors to natural areas, without enhanced tick awareness, could create circumstances where high levels of tick bites are reported.

75.2 Future Predictions

It is not possible to directly correlate increased temperatures, milder winters or extreme events to changes in tick populations or TBDs without taking other driving factors into consideration, particularly at a local scale (Ehrmann *et al.*, 2017). There are many factors that impact the habitat, microclimate and hosts (Li *et al.*, 2016; Millins *et al.*, 2016), as well as the raft of environmental changes that are taking place. The way in which humans are interacting with natural habitats in a changing climate also needs consideration. Preparing for and responding to these changing risks requires a multi-agency response, with wide stakeholder engagement. There are generic lessons, but action must be local and bespoke. Research needs to focus on the interaction of these many drivers, incorporating long-term data sets on ticks, hosts, habitat and climate, if we are to better understand the potential public health impact of TBDs in the future. In the future, towns and cities will continue to expand into green space and natural habitats, with a greater drive towards urban greening and access to green space. In more rural areas, rewilding is now a phenomenon that is gathering pace, whether this is at a landscape scale through natural regeneration or widescale tree planting, or through a paradigm shift in farming practices that promote and reward management for biodiversity. Coupled with the current trend in increasing deer abundance, the potential increased human exposure to ticks is a concern. If we are to minimize the increasing risks from ticks and TBDs over the next 10–50 years, while promoting biodiversity strategies, a raft of interventions and guidance on countryside and urban green space management may be needed. These will need to be geared towards implementation at a local scale, with ticks and TBDs considered in the planning process as we change our environment in a changing climate.

References

Anderson, J.F. and Magnarelli, L.A. (2008) Biology of ticks. *Infectious Disease Clinics of North America* 22, 195–215. https://doi.org/10.1016/j.idc.2007.12.006

Baylis, M. (2017) Potential impact of climate change on emerging vector-borne and other infections in the UK. *Environmental Health* 16, 112. https://doi.org/10.1186/s12940-017-0326-1

Becker, D. and Han, B. (2020) The macroecology and evolution of avian competence for *Borrelia burgdorferi*. *bioRxiv* 2020.04.15.040352. https://doi.org/10.1101/2020.04.15.040352

Bregnard, C., Rais, O. and Voordouw, M.J. (2020) Climate and tree seed production predict the abundance of the European Lyme disease vector over a 15-year period. *Parasites & Vectors* 13, 408. https://doi.org/10.1186/s13071-020-04291-z

Černý, J., Lynn, G., Hrnková, J., Golovchenko, M., Rudenko, N. and Grubhoffer, L. (2020) Management options for *Ixodes ricinus*-associated pathogens: a review of prevention strategies. *International Journal of Environmental Research and Public Health* 17, 1830. https://doi.org/10.3390/ijerph17061830

Cull, B., Pietzsch, M.E., Hansford, K.M., Gillingham, E.L. and Medlock, J.M. (2018) Surveillance of British ticks: an overview of species records, host associations, and new records of *Ixodes ricinus* distribution. *Ticks and Tick-borne Diseases* 9, 605–614. https://doi.org/10.1016/j.ttbdis.2018.01.011

Cull, B., Pietzsch, M.E., Gillingham, E.L., McGinley, L., Medlock, J.M. and Hansford, K.M. (2020) Seasonality and anatomical location of human tick bites in the United Kingdom. *Zoonoses and Public Health* 67, 112–121. https://doi.org/10.1111/zph.12659

Danielová, V., Daniel, M., Rudenko, N. and Golovchenko, M. (2004) Prevalence of *Borrelia burgdorferi* sensu lato genospecies in host-seeking *Ixodes ricinus* ticks in selected South Bohemian locations (Czech Republic). *Central European Journal of Public Health* 12, 151–156.

Dantas-Torres, F. (2010) Biology and ecology of the brown dog tick, *Rhipicephalus sanguineus*. *Parasites & Vectors* 3, 26. https://doi.org/10.1186/1756-3305-3-26

de Marco, M.D.M.F., Hernández-Triana, L.M., Phipps, L.P., Hansford, K., Mitchell, E.S., *et al.* (2017) Emergence of *Babesia canis* in southern England. *Parasites & Vectors* 10, 241. https://doi.org/10.1186/s13071-017-2178-5

Drehmann, M., Springer, A., Lindau, A., Fachet, K., Mai, S., *et al.* (2020) The spatial distribution of *Dermacentor* ticks (Ixodidae) in Germany – evidence of a continuing spread of *Dermacentor reticulatus*. *Frontiers in Veterinary Science* 7, 578220. https://doi.org/10.3389/fvets.2020.578220

Ehrmann, S., Liira, J., Gärtner, S., Hansen, K., Brunet, J., *et al.* (2017) Environmental drivers of *Ixodes ricinus* abundance in forest fragments of rural European landscapes. *BMC Ecology* 17, 31. https://doi.org/10.1186/s12898-017-0141-0

Ehrmann, S., Ruyts, S.C., Scherer-Lorenzen, M., Bauhus, J., Brunet, J., *et al.* (2018) Habitat properties are key drivers of *Borrelia burgdorferi* (s.l.) prevalence in *Ixodes ricinus* populations of deciduous forest fragments. *Parasites & Vectors* 11, 23. https://doi.org/10.1186/s13071-017-2590-x

England, M.E., Phipps, P., Medlock, J.M., Atkinson, P.M., Atkinson, B., *et al.* (2016) *Hyalomma* ticks on northward migrating birds in southern Spain: implications for the risk of entry of Crimean–Congo haemorrhagic fever virus to Great Britain. *Journal of Vector Ecology* 41, 128–134. https://doi.org/10.1111/jvec.12204

Estrada-Peña, A. (2003) The relationships between habitat topology, critical scales of connectivity and tick abundance *Ixodes ricinus* in a heterogeneous landscape in northern Spain. *Ecography* 26, 661–671. https://doi.org/10.1034/j.1600-0587.2003.03530.x

Estrada-Peña, A. (2015) Ticks as vectors: taxonomy, biology and ecology. *Revue Scientifique et Technique* 34, 53–65. https://doi.org/10.20506/rst.34.1.2345

Estrada-Peña, A., Martínez Avilés, M. and Muñoz Reoyo, M.J. (2011) A population model to describe the distribution and seasonal dynamics of the tick *Hyalomma marginatum* in the Mediterranean basin. *Transboundary and Emerging Diseases* 58, 213–223. https://doi.org/10.1111/j.1865-1682.2010.01198.x

Estrada-Peña, A., Palomar, A.M., Santibáñez, P., Sánchez, N., Habela, M.A., *et al.* (2012) Crimean–Congo hemorrhagic fever virus in ticks, southwestern Europe, 2010. *Emerging Infectious Diseases* 18, 179–180. https://doi.org/10.3201/eid1801.111040

Fernández-Ruiz, N. and Estrada-Peña, A. (2020) Could climate trends disrupt the contact rates between *Ixodes ricinus* (Acari, Ixodidae) and the reservoirs of *Borrelia burgdorferi* s.l.? *PLoS ONE* 15, 0233771. https://doi.org/10.1371/journal.pone.0233771

Földvári, G., Široký, P., Szekeres, S., Majoros, G. and Sprong, H. (2016) *Dermacentor reticulatus*: a vector on the rise. *Parasites & Vectors* 9, 314. https://doi.org/10.1186/s13071-016-1599-x

Gilbert, L. (2010) Altitudinal patterns of tick and host abundance: a potential role for climate change in regulating tick-borne diseases? *Oecologia* 162, 217–225. https://doi.org/10.1007/s00442-009-1430-x

Gilbert, L., Aungier, J. and Tomkins, J.L. (2014) Climate of origin affects tick (*Ixodes ricinus*) host-seeking behavior in response to temperature: implications for resilience to climate change? *Ecology and Evolution* 4, 1186–1198. https://doi.org/10.1002/ece3.1014

Grandi, G., Chitimia-Dobler, L., Choklikitumnuey, P., Strube, C., Springer, A., *et al.* (2020) First records of adult *Hyalomma marginatum* and *H. rufipes* ticks (Acari: Ixodidae) in Sweden. *Ticks and Tick-borne Diseases* 11, 101403. https://doi.org/10.1016/j.ttbdis.2020.101403

Gray, J.S., Dautel, H., Estrada-Peña, A., Kahl, O. and Lindgren, E. (2009) Effects of climate change on ticks and tick-borne diseases in Europe. *Interdisciplinary Perspectives on Infectious Diseases* 2009, 593232. https://doi.org/10.1155/2009/593232

Hansford, K.M., Pietzsch, M., Cull, B. and Medlock, J.M. (2015) Brown dog tick infestation of a home in England. *Veterinary Record* 176, 129–130. https://doi.org/10.1136/vr.h496

Hansford, K.M., Medlock, J.M., Atkinson, B. and Santos-Silva, M.M. (2016) Importation of a *Hyalomma lusitanicum* tick into the UK on a dog. *Veterinary Record* 179, 415–415. https://doi.org/10.1136/vr.i5645

Hansford, K.M., Phipps, L.P., Cull, B., Pietzsch, M.E. and Medlock, J.M. (2017a) *Rhipicephalus sanguineus* importation into the UK: surveillance, risk, public health awareness and One Health response. *Veterinary Record* 180, 119. https://doi.org/10.1136/vr.104061

Hansford, K.M., Fonville, M., Gillingham, E.L., Coipan, E.C., *et al.* (2017b) Ticks and *Borrelia* in urban and peri-urban green space habitats in a city in southern England. *Ticks and Tick-borne Diseases* 8, 353–361. https://doi.org/10.1016/j.ttbdis.2016.12.009

Hansford, K.M., Pietzsch, M.E., Cull, B., Gillingham, E.L. and Medlock, J.M. (2018) Potential risk posed by the importation of ticks into the UK on animals: records from the Tick Surveillance Scheme. *Veterinary Record* 182, 107. https://doi.org/10.1136/vr.104263

Hansford, K.M., Carter, D., Gillingham, E.L., Hernandez-Triana, L.M., Chamberlain, J., *et al.* (2019) *Hyalomma rufipes* on an untraveled horse: is this the first evidence of *Hyalomma* nymphs successfully moulting in the United Kingdom? *Ticks and Tick-borne Diseases* 10, 704–708. https://doi.org/10.1016/j.ttbdis.2019.03.003

Hawes, P. (2015) Sheep grazing and the management of chalk grassland. *British Wildlife* 21, 25–30.

Holding, M., Dowall, S.D., Medlock, J.M., Carter, D.P., McGinley, L., *et al.* (2019) Detection of new endemic focus of tick-borne encephalitis virus (TBEV), Hampshire/Dorset border, England, September 2019. *Eurosurveillance* 24, 1900658. https://doi.org/10.2807/1560-7917.ES.2019.24.47.1900658

Hubálek, Z., Sedláček, P., Estrada-Peña, A., Vojtíšek, J. and Rudolf, I. (2020) First record of *Hyalomma rufipes* in the Czech Republic, with a review of relevant cases in other parts of Europe. *Ticks and Tick-borne Diseases* 11, 101421. https://doi.org/10.1016/j.ttbdis.2020.101421

Hvidsten, D., Frafjord, K., Gray, J.S., Henningsson, A.J., Jenkins, A., *et al.* (2020) The distribution limit of the common tick, *Ixodes ricinus*, and some associated pathogens in north-western Europe. *Ticks and Tick-borne Diseases* 11, 101388. https://doi.org/10.1016/j.ttbdis.2020.101388

Jaenson, T.G.T., Jaenson, D.G.E., Eisen, L., Petersson, E. and Lindgren, E. (2012) Changes in the geographical distribution and abundance of the tick *Ixodes ricinus* during the past 30 years in Sweden. *Parasites & Vectors* 5, 8. https://doi.org/10.1186/1756-3305-5-8

James, M.C., Bowman, A.S., Forbes, K.J., Lewis, F., McLeod, J.E. and Gilbert, L. (2013) Environmental determinants of *Ixodes ricinus* ticks and the incidence of *Borrelia burgdorferi* sensu lato, the agent of Lyme borreliosis, in Scotland. *Parasitology* 140, 237–246. https://doi.org/10.1017/S003118201200145X

Jameson, L.J. and Medlock, J.M. (2011) Tick surveillance in Great Britain. *Vector-Borne and Zoonotic Diseases* 11, 403–412. https://doi.org/10.1089/vbz.2010.0079

Jameson, L.J., Morgan, P.J., Medlock, J.M., Watola, G. and Vaux, A.G.C. (2012) Importation of *Hyalomma marginatum*, vector of Crimean–Congo haemorrhagic fever virus, into the United Kingdom by migratory birds. *Ticks and Tick-borne Diseases* 3, 95–99. https://doi.org/10.1016/j.ttbdis.2011.12.002

Kiewra, D., Szymanowski, M., Czułowska, A. and Kolanek, A. (2021) The local-scale expansion of *Dermacentor reticulatus* ticks in Lower Silesia, SW Poland. *Ticks and Tick-borne Diseases* 12, 101599. https://doi.org/10.1016/j.ttbdis.2020.101599

Krawczyk, A.I., van Duijvendijk, G.L.A., Swart, A., Heylen, D., Jaarsma, R.I., *et al.* (2020) Effect of rodent density on tick and tick-borne pathogen populations: consequences for infectious disease risk. *Parasites & Vectors* 13, 34. https://doi.org/10.1186/s13071-020-3902-0

Kurtenbach, K., Peacey, M., Rijpkema, S.G., Hoodless, A.N., Nuttall, P.A. and Randolph, S.E. (1998) Differential transmission of the genospecies of *Borrelia burgdorferi* sensu lato by game birds and small rodents in England. *Applied Environmental Microbiology* 64, 1169–1174. https://doi.org/10.1128/AEM.64.4.1169-1174.1998

Li, S., Gilbert, L., Harrison, P.A. and Rounsevell, M.D.A. (2016) Modelling the seasonality of Lyme disease risk and the potential impacts of a warming climate within the heterogeneous landscapes of Scotland. *Journal of the Royal Society Interface* 13, 20160140. https://doi.org/10.1098/rsif.2016.0140

Macrelli, M., Phipps, P.L., McGinley, L., Medlock, J.M. and Johnson, N. (2020) First report of fatal tick pyaemia caused by heavy infestation with the red sheep tick, *Haemaphysalis punctata* and co-infection with *Babesia* and *Theileria* species. *Veterinary Record Case Reports* 8, e001267. https://doi.org/10.1136/vetreccr-2020-001267

Mannelli, A., Bertolotti, L., Gern, L. and Gray, J. (2012) Ecology of *Borrelia burgdorferi* sensu lato in Europe: transmission dynamics in multi-host systems, influence of molecular processes and effects of climate change. *FEMS Microbiology Reviews* 36, 837–861. https://doi.org/10.1111/j.1574-6976.2011.00312.x

McGinley, L., Hansford, K.M., Cull, B., Gillingham, E.L., Carter, D.P., *et al.* (2021) First report of human exposure to *Hyalomma marginatum* in England: further evidence of a *Hyalomma* moulting event in north-western Europe? *Ticks and Tick-borne Diseases* 12, 101541. https://doi.org/10.1016/j.ttbdis.2020.101541

Medlock, J.M. and Leach, S.A. (2015) Effect of climate change on vector-borne disease risk in the UK. *Lancet Infectious Diseases* 15, 721–730. https://doi.org/10.1016/S1473-3099(15)70091-5

Medlock, J.M., Shuttleworth, H., Copley, V., Hansford, K.M. and Leach, S. (2012) Woodland biodiversity management as a tool for reducing human exposure to *Ixodes ricinus* ticks: a preliminary study in an English woodland. *Journal of Vector Ecology* 37, 307–315. https://doi.org/10.1111/j.1948-7134.2012.00232.x

Medlock, J.M., Hansford, K.M., Bormane, A., Derdakova, M., Estrada-Peña, A., *et al.* (2013) Driving forces for changes in geographical distribution of *Ixodes ricinus* ticks in Europe. *Parasites & Vectors* 6, 1. https://doi.org/10.1186/1756-3305-6-1

Medlock, J.M., Hansford, K.M., Vaux, A.G.C., Cull, B., Abdullah, S., *et al.* (2017) Distribution of the tick *Dermacentor reticulatus* in the United Kingdom. *Medical and Veterinary Entomology* 31, 281–288. https://doi.org/10.1111/mve.12235d

Medlock, J.M., Hansford, K.M., Vaux, A.G.C., Cull, B., Pietzsch, M.E., *et al.* (2018) Has the red sheep tick, *Haemaphysalis punctata*, recently expanded its range in England? *Medical and Veterinary Entomology* 32, 473–480. https://doi.org/10.1111/mve.12335

Medlock, J.M., Vaux, A.G.C., Hansford, K.M., Pietzsch, M.E. and Gillingham, E.L. (2020) Ticks in the ecotone: the impact of agri-environment field margins on the presence and intensity of *Ixodes ricinus* ticks (Acari: Ixodidae) in farmland in southern England. *Medical and Veterinary Entomology* 34, 175–183. https://doi.org/10.1111/mve.12428

Met Office (2019) *UK Climate Projections: Headline Findings*. Met Office, Exeter, UK. Available at: https://www.metoffice.gov.uk/binaries/content/assets/metofficegovuk/pdf/research/ukcp/ukcp-headline-findings-v2.pdf (accessed 21 June 2021).

Millins, C., Gilbert, L., Johnson, P., James, M., Kilbride, E., *et al.* (2016) Heterogeneity in the abundance and distribution of *Ixodes ricinus* and *Borrelia burgdorferi* (sensu lato) in Scotland: implications for risk prediction. *Parasites & Vectors* 9, 595. https://doi.org/10.1186/s13071-016-1875-9

Mysterud, A., Stigum, V.M., Jaarsma, R.I. and Sprong, H. (2019) Genospecies of *Borrelia burgdorferi* sensu lato detected in 16 mammal species and questing ticks from northern Europe. *Scientific Reports* 9, 5088. https://doi.org/10.1038/s41598-019-41686-0

Negredo, A., de la Calle-Prieto, F., Palencia-Herrejón, E., Mora-Rillo, M., Astray-Mochales, J., *et al.* (2017) Autochthonous Crimean–Congo hemorrhagic fever in Spain. *New England Journal of Medicine* 377, 154–161. https://doi.org/10.1056/nejmoa1615162

Ostfeld, R.S. and Keesing, F. (2000) Biodiversity and disease risk: the case of Lyme disease. *Conservation Biology* 14, 722–728. https://doi.org/10.1046/j.1523-1739.2000.99014.x

Perez, G., Bastian, S., Agoulon, A., Bouju, A., Durand, A., *et al.* (2016) Effect of landscape features on the relationship between *Ixodes ricinus* ticks and their small mammal hosts. *Parasites & Vectors* 9, 20. https://doi.org/10.1186/s13071-016-1296-9

Qviller, L., Viljugrein, H., Loe, L.E., Meisingset, E.L. and Mysterud, A. (2016) The influence of red deer space use on the distribution of *Ixodes ricinus* ticks in the landscape. *Parasites & Vectors* 9, 545. https://doi.org/10.1186/s13071-016-1825-6

Randolph, S.E. (2000) Ticks and tick-borne disease systems in space and from space. *Advances in Parasitology* 47, 217–243. https://doi.org/10.1016/S0065-308X(00)47010-7

Randolph, S.E. (2004) Tick ecology: processes and patterns behind the epidemiological risk posed by ixodid ticks as vectors. *Parasitology* 129(Suppl.), S37–S65. https://doi.org/10.1017/S0031182004004925

Randolph, S.E. and Dobson, A.D.M. (2012) Pangloss revisited: a critique of the dilution effect and the biodiversity–buffers–disease paradigm. *Parasitology* 139, 847–863. https://doi.org/10.1017/S0031182012000200

Rizzoli, A., Silaghi, C., Obiegala, A., Rudolf, I., Hubálek, Z., *et al.* (2014) *Ixodes ricinus* and its transmitted pathogens in urban and peri-urban areas in Europe: new hazards and relevance for public health. *Frontiers in Public Health* 2, 251. https://doi.org/10.3389/fpubh.2014.00251

Rizzoli, A., Tagliapietra, V., Cagnacci, F., Marini, G., Arnoldi, D., *et al.* (2019) Parasites and wildlife in a changing world: the vector–host–pathogen interaction as a learning case. *International Journal for Parasitology: Parasites and Wildlife* 9, 394–401. https://doi.org/10.1016/j.ijppaw.2019.05.011

Rosà, R., Andreo, V., Tagliapietra, V., Baráková, I., Arnoldi, D., *et al.* (2018) Effect of climate and land use on the spatio-temporal variability of tick-borne bacteria in Europe. *International Journal of Environmental Research and Public Health* 15, 732. https://doi.org/10.3390/ijerph15040732

Rubel, F., Brugger, K., Belova, O.A., Kholodilov, I.S., Didyk, Y.M., *et al.* (2020a) Vectors of disease at the northern distribution limit of the genus *Dermacentor* in Eurasia: *D. reticulatus* and *D. silvarum*. *Experimental and Applied Acarology* 82, 95–123. https://doi.org/10.1007/s10493-020-00533-y

Rubel, F., Walter, M., Vogelgesang, J.R. and Brugger, K. (2020b) Tick-borne encephalitis (TBE) cases are not random: explaining trend, low- and high-frequency oscillations based on the Austrian TBE time series. *BMC Infectious Diseases* 20, 448. https://doi.org/10.1186/s12879-020-05156-7

Ruyts, S.C., Landuyt, D., Ampoorter, E., Heylen, D., Ehrmann, S., *et al.* (2018) Low probability of a dilution effect for Lyme borreliosis in Belgian forests. *Ticks and Tick-borne Diseases* 9, 1143–1152. https://doi.org/10.1016/j.ttbdis.2018.04.016

Semenza, J.C. and Suk, J.E. (2018) Vector-borne diseases and climate change: a European perspective. *FEMS Microbiology Letters* 365, fnx244. https://doi.org/10.1093/femsle/fnx244

Sprong, H., Azagi, T., Hoornstra, D., Nijhof, A.M., Knorr, S., *et al.* (2018) Control of Lyme borreliosis and other *Ixodes ricinus*-borne diseases. *Parasites & Vectors* 11, 145. https://doi.org/10.1186/s13071-018-2744-5

Strnad, M., Hönig, V., Ruzek, D., Grubhoffer, L. and Rego, R.O.M. (2017) Europe-wide meta-analysis of *Borrelia burgdorferi* sensu lato prevalence in questing *Ixodes ricinus* ticks. *Applied Environmental Microbiology* 83, e00609-17. https://doi.org/10.1128/AEM.00609-17

Telford, S.R. (2017) Deer reduction is a cornerstone of integrated deer tick management. *Journal of Integrated Pest Management* 8, 25. https://doi.org/10.1093/jipm/pmx024

Valcárcel, F., González, J., González, M.G., Sánchez, M., Tercero, J.M., *et al.* (2020) Comparative ecology of *Hyalomma lusitanicum* and *Hyalomma marginatum* Koch, 1844 (Acarina: Ixodidae). *Insects* 11, 303. https://doi.org/10.3390/insects11050303

van Oort, B.E.H., Hovelsrud, G.K., Risvoll, C., Mohr, C.W., and Jore, S. (2020) A mini-review of *Ixodes* ticks climate sensitive infection dispersion risk in the Nordic region. *International Journal of Environmental Research and Public Health* 17, 5387. https://doi.org/10.3390/ijerph17155387

Wierzbicka, A., Rączka, G., Skorupski, M., Michalik, J. and Lane, R.S. (2016) Human behaviors elevating the risk of exposure to *Ixodes ricinus* larvae and nymphs in two types of lowland coniferous forests in west-central Poland. *Ticks and Tick-borne Diseases* 7, 1180–1185. https://doi.org/10.1016/j.ttbdis.2016.07.018

Wilhelmsson, P., Lindblom, P., Fryland, L., Nyman, D., Jaenson, T.G., *et al.* (2013) *Ixodes ricinus* ticks removed from humans in Northern Europe: seasonal pattern of infestation, attachment sites and duration of feeding. *Parasites & Vectors* 6, 362. https://doi.org/10.1186/1756-3305-6-362

Wright, I., Wildgoose, W.H., Wall, R. and McGarry, J. (2018) *Rhipicephalus sanguineus* in an imported dog. *Veterinary Record* 182, 25–26. https://doi.org/10.1136/vr.k3

eo76 Climate Change, Ticks and Tick-Borne Pathogens in Northern Europe

John H.O. Pettersson[1,2*]
[1]*University of Sydney, Sydney, Australia;* [2]*Uppsala University, Uppsala, Sweden*

76.1 Expert Opinion

76.1.1 Introduction

Ticks (Ixodida) are some of the most biologically and evolutionarily successful organisms. Since their evolutionary origin more than 100 million years ago, they have diversified, colonized and adapted to most of the Earth's different continents, climate zones, local environments and the animal hosts that are present there. They have evolved to exist in complex host–parasite relationships, affected by numerous biotic and abiotic ecological factors, and are known vectors for multiple different pathogenic viruses, bacteria, parasites and nematodes, many of which significantly affect human and animal health and economy. Like all other organisms, ticks are subjected to the ongoing climate change. Long-term predictions of climate change for Northern Europe suggest that average temperatures will have increased *ca.*2–4°C by the year 2100 (Collins *et al.*, 2013). One of the most important consequences of this climate change for Northern Europe is that winters are expected to be less cold with fewer cold temperature extremes while precipitation will increase. Conversely, summers will tend to be drier with more frequent heatwaves. The question is, what will a changing climate in Northern Europe mean in the near and distant future for ticks and the disease-causing microorganisms they carry?

The potential effect a warmer climate may have on ticks and the pathogens they vector is one of the most complex research topics within the field of science and highlights the problem of drawing firm conclusions from correlated patterns when the specific causative factors of said components are difficult to identify (Estrada-Peña *et al.*, 2012). There is no question that climate and temperature affect ticks. For example, the nymphal stage of the most common tick in Northern Europe, *Ixodes ricinus*, does not commonly search for hosts below 5°C, development time to new tick stages following a blood meal increases with decreasing temperature and exposure to cold temperatures increases tick mortality. Conversely, although ticks are resilient and relatively temperature-tolerant organisms, they are also sensitive to changes in ground humidity, where areas with a relative ground-level humidity below 85% prohibit the establishment of a sustainable *I. ricinus* tick population (Randolph *et al.*, 2002). In addition to dependence on abiotic variables for their survival, ticks are just as dependent on multiple biotic variables. To be able to establish a tick population, not only are there physical attributes to be fulfilled;

*Email: john.pettersson@gmail.com

DOI: 10.1079/9781789249637.0076

it is also necessary to have suitable hosts present for all life stages requiring a blood meal prior to the next developmental stage and for the tick female to gain energy and nutrients to produce eggs following mating. For the common tick, *I. ricinus*, this translates to having a habitat that includes both small and large mammals. Small mammals, such as rodents and shrews, are the main hosts for the larval stage of *I. ricinus*, whereas nymphs and adult ticks feed off larger mammals, for example hares and roe deer. For adult ticks in particular, the presence of larger mammals in the habitat is a necessity as they rarely quest for, feed from and mate on smaller mammals. Thus, a particular habitat needs to accommodate the abiotic and biotic requirements of multiple different organisms in order to maintain a healthy tick population. Clearly, there are many interrelated forces and factors that determine which past, current and potential future distribution of ticks (Jaenson *et al.*, 2012; Medlock *et al.*, 2013) should be taken into account when considering the potential impacts climate change might have on ticks and the pathogens they vector in Northern Europe.

76.1.2 What may be the consequences of a warmer climate in Northern Europe?

Tick survival, reproduction and distribution

Perhaps the most obvious and direct consequence of a warmer climate on ticks would be increased survival and reproductive rates. In Northern Europe, such a climate change-induced effect has most likely already been observed with regard to *I. ricinus*. Due to an apparent increase in the length of the vegetation period, where spring occurs earlier and autumn happens later, ticks may become active earlier and begin winter hibernation later. As a consequence, the extended period of activity will increase the chances of finding both a host and a mate. In addition to increasing the chances for survival and reproduction, a warmer climate, and milder winters in particular, may also increase the chances for winter survival of tick hosts.

While milder winters and a longer vegetation period are likely to have a positive impact on the survival and reproduction of both ticks and their hosts in areas where they already occur, perhaps the main effect will be seen at the border zones of the ticks' northernmost distributional range. Still, it is important to keep in mind that the absence or low abundance of ticks in these border regions may also partly be explained by the absence or too low abundance of tick hosts. Hence, ticks would not be able to expand their distributional range until the various tick hosts needed for a sustainable long-term tick population had established and/or increased in abundance in the same region. The northward migration and establishment of tick hosts may also be dependent on the migration of plants in order to create the habitats needed for small and large mammals to expand their own distributional range and population size in those areas. Thus, while we can expect ticks to expand their northern distributional range as climate becomes warmer with time (Kjær *et al.*, 2019), there will likely be a time-lag effect for this expansion related to several ecological factors.

If the survival, reproduction and distribution of ticks and tick hosts increase with climate change, is it likely that the prevalence and distribution of tick-borne pathogens also will increase in the same manner? For the tick-borne encephalitis virus, it has been argued that the distribution is patchy whereby some areas, so-called foci, have significantly higher prevalence than other areas. The characteristics of such areas appear to be a favourable microclimate with warmer than average temperatures and accessibility to mainly small mammalian hosts that allow for the density of ticks to be higher than in other areas. A possible scenario is that such endemic hotspots, which favour the virus, would increase in frequency such that the overall virus prevalence would also increase. If so, it is likely that the number of human disease cases also would increase due to a higher overall risk of exposure to a tick-borne encephalitis virus from an infected tick.

Introduction of 'exotic' viruses and ticks

Since ticks feed off a host for several days, sometimes more than 10 days, they can be transported extensive distances, hundreds to thousands of kilometres, depending on the host and the time it takes for the tick to complete a blood meal.

Importantly, during the spring and early summer, millions of migratory birds fly to Northern Europe, many of which are infested with different species of ticks and other ectoparasites. Some of these birds migrate from latitudes well below the equator before they reach their destinations in Northern Europe. Studies on ticks from migratory birds have identified several human pathogens and various tick-associated viruses of unknown association to human and veterinary health (Pettersson *et al.*, 2017; Pedersen *et al.*, 2020; Wilhelmsson *et al.*, 2020). However, it should also be pointed out that not only ticks function as long-distance dispersal vectors of pathogens and other microbes. It has been shown that many migratory birds are likely to be infected with various human pathogenic viruses, which may be transported between continents or even the polar regions (Pettersson *et al.*, 2020). Consequently, upon reaching their migratory end destination, these viruses and other microorganisms may become established in the local tick population.

Thus, there is a regular and steady 'flow' of different species of ticks, tick-borne pathogens, potential pathogens and other microorganisms that are being imported and disseminated to and across northern and other parts of Europe every year. Clearly, not all types of ticks, viruses and other microorganisms are capable of surviving and establishing themselves in the type of environment, vectors and hosts that are present in Northern Europe. At the moment, 'exotic' ticks and microbes that are imported to Northern Europe may establish themselves temporarily during the summer period (Grandi *et al.*, 2020), when temperatures and other physical properties are not deviating too much from their usual habitat. But as winter arrives, survival until spring is most likely improbable for most exotic ticks and thereby any microorganisms they carry. On the other hand, the core of host–vector and virus evolution are opportunity and repetition. Eventually, a variant of, for example, a particular virus might have acquired just through chance the appropriate mutation or mutations that increase its chances of survival in a new tick vector or host, whereafter it could become established and spread in the new geographical area. Furthermore, warmer winters and temperatures that approach the tolerance limits of other tick species imply it is possible that this would create opportunities for the temporary period of establishment to increase to more than two seasons and, if winter and summer temperatures allow, permit development through the moulting and diapausing life stages. Eventually, if climate change continues and annual temperatures become warmer, it is to be expected that the diversity of both ticks and pathogenic and non-pathogenic microorganisms in Northern Europe would increase.

76.2 Future Predictions

The potential impacts of climate change in relation to ticks and tick-borne pathogens are a multifactorial and difficult topic to address. In the long term, given the current future projections for climate change to a warmer climate, this will undoubtedly have an effect on ticks. The question is, to what extent will that effect be short-term and long-term? There are at least four potential consequences with regard to ticks and tick-borne pathogens in relation to climate change:

1. It is likely that warmer winters and an extended activity period will increase overall tick survival and push the northernmost edges of the *I. ricinus* distributional range even further north and increase the abundance in these edge zones.
2. With increased survival and reproduction, some tick-borne pathogens may become more common with time and/or become more prevalent in areas where they previously have not been reported.
3. The number of successful introductions and periods of temporary establishment of 'exotic' tick species and their pathogens to Northern Europe might become increasingly frequent and longer with a warmer climate, particularly during years of unusually mild winters.
4. Given the appropriate environmental conditions and regular and continuous importation of ticks and tick-associated microorganisms, it is not impossible to imagine that eventually the 'right' variant of a novel pathogen for Northern Europe could emerge with interesting consequences.

Another related issue is that of research and research funding. Given the relatively long time periods over which climate change will occur and the complex ecosystem dependencies that are in play, of which climate and weather are one, it will also require long-term continuous research that bridges multiple fields of study and scientific disciplines. In view of that, it is not fair to ask scientists to answer or study questions related to climate change when long-term and dedicated funding programmes are non-existent, be it at the national or EU level. In contrast, scientific funding favours short-term projects and high-impact research, none of which ecologically oriented studies tend to fall within. Thus, given the current funding strategies for research in general and on ticks and tick-borne pathogen-focused research, we will have to rely on an ever so important 'patchwork' of short-term projects to understand something that will go on and affect us all for decades to come. Indeed, for 'climate change' simply not to become a buzzword to increase one's chances of attracting funding, science funding strategies also need to change more quickly than the consequences we all will experience from climate change.

References

Collins, M., Knutti, R., Arblaster, J., Dufresne, J.-L., Fichefet, T., *et al.* (2013) Long-term climate change: projections, commitments and irreversibility. In: *Climate Change 2013 – The Physical Science Basis. Contribution of Working Group I to the Fifth Assessment Report of the Intergovernmental Panel on Climate Change*. Cambridge University Press, Cambridge, pp. 1029–1136. https://doi.org/10.1017/CBO9781107415324.024

Estrada-Peña, A., Ayllón, N. and de la Fuente, J. (2012) Impact of climate trends on tick-borne pathogen transmission. *Frontiers in Physiology* 3, 64. https://doi.org/10.3389/fphys.2012.00064

Grandi, G., Chitimia-Dobler, L., Choklikitumnuey, P., Strube, C., Springer, A., *et al.* (2020) First records of adult *Hyalomma marginatum* and *H. rufipes* ticks (Acari: Ixodidae) in Sweden. *Ticks and Tick-borne Diseases* 11, 101403. https://doi.org/10.1016/j.ttbdis.2020.101403

Jaenson, T.G., Hjertqvist, M., Bergstrom, T. and Lundkvist, A. (2012) Why is tick-borne encephalitis increasing? A review of the key factors causing the increasing incidence of human TBE in Sweden. *Parasites & Vectors* 5, 184. https://doi.org/10.1186/1756-3305-5-184

Kjær, L.J., Soleng, A., Edgar, K.S., Lindstedt, H.E.H., Paulsen, K.M., *et al.* (2019) Predicting and mapping human risk of exposure to *Ixodes ricinus* nymphs using climatic and environmental data, Denmark, Norway and Sweden, 2016. *Eurosurveillance* 24, 1800101. https://doi.org/10.2807/1560-7917.ES.2019.24.9.1800101

Medlock, J.M., Hansford, K.M., Bormane, A., Derdakova, M., Estrada-Peña, A., *et al.* (2013) Driving forces for changes in geographical distribution of *Ixodes ricinus* ticks in Europe. *Parasites & Vectors* 6, 1. https://doi.org/10.1186/1756-3305-6-1

Pedersen, B.N., Jenkins, A. and Kjelland, V. (2020) Tick-borne pathogens in *Ixodes ricinus* ticks collected from migratory birds in southern Norway. *PLoS ONE* 15, e0230579. https://doi.org/10.1371/journal.pone.0230579

Pettersson, J.H.-O., Shi, M., Bohlin, J., Eldholm, V., Brynildsrud, O.B., *et al.* (2017) Characterizing the virome of *Ixodes ricinus* ticks from northern Europe. *Scientific Reports* 7, 10870. https://doi.org/10.1038/s41598-017-11439-y

Pettersson, J.H.-O., Ellström, P., Ling, J., Nilsson, I., Bergström, S., *et al.* (2020) Circumpolar diversification of the *Ixodes uriae* tick virome. *PLoS Pathogens* 16, e1008759. https://doi.org/10.1371/journal.ppat.1008759

Randolph, S.E., Green, R.M., Hoodless, A.N. and Peacey, M.F. (2002) An empirical quantitative framework for the seasonal population dynamics of the tick *Ixodes ricinus*. *International Journal for Parasitology* 32, 979–989. https://doi.org/10.1016/S0020-7519(02)00030-9

Wilhelmsson, P., Jaenson, T.G.T., Olsen, B., Waldenström, J. and Lindgren, P.-E. (2020) Migratory birds as disseminators of ticks and the tick-borne pathogens *Borrelia* bacteria and tick-borne encephalitis (TBE) virus: a seasonal study at Ottenby Bird Observatory in south-eastern Sweden. *Parasites & Vectors* 13, 607. https://doi.org/10.1186/s13071-020-04493-5

eo77 Tick and Tick-Borne Disease Circulation in a Changing Marine Ecosystem

Karen D. McCoy*
CNRS, Montpellier, France

77.1 Expert Opinion

One may wonder how ticks and tick-borne disease could be present in marine ecosystems. It is true that ticks do not live underwater (although they can survive immersion) and do not exploit fish. However, ticks do feed on other vertebrates that are marine animals, in particular those that spend part of their lives on land, predominantly seabirds. Indeed, at least 30 tick species occur regularly in this ecosystem and can be found across the globe, including the most extreme areas of both hemispheres (Table 77.1).

As in terrestrial systems, high infestations by ticks can reduce seabird body condition and lower reproductive success, which in turn can affect seabird population dynamics (Boulinier and Danchin, 1996; Gauthier-Clerc *et al.*, 1998; Mangin *et al.*, 2003). The impact of tick exploitation can be especially important under marine conditions, where thermal stress and the physiological demands of foraging can be arduous. Hyper-infestations by *Ixodes uriae*, resulting in blood loss and hypothermia, was, for example, implicated as a direct cause of death of breeding adult king penguins, *Aptenodytes patagonicus* (Gauthier-Clerc *et al.*, 1998). According to attachment site preferences, heavy infestations can also directly impair locomotion, leading to starvation or predation (Fig. 77.1a). Even lighter tick infestations have been associated with lower reproductive success and colony abandonment (Bergstrom *et al.*, 1999; Monticelli *et al.*, 2008; Sanz-Aguilar *et al.*, 2020); this can be due to direct impacts on young birds or irritation that perturbs normal feeding and vigilance behaviours in adults. Tropical seabirds are reported to shift colony locations as a means of avoiding the build-up of large numbers of ticks, such as *Ornithodoros capensis* (Fig. 77.1b) (Feare, 1976; Duffy, 1983).

In addition to these direct effects, general surveys of marine ticks have shown that a range of infectious agents are present and circulating in these systems, including diverse bacteria, viruses and protozoan parasites (Table 77.1). Some of these agents are known pathogens for humans and/or their domestic animals, such as the bacterial agents of Lyme disease, carried by *I. uriae*, or of African tick-bite fever (*Rickettsia africae*) found in *Amblyomma loculosum*. The impact of these agents on seabirds (and on ticks, for that matter) remain largely unknown. However, it is likely that the presence of these organisms is of some consequence for seabird health and population ecology. For example, the Soldado virus (Hughes serogroup) has been isolated from sick birds and shown experimentally to kill chickens (Converse *et al.*, 1975). Infection by *Babesia peircei* has been associated with higher mortality

*Email: karen.mccoy@ird.fr

 Climate, Ticks and Disease (ed. P. Nuttall)
DOI: 10.1079/9781789249637.0077

Table 77.1. Known tick species from the marine ecosystem and associated infectious agents. (Updated from Dietrich *et al*., 2011 with only new references included. Author's own table.)

Tick family/genus	Species	Recorded hosts	Known distribution	Infectious agents
Argasidae				
Argas	*cucumerinus*	*Pelecanus thagus*	Peru	Unknown
	macrostigmatus	*Phalacrocorax aristotelis*	Corsica, Crimea	Unknown
	monolakensis	*Larus californicus*	California	Mono Lake virus (*Orbivirus*, Kemerovo serogroup)
Ornithodoros (*Carios*/ *Alectorobius*)	*amblus*	<10 species: Suliformes, Charadriiformes, Sphenisciformes[a]	Peru, Chile	Huacho virus (*Orbivirus*, Kemerovo serogroup); Punta Salinas virus (*Nairovirus*, Hughes serogroup)
	capensis	>35 species: diverse Procellariiformes, Suliformes, Charadriiformes, Sphenisciformes, Phaethontiformes, Pelecaniformes	Globally distributed across tropical/subtropical regions	Soldado, Farallon viruses (*Nairovirus*, Hughes serogroup); Upolu, Aransas Bay viruses (*Thogotovirus*); Saumarez Reef virus (*Flavivirus*); Midway virus (*Nyamivirus*); Johnson Atoll virus (*Quaranjavirus*); adenoviruses (group B); *Rickettsia bellii*; *Rickettsia hoogstraalii*[d]
	cheikhi	*Sterna* spp.	Mauritania	Unknown
	denmarki	>10 species: diverse Procellariiformes, Suliformes, Charadriiformes, Sphenisciformes	Hawaii islands, Peru, Mexico, Jamaica, Martinique, Florida, Australia, New Zealand	Soldado virus (*Nairovirus*, Hughes serogroup); Farallon-like (*Nairovirus*, Hughes serogroup); Frazer Point and Raza viruses; Midway virus (*Nyamivirus*); Johnson Atoll virus (*Quaranjavirus*)
	maritimus	>35 species: diverse Procellariiformes, Suliformes, Charadriiformes, Phaethontiformes, Pelecaniformes, Sphenisciformes	British Isles, France, Spain, Portugal, Morocco	Essaouira, Kala Iris, Baku, Moellez viruses (*Orbivirus*, Kemerovo serogroup); Soldado, Great Saltee, Pettico Wick, Puffin Island viruses (*Nairovirus*, Hughes serogroup); West Nile, Caspiy, Meaban viruses (*Flavivirus*); *Rickettsia* spp.; *Borrelia* spp.; *Babesia* spp.[e]
	muesebecki	<10 species: Charadriiformes, Suliformes	Arabian Sea	Zirqa virus (*Nairovirus*, Hughes serogroup)
	sawaii	2 species: Procellariiformes	Japan	*Rickettsia* sp.; *Borrelia* sp.
	spheniscus	*Spheniscus humboldti*	Peru	Unknown
	yunkeri	<10 species: certain Procellariiformes, Suliformes, Charadriiformes, Sphenisciformes	Galapagos	Unknown

Continued

Table 77.1. Continued.

Tick family/genus	Species	Recorded hosts	Known distribution	Infectious agents
Ixodidae				
Ixodes	*amersoni*	>5 species: Procellariiformes (Diomedeidae), Suliformes (Sulidae), Charadriiformes (Laridae)	Central Pacific Ocean	Unknown
	auritulus	>35 species: diverse Procellariiformes, Suliformes, Charadriiformes, Sphenisciformes	Southern circumpolar region	Unknown
	diomedea	1 species: Procellariiformes (Diomedeidae)	Tristan da Cunha	Unknown
	eudyptidus	>10 species: diverse Procellariiformes, Suliformes, Charadriiformes, Sphenisciformes	Australia, New Zealand	Saumarez Reef virus (*Flavivirus*)
	jacksoni	1 species: Suliformes	Australia, New Zealand	Unknown
	kerguelenensis	>10 species: diverse Procellariiformes (Diomedeidae, Procellariidae)	Southern circumpolar region	Unknown
	kohlsi	>5 species: Procellariiformes, Suliformes, Charadriiformes, Sphenisciformes	Australia	Unknown
	layanensis	10 species: Procellariiformes, Suliformes, Charadriiformes, Phaethontiformes	Central Pacific Ocean	Unknown
	murreleti	1 species: *Synthliboramphus hypoleucus*	Mexico	Unknown
	percavatus	2 species: Procellariiformes (Diomedeidae), Sphenisciformes	Tristan da Cunha	Unknown
	philippi	2 species: Procellariiformes	Japan	Unknown
	rothschildi	>10 species: diverse Procellariiformes, Suliformes, Charadriiformes	North-eastern Atlantic Ocean	*Phlebovirus* (Uukuniemi group)
	signatus	~4 species: Suliformes, Charadriiformes	North Pacific Ocean	*Orbivirus* (Kemerovo group); *Orthonairovirus* (Sakhalin group); *Phlebovirus* (Uukuniemi group)
	unicavatus	2 *Phalacrocorax* spp.	Eastern North Atlantic Ocean, Baltic Sea	Unknown

	uriae	>60 species: diverse Procellariiformes, Suliformes, Charadriiformes, Sphenisciformes	Circumpolar (both hemispheres)	*Orbivirus* (Kemerovo group); *Orthonairovirus* (Sakhalin and Hughes groups); *Phlebovirus* (Uukuniemi group); *Flaviviruses*; *Borrelia burgdorferi* sensu lato (*Borrelia garinii*, *Borrelia bavariensis*, *Borrelia burgdorferi* sensu stricto, *Borrelia lusitaniae*)[f]
Amblyomma	*darwini*	1 species: *Amblyrhynchus cristatus* (iguana)[b]	Galapagos	Unknown
	loculosum	>10 species: diverse Procellariiformes, Suliformes, Charadriiformes, Phaethontiformes	Australia, Indian Ocean	Soldado virus (*Nairovirus*, Hughes serogroup); *Rickettsia africae*[d]
	nitidum	Sea snakes (*Laticauda*)[c]	South-western Pacific Ocean	Unknown

[a]Llanos-Soto *et al.* (2019).
[b]Schulze (1936).
[c]Kwak *et al.* (2020).
[d]Dietrich *et al.* (2014a).
[e]Dupraz *et al.* (2017); Gomard *et al.* (2020); Sanz-Aguilar *et al.* (2020).
[f]Munro *et al.* (2017).

Fig. 77.1. Marine ticks can pose significant problems for breeding seabirds in cases of heavy infestations and via the transmission of infectious agents. While infestations by ixodid ticks are relatively easy to quantify, as seen by this massively infested Atlantic puffin *Fratercula arctica* (a), those by argasid ticks are much more difficult. Indeed, soft ticks can be present in significant numbers within tropical and temperate seabird colonies, such as in this sooty tern, *Onychoprion fuscatus*, colony on the Ile de Lys, Iles Eparses (b), but only leave shelters in the vegetation for their short blood meal during the night (insets: sampling of *Ornithodoros capensis*). (Credits: (a) Photograph by D. Playle used with permission from A. Spellman (www.mullbirds.com); (b) author's own photographs.)

rates in certain seabirds brought to rehabilitation centres, particularly penguins (Parsons *et al.*, 2017, 2018). More generally, quantifying the impact of infection on seabirds, as in other wild populations, is extremely difficult when pathogenicity is chronic and acts in concert with other sources of environmental stress (e.g. food availability, predators, etc).

It is evident that current global changes will alter interactions between ticks and their marine hosts, with subsequent consequences for the circulation of infectious agents. This may occur via several possible mechanisms:

1. Marine ticks can only exist where their hosts breed. The presence of breeding colonies is dependent on both suitable nesting habitat and adequate food resources for nestlings. The distribution of seabird species has been dynamic over evolutionary time, shifting with changes in climate. For example, the formation of the Antarctic Polar Front during the Miocene led to the formation of a region of extremely high biodiversity, favouring its use by a large diversity of seabirds (Bost *et al.*, 2009). The discovery of fossilized nests on Bermuda showed that albatross bred within the North Atlantic up until approximately 400,000 years ago, likely going extinct during an interglacial period when seawater levels and storm surge were maximized (Olson and Hearty, 2003). Climate-driven changes in gyres where seabirds feed are regularly noted and known to affect adult survival (Fluhr *et al.*, 2017). Current rapid changes will therefore certainly result in the displacement or extinction of certain colonies. Indeed, as warming waters cause thermal expansion and ice sheets to melt, sea levels will rise and will eliminate low-lying colonies. Increases in water column stratification, deoxygenation and salinity, along with shifts in ocean circulation patterns, will accompany increasing sea-surface temperatures and affect the availability of local food sources for birds (Hoegh-Guldberg *et al.*, 2018). Increasing distances between established nesting sites and food sources will put hatchlings at risk (Jenouvrier *et al.*, 2015) and may lead to colony abandonment. Some seabird species may respond rapidly to such changes, particularly if replacement zones for reproduction and foraging can be found. For example, warmer conditions in high latitudes may allow some species to expand their ranges poleward (Péron *et al.*, 2010). However, some species will not have this opportunity and will need to adapt to a warmer world. Given the 'slow pace' of life of seabirds and the relative speed of current changes, many of these species will likely be driven to extinction, along with their parasites and pathogens (Grose *et al.*, 2020).

2. Climate changes may directly impact the presence and abundance of ticks in coastal marine environments. Marine ticks are only found on their hosts for the time of the blood meal and have adapted to a range of climates, with only the very coldest and driest polar regions seemingly unoccupied at present. Increasing temperatures and modified humidity regimes will change this distribution. Indeed, studies have shown that many terrestrial tick species are currently expanding their distributions northward due to milder winters that enable tick establishment (see references in Léger *et al.*, 2013). Surveys now suggest the same pattern holds true for marine ticks. For example, *I. uriae* seems to be increasingly present in high Arctic seabird colonies due to milder temperatures (Coulson *et al.*, 2009; Descamps, 2013). However, as the current distribution of marine ticks is still only partially known, such changes will be difficult to quantify. More importantly, if ticks expand in some ranges, will they actually retract in others? How plastic are ticks to changing conditions and is the current rate of change too fast for adaptation to follow? We know that marine ticks can survive under extreme conditions. For example, *I. uriae* has the largest temperature tolerance range known in ticks, from −30°C to +30°C, but is almost absent from seabird colonies in latitudes between 43°N and 45°S. Is this due to temperature–humidity interactions (Benoit *et al.*, 2007) or to some other environmental or ecological factor (e.g. competition)? Ticks seem to have large, highly repetitive genomes (Ullmann *et al.*, 2005). This may enable rapid adaptation to changing environmental conditions and shifts in host species use, ensuring their survival. Such a mechanism has been hypothesized to explain the worldwide distribution and broad host range of *I. uriae* (Dietrich *et al.*, 2014b).

3. The impact of ticks and their infectious agents can be exacerbated by the concurrent influence of other environmental factors, such as extreme weather conditions, reduced food resources, pollution and the presence of predators or co-infecting parasites/pathogens. It is the combination of these different individual factors which may have the greatest impact on marine vertebrate populations (Dias *et al.*, 2019) and consequently on their interactions with ticks and tick-borne infectious agents. These additional sources of stress are augmenting rapidly. For example, by-catch and competition with fisheries for depleting stocks have increased over time, leading to seabird population declines (Grémillet *et al.*, 2018). Pollutants are ubiquitous in the marine environment and have reached even the most untouched zones of the planet via ocean currents (Cózar *et al.*, 2017). Indeed, massive accumulations of plastics have formed, destroying foraging and breeding habitats for marine animals and bioaccumulating along trophic food chains with potential chronic toxicity effects (Auta *et al.*, 2017). Climate change is rapidly modifying access to habitats and food resources used by marine animals and destabilizing populations via an increasing number of extreme weather events (Jenouvrier *et al.*, 2015). Although seabirds are among the most vagile creatures on Earth, most show strong migratory and breeding site fidelity, limiting their movements to within well-defined limits. Exploratory movements only tend to occur when individuals prospect for new locations to breed. This is seen in young birds during the period between fledgling and first reproduction, but also in adult birds that have failed a reproductive attempt. As the proportion of local nest success declines within a colony, the probability that a given breeding pair will leave to prospect increases (Boulinier *et al.*, 2008). Failed birds in colonies with widespread failure will thus show reduced site fidelity the following year and will have a higher probability of changing breeding partners. This behaviour sets up a feedback loop in which breeding failures (caused by a combination of potential factors) lead to unstable spatiotemporal population dynamics and a higher probability of dispersal. This, in turn, increases the probability of dispersing parasites (ticks) and infectious agents harboured by the birds, allowing the invasion of new locations and novel hosts. Therefore, the combined effect of different environmental stresses can alter the nature and outcome of host–parasite interactions and result in large-scale changes to the ecological and evolutionary pathways taken by these species. As global changes increase pressure on marine birds, this phenomenon may become more frequent, resulting in the large-scale homogenization of ticks and infectious agents across ranges of suitable ecological conditions.

77.2 Future Predictions

So can we make any concrete predictions? This is a difficult question to answer in a complex and rapidly changing marine ecosystem. However, I expect that certain opportunistic tick species, such as *A. loculosum* and *O. capensis* sensu lato, will expand their distributions and explode in numbers. Other, more specialist species, and particularly polar species such as *Ixodes signatus* and *Ixodes kerguelenensis*, will likely follow the dynamics of their hosts and decline in prevalence and abundance. Still other, more plastic, species like *I. uriae* may be able to maintain the status quo, shifting their distributions to the limits of their physiological capacities and adapting to those host species available locally. What is sure is that these changes will interplay with other environmental factors that affect seabird community dynamics, patterns of colony use and dispersal. This, in turn, will feed back on the interaction to alter the spread of ticks and the infectious agents these ticks and their host birds carry. As competition for limiting marine resources continues to intensify, contact between humans and wildlife are likely to increase, opening new links between marine and terrestrial tick and tick-borne disease systems. In this light, a better understanding of what limits the spatial distribution of marine ticks, along with more basic information on the infectious agents they carry, seems an essential first step for predicting future impacts of ticks and tick-borne disease on marine ecosystems and beyond.

References

Auta, H.S., Emenike, C. and Fauziah, S. (2017) Distribution and importance of microplastics in the marine environment: a review of the sources, fate, effects, and potential solutions. *Environment International* 102, 165–176.

Benoit, J.B., Yoder, J.A., Lopez-Martinez, G., Elnitsky, M.A., Lee, R.E. and Denlinger, D.L. (2007) Habitat requirements of the seabird tick, *Ixodes uriae* (Acari: Ixodidae), from the Antarctic Peninsula in relation to water balance characteristics of eggs, nonfed and engorged stages. *Journal of Comparative Physiology B* 177, 205–215.

Bergstrom, S., Haemig, P.D. and Olsen, B. (1999) Increased mortality of black-browed albatross chicks at a colony heavily infested with the tick *Ixodes uriae*. *International Journal for Parasitology* 29, 1359–1361.

Bost, C.A., Cotté, C., Bailleul, F., Cherel, Y., Charrassin, J.B., *et al.* (2009) The importance of oceanographic fronts to marine birds and mammals of the southern oceans. *Journal of Marine Systems* 78, 363–376.

Boulinier, T. and Danchin, E. (1996) Population trends in kittiwake *Rissa tridactyla* colonies in relation to tick infestation. *Ibis* 138, 326–334.

Boulinier, T., McCoy, K.D., Yoccoz, N.G., Gasparini, J. and Tveraa, T. (2008) Public information affects breeding dispersal in a colonial bird: kittiwakes cue on neighbours. *Biology Letters* 4, 538–540.

Converse, J.D., Hoogstraal, H., Moussa, M.I., Feare, C.J. and Kaiser, M.N. (1975) Soldado virus (Hughes group) from *Ornithodoros* (*Alectorobius*) *capensis* (Ixodoidea – Argasidae) infesting sooty tern colonies in Seychelles, Indian Ocean. *American Journal of Tropical Medicine and Hygiene* 24, 1010–1018.

Coulson, S.J., Lorentzen, E., Strom, H. and Gabrielsen, G.W. (2009) The parasitic tick *Ixodes uriae* (Acari: Ixodidae) on seabirds from Spitsbergen, Svalbard. *Polar Research* 28, 399–402.

Cózar, A., Martí, E., Duarte, C.M., García-de-Lomas, J., van Sebille, E., *et al.* (2017) The Arctic Ocean as a dead end for floating plastics in the North Atlantic branch of the Thermohaline Circulation. *Science Advances* 3, e1600582.

Descamps, S. (2013) Winter temperature affects the prevalence of ticks in an Arctic seabird. *PLoS ONE* 8, e65374.
Dias, M.P., Martin, R., Pearmain, E.J., Burfield, I.J., Small, C., *et al.* (2019) Threats to seabirds: a global assessment. *Biological Conservation* 237, 525–537.
Dietrich, M., Gomez-Diaz, E. and McCoy, K.D. (2011) Worldwide distribution and diversity of seabird ticks: implications for the ecology and epidemiology of tick-borne pathogens. *Vector-Borne and Zoonotic Diseases* 11, 453–470.
Dietrich, M., Lebarbenchon, C., Jaeger, A., Le Rouzic, C., Bastien, M., *et al.* (2014a) *Rickettsia* spp. in seabird ticks from western Indian Ocean islands, 2011–2012. *Emerging Infectious Diseases* 20, 838–842.
Dietrich, M., Lobato, E., Boulinier, T. and McCoy, K.D. (2014b) An experimental test of host specialization in a ubiquitous polar ectoparasite: a role for adaptation? *Journal of Animal Ecology* 83, 576–587.
Duffy, D.C. (1983) The ecology of tick parasitism on densely nesting Peruvian seabirds. *Ecology* 64, 110–119.
Dupraz, M., Toty, C., Devillers, E., Blanchon, T., Elguero, E., *et al.* (2017) Population structure of the soft tick *Ornithodoros maritimus* and its associated infectious agents within a colony of its seabird host *Larus michahellis*. *International Journal for Parasitology: Parasites and Wildlife* 6, 122–130.
Feare, C.J. (1976) Desertion and abnormal development in a colony of sooty terns *Sterna fuscata* infested by virus-infected ticks. *Ibis* 118, 112–115.
Fluhr, J., Strøm, H., Pradel, R., Duriez, O., Beaugrand, G. and Descamps, S. (2017) Weakening of the subpolar gyre as a key driver of North Atlantic seabird demography: a case study with Brünnich's guillemots in Svalbard. *Marine Ecology Progress Series* 563, 1–11.
Gauthier-Clerc, M., Clerquin, Y. and Handrich, Y. (1998) Hyperinfestation by ticks *Ixodes uriae*: a possible cause of death in adult king penguins, a long-lived seabird. *Colonial Waterbirds* 21, 229–233.
Gomard, Y., Flores, O., Vittecoq, M., Blanchon, T., Toty, C., *et al.* (2020) Changes in bacterial diversity, composition and interactions during the development of the seabird tick *Ornithodoros maritimus* (Argasidae). *Microbial Ecology* 81, 770–783.
Grémillet, D., Ponchon, A., Paleczny, M., Palomares, M.-L.D., Karpouzi, V. and Pauly, D. (2018) Persisting worldwide seabird–fishery competition despite seabird community decline. *Current Biology* 28, 4009–4013.e2.
Grose, S.O., Pendleton, L., Leathers, A., Cornish, A. and Waitai, S. (2020) Climate change will re-draw the map for marine megafauna and the people who depend on them. *Frontiers in Marine Science* 7, 547.
Hoegh-Guldberg, O., Jacob, D., Taylor, M., Bindi, M., Brown, S., *et al.* (2018) Impacts of 1.5°C global warming on natural and human systems. In: Masson-Delmotte, V., Zhai, P., Pörtner, H.-O., Roberts, D., Skea, J., *et al.* (eds) *Global Warming of 1.5°C. An IPCC Special Report on the impacts of global warming of 1.5°C above pre-industrial levels and related global greenhouse gas emission pathways, in the context of strengthening the global response to the threat of climate change, sustainable development, and efforts to eradicate poverty*. World Meteorological Organization, Geneva, Switzerland, pp. 175–312.
Jenouvrier, S., Péron, C. and Weimerskirch, H. (2015) Extreme climate events and individual heterogeneity shape life-history traits and population dynamics. *Ecological Monographs* 85, 605–624.
Kwak, M.L., Kuo, C.-C. and Chu, H.-T. (2020) First record of the sea snake tick *Amblyomma nitidum* Hirst and Hirst, 1910 (Acari: Ixodidae) from Taiwan. *Ticks and Tick-borne Diseases* 11, 101383.
Léger, E., Vourc'h, G., Vial, L., Chevillon, C. and McCoy, K.D. (2013) Changing distributions of ticks: causes and consequences. *Experimental and Applied Acarology* 59, 219–244.
Llanos-Soto, S., Muñoz-Leal, S. and González-Acuña, D. (2019) Hyperparasitism in the seabird tick *Ornithodoros amblus* (Acari: Argasidae). *Systematic and Applied Acarology* 24, 525–528.
Mangin, S., Gauthier-Clerc, M., Frenot, Y., Gendner, J.-P. and Le Maho, Y. (2003) Ticks *Ixodes uriae* and the breeding performance of a colonial seabird, king penguin *Aptenodytes patagonicus*. *Journal of Avian Biology* 34, 30–34.
Munro, H.J., Ogden, N.H., Lindsay, L.R., Robertson, G.J., Whitney, H. and Lang, A.S. (2017) Evidence for *Borrelia bavariensis* infections of *Ixodes uriae* within seabird colonies of the North Atlantic Ocean. *Applied and Environmental Microbiology* 83, 20.
Monticelli, D., Ramos, J.A., Hines, J.E., Nichols, J.D. and Spendelow, J.A. (2008) Juvenile survival in a tropical population of roseate terns: interannual variation and effect of tick parasitism. *Marine Ecology Progress Series* 365, 277–287.

Olson, S.L. and Hearty, P.J. (2003) Probable extirpation of a breeding colony of short-tailed albatross (*Phoebastria albatrus*) on Bermuda by Pleistocene sea-level rise. *Proceedings of the National Academy of Sciences USA* 100, 12825–12829.
Parsons, N.J., Voogt, N.M., Schaefer, A.M., Peirce, M.A. and Vanstreels, R.E.T. (2017) Occurrence of blood parasites in seabirds admitted for rehabilitation in the Western Cape, South Africa, 2001–2013. *Veterinary Parasitology* 233, 52–61.
Parsons, N.J., Gous, T.A., Cranfield, M.R., Cheng, L.I., Schultz, A., *et al.* (2018) Novel vagrant records and occurrence of vector-borne pathogens in king penguins (*Aptenodytes patagonicus*) in South Africa. *Polar Biology* 41, 79–86.
Péron, C., Authier, M., Barbraud, C., Delord, K., Besson, D. and Weimerskirch, H. (2010) Interdecadal changes in at-sea distribution and abundance of subantarctic seabirds along a latitudinal gradient in the southern Indian Ocean. *Global Change Biology* 16, 1895–1909.
Sanz-Aguilar, A., Payo-Payo, A., Rotger, A., Yousfi, L., Moutailler, S., *et al.* (2020) Infestation of small seabirds by *Ornithodoros maritimus* ticks: effects on chick body condition, reproduction and associated infectious agents. *Ticks and Tick-borne Diseases* 11, 101281.
Schulze, P. (1936) The Norwegian zoological expedition to the Galapagos Island. *Die Ixodoidea der Galapagos-Inseln. Meddelelser fra det Zoologiske Museum* 45, 155–162.
Ullmann, A.J., Lima, C.M.R., Guerrero, F.D., Piesman, J. and Black, W.C. (2005) Genome size and organization in the blacklegged tick, *Ixodes scapularis* and the Southern cattle tick, *Boophilus microplus*. *Insect Molecular Biology* 14, 217–222.

Synopsis: Disease

Pat Nuttall*
University of Oxford, Oxford, UK

Expert opinions related to disease consider whether or not changes in climate affect various aspects of the tick–host–pathogen biosphere (eo45: Munderloh and Kurtti, Fig. 45.1), ranging from the triangle of tick–host–pathogen molecular interactions to disease ecology in various regions and ecosystems of the world. This synopsis provides a brief overview.

Tick-borne infectious agents (mainly viruses, bacteria and protozoa) survive through alternate replication in a vertebrate host and a tick vector. When the agent causes disease in the vertebrate host, a complex triangle of parasitic interactions ensues: tick–host, pathogen–tick, pathogen–host. Added to this complexity are opportunities for multiple pathogens to co-infect the tick vector and/or vertebrate host. Conceivably, the incidence of co-infections may increase as environmental conditions become more benevolent for endemic and exotic tick and host species (eo38: Pfeffer *et al.*). de la Fuente and Villar (eo33) propose that conflict and cooperation between ticks, their hosts and tick-borne pathogens have evolved to increase tick (and pathogen) fitness and survival under extreme environmental conditions, e.g. pathogen-induced changes in heat-shock protein and subolesin expression affecting tick behaviour, in epigenetic regulatory mechanisms favouring survival under extreme conditions and in exoskeleton chemical composition. As yet, none of these pathogen-induced changes are reflected in the geographical expansion of infected (compared with uninfected) tick vectors (eo42: Sonenshine), although they may help identify new control strategies (eo33: de la Fuente and Villar). Nevertheless, tick-borne pathogens have undoubtedly experienced climate extremes during their evolution, illustrated by the phylogenetic tree of tick-borne flaviviruses showing their emergence from Africa more than 16,000 years ago and dispersal northwards, experiencing notable climate extremes along the way (eo36: Gould and de Lamballerie).

One factor often overlooked in considering tick–host–pathogen interactions is the impact of seasonality changes, which may create mismatches in e.g. timing of tick contact with the pathogen reservoir host (eo34: Estrada-Peña *et al.*). Seasonal activity patterns of ticks depend on the combined effects of climate on stage-to-stage development rates and weather conditions suitable for host-seeking activity. Changes in climate may influence feeding synchrony between larvae and nymphs, with significant impact on pathogen transmission especially for pathogens with relatively short infectious periods in their vertebrate host that rely on co-feeding transmission for survival (e.g. tick-borne encephalitis

*Email: pat.nuttall@zoo.ox.ac.uk

DOI: 10.1079/9781789249637.syn3

(TBE) virus, Powassan virus, *Ehrlichia muris eauclairensis*) (eo41: Diuk-Wasser). Co-feeding transmission is a non-systemic transmission mode in which the host provides an infective bridge between co-feeding susceptible and infected ticks, facilitating pathogen transmission (eo39: Wu and Zhang). As non-systemic transmission requires co-feeding of infected and uninfected ticks in close (both spatially and temporally) proximity, environmental conditions affecting the sylvatic cycle are critical. Studies of TBE and the sylvatic cycle provide compelling evidence of the importance of co-feeding transmission in models of disease risk and of the inclusion of region-specific climate change patterns in projections of future risk (eo39: Wu and Zhang).

A second factor often overlooked in attempts to predict climate change impacts is the need for a critical threshold of reservoirs necessary to support the circulation of a pathogen (eo35: Mysterud; eo34: Estrada-Peña *et al.*). Evaluating how R_0 (the basic reproduction number of an infection) changes under different climate scenarios is one approach that can take account of seasonality and host thresholds. Models for R_0 that include phenologies of immature *Ixodes* tick vectors and their interactions with pathogens have generated insights into how shifts in tick feeding seasonality might affect the persistence of tick-borne pathogens (eo41: Diuk-Wasser). However, R_0 does not take account of human actions (e.g. agricultural practices, urban encroachment of habitats, recreational activities), which can overwhelm any impacts of climate change (eo45: Munderloh and Kurtti). Telford (eo40) argues that human aspects of risk, particularly human behaviour, are poorly studied and need to be considered in any discussion of the role of global climate change on risk for tick-borne infections.

Whatever the reasons, tick-borne diseases are emerging worldwide, but more importantly, their distributions are shifting (eo37: Foley; eo44: Samy *et al.*). Modelling how climate change might impact the prevalence and incidence of tick-borne infections is usually restricted to the distribution of the principal vector species. For example, physiological models used in hindsight show that almost every site in Sweden under the latitude 67°N experienced a theoretical boosting of the *Ixodes ricinus* life cycle during the past 30 years (eo34: Estrada-Peña *et al.*). As the pace of change accelerates, data we have from past records become less directly applicable to the present and near future (eo37: Foley). Hence it is important to assess uncertainty, which may be introduced by transferring models to future conditions and provide a guide to where model predictions are less robust; hierarchical partitioning of variance is one approach (eo44: Samy *et al.*). Fine-scale, real-time geographical data on disease risk are needed as the public becomes increasingly aware of the risks of tick-borne disease and emerging infectious disease in general (eo37: Foley).

Several tick-borne pathogens cause disease in humans, of which Lyme borreliosis (Lyme disease) is the most frequently recorded, with >1 million cases annually worldwide (eo49: Rudenko *et al.*). Caused by some bacterial species in the *Borrelia burgdorferi* sensu lato (s.l.) complex, the disease is increasing in incidence and spreading geospatially. Drivers of this growth include those considered for the tick vector species (Section 2) with additional factors such as gene transfer between *B. burgdorferi* s.l. strains creating recombinant genotypes new to a region (eo49: Rudenko *et al.*). Based on a 5°C lower thermal threshold for the Lyme borreliosis season, Trájer (eo54) compares modelled maximum seasonal duration of the disease across Eastern and Central Europe for 1960–1990 with 2040–2060. The worse-case scenario predicts a 2-month increase in seasonal duration of Lyme borreliosis risk for some regions.

The second most frequently recorded human tick-borne infection is TBE caused by a group of viruses that evolved some 3000 years ago (eo36: Gould and de Lamballerie). In Europe, TBE virus was first isolated from patients in 1948 in former Czechoslovakia (eo47: Danielová and Daniel). Despite the availability of vaccines, TBE cases in Europe have risen in recent decades. Rubel (eo50) considers three effects of climate change as contributing factors to increased TBE cases in the Greater Alpine region: (i) increasing tick vector densities reflecting replacement of spruce forests by deciduous or mixed forests, a human-induced climate change adaptation strategy; (ii) more years with conditions favouring the TBE sylvatic cycle (more beechnuts resulting in more rodents on which immature ticks feed); and (iii) extension of endemic areas as temperatures increase in the mountains. In the 1980s, the upper limit of *I. ricinus* distribution in the Krkonoše Mountains of Central Europe was 700 m above sea level; two decades on, this upper limit shifted to ~1100 m, coincident with a 1.4°C rise in air temperature at 1000 m and TBE cases recorded at higher altitudes (eo47: Danielová and Daniel; eo50: Rubel).

Kyasanur forest disease (KFD), caused by a relative of TBE virus, was first reported in 1957 when human cases of severe haemorrhagic fever were recorded in the forested Shimoga district of India's Karnataka state. Locally known as monkey fever or monkey disease, the virus is transmitted (horizontally and transovarially) mainly by *Haemaphysalis spinigera* in a seasonal cycle dictated by the monsoon period (eo52: Pandey and Singh). Deforestation impacts the microclimate, changing local rain patterns which affect the infective period of larvae and, potentially, the emergence and re-emergence of KFD through effects on the vector–host–virus triangle. Although KFD has become more widespread in India, data are lacking that relate seasonal changes in temperature and rainfall to KFD virus emergence (eo52: Pandey and Singh).

Beside KFD virus, several other tick-borne viruses cause haemorrhagic fever in humans, of which Crimean–Congo haemorrhagic fever (CCHF) virus is the most notorious, primarily because of its wide geographical distribution and ability to be transmitted directly to humans via infected blood and other infectious products (eo48: Bente). CCHF virus is also notable for the high number (>30) of tick species from which it has been isolated, although the vector capacity of only a few species has been demonstrated. Bente (eo48) questions whether vector switching might occur as temperature and environmental conditions change, allowing other tick species to play a more prominent role in virus transmission and disease dynamics.

Another severe human disease with increasing case numbers and an expanding geographical distribution is severe fever with thrombocytopaenia syndrome (SFTS) caused by a phlebovirus (eo55: Liu *et al.*). The disease was first identified in China, in 2010, and subsequently in South Korea and Japan. In China, there are four hotspots, characterized by temperate humid or subtropical climates but with no data on climate impacts. The sylvatic cycle is poorly understood, although transmission via ticks is deemed the most likely source of human infections. Although *Haemaphysalis longicornis* is thought to be the principal vector, SFTS virus has been detected in several other tick species, raising the question of vector switching posed by Bente (eo48).

Case numbers of Japanese spotted fever (JSF) have also been increasing in Japan since the rickettsial disease was first described there in 1984. The aetiological agent, *Rickettsia japonica*, is one of 11 spotted fever group rickettsiae identified in ixodid tick species collected in Japan (eo51: Sato *et al.*). The increase in human cases of JSF is probably caused by exposure of more people to ticks as populations of natural tick hosts (e.g. deer and wild pigs) expand into previously non-endemic regions (eo51: Sato *et al.*). Again, there are no solid data on climate impacts.

All tick-borne relapsing fevers (TBRFs) are transmitted by *Ornithodoros* spp., argasid ticks that usually engorge within 30 min. The most severe form of TBRF is caused by *Borrelia duttoni* transmitted by *Ornithodoros moubata* which feeds entirely on humans in parts of Africa, living within the protection of human dwellings. In western North America, the distribution of *Ornithodoros hermsi*, the vector of *Borrelia hermsii* (the most common cause of TBRF in North America), will likely shrink to higher elevations as winters become warmer and drier (eo53: Teglas). Although geographic regions considered endemic for TBRFs will change with changing climate, Teglas (eo53) considers TBRFs will continue to be a global public health concern in the next century while Stuen (eo43) believes improved diagnostics and more active surveillance will likely detect increasing cases of zoonotic infections with *Ehrlichia* and *Anaplasma* pathogens.

As with most human diseases, long-term data on the incidence, prevalence and geographical distribution of diseases affecting domesticated animals are lacking, both in developed (e.g. UK) (eo59: Johnson and Phipps) and developing (e.g. Pakistan) (eo60: Ghafar *et al.*) parts of the world, and the impacts of climate change are unknown. African swine fever (ASF) virus is probably the most notorious tick-borne infection of domesticated animals, having caused national pig populations to be culled in efforts to control disease outbreaks. However, ticks are generally not the culprit as transmission occurs readily from pig to pig and from contaminated products to domestic pigs (eo58: Bastos and Boinas). In eastern and southern Africa, the vector species, *O. moubata*, enjoys a buffered microclimate living in warthog (*Phacochoerus africanus*) burrows where ASF virus is maintained in a sylvatic cycle. Bastos and Boinas (eo58) predict climate change will drive this sylvatic cycle to extinction because aardvarks (*Orycteropus afer*), which excavate burrows upon which many species depend for survival (including neonatal warthogs), are particularly vulnerable to climate change impacts.

Most livestock diseases in the UK associated with ticks involve *I. ricinus* (the UK's sheep tick), e.g. louping ill encephalomyelitis in sheep, anaplasmosis (aka, tick-borne fever), babesiosis (aka, redwater fever) and tick pyaemia (caused by *Staphylococcus aureus*); the exception being ovine theileriosis caused by *Theileria luwenshuni* and transmitted by *Haemaphysalis punctata*. Recently, an outbreak of bovine babesiosis in cattle was linked to abnormally high winter temperatures, demonstrating the potential for climate change impacts (eo59: Johnson and Phipps).

Pakistan faces a daunting task tackling ticks and associated diseases affecting its economically critical livestock industry (eo60: Ghafar *et al.*). In tropical to temperate climates, and with millions of ruminants, poor husbandry practices and lack of proper control measures, *Haemaphysalis*, *Hyalomma* and *Rhipicephalus* spp. ticks thrive. Recent studies suggest more than half the ruminant population is affected by at least one of three major diseases – theileriosis, babesiosis and anaplasmosis – of which theileriosis is considered the most important. Ghafar *et al.* (eo60) speculate climate change is likely to increase the prevalence and distribution of *Hyalomma* and *Rhipicephalus* spp., which mainly infest ruminants in Pakistan.

Although Australia has an abundance of tick species, the disease challenge comes from five species introduced with domesticated animals: poultry, *Argas persicus*; horses, *Otobius megnini*; cattle, *H. longicornis* and *Rhipicephalus australis* (syn. *Boophilus microplus*); and dogs, *Rhipicephalus sanguineus*. Oskam *et al.* (eo61) consider two case studies to illustrate how climate change can potentially influence the spread of introduced tick species and the pathogens they transmit: (i) the recent incursion of canine monocytic ehrlichiosis (CME) into northern regions of Australia in April/June 2020; and (ii) the geographical expansion of bovine anaemia due to *Theileria orientalis* group (BATOG) across Australia since 2006. To date, the respective tick vector species, *R. sanguineus* and *H. longicornis*, are unlikely to have reached their full geographical range in Australia; hence there is concern that CME and BATOG will become increasingly widespread, especially if climate change leads to increasingly favourable environmental conditions for these introduced species (eo61: Oskam *et al.*). The situation for humans is less clear in Australia regarding tick-borne infections, although the potently toxic bite of *Ixodes holocyclus* is a recognized threat for humans and dogs (eo56: Dehhaghi *et al.*).

Looking at how climate is affecting different parts of the world, the Arctic region is clearly undergoing some of the most profound changes. However, the lack of baseline data makes it difficult to assess whether increased abundance of established tick species, and records of new species, are real phenomena or simply the consequence of increased research and public awareness (eo67: Černý *et al.*). For example, the numbers of *Ixodes uriae* feeding on seabirds in Svalbard have increased compared with the beginning of this century, and *I. ricinus* and *Ixodes persulcatus* were recently found at the southern borders of the Arctic. As yet, no tick-borne infections have been detected in Svalbard's *I. uriae* although this species is a known vector of *B. burgdorferi* s.l. and numerous arboviruses (eo67: Černý *et al.*). The Arctic is virgin territory as far as tick-borne pathogens are concerned. If the experience in Canada is anything to go by, the Arctic is surely at risk.

In response to the emerging threat of Lyme borreliosis in the 1990s, Canada put in place a tick surveillance system, providing a unique view of disease emergence and pointing to climate warming as a significant driver (eo73: Leighton *et al.*). In the early 1990s, *Ixodes scapularis* invaded eastern Canada and spread rapidly; Lyme borreliosis followed and is now endemic in most Canadian provinces. Native Canadian tick vector species are more abundant, and anaplasmosis, babesiosis and Powassan virus infections are more common, as is the case in the USA (eo69: Hart and Thangamani). Although physiological constraints impose a theoretical northern range limit for ticks in Canada, this limit is expected to continue shifting northwards with further climate warming, expanding areas for tick population establishment (where habitats are suitable) with opportunities for tick-borne pathogens to follow (eo73: Leighton *et al.*). The dramatic northern migration of *Amblyomma americanum* in the USA, followed by an expanded range of *Ehrlichia chaffeensis*, the cause of human monocytic ehrlichiosis, may also be driven by milder winters (eo69: Hart and Thangamani). The futures for *H. longicornis*, a newly established species in the USA and not yet recorded as a significant disease vector (unlike in its native Asia), and *Rhipicephalus microplus* (the scourge of the cattle industry) under a warming climate, remain to be seen.

For Northern Europe, data for the last three to four decades reveal the emergence of TBE and Lyme borreliosis in Norway and Finland. In Norway, records show an increase in endemic areas and spatial expansion for babesiosis in cattle, anaplasmosis in sheep and cattle, and Lyme borreliosis in humans (eo35: Mysterud). For TBE, the distribution is patchy, possibly reflecting the prevalence of natural sylvatic foci. Such foci appear to have favourable microclimates with warmer than average temperatures and small mammalian hosts supporting higher densities of ticks compared with other areas. Pettersson (eo76) considers these endemic hotspots may increase in frequency with increasing temperatures in Northern Europe, increasing the risk of human exposure to TBE virus-infected ticks. It will be interesting to see if this pans out in England where TBE virus-infected ticks were recently discovered in two forest foci, possibly linked to a temperature driver (eo75: Medlock and Hansford).

Scotland suffers the highest UK incidence of Lyme borreliosis, with >37 cases per 100,000 people (eo74: Gilbert). The prevalence and density of *B. burgdorferi*-infected questing nymphs are higher in warmer areas of Scotland and at lower altitude, and Lyme disease cases increase in years with warmer winters and springs. However, numbers of questing *I. ricinus*, and the likelihood of their infection with *B. burgdorferi*, are negatively associated with rainfall and relative humidity. Hence, in a cool, wet country such as Scotland, it can be too wet for *I. ricinus* and Lyme borreliosis (and for humans too) (eo74: Gilbert). Other factors, particularly the increased distribution and abundance of red and roe deer, have helped grow the UK's *I. ricinus* population over the past few decades, aided by milder winters (eo75: Medlock and Hansford).

The year 2019 was the warmest on record in Russia, although the effect of climate warming varies over this vast territory. Despite the warming climate, northern distribution limits of *I. persulcatus* (vector of *B. burgdorferi* and TBE virus) in several Russian regions are unchanged compared with the 1960s and 1970s (eo63: Korenberg). Similarly, changes in the abundance of unfed adult *I. persulcatus* ticks show no definite trend attributable to global climate change over at least the past 40 years. In fact, evidence for the directional effect of climate change on the morbidity of tick-borne infections in Russia has not been obtained to date. The highest TBE morbidity rate recorded in Russia, 6.8 cases per 100,000 in 1999, followed legislative changes which allowed undeveloped land to be used for housing and gardens. This resulted in unprecedented mass contacts of city dwellers with ticks. Subsequently, TBE morbidity in Russia has been falling (eo63: Korenberg).

In Central Europe, where Lyme borreliosis and TBE are the most important tick-borne infections in humans, there is growing evidence of: (i) geographical expansion of indigenous tick species northward and to higher altitudes; (ii) increasing tick survival and abundance in areas such as urban parks and gardens; (iii) prolonged seasonal tick questing activities; (iv) emergence of new disease foci; and (v) changing epidemiological patterns and increasing incidence of tick-borne diseases in endemic foci (eo62: Kazimirova).

Turkey is a haven for ticks and tick-borne diseases with its rich diversity of ecological niches, vegetation and wildlife, and different climatic zones (eo66: Kar and Ergünay). Since the emergence of CCHF virus infections in the early 2000s, Turkey has become the epicentre of the disease. However, within the climatic range of Turkey, higher temperatures and milder conditions are insufficient to explain ongoing regional variations in disease prevalence. Even the Black Sea Flyway of migrating birds appears to have had negligible impact, at least over the past

20 years in Turkey (eo66: Kar and Ergünay). Indeed, in the neighbouring Middle East, where a total of 136 tick species have been recorded, extraneous factors may have more bearing on ticks and tick-borne infections, such as political unrest changing human demographics, human migration and movement of livestock (eo68: Chavshin and Seyyed-Zadeh).

In the Caribbean region of tropical and subtropical islands, *R. microplus* (the cattle tick) is the most widespread tick species and the vector of *Anaplasma marginale* as well as the protozoan parasites, *Babesia bovis* and *Babesia bigemina*, which cause significant losses of cattle and buffalo (eo65: Obregón Alvarez *et al.*). Climate change impacts are not apparent in the region although the potential is recognized for *R. microplus* to expand its known range and/or speed up its development cycle under a warming climate. By contrast, in tropical Africa, tick-borne disease epizootiology is complex owing to co-infestations with multiple tick species resulting in co-infection of livestock and wildlife with multiple tick-borne pathogens (eo70: Githaka *et al.*). *R. microplus* (locally called the Asian blue tick) is rapidly expanding its geographical range within the continent. Uncontrolled transboundary cattle trade in regions with mostly porous international boundaries is believed to be the main driver, with civil unrest catalysing the spread of animal diseases including those transmitted by *R. microplus*. The indigenous species, *Rhipicephalus appendiculatus* (vector of *Theileria parva*, the cause of East Coast fever in cattle), is limited to relatively higher, cooler and wetter regions of sub-Saharan Africa. Although predicted to expand its distribution in Central Africa, *R. appendiculatus* is absent from some ecologically suitable habitats where there is also an absence of *T. parva* in Cape buffalo (*Syncerus caffer*), the reservoir host of this important pathogen. Competition with other tick species and predation are considered possible reasons for its absence (eo70: Githaka *et al.*). Cross-bred cattle, reared in Africa to enhance productivity and increase resilience to ticks and tick-borne infections, are more susceptible to tick infestation and parasitic diseases during periods of nutritional restriction, as occurs increasingly during droughts (eo70: Githaka *et al.*). Global warming is linked to a rise in human attacks by the brown dog tick, *R. sanguineus*, and increased transmission of life-threatening rickettsioses. In south-western Tanzania, high seropositivity to spotted fever group rickettsiae was found in localities with higher temperatures. Nevertheless, there are insufficient data on whether climate change is having an impact on the incidence, prevalence and distribution of rickettsial infections in Africa (eo72: Kolo).

In China, more than 120 tick species have been documented together with 103 tick-borne agents, the most important being SFTS virus and TBE virus in terms of disease burden (eo71: Li *et al.*). Although an increasing number of tick-borne pathogens have been reported in the past 30 years, this is largely attributed to the application of new diagnostic technologies. Changes in land use, animal host communities, human living conditions and societal factors are considered more important than climate change in the emergence of diseases associated with ticks in China (eo71: Li *et al.*).

Like many parts of the world, in Taiwan, the effects of climate change (including temperature and precipitation) on ticks have not been investigated explicitly (eo64: Kuo). SFTS and Lyme borreliosis are notifiable tick-borne diseases in Taiwan although only one case of SFTS has been recorded (in 2019) to date and no autochthonous cases of Lyme borreliosis have been documented despite *B. burgdorferi* having been identified in ticks and rodents.

Marine ecosystems provide an insight into ticks and tick-borne infections globally, including polar regions where seabirds are hosts to the remarkably resilient *I. uriae* (eo77: McCoy). At least 30 tick species have been recorded from marine ecosystems together with viruses, bacteria and protozoan parasites, some of which are human pathogens (e.g. *B. burgdorferi* transmitted by *I. uriae*, *Rickettsia africae* transmitted by *Amblyomma loculosum*). Climate-driven changes in gyres, where seabirds feed, are regularly noted and affect seabird survival. Such effects may result in displacement or extinction of certain seabird colonies (eo77: McCoy). Coastal marine environments appear to be experiencing the same northward expansion observed for terrestrial tick species, facilitated by milder winters.

The influence of climate on mosquito-borne pathogens gives pointers for tick-borne pathogens. Environmental temperature is known to affect the ability of mosquitoes to transmit a variety of pathogens (knowledge that is generally lacking for tick-borne pathogens, especially tick-borne viruses) (eo46: Huang *et al.*). Various studies have shown an association between increased temperature and increased case numbers of disease caused by mosquito-borne pathogens (eo57: Turell). Warmer temperatures increase the severity and risk of introduction into new regions of zoonotic pathogens such as West Nile virus, Japanese encephalitis virus and Rift Valley fever virus, whereas anthroponotic pathogens such as malaria parasites and dengue viruses are constrained by socio-economic factors (e.g. air conditioning, household screening, piped water and television) (eo57: Turell). Rapid transportation of people and merchandise introduces novel pathogens into virgin territory. Resulting outbreaks of mosquito-borne infections (e.g. chikungunya and Zika) exploited the lack of protective immunity in susceptible populations rather than a warming climate (eo57: Turell). Fortunately, ticks cannot fly (unless as hitchhikers) and must moult between feeds, hence the rapid rate of pathogen transmission achieved by mosquito vectors is not such an issue for tick vectors.

Final Synopsis and Future Predictions

Pat Nuttall*
University of Oxford, Oxford, UK

In general, experts believe changes in climate must impact ticks and ergo tick-transmitted infections. Most ticks spend most of their life off a host, sheltering under leaf litter or stones, in cracks in rocks or walls of buildings, in nests or burrows of their hosts – anywhere in their ecological niche where they can digest a blood meal, moult to their next developmental stage, lay eggs, avoid predators and parasites, and conserve water and energy. The off-host phase of their existence is exposed to environmental conditions (to a greater or lesser degree, depending on the nature of their shelter), particularly during host-seeking (questing). Even one-host ticks, which spend prolonged periods on their host, can experience challenging heat stress conditions (eo29: Santos *et al.*). Despite extensive exposure to the elements, clear-cut evidence ticks and tick-borne pathogens are being directly affected by a warming climate is scarce. This final synopsis attempts to summarize evidence provided by expert opinions of tangible impacts and provide a glimpse of the experts' thoughts on the future.

Some of the most compelling evidence of climate affecting ticks and tick-borne pathogens is from Canada where climate warming is occurring at over twice the rate experienced in most other parts of the world. Thirty years of sustained tick surveillance has tracked the encroachment of *Ixodes scapularis* from the USA and its expansion across southern parts of eastern and central Canada (eo73: Leighton *et al.*). Sylvatic cycles of *Borrelia burgdorferi* appeared some 5 years after the tick vector established, with Lyme borreliosis now endemic in most Canadian provinces. Native tick species have also expanded north in Canada and cases of anaplasmosis, babesiosis and Powassan virus encephalitis are more frequent than when surveillance first began (eo73: Leighton *et al.*). Similar evidence of a northward expansion of *Ixodes ricinus* has been recorded in Scandinavia, together with an increase in tick abundance in already endemic areas and rises in incidence of Lyme borreliosis and tick-borne encephalitis (TBE) in humans, as well as babesiosis and anaplasmosis in livestock (eo14: Kjær and Bødker; eo35: Mysterud). Other examples of range expansion include *Amblyomma americanum* and *Dermacentor variabilis* in North America (eo21: Lynn *et al.*; eo28: Eisen and Eisen; eo69: Hart and Thangamani; eo73; Leighton *et al.*) and *Dermacentor reticulatus* in Northern Europe (eo14: Kjær and Bødker; eo24: Zając). Just as some tick species have expanded their latitudinal range, so they have climbed to higher altitudes. For example, the progression of *I. ricinus* (and TBE virus) to increasing heights has been monitored in the Czech mountain regions where mean annual air temperatures have

*Email: pat.nuttall@zoo.ox.ac.uk

 Climate, Ticks and Disease (ed. P. Nuttall)
DOI: 10.1079/9781789249637.syn4

risen (eo47: Danielová and Daniel). In addition, there are increasing examples of exotic tick species establishing in new locations, seemingly facilitated by a warming climate, e.g. *Hyalomma marginatum* in France (eo18: Stachurski *et al.*).

Although there are compelling examples of climate change creating new locations in which ticks and tick-borne pathogens are flourishing, most other regions/countries have not recorded any influence of a warming climate on ticks and/or associated diseases. In many cases (e.g. Africa, the Arctic, Australia, China, India, Japan, the Middle East, Pakistan, Taiwan) long-term replicated data on tick distribution, prevalence and abundance are not available (eo37: Foley), diseases are often not notifiable and fundamental data on tick–host–pathogen combinations are lacking (eo48: Bente; eo52: Pandey and Singh; eo55: Liu *et al.*; eo60: Ghafar *et al.*; eo61: Oskam *et al.*; eo64: Kuo; eo67: Černý *et al.*; eo68: Chavshin and Seyyed-Zadeh; eo71: Li *et al.*; eo72: Kolo). Arguably, there have not yet been enough years of rapid global climate change to detect an impact (eo37: Foley), particularly when socio-economic changes have overwhelmed any possible climate signal. For example, Russia has a long and distinguished record in studies of ticks and tick-borne infections. Records show the greatest influence on the risk of contracting TBE in Russia has been socio-economic (eo63: Korenberg). A similar influence appears to be the root cause of increased cases of Kyasanur forest disease in India (eo52: Pandey and Singh). In the USA, the rising numbers of Lyme borreliosis and other infections transmitted by *I. scapularis* can be explained by this vector reclaiming territory it once inhabited before settlers destroyed forests and decimated white-tailed deer populations (eo26: Fish).

What then of the future? In constructing their expert opinions, contributors were asked to give their views on what the future might hold. Here is a snapshot of their expert thoughts.

At the physiological level, the tick microbiome will adapt faster to environmental change than the rest of the tick, causing an imbalance with tick-borne pathogens and potential knock-on effects for disease ecology (eo7: Cabezas-Cruz). However, rapid and radical changes in tick endosymbiont populations will likely enhance tick survival through selection of supportive symbiont genes, including those acquired by horizontal transfer of symbionts from other ticks (and even other arthropods) encountered in new ecological niches (eo10: Gottlieb and Duron). Tick-borne pathogen prevalence and transmission dynamics may be impacted by such changes and/or by increased temperatures (eo21: Lynn *et al.*). Behavioural plasticity will enable ticks to adjust their questing behaviour to new environmental conditions, facilitated by the extreme ranges of tolerance and adaptability of their nervous system (eo8: Šimo). If temporal and spatial occurrence of questing changes, tick–host interactions will change with further knock-on effects on tick-borne infections (eo9: Ronai). For example, game birds released for shooting are likely to feed more ticks (especially larvae) in future owing to greater seasonal synchrony in autumn, affecting local circulation of *B. burgdorferi* genospecies (eo19: Hoodless and Sage). Increased drought will increase the redistribution of dry- and heat-adapted tick species (e.g. *Amblyomma maculatum*) and increase tick abundances near rivers and lakes (eo11: Benoit and Oyen). Where humidity is not physiologically limiting, as in Scandinavia, further changes in tick distribution and abundance are inevitable (eo14: Kjær and Bødker; eo76: Pettersson). Expansion into sparsely populated regions will have minimal impact; however, establishment of new species, e.g. *Ixodes persulcatus* in Sweden, creates new threats through the introduction of new pathogens, e.g. Siberian and Far Eastern subtypes of TBE virus. The seeding of new regions with exotic tick species will change as patterns of bird migration respond to climate change (eo15: Kelly *et al.*). Permanent establishment of *Hyalomma rufipes* in Europe and Turkey, facilitated by long- and short-distance dispersal by birds, will be a major change to the resident ixodid tick fauna. Likewise, *H. marginatum* will become permanently established in southern France and other Mediterranean regions of Europe (eo18: Stachurski *et al.*). However, *H. marginatum* will probably lose out in Turkey as livestock management practices adapt to climate change and other pressures (eo17: Kar and Gargili Keles). In the Caribbean region, *Rhipicephalus microplus* could either expand its known range and/or speed up its developmental cycle (eo65: Obregón Alvarez *et al.*). Increasing temperatures are predicted to increase the chances for re-introducing *R. microplus* in major cattle-producing areas of the USA (eo29: Santos *et al.*). In East and Central Africa, shifts from cattle to more climate-tolerant livestock species, together with growing wildlife losses, will change tick communities and pathogen prevalence (eo20: Titcomb). Intermingling of endemic with exotic species may lead to unpredictable co-infection combinations in already established tick species populations (especially if endemic species increase in density), creating new challenges for clinicians and veterinarians alike (eo38: Pfeffer *et al.*; eo69: Hart and Thangamani). Warming climates may increase the fitness of pathogens transmitted transovarially in ticks or transplacentally in vertebrate hosts (eo41: Diuk-Wasser). There is some evidence a warming climate will increase tick feeding synchrony, thus favouring pathogens with short infectious periods and/or those relying on co-feeding for transmission (eo41: Diuk-Wasser).

Worldwide, the future is one of tick population expansion rather than contraction. Predicted changes in climate are expected to further affect *I. ricinus*' spatial distribution and its seasonal dynamics in the next decades (eo25: Rizzoli). Changes in seasonality could extend the Lyme borreliosis risk period by up to 2 months in some parts of Central and Eastern Europe (eo54: Trájer), and similarly for louping ill in Scotland (eo59: Johnson and Phipps). Northward expansion of *I. persulcatus* will broaden the zone of sympatry between this species and *I. ricinus* (eo63: Korenberg), with implications for the guild of pathogens they transmit. Continuing northward expansion of *D. reticulatus* is likely and the so-called Eastern and Western European populations of this species may merge while local abundance increases, enhancing infection risk (eo24: Zając). The occurrence of *D. reticulatus* and *Haemaphysalis punctata* in more European locations will further spread *Babesia* infections (eo32: Gray). In the USA, rapid range expansion of the northern form of *I. scapularis* (and its associated pathogens) is likely to continue until it has regained its precolonial distribution; the influence of climate change on further expansion can then be more readily determined (eo26: Fish). As temperatures increase and conditions further support tick development and questing, other species (e.g. *A. americanum*, *D. variabilis*) in temperate North America will continue to expand northwards and to higher altitudes (eo28: Eisen and Eisen; eo42: Sonenshine); tick-borne pathogens will generally follow (eo41: Diuk-Wasser; eo45: Munderloh and Kurtti; eo69: Hart and Thangamani). In Brazil, *Rhipicephalus sanguineus* of the tropical lineage may expand to higher latitudes, occupying areas previously inhabited only by its temperate lineage (eo30: Bechara and Sanches). In Japan and neighbouring countries, there is concern about possible range expansion of *Haemaphysalis longicornis* and recognition of insufficient knowledge of the effects of climate variability and global warming in the region (eo27: Umemiya-Shirafuji). Similar concern surrounds the continued establishment and spread of *H. longicornis* throughout the USA and its potential to transmit endemic and exotic pathogens (eo69: Hart and Thangamani).

While tick populations expand northwards, it is less clear what the future holds at the southern limits of their range (eo28: Eisen and Eisen). Models restricted to climatically analogous areas estimate a reduction in suitable areas for *Amblyomma cajennense* sensu stricto and *Amblyomma sculptum* in Brazil (eo23: Oliveira). In the southern part of *I. ricinus*' range, where lack of humidity rather than low temperature is the limiting condition, ecosystem niche models have a high level of uncertainty (eo32: Gray). Nevertheless, spatial distribution models of *I. ricinus*, and a climate envelope model for Lyme borreliosis at 2°C of global warming, show a projected retreat of the vector and a reduced incidence of the disease in Southern and Central Europe (eo3: Cox *et al.*; eo5: Wint and Alexander). Indeed, there is a notable gap in understanding species that may be limited by drought rather than temperature, such as species distributed across humid-to-arid gradients in tropical regions (eo44: Samy *et al.*). If southern parts of Europe are hotter and drier, the distribution of species dependent on high humidity and soil moisture (e.g. *I. ricinus*, *Haemaphysalis concinna*) may diminish, while *R. sanguineus* and the pathogens it transmits may flourish (eo38: Pfeffer *et al.*). In southern and lowland areas of *I. ricinus*' distribution, TBE incidence will most likely reduce (eo47: Danielová and Daniel). Climate impacts on marine ecosystems will follow seabird community dynamics, patterns of colony use and bird dispersal which, in turn, will feed back on the ticks and infections these ticks and their avian hosts carry (eo77: McCoy).

Speculation on future effects of climate change on argasid tick distributions and vectorial capacities is difficult as relatively few species have been studied. Some argasid ticks may benefit from their ability to adapt to human-made habitats while others may become extinct in parts of their range owing to anthropogenic pressures (eo31: Filatov and Rego). Although regions considered endemic for tick-borne relapsing fevers associated with argasid *Ornithodoros* spp. will change under global warming, it is highly likely tick-borne relapsing fever will remain a global public health concern in the next century (eo53: Teglas).

A common framework for predictive projections has not been established. Ideally, the framework should combine every single partner involved in tick dynamics (climate, land use, vegetation, communities of vertebrates, tick physiology, stress, mortality, fecundity and activity) (eo34: Estrada-Peña *et al.*). However, there are still major gaps in basic understanding of what specifically limits distribution and activity of many important tick species (eo37: Foley). This also applies to tick-borne pathogens; for example, future predictions of climate change impacts on co-feeding transmission dynamics require further refinement of model formulations to account for diapause (eo39: Wu and Zhang). Notwithstanding, Telford (eo40) believes the major flaw in speculation on the effect of global warming on human vector-borne disease risk is a failure to recognize the compensatory behaviours people adopt when temperatures rise. This extends to roles played by socioeconomic factors and rewilding initiatives that modify habitats through changes in land use, and livestock and wildlife management, increasing the frequency of human contact with natural foci of infection (eo49: Rudenko *et al.*; eo58: Bastos and Boinas; eo63: Korenberg; eo74: Gilbert; eo75: Medlock and Hansford). In much of Africa, future climatic variability may amplify anthropogenic factors, such as transboundary cattle movement for trade or to find grazing (eo70: Githaka *et al.*). The challenge in predicting the future is illustrated by the prediction (made in 2000) that in the 2020s, France, Switzerland, Croatia and much of Austria would be cleared of TBE virus, while in reality TBE cases in these countries have risen sharply (eo50: Rubel). For the short-term future, using the previous decade to predict the next decade will give a fair approximation (eo35: Mysterud). However, as the pace of change accelerates, data we have now become less directly applicable to the future (eo37: Foley).

What about solutions? Collaborative, predictive and synthetic data sharing and analysis, with adaptive management of risk, will best position public health authorities to respond when ticks pose a threat (eo37: Foley; eo51: Sato *et al.*; eo71: Li *et al.*). Action must be local and bespoke (eo1: Cornforth and Plumpton; eo75: Medlock and Hansford). More effort should be made in studying humans and their interactions with the environment ticks inhabit (eo40: Telford; eo45: Munderloh and Kurtti). The Czech Republic already has a publicized weather-based forecasts of human risk of tick bite (TICKPRO) with proposals for upgrading to give risk warnings of TBE virus infection (eo47: Danielová and Daniel). In countries such as Pakistan and Turkey, where tick-borne infections have a high socioeconomic impact, a One Health approach is advocated (eo60: Ghafar *et al.*; eo66: Kar and Ergünay). Australia faces challenges from exotic (introduced) species (*R. sanguineus* and *Haemaphysalis* spp.) that will likely expand their geographical ranges as climate change creates more favourable environmental conditions for these species; a system of real-time updates on the presence and geographical distribution of pathogens and associated vectors is called for (eo56: Dehhaghi *et al.*; eo61: Oskam *et al.*). In Brazil, options for pasture grasses must now be considered in the context of climate change as some grasses repel ticks but are not compatible with drought and fire resistance (eo29: Santos *et al.*). Genomic selection for heat tolerance in cattle needs to consider tick resistance, in addition to production and product quality traits, to provide reliable solutions to challenges brought on by climate change (eo29: Santos *et al.*). Integrated tick-control management, linked with livestock movement control and efficient monitoring of tick spread, is recommended to slow or prevent the spread of ticks into new regions (eo16: Wilke *et al.*; eo22: Mans). Developing effective control strategies will benefit from multifactorial approaches, taking account of tick and pathogen genomic resources, tick midgut microbiome, glycoproteins, exoskeleton chemical composition, paratransgenesis, quantum vaccinomics and modelling tick vaccines (eo33: de la Fuente and Villar). Other exciting prospects are diagnostic techniques allowing holistic tick analyses (i.e. detection of all putatively pathogenic organisms), by anyone in almost no time, providing public health alerts and citizen science initiatives (eo38: Pfeffer *et al.*). Improved diagnostics and more active surveillance may detect an increasing number of *Ehrlichia* and *Anaplasma* infections in people with non-specific flu-like symptoms in areas where ticks occur (eo43: Stuen). However, it must be remembered that finding a new tick-borne pathogen is not necessarily the consequence of climate change (or globalization) but may be the result of improved diagnostics and more intensive research (eo62: Kazimirova). The best solution of all to prevent a future of more ticks and more accompanying disease is to stop the anthropogenic emissions of greenhouse gases that drive climate change and to focus on cost-effective surveillance and control measures.

Index

Note: Page numbers in bold and italics represent tables and figures respectively.

acaricides 142
active/passive surveillance programmes 241
acute respiratory distress symptoms 301
acute-phase proteins 201
Acyrthosiphon pisum 62
adaptation planning 5, 139, 245
Aedes aegypti 402, 447
Aedes albopictus 266
African climate
  agricultural decision making 4
  agriculture and people's livelihoods 3–4
  collecting water from community borehole 4
  convection-permitting model (CPM) 3
  flood-inducing events 3
  flooding and waterlogging 5
  harvesting sesame seed 2
  heavy rainfall events 2
  introduction 1
  livestock comfort temperature 2
  livestock fodder 5
  precipitation regime 2, 3
  rainfall intensity 3
  timing, rainy seasons 2
  uncertainty 1, 4, 5
  water resources 5
  wet-season rainfall 3
African swine fever (ASF) 78–79, 497
  biological cycle, *O. erraticus* 408
  definition 405
  infected pork 405
  infected tick population 405
  Ornithodoros erraticus 406, 408
  Ornithodoros moubata 406, 410, 411
  Ornithodoros tick populations, Africa **407**
  Ornithodoros tick species 406, **409**
  suid-associated tick species 408
  sylvatic cycle 405, *406*
African tick-bite fever (ATBF) 450, 500
agent-based modelling 350
agricultural decision making 4
agricultural development programmes 1
agricultural intensification 140
agroforestry systems 197
Akaike Information Criterion (AIC) 41
Alkhumra haemorrhagic fever (AHF) 341
Alkhumra haemorrhagic fever virus (AHFV) 342
Alongshan virus (ALSV) 493
alpha-gal syndrome (AGS) 233
ambient temperature 57, 168, 195, 200, 211, 212
Amblyomma americanum 147, 159, 189–190, 241, 302, 309, 318, 370
Amblyomma cajennense 162
Amblyomma hebraeum 111, 154, 302
Amblyomma loculosum 532
Amblyomma maculatum 70, 147–148, 159, 241, 261, 310–311
Amblyomma mericanum 146
Amblyomma mixtum 450
Amblyomma sculptum 163
Amblyomma ticks
  climate change and tick-borne diseases 160
  climate suitability 162
  distribution 160–162
  ecological niche models 163–164
  hydroclimate projections 163
  introduction 159–160
  public health information campaigns 161
  uncertainty 163

Amblyomma tigrinum 261
Amblyomma triguttatum 159
Amblyomma triste 261
Amblyomma variegatum 125, 154, 302, 449
ambush strategy 56
American dog tick 190
Anaplasma capra 300
Anaplasma marginale 166
Anaplasma ovis 300
Anaplasma phagocytophilum 46, 146, 149, 179, 232, 233, 265, 279, 294, 300
Anaplasma platys 261, 301
Anaplasma species
  evidence against climate change impacts 296–297
  evidence for climate change impacts 294–296
  introduction 293–294
  tick densities 297
anaplasmosis 132, 159, 293
Animal and Plant Health Agency (APHA) 415
anthropocene 154
anthropogenic activities 54
anthroponotic pathogens 402
anti-tick grasses 197
anti-tick immunity 47
antifreeze glycoproteins 233
antifreeze protein 295
Aptenodytes patagonicus 532
arboviruses 26, 326, 327, 328, 329
Arctic amplification 21
arctic ecosystem
  arthropods 461, 465
  climate change 462
  climatological models 464
  colonization 463
  distribution of species *465*
  dominant tick species 462
  ecological conditions 461
  expert opinion 466
  geographic ranges *461*
  Ixodes persulcatus 463, 464
  Ixodes uriae 462, 463
  low diversity 460
  tick species 460, **463**
  warming 460
Argas persicus 219
Argas polonicus 101
Argas vulgaris 101
Argasidae
  anthropogenic pressures 220
  ecological niches 218–219
  global environmental change and anthropogenic influence 219
  introduction 216–217
  physiological limits 217–218
  soft tick ecology and distributions 217
arid environment ticks 118
arthropod vectors 110
arthropods 38, 61–63, 148, 261
ASF virus (ASFV) 497
atmospheric/planetary boundary layer (ABL) 11
autumn-recruitment model 133

Babesia bigemina 224
Babesia bovis 224
Babesia canis 91
Babesia divergens 223
Babesia microti 223, 286
Babesia ovata 185
Babesia vogeli 225
babesiosis 132, 159, 278
babesiosis vectors
  blood transfusions 226
  Dermacentor reticulatus 224
  ecological niche models 225
  European distribution 225
  Hyalomma punctata 224
  Hyalomma spp. 224
  Ixodes ricinus 223–224
  Ixodes scapularis 225
  introduction 223
  Rhipicephalus spp. 224–225
  small mammalian reservoir hosts 225
  temperature-driven 226
  uncertainty 225
  vegetation growth season (VGS) 226
Bambi effect 102
barn swallow *(Hirundo rustica)* 99
barns 120
Bartonella henselae 266
Bartonella spp. 166
beech climate (Cfb) 358
behavioural plasticity 58
Bemisia tabaci 64
Bicyclus anynana 196
BIOCLIM 40
biodiversity 85, 216
biogeochemical cycles 8
biomass 97
biomes 11
biomineralization 233
biosafety level 4 (BSL4) laboratories 365
biotelemetry 99
birds
  climate change impacts 97–99
  distributional ranges 103–104
  Ixodes ricinus 102
  introduction 96
  migration 97–103
  populations 103
  vectors 96–97
black-legged tick 188–189
black-throated thrush (*Turdus astrogularis*) 104

blackbird (*Turdus merula*) 98
blood feeding 76
blood-feeding process 254
Boophilus (Rhipicephalus) annulatus 121
Boophilus decoloratus 111
boosted regression trees (BRT) 32, 34
Borrelia anserina 393
Borrelia burgdorferi 18, 46, 80, 84, 97, 110, 134, 146, 166, 178, 264, 277, 284, 319, 393, 444, 509, 513, 542
Borrelia duttonii 78
Borrelia garinii 134
Borrelia hermsii 372
Borrelia miyamotoi 188–189
Borrelia valaisiana 134
Bourbon virus 159
Brachiaria brizantha 197
Brazilian Legal Amazon 194
brown dog tick 265
Buchnera aphidicola 62
bunyaviruses 326

Canada 146–148
  climate 508–509, 510
  climate warming 507, 509
  geographic range limits 509
  invasion speeds 509
  Lyme disease 507, 508
  rocky mountain spotted fever (RMSF) 507
  temperature/tick population growth 508, 509
  tick physiology/life cycle 508
  tick-friendly climate 508
canine babesiosis 166
canine monocytic ehrlichiosis (CME) 211, 212, 424
Canis aureus 91
Capreolus capreolus 85
carbonic anhydrase 10
Caribbean animal health network (CaribVET) 449
Catharus minima 99
cattle tick fever complex (CTF) 450
Centers for Disease Control and Prevention (CDC) 307, 349
Central Europe
  biostatistics 338
  climate warming 337
  extraordinary meteorological situations 337
  Ixodes ricinus 332–333, 336, 337
  introduction 331–332
  tick-borne encephalitis (TBE) 332, 333–335, 336
  weather-based forecasts 336
Central European encephalitis 332
central nervous system (CNS) 331
chemosensilla 50
chikungunya virus (CHIKV) 400
China
  Anaplasma 494
  Babesia 494, 495
  biotopes 492
  Borrelia 496
  infectious diseases 492, 493, 494
  microbiome 497
  natural habitat types 492
  spotted fever group rickettsia (SFGR) 495, 496
  tick distributions 492, 493
  viruses 496, 497
chlorofluorocarbons 253
Chromolaena odorata 368
climate change
  babesiosis vectors 223–226
  birds 96–104, 97–99
  drought and tick dynamics 67–71
  evolutionary periods 153–154
  food production challenges 1
  influences 3, 78
  Lyme disease 18–24
  microbial communities 48
  R. microplus 195–197, 198–199
  regional and national scales 4
  species extinction 9
  temperatures 1
  tick life history 38
  tick-borne diseases 4
  tick–endosymbiont interactions 63
  vector-borne diseases (VBDs) 26–29
  water resources 5
climate envelope models (CEMs) 11
climate instability/emergence event
  active and passive surveillance 262
  collaborative, predictive and synthetic data sharing 262
  coping 260
  data-driven 260
  ecological and physiological characteristics 260
  ecological change 260
  ecological synthesis 261
  exotic ticks 259
  freeze–thaw cycles 260
  health management 261–262
  lineages 261
  mean annual temperatures 260
  public health 261
  societal disruption 259
  tipping points 260
  unstable weather 260
  virulent strains 259
climate-driven livestock management shifts
  climate change affects 139–140
  climate futures and near-term projections 143
  humans affect ticks and pathogens 140–142
  intertwining effects, hosts and ticks 140
  introduction 139
  questions and considerations 142–143

clock ticking
beyond 50 years 256–257
global catastrophes 257
global warming 257
introduction 253
next 50 years (2021–2071) 255
survival strategy 254–255
tick-borne encephalitis virus (TBEV) dispersal 255–256
virus evolution, dispersal and transmission characteristics 255
CNS *see* central nervous system (CNS)
co-feeding transmission 284–286
climate change impact 271–272
diapause 273
infestation dynamics and Allee effects 272–273
introduction 270–271
larval tick 272
co-infections
climate change impacts 266
introduction 264–266
next 10 years 266–267
next 20 years 267
next 50 years 267
cold hardiness phenomenon 295
Community Climate System Model 4 (CCSM4) 376, *380*
convection-permitting models (CPMs) 3
Coupled Model Inter-Comparison Project (CMIP) 29
Coxiella burnetii 166
Crimean-Congo haemorrhagic fever (CCHF) 26–27, 265, 308, 310, 519, 520, 543, 545
Crimean–Congo haemorrhagic fever virus (CCHFV) 326, 341, 455, 493, 519, 520
critical equilibrium humidity (CEH) 68, 217
critical transition temperature (CTT) 218
cuckoo *(Cuculus canorus)* 99
cytokines 200
Czech Hydrometeorological Institute (CHMI) 334–335
Czechoslovak encephalitis 332

Dabie bandavirus 342
data reduction techniques 34
dead clade walking 153
Debilitating Symptom Complexes Attributed to Ticks (DSCATT) 393
deforestation 177
dehydration 67–70
dengue virus (DENV) 401
density of infected nymphs (DIN) 172, 276
Dermacentor andersoni 218
Dermacentor marginatus 122
Dermacentor nitens 195
Dermacentor reticulatum 33
Dermacentor reticulatus 90, 224, 264, 309, 310, 354, 430
adult 167
Europe *vs.* progressive climate change 166–168
introduction 166
Northward expansion 169
populations 168, 169
Dermacentor spp. 122–123
Dermacentor variabilis 39, 41, 69, 147, 190, 241, 294, 302
Dermacentor-borne necrosis erythema and lymphadenopathy (DEBONEL) 166
Dermatophilum congolensis 450
desiccation-tolerant species 85
distribution models 33
domestic animals, Australia
Bovine Anaemia due to *Theileria orientalis* Group (BATOG) 426
climate 424
climate change 427
canine monocytic ehrlichiosis (CME) 425
Ehrlichia canis 425
Haemaphysalis longicornis 426, 427
Rhipicephalus sanguineus 425, 427
Theileria orientalis 426
tick species 424
Drosophila melanogaster 202
drought 69–70
Dugbe virus 342
dunnock (*Prunella modularis*) 102

earth system models (ESMs) 12
ecdysteroids 76
ecological change 11–13
ecological informatics 13
ecological niche models 163–164, 171, 218–219, 225, 226, 308
ecological processes 9
ecosystems 11
ectothermic organisms 145, 240, 247
egg protein 75
Ehrlichia canis 261, 301
Ehrlichia chaffeensis 159, 302
Ehrlichia ewingii 159, 302
Ehrlichia muris 302
Ehrlichia ruminantium 302
ehrlichiosis 132
embryonic development 167
endemic stability 344–345
endosymbionts
acquisition 63
arthropods 61
evolutionary innovations 61
facultative 61, 62, 64
genera 61
heritable 62–64

high metabolic demand 62
insect symbiosis systems 63
mechanisms 63
pea aphid 64
selection pressures 62
England
changing temperature and disease 520
human exposure 522, 523
increased temperatures/heatwaves/mild winters 518
indirect effects, climate change 520, 521
landscape management 521, 522
microclimate/hosts 523
native ticks, temperature 519
non-native ticks/climatic change 519
survival/density/distribution 518
tick hosts 522
tick species 518
entomological risk index (ERI) 276
entomopathogenic fungi 260
entomopathogens 197
environmental variables 11
enzootic cycles 271, 276, 280
enzootic pathogen transmission 188, 189
enzootic spirochaete transmission 351
enzootic transmission cycles 190, 283
epigenetic regulatory mechanisms 541
erythrocytes 223
Eurasian reed warbler (*Acrocephalus scirpaceus*) 99
Europe
endemic populations 126–127
engorged nymphs 129
introduction 125–126
invasive populations 127–128
tick survival and activity thresholds 129
vector systems 129
evaporation 10
exophilic *vs.* endophilic ticks 56, 97, 240
extreme meteorological events 117
eyebrowed thrush *(Turdus obscurus)* 104

fallow deer (*Dama dama*) 102
fat body-stimulating factor (FSF) 75
FCFA *see* Future Climate for Africa (FCFA)
feedback process 257
filoviruses 341
flaviviruses 326
flood-inducing events 3
FLUXNET 11
fossil fuel resources 322
fragmentation 140
Francisella tularensis 159, 166
Free-Air CO_2 Enrichment (FACE) 12
freeze–thaw cycles 260
Future Climate for Africa (FCFA) 1–2
fuzzy logic 262

Gadgets Gully virus (GGYV) 255
game management 135, 278
gaps/caveats
climate effects 241–242
confounding effects 242
exophilic and endophilic ticks 240
Ixodes ricinus 242–244
multivariate statistics/interpolated climatology 245
predictive projections 244
temperature, Europe 243
tick-borne pathogens spreading 240–241
garden warbler (*Sylvia borin*) 99
general circulation models (GCM) 41, 173, 312
generalized linear models (GLMs) 357
genomic selection technologies 202
Global Land–Atmosphere Climate Experiment (GLACE) 12
global warming 18, 51, 97, 99, 245, 253, 257, 280, 309, 352
challenges 111–112
effects 111
introduction 110
range of tick species 111
tick-borne diseases (TBDs) 110–111
ticks 110
uncertainties 112
globalization 303
glycoproteins 235
great tit (*Parus major*) 102
Greater Alpine Region (GAR)
evidence for climate change impacts 356–357
introduction 354
long-range TBE forecast (next 80 years) 357–358
medium-range TBE forecast (next 10 years) 357
short-range TBE forecast (next 2 years) 357
tick-borne encephalitis (TBE) 354–356
greenhouse gas emissions 54, 136
grey squirrels (*Sciurus carolinensis*) 134
grey-cheeked thrush (*Hylocichla minima*) 99

habitat 296
Haemaphysalis bremneri 426
Haemaphysalis concinna 264, 430
Haemaphysalis cornigera 361
Haemaphysalis inermis 122
Haemaphysalis japonica 184
Haemaphysalis leporispalustris 99
Haemaphysalis longicornis 146, 148, 154, 311, 328, 361, 385, 543
geographical distribution 184
introduction 183
Japan
distribution 183–184
Protozoan parasites 185
seasonal occurrence 184–185

Haemaphysalis megaspinosa 184
Haemaphysalis parva 122
Haemaphysalis punctata 117, 122, 154, 224, 413
Haemaphysalis spinigera 345, 543
Haemaphysalis spp. 122–123
Haemaphysalis sulcata 122
haematophagous arthropods 200
haematophagous ectoparasites 196
haemoglobin–haptoglobin (Hb–Hp) complex 201
haemolytic anaemia 300
haemorrhagic fever infections
  host 344–345
  host abundance 345
  host diversity 346
  introduction 341–342
  research efforts 347
  role of host 345
  tick vector 344
  virus 346
  virus dynamics 342–344
haemorrhagic syndrome 341
hair-shedding ability 202
Hamiltonella defensa 63, 64
harmonic regression models 357
harmonized collection procedures 245
Heartland virus (HRTV) 159, 190, 328
heat-shock proteins 196, 233, 295
Hepatozoon canis 166
HGA *see* human granulocytic anaplasmosis (HGA)
high-volume data transfer streams 13
Holdridge life zone classification 11
holobiont 46, 47
host-seeking behaviour
  abiotic environmental stimuli 57–58
  behavioural plasticity 58
  desiccation stress 58
  introduction 56
  pathogen transmission dynamics 58
human behaviour trumps entomological risk
  acute Lyme disease seasonality 277
  atmospheric moisture 279
  enzootic parameters 280
  entomological risk index (ERI)-modulating factors 280
  game management, suburbanization, lack of natural predators 278
  hosts and microbial assemblages 278
  human demography 277
  longitudinal measurements 279
  'mad dogs and Englishmen' hypothesis 280
  microbial guild 276
  Nantucket Island (2000–2015) *279*
  nymphal anthropophily 280
  suburbanization and increased recreation 278
  transmission cycles 276
  vector-borne agents 280
  zoonotic transmission 276
human granulocytic anaplasmosis (HGA) 278, 284, 293
human-assisted movement 322
human-infectious agents 321
human-mediated impacts 117
human-triggered land changes 352
hunting strategy 56
Hyalomma aegyptium 117, 119
Hyalomma anatolicum 120, 154
Hyalomma dromedarii 309
Hyalomma excavatum 119
Hyalomma hussaini 418
Hyalomma lusitanicum 242, 520
Hyalomma marginatum 33, 91, 96, 100, 101, 103, 117–119, 126–128, 225, 250, 265, 310, 327, 344, 455, 519
Hyalomma rufipes 91, 96, 100, 101, 104, 117, 119, 154, 265
Hyalomma scupense 118, 120–121, 128
Hyalomma spp. 224
Hyalomma truncatum 111

Ixodes ricinus serine protease inhibitor (IrSPI) 266
immune responses
  antimicrobial peptides 79
  African swine fever (ASF) 78–79
  defensin gene 79
  gonotrophic cycles 80
  influences 79–80
  innate and adaptive 97
  tick-borne relapsing fever 78
  transovarial pathogen transmission 79
indigenous Amerindian population 177
indirect immunofluorescence/indirect immunoperoxidase assay 360
inferential approaches 217
innate lymphoid cells 200
integrated vector management 112
Inter-Sectoral Impact Model Intercomparison Project 29
Intergovernmental Panel on Climate Change (IPCC) 2, 20, 125, 160, 312, 413
International Long-Term Ecological Research Network (ILTER/LTER) 11
interstadial development 286
intertwined pathogen–vector cycles 321
invertebrate host 346
Israeli spotted fever (ISF) 502
Ixodes arboricola 97
Ixodes frontalis 101
Ixodes inopinatus 265, 431
Ixodes kerguelenensis 538
Ixodes lividus 97
Ixodes ovatus 184
Ixodes pacificus 18, 27, 261, 508
Ixodes persulcatus 18, 33, 36, 91, 92, 103–104, 111, 184, 253, 270, 308

Ixodes ricinus 18, 27, 33, 35, 36, 52, 69, 79–80, 84, 90, 93, 96, 102, 103, 117, 122, 126–127, 210, 223–224, 226, 241–244, 247, 253, 254, 264, 270, 309–310, 332–333, 413
- climate change 514
- climate change, resilience 516
- climate warming, Scotland *515*, 516
- contact disruption 174
- green recovery impacts 516
- indirect climate change impacts 514
- introduction 171–172
- land vertebrate 513
- macroclimatic conditions 173
- non-systemic transmission 174
- Scotland's changing climate 513
- spatial and temporal distribution 172–173
- seasonality 515
- temperature 516

Ixodes scapularis 18, 27, 52, 69, 96, 111, 188–189, 210, 225, 234, 241, 247, 259, *285*, 311, 318, 516, 548
- description 176
- distribution 176
- establishment 178–179
- expansion 180
- forest history, USA 177
- history 178
- mode of range expansion 178
- nymphal-stage ticks 176
- pathways 179
- population density 180
- relict populations 177
- theories 179
- white-tailed deer 177–178

Ixodes uriae 69
Ixodidae 216

Japanese encephalitis virus (JEV) 402
Japanese spotted fever (JSF) 360, 543
Jingmen tick virus (JMTV) 455, 493
juvenile hormone (JH) 76

K–T extinction event 153–154
Kadam virus (KADV) 255
Köppen–Geiger climate classification 11, 19, 115, 189, 190
Kyasanur forest disease (KFD) 341, 543
Kyasanur forest disease virus (KFDV) 342
- affected areas *367*
- climate change 366–368, *368*
- definition 365
- deforestation 368
- epidemiology and transmission 366
- histopathological findings 365
- parenchymal degeneration 365
- tick vector life cycle 368

Lantana camara 368
leaf water status 10
leishmaniasis 321
leukopaenia 302
light intensity 58
lipopolysaccharide 201
Lithuanian woodland 86
livestock 345
- anecdotal observations 416
- babesiosis 413
- climate change 413
- cross-sector coordinated approach 416
- endemic tick-borne diseases **414**
- exotic vectors/infectious hazards 415
- *Hyalomma punctata* 415
- human-animal interface 415
- *Ixodes ricinus* 414, 415
- IPCC 413
- Pakistan
  - annual temperature *420*
  - climate 418–419
  - *Hyalomma* ticks 420
  - longitudinal data sets 419
  - ruminants 419, *421*
  - *in silico* models 421
  - temperature and precipitation conditions 418
  - ticks and tick-borne diseases (TTBD) 421
- transmission 413, 415

livestock diseases 544
livestock management 139
- altered host composition 140–141
- altered tick mortality 142
- habitat modification 142
- non-metric multidimensional scaling 141

livestock production 159
local host demography 117
lone star tick 189–190
louping ill virus (LIV) 414
Lyme borreliosis (LB) 26, 85, 90, 102, 171, *249*, 334, 430, 542
- East-Central Europe
  - climate change 375, 376
  - climatic conditions 375
  - forecasting models 376
  - forest and forest–steppe biomes 376
  - geographical distribution 376
  - longest LB seasons 376
  - modelled relative incidence *380*
  - modelled relative incidence *379*
  - patterns 376
  - RCP-scenario-based CCSM4 climate models 379
  - seasonal duration *377*, *378*
  - temperature-based component 377
- evidence for climate change impacts 350–352
- introduction 349–350

Lyme borreliosis (LB) (*Continued*)
One Health concept 352
prevalence, overview 544–546
socio-economic factors 352
Lyme disease (LD) 26, 132, 176, 260, 264, 277, 280
climate change 20–21
climate variables 22, 23
climate zone 21
correspondence 19
geographical distribution 24
humidity deficit 23
incidence 23
introduction 18–19
polar amplification 22
prevalence 27
threshold model 22, 23
tick vector life cycle 19–20
Lyme neuroborreliosis 171
LYMERISK model 350, 353
Lymphangitis-associated rickettsiosis (LAR) 502
lymphocytopaenia 302
lymphoid tissue 200
lynx *(Lynx lynx)* 248

machine learning (ML) 13, 28, 38, 262
macroclimates 172
marine ecosystem
climate conditions 537
climate-driven changes 536
hyper-infestations 532
Ixodes uriae 538
infectious agents 536
ixodid ticks *536*
seabirds 537
terrestrial systems 532
tick species **533–535**, 538
vertebrates 532
Markov transition process 271
maternal antibodies 97
mating-dependent regulation 76, 77
Maxent method 28, 32
maximum entropy 28, 32, 39
measuring impacts
animal welfare 88
climate impact assessments 88
evidence and limitations 85–86
introduction 84–85
metabolic rate 86–87
mortality/morbidity 87
temperature- and moisture-dependent 88
mechanistic agent-based models 514
mechanosensilla 50
Mediterranean spotted fever (MSF) 501
meningoencephalitis 331, 332
metamorphological development 147
micro-scale monitoring 12
microbial metabolic pathways 47
microbiome 212
microbiome resistance 47
microclimates 8, 9, 38, 70, 101
microclimatic conditions/RNA viruses
arboviruses 326, 327, 328, 329
detection and autochthonous transmission 326
ecological and behavioural factors 327
farmlands 327
genomes 326
hard and soft ticks 328
human exposure 327
metamorphosis 328
meteorological factors 326
molecular virology 329
predictive models 327
severe fever with thrombocytopaenia syndrome virus (SFTSV) 326
transstadial transmission 327, 328
viral loads 329
Middle East
climatic diversity 473
influential factors 473
ixodidae 469
Köppen–Geiger climate classification *470*
medically important ticks *471*
migratory birds 473
ticks 469–473
vertebrate hosts 473
migration
behavioural and ecological events 98
definition 98
ecological consequences 97
economic resources 99
environment and climate change 101
journey 100–101
long-distance migrants 97–98
mass movements 98
passerine birds 98
pattern 102
processes 98, 103
radar discovery 98
seasonal environments 98
short-distance migrants 98
sub-Sahara 99–100
mitochondrial DNA (mtDNA) 62
mobility-oriented parity (MOP) 42
moisture-sensitive neurons 52
molecular clock analysis 152, 155
mosquito-borne arboviruses 211
mosquito-borne pathogens
environmental temperature 400
exotic virus 401, 402
global warming 400
malaria/yellow fever 400, 401
zoonotic *vs.* anthroponotic diseases 401
zoonotic viruses 403

multi-omics data integration 235
multifactorial approaches 551

National Health and Medical Research Council (NHMRC) 397
National Institute of Allergy and Infectious Diseases (NIAID) 365
natural ecosystems 219
Naumann's thrush *(Turdus naumanni)* 104
network analysis 235
neural plasticity 51
neural/sensory systems 50–52
neurobiology
  arthropods 54
  climate change-mediated desiccation stress 52–53
  climate influence 53
  directional thermal stress 51–52
  human-induced environmental factors 53–54
  introduction 50
  molecular determinants 54
  nervous and sensory systems 50–52
neuroborreliosis 250
neuroendocrine system 53
nidicolous ticks 97
non-systemic transmission 271, 346
non-viraemic transmission 255, 331, 346
North America
  American dog tick, *Dermacentor variabilis* 190
  black-legged tick, *Ixodes scapularis* 188–189
  hard tick species 146–148
  host immunity 149
  human-biting tick species 188
  introduction 145, 188
  land use and anthropogenic factors 148
  lone star tick, *Amblyomma americanum* 189–190
  molecular changes 148
  pathogen transmission 149
  temperatures 190
  tick biology 145–146
  tick species distributions 191
  trypanosome parasites 149
nutrient-dependent regulation 76, 77
Nuttalliella namaqua 154

off-host ecosystem 193
oligoxenous 97
omics technologies 232
OmRelish gene 79
Omsk haemorrhagic fever (OHF) 341
Omsk haemorrhagic fever virus (OHFV) 342
on-host ecosystem 193
'One Health' approach 422
oocytes 76
operational taxonomic units (OTUs) 46
ornate cow/meadow tick *see Dermacentor reticulatus*
Ornithodoros erraticus 77, 371
Ornithodoros hermsi 372
Ornithodoros lahorensis 75
Ornithodoros moubata 371, 405, 543
  immune responses 78–80
  influences, climate reproduction 77–78
  introduction 74
  life cycle 74–75
  reproduction 80
  reproduction regulation 75–76
  tick-control methods 80
Ornithodoros papillipes 219
Ornithodoros savignyi 371
Ornithodoros turicata 75
Ornithodoros verrucosus 219
Oryzaephilus surinamensis 62
osmoregulatory organs 53
overwintering 58

Palaearctic–African system 98
Panola Mountain *Ehrlichia* (PME) 302
paratransgenesis 235
pathogen transmission dynamics
  canonical seasonal feeding inversion 284
  climate and weather 286–287
  co-feeding transmission 284–286
  integration 287–288
  introduction 283
  non-horizontal transmission pathways 286
  synchrony, immature activity 284
  system resilience 288
  warming climates 288
pathogen–tick–host triad, Northern latitudes
  climate change impacts 247–248
  complexity 249
  denser populations 250
  disease statistics 250
  drivers of change 248
  host drivers 250
  introduction 247
  life cycle 248
  limitations 248, 250
  non-linearities 250
  qualitative prediction 250
  scenario 248
  tick species 250
  vertebrate hosts 249
pattern scaling 21
Pediculus humanus humanus 370
peripheral sensory system 50
Peromyscus leucopus 286
Phacochoerus africanus 405
phagolysosomes 294
pheasants (*Phasianus colchicus*) 134

phenology 11
photosensilla 50
photosynthetically active radiation (PAR) 9
piroplasmosis 183, 184
population vulnerability 1
Powassan encephalitis 278
Powassan virus (POWV) 179, 255, 284–286, 542
principal components analysis (PCA) 40
proteomics analyses 233
Pseudomonas syringae 64
Public Health England Tick Surveillance Scheme (PHE-TSS) 415

quantum vaccinomics 235–236
questing 56–58, 67, 69, 247

R_0 models 287–288
radar ornithology 98
random forest 32, 34
RCP8.5-based seasonality models 376
receiver operator characteristic (ROC) 41
red deer (*Cervus elaphus*) 102
red fox (*Vulpes vulpes*) 248
red grouse (*Lagopus lagopus scoticus*) 134
red-flanked bluetail (*Tarsiger cyanurus*) 104
relapsing fever
 argasid ticks 371
 definition 370
 distribution and habitat requirements 371, 373
 geographic regions 373
 infected human cases 372
 ixodid tick species 371
 Lyme disease 372
 microclimatic factors 371
 Ornithodoros moubata 373
 Ornithodoros turicata 373
 spirochaetaemia 370
 ticks 371
relative humidity (RH) 57, 68, 74, 196
relict tick 264
Representative Concentration Pathway (RCP) 29, 34, 41, 162, 173, 312, 376, 376, *379*
reverse migration 104
Rhipicephalus annulatus 154, 225
Rhipicephalus appendiculatus 111, 142, 154
Rhipicephalus bursa 122
Rhipicephalus decoloratus 154
Rhipicephalus evertsi 154
Rhipicephalus lunulatus 503
Rhipicephalus microplus 125, 154, 210, 224, 225, 241, 545
 biogeography and ecology 194
 biological control 197–198
 distribution 194
 host specificity 199–201
 host-questing 198
 introduction 193–195
 off-host ecology 195–197
 on-host ecology 198–199
 vectorial capacity 201–202
 warmer temperatures 202
Rhipicephalus sanguineus 33, 85, 121–122, 154, 199, 225, 265, 301, 309, 344, 431, 500
 introduction 209
 lineages
  distribution 209–210
  distribution and abundance 210–211
  vector competence 211–212
Rhipicephalus spp. 224–225
Rickettsia aeschlimannii 91, 421, 502
Rickettsia africae 500
Rickettsia bellii 64
Rickettsia conorii 501, 502
Rickettsia heilongjiangensis 360
Rickettsia helvetica 360, 502
Rickettsia massiliae 421
Rickettsia monacensis 362
Rickettsia raoultii 91, 166, 264
Rickettsia rickettsii 110, 147, 159, 261
Rickettsia slovaca 166
Rickettsiella 64
Rift Valley fever virus (RVFV) 402
Riparia riparia 97
risk mitigation 323
robin (*Erithacus rubecula*) 98
Rocky Mountain spotted fever (RMSF) 477
roe deer (*Capreolus capreolus*) 102, 133, 134

scalp eschar and neck lymphadenopathy (SENLAT) syndrome 166, 502
scaly thrush (*Zoothera daumi*) 104
Scandinavia
 disease-causing tick vectors 92
 endemic *Ixodes ricinus* ticks 90, 92
 exotic ticks 90–93
 introduction 90
 non-climatic factors 93
Scandinavian index 356
sea-surface temperature (SST) 195
sedge warbler (*Acrocephalus schoenobaenus*) 99
sedimentary fossil records 153
semiochemical conditions 199
septic shock-like syndrome 301
Serratia symbiotica 62
severe fever with thrombocytopaenia syndrome (SFTS) 493, 543
 animal vectors 386, 387
 cluster II 383
 confirmed case 382
 definition 382

demographic characteristics 383, **385**
endemic region 382
enzootic cycles 387
epidemiological risk factors 383
geographical clusters *384*
geographical distribution 382
Haemaphysalis longicornis ticks 387
immunoglobulin G antibody 382
laboratory-confirmed cases 382
laboratory-identified abnormalities 382
natural reservoir hosts 386
person-to-person transmission 385
seasonal distribution 383
spatial-temporal distribution/spread 386, 387
tick growth and reproduction 387
severe fever with thrombocytopaenia syndrome virus (SFTSV) 326, 328, 482
severe haemorrhagic syndrome disease (SFTS) 496
shared socio-economic pathway (SSP) 29
Siberian thrush (*Geokichla sibirica*) 104
sika deer (*Cervus nippon*) 102
silvopastoral systems 202, 203
single nucleotide polymorphism (SNP) 202
sixth mass extinction event 154
skin microbiota 200
small-scale developments 242
socio-ecological systems 12
socio-economic development 125
socio-economic transitions 311
solar radiation 10
song thrush (*Turdus philomelos*) 98
southern tick-associated rash illness (STARI) 159
spatial distribution models (SDMs) 38, 42
spatial statistics 262
species distribution models (SDMs) 11
spectrophotometric techniques 86
spirochaetes 350
Spiroplasma ixodetis 63
spotted fever group (SFG) 500
spotted fever group rickettsiae (SFGR) 450, 492
aetiological agent 360
infectious disease 360
Japan coastal region 362
Japanese spotted fever (JSF) 361, 362
pathogens 363
rickettsiosis and tick species 362
symptoms 360
ticks and spotted fever **361**
Staphylococcus aureus 414
starling (*Sturnus vulgaris*) 102
stepping stone model 103
stomatal conductance 10
subolesin 233
suburbanization 278
Šumava Mountains 337
susceptible, exposed, infectious, recovered (SEIR) model 28
sylvatic cycle 542
synganglion 50, 53, 54
synganglion cholinergic neurons 53
systems biology 232, 235

Taiwan
Aedes aegypti 447
climate change impacts 446
climate change/tick-borne pathogens *445*, *446*
definition 444
temperature 445
tick species 444
Tamdy virus (TAMV) 455
target of rapamycin (TOR) 76
terrestrial biosphere models (TBMs) 12
Theileria annulata 419
Theileria equi 166
Theileria luwenshuni 544
Theileria ovis 419
Theileria parva 111, 142
theileriosis 159
thermal stress 51–52
thermoneutral zone (TNZ) 198
thermosensilla 50
threshold viraemia 345, 346
thrombocytopaenia 302
thrombocytopaenia syndrome 267, 327
thrombocytopaenia syndrome virus 111
tick density 352
tick distributions
baseline setting 32–34
caveats 36
contemporary models 34, 36
continent vector database 36
covariate model 36
human population densities 34
inherent faults 34
introduction 32
tick eradication quarantine area (TEQA) 194
tick evolution
anthropocene and sixth mass extinction event 154
climate change 153–154
habitat suitability 155
introduction 152
origins 152–153
tick feeding synchrony 550
tick microbiome
antibiotics 49
biotic disturbance 47
community disruption 49
contributions 46
microbial metabolic pathways 47
taxonomic composition 46
temperature 48

tick species
abiotic conditions 42
correlation estimate 40
ecological modelling 39
environmental variables 40
functional forms 40
growth rates 38
potential distribution 41
predicted distribution 39
reliable models 39
spatial distribution models (SDMs) 38, 41, 42
spatial distributional extent 39
spatial filters 40
spatiotemporal distributions 38
thresholding models 41
tick-borne diseases (TBDs) 4, 26, 29, 58, 80, 84, 86
Canada (*see* Canada)
characteristics 312
China (*see* China)
endemic foci 308
England (*see* England)
epidemiology 110–111
evidence against climate change impact 311
evidence for climate change impacts 308–311
factors 307, 312
life-cycle stage 307
livestock, UK (*see* livestock)
macro- and microclimatic requirements 307
marine ecosystem (*see* marine ecosystem)
Middle East (*see* Middle East)
Taiwan (*see* Taiwan)
ticks 110
uncertainty 312
USA (*see* USA)
vegetative productivity 308
tick-borne encephalitis (TBE) 27, 84, 90, 102, 111, 132, 171, 248, 253, 264, 270, 279, 283, 333–335, 354–356, 456
Greater Alpine Region 542
human 548
natural foci 331
Russia 549
tick-borne encephalitis virus (TBEV) 91, 253, 255–256, 264, 430, 456, 493, 520
tick-borne fever (TBF) 293, 300
tick-borne haemorrhagic fever viruses (TBH-FVs) 341, 343
Tick-borne Illness Community Network Australia Inc. (TICNA) 397
tick-borne infections, Central Europe
Babesia genotypes 433
clusters 433
competent vector 430
Dermacentor marginatus 432
Dermacentor reticulatus 432, 433
Ixodes inopinatus 433
Ixodes ricinus 430
indigenous tick species 431, 432
Lyme borreliosis 430
non-endemic areas 433, 434
relict species 433
Russia
abiotic factors 440–441
climatological data 438
fed larvae and nymphs 439
forest tick 439
habitation 438–439
Ixodes persulcatus 442
ixodid ticks 439–440
obligatory transmissible infections 441–442
species-specific requirements 438
socio-economic factors 434
tick-borne infectious agents
co-feeding transmission 542
epigenetic regulatory mechanisms 541
non-systemic transmission 542
vertebrate host 541
tick-borne pathogen (TBP)
anthroponotic pathogens 547
Arctic (*see* Arctic ecosystem)
climate change 552
climate warming, Canada 548
environmental temperature 546
indigenous species 546
Lyme borreliosis (*see* Lyme borreliosis)
mosquito-borne pathogens 547
Northern Europe
'exotic' viruses and ticks 529, 530
climate change 530, 531
climate zones 528
mammals 529
tick reproduction/distribution 529
warmer climate 528
winters 528
Rhipicephalus microplus 546
transmission dynamics 549
tick-borne pathogens (TBPs) 264, 265
Caribbean region
Amblyomma mixtum 452
Amblyomma variegatum 450
biodiversity hotspot 449
brown dog tick 451
climatic change 451
cattle tick fever (CTF) 450
Dermacentor nitens 451
Ehrlichia ruminantium 452
exotic tick species 449
laying eggs 453
Rhipicephalus microplus 452
temperature and precipitation 452
tick fauna 449
tick species diversity 450
tick-borne relapsing fevers (TBRFs) 370, 543

tick-borne rickettsioses
definition 500
Rickettsia aeschlimannii 502
Rickettsia africae 500, 501, *501*
Rickettsia conorii 501, 502
Spotted fever group (SFG) 500
tick bite *501*
tick-borne detected, Africa 502, 503
tick-borne viruses
Black Sea flyway 455
climate change 458
Crimean Congo haemorrhagic fever virus (CCHFV) 457
metagenome-based assessments 456
tick-borne encephalitis (TBE) 456–457
tick-borne zoonotic pathogens 421
tick-transmitted pathogen populations/distribution
African continent 486
animal hosts 486
anthropogenic factors 488
climate 486
climatic and non-climatic drivers *487*
climatic variability 490
human hosts 487
indigenous livestock 490
mammalian host 487
Rhipicephalus microplus 488
Rhipicephalus appendiculatus 488, 489
vector-borne diseases, livestock 486
tick–game animal dynamics
bird densities 136
climate, habitat and management interactions 134–136
distribution and behaviour 133
endemic zones 136
greenhouse gas emissions 136
human population growth 136
introduction 132
pheasant densities 135
tick hosts 133–134
tick–host enzootic cycle 270
tick–host–disease dynamics 132
tick–host–pathogen interactions
Anaplasma phagocytophilum 233
conflict *vs.* cooperation 233–234
epigenetic regulatory mechanisms 235
exoskeleton chemical composition 235
genomic resources 234–235
glycoproteins 235
Ixodes scapularis proteome 234
mechanisms 234
midgut microbiome 235
modelling tick vaccines 236
multifactorial approach 234
paratransgenesis 235
quantum vaccinomics 235–236
study 232–233
ticks
Australia
borreliosis infections 393
climate change/human diseases 394–396
human bite 394
human-biting ticks *393*
Lyme disease 393
mean annual temperature anomaly *392*
non-infectious illnesses 394
population density *395*, *396*
Q fever 391
subfamilies 391
tick typhus 393
tick-borne illnesses 397
tick-borne infections 396
transcriptomes 196
transcriptomics 233
transient receptor potential (TRP) 52
transovarial transmission 331
transpiration 9
transstadial transmission 293, 331
tree pipit (*Anthus trivialis*) 99
tropical canine pancytopaenia 301
Turkey
Boophilus annulatus 121
climate characteristics 116
geographical, ecological and climate characteristics 115
Hyalomma aegyptium 119
Hyalomma anatolicum 120
Hyalomma excavatum 119
Hyalomma marginatum 118–119
Hyalomma rufipes 119
Hyalomma scupense 120–121
Haemaphysalis spp. and *Dermacentor* spp. 122–123
Ixodes ricinus 122
Rhipicephalus bursa 122
Rhipicephalus sanguineus 121–122
Rhipicephalus turanicus 121
tick fauna 117–118

uncertainty 1, 4, 5, 112, 163, 225
Urochloa brizantha 197
USA
Amblyomma spp. 478, 480, 481
Dermacentor spp. and transmitted pathogens 477, 478, 480
Ixodes spp. 478, 479
Ixodes ticks, changes in ranges 481, 482
native tick species 477
Ornithodoros, influential factors 477
Ornithodoros spp. 479
Rocky Mountain spotted fever (RMSF) 477
sylvatic cycles 479, 480
threats 482, 483
tick-endemic regions 480

vaccine development 303
vaccinomics 235
vector competence 79
vector distribution modelling 161
vector diversity 344
vector-borne diseases (VBDs) 86, 93, 161, 171, 297, 351, 391
  introduction 26–27
  mathematical models 27–29
  modelling guidelines 29
vector-borne pathogens 322
vector-borne zoonoses 247
vector–host population dynamics 272
vectorial capacity 216–217, 219, 276, 279
VectorNET project 28
vegetation growing season (VGS) 226
vegetation–climate interactions
  across scales 9–11
  analytical methods 13
  ecological change 11–13
  interdisciplinary collaborations 13
  introduction 8–9
  scale and pattern 9
viraemic transmission 255
virus-borne infections 250
vitamin B 47
vitellogenin (Vg) 75–77
vitellogenin-inducing factor (VIF) 75

Walter classification 11
warming climate 147
water balance
  behavioural phenotypes 69
  blood-feeding arthropods 67
  critical equilibrium humidity (CEH) 68
  dehydration 67–69
  glycerol production 69
  larvae and nymphs 67
  longevity/starvation resistance 68
  RNA-seq and metabolomic studies 69
  seasonal quiescence 69
  stress exposure 68
  tick biology 67
water loss rates (WLRs) 218
water-borne diseases 5
weather forecasting models 12
West Nile virus (WNV) 400
white-footed mouse (*Peromyscus leucopus*) 146
white-tailed deer (*Odocoileus virginianus*) 146, 177–178
whole-genome sequencing techniques 265
willow warbler (*Phylloscopus trochilus*) 99
wolf (*Canis lupus*) 248
wood mice (*Apodemus sylvaticus*) 134
World Health Organization (WHO) 307
WorldClim project 162

xerophiles 217
Xyllela fastidiosa 26

yellow fever virus (YFV) 401

zoonotic interface
  biotic elements *319*
  control and treatment measures 318
  disease incidence 320–321
  fever spirochaetes and arboviruses 318
  fossil fuel resources 322
  human disease risk 323
  human-assisted movement 322
  pathogen reservoirs and tick maintenance hosts 320
  risk mitigation 323
  spatial and temporal changes 318
  tick host availability, distribution and abundance 321–322
  vertebrates 318
zoonotic pathogens 171
zoonotic potential
  clinical symptoms 300
  diagnostics 303
  distribution 302–303
  human infection 302
  microorganisms 300
  pathogens 303
  vectors and hosts 301